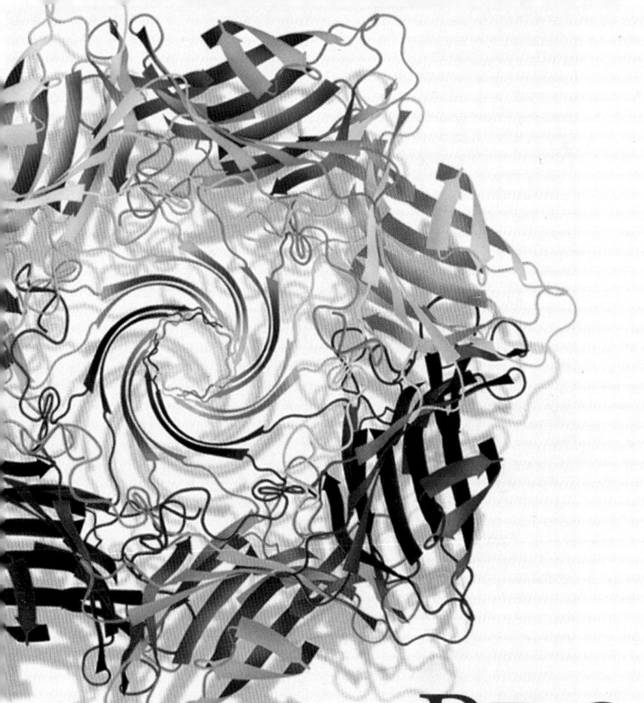

PROTEINS

STRUCTURE AND FUNCTION

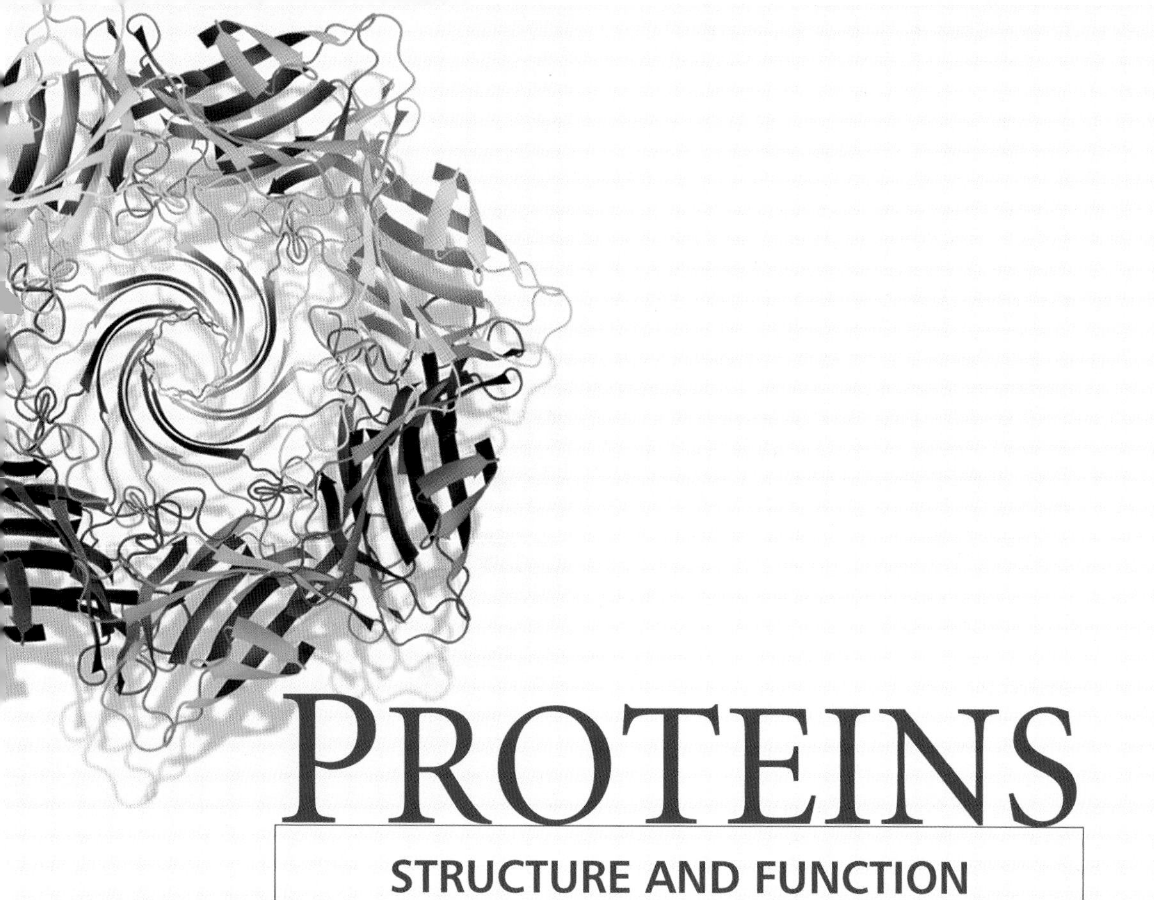

PROTEINS

STRUCTURE AND FUNCTION

David Whitford

John Wiley & Sons, Ltd

Telephone (+44) 1243 779777

Email (for orders and customer service enquiries): cs-books@wiley.co.uk
Visit our Home Page on www.wiley.com

Other Wiley Editorial Offices

John Wiley & Sons Inc., 111 River Street, Hoboken, NJ 07030, USA

Jossey-Bass, 989 Market Street, San Francisco, CA 94103-1741, USA

Wiley-VCH Verlag GmbH, Boschstr. 12, D-69469 Weinheim, Germany

John Wiley & Sons Australia Ltd, 33 Park Road, Milton, Queensland 4064, Australia

John Wiley & Sons (Asia) Pte Ltd, 2 Clementi Loop #02-01, Jin Xing Distripark, Singapore 129809

John Wiley & Sons Canada Ltd, 22 Worcester Road, Etobicoke, Ontario, Canada M9W 1L1

Wiley also publishes its books in a variety of electronic formats. Some content that appears
in print may not be available in electronic books.

British Library Cataloguing in Publication Data

A catalogue record for this book is available from the British Library

ISBN 0-471-49893-9 HB
ISBN 0-471-49894-7 PB

Typeset in 10/12pt Times by Laserwords Private Limited, Chennai, India
Printed and bound by Graphos SpA, Barcelona, Spain
This book is printed on acid-free paper responsibly manufactured from sustainable forestry
in which at least two trees are planted for each one used for paper production.

For my parents,
Elizabeth and Percy Whitford,
to whom I owe everything

Contents

Preface

When I first started studying proteins as an undergraduate I encountered for the first time complex areas of biochemistry arising from the pioneering work of Pauling, Sumner, Kendrew, Perutz, Anfinsen, together with other scientific 'giants' too numerous to describe at length in this text. The area seemed complete. How wrong I was and how wrong an undergraduate's perception can be! The last 30 years have seen an explosion in the area of protein biochemistry so that my 1975 edition of *Biochemistry* by Albert Lehninger remains, perhaps, of historical interest only. The greatest change has occurred through the development of molecular biology where fragments of DNA are manipulated in ways previously unimagined. This has enabled DNA to be sequenced, cloned, manipulated and expressed in many different cells. As a result areas of recombinant DNA technology and protein engineering have evolved rapidly to become specialist disciplines in their own right. Almost any protein whose primary sequence is known can be produced in large quantity via the expression of cloned or synthetic genes in recombinant host cells. Not only is the method allowing scientists to study some proteins for the first time but the increased amount of protein derived from recombinant DNA technology is also allowing the application of new and continually advancing structural techniques. In this area X-ray crystallography has remained at the forefront for over 40 years as a method of determining protein structure but it is now joined by nuclear magnetic resonance (NMR) spectroscopy and more recently by cryoelectron microscopy whilst other methods such as circular dichroism, infrared and Raman spectroscopy, electron spin resonance spectroscopy, mass spectrometry and fluorescence provide more limited, yet often vital and complementary, structural data. In many instances these methods have become established techniques only in the last 20 years and are consequently absent in many of those familiar textbooks occupying the shelves of university libraries.

An even greater impact on biochemistry has occurred with the rapid development of cost-effective, powerful, desktop computers with performance equivalent to the previous generation of supercomputers. Many experimental techniques relied on the co-development of computer hardware but software has also played a vital role in protein biochemistry. We can now search databases comparing proteins at the level of DNA or amino acid sequences, building up patterns of homology and relationships that provide insight into origin and possible function. In addition we use computers routinely to calculate properties such as isoelectric point, number of hydrophobic residues or secondary structure – something that would have been extraordinarily tedious, time consuming and problematic 20 years ago. Computers have revolutionized all aspects of protein biochemistry and there is little doubt that their influence will continue to increase in the forthcoming decades. The new area of bioinformatics reflects these advances in computing.

In my attempt to construct an introductory yet extensive text on proteins I have, of necessity, been circumspect in my description of the subject area. I have often relied on qualitative rather than quantitative descriptions and I have attempted to minimise the introduction of unwieldy equations or formulae. This does not reflect my own interests in physical biochemistry because my research, I hope, was often quantitative. In some cases particularly the chapters on enzymes and physical methods the introduction of equations is unavoidable but also necessary to an initial description of the content of these chapters. I would be failing in my duty as an educator if I omitted some of these equations and I hope students will keep going at these 'difficult' points or failing that just omit them entirely on first reading this book. However, in general I wish to introduce students to proteins by describing principles governing their structure and function and to avoid over-complication in this presentation through rigorous and quantitative treatment. This book is firmly intended to be a broad introductory text suitable for undergraduate and postgraduate study, perhaps after an initial exposure to the subject of protein biochemistry, whilst at the same time introducing specialist areas prior to future advanced study. I hope the following chapters will help to direct students to the amazing beauty and complexity of protein systems.

Target audience

The present text should be suitable for all introductory modes of biochemistry, molecular biology, chemistry, medicine and dentistry. In the UK this generally means the book is suitable for all undergraduates between years 1 and 3 and this book has stemmed from lectures given as parts of biochemistry courses to students of biochemistry, chemistry, medicine and dentistry in all 3 years. Where possible each chapter is structured to increase progressively in complexity. For purely introductory courses as would occur in years 1 or 2 it is sufficient to read only the first parts, or selected sections, of each chapter. More advanced courses may require thorough reading of each chapter together with consultation of the bibliography and secondly the list of references given at the end of the book.

The world wide web

In the last ten years the world wide web (WWW) has transformed information available to students. It provides a new and useful medium with which to deliver lecture notes and an exciting and new teaching resource for all. Consequently within this book URLs direct students to learning resources and a list of important addresses is included in the appendix. In an effort to exploit the power of the internet this book is associated with 'web-based' tutorials, problems and content and is accessed from the following URL http://www.wiley.com/go/whitfordproteins. These 'pages' are continually updated and point the interested reader towards new areas as they emerge. The Bibliography points interested readers towards further study material suitable for a first introduction to a subject whilst the list of references provides original sources for many areas covered in each of the twelve chapters.

For the problems included at the end of each chapter there are approximately 10 questions that aim to build on the subject matter discussed in the preceding text. Often the questions will increase in difficulty although this is not always the case. In this book I have limited the bibliography to broad reviews or accessible journal papers and I have deliberately restricted the number of 'high-powered' (difficult!) articles since I believe this organization is of greater use to students studying these subjects for the first time. To aid the learning process the web edition has multiple-choice questions for use as a formative assessment exercise. I should certainly like to hear of all mistakes or omissions encountered in this text and my hope is that educators and students will let me know via the e-mail address at the end of this section of any required corrections or additions.

Proteins are three-dimensional (3D) objects that are inadequately represented on book pages. Consequently many proteins are best viewed as molecular images using freely available software. Here, real-time manipulation of coordinate files is possible and will prove helpful to understanding aspects of structure and function. The importance of viewing, manipulating and even changing the representation of proteins to comprehending structure and function cannot be underestimated. Experience has suggested that the use of computers in this area can have a dramatic effect on student's understanding of protein structures. The ability to visualize in 3D conveys so much information – far more than any simple 2D picture in this book could ever hope to portray. Alongside many figures I have written the Protein DataBank files (e.g. PDB: 1HKO) used to produce diagrams. These files can be obtained from databases at several permanent sites based around the world such as http://www.rscb.org/pdb or one of the many 'mirrors' that exist (for example, in the UK this data is found at http://pdb.ccdc.cam.ac.uk). For students with Internet access each PDB file can be retrieved and manipulated independently to produce comparable images to those shown in the text. To explore these macromolecular images with reasonable efficiency does not require the latest 'all-powerful' desktop computer. A computer with a Pentium III (or later) based processor, a clock speed of 200 MHz or greater, 32–64 MB RAM, hard disks of 10 GB, a graphics video card with at least 8 MB memory and a connection to the internet are sufficient to view and store a significant number of files together with representative images. Of course things are easier with a computer with a surfeit of memory (>256 MB) and a high 'clock' speed (>2 GHz) but it is not obligatory to see 'on-line' content or to manipulate molecular images. This book was started on a 700 MHz Pentium III based processor equipped with 256 MB RAM and 16 MB graphics card.

Organization of this book

This book will address the structure and function of proteins in 12 subsequent chapters each with a definitive theme. After an initial chapter describing why one would wish to study proteins and a brief historical background the second chapter deals with the 'building blocks' of proteins, namely the amino acids together with their respective chemical and physical properties. No attempt is made at any point to describe the metabolism connected with these amino acids and the reader should consult general textbooks for descriptions of the synthesis and degradation of amino acids. This is a major area in its own right and would have lengthened the present book too much. However, I would like to think that students will not avoid these areas because they remain an equally important subject that should be covered at some point within the undergraduate curriculum. Chapter 3 covers the assembly of amino acids into polypeptide chains and levels of organizational structure found within proteins. Almost all detailed knowledge of protein structure and function has arisen through studies of globular proteins but the presence of fibrous proteins with different structures and functional properties necessitated a separate chapter devoted to this area (Chapter 4). Within this class the best understood structures are those belonging to the collagen class of proteins, the keratins and the extended β sheet structures such as silk fibroin. The division between globular proteins and fibrous proteins was made at a time when the only properties one could compare readily were a protein's amino acid composition and hydrodynamic radius. It is now apparent that other proteins exist with properties intermediate between globular and fibrous proteins that do not lend themselves to simple classification. However, the 'old' schemes of identification retain their value and serve to emphasize differences in proteins.

Membrane proteins represent a third group with different composition and properties. Most of these proteins are poorly understood, but there have been spectacular successes from the initial low-resolution structure of bacteriorhodopsin to the highly defined structure of bacterial photosynthetic reaction centres. These advances paved the way towards structural studies of G proteins and G-protein coupled receptors, the respiratory complexes from aerobic bacteria and the structure of ATP synthetases.

Chapter 6 focuses both on experimental and computational methods of comparing proteins where *in silico* methods have become increasingly important as a vital tool to assist with modern protein biochemistry. Chapter 7 focuses on enzymes and by discussing basic reaction rate theories and kinetics the chapter leads to a discussion of enzyme-catalysed reactions. Enzymes catalyse reactions through a variety of mechanisms including acid–base catalysis, nucleophilic driven chemistry and transition state stabilization. These and other mechanisms are described along with the principles of regulation, active site chemistry and binding.

The involvement of proteins in the cell cycle, transcription, translation, sorting and degradation of proteins is described in Chapter 8. In 50 years we have progressed from elucidating the structure of DNA to uncovering how this information is converted into proteins. The chapter is based around the structure of two macromolecular systems: the ribosome devoted towards accurate and efficient synthesis and the proteasome designed to catalyse specific proteolysis. Chapter 9 deals with the methods of protein purification. Very often, biochemistry textbooks describe techniques without placing the technique in the correct context. As a result, in Chapter 9 I have attempted to describe equipment as well as techniques so that students may obtain a proper impression of this area.

Structural methods determine the topology or fold of proteins. With an elucidation of structure at atomic levels of resolution comes an understanding of biological function. Chapter 10 addresses this area by describing different techniques. X-ray crystallography remains at the forefront of research with new variations of the basic principle allowing faster determination of structure at improved resolution. NMR methods yield structures of comparable resolution to crystallography for small soluble proteins. In ideal situations these methods provide complete structural determination of all heavy atoms but they are complemented by other spectroscopic methods such as absorbance and fluorescence methods, mass spectrometry and infrared spectroscopy. These techniques provide important ancillary information on tertiary structure such as the helical content of the protein, the proportion and environment of aromatic residues within a protein as well as secondary structure content.

Chapter 11 describes protein folding and stability – a subject that has generated intense research interest with the recognition that disease states arise from aberrant folding or stability. The mechanism of protein folding is illustrated by *in vitro* and *in vivo* studies. Whilst the broad concepts underlying protein folding were deduced from studies of 'model' proteins such as ribonuclease, analysis of cell folding pathways has highlighted specialised proteins, chaperones, with a critical function to the overall process. The GroES–GroEL complex is discussed to highlight the integrated process of synthesis and folding *in vivo*.

The final chapter builds on the preceding 11 chapters using a restricted set of well-studied proteins (case studies) with significant impact on molecular medicine. These proteins include haemoglobin, viral proteins, p53, prions and α_1-antitrypsin. Although still a young subject area this branch of protein science will expand in the next few years and will rely on the techniques, knowledge and principles elucidated in Chapters 1–11. The examples emphasize the impact of protein science and molecular medicine on the quality of human life.

Acknowledgements

I am indebted to all research students and post-docs who shared my laboratories at the Universities of London and Oxford during the last 15 years in many cases acting as 'test subjects' for teaching ideas. I should like to thank Drs Roger Hewson, Richard Newbold and Susan Manyusa whose comments throughout my research and teaching career were always valued. I would also like to thank individuals, too numerous to name, with whom I interacted at King's College London, Imperial College of Science, Technology and Medicine and the University of Oxford. In this context I should like to thank Dr John Russell, formerly of Imperial College London whose goodwill, humour and fantastic insight into the history of science, the scientific method and 'day to day' experimentation prevented absolute despair.

During preparation of this book many individuals read and contributed valuable comments to the manuscript's content, phrasing and ideas. In particular I wish to thank these unnamed and some times unknown individuals who read one or more of the chapters of this book. As is often said by most authors at this point despite their valuable contributions all of the remaining errors and deficiencies in the current text are my responsibility. In this context I could easily have spent more months attempting to perfect the current text. I am very aware that this text has deficiencies but I hope these defects will not detract from its value. In addition my wish to try other avenues, other roads not taken, dictates that this manuscript is completed without delay.

Writing and producing a textbook would not be possible without the support of a good publisher. I should like to thank all the staff at John Wiley & Sons, Chichester, UK. This exhaustive list includes particularly Andrew Slade as senior Publishing Editor who helped smooth the bumpy route towards production of this book, Lisa Tickner who first initiated events leading to commissioning this book, Rachel Ballard who supervised day to day business on this book, replacing every form I lost without complaint and monitoring tactfully and gently about possible completion dates, Robert Hambrook who translated my text and diagrams into a beautiful book, and the remainder of the production team of John Wiley and Sons. Together we inched our way towards the painfully slow production of this text, although the pace was entirely attributable to the author.

Lastly I must also thank Susan who tolerated the protracted completion of this book, reading chapters and offering support for this project throughout whilst coping with the arrival of Alexandra and Ethan effortlessly (unlike their father).

David Whitford
April 2004
david.whitford@ntlworld.com

1 An Introduction to protein structure and function

Biochemistry has exploded as a major scientific endeavour over the last one hundred years to rival previously established disciplines such as chemistry and physics. This occurred with the recognition that living systems are based on the familiar elements of organic chemistry (carbon, oxygen, nitrogen and hydrogen) together with the occasional involvement of inorganic chemistry and elements such as iron, copper, sodium, potassium and magnesium. More importantly the laws of physics including those concerning thermodynamics, electricity and quantum physics are applicable to biochemical systems and no 'vital' force distinguishes living from non-living systems. As a result the laws of chemistry and physics are successfully applied to biochemistry and ideas from physics and chemistry have found widespread application, frequently revolutionizing our understanding of complex systems such as cells.

This book focuses on one major component of all living systems – the proteins. Proteins are found in all living systems ranging from bacteria and viruses through the unicellular and simple eukaryotes to vertebrates and higher mammals such as humans. Proteins make up over 50 percent of the dry weight of cells and are present in greater amounts than any other biomolecule. Proteins are unique amongst the macromolecules in underpinning every reaction occurring in biological systems. It goes without saying that one should not ignore the other components of living systems since they have indispensable roles, but in this text we will consider only proteins.

A brief and very selective historical perspective

With the vast accumulation of knowledge about proteins over the last 50 years it is perhaps surprising to discover that the term *protein* was introduced nearly 170 years ago. One early description was by Gerhardus Johannes Mulder in 1839 where his studies on the composition of animal substances, chiefly fibrin, albumin and gelatin, showed the presence of carbon, hydrogen, oxygen and nitrogen. In addition he recognized that sulfur and phosphorus were present sometimes in 'animal substances' that contained large numbers of atoms. In other words, he established that these 'substances' were macromolecules. Mulder communicated his results to Jöns Jakob Berzelius and it is suggested the term protein arose from this interaction where the origin of the word protein has been variously ascribed to derivation from the Latin word *primarius* or from the Greek god *Proteus*. The definition of proteins was timely since in 1828 Friedrich Wöhler had shown that

Proteins: Structure and Function by David Whitford

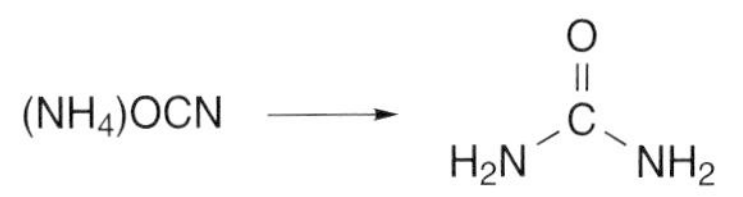

$(NH_4)OCN \longrightarrow H_2N{-}C({=}O){-}NH_2$

Figure 1.1 The decomposition of ammonium cyanate yields urea

heating ammonium cyanate resulted in isomerism and the formation of urea (Figure 1.1). Organic compounds characteristic of living systems, such as urea, could be derived from simple inorganic chemicals. For many historians this marks the beginning of biochemistry and it is appropriate that the discovery of proteins occurred at the same period.

The development of biochemistry and the study of proteins was assisted by analysis of their composition and structure by Heinrich Hlasiwetz and Josef Habermann around 1873 and the recognition that proteins were made up of smaller units called amino acids. They established that hydrolysis of casein with strong acids or alkali yielded glutamic acid, aspartic acid, leucine, tyrosine and ammonia whilst the hydrolysis of other proteins yielded a different group of products. Importantly their work suggested that the properties of proteins depended uniquely on the constituent parts – a theme that is equally relevant today in modern biochemical study.

Another landmark in the study of proteins occurred in 1902 with Franz Hofmeister establishing the constituent atoms of the peptide bond with the polypeptide backbone derived from the condensation of free amino acids. Five years earlier Eduard Buchner revolutionized views of protein function by demonstrating that yeast cell *extracts* catalysed fermentation of sugar into ethanol and carbon dioxide. Previously it was believed that only living systems performed this catalytic function. Emil Fischer further studied biological catalysis and proposed that components of yeast, which he called enzymes, combined with sugar to produce an intermediate compound. With the realization that cells were full of enzymes 100 years of research has developed and refined these discoveries. Further landmarks in the study of proteins could include Sumner's crystallization of the first enzyme (urease) in 1926 and Pauling's description of the geometry of the peptide bond; however, extensive discussion of these advances and many other important discoveries in protein biochemistry are best left to history of science textbooks.

A brief look at the award of the Nobel Prizes for Chemistry, Physiology and Medicine since 1900 highlighted in Table 1.1 reveals the involvement of many diverse areas of science in protein biochemistry. At first glance it is not obvious why William and Lawrence Bragg's discovery of the diffraction of X-rays by sodium chloride crystals is relevant, but diffraction by protein crystals is the main route towards biological structure determination. Their discovery was the first step in the development of this technique. Discoveries in chemistry and physics have been implemented rapidly in the study of proteins. By 1958 Max Perutz and John Kendrew had determined the first protein structure and this was soon followed by the larger, multiple subunit, structure of haemoglobin and the first enzyme, lysozyme. This remarkable advance in knowledge extended from initial understanding of the atomic composition of proteins around 1900 to the determination of the three-dimensional structure of proteins in the 1960s and represents a major chapter of modern biochemistry. However, advances have continued with new areas of molecular biology proving equally important to understanding protein structure and function.

Life may be defined as the ordered interaction of proteins and all forms of life from viruses to complex, specialized, mammalian cells are based on proteins made up of the same building blocks or amino acids. Proteins found in simple unicellular organisms such as bacteria are identical in structure and function to those found in human cells illustrating the evolutionary lineage from simple to complex organisms.

Molecular biology starts with the dramatic elucidation of the structure of the DNA double helix by James Watson, Francis Crick, Rosalind Franklin and Maurice Wilkins in 1953. Today, details of DNA replication, transcription into RNA and the synthesis of proteins (translation) are extensive. This has established an enormous body of knowledge representing a whole new subject area. All cells encode the information content of proteins within genes, or more accurately the order of bases along the DNA strand, yet it is the

Table 1.1 Selected landmarks in the study of protein structure and function from 1900–2002 as seen by the award of the Nobel Prize for Chemistry, Physiology or Medicine

Date	Discoverer + Discovery
1901	Wilhelm Conrad Röntgen 'in recognition of the ... discovery of the remarkable rays subsequently named after him'
1907	Eduard Buchner 'cell-free fermentation'
1914	Max von Laue 'for his discovery of the diffraction of X-rays by crystals'
1915	William Henry Bragg and William Lawrence Bragg 'for their services in the analysis of crystal structure by ... X-rays'
1923	Frederick Grant Banting and John James Richard Macleod 'for the discovery of insulin'
1930	Karl Landsteiner 'for his discovery of human blood groups'
1946	James Batcheller Sumner 'for his discovery that enzymes can be crystallized'.
	John Howard Northrop and Wendell Meredith Stanley 'for their preparation of enzymes and virus proteins in a pure form'
1948	Arne Wilhelm Kaurin Tiselius 'for his research on electrophoresis and adsorption analysis, especially for his discoveries concerning the complex nature of the serum proteins'
1952	Archer John Porter Martin and Richard Laurence Millington Synge 'for their invention of partition chromatography'
1952	Felix Bloch and Edward Mills Purcell 'for their development of new methods for nuclear magnetic precision measurements and discoveries in connection therewith'
1954	Linus Carl Pauling 'for his research into the nature of the chemical bond and ... to the elucidation of ... complex substances'
1958	Frederick Sanger 'for his work on the structure of proteins, especially that of insulin'
1959	Severo Ochoa and Arthur Kornberg 'for their discovery of the mechanisms in the biological synthesis of ribonucleic acid and deoxyribonucleic acid'
1962	Max Ferdinand Perutz and John Cowdery Kendrew 'for their studies of the structures of globular proteins'
1962	Francis Harry Compton Crick, James Dewey Watson and Maurice Hugh Frederick Wilkins 'for their discoveries concerning the molecular structure of nucleic acids and its significance for information transfer in living material'
1964	Dorothy Crowfoot Hodgkin 'for her determinations by X-ray techniques of the structures of important biochemical substances'
1965	François Jacob, André Lwoff and Jacques Monod 'for discoveries concerning genetic control of enzyme and virus synthesis'
1968	Robert W. Holley, Har Gobind Khorana and Marshall W. Nirenberg 'for ... the genetic code and its function in protein synthesis'
1969	Max Delbrück, Alfred D. Hershey and Salvador E. Luria 'for their discoveries concerning the replication mechanism and the genetic structure of viruses'

(*continued overleaf*)

Table 1.1 (*continued*)

Date	Discoverer + Discovery
1972	Christian B. Anfinsen 'for his work on ribonuclease, especially concerning the connection between the amino acid sequence and the biologically active conformation' Stanford Moore and William H. Stein 'for their contribution to the understanding of the connection between chemical structure and catalytic activity of ... ribonuclease molecule'
1972	Gerald M. Edelman and Rodney R. Porter 'for their discoveries concerning the chemical structure of antibodies'
1975	John Warcup Cornforth 'for his work on the stereochemistry of enzyme-catalyzed reactions'. Vladimir Prelog 'for his research into the stereochemistry of organic molecules and reactions'
1975	David Baltimore, Renato Dulbecco and Howard Martin Temin 'for their discoveries concerning the interaction between tumour viruses and the genetic material of the cell'
1978	Werner Arber, Daniel Nathans and Hamilton O. Smith 'for the discovery of restriction enzymes and their application to problems of molecular genetics'
1980	Paul Berg 'for his fundamental studies of the biochemistry of nucleic acids, with particular regard to recombinant-DNA' Walter Gilbert and Frederick Sanger 'for their contributions concerning the determination of base sequences in nucleic acids'
1982	Aaron Klug 'development of crystallographic electron microscopy and structural elucidation of nucleic acid–protein complexes'
1984	Robert Bruce Merrifield 'for his development of methodology for chemical synthesis on a solid matrix'
1984	Niels K. Jerne, Georges J.F. Köhler and César Milstein 'for theories concerning the specificity in development and control of the immune system and the discovery of the principle for production of monoclonal antibodies'
1988	Johann Deisenhofer, Robert Huber and Hartmut Michel 'for the determination of the structure of a photosynthetic reaction centre'
1989	J. Michael Bishop and Harold E. Varmus 'for their discovery of the cellular origin of retroviral oncogenes'
1991	Richard R. Ernst 'for ... the methodology of high resolution nuclear magnetic resonance spectroscopy'
1992	Edmond H. Fischer and Edwin G. Krebs 'for their discoveries concerning reversible protein phosphorylation as a biological regulatory mechanism'
1993	Kary B. Mullis 'for his invention of the polymerase chain reaction (PCR) method' and Michael Smith 'for his fundamental contributions to the establishment of oligonucleotide-based, site-directed mutagenesis'
1994	Alfred G. Gilman and Martin Rodbell 'for their discovery of G-proteins and the role of these proteins in signal transduction'

Table 1.1 *(continued)*

Date	Discoverer + Discovery
1997	Paul D. Boyer and John E. Walker 'for their elucidation of the enzymatic mechanism underlying the synthesis of adenosine triphosphate (ATP)'. Jens C. Skou 'for the first discovery of an ion-transporting enzyme, Na^+, K^+-ATPase'
1997	Stanley B. Prusiner 'for his discovery of prions – a new biological principle of infection'
1999	Günter Blobel 'for the discovery that proteins have intrinsic signals that govern their transport and localization in the cell'
2000	Arvid Carlsson, Paul Greengard and Eric R Kandel 'signal transduction in the nervous system'
2001	Paul Nurse, Tim Hunt and Leland Hartwill 'for discoveries of key regulators of the cell cycle'
2002	Kurt Wuthrich, 'for development of NMR spectroscopy as a method of determining biological macromolecules structure in solution.' John B. Fenn and Koichi Tanaka 'for their development of soft desorption ionization methods for mass spectrometric analyses of biological macromolecules'. Sydney Brenner, H. Robert Horvitz and John E. Sulston 'for their discoveries concerning genetic regulation of organ development and programmed cell death'

conversion of this information or expression into proteins that represents the tangible evidence of a living system or life.

$$\text{DNA} \longrightarrow \text{RNA} \longrightarrow \text{protein}$$

Cells divide, synthesize new products, secrete unwanted products, generate chemical energy to sustain these processes via specific chemical reactions, and in all of these examples the common theme is the mediation of proteins.

In 1944 the physicist Erwin Schrödinger posed the question 'What is Life?' in an attempt to understand the physical properties of a living cell. Schrödinger suggested that living systems obeyed all laws of physics and should not be viewed as exceptional but instead reflected the statistical nature of these laws. More importantly, living systems are amenable to study using many of the techniques familiar to chemistry and physics. The last 50 years of biochemistry have demonstrated this hypothesis emphatically with tools developed by physicists and chemists rapidly employed in biological studies. A casual perusal of Table 1.1 shows how quickly methodologies progress from discovery to application.

The biological diversity of proteins

Proteins have diverse biological functions ranging from DNA replication, forming cytoskeletal structures, transporting oxygen around the bodies of multicellular organisms to converting one molecule into another. The types of functional properties are almost endless and are continually being increased as we learn more about proteins. Some important biological functions are outlined in Table 1.2 but it is to be expected that this rudimentary list of properties will expand each year as new proteins are characterized. A formal demarcation of proteins into one class should not be pursued too far since proteins can have multiple roles or functions; many proteins do not lend themselves easily to classification schemes. However, for all chemical reactions occurring in cells a protein is involved intimately in the biological process. These proteins are united through their composition based on the same group of 20 amino acids. Although all proteins are composed of the same group of 20 amino acids they differ in their composition – some contain a surfeit of one amino acid whilst others may lack one or two members of the group of 20 entirely. It was realized early in the study of proteins that

Table 1.2 A selective list of some functional roles for proteins within cells

Function	Examples
Enzymes or catalytic proteins	Trypsin, DNA polymerases and ligases,
Contractile proteins	Actin, myosin, tubulin, dynein,
Structural or cytoskeletal proteins	Tropocollagen, keratin,
Transport proteins	Haemoglobin, myoglobin, serum albumin, ceruloplasmin, transthyretin
Effector proteins	Insulin, epidermal growth factor, thyroid stimulating hormone,
Defence proteins	Ricin, immunoglobulins, venoms and toxins, thrombin,
Electron transfer proteins	Cytochrome oxidase, bacterial photosynthetic reaction centre, plastocyanin, ferredoxin
Receptors	CD4, acetycholine receptor,
Repressor proteins	Jun, Fos, Cro,
Chaperones (accessory folding proteins)	GroEL, DnaK
Storage proteins	Ferritin, gliadin,

variation in size and complexity is common and the molecular weight and number of subunits (polypeptide chains) show tremendous diversity. There is no correlation between size and number of polypeptide chains. For example, insulin has a relative molecular mass of 5700 and contains two polypeptide chains, haemoglobin has a mass of approximately 65 000 and contains four polypeptide chains, and hexokinase is a single polypeptide chain with an overall mass of ~100 000 (see Table 1.3).

The molecular weight is more properly referred to as the relative molecular mass (symbol M_r). This is defined as the mass of a molecule relative to 1/12th the mass of the carbon (^{12}C) isotope. The mass of this isotope is defined as exactly 12 atomic mass units. Consequently the term molecular weight or relative molecular mass is a dimensionless quantity and should not possess any units. Frequently in this and many other textbooks the unit Dalton (equivalent to 1 atomic mass unit, i.e. 1 Dalton = 1 amu) is used and proteins are described with molecular weights of 5.5 kDa (5500 Daltons). More accurately, this is the absolute molecular weight representing the mass in grams of 1 mole of protein. For most purposes this becomes of little relevance and the term 'molecular weight' is used freely in protein biochemistry and in this book.

Table 1.3 The molecular masses of proteins together with the number of subunits. The term 'subunit' is synonymous with the number of polypeptide chains and is used interchangeably

Protein	Molecular mass	Subunits
Insulin	5700	2
Haemoglobin	64 500	4
Tropocollagen	285 000	3
Subtilisin	27 500	1
Ribonuclease	12 600	1
Aspartate transcarbamoylase	310 000	12
Bacteriorhodopsin	26 800	1
Hexokinase	102 000	1

Proteins are joined covalently and non-covalently with other biomolecules including lipids, carbohydrates, nucleic acids, phosphate groups, flavins, heme groups and metal ions. Components such as hemes or metal ions are often called prosthetic groups. Complexes formed between lipids and proteins are lipoproteins, those with carbohydrates are called glycoproteins, whilst complexes with metal ions lead to metalloproteins, and so on. The complexes formed between metal ions and proteins increases the involvement of elements of the periodic table beyond that expected of typical organic molecules (namely carbon, hydrogen, nitrogen and oxygen). Inspection of the periodic table (Figure 1.2) shows that at least 20 elements have been implicated directly in the structure and function of proteins (Table 1.4). Surprisingly elements such as aluminium and silicon that are very abundant in the Earth's crust (8.1 and 25.7 percent by weight, respectively) do not occur in high concentration within cells. Aluminium is rarely, if ever, found as part of proteins whilst the role of silicon is confined to biomineralization where it is the core component of shells. The involvement of carbon, hydrogen, oxygen, nitrogen, phosphorus and sulfur is clear although the role of other elements, particularly transition metals, has been difficult to establish. Where transition metals occur in proteins there is frequently only one metal atom per mole of protein and led in the past to a failure to detect metal. Other elements have an inferred involvement from growth studies showing that depletion from the diet leads to an inhibition of normal cellular function. For metalloproteins the absence of the metal can lead to a loss of structure and function.

Metals such as Mo, Co and Fe are often found associated with organic co-factors such as pterin, flavins, cobalamin and porphyrin (Figure 1.3). These organic ligands hold metal centres and are often tightly associated to proteins.

Table 1.4 The involvement of trace elements in the structure and function of proteins

Element	Functional role
Sodium	Principal intracellular ion, osmotic balance
Potassium	Principal intracellular ion, osmotic balance
Magnesium	Bound to ATP/GTP in nucleotide binding proteins, found as structural component of hydrolase and isomerase enzymes
Calcium	Activator of calcium binding proteins such as calmodulin
Vanadium	Bound to enzymes such as chloroperoxidase.
Manganese	Bound to pterin co-factor in enzymes such as xanthine oxidase or sulphite oxidase. Also found in nitrogenase and as component of water splitting enzyme in higher plants.
Iron	Important catalytic component of heme enzymes involved in oxygen transport as well as electron transfer. Important examples are haemoglobin, cytochrome oxidase and catalase.
Cobalt	Metal component of vitamin B_{12} found in many enzymes.
Nickel	Co-factor found in hydrogenase enzymes
Copper	Involved as co-factor in oxygen transport systems and electron transfer proteins such as haemocyanin and plastocyanin.
Zinc	Catalytic component of enzymes such as carbonic anhydrase and superoxide dismutase.
Chlorine	Principal intracellular anion, osmotic balance
Iodine	Iodinated tyrosine residues form part of hormone thyroxine and bound to proteins
Selenium	Bound at active centre of glutathione peroxidase

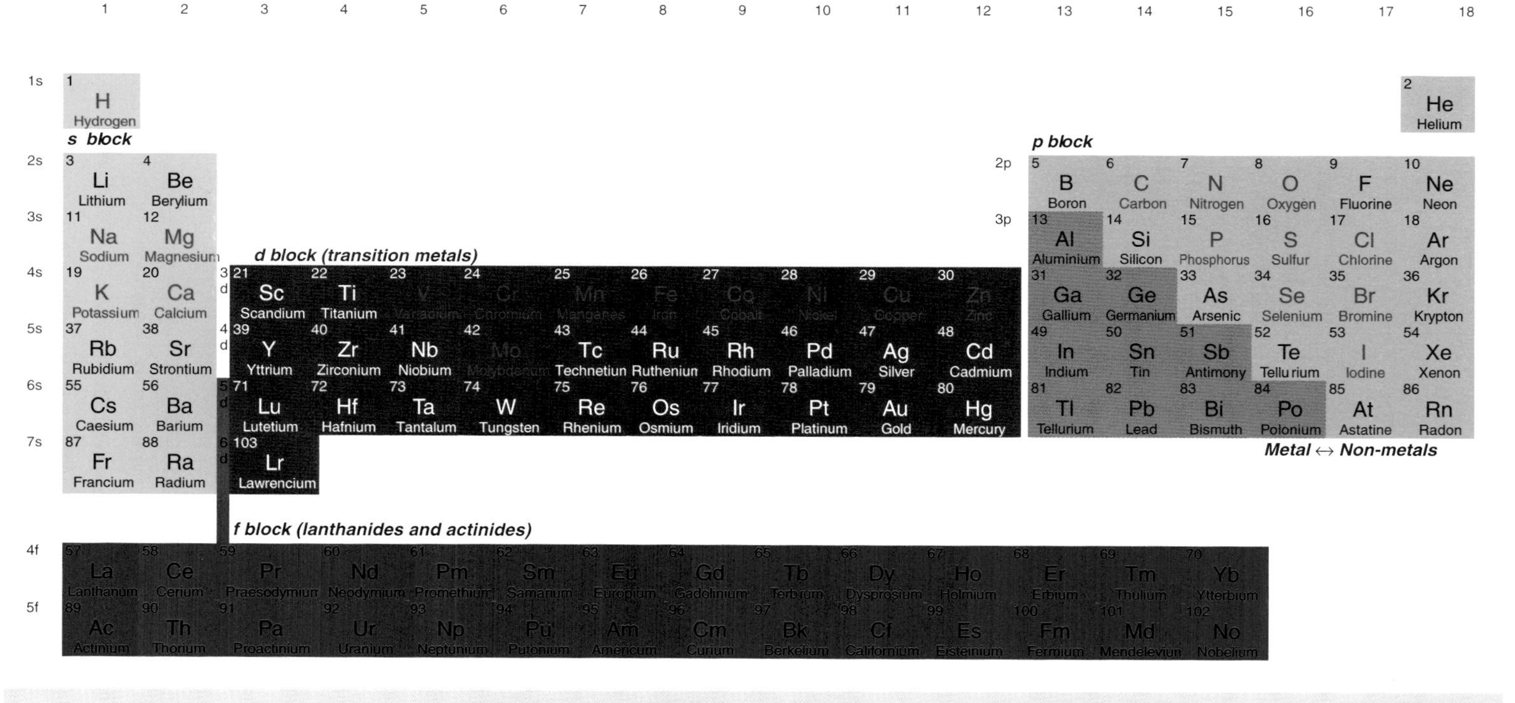

Figure 1.2 The periodic table showing the elements highlighted in red known to have involvement in the structure and/or function of proteins. The involvement of some elements is contentious tungsten and cadmium are claimed to be associated with proteins yet these elements are also known to be toxic

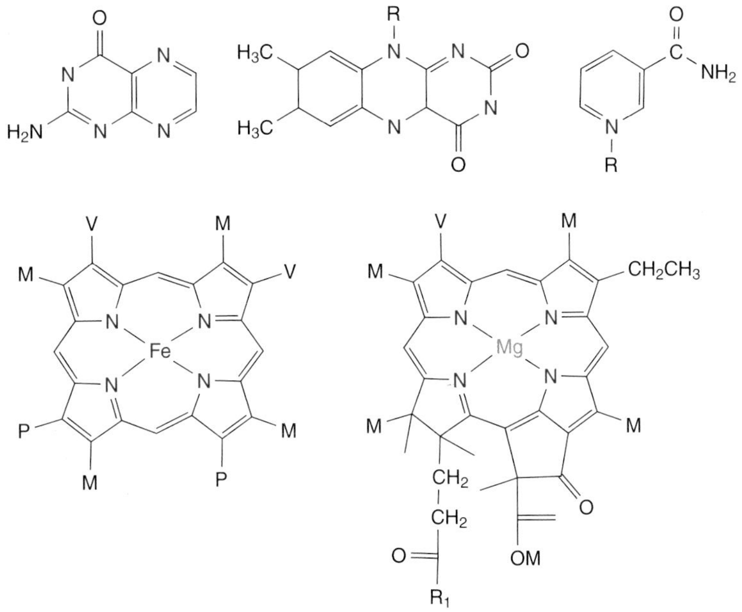

Figure 1.3 Organic co-factors found in proteins. These co-factors are pterin, the isoalloxine ring found as part of flavin in FAD and FMN, the pyridine ring of NAD and its close analogue NADP and the porphyrin skeletons of heme and chlorophyll. R represents the remaining part of the co-factor whilst M and V signify methyl and vinyl side chains

Proteins and the sequencing of the human and other genomes

Recognition of the diverse roles of proteins in biological systems increased largely as a result of the enormous amount of sequencing information generated via the Human Genome Mapping project. Similar schemes aimed at deciphering the genomes of *Escherichia coli*, yeast (*Sacharromyces cerevisiae*), and mouse provided related information. With the completion of the first draft of the human genome mapping project in 2001 human chromosomes contain approximately 25–30 000 genes. This allows a conservative estimate of the number of polypeptides making up most human cells as ~25 000, although alternative splicing of genes and variations in subunit composition increase the number of proteins further. Despite sequencing the human genome it is an unfortunate fact that we do not know the role performed by most proteins. Of those thousands of polypeptides we know the structures of only a small number, emphasizing a large imbalance between the abundance of sequence data and the presence of structure/function information. An analysis of protein databases suggests about 1000 distinct structures or folds have been determined for globular proteins. Many proteins are retained within cell membranes and we know virtually nothing about the structures of these proteins and only slightly more about their functional roles. This observation has enormous consequences for understanding protein structure and function.

Why study proteins?

This question is often asked not entirely without reason by many undergraduates during their first introduction to the subject. Perhaps the best reply that can be given is that proteins underpin every aspect of biological activity. This is particularly important in areas where protein structure and function have an impact on human endeavour such as medicine. Advances in molecular genetics reveal that many diseases stem from specific protein defects. A classic example is cystic

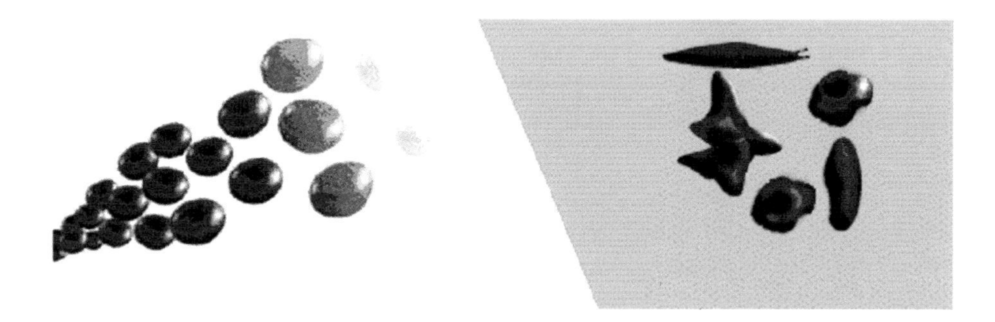

Figure 1.4 The shape of erythrocytes in normal and sickle cell anemia arises from mutations to haemoglobin found within the red blood cell. (Reproduced with permission from Voet, D, Voet, J.G and Pratt, C.W. *Fundamentals of Biochemistry*. John Wiley & Sons Inc.)

fibrosis, an inherited condition that alters a protein, called the cystic fibrosis transmembrane conductance regulator (CFTR), involved in the transport of sodium and chloride across epithelial cell membranes. This defect is found in Caucasian populations at a ratio of ~1 in 20, a surprisingly high frequency. With 1 in 20 of the population 'carrying' a single defective copy of the gene individuals who inherit defective copies of the gene from each parent suffer from the disease. In the UK the incidence of cystic fibrosis is approximately 1 in 2000 live births, making it one of the most common inherited disorders. The disease results in the body producing a thick, sticky mucus that blocks the lungs, leading to serious infection, and inhibits the pancreas, stopping digestive enzymes from reaching the intestines where they are required to digest food. The severity of cystic fibrosis is related to *CFTR* gene mutation, and the most common mutation, found in approximately 65 percent of all cases, involves the deletion of a single amino acid residue from the protein at position 508. A loss of one residue out of a total of nearly 1500 amino acid residues results in a severe decrease in the quality of life with individuals suffering from this disease requiring constant medical care and supervision.

Further examples emphasize the need to understand more about proteins. The pioneering studies of Vernon Ingram in the 1950s showed that sickle cell anemia arose from a mutation in the β chain of haemoglobin. Haemoglobin is a tetrameric protein containing 2α and 2β chains. In each of the β chains a mutation is found that involves the change of the sixth amino acid residue from a glutamic acid to a valine. The alteration of two residues out of 574 leads to a drastic change in the appearance of red blood cells from their normal biconcave disks to an elongated sickle shape (Figure 1.4).

As the name of the disease suggests individuals are anaemic showing decreased haemoglobin content in red blood cells from approximately 15 g per 100 ml to under half that figure, and show frequent illness. Our understanding of cystic fibrosis and of sickle cell anaemia has advanced in parallel with our understanding of protein structure and function although at best we have very limited and crude means of treating these diseases.

However, perhaps the greatest impetus to understand protein structure and function lies in the hope of overcoming two major health issues confronting the world in the 21st century. The first of these is cancer. Cancer is the uncontrolled proliferation of cells that have lost their normal regulated cell division often in response to a genetic or environmental trigger. The development of cancer is a multistep, multifactorial process often occurring over decades but the precise involvement of specific proteins has been demonstrated in some instances. One of the best examples is a protein called p53, normally present at low levels in cells, that 'switches on' in response to cellular damage and as a transcription factor controls the cell cycle process. Mutations in p53 alter the normal cycle of events leading eventually to cancer and several tumours including lung, colorectal and skin carcinomas are attributed to molecular defects in p53. Future research on p53 will enable its physicochemical properties to be thoroughly appreciated and by understanding the link between structure, folding, function and regulation comes the prospect of unravelling its role in tumour formation and manipulating its activity via therapeutic intervention. Already some success is being achieved in this area and the future holds great promise for 'halting' cancer by controlling the properties of p53 and similar proteins.

A second major problem facing the world today is the estimated number of people infected with the human immunodeficiency virus (HIV). In 2003 the World Health Organization (WHO) estimated that over 40 million individuals are infected with this virus in the world today. For many individuals, particularly those in the 'Third World', the prospect of prolonged good health is unlikely as the virus slowly degrades the body's ability to fight infection through damage to the immune response mechanism and in particular to a group of cells called cytotoxic T cells. HIV infection encompasses many aspects of protein structure and function, as the virus enters cells through the interaction of specific viral coat proteins with receptors on the surface of white blood cells. Once inside cells the virus 'hides' but is secretly replicating and integrating genetic material into host DNA through the action of specific enzymes (proteins). Halting the destructive influence of HIV relies on understanding many different, yet inter-related, aspects of protein structure and function. Again, considerable progress has been made since the 1980s when the causative agent of the disease was recognized as a retrovirus. These advances have focussed on understanding the structure of HIV proteins and in designing specific inhibitors of, for example, the reverse transcriptase enzyme. Although in advanced health care systems these drugs (inhibitors) prolong life expectancy, the eradication of HIV's destructive action within the body and hence an effective cure remains unachieved. Achieving this goal should act as a timely reminder for all students of biology, chemistry and medicine that success in this field will have a dramatic impact on the quality of human life in the forthcoming decades.

Central to success in treating any of the above diseases are the development of new medicines, many based on proteins. The development of new therapies has been rapid during the last 20 years with the list of new treatments steadily increasing and including minimizing serious effects of different forms of cancer via the use of specific proteins including monoclonal antibodies, alleviating problems associated with diabetes by the development of improved recombinant 'insulins' and developing 'clot-busting' drugs (proteins) for the management of strokes and heart attacks. This highly selective list is the productive result of understanding protein structure and function and has contributed to a marked improvement in disease management. For the future these advances will need to be extended to other diseases and will rely on an extensive and thorough knowledge of proteins of increasing size and complexity. We will need to understand the structure of proteins, their interaction with other biomolecules, their roles within different biological systems and their potential manipulation by genetic or chemical methods. The remaining chapters in this book represent an attempt to introduce and address some of these issues in a fundamental manner helpful to students.

2

Amino acids: the building blocks of proteins

Despite enormous functional diversity all proteins consist of a linear arrangement of amino acid residues assembled together into a polypeptide chain. Amino acids are the 'building blocks' of proteins and in order to understand the properties of proteins we must first describe the properties of the constituent 20 amino acids. All amino acids contain carbon, hydrogen, nitrogen and oxygen with two of the 20 amino acids also containing sulfur. Throughout this book a colour scheme based on the CPK model (after Corey, Pauling and Kultun, pioneers of 'space-filling' representations of molecules) is used. This colouring scheme shows nitrogen atoms in blue, oxygen atoms in red, carbon atoms are shown in light grey (occasionally black), sulfur is shown in yellow, and hydrogen, when shown, is either white, or to enhance viewing on a white background, a lighter shade of grey. To avoid unnecessary complexity 'ball and stick' representations of molecular structures are often shown instead of space-filling models. In other instances cartoon representations of structure are shown since they enhance visualization of organization whilst maintaining clarity of presentation.

The 20 amino acids found in proteins

In their isolated state amino acids are white crystalline solids. It is surprising that crystalline materials form the building blocks for proteins since these latter molecules are generally viewed as 'organic'. The crystalline nature of amino acids is further emphasized by their high melting and boiling points and together these properties are atypical of most organic molecules. Organic molecules are not commonly crystalline nor do they have high melting and boiling points. Compare, for example, alanine and propionic acid – the former is a crystalline amino acid and the other is a volatile organic acid. Despite similar molecular weights (89 and 74) their respective melting points are 314 °C and −20.8 °C. The origin of these differences and the unique properties of amino acids resides in their ionic and dipolar nature.

Amino acids are held together in a crystalline lattice by charged interactions and these relatively strong forces contribute to high melting and boiling points. Charge groups are also responsible for electrical conductivity in aqueous solutions (amino acids are electrolytes), their relatively high solubility in water and the large dipole moment associated with crystalline material. Consequently amino acids are best viewed as charged molecules that crystallize from solutions containing dipolar ions. These dipolar ions are called zwitterions. A proper representation of amino acids reflects amphoteric behaviour and amino acids are always represented as the zwitterionic state in this

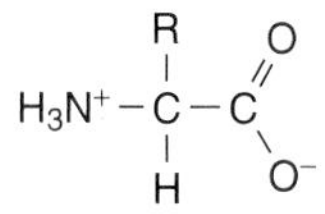

R O
H3N+ – C – C
H O–

Figure 2.1 A skeletal model of a generalized amino acid showing the amino (blue) carboxyl (red) and R groups attached to a central or α carbon

textbook as opposed to the undissociated form. For 19 of the twenty amino acids commonly found in proteins a general structure for the zwitterionic state has charged amino (NH_3^+) and carboxyl (COO^-) groups attached to a central carbon atom called the α carbon. The remaining atoms connected to the α carbon are a single hydrogen atom and the R group or side chain (Figure 2.1).

The acid–base properties of amino acids

At pH 7 the amino and carboxyl groups are charged but over a pH range from 1 to 14 these groups exhibit a series of equilibria involving binding and dissociation of a proton. The binding and dissociation of a proton reflects the role of these groups as weak acids or weak bases. The acid–base behaviour of amino acids is important since it influences the eventual properties of proteins, permits methods of identification for different amino acids and dictates their reactivity. The amino group, characterized by a basic pK value of approximately 9, is a weak base. Whilst the amino group ionizes around pH 9.0 the carboxyl group remains charged until a pH of ∼2.0 is reached. At this pH a proton binds neutralizing the charge of the carboxyl group. In each case the carboxyl and amino groups ionize according to the equilibrium

$$HA + H_2O \longrightarrow H_3O^+ + A^- \qquad (2.1)$$

where HA, the proton donor, is either –COOH or $-NH_3^+$ and A^- the proton acceptor is either $-COO^-$ or $-NH_2$. The extent of ionization depends on the equilibrium constant

$$K = [H^+][A^-]/[HA] \qquad (2.2)$$

and it becomes straightforward to derive the relationship

$$pH = pK + \log[A^-]/[HA] \qquad (2.3)$$

known as the Henderson–Hasselbalch equation (see appendix). For a simple amino acid such as alanine a biphasic titration curve is observed when a solution of the amino acid (a weak acid) is titrated with sodium hydroxide (a strong base). The titration curve shows two zones where the pH changes very slowly after additions of small amounts of acid or alkali (Figure 2.2). Each phase reflects different pK values associated with ionizable groups.

During the titration of alanine different ionic species predominate in solution (Figure 2.3). At low pH (<2.0) the equilibrium lies in favour of the positively charged form of the amino acid. This species contains a charged amino group and an uncharged carboxyl group leading to the overall or *net* charge of +1. Increasing the pH will lead to a point where the concentration of each species is equal. This pH is equivalent to the first pK value (∼pH 2.3) and further increases in pH lead to point of inflection, where the dominant

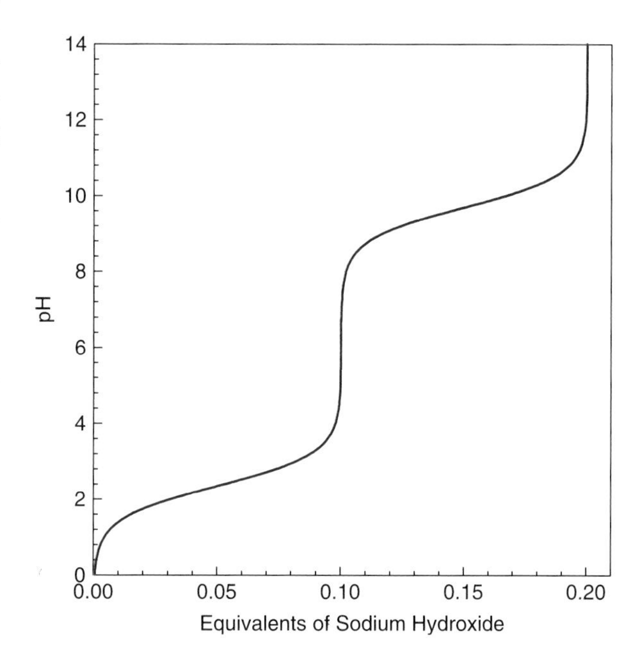

Figure 2.2 Titration curve for alanine showing changes in pH with addition of sodium hydroxide

low pH ~ 2 $H_3N^+{-}C(CH_3)(H){-}C({=}O)OH$ +1

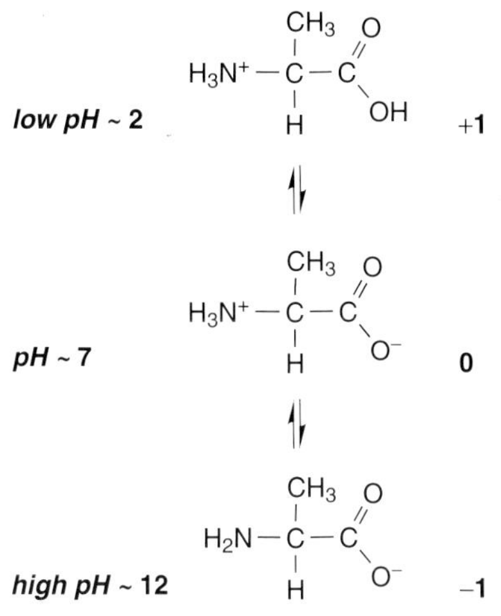

⇅

pH ~ 7 $H_3N^+{-}C(CH_3)(H){-}C({=}O)O^-$ 0

⇅

high pH ~ 12 $H_2N{-}C(CH_3)(H){-}C({=}O)O^-$ −1

Figure 2.3 The three major forms of alanine occurring in titrations between pH 1 and 14

species in solution is the zwitterion. The zwitterion, although dipolar, has no overall charge and at this pH the amino acid will not migrate towards either the anode or cathode when placed in an electric field. This pH is called the isoelectric point or pI and for alanine reflects the arithmetic mean of the two pK values $pI = (pK_1 + pK_2)/2$. Continuing the pH titration still further into alkaline conditions leads to the loss of a proton from the amino group and the formation of a species containing an overall charge of −1. The R group may contain functional groups that donate or accept protons and this leads to more complex titration curves. Amino acids showing additional pK values include aspartate, glutamate, histidine, argininine, lysine, cysteine and tyrosine (see Table 2.1).

Amino acids lacking charged side chains show similar values for pK_1 of about 2.3 that are significantly lower than the corresponding values seen in simple organic acids such as acetic acid (pK_1 ~4.7). Amino acids are stronger acids than acetic acid as a result of the electrophilic properties of the α amino group that increase the tendency for the carboxyl hydrogen to dissociate.

Stereochemical representations of amino acids

Although an amino acid is represented by the skeletal diagram of Figure 2.1 it is more revealing, and certainly more informative, to impose a stereochemical view on the arrangement of atoms. In these views an attempt is made to represent the positions in space of each atom. The amino, carboxyl, hydrogen and R groups are arranged tetrahedrally around the central α carbon (Figure 2.4).

Table 2.1 The pK values for the α-carboxyl, α-amino groups and side chains found in the individual amino acids

Amino acid	pK_1	pK_2	pK_R	Amino acid	pK_1	pK_2	pK_R
Alanine	2.4	9.9	–	Leucine	2.3	9.7	–
Arginine	1.8	9.0	12.5	Lysine	2.2	9.1	10.5
Asparagine	2.1	8.7	–	Methionine	2.1	9.3	–
Aspartic Acid	2.0	9.9	3.9	Phenylalanine	2.2	9.3	–
Cysteine	1.9	10.7	8.4	Proline	2.0	10.6	–
Glutamic Acid	2.1	9.5	4.1	Serine	2.2	9.2	–
Glutamine	2.2	9.1	–	Threonine	2.1	9.1	–
Glycine	2.4	9.8	–	Tyrosine	2.2	9.2	10.5
Histidine	1.8	9.3	6.0	Tryptophan	2.5	9.4	–
Isoleucine	2.3	9.8	–	Valine	2.3	9.7	–

Adapted from Dawson, R.M.C, Elliot, W.H., & Jones, K.M. 1986 *Data for Biochemical Research*, 3rd edn. Clarendon Press Oxford.

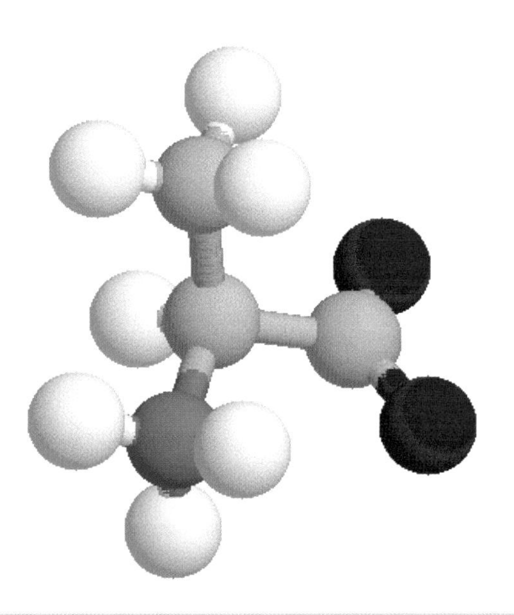

Figure 2.4 The spatial arrangement of atoms in the amino acid alanine

The nitrogen atom (blue) is part of the amino ($-NH_3^+$) group, the oxygen atoms (red) are part of the carboxyl ($-COO^-$) group. The remaining groups joined to the α carbon are one hydrogen atom and the R group.

The R group is responsible for the different properties of individual amino acids. As amino acids make up proteins the properties of the R group contribute considerably to the physical properties of proteins. Nineteen of the 20 amino acids found in proteins have the arrangement shown by Figure 2.4 but for the remaining amino acid, proline, an unusual cyclic ring is formed by the side chain bonding directly to the amide nitrogen (Figure 2.5).

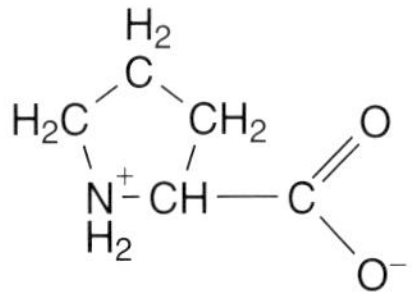

H_2
C
H_2C CH_2 O
$\overset{+}{N}$–CH—C
H_2
O^-

Figure 2.5 The structure of proline – an unusual amino acid containing a five-membered pyrrolidine ring

A glance at the structures of the 20 different side chains reveals major differences in, for example, size, charge and hydrophobicity although the R group is always attached to the α carbon (C2 carbon). From the α carbon subsequent carbon atoms in the side chains are designated as β, γ, δ, ε and ζ. In some databases of protein structures the C_β is written as CB, the C_δ as CD, C_ζ as CZ, etc. Both nomenclatures are widely used. The nomenclature is generally unambiguous but care needs to be exercised when describing the atoms of the side chain of isoleucine. Isoleucine has a branched side chain in which the C_γ or CG is either a methyl group or a methylene group. In this instance the two groups are distinguished by the use of a subscript 1 and 2, i.e. CG1 and CG2. A similar line of reasoning applies to the carbon atoms of aromatic rings. In phenylalanine, for example, the aromatic ring is linked to the C_β atom by the C_γ atom and contains two C_δ and C_ε atoms ($C_{\delta 1}$ and $C_{\delta 2}$, $C_{\varepsilon 1}$ and $C_{\varepsilon 2}$) before completing ring at the C_ζ (or CZ) atom.

Peptide bonds

Amino acids are joined together by the formation of a peptide bond where the amino group of one molecule reacts with the carboxyl group of the other. The reaction is described as a condensation resulting in the elimination of water and the formation of a dipeptide (Figure 2.6).

Three amino acids are joined together by two peptide bonds to form a tripeptide and the sequence

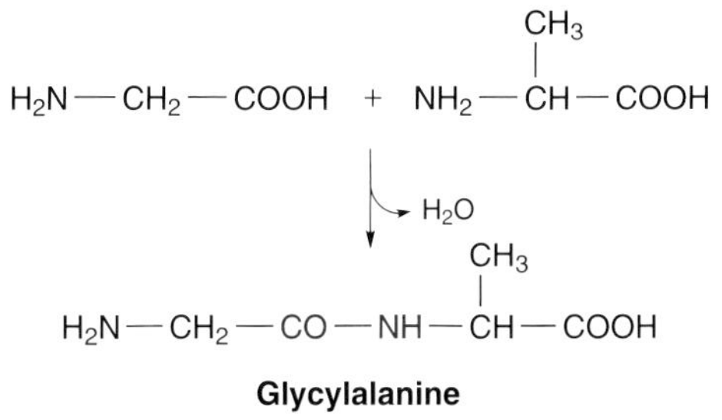

$$H_2N{-}CH_2{-}COOH + NH_2{-}\underset{}{CH}(CH_3){-}COOH \xrightarrow{-H_2O} H_2N{-}CH_2{-}CO{-}NH{-}CH(CH_3){-}COOH$$

Glycylalanine

Figure 2.6 Glycine and alanine react together to form the dipeptide glycylalanine. The important peptide bond is shown in red

continues with the formation of tetrapeptides, pentapeptides, and so on. When joined in a series of peptide bonds amino acids are called residues to distinguish between the free form and the form found in proteins. A short sequence of residues is a peptide with the term polypeptide applied to longer chains of residues usually of known sequence and length. Within the cell protein synthesis occurs on the ribosome but today peptide synthesis is possible *in vitro* via complex organic chemistry. However, whilst the organic chemist struggles to synthesize a peptide containing more than 50 residues the ribosome routinely makes proteins with over 1000 residues.

All proteins are made up of amino acid residues linked together in an order that is ultimately derived from the information residing within our genes. Some proteins are clearly related to each other in that they have similar sequences whilst most proteins exhibit a very different composition of residues and a very different order of residues along the polypeptide chain. From the variety of side chains a single amino acid can link to 19 others to create a total of 39 different dipeptides. Repeating this for the other residues leads to a total of 780 possible dipeptide permutations. If tripeptides and tetrapeptides are considered the number of possible combinations rapidly reaches a very large figure. However, when databases of protein sequences are studied it is clear that amino acid residues do not occur with equal frequency in proteins and sequences do not reflect even a small percentage of all possible combinations. Tryptophan and cysteine are rare residues (less than 2 percent of all residues) in proteins whilst alanine, glycine and leucine occur with frequencies between 7 and 9 percent (see Table 2.2).

Amino acid sequences of proteins are read from left to right. This is from the amino or N terminal to the carboxyl or C terminal. The individual amino acids have three-letter codes, but increasingly, in order to save space in the presentation of long protein sequences, a single-letter code is used for each amino acid residue. Both single- and three-letter codes are shown alongside the R groups in Table 2.2 together with some of the relevant properties of each side chain. Where possible the three-letter codes for amino acids will be used but it should be stressed that single letter codes avoid potential confusion. For example Gly, Glu and Gln are easily mistaken when rapidly reading protein sequences but their single letter codes of G, E and Q are less likely to be misunderstood.

Joining together residues establishes a protein sequence that is conveniently divided into main chain and side chain components. The main chain, or polypeptide backbone, has the same composition in all proteins although it may differ in extent – that is the number of residues found in the polypeptide chain. The backbone represents the effective repetition of peptide bonds made up of the N, C_α and C atoms, with proteins such as insulin having approximately 50 residues whilst other proteins contain over 1000 residues and more than one polypeptide chain (Figure 2.7). Whilst all proteins link atoms of the polypeptide backbone similarly the side chains present a variable component in each protein.

Properties of the peptide bond

The main chain or backbone of the polypeptide chain is established by the formation of peptide bonds between amino acids. The backbone consists of the amide N, the α-carbon and the carbonyl C linked together (Figure 2.8).

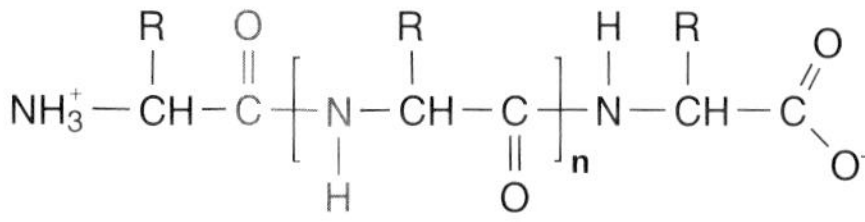

Figure 2.7 Part of a polypeptide chain formed by the covalent bonding of amino acids where n is often 50–300, although values above and below these limits are known.

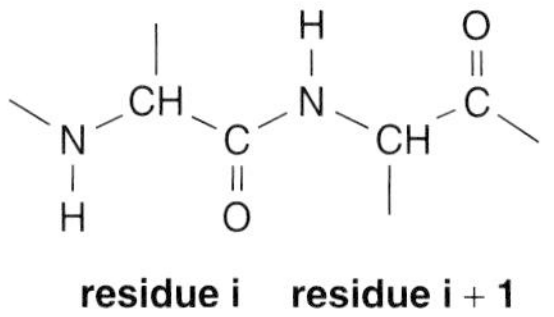

Figure 2.8 The polypeptide backbone showing arrangement of i, i + 1 residues within a chain.

Table 2.2 The frequencies with which amino acid residues occur in proteins

Amino acid	Property of individual amino acid residues	Ball and stick representation of each amino acid
Alanine A Ala M_r 71.09	Non-polar side chain. Small side chain volume. Van der Waals volume = 67 $Å^{3*}$ Frequency in proteins = 7.7 % Surface area = 115 $Å^2$ Unreactive side chain	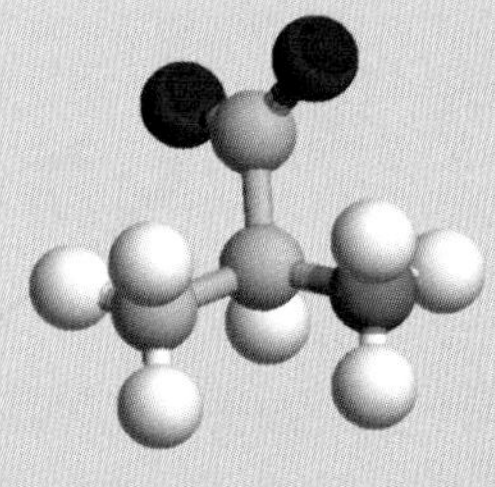
Arginine R Arg M_r 156.19	Positively charged side chain at pH 7.0. pK for guanidino group in proteins ~12.0 Van der Waals volume = 167 $Å^3$ Frequency in proteins = 5.1 % Surface area = 225 $Å^2$ Participates in ionic interactions with negatively charged groups	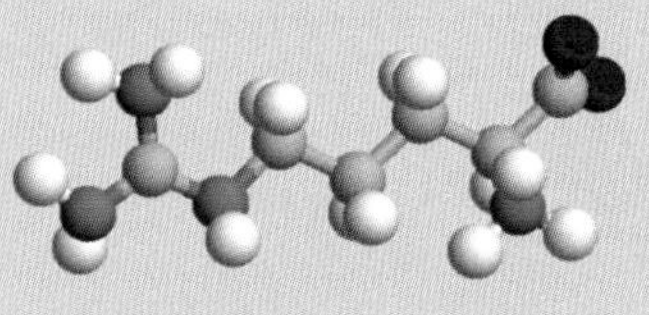
Asparagine N Asn M_r 114.11	Polar, but uncharged, side chain Van der Waals volume = 148 $Å^3$ Frequency in proteins = 4.3 % Surface area = 160 $Å^2$ Polar side chain will hydrogen bond Relatively small side chain volume leads to this residue being found relatively frequently in turns	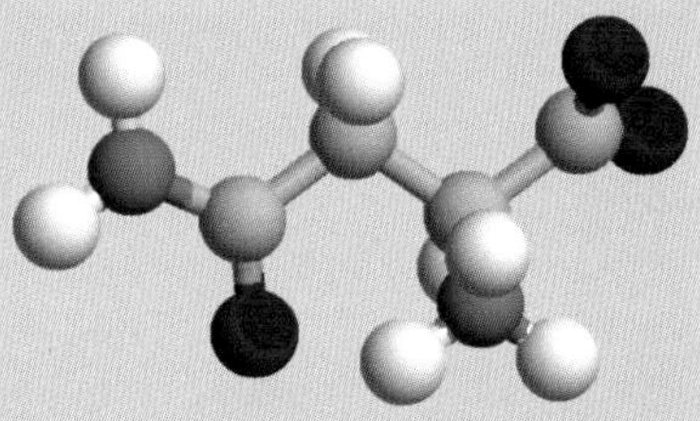
Aspartate D Asp M_r 115.09	Negatively charged side chain pK for side chain of ~4.0 Van der Waals volume = 67 $Å^3$ Frequency in proteins = 5.2 % Surface area = 150 $Å^2$ Charged side chain exhibits electrostatic interactions with positively charged groups.	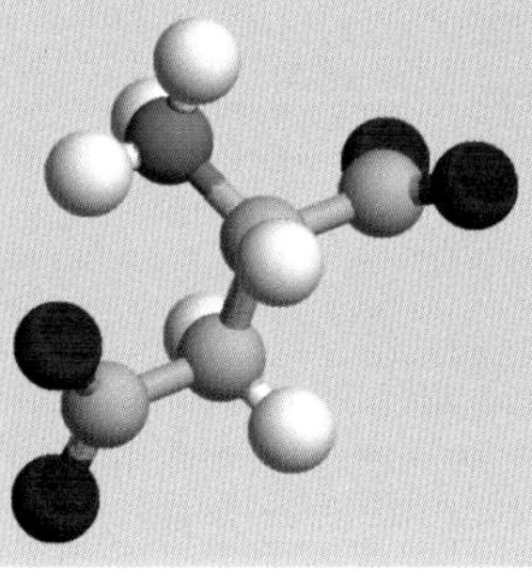

Table 2.2 (*continued*)

Amino acid	Property of individual amino acid residues	Ball and stick representation of each amino acid
Cysteine C Cys M_r 103.15	Side chain contains thiol (SH) group. Van der Waals volume = 86 Å^3 Frequency in proteins = 2.0 % Surface area = 135 Å^2 Thiol side chain has pK in isolated amino acid of ~8.5 but in proteins varies 5–10 Thiol group is very reactive	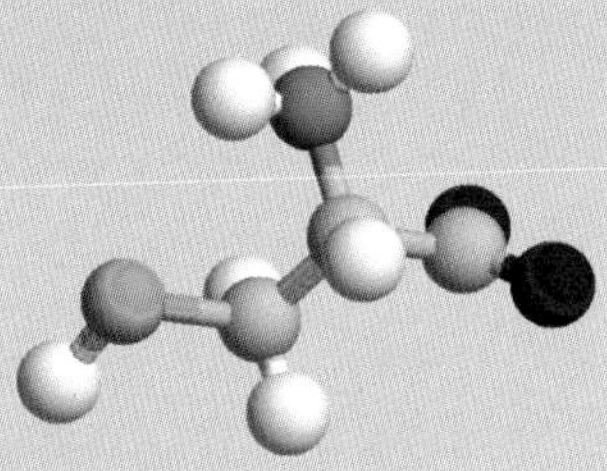
Glutamine Q Gln M_r 128.12	Polar but uncharged side chain Van der Waals volume = 114 Å^3 Frequency in proteins = 4.1 % Surface area = 180 Å^2 Polar side chain can hydrogen bond	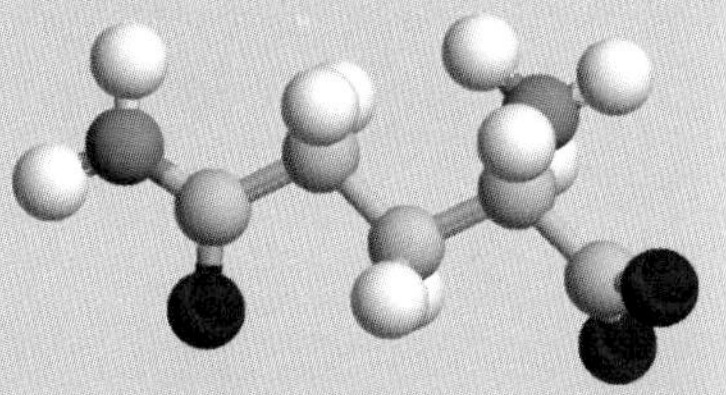
Glutamate E Glu M_r 129.12	Negatively charged side chain. Van der Waals volume = 109 Å^3 Frequency in proteins = 6.2 % Surface area = 190 Å^2 Side chain has pK of ~4.5.	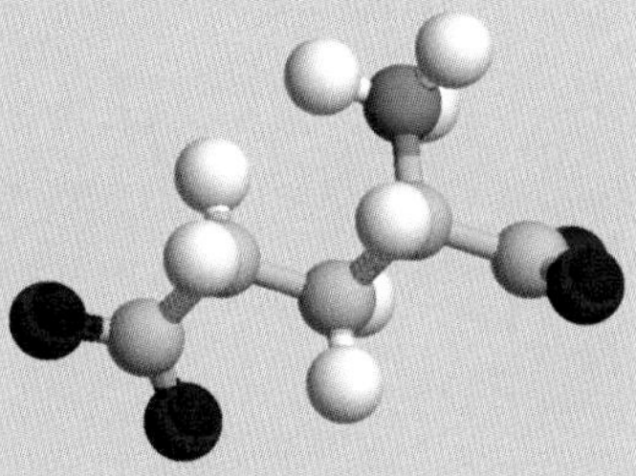
Glycine G Gly M_r 57.05	Uncharged, small side chain. Often found in turn regions of proteins or regions of conformational flexibility No chiral centre; due to two hydrogens attached to C_α centre Van der Waals volume = 48 Å^3 Frequency in proteins = 7.4 % Surface area = 75 Å^2	

(*continued overleaf*)

Table 2.2 (*continued*)

Amino acid	Property of individual amino acid residues	Ball and stick representation of each amino acid
Histidine H His M_r 137.14	Imidazole side chain Van der Waals volume = 118 $Å^3$ Frequency in proteins = 2.3 % Surface area = 195 $Å^2$ The side chain exhibits a pK ~ 6.0 in model peptides but in proteins can vary from 4–10	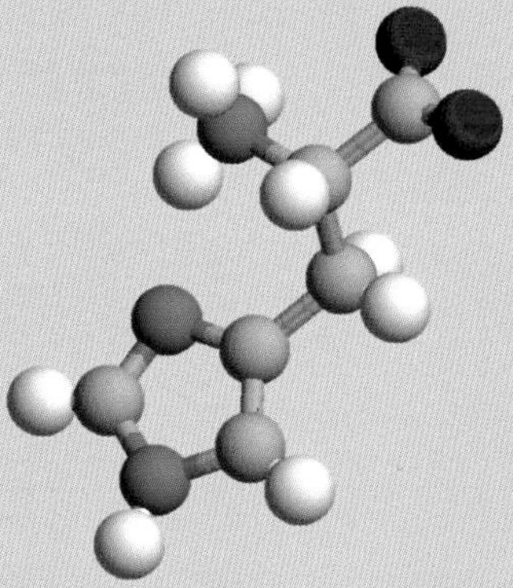
Isoleucine I Ile M_r 113.16	Hydrophobic side chain exhibiting non-polar based interactions but generally unreactive Van der Waals volume = 124 $Å^3$ Frequency in proteins = 5.3 % Surface area = 175 $Å^2$	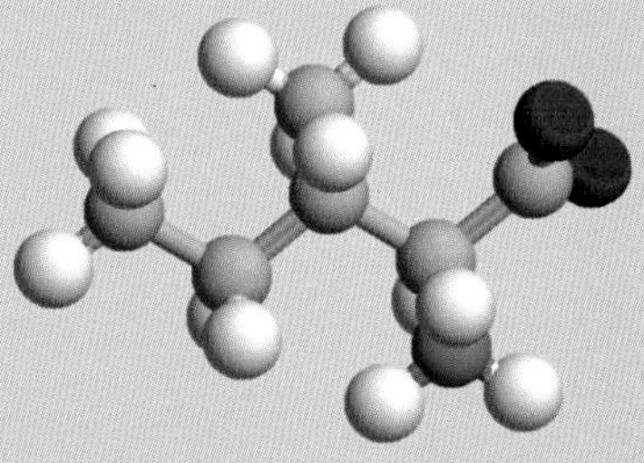
Leucine L Leu M_r 113.16	Hydrophobic side chain Van der Waals volume = 124 $Å^3$ Frequency in proteins = 8.5 % Surface area = 170 $Å^2$	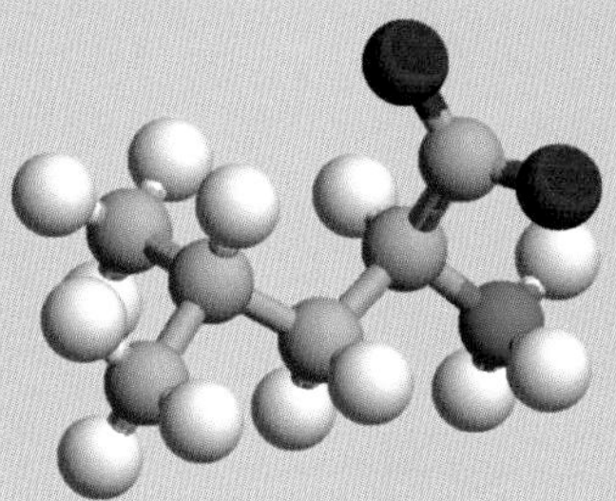
Lysine K Lys M_r 128.17	Positively charged side chain Van der Waals volume = 135 $Å^3$ Frequency in proteins = 5.9 % Surface area = 200 $Å^2$ Side chain is basic with pK of ~10.5. Shows ionic interactions	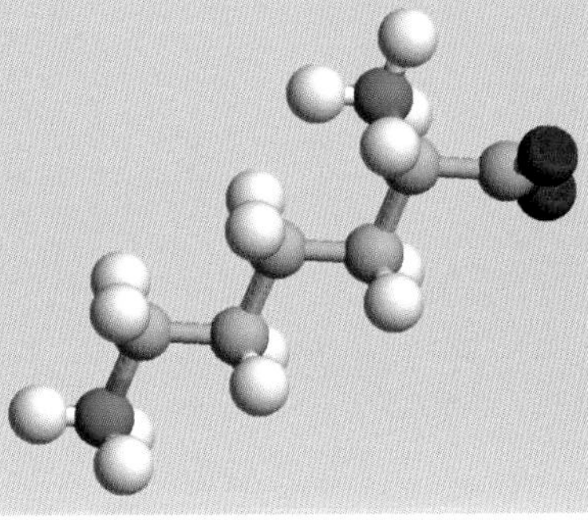

Table 2.2 (*continued*)

Amino acid	Property of individual amino acid residues	Ball and stick representation of each amino acid
Methionine M Met M_r 131.19	Sulfur containing hydrophobic side chain The sulfur is unreactive especially when compared with thiol group of cysteine Van der Waals volume = 124 Å^3 Frequency in proteins = 2.4 % Surface area = 185 Å^2	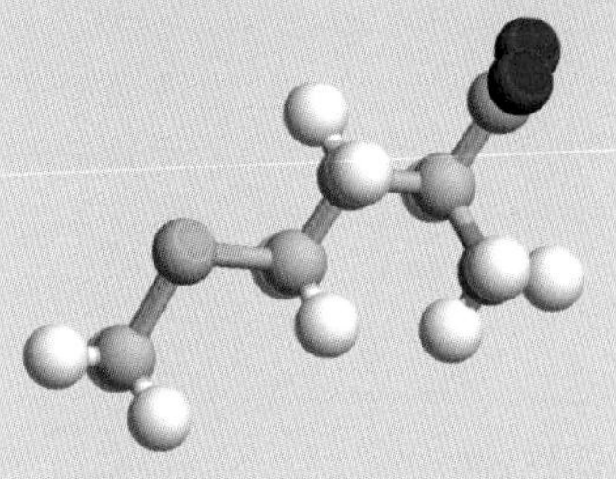
Phenylalanine F Phe M_r 147.18	Hydrophobic, aromatic side chain Phenyl ring is chemically unreactive in proteins. Exhibits weak optical absorbance around 280 nm Van der Waals volume = 135 Å^3 Frequency in proteins = 4.0 % Surface area = 210 Å^2	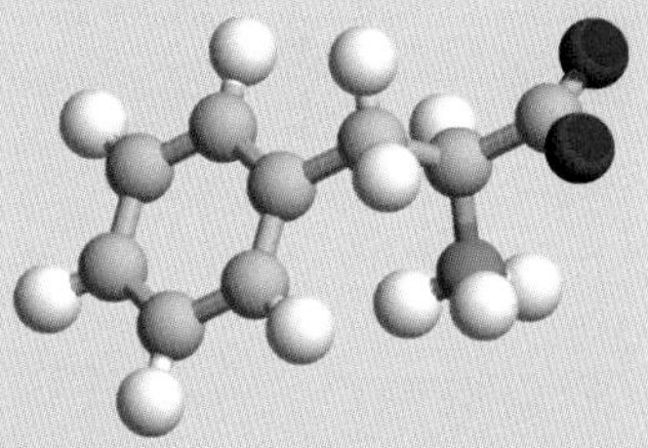
Proline P Pro M_r 97.12	Cyclic ring forming hydrophobic side chain The cyclic ring limits conformational flexibility around N-C_α bond In a polypeptide chain lacks amide hydrogen and cannot form backbone hydrogen bonds Van der Waals volume = 90 Å^3 Frequency in proteins = 5.1 % Surface area = 145 Å^2	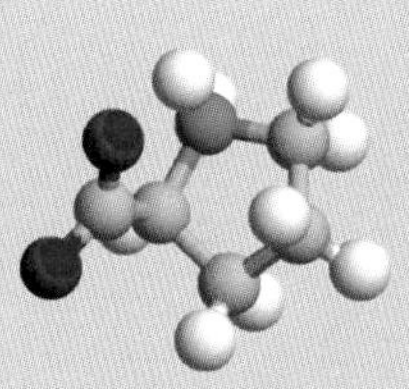
Serine S Ser M_r 87.08	Polar but uncharged side chain. Contains hydroxyl group (–OH) that hydrogen bonds Oxygen atom can act as potent nucleophile in some enzymes Van der Waals volume = 73 Å^3 Frequency in proteins = 6.9 % Surface area = 115 Å^2	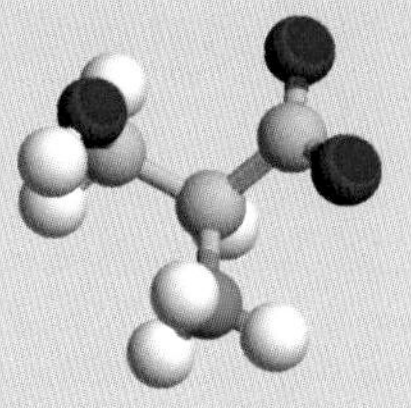

(*continued overleaf*)

Table 2.2 *(continued)*

Amino acid	Property of individual amino acid residues	Ball and stick representation of each amino acid
Threonine T Thr M_r 101.11	Polar but uncharged side chain. Contains hydroxyl group (–OH) Hydrogen bonding side chain Van der Waals volume = 93 Å^3 Frequency in proteins = 5.9 % Surface area = 140 Å^2	
Tryptophan W Trp M_r 186.21	Large, hydrophobic and aromatic side chain Almost all reactivity is based around the indole ring nitrogen Responsible for majority of near uv absorbance in proteins at 280 nm Van der Waals volume = 163 Å^3 Frequency in proteins = 1.4 % Surface area = 255 Å^2	
Tyrosine Y Tyr M_r 163.18	Aromatic side chain Van der Waals volume = 141 Å^3 Frequency in proteins = 3.2 % Surface area = 230 Å^2 Phenolic hydroxyl group ionizes at pH values around pH 10 Aromatic ring more easily substituted than that of phenylalanine	
Valine V Val M_r 99.14	Hydrophobic side chain Van der Waals volume = 105 Å^3 Frequency in proteins = 6.6 % Surface area = 155 Å^2	

From Jones, D.T. Taylor, W.R. & Thornton, J.M. (1991) *CABIOS* **8**, 275–282. Databases of protein sequences are weighted towards globular proteins but with the addition of membrane proteins to databases a gradual increase in the relative abundance of hydrophobic residues such as Leu, Val, Ile, Phe, Trp is expected. The surface area was calculated for an accessible surface of residue X in the tripeptide G-X-G (Chothia, C. (1975) *J. Mol. Biol.*, **105**, 1–14). Volumes enclosed by the van der Waals radii of atoms as described by Richards, F.M. (1974) *J. Mol. Biol.* **82**, 1–14.

*1 Å = 0.1 nm.

The linear representation of the polypeptide chain does not convey the intricacy associated with the bond lengths and angles of the atoms making up the peptide bond. The peptide bond formed between the carboxyl and amino groups of two amino acids is a unique bond that possesses little intrinsic mobility. This occurs because of the partial double bond character (Figure 2.9)–a feature associated with the peptide bond and resonance between two closely related states.

One of the most important consequences of resonance is that the peptide bond length is shorter than expected for a simple C–N bond. On average a peptide bond length is 1.32 Å compared to 1.45 Å for an ordinary C–N bond. In comparison the average bond length associated with a C=N double bond is 1.25 Å, emphasizing the intermediate character of the peptide bond. More importantly the partial double bond between carbon and nitrogen atoms restricts rotation about this bond. This leads to the six atoms shown in Figure 2.9 being coplanar; that is all six atoms are found within a single imaginary plane. For any polypeptide backbone represented by the sequence –N–C_α–C–N–C_α–C–only the C_α–C and N–C_α bonds exhibit rotational mobility. As a result of restricted motion about the peptide bond two conformations related by an angle of 180° are possible. The first occurs when the C_α atoms are *trans* to the peptide bond whilst the second and less favourable orientation occurs when the C_α atoms are *cis* (Figure 2.10).

Figure 2.9 The peptide bond may be viewed as a partial double bond as a result of resonance

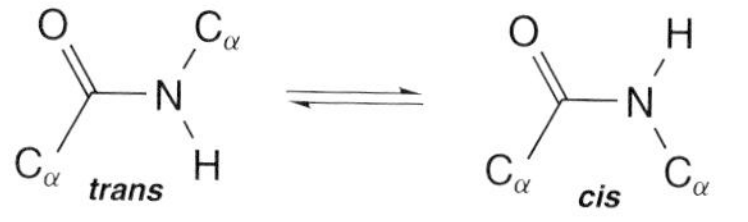

Figure 2.10 *Cis* and *trans* configurations are possible about the rigid peptide bond

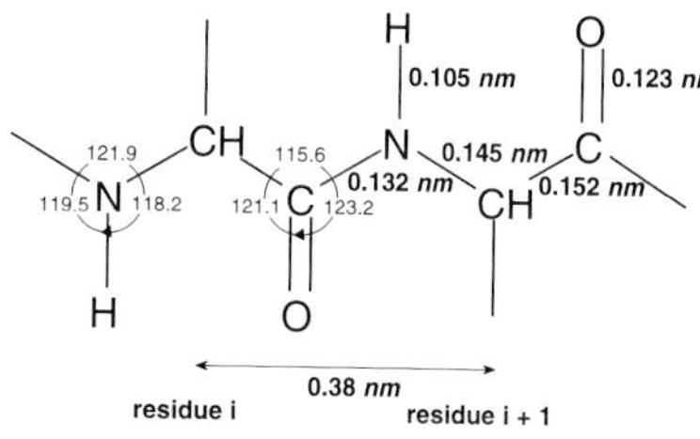

Figure 2.11 Detailed bond lengths and angles for atoms of the polypeptide backbone

The *trans* form is the more favoured state because in this arrangement repulsion between non-bonded atoms connected to the C_α centre are minimized. For most peptide bonds the ratio of *cis* to *trans* configurations is approximately 1:1000. However, one exception to this rule is found in peptide bonds where the following residue is proline. Proline, unusual in having a cyclic side chain that bonds to the backbone amide nitrogen, has less repulsion between side chain atoms. This leads to an increase in the relative stability of the *cis* peptide bond when compared with the *trans* state, and for peptide bonds formed between Xaa and proline (Xaa is any amino acid) the *cis* to *trans* ratio is 1:4.

The dimensions associated with the peptide bond have generally been obtained from crystallographic studies of small peptides and besides the peptide bond length of 0.132 nm other characteristic bond lengths and angles have been identified (Figure 2.11). One of the most important dimensions is the maximum distance between corresponding atoms in sequential residues. In the *trans* peptide bond and a fully extended conformation this distance is maximally 0.38 nm although in proteins it is often much less.

The chemical and physical properties of amino acids

Chemical reactions exhibited by the 20 amino acids are extensive and revolve around the reactivity of amino and carboxyl groups. In proteins, however, these groups are involved in peptide bonds and the defining properties of amino acids are those associated with side chains.

Glycine

Glycine is the simplest amino acid containing two hydrogen atoms attached to the central α carbon. Consequently it lacks an asymmetric centre and does not occur as R/S isomers (see below). More importantly the absence of any significant functional group means that glycine possesses little intrinsic chemical reactivity. However, the absence of a large side chain results in conformational flexibility about the N–C_α and C_α–C bonds in a polypeptide chain. This fact alone has important consequences for the overall structure of proteins containing significant numbers of glycine residues.

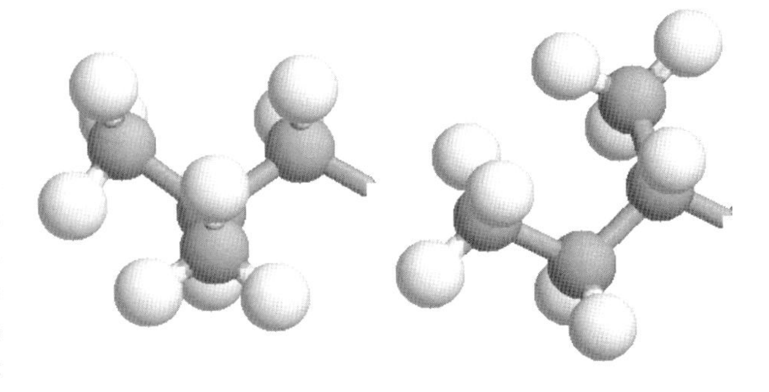

Figure 2.13 The structures of the side chains of leucine (left) and isoleucine (right)

Aliphatic side chains: alanine, valine, isoleucine and leucine

Alanine, like glycine, has a very simple side chain – a single methyl group (CH_3) that is chemically inert and inactive (Figure 2.12). It is joined in this remarkable non-reactivity by the side chains of valine, leucine and isoleucine (Figure 2.13). The side chains of these amino acids are related in containing methine (CH) or methylene (–CH_2) groups in branched aliphatic chains terminated by methyl groups. However, one very important property possessed by these side chains is their unwillingness to interact with water – their hydrophobicity. The side chains interact far more readily with each other and the other non-polar side chains of amino acids such as tryptophan or phenylalanine. In later chapters on tertiary structure, protein stability and folding the important role of the weak hydrophobic interaction in maintaining the native state of proteins is described. Alanine, valine, leucine and isoleucine will all exhibit hydrophobic interactions. Isoleucine and leucine as their names suggest are isomers differing in the arrangement of methylene and methyl groups. The first carbon atom in the side chain (CB) is a methylene (CH_2) group in leucine whilst in isoleucine it is a methine (–CH–) moiety. The CB of leucine is linked to a methine group (CG) whilst in isoleucine the CB atom bonds to a methyl group *and* a methylene group. The side chain of valine (Figure 2.14) has one methylene group less than leucine but is otherwise very similar in property.

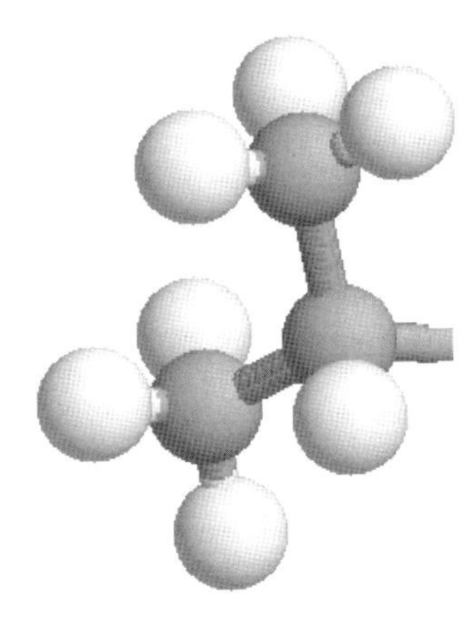

Figure 2.14 The side chain of valine is branched at the CB (–CH or methine) group and leads to two methyl groups

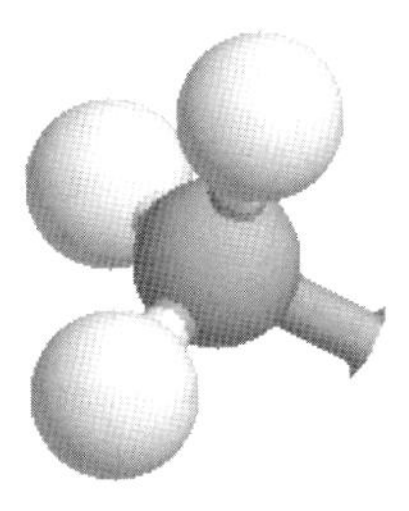

Figure 2.12 The side chain of alanine is the unreactive methyl group (–CH_3)

The hydroxyl-containing residues: serine and threonine

The side chains of serine and threonine are characterized by the presence of a polar hydroxyl (–OH) group (Figure 2.15). Their side chains are generally small (Table 2.2). Although small the side groups of both serine and threonine hydrogen bond with other residues in proteins whilst in an isolated state their reactivity is confined to that expected for a primary alcohol, with esterification representing a common reaction with organic acids. One of the most important reactions to a serine or threonine side chain is the addition of a phosphate group to create phosphoserine or phosphothreonine. This occurs as a post-translational modification in the cell after protein synthesis on the ribosome and is important in protein–protein interactions and intracellular signalling pathways.

In one group of enzymes – the serine proteases – the activity of a single specific serine side chain is enhanced by its proximity to histidine and aspartyl side chains. By itself the side chain of Ser is a weak nucleophile but combined with His and Asp this triad of residues becomes a potent catalytic group capable of splitting peptide bonds.

The acidic residues: aspartate and glutamate

Aspartic acid and glutamic acid have side chains with a carboxyl group (Figure 2.16). This leads to the side chain having a negative charge under physiological conditions and hence to their description as acidic side chains. For this reason the residues are normally referred to as aspartate and glutamate reflecting the ionized and charged status under most cellular conditions. The two side chains differ by one methylene group (CH_2) but are characterized by similar pK values (range 3.8–4.5).

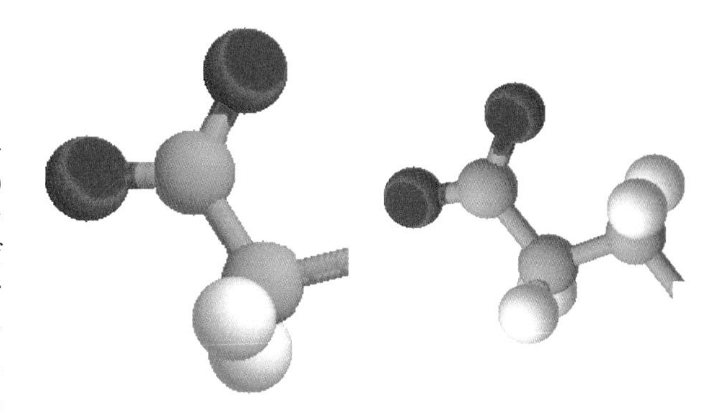

Figure 2.16 The acidic side chains of aspartate and glutamate

The side chains behave as typical organic acids and exhibit a wide range of chemical reactions including esterification with alcohols or coupling with amines. Both of these reactions have been exploited in chemical modification studies of proteins to alter one or more aspartate/glutamate side chains with a loss of the negatively charged carboxyl group. As expected the side chains of aspartate and glutamate are potent chelators of divalent metal ions and biology exploits this property to bind important ions such as Ca^{2+} in proteins such as calmodulin or Zn^{2+} in enzymes such as carboxypeptidase.

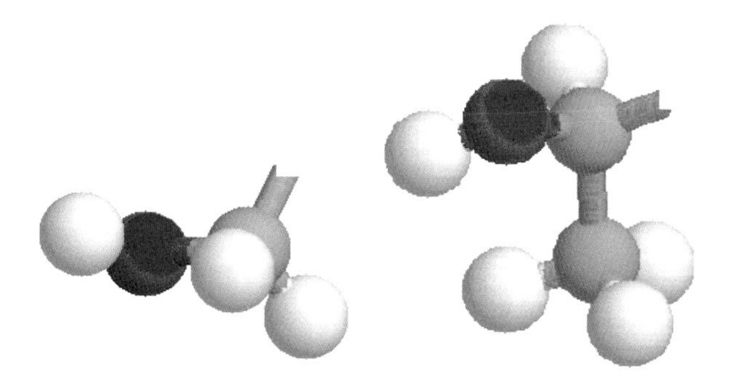

Figure 2.15 The side chains of serine and threonine are aliphatic containing hydroxyl groups

The amide-containing residues: asparagine and glutamine

Asparagine and glutamine residues are often confused with their acidic counterparts, the side chains of aspartate and glutamate. Unlike the acidic side chains the functional group is an amide – a generally unreactive group that is polar and acts as hydrogen bond donor and acceptor (see Figure 2.17). The amide group is labile at alkaline pH values, or extremes of temperature, being deamidated to form the corresponding acidic side

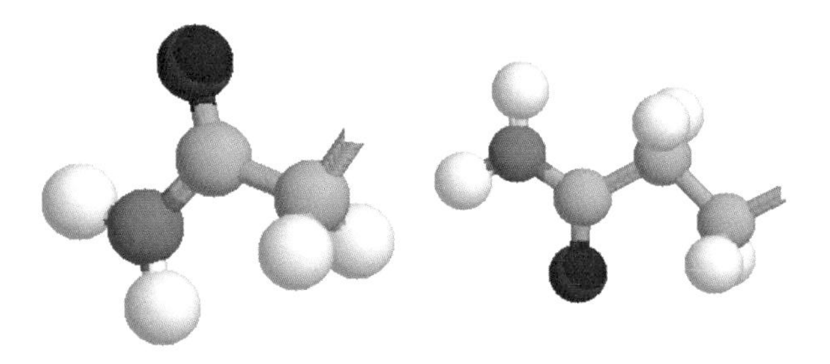

Figure 2.17 The uncharged side chains of asparagine and glutamine

chain. This reaction can occur during protein isolation and occasionally leads to protein sequences containing the three letter code Glx or Asx where the identity of a residue, either Gln or Glu and Asp or Asn, is unclear.

The sulfur-containing residues: cysteine and methionine

Sulfur occurs in two of the 20 amino acids: cysteine and methionine. In cysteine the sulfur is part of a reactive thiol group whilst in methionine sulfur is found as part of a long, generally unreactive, side chain. The side chain of methionine is non-polar (Figure 2.18) and is larger than that of valine or leucine although it is unbranched. Its properties are dominated by the presence of the sulfur atom, a potent nucleophile under certain conditions.

The sulfur atom is readily methylated using methyl iodide in a reaction that is often used to introduce a 'label' onto methionine residue via the use of ^{13}C labelled reactant. In addition the sulfur of methionine interacts with heavy metal complexes particularly those involving mercury and platinum such as K_2PtCl_4 or

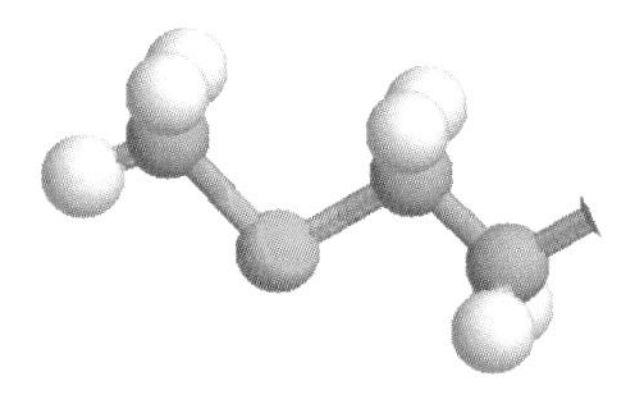

Figure 2.18 The side chain of methionine

$$-CH_2-CH_2-S-CH_3$$

$$\downarrow$$

$$-CH_2-CH_2-\overset{\displaystyle O}{\overset{\|}{S}}-CH_3$$

$$\downarrow$$

$$-CH_2-CH_2-\overset{\displaystyle O}{\overset{\|}{\underset{\underset{\displaystyle O}{\|}}{S}}}-CH_3$$

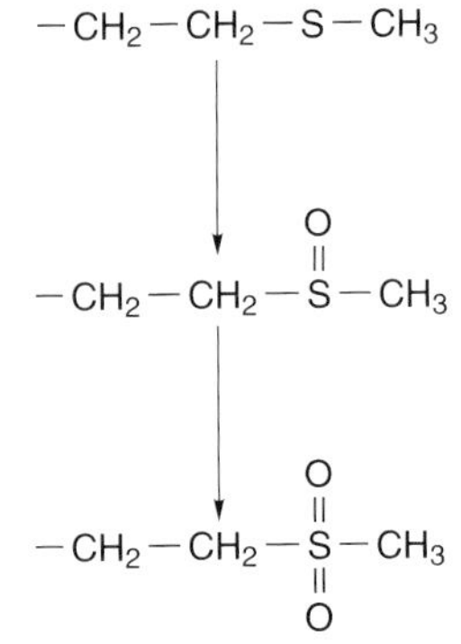

Figure 2.19 Oxidation of methionine side chains by strong oxidizing agents

$HgCl_2$ and these have proved extremely useful in the formation of isomorphous heavy atom derivatives in protein crystallography (see Chapter 10). The sulfur atom of methionine can be oxidized to form first a sulfoxide and finally a sulfone derivative (Figure 2.19). This form of oxidative damage is known to occur in proteins and the reaction scheme involves progressive addition of oxygen atoms.

One of the most important reactions of methionine involves cyanogen bromide – a reagent that breaks the polypeptide chain on the C-terminal side of methionine residues by sequestering the carbonyl group of the next peptide bond in a reaction involving water and leading to formation of a homoserine lactone (Figure 2.20). This reaction is used to split polypeptide chains into smaller fragments for protein sequencing. When compared with cysteine, however, the side chain of methionine is less reactive undergoing comparatively few important chemical reactions.

The functional group of cysteine is the thiol (–SH) group, sometimes called the sulfhydryl or mercapto group. It is the most reactive side chain found amongst the 20 naturally occurring amino acids undergoing many chemical reactions with diverse reagents (Figure 2.21). Some enzymes exploit reactivity by using a conserved cysteine residue at their active sites that participates directly in enzyme-catalysed reactions.

The large sulfur atom as part of the thiol group influences side chain properties significantly, with disulfide bonds forming between cysteine residues that are close

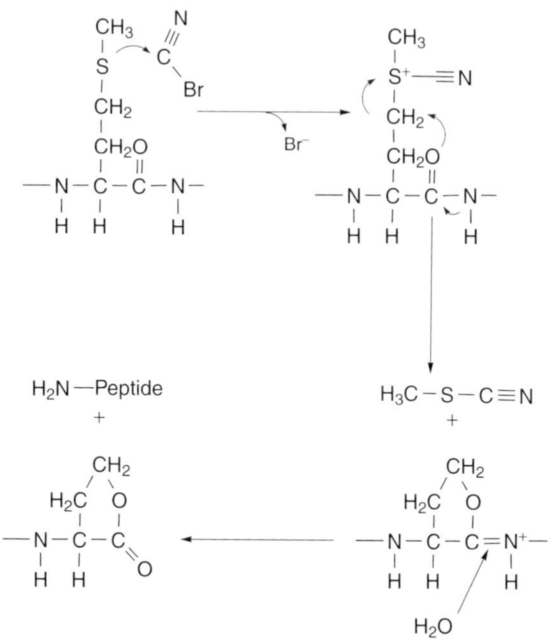

Figure 2.20 The reaction of cyanogen bromide with methionine residues

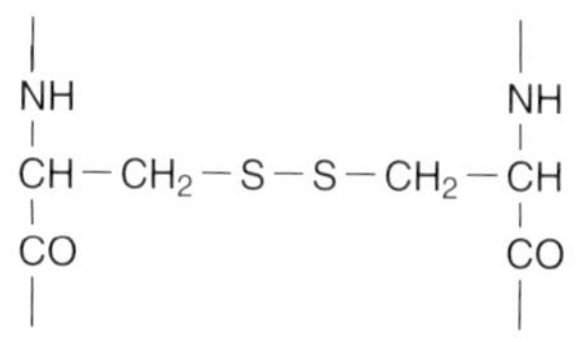

Figure 2.22 Formation of a disulfide bridge between two thiol side chains

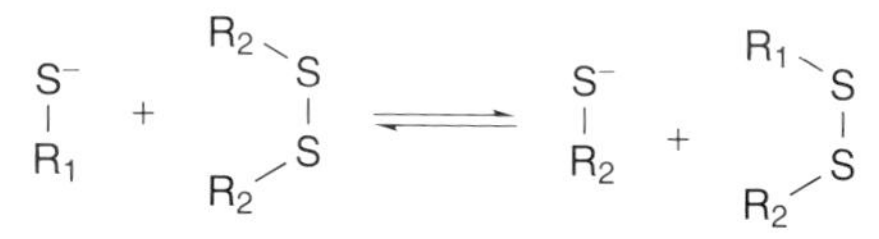

Figure 2.23 The formation of mixed disulfides via the interaction of thiolate groups. A thiolate anion reacts with other symmetrical disulfides to form a mixed disulfide formed between R_1 and R_2

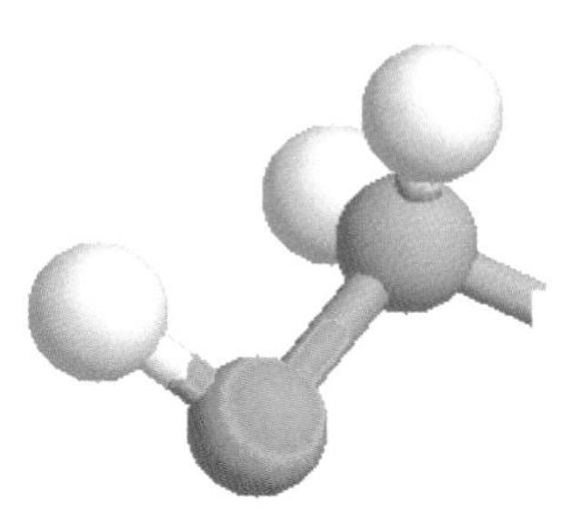

Figure 2.21 The thiol group in the side chain of cysteine is the most reactive functional group found in amino acid residues

together in space. This forms a strong covalent bond and exercises considerable conformational restraint on the structure adopted in solution by polypeptides. Formation of a disulfide bridge is another example of post-translational modification (Figure 2.22).

The thiol group ionizes at alkaline pH values (~pH 8.5) to form a reactive thiolate anion (S^-). The thiolate anion reacts rapidly with many compounds but the most important includes other thiols or disulfides in exchange type reactions occurring at neutral to alkaline pH values. A general reaction scheme for thiol-disulfide exchange is given in Figure 2.23.

A common reaction of this type is the reaction between cysteine and Ellman's reagent (dithionitrobenzoic acid, DTNB; Figure 2.24). The aromatic disulfide undergoes exchange with reactive thiolate anions forming a coloured aromatic thiol – nitrothiobenzoate. The benzoate anion absorbs intensely at 416 nm allowing the concentration of free thiol groups to be accurately estimated in biological systems.

Thiols are also oxidized by molecular oxygen in reactions catalysed by trace amounts of transition metals, including Cu and Fe. More potent oxidants such as performic acid oxidize the thiol group to a sulfonate (SO_3^{2-}) and this reaction has been exploited as a method of irreversibly breaking disulfide bridges to form two cysteic acid residues. More frequently the disulfide bridge between cysteine residues is broken by reducing agents that include other thiols such as mercaptoethanol or dithiothreitol (Figure 2.25) as well as more conventional reductants such as sodium borohydride or molecular hydrogen. Dithiothreitol,

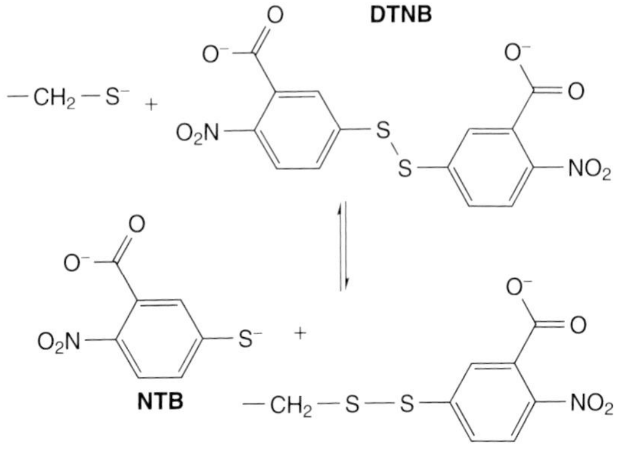

Figure 2.24 The reaction between the thiolate anion of cysteine and Ellman's reagent

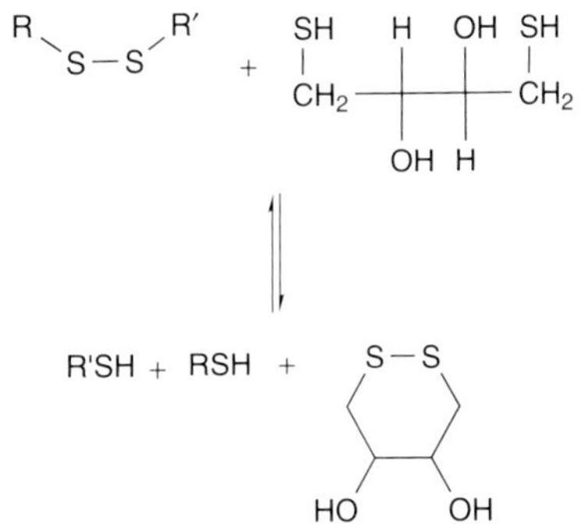

Figure 2.25 The reaction between dithiothreitol and disulfide groups leads to reduction of the disulfide and the formation of two thiols

sometimes called Cleland's reagent, reacts with disulfide groups to form initially a mixed disulfide intermediate but this rapidly rearranges to yield a stable six-membered ring and free thiol groups (–SH). The equilibrium constant for the reaction lies over to the right and is largely driven by the rapid formation of cyclic disulfide and its inherent stability.

The basic residues: lysine and arginine

The arginine side chain (Figure 2.26) contains three methylene groups followed by the basic guanidino (sometimes called guanadinium) group, which is usually protonated, planar and with the carbon atom exhibiting sp^2 hybridization.

The guanidino group is the most basic of the side chains found in amino acids with a pK of 12 and under almost all conditions the side chain retains a net positive charge (Figure 2.27). The positive charge is distributed over the entire guanidino group as a result of resonance between related structures

Lysine possesses a long side chain of four methylene groups terminated by a single ε amino group (Figure 2.28). The amino group ionizes with a pK of approximately 10.5–11.0 and is very basic. As expected the side chain interacts strongly in proteins with oppositely charged side chains but will also undergo methylation, acetylation, arylation and acylation. Many of these reactions are performed at high pH (above pH 9.0) since the unprotonated nitrogen is a potent nucleophile reacting rapidly with suitable reagents. One of the most popular lysine modifications

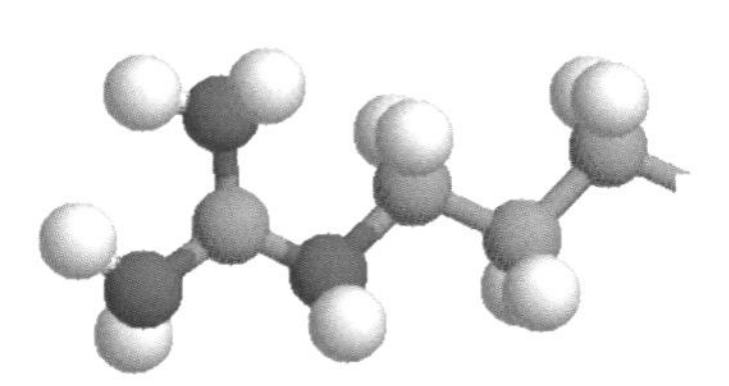

Figure 2.26 The side chain of arginine is very basic and contains a guanidino group

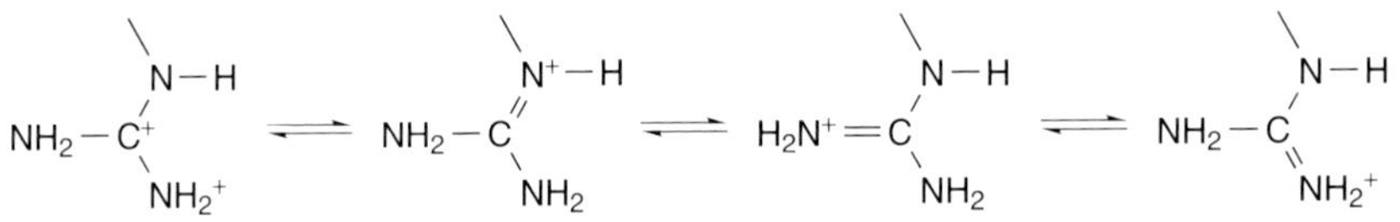

Figure 2.27 Charge delocalization and isomerization within the guanidino group

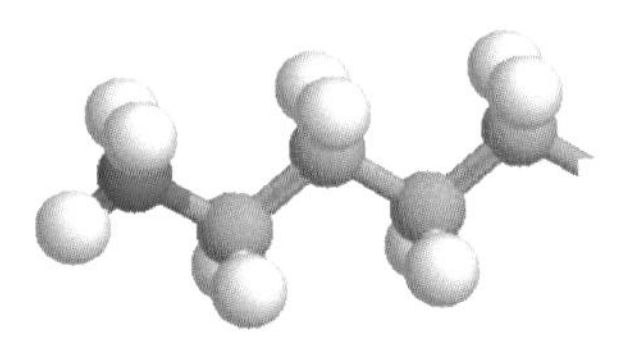

Figure 2.28 The lysine side chain is basic

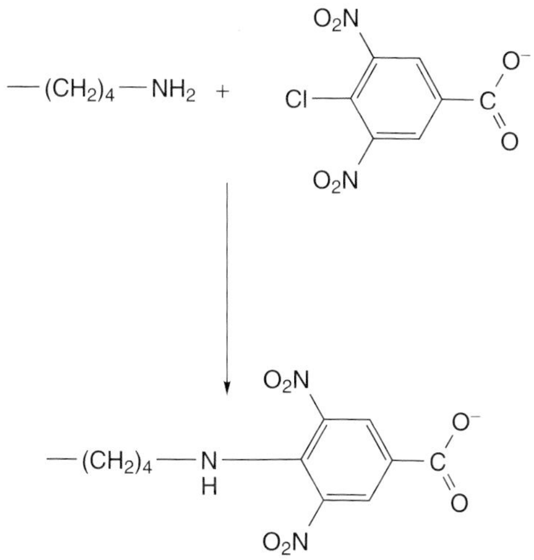

Figure 2.29 Reaction of lysine side chains with dinitrobenzene derivatives

involves adding a nitrobenzene derivative to the ε-amino group. The nitrobenzene group is coloured and the rate of the reaction is therefore comparatively easily followed spectrophotometrically. In the example shown in Figure 2.29 4-chloro-3, 5-dinitrobenzoic acid reacts with lysine side chains to form a negatively charged dinitrophenol derivative. Methylation, unlike the arylation reaction described above, preserves the positive charge on the side chain and in some systems, particularly fungi, trimethylated lysine residues are found as natural components of proteins.

One of the most important reactions occurring with lysine side chains is the reaction with aldehydes to form a Schiff base (see Figure 2.30). The reaction is important within the cell because pyridoxal phosphate, a co-factor derived from vitamin B_6, reacts with the ε amino group of lysine and is found in many enzyme active sites. Pyridoxal phosphate is related to vitamin B_6, pyridoxine.

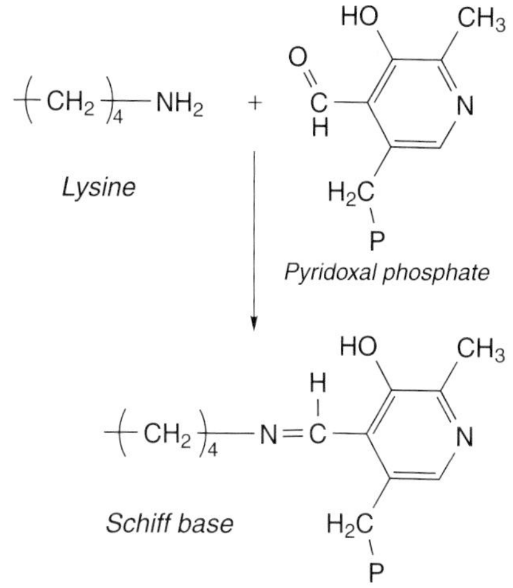

Figure 2.30 Reaction between pyridoxal phosphate and lysine side chains results in formation of a covalent Schiff base intermediate. The aldehyde group of pyridoxal phosphate links with the ε amino group of a specific lysine residue at the active site of many enzymes

Proline

The side chain of proline is unique in possessing a side chain that covalently bonds with the backbone nitrogen atom to form a cyclic pyrrolidine ring with groups lacking reactivity. One of the few reactions involving prolyl side chains is enzyme-catalysed hydroxylation. The cyclic ring imposes rigid constraints on the N-C_α bond leading to pronounced effects on the configuration of peptide bonds preceding proline. In addition the ring is puckered with the C_γ atom displaced by 0.5 Å from the remaining atoms of the ring which show approximate co-planarity.

Histidine

The imidazole side chain of histidine (Figure 2.31) is unusual and alone amongst the side chains of amino acids in exhibiting a pK around 7.0. In its ionized

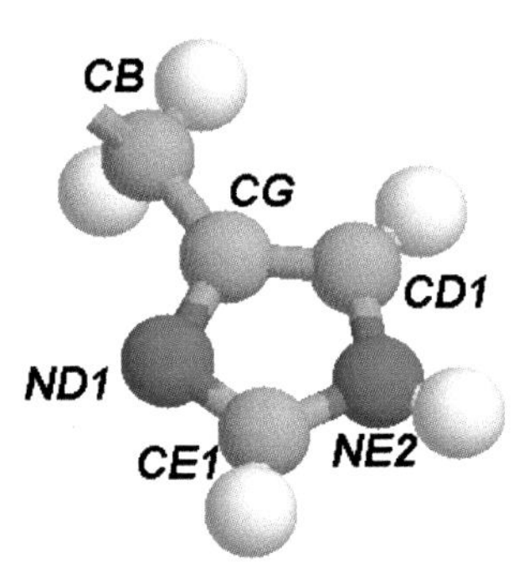

Figure 2.31 The side chain of histidine showing nomenclature for side chain atoms. The IUPAC scheme encourages the use of *pros* (meaning 'near' and abbreviated with the symbol π) and *tele* (meaning 'far', abbreviated with the symbol τ) to show their position relative to the main chain

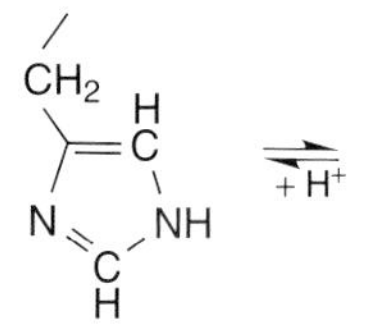

Figure 2.32 Protonation of the imidazole side chain leads to a positively charge

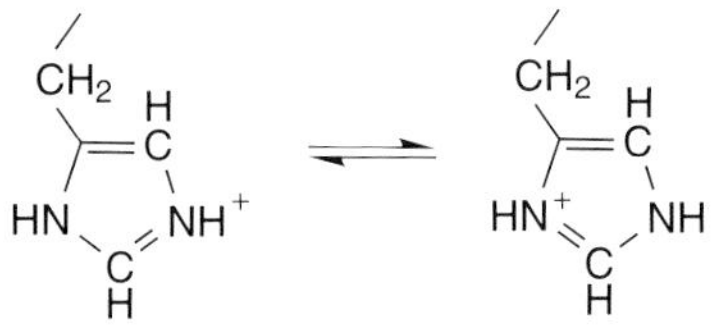

Figure 2.33 Delocalization of charge in protonated imidazole side chains of histidine is possible

state the side chain has a positive charge whilst in the unionized state the side chain remains neutral. However, protonation not only modulates the charge and acid–base behaviour of the side chain but also alters nucleophilic and electrophilic properties of the ring (Figure 2.32).

The protonated nitrogen, shown in red in Figure 2.32, is called the NE2 or ε2. Experimental evidence suggests that the hydrogen atom is usually located on the NE2 nitrogen but upon further protonation the structure on the right is formed with the ND1 nitrogen now binding a proton. The charge cannot be accurately assigned to one of the nitrogen atoms and resonance structures exist (Figure 2.33).

The unprotonated nitrogen of the uncharged imidazole ring is a potent nucleophile and has a capacity for hydrogen bonding.

The aromatic residues: phenylalanine, tyrosine and tryptophan

The aromatic side chains have a common property of absorbing in the ultraviolet region of the electromagnetic spectrum. Table 2.3 shows the spectroscopic properties of Phe, Tyr and Trp. As a result Phe, Tyr and Trp are responsible for the absorbance and fluorescence of proteins frequently measured between

Table 2.3 Spectroscopic properties of the aromatic amino acids

Amino acid	Absorbance		Fluorescence	
	λ_{max} (nm)	ε ($M^{-1}\ cm^{-1}$)	λ_{max} (nm)	Quantum yield
Phenylalanine	257.4	197	282	0.04
Tyrosine	274.6	1420	303	0.21
Tryptophan	279.8	5600	348	0.20

λ, wavelength; ε, molar absorptivity coefficient.

250 and 350 nm. In this region the molar extinction coefficients of Phe, Tyr and Trp, an indication of how much light is absorbed at a given wavelength, are not equal. Tryptophan exhibits a molar extinction coefficient approximately four times that of tyrosine and over 28 times greater than phenylalanine. Almost all spectrophotometric measurements of a protein's absorbance at 280 nm reflect the intrinsic Trp content of that protein. At equivalent molar concentrations proteins with a high number of tryptophan residues will give a much larger absorbance at 280 nm when compared with proteins possessing a lower Trp content.

Phenylalanine

The aromatic ring of phenylalanine is chemically inert and has generally proved resistant to chemical modification (Figure 2.34). Only with genetic modification of proteins have Phe residues become routinely altered. The numbering scheme of atoms around the aromatic benzene ring is shown for carbon and hydrogens. Alternative nomenclatures exist for the ring proteins with labels of HD, HE or HZ (H_δ, H_ε or H_ζ in some schemes) being the most common. In many proteins rotation of the aromatic ring around the CB–CG axis leads to the protons at the 2,6 positions (HD) and the 3,5 positions (HE) being indistinguishable.

Although the aromatic ring is inert chemically it remains hydrophobic and prefers interactions in proteins with other non-polar residues. With other aromatic rings Phe forms important $\pi-\pi$ interactions derived from the delocalized array of electrons that arises from bonding between the ring carbon atoms. In a simple alkene such as ethene each carbon links to two hydrogens and a single carbon. This is achieved by sp^2 hybridization where the normal ground state of carbon $1s^22s^22p_x{}^12p_y{}^1$ is transformed by elevating one of the 2s electrons into the vacant p_z orbital yielding $1s^22s^12p_x{}^12p_y{}^12p_z{}^1$. The sp^2 hybrids, made from rearrangement of the 2s orbital and two of the three 2p orbitals, are orientated at an angle of 120° to each other within a plane. The remaining $2p_z$ orbital is at right angles to these hybrids. Formation of three bonds leaves the p_z orbital vacant and a similar situation exists in benzene or the aromatic ring of phenylalanine; each sp^2 hybridized carbon is bonded to two carbon atoms and a single hydrogen and has a single unpaired electron. This electron lies in the vacant p_z orbital proximal to and overlapping with another carbon atom possessing an identical configuration. The extensive overlap produces a system of π bonds where the electrons are not confined between two carbon atoms but delocalize around the entire ring. Six delocalized electrons go into three π molecular orbitals – two in each. The interaction of these π molecular orbitals by the close approach of two aromatic rings leads to further delocalization. The coplanar orientation is the basis for $\pi-\pi$ interactions and may account for the observation that aromatic side chains are found in proteins in close proximity to other aromatic groups. $\pi-\pi$ interactions are expected to make favourable and significant contributions to the overall protein stability. Studies with model compounds suggest that an optimal geometry exists with the aromatic rings perpendicular to each other and leading to positively charged hydrogens on the edge of one ring interacting favourably with the π electrons and partially negatively charged carbons of the other. For proteins, as opposed to model compounds, this type of interaction is less common than a co-planar orientation of aromatic rings.

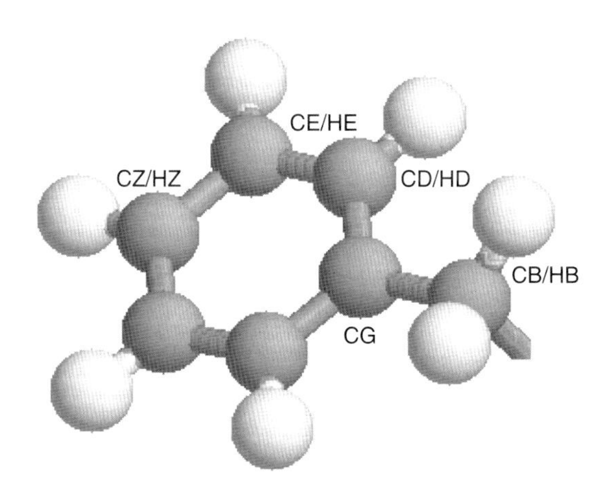

Figure 2.34 The aromatic side chain of phenylalanine

Tyrosine

Tyrosine is more reactive than phenylalanine due to the hydroxyl group (Figure 2.35) substituted at the fourth (C_Z) aromatic carbon, often called the *para*

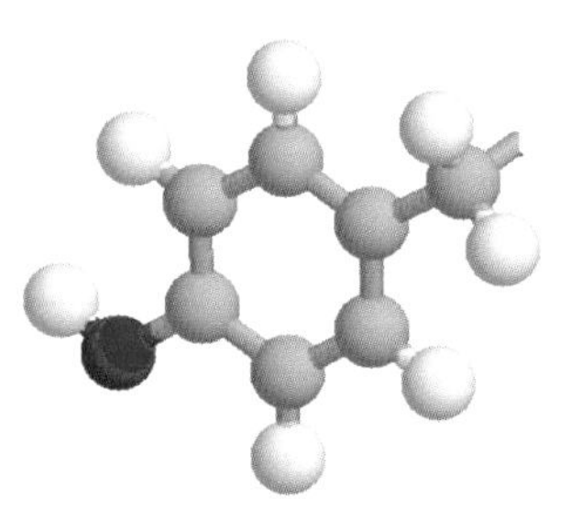

Figure 2.35 The side chain of tyrosine contains a reactive hydroxyl group

position. The hydroxyl group leaves the aromatic ring susceptible to substitution reactions.

Nucleophiles such as nitrating agents or activated forms of iodide react with tyrosine side chains in proteins and change the acid–base properties of the ring. Reaction with nitrating agents yields mono nitrotyrosine derivatives at the 3,5 position followed by dinitro derivatives. Normally the hydroxyl group is a weak acid with a p$K \sim 11.0$ but the addition of one or more nitro groups causes the pK to drop by up to 4 pH units. The tyrosine interacts with other aromatic rings but the presence of the OH group also allows hydrogen bonding.

Tryptophan

The indole side chain (Figure 2.36) is the largest side chain occurring in proteins and is responsible for most of the intrinsic absorbance and fluorescence. As a crude approximation the molar extinction coefficient of a protein at 280 nm may be estimated by adding up the number of Trp residues found in the sequence and multiplying by 5800 (see Table 2.3). However, the relatively low frequency of Trp residues in proteins means that this approach is not always accurate and in some cases (for proteins lacking Trp) will be impossible.

The side chain is hydrophobic and does not undergo extensive chemical reactions. Exceptions to the rule of unreactivity include modifying reagents such as iodine, N-bromosuccinimide and ozone although their use in chemical modification studies of proteins is limited. Universally the weakest or most sensitive part of the indole ring is the pyrrole nitrogen atom.

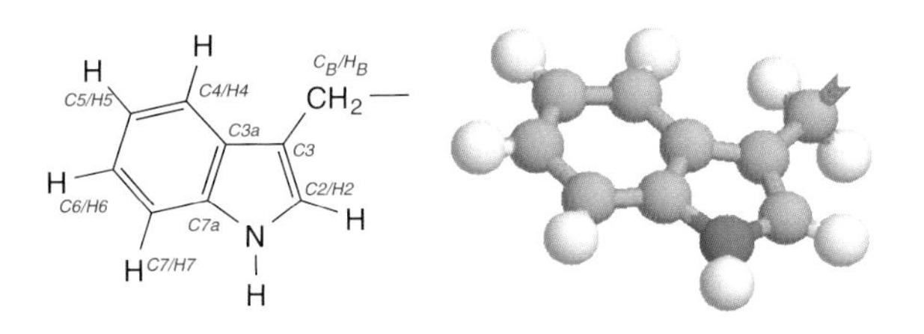

Figure 2.36 The indole ring of tryptophan. The nomenclature of the ring protons names the ring protons as H2, H4, H5, H6 and H7. The carbons with protons attached are similarly named i.e C2, C3 ... C7. The carbon centres lacking protons are C3, C3a and C7a. The nitrogen is properly termed the N1 centre but is frequently called the indole, or NE1, nitrogen. In addition Greek symbols are often applied to these carbon and protons with the following symbols used $C_{\xi 2} = C7$, $C_{\eta 2} = C6$, $C_{\xi 3} = C5$, $C_{\eta 3} = C4$

Detection, identification and quantification of amino acids and proteins

To quantify protein isolation requires a means of estimating protein concentration and most methods use the properties of amino acid side chains. Proteins absorb around 280 nm principally because of the relative contributions of tryptophan, phenylalanine and tyrosine. If their occurrence in proteins is known then, in theory at least, the concentration of a protein solution is calculated using the data of Table 2.4 and Beer–Lamberts law where

$$A_{280} = \varepsilon_{280} c \, l \qquad (2.4)$$

(where A is absorbance, ε is molar absorptivity coefficient, c is the concentration in moles dm^{-3} and l is the light path length normally 1 cm). In practice this turns out to be a very crude method of measuring protein concentration and of limited accuracy. The number of Trp, Tyr or Phe residues is not always known and more importantly the absorbance of proteins in their native state does not usually equate with a simple summation of their individual contributions. Moreover the presence of any additional components absorbing at 280 nm will lead to inaccuracies and this could include the presence of impurities, disulfide bridges, or additional co-factors such as heme or flavin. Performing spectrophotometric measurements under denaturing conditions allows improved estimation of protein concentration from the sum of the Trp, Tyr and Phe components. The contribution of disulfide bridges, although weak around 280 nm, can be eliminated by the addition of reducing agents, and under denaturing conditions co-factors are frequently lost.

An alternative method of determining protein concentration exploits the reaction between the thiol group of cysteine and Ellman's reagent. The reaction produces the nitrothiobenzoate anion and since the molar absorptivity coefficient for this product at 410 nm is accurately known (13 600 $\text{M}^{-1}\,\text{cm}^{-1}$) the reagent offers one route of determining protein concentration if the number of thiol groups is known beforehand. In general, protein concentration may be determined by any reaction with side chains that lead to coloured products that can be quantified by independent methods. Other reagents with applications in estimating protein, peptide or amino acid concentration include ninhydrin, fluorescamine, dansyl chloride, nitrophenols and fluorodinitrobenzene. They have the common theme of possessing a reactive group towards certain side chains combined with a strong chromophore. A common weakness is that all rely on reactions with a restricted number of side chains to provide estimates of total protein concentration. A better approach is to devise assays based on the contribution of all amino acids.

Historically the method of choice has been the biuret reaction (Figure 2.37) where a solution of copper(II) sulfate in alkaline tartrate solution reacts with peptide bonds to form a coloured (purple) complex absorbing around 540 nm. The structure of the complex formed is unclear but involves the coordination of copper ions by the amide hydrogen of peptide bonds. An important aspect of this assay is that it is based around the properties of the polypeptide backbone. Reduction of Cu(II) to Cu(I) is accompanied by a colour change and if the absorbance of the unknown protein is compared with a calibration curve then concentration is accurately determined. Ideally the standard protein used to construct a calibration curve will have similar properties, although frequently bovine serum albumin (BSA) is used.

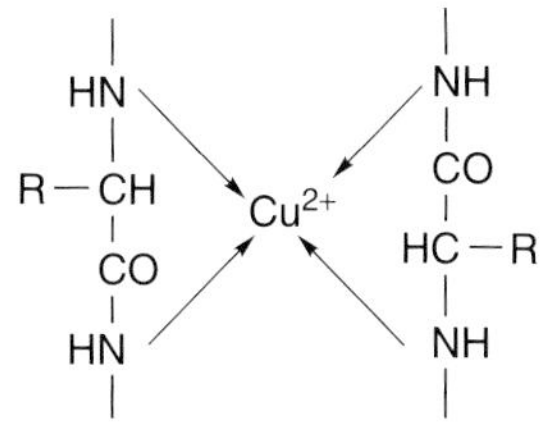

Another commonly used method of estimating protein concentration is the Folin–Lowry method. The colour reaction is enhanced by the addition of Folin–Ciocalteau's reagent to a protein solution containing copper ions. The active ingredient of Folin–Ciocalteau's reagent is a complex mixture of phosphotungstic and phosphomolybdic acids. The reduced Cu(I) generated in the biuret reaction forms a number of reduced acid species in solution each absorbing at a wavelength maximum between 720 and 750 nm and leading to an obvious blue colour. Estimation of protein concentration arises by comparing the intensity of this blue colour to a 'standard' protein of known concentration at wavelengths between 650 and 750 nm.

Coomassie Brilliant Blue (Figure 2.38) does not undergo chemical reactions with proteins but forms

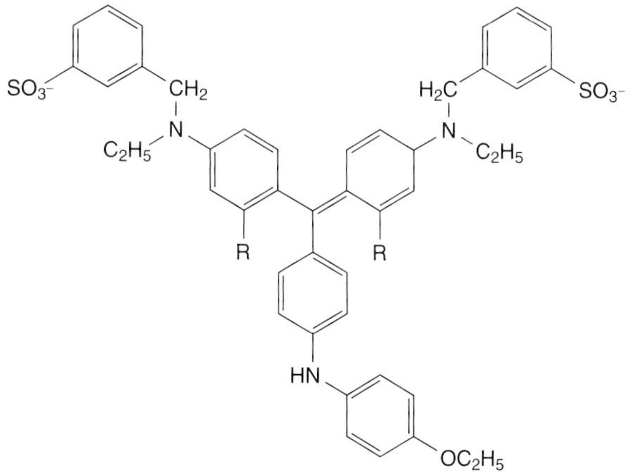

Figure 2.37 The complexation of cupric ions by peptide bonds in the biuret method

Figure 2.38 Coomassie blue dyes are commonly used to estimate protein concentration (R250 = H; G250 = CH_3)

stable complexes via non-covalent interactions. Coomassie blue (G250) is used to estimate the amount of protein in solution because complex formation is accompanied by a shift in absorbance maxima from 465 to 595 nm. The mechanism of complex formation between dye and protein is unclear but from the structure of the dye involves non-polar interactions.

Stereoisomerism

Up to this point the α carbon of amino acids has been described as an asymmetric centre and little attention has been paid to this characteristic property of tetrahedral carbon centres. One of the most important consequences of the asymmetric α carbon is that it gives rise to a chiral centre and the presence of two isomers. The two potential arrangements of atoms about the central carbon are shown for alanine (Figure 2.39) and whilst superficially appearing identical a closer inspection of the *arrangement* of atoms and the relative positions of the amino, carboxyl, hydrogen and methyl groups reveals that the two molecules can never be exactly superimposed.

The molecules are mirror images of each other or stereoisomers. Most textbooks state that these isomers are termed the L and D isomers without resorting to further explanation. All naturally occurring amino acids found in proteins belong to the L absolute configuration. However, at the same time the asymmetric or chiral centre is also described as an optically active centre. An optically active centre is one that rotates plane-polarized light. Amino acids derived from the hydrolysis of proteins under *mild* conditions will, with the exception of glycine, rotate plane-polarized light in a single direction. If this direction is to the right the amino acids are termed *dextrorotatory* (+) whilst rotation to the left is *laevorotatory* (−). Values for isolated amino acids are known, with L-amino acids rotating plane-polarized light in different directions and to differing extents (Table 2.4).

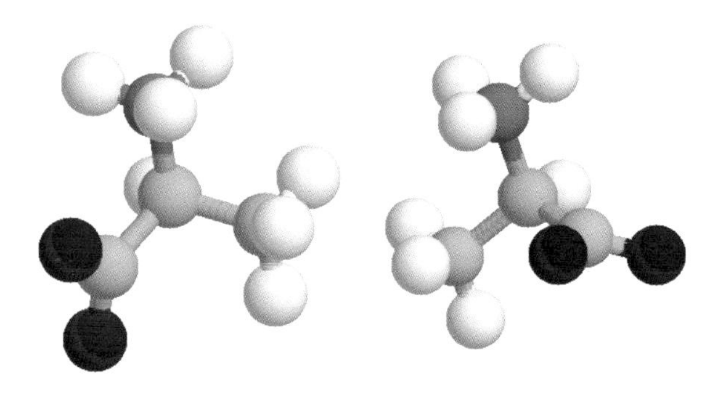

Figure 2.39 Two stereoisomers of alanine. As viewed the C-H bond of the α-carbon is pointing away from the viewer (down into the page) and superposition of the remaining groups attached this chiral centre is impossible. Each molecule is a mirror image of the other known variously as enantiomers, optical isomers or stereoisomers.

Table 2.4 The specific optical rotation of selected amino acids

L-Amino Acid	$[\alpha]_D(H_2O)$	L-Amino Acid	$[\alpha]_D(H_2O)$
Alanine	+1.8	Isoleucine	+12.4
Arginine	+12.5	Leucine	−11.0
Cysteine	−16.5	Phenylalanine	−34.5
Glutamic Acid	+12.0	Threonine	−28.5
Histidine	−38.5	Tryptophan	−33.7

The terms dextro- and levorotatary must not be confused with the L and D isomers, which for historical reasons are related to the arrangement of atoms in L- and D-glyceraldehyde. L-Trp is laevorotatory whilst L-Glu is dextrorotatory. The D and L stereoisomers of any amino acid have identical properties with two exceptions: they rotate plane-polarized light in opposite directions and they exhibit different reactivity with asymmetric reagents. This latter point is important in protein synthesis where D amino acids are effective inhibitors and in enzymes where asymmetric active sites discriminate effectively between stereoisomers.

The Cahn–Ingold–Prelog scheme has been devised to avoid confusion and provides an unambiguous method of assigning absolute configurations particularly for molecules containing more than one asymmetric centre. The amino acids isoleucine and threonine each contain two asymmetric centres, four stereoisomers, and potential confusion in naming isomers using the DL system. Using threonine as an example we can see that the α carbon and the β carbon are each attached to four different substituents. For the β carbon these are the hydroxyl group, the methyl group, the hydrogen atom and the α carbon. The RS system is based on ranking the substituents attached to each chiral centre according to atomic number. The smallest or lowest ranked group is arranged to point away from the viewer. This is always the C–H bond for the C_α carbon and the remaining three groups are then viewed in the direction of decreasing priority. For biomolecules the functional groups commonly found in proteins are ranked in order of decreasing priority:

$$SH > P > OR > OH > NH_3 > COO > CHO$$
$$> CH_2OH > C_6H_5 > CH_3 > H$$

The priority is established initially on the basis of atomic number so H (atomic no. = 1) has the lowest priority whilst groups containing sulfur (atomic no. = 16) are amongst the highest. To identify RS isomers one first identifies the asymmetric or stereogenic centre and in proteins this involves an sp^3 carbon centre with a tetrahedral arrangement containing four different groups. With the C–H bond (lowest priority) pointing away from the viewer the three remaining groups are ranked according to the scheme shown above. In the case where two carbon atoms are attached to an asymmetric centre then one moves out to the next atom and applies the same selection rule at this point. This leads to CH_3 being of lower priority than C_2H_5 which in turn is lower than COO^-. Where two isotopes are present and this is most commonly deuterium and hydrogen the atomic number rule is applied leading to D having a greater priority than H. Having ranked the groups we now assess if the priority (from high to low) is in a clockwise or anticlockwise direction. A clockwise direction leads to the R isomer (Latin: *rectus* = right) whilst an anticlockwise direction leads to the S isomer (Latin: *sinister* = left).

In L-threonine at the first asymmetric centre (C_α) when looking down the C_α–H bond (pointing away from the viewer in the arrangement shown in Figure 2.40) the direction of decreasing order of priority (NH_3^+, COO^- and $CH–OH–CH_3$) is counter clockwise. The C_α, the second carbon, is therefore configuration S. A similar line of reasoning for the asymmetric centre located at the third carbon (C_G) yields a decreasing order of priority in a clockwise direction and hence the R configuration. Combining these two schemes yields for L-threonine the nomenclature [2S-3R] threonine (Figure 2.41).

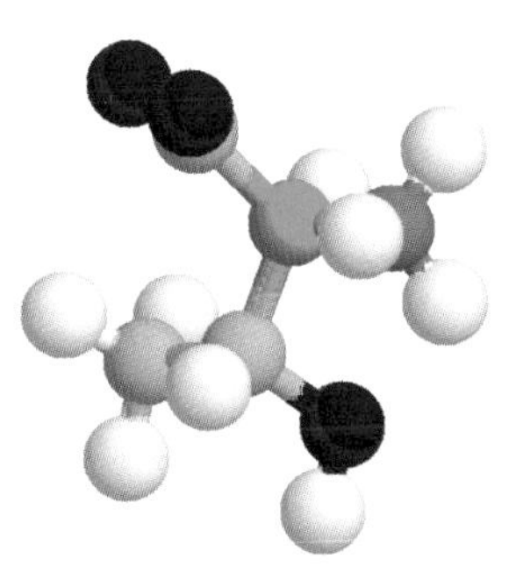

Figure 2.40 The asymmetric C_α centre of Thr is in the S configuration. The C_α is shown on green whilst the other atoms have their normal CPK colours

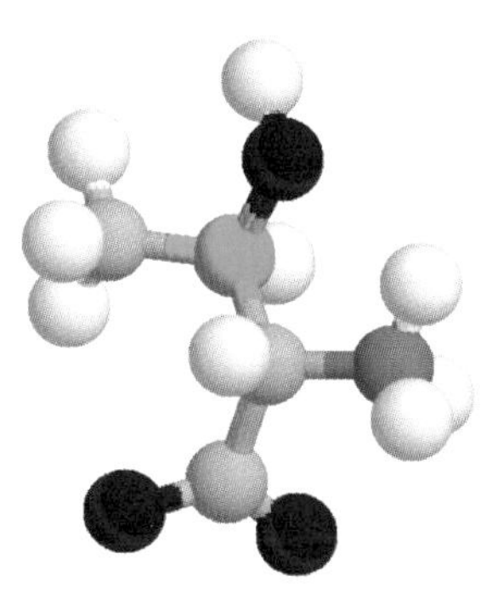

Figure 2.41 The asymmetric C_G centre of threonine is in the R configuration. The C_G atom is shown in green and the C–H bond is pointing away from the viewer

Non-standard amino acids

The 20 amino acids are the building blocks of proteins but are also precursors for further reactions that produce additional amino acids. Amino acids with unusual stereochemistry about the C_α carbon often called D amino acids are comparatively common in micro-organisms. Prominent examples are D-alanine and D-isoglutamate in the cell wall of the Gram-positive bacterium *Staphylococcus aureus*. In other bacteria

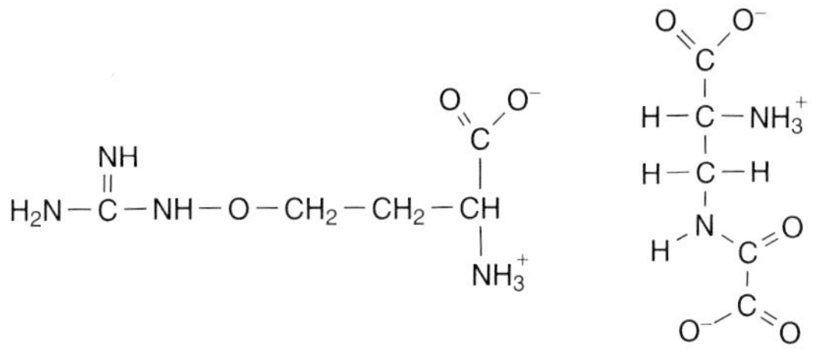

Figure 2.42 Canavanine and β-*N*-oxalyl L-α, β-diamino propionic acid (ODAP)

small peptide molecules known as ionophores form channels in membranes through the use of proteins containing D amino acid residues. A well-known example is the 15 residue peptide Gramicidin A containing a series of alternating D and L amino acids. In each case the use of these unusual or non-standard amino acids may be viewed as a defensive mechanism. Plants and microorganism use non-standard amino acids as protective weapons. Canavanine, a homologue of arginine containing an oxygen instead of a methylene group at the δ position, accumulates as a storage protein in alfafa seeds where it acts as a natural defence against insect predators (Figure 2.42).

A serious occurrence of an unusual amino acid is the presence of β-*N*-oxalyl L-α, β-diamino propionic acid (ODAP) in the legume *Lathyrus sativus*. Ingestion of seeds containing this amino acid is harmless to humans in small quantities but increased consumption leads to a neurological disorder (lathyrism) resulting in irreversible paralysis of leg muscles. ODAP is the culprit and acts by mimicking glutamate in biological systems.

Amino acids undergo a wide range of metabolic conversions as part of the cell's normal synthetic and degradative pathways. However, in a few cases amino acids are converted to specific products. Instead of undergoing further metabolism these modified amino acids are used to elicit biological response. The most important examples of the conversion of amino acids into other modified molecules include histamine, dopamine, thyroxine and γ-amino butyric acid (GABA) (Figure 2.43). GABA is formed from the amino acid glutamate as a result of decarboxylation and is an important neurotransmitter; thyroxine is an iodine-containing hormone stimulating basal metabolic rate via increases in carbohydrate metabolism in vertebrates, whilst in amphibians it plays a role in metamorphosis; histamine is a mediator of allergic response reactions and is produced by mast cells as part of the body's normal response to allergens.

Summary

Twenty different amino acids act as the building blocks of proteins. Amino acids are found as dipolar ions in solutions. The charged properties result from the presence of amino and carboxyl groups and lead to solubility in water, an ability to act as electrolytes, a crystalline appearance and high melting points.

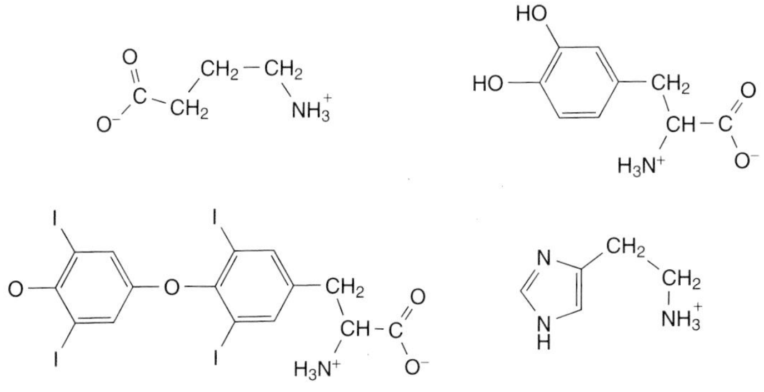

Figure 2.43 Biologically active amino acid derivatives. Clockwise from top left are GABA, dopamine, histamine and thyroxine

Of the 20 amino acids found in proteins 19 have a common structure based around a central carbon, the C_α carbon, in which the amino, carboxyl, hydrogen and R group are arranged tetrahedrally. An exception to this arrangement of atoms is proline. The C_α carbon is asymmetric with the exception of glycine and leads to at least two stereoisomeric forms.

Amino acids form peptide bonds via a condensation reaction and the elimination of water in a process that normally occurs on the ribosomes found in cells. The formation of one peptide bond covalently links two amino acids forming a dipeptide. Polypeptides or proteins are built up by the repetitive formation of peptide bonds and an average sized protein may contain 1000 peptide bonds.

The peptide bond possesses hybrid characteristics with properties between that of a C–N single bond and those of a C=N double bond. These properties result in decreased peptide bond lengths compared to a C–N single bond, a lack of rotation about the peptide bond and a preferred orientation of atoms in a *trans* configuration for most peptide bonds. One exception to this rule is the peptide bond preceding proline residues where the *cis* configuration is increased in stability relative to the *trans* configuration.

The side chains dictate the chemical and physical properties of proteins. Side chain properties include charge, hydrophobicity and polarity and underpin many aspects of the structure and function of proteins.

Problems

1. Why are amino acids white crystalline solids and how does this account for their physicochemical properties?
2. Using Figures 2.13 and 2.14 identify methine, methene and methyl groups in the side chains of leucine, isoleucine and valine. Label each carbon atom according to the usual nomenclature. Repeat the exercise of nomenclature for Figure 2.18.
3. Translate the following sequence into a sequence based on the single letter codes.
   Ala-Phe-Phe-Lys-Arg-Ser-Ser-Ser-Ala-Thr-Leu-Ile-Val-Thr-Lys-Lys-Gln-Gln-Phe-Asn-Gly-Gly-Pro-Asp-Glu-Val-Leu-Arg-Thr-Ala-Ser-Thr-Lys-Ala-Thr-Asp.
4. What is the average mass of an amino acid residue? Why is such information useful yet at the same time limited?
5. Which of the following peptides might be expected to be positively charged, which are negatively charged and which carry no net charge?
   Ala-Phe-Phe-Lys-Arg-Ser-Ser-Ser-Ala-Thr-Leu-Ile-Val-
   Ala-Phe-Phe-Lys-Arg-Ser-Glu-Asp-Ala-Thr-Leu-Ile-Val-
   Ala-Phe-Phe-Lys-Asp-Ser-Ser-Asp-Ala-Thr-Leu-Ile-Val-
   Ala-Phe-Phe-Lys-Asp-Ser-Ser-His-Ala-Thr-Leu-Ile-Val-
   Ala-Phe-Phe-Lys-Asp-Ser-Glu-His-Ala-Thr-Leu-Ile-Val-
   Are there any contentious issues?
6. The reagent 1-fluoro 2,4-dinitrobenzene has been used to identify amino acids. Describe the groups and residues you would expect this reagent to react with preferentially. Under what conditions would you perform the reactions.
7. Histidine has three ionizable functional groups. Draw the structures of the major ionized forms of histidine at pH 1.0, 5.0, 8.0 and 13.0. What are the respective charges at each pH?
8. The pK_1 and pK_2 values for alanine are 2.34 and 9.69. In the dipeptide Ala-Ala these values are 3.12 and 8.30 whilst the tripeptide Ala-Ala-Ala has values of 3.39 and 8.00. Explain the trends for the values of pK_1 and pK_2 in each peptide.
9. A peptide contains the following aromatic residues 3 Trp, 6 Tyr and 1 Phe and gives an absorbance at 280 nm of 0.8. Having added 1 ml of a solution of this peptide to a cuvette of path length 1 cm light calculate from the data provided in Table 2.4 the approximate concentration of the peptide (you can assume the differences in ε between 280 nm and those reported in Table 2.4 are negligible). The molecular mass of the peptide is 2670 what was the concentration in mg ml^{-1}.

10. Citrulline is an amino acid first isolated from the watermelon (*Citrullus vulgaris*). Determine the absolute (RS) configuration of citrulline for any asymmetric centre. Isoleucine has two asymmetric centres. Identify these centres and determine the configuration at each centre. How many potential optical isomers exist for isoleucine?

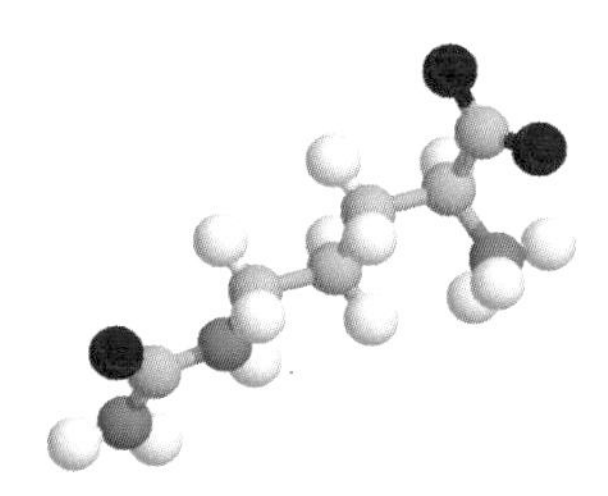

3

The three-dimensional structure of proteins

Amino acids linked together in a flat 'two-dimensional' representation of the polypeptide chain fail to convey the beautiful three-dimensional arrangement of proteins. It is the formation of regular secondary structure into complicated patterns of protein folding that ultimately leads to the characteristic functional properties of proteins.

Primary structure or sequence

The primary structure is the linear order of amino acid residues along the polypeptide chain. It arises from covalent linkage of individual amino acids via peptide bonds. Thus, asking the question 'What is the primary structure of a protein?' is simply another way of asking 'What is the amino acid sequence from the N to C terminals?' To read the primary sequence we simply translate the three or single letter codes from left to right, from amino to carboxyl terminals. Thus in the sequence below two alternative representations of the same part of the polypeptide chain are given, starting with alanine at residue 1, glutamate at position 2 and extending to threonine as the 12th residue (Figure 3.1).

Every protein is defined by a unique sequence of residues and all subsequent levels of organization (secondary, super secondary, tertiary and quaternary) rely on this primary level of structure. Some proteins are related to one another leading to varying degrees of similarity in primary sequences. So myoglobin, an oxygen storage protein found in a wide range of organisms, shows similarities in human and whales in the 153 residue sequence (Figure 3.2). Most of the sequence is identical and it is easier to spot the differences. When a change occurs in the primary sequence it frequently involves two closely related residues. For example, at position 118 the human variant has a lysine residue whilst whale myoglobin has an arginine residue. Reference to Table 2.2 will show that arginine and lysine are amino acids that contain a positively charged side chain and this change is called a conservative transition. In contrast in a few positions there are very different amino acid residues. Consider position 145 where asparagine (N) is replaced by lysine (K). This transition is not conservative; the small, polar, side chain of asparagine is replaced by the larger, charged, lysine. Regions, or residues, that never change are called invariant.

Secondary structure

Primary structure leads to secondary structure; the local conformation of the polypeptide chain or the spatial relationship of amino acid residues that are close together in the primary sequence. In globular proteins

Proteins: Structure and Function by David Whitford

NH_3-Ala-Glu-Glu-Ser-Ser-Lys-Ala-Val-Lys-Tyr-Tyr-Thr-.........

NH_3---A---E---E---S----S----K----A---V---K---Y----Y----T........

Figure 3.1 Single- and three-letter codes for amino terminal of a primary sequence

the three basic units of secondary structure are the α helix, the β strand and turns. All other structures represent variations on one of these basic themes. This chapter will focus on the chemical and physical properties of polypeptides that permit the transition from randomly oriented polymers of amino acid residues to regular repeating secondary structure. With 20 different amino acid residues found in proteins there are 780 possible permutations for a dipeptide. In an average size polypeptide of ~100 residues the number of potential sequences is astronomical. In view of the enormous number of possible conformations many fundamental studies of protein secondary structure have been performed on homopolymers of amino acids for example poly-alanine, poly-glutamate, poly-proline or poly-lysine.

Homopolymers have the advantage that either all the residues are identical or in some cases simple repeating units such as poly Ala-Gly, where Ala-Gly dipeptides are repeated along the length of the polymer. In comparison with polypeptides derived from proteins these polymers have the advantage of consistent conformations. This is not to imply that the conformation of homopolymers is regular and ordered. In some instances these polymers are unstructured, however, the common theme is that the polymers are uniform in their conformations and hence are much more attractive candidates for initial studies of structure.

Studies of homopolymers have been advanced by host–guest studies where a polymer of alanine residues is modified by an introduction of a single different residue in the middle of the polypeptide. By measuring changes in stability, solubility, or helical properties these studies allow the effect of the new or guest residue to be accurately defined. Indeed, much of the data reported in tables throughout this book were obtained by the introduction of a single amino acid residue into a polymer containing just one type of residue. The host peptide is usually designed to be monomeric (i.e. non-aggregating), to be soluble and not more than 15 residues in length. By systematically replacing residue 8 with any of the other 19 amino acids these studies have elucidated many of the properties of residues within polypeptide chains. One property that has been extensively studied using this approach is the relative helical tendency of amino acid residues in a peptide of poly-alanine. Results suggest, unsurprisingly, that alanine is the most stable residue to substitute into a poly-alanine peptide whilst proline is the most destabilizing. More revealing is the relative

```
                    *        20         *        40
Human Mb : GLSDGEWQLVLNVWGKVEADIPGHGQEVLIRLFKGHPETL :  40
Whale Mb : VLSEGEWQLVLHVWAKVEADVAGHGQDILIRLFKSHPETL :  40

                    *        60         *        80
Human Mb : EKFDKFKHLKSEDEMKASEDLKKHGATVLTALGGILKKKG :  80
Whale Mb : EKFDRFKHLKTEAEMKASEDLKKHGVTVLTALGAILKKKG :  80

                    *       100         *       120
Human Mb : HHEAEIKPLAQSHATKHKIPVKYLEFISECIIQVLQSKHP : 120
Whale Mb : HHEAELKPLAQSHATKHKIPIKYLEFISEAIIHVLHSRHP : 120

                    *       140         *
Human Mb : GDFGADAQGAMNKALELFRKDMASNYKELGFQG : 153
Whale Mb : GDFGADAQGAMNKALELFRKDIAAKYKELGYQG : 153
```

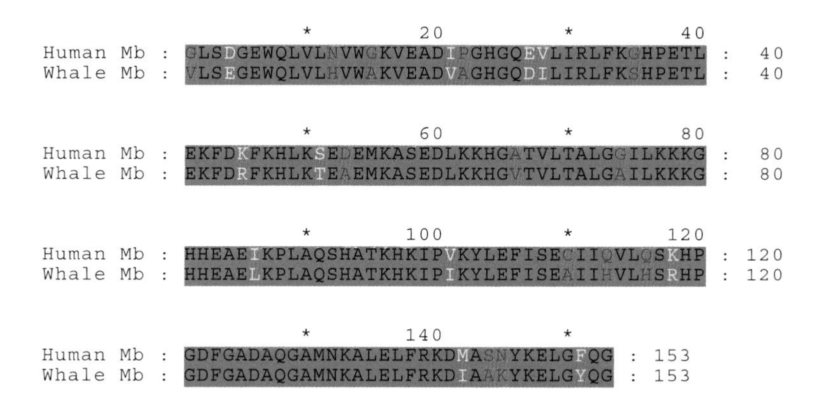

Figure 3.2 The primary sequences of human and sperm whale myoglobin. Identical residues are shown in blue, the regions in yellow show conserved substitutions whilst the red regions show non-conservative changes. Astericks indicate every tenth residue. Single letter codes for residues are used

Table 3.1 The helical propensity of amino acid residues substituted into alanine polymers

Residue	Helix propensity, ΔG (kJ mol^{-1})	Residue	Helix propensity, ΔG (kJ mol^{-1})
Ala	0	Ile	0.41
Arg	0.21	Leu	0.21
Asn	0.65	Lys	0.26
Asp^0	0.43	Met	0.24
Asp$^-$	0.69	Phe	0.54
Cys	0.68	Pro	3.16
Gln	0.39	Ser	0.50
Glu^0	0.16	Thr	0.66
Glu$^-$	0.40	Tyr	0.53
Gly	1.00	Trp	0.49
His^0	0.56	Val	0.61
His$^+$	0.66		

Derived from Pace, C.N. & Scholtz, J.M. *Biophys. J.* 1998, 75, 422–427. The data includes uncharged Glu, Asp and His residues (superscript ^{0}). All residues form helices with less propensity than poly-Ala hence the positive values for ΔG.

helical tendencies or propensities of the other residues when measured relative to alanine (Table 3.1).

When the first crystal structures of proteins became available it allowed a comparison between residue identity and secondary structure. Ala, Leu and Glu are found more frequently in α helices whilst Pro, Gly and Asp were found less frequently than average. Using this analysis of primary sequence a helix propensity scale was derived, and is still used in predicting the occurrence of helices and sheets in folded soluble proteins.

The α helix

The right-handed α helix is probably the best known and most identifiable unit of secondary structure. The structure of the α helix was derived from model-building studies until the publication of the crystallographic structure of myoglobin. This demonstrated that α helices occurred in proteins and were largely as predicted from theoretical studies by Linus Pauling. The α helix is the most common structural motif found in proteins; in globular proteins over 30 percent of all residues are found in helices.

The regular α helix (Figure 3.3) has 3.6 residues per turn with each residue offset from the preceding residue by 0.15 nm. This parameter is called the translation per residue distance. With a translation distance of 0.15 nm and 3.6 residues per turn the pitch of the α helix is simply 0.54 nm (i.e. 3.6×0.15 nm). The pitch is the translation distance between any two corresponding atoms on the helix. One of the major results of model building studies was the realization that the α helix arises from regular values adopted for ϕ (phi) and ψ (psi), the torsion or dihedral angles (Figure 3.4).

The values of ϕ and ψ formed in the α helix allow the backbone atoms to pack close together with few unfavourable contacts. More importantly this arrangement allows some of the backbone atoms to form hydrogen bonds. The hydrogen bonds occur between the backbone carbonyl oxygen (acceptor) of one residue and the amide hydrogen (donor) of a residue four ahead in the polypeptide chain. The hydrogen bonds are 0.286 nm long from oxygen to nitrogen atoms, linear and lie (in a regular helix) parallel to the helical axis. It is worth noting that in 'real' proteins the arrangement of hydrogen bonds shows variation in length and angle with respect to helix axes (Figure 3.5).

Hydrogen bonds have directionality that reflects the intrinsic polarization of the hydrogen bond due to the electronegative oxygen atom. In a similar fashion the peptide bond also has polarity and the combined effect of these two parameters give α helices pronounced dipole moments. On average the amino end of the α helix is positive whilst the carboxyl end is negative. In the α helix, the first four NH groups and last four CO groups will normally lack backbone hydrogen bonds. For this reason very short helices often have distorted conformations and form alternative hydrogen bond partners. The distortion of hydrogen bonds and lengths that occur in real helices are accompanied by the dihedral angles (ϕ and ψ) that deviate significantly from the ideal values of $-57°$ and $-47°$ (see Table 3.2).

Visualization of the α helix frequently neglects the side chains but an ideal arrangement involves the

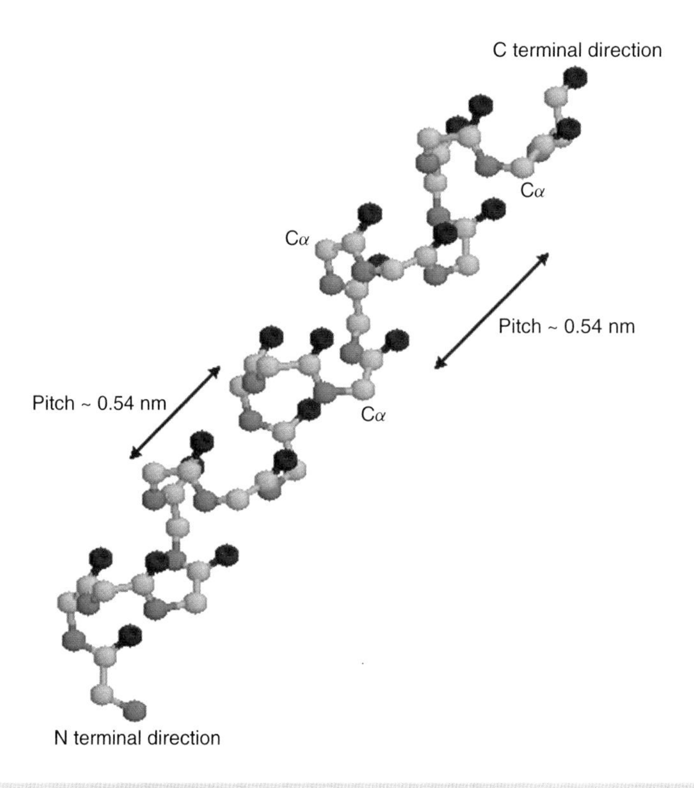

Figure 3.3 A regular α helix. Only heavy atoms (C, N and O, but not hydrogen) are shown and the side chains are omitted for clarity

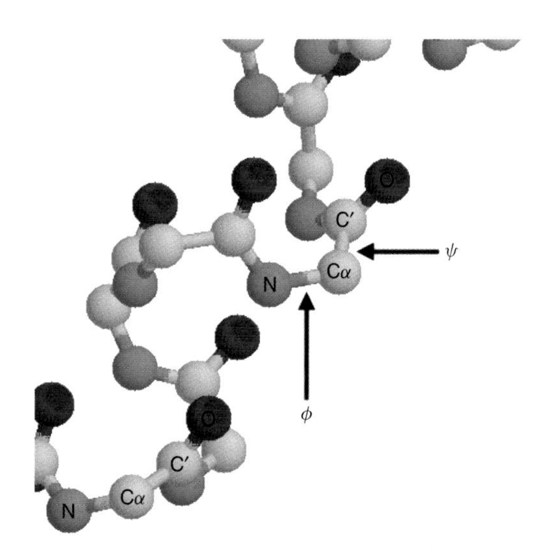

Figure 3.4 φ is defined by the angle between the C′–N–Cα –C′ atoms whilst ψ is defined by the atoms N–Cα –C′–N

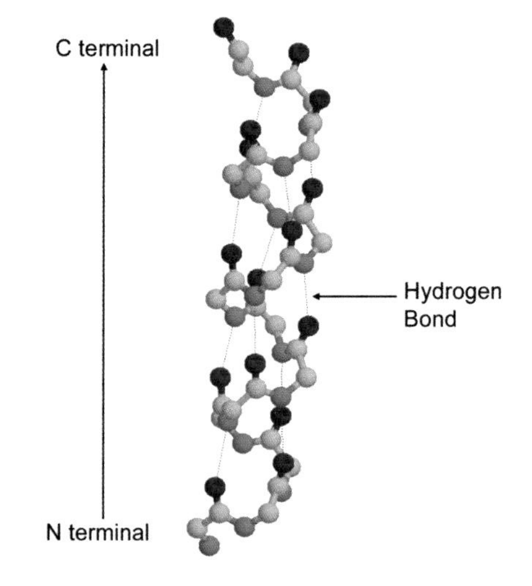

Figure 3.5 Arrangement of backbone hydrogen bonds in a *real* α helix from myoglobin, shows deviations from ideal geometry

Table 3.2 Dihedral angles, translation distances and number of residues per turn for regular secondary structure conformations. In poly(Pro) I ω is 0° whilst in poly(Pro) II ω is 180°

Secondary structure element	Dihedral angle (°) ϕ	ψ	Residues/turn	Translation distance per residue (nm)
α helix	−57	−47	3.6	0.150
3_{10} helix	−49	−26	3.0	0.200
π helix	−57	−70	4.4	0.115
Parallel β strand	−139	+135	2.0	0.320
Antiparallel β strand	−119	+113	2.0	0.340
Poly(Pro) I	−83	+158	3.3	0.190
Poly(Pro) II	−78	+149	3.0	0.312

atoms projecting outwards into solution. Although the side chains radiate outwards there are conformational restrictions because of potential overlap with atoms in neighbouring residues. This frequently applies to branched side chains such as valine, isoleucine and threonine where the branch occurs at the CB atom and is closest to the helix. Steric restriction about the CA–CB bond leads to discrete populations or rotamers (Figure 3.6). The symbol χ_1 (pronounced "KI one") is used to define this angle and is best appreciated by viewing projections along the CA–CB bond.

Alanine, glycine and proline do not have χ_1 angles whilst the χ_2 angle for serine, threonine and cysteine is difficult to measure because it involves determining the position of a single H atom accurately. However, in databases of protein structures the χ_1 angles adopted for all residues (except Ala, Gly and Pro) have been documented, whilst the χ_2 distribution of Arg, Glu, Gln, Ile, Leu, Lys and Met are also well known. This has lead to the idea of rotamer libraries that reflect the most probable side chain conformations in elements of secondary structure (see Table 3.3).

Proline does not form helical structure for the obvious reason that the absence of an amide proton (NH) precludes hydrogen bonding whilst the side chain covalently bonded to the N atom restricts backbone rotation. The result is that proline often locates at the beginning of helices or in turns between two α helical units. Occasionally proline is found in a long helical region but invariably a major effect is to distort the helix causing a kink or change of direction of the polypeptide backbone.

Table 3.3 The range or distribution of χ_1 bond angles in different rotamer populations

Rotamer	χ_1 angle
g^+ (gauche $^+$)	−120° – 0°
trans	120° −240°
g^- (gauche $^-$)	0° – 120°

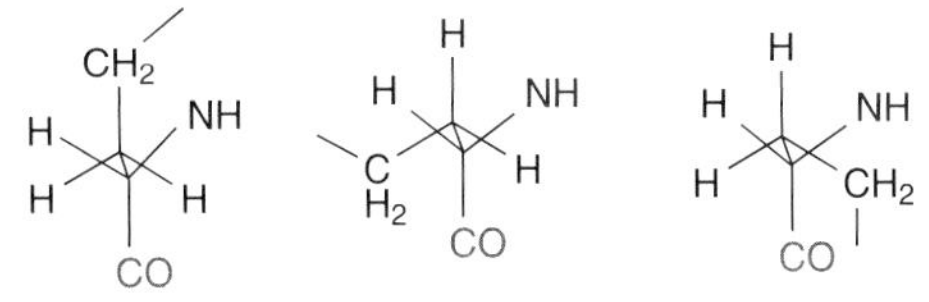

Figure 3.6 The CA–CB bond and different rotamer populations. Instead of looking directly along the CA–CB bond the bond is offset slightly to aid viewing. The front face represents the 'backbone' portion of the molecule. A clockwise rotation is defined as positive (+) whilst a counter clockwise rotation is negative (−) and leads to angles between methylene and CO group of ~180°, −60° and +60°

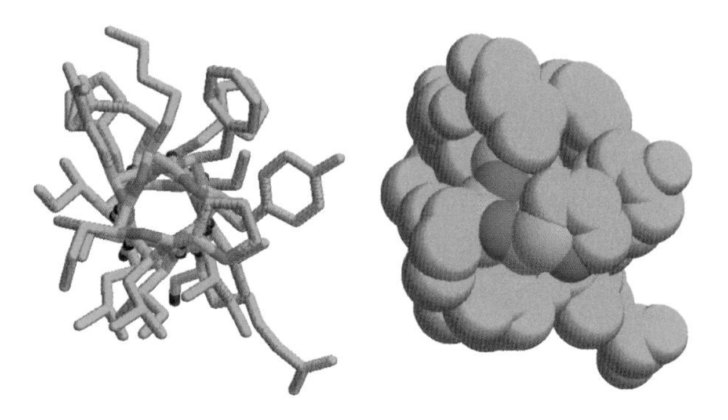

Figure 3.7 Wireframe and space-filling representations of one end of a regular α helix

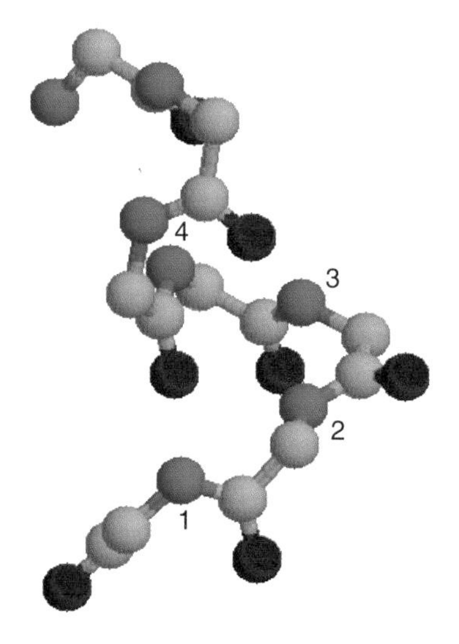

Figure 3.8 A 3_{10} helix shows many similarities to a regular α helix. Closer inspection of the hydrogen bond donor and acceptor reveals i, $i + 3$ connectivity, three residues per turn and dihedral angle values of ∼−50 and −25 reflect a more tightly coiled helix (see Table 3.2)

The α helix is frequently portrayed in textbooks including this one with a 'ball and stick' or 'wireframe' representation (Figure 3.7). A better picture is provided by 'space-filling' representations (Figure 3.7) where atoms are shown with their van der Waals radii. In the 'end-on' view of the α helix the wireframe representation suggests a hollow α helix. In contrast, the space filling representation emphasizes that little space exists anywhere along the helix backbone. Other representations of helices include cylinders showing the length and orientation of each helix or a ribbon representation that threads through the polypeptide chain. The preceding section views the helix as a stable structure but experiments with synthetic poly-amino acids suggest that very few polymers fold into regular helical conformation.

Other helical conformations

The 3_{10} helix is a structural variation of the α helix found in proteins (Figure 3.8). The 3_{10} helices are often found in proteins when a regular α helix is distorted by the presence of unfavourable residues, near a turn region or when short sequences fold into helical conformation. In the 3_{10} helix the dominant hydrogen bonds are formed between residues i, $i + 3$ in contrast to (i, $i + 4$) bonds seen in the regular α helix. The designation 3_{10} refers to the number of backbone atoms located between the donor and acceptor atoms (10) and the fact that there are three residues per turn. With three residues per turn the 3_{10} helix is a tighter, narrower structure in which the potential for unfavourable contacts between backbone or side chain atoms is increased.

Whilst the 3_{10} helix is a narrower structure than the α helix a third possibility is a more loosely coiled helix with hydrogen bonds formed between the CO and NH groups separated by five residues (i, $i + 5$). This structure is the π helix and at one stage it was thought not to occur naturally. However, examples of this helix structure have been described in proteins. Soyabean lipoxygenase has a 43-residue helix containing regions essential to enzyme function and stability running through the centre of the molecule. Three turns of an expanded helix with eight (i, $i + 5$) hydrogen bonds and 4.4 residues per turn make it an example of a π helix containing more than one turn found in a protein (Figure 3.9).

The rarity of this form of secondary structure arises for a number of reasons. One major limitation is that the ϕ/ψ angles of a π helix lie at the edge of the allowed, minimum energy, region of the Ramachandran

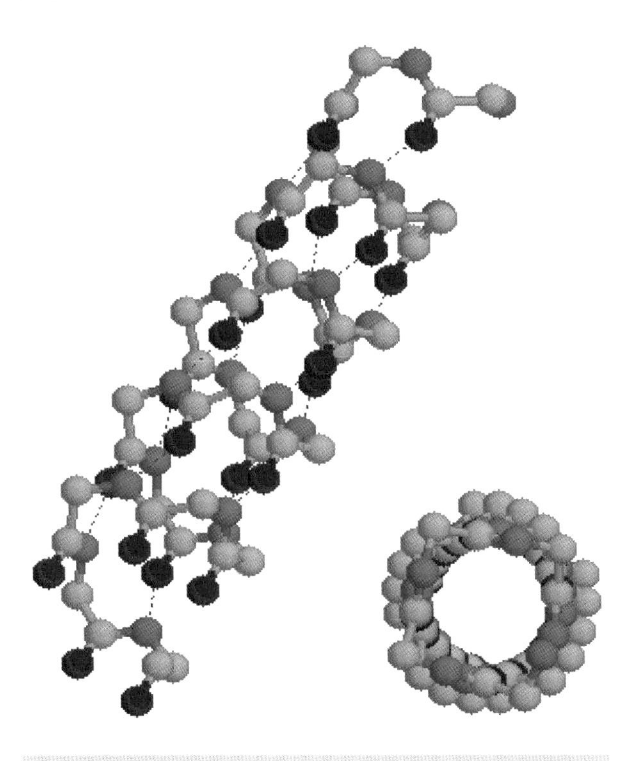

Figure 3.9 The π helix showing i, $i + 5$ hydrogen bonds and the end on view of the helix

plot. The large radius of the π helix means that backbone atoms do not make van der Waals contact across the helix axis leading to the formation of a hole down the middle of the helix that is too small for solvent occupation.

The β strand

The β strand, so called because it was the second unit of secondary structure predicted from the model-building studies of Pauling and Corey, is an extended conformation when compared with the α helix. Despite its name the β strand is a helical arrangement although an extremely elongated form with two residues per turn and a translation distance of 0.34 nm between similar atoms in neighbouring residues. Although less easy to recognize, this leads to a pitch or repeat distance of nearly 0.7 nm in a regular β strand (Figure 3.10). A single β strand is not stable largely because of the limited number of local stabilizing interactions. However, when two or more β strands form additional hydrogen bonding interactions a stable sheet-like arrangement is created (Figure 3.11). These β sheets result in significant increases in overall stability and

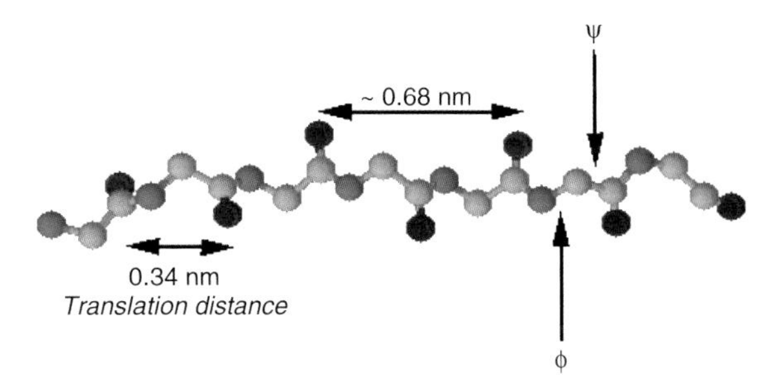

Figure 3.10 The polypeptide backbone of a single β strand showing only the heavy atoms

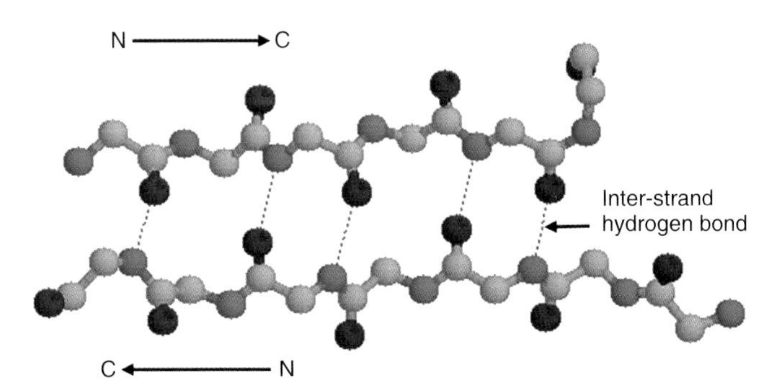

Figure 3.11 Two adjacent β strands are hydrogen bonded to form a small element of β sheet. The hydrogen bonds are inter-strand between neighbouring CO and NH groups. Only the heavy atoms are shown in this diagram for clarity

are stabilized by the formation of backbone hydrogen bonds between adjacent strands that may involve residues widely separated in the primary sequence.

Adjacent strands can align in parallel or antiparallel arrangements with the orientation established by determining the direction of the polypeptide chain from the N- to the C-terminal. 'Cartoon' representations of β strands make establishing directions in molecular structures easy since β strands are often shown as arrows; the arrowheads indicate the direction towards the C terminal. These cartoons take different forms but all 'trace' the arrangement of the polypeptide backbone and summarize secondary structural elements found in proteins without the need to show large numbers of atoms.

Polyamino acid chains do not form β sheets when dispersed in solution and this has hindered study of the formation of such structures. However, despite this observation many proteins are based predominantly on β strands, with chymotrypsin a proteolytic enzyme being one example (see Figure 3.12).

Unlike ideal representations strands found in proteins are often distorted by twisting that arises from a systematic variation of dihedral angles (ϕ and ψ) towards more positive values. The result is a slight, but discernable, right hand twists in the polypeptide chain. In addition when strands hydrogen bond together to form sheets further distortions occur especially with mixtures of anti-parallel and parallel β strands. On average β sheets containing antiparallel strands are more common than sheets made up entirely of parallel strands. Anti-parallel sheets often form from just two β strands running in opposite directions whilst it is observed that at least four β strands are required to form parallel sheets. β strands associate spectacularly into extensive curved sheets known as β barrels. The β barrel is found in many proteins (Figure 3.13) and consists of eight parallel β strands linked together by helical segments.

Turns as elements of secondary structure

As more high-resolution structures are deposited in protein databases it has allowed turns from different proteins to be defined and compared in terms of

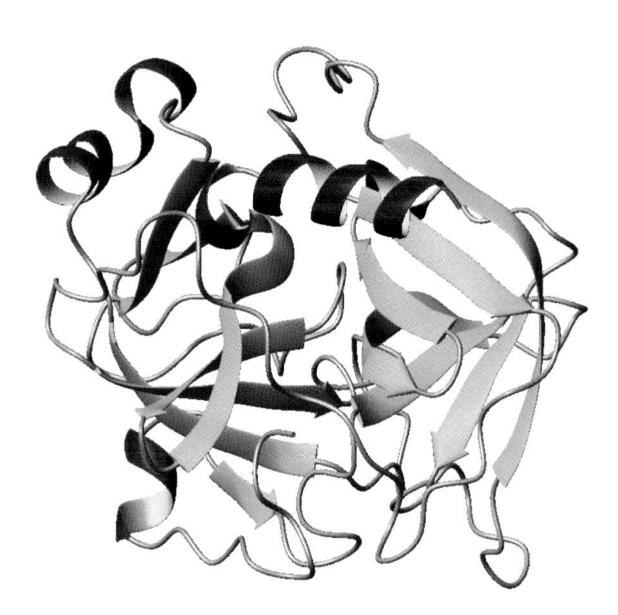

Figure 3.12 Representation of the elements of β strands and α helices found in the serine protease chymotrypsin. (PDB:2CGA). The strands shown in cyan have arrows indicating a direction leading from the N > C terminus, the helices are shown in red and yellow with turns in grey. There is no convention describing the use of colours to a particular element of secondary structure

Figure 3.13 The β barrel of triose phosphate isomerase seen in three different views. The eight β strands encompassed by helices are shown in the centre with two different views of the eight parallel β strands shown in the absence of helical elements. Eight strands are arranged at an angle of approximately 36° to the barrel axis (running from top to bottom in the right most picture) with each strand offset from the previous one by a constant amount

residue composition, angles and bond distances. In some proteins the proportion of residues found in turns can exceed 30 percent and in view of this high value it is unlikely that turns represent random structures. Turns have the universal role of enabling the polypeptide to change direction and in some cases to reverse back on itself. The reverse turns or bends arise from the geometric properties associated with these elements of protein structure.

Analysis of the amino acid composition of turns reveals that bulky or branched side chains occur at very low frequencies. Instead, residues with small side chains such as glycine, aspartate, asparagine, serine, cysteine and proline are found preferentially. An analysis of the different types of turns has established that perhaps as many as 10 different conformations exist on the basis of the number of residues making up the turn and the angles ϕ and ψ associated with the central residues. Turns can generally be classified according to the number of residues they contain with the most common number being three or four residues.

A γ turn contains three residues and frequently links adjacent strands of antiparallel β sheet (Figure 3.14). The γ turn is characterized by the residue in the middle of the turn ($i + 1$) not participating in hydrogen bonding whilst the first and third residues can form the final and initial hydrogen bonds of the antiparallel β strands. The change in direction of the polypeptide chain caused by a γ turn is reflected in the values of ϕ and ψ for the central residue. As a result of its size and conformational flexibility glycine is a favoured residue in this position although others are found.

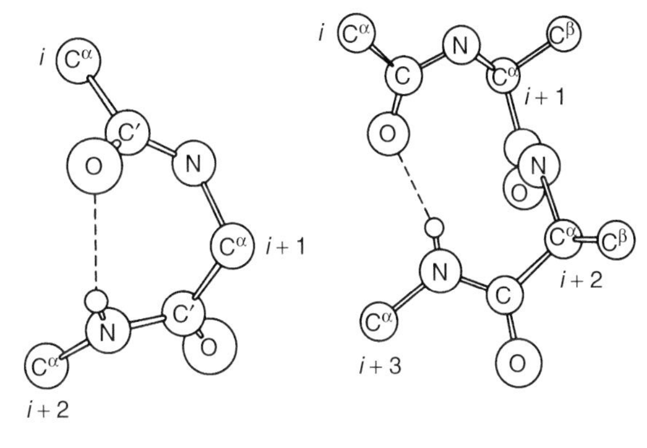

Figure 3.14 Arrangement of atoms in γ and β turns connected to strands of an antiparallel β sheet. γ turns contain three residues whilst β turns have four residues. A large number of variations on this basic theme exist in proteins. Hydrogen bonds are shown by a dotted line whilst only heavy atoms of the polypeptide backbone together with the CB atom are shown

More commonly found in protein structures are four residue turns (β turns). Here the middle two residues ($i + 1, i + 2$) are never involved in hydrogen bonding whilst residues i and $i + 3$ will participate in hydrogen bonding if a favourable arrangement forms between donor and acceptor. Analysis of structures deposited in protein databases reveals strong residue preferences. In the relatively common type 1β turn any residue can occur at position i, $i + 3$ with the exception that Pro is never found. More significantly glycine is often found at position $i + 3$ whilst proline predominates at $i + 1$. Asn, Asp, Cys and Ser are also found frequently in β-turns as the first residue.

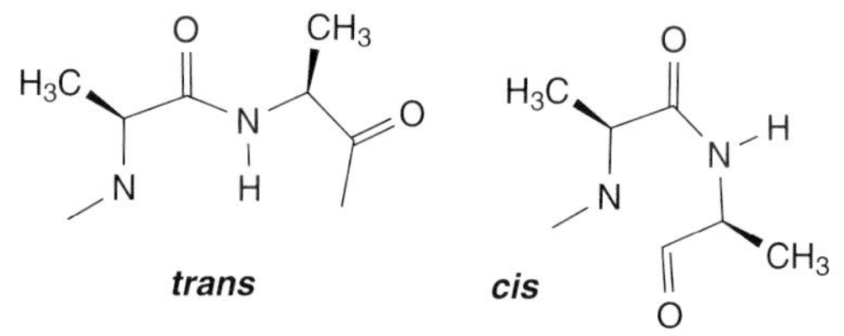

Figure 3.15 *Cis* and *trans* peptide bonds

Additional secondary structure

Both glycine and proline are associated with unusual conformational flexibility when compared with the remaining 18 residues. In the case of glycine there is very little restriction to either ϕ/ψ as a result of the small side chain (H). In the case of proline the opposite situation applies with ϕ, the angle defined by the N–Cα bond, restricted by the five-membered cyclic (pyrrolidine) ring. Proline residues are not suited to either helical or strand arrangements but are found at high frequency in turns or bends.

Additionally polyproline chains adopt unique and regular conformations distinct from helices, turns or strands. Two recognized conformations are called poly(Pro) I and poly(Pro) II. Proline is unique amongst the twenty residues in showing a much higher proportion of *cis* peptide bonds. The two forms of poly(Pro) therefore contain all *cis* (I) or all *trans* (II) peptide bonds. The value of ϕ is restricted by bond geometry to $-83°$ (I) and $-78°$ (II) and the torsion angle restrictions create poly(Pro) I as a right-handed helix with 3.3 residues per turn, whilst poly(Pro) II is a left-handed helix with three residues per turn. Poly(Gly) chains also adopt regular conformation and presents the opposite situation to poly(Pro) chains. Glycine has extreme conformational flexibility. In the solid state poly(Gly) has been shown to adopt two regular conformations designated I and II. State I has an extended conformation like a β strand whilst state II has three residues per turn and is similar to poly(Pro).

The Ramachandran plot

The peptide bond is planar as a result of resonance and its bond angle, **ω**, has a value of 0 or 180°. A peptide bond in the *trans* (Figure 3.15) conformation ($\omega = 180°$) is favoured over the *cis* (Figure 3.15) arrangement ($\omega = 0°$) by a factor of ~1000 because the preferential arrangement of non-bonded atoms leads to fewer repulsive interactions that otherwise decrease stability. In the *cis* peptide bond these non-bonded interactions increase due to the close proximity of side chains and C_α atoms with the preceding residue and hence results in decreased stability relative to the *trans* state. Peptide bonds preceding proline are an exception to this trend with a *trans/cis* ratio of approximately 4.

The peptide bond is relatively rigid, but far greater motion is possible about the remaining backbone torsion angles. In the polypeptide backbone C–N–C_α–C defines the torsion angle ϕ whilst N–C_α–C–N defines ψ. In practice these angles are limited by unfavourable close contacts with neighbouring atoms and these steric constraints limit the conformational space that is sampled by polypeptide chains. The allowed values for ϕ and ψ were first determined by G.N. Ramachandran using a 'hard sphere model' for the atoms and these values are indicated on a two-dimensional plot of ϕ against ψ that is now called a Ramachandran plot (Figure 3.16).

In the Ramachandran plot shown in Figure 3.16 the freely available conformational space is shaded in green. This represents ideal geometry and is exhibited by regular strands or helices. Analysis of crystal structures determined to a resolution of <2.5 Å showed that over 80 percent of all residues are found in this region of the Ramachandran plot. The yellow region indicates areas that although less favourable can be formed with small deviations from the ideal angular values for ϕ or ψ. The yellow and green regions include 95 percent of all residues within a protein. Finally, the purple coloured region, although much less favourable,

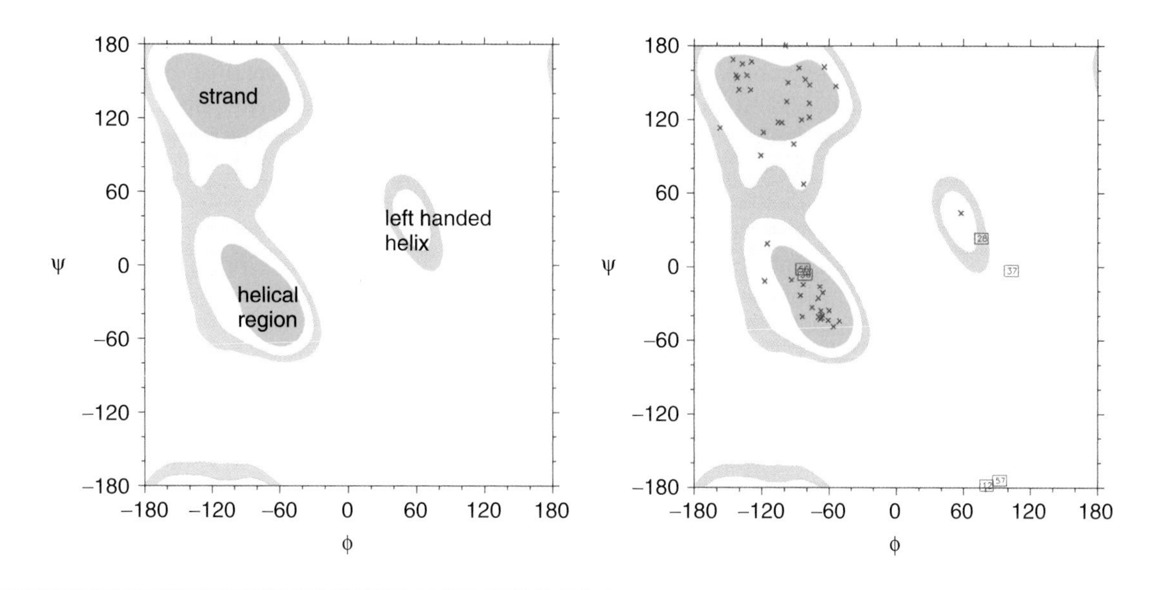

Figure 3.16 Ramachandran plots showing favourable conformational parameters for different ϕ/ψ values. Left: Ramachandran plot showing the ϕ/ψ angles exhibited by regular α helix and β strands. In addition the left-handed helix has a region of limited stability and although not widely shown in protein structures isolated residues do adopt this conformation. Right: Ramachandran plot derived from the crystal structure of bovine pancreatic trypsin inhibitor PDB: 1BPI. The residue numbers for glycine are shown. In BPTI some of these residues do not exhibit typical ϕ/ψ angles

will account for 98 percent of all residues in proteins. All other regions are effectively disallowed with the minor exception of a small region representing left-handed helical structure. In total only 30 percent of the total conformational space is available suggesting that the polypeptide chain itself imposes severe restrictions.

One exception to this rule is glycine. Glycine lacks a C_β atom and with just two hydrogen atoms attached to the C_α centre this residue is able to sample a far greater proportion of the space represented in the Ramachandran plot (Figure 3.17). For glycine this leads to a symmetric appearance for the allowed regions. As expected residues with large side chains are more likely to exhibit unfavourable, non-bonded, interactions that limit the possible values of ϕ and ψ. In the Ramachandran plot the allowed regions are smaller for residues with large side chains such as phenylalanine, tryptophan, isoleucine and leucine when compared with, for example, the allowed regions for alanine.

Similar plots of χ_1 versus χ_2 describe their distribution in proteins and it was found that the distribution

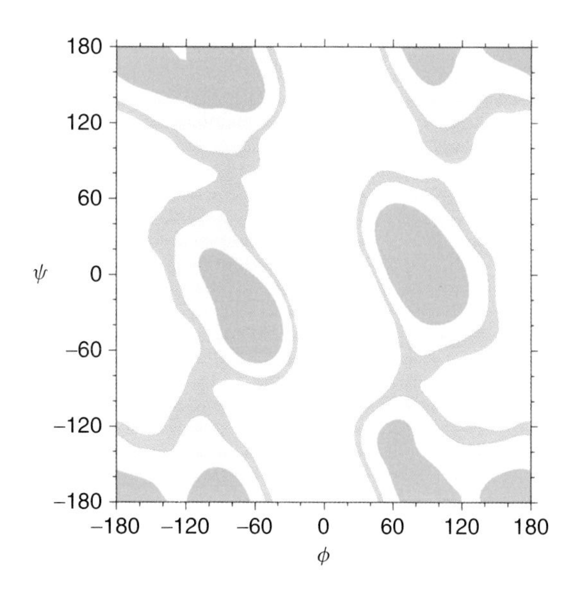

Figure 3.17 Ramachandran plot showing conformational space 'sampled' by glycine residues

of χ_1 and χ_2 side-chain torsion angles followed simple energy-based calculations with preferences for values of approximately $+60^\circ$, 180° and -60° for χ_1 and χ_2 in aliphatic side chains, and $+90^\circ$ and -90° for the χ_2 torsion angle of aromatic residues. Inspection of χ_1/χ_2 distribution reveals that several residues display preferences for certain combinations of torsion angles. Leucine residues prefer the combinations $-60^\circ/180^\circ$ and $180^\circ/+60^\circ$. This approach has led to the derivation of a library of preferred, but not obligatory, side-chain rotamers. These libraries are useful in the refinement of protein structures and include residue-specific preferences for combinations of side-chain torsion angles. Valine, for example, shows the greatest preference for one rotamer (χ_1 is predominantly t) and is unique in this respect amongst the side chains.

Tertiary structure

At the beginning of 2004 ~22 000 sets of atomic coordinates were deposited in databases such as the Protein Data Bank (PDB) and this value increases daily with the deposition of new structures from increasingly diverse organisms (see Figure 3.18). This database is currently maintained at Rutgers University with several mirrors (identical sites) found at other sites across the world (http://www.rcsb.org/pdb). Whilst some of these proteins are duplicated or related (there are over 50 structures of T4 lysozyme and over 200 globin structures) the majority represent the determination of novel structures using mainly X-ray diffraction and nuclear magnetic resonance (NMR) spectroscopy. Over 1000 different protein folds have been discovered to date with undoubtedly more to follow.

PDB files contain information of the positions in space of the vast majority of atoms making up a protein. By identifying the x, y and z coordinate of each atom we define the whole molecule. However, besides the x, y and z coordinates of atoms PDB files also contain header information describing the primary sequence, the method used to determine the structure, the organism from which this protein was derived, the elements of secondary structure, any post-translational modifications as well as the authors. As a consequence of the great development in bioinformatics PDB files are undergoing homologization to enable easy comparison between structures and to permit widespread use with different computer software packages. Figure 3.19 shows a representative PDB file.

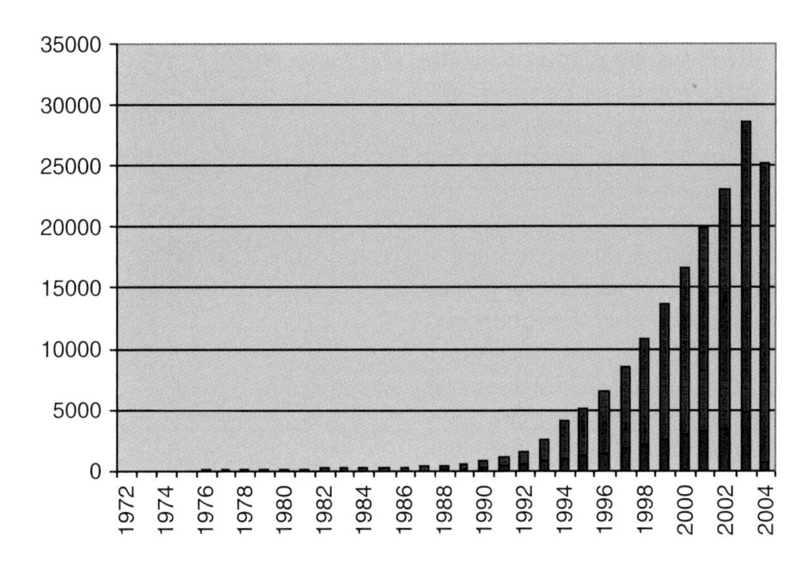

Figure 3.18 Data produced by the Protein Data Bank on the deposition of structures. From Berman, H.M. *et al.* The Protein Data Bank. *Nucl. Acids Res.* 2000 **28**, 235–242. See http://www.rcsb.org/pdb for latest details. Red blocks indicate total structures, purple blocks the number of submissions per year

```
HEADER    HYDROLASE(ZYMOGEN)                          16-JAN-87   2CGA      2CGA   3
COMPND    CHYMOTRYPSINOGEN *A                                               2CGA   4
SOURCE    BOVINE (BOS $TAURUS) PANCREAS                                     2CGA   5
AUTHOR    D.WANG,W.BODE,R.HUBER                                             2CGA   6
>>>>>>
>>>>>>
JRNL        AUTH   D.WANG,W.BODE,R.HUBER                                    2CGA   8
JRNL        TITL   BOVINE CHYMOTRYPSINOGEN *A. X-RAY CRYSTAL STRUCTURE      2CGA   9
>>>>>>
REMARK   1                                                                  2CGA  14
REMARK   1 REFERENCE 1                                                      2CGA  15
>>>>>>
SEQRES   1 A  245   CYS GLY VAL PRO ALA ILE GLN PRO VAL LEU SER GLY LEU      2CGA  71
SEQRES   2 A  245   SER ARG ILE VAL ASN GLY GLU GLU ALA VAL PRO GLY SER      2CGA  72
SEQRES   3 A  245   TRP PRO TRP GLN VAL SER LEU GLN ASP LYS THR GLY PHE      2CGA  73
>>>>>>
>>>>>>
SEQRES  17 B  245   LEU VAL GLY ILE VAL SER TRP GLY SER SER THR CYS SER      2CGA 106
SEQRES  18 B  245   THR SER THR PRO GLY VAL TYR ALA ARG VAL THR ALA LEU      2CGA 107
SEQRES  19 B  245   VAL ASN TRP VAL GLN GLN THR LEU ALA ALA ASN              2CGA 108
>>>>>>
CRYST1   59.300   77.100  110.100  90.00  90.00  90.00 P 21 21 21    8      2CGA 113
ORIGX1      1.000000  0.000000  0.000000        0.00000                     2CGA 114
ORIGX2      0.000000  1.000000  0.000000        0.00000                     2CGA 115
ORIGX3      0.000000  0.000000  1.000000        0.00000                     2CGA 116
SCALE1       .016863  0.000000  0.000000        0.00000                     2CGA 117
SCALE2      0.000000   .012970  0.000000        0.00000                     2CGA 118
SCALE3      0.000000  0.000000   .009083        0.00000                     2CGA 119
MTRIX1   1   .987700   .155000   .017700        6.21700    1                2CGA 120
MTRIX2   1   .022800  -.031400  -.999200      115.61600    1                2CGA 121
MTRIX3   1  -.154300   .987400  -.034600       -3.74800    1                2CGA 122
ATOM      1  N    CYS A   1     -10.656  55.938  41.808  1.00 11.66         2CGA 123
ATOM      2  CA   CYS A   1     -10.044  57.246  41.343  1.00 11.66         2CGA 124
ATOM      3  C    CYS A   1     -10.076  58.323  42.431  1.00 11.66         2CGA 125
ATOM      4  O    CYS A   1     -10.772  58.097  43.448  1.00 11.66         2CGA 126
ATOM      5  CB   CYS A   1     -10.807  57.718  40.066  1.00 11.66         2CGA 127
>>>>>>
>>>>>>
>>>>>>
ATOM    744  N    ASN A 100     -13.152  77.724  22.378  1.00  8.65         2CGA 866
ATOM    745  CA   ASN A 100     -14.213  76.940  23.011  1.00  8.65         2CGA 867
ATOM    746  C    ASN A 100     -14.134  75.441  22.693  1.00  8.65         2CGA 868
ATOM    747  O    ASN A 100     -13.706  75.062  21.563  1.00  8.65         2CGA 869
>>>>>>
>>>>>>
ATOM   1461  N    VAL A 200      -9.212  70.793  39.923  1.00  9.30         2CGA1583
ATOM   1462  CA   VAL A 200      -9.875  69.689  40.639  1.00  9.30         2CGA1584
ATOM   1463  C    VAL A 200     -10.634  70.148  41.868  1.00  9.30         2CGA1585
ATOM   1464  O    VAL A 200     -10.151  70.985  42.657  1.00  9.30         2CGA1586
>>>>>>
HETATM 3601  O    HOH     601     -20.008  66.224  26.138  1.00 26.69         2CGA3723
HETATM 3602  O    HOH     602     -21.333  66.182  28.756  1.00 18.10         2CGA3724
HETATM 3603  O    HOH     603     -18.000  68.022  22.774  1.00 34.03         2CGA3725
MASTER         60    3    0    0    0    0    0    9 3927    2    0   38      2CGAA  6
END                                                                         2CGA4053
```

Figure 3.19 An abbreviated version of a representative PDB file. The file 2CGA refers to the chymotrypsin. In the above example the structure was determined by X-ray crystallography and the initial lines (header and remarks) of a PDB file give the authors, important citations, source of the protein and other useful information. Later the primary sequence is described and this is followed by crystallographic data on the unit cell dimensions and space group. The important lines are those beginning with ATOM since collectively these lines list all heavy atoms of the protein together with their respective *x*, *y* and *z* coordinates. A line beginning HETATM lists the position of hetero atoms which might include co-factors but more frequently lists water molecules found in the crystal structure. The symbol ≫≫≫≫≫≫ is used here to denote the omission of many lines of text. Where more than one chain exists the coordinates will be listed as chain A, chain B, etc.

Ten years ago the representation of tertiary structures on flat pages of a book represented a major problem as well as creating conceptual difficulty for students. Although tertiary structures remain complex to visualize the widespread development of desktop computers capable of handling these relatively large files coupled with graphical software to represent the structures of proteins in different fashions has revolutionized this area. Many of the images of proteins shown in this book can be viewed as true 'three-dimensional' structures by consulting the on-line version of this book or by downloading the PDB file from one of the many available databases and using this file in a suitable molecular graphics package. An alternative sometimes used is to represent molecules as stereo images that can be viewed with either special glasses or more easily on computers equipped to display such structures again using special glasses. In this book stereo images have not been widely used purely for simplicity but students should endeavour to view such presentations because of their ability to convey space, folding and depth in highly complex structures.

Numerous software packages have been developed for viewing PDB files and each has its own advantages and disadvantages as well as its own supporters and detractors within the scientific community. Many of these packages are public domain software whilst others have been developed as commercial entities. However, there is little doubt that as a result of this important, yet often under-rated, development it has become easier to portray tertiary structures conveying their complexity and beauty.

Detailed tertiary structure

The tertiary structure represents the folded polypeptide chain. It is defined as the spatial arrangement of amino acid residues that are widely separated in the primary sequence or more succinctly as the overall topology formed by the polypeptide (see Figure 3.20). For small globular proteins of 150 residues or less the folded

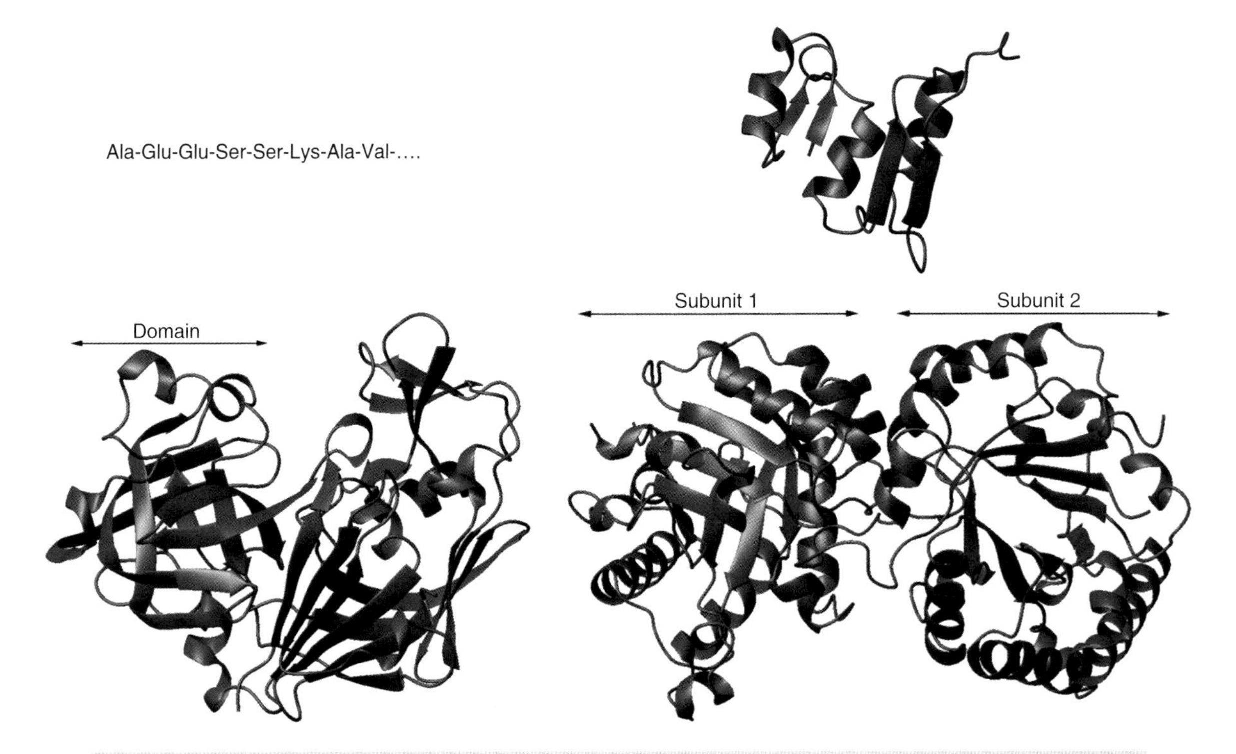

Figure 3.20 Four levels of organization within proteins; primary, secondary, tertiary and quaternary

structure involves a spherical compact molecule composed of secondary structural motifs with little irregular structure. Disordered or irregular structure in proteins is normally confined to the N and C terminals or more rarely to loop regions within a protein or linker regions connecting one or more domains. Asking the question 'what is the tertiary structure of a protein?' is synonymous with asking 'what is the protein fold?.' The fold arises from linking together secondary structures forming a compact globular molecule. Elements of secondary structure interact via hydrogen bonds, as in β sheets, but also depend on disulfide bridges, electrostatic interactions, van der Waals interactions, hydrophobic contacts and hydrogen bonds between non-backbone groups.

Interactions stabilizing tertiary structure

To form stable tertiary structure proteins must clearly form *more* attractive interactions than unfavourable or repulsive ones. The formation of stable tertiary folds relies on interactions that differ in their relative strengths and frequency in proteins.

Disulfide bridges

Disulfide bridges dictate a protein fold by forming strong covalent links between cysteine side chains that are often widely separated in the primary sequence. A disulfide bridge cannot form between consecutive cysteine residues and it is normal for each cysteine to be separated by at least five other residues. The formation of a disulfide bridge restrains the overall conformation of the polypeptide and in bovine pancreatic trypsin inhibitor (BPTI; Figure 3.21), a small protein of 58 residues, there are three disulfide bridges formed between residues 5–55, 14–38 and 30–51. The effect is to bring the secondary structural elements closer together. Disulfide bonds are only broken at high temperatures, acidic pH or in the presence of reductants. In BPTI reduction of the disulfide bonds leads to decreased protein stability that is mirrored by other disulfide-rich proteins.

The hydrophobic effect

The importance of the hydrophobic effect was for a long time underestimated. Charged interactions and hydrogen bonds are not strong *intra*molecular forces because water molecules compete significantly with these effects. However, water is a very poor solvent for many non-polar molecules and this is exemplified by dissolving an organic solute such as cyclohexane in water. Non-polar molecules cannot form hydrogen bonds with water and this prevents molecules such as cyclohexane dissolving extensively in aqueous solutions. As a consequence interactions between water and non-polar molecules are weakened and may be virtually non-existent. The result is an enhancement of interactions between non-polar molecules and the formation of hydrophobic clusters within water. The enhanced interactions between non-polar molecules in the presence of water are the basis for the hydrophobic effect. Since the side chains of many amino acid residues are hydrophobic it is clear that the hydrophobic effect may contribute significantly to intramolecular interactions. The hydrophobic effect can be restated as the preference of non-polar atoms for non-aqueous environments.

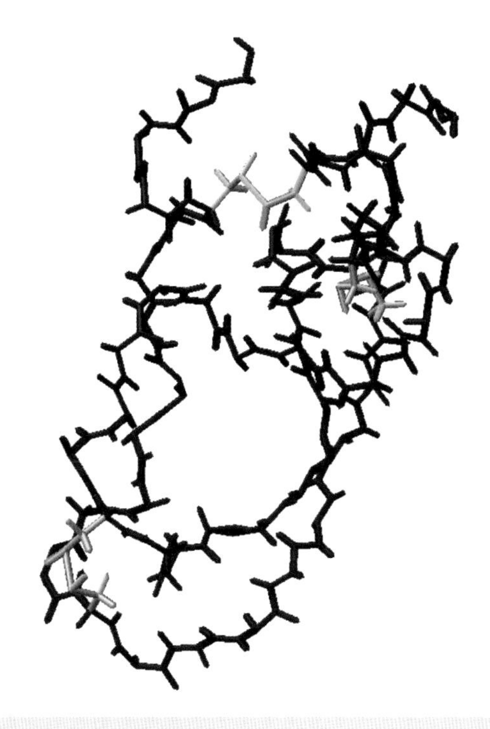

Figure 3.21 Structure of BPTI (PDB: 5PTI) showing the protein fold held by three disulfide bridges (5–55, 14–38 and 30–51)

The magnitude of the hydrophobic effect has proved difficult to estimate but has been accomplished by measuring the free energy associated with transfer of a non-polar solvent into water from the gaseous, liquid or solid states. The thermodynamics governing the transfer of non-polar molecules between phases are complicated but it is worth remembering that the enthalpy change, ΔH, represents changes in non-covalent interactions in going between the two phases whilst the entropy change (ΔS) reflects differences in the order of each system. A summation of these two terms gives ΔG_{tr}, the free energy of transfer of a non-polar or hydrophobic molecule from one phase to another, via the relationship $\Delta G = \Delta H - T\Delta S$, where T is temperature. The overall process involves the transfer of a solute molecule into the aqueous phase by (i) creating a cavity in the water, (ii) adding solute to the cavity, and (iii) maximizing favourable interactions between solute molecules and between solvent molecules.

In ice water molecules are arranged in a regular crystal lattice that maximizes hydrogen bonding with other water molecules, forming on average four hydrogen bonds. In the liquid phase these hydrogen bonds break and form rapidly with an estimated half-life of less than 1 ns. This leads to each water molecule forming an average of 3.4 hydrogen bonds. The capacity for hydrogen bonding and intermolecular attraction accounts for the high boiling point of water and the relatively large amounts of energy required to break these interactions, especially when compared with the interactions between non-polar liquids such as cyclohexane (Figure 3.22). Here the C–H bonds show little tendency to hydrogen bond and this is true for most of the non-polar side chains of amino acids in proteins. The hydrophobic interaction does not derive from the interaction between non-polar molecules or from the interaction between water and non-polar solutes because hydrogen bonds do not form. The driving force for the formation of hydrophobic clusters is the tendency for water molecules to hydrogen bond with each other. Water forms hydrogen bonded networks around non-polar solutes to become more ordered and one extreme example of this effect is the observation of clathrates where water forms an ordered cage around non-polar solutes. Around the cage water forms with maximum hydrogen bonding but each bond has less than optimal geometry.

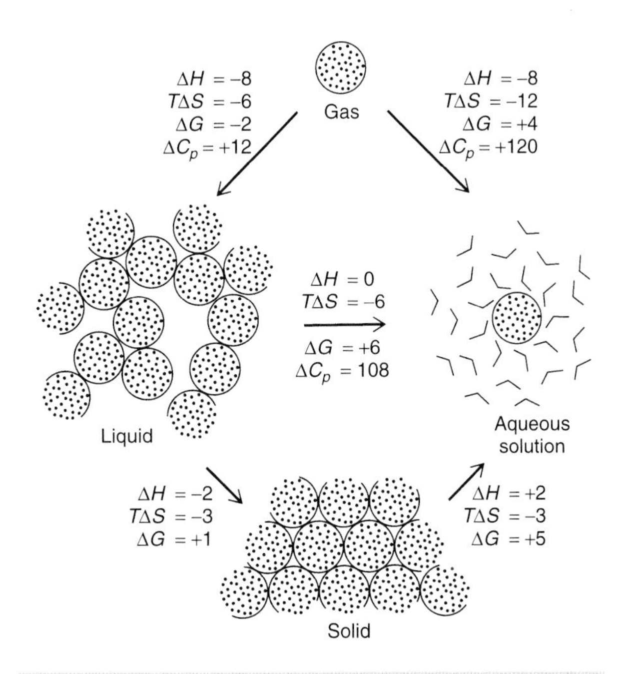

Figure 3.22 Transfer of a non-polar solute (cyclohexane) from gas, liquid and solid phases to aqueous solution around room temperature. The hydrophobic interaction is temperature dependent and the respective enthalpy (ΔH), entropy ($T\Delta S$) and free energy (ΔG) of transfer are given in kcal mol^{-1}. (Reproduced courtesy of Creighton, T.E. *Proteins Structure & Molecular Properties*, 2nd edn. W.H. Freeman, 1993)

The energetics of the hydrophobic interaction can now be explored in more detail to provide a physical basis for the phenomena in protein structure. The most interesting transitions are those occurring from liquid phases involving the transfer of a non-polar solute from a non-polar liquid to water. The adverse ordering of water around the solute leads to an unfavourable decrease in entropy, whilst the ΔH_{tr} term is approximately zero. It is the entropic factor that dominates in the transfer of a non-polar solute into aqueous solutions with enthalpy terms reflecting increased hydrogen bonding.

The temperature dependence of the hydrophobic interaction provides still more clues concerning the process. As the temperature is increased water around non-polar solutes is disrupted by breaking hydrogen bonds and becomes more like bulk water. This process is reflected by a large heat capacity (ΔC_p) that accompanies the hydrophobic interaction and is characteristic of this type of interaction. The large change in heat capacity underpins the temperature dependency of the hydrophobic interaction through its effect on both enthalpic and entropic terms. The magnitude of C_p is related to the non-polar surface area of the solute exposed to water along with many other thermodynamic parameters. As a result a large number of correlations have been made between accessible surface area, solubility of non-polar solutes in aqueous solutions and the energetics associated with the hydrophobic interaction. The hydropathy index established for amino acids is just one manifestation of the hydrophobic interaction described here.

Charge–charge interactions

These interactions occur between the side chains of oppositely charged residues as well as between the NH_3^+ and COO^- groups at the ends of polypeptide chains (Figure 3.23). Of importance to charge–charge interactions[1] are the side chains of lysine, arginine and histidine together with the side chains of aspartate, glutamate and to a lesser extent tyrosine and cysteine.

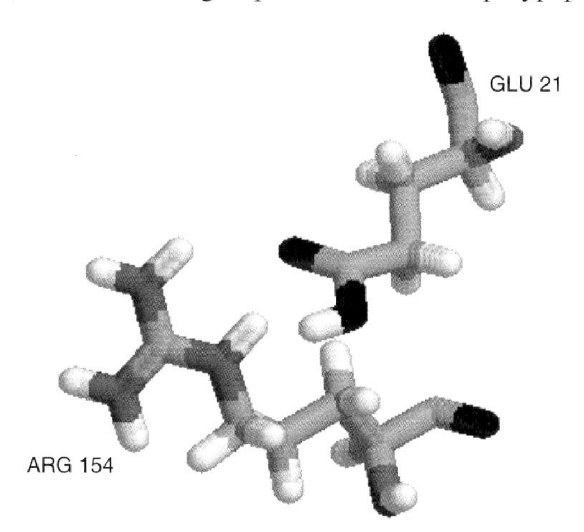

Figure 3.23 An example of electrostatic or charge-charge interactions occurring in proteins. The example shown is from the protease chymotrypsin (PDB: 2CGA). Here the interaction is between arginine and glutamate side chains. The donor atom is the hydrogen of the NE nitrogen whilst the acceptor groups are both oxygen atoms, OE1 and OE2. Reproduced courtesy Creighton, T.E. Proteins Structure & Molecular Properties, 2nd edn. W.H. Freeman, 1993

As a result of their charge the side chains of these residues are found on the protein surface where interactions with water or solvent molecules dramatically weaken these forces. In view of their low frequency and solvated status these interactions do not usually contribute significantly to the overall stability of a protein fold. Coulomb's law describes the potential energy (V) between two separated charges (in a perfect vacuum) according to the relationship

$$V = q_1\ q_2/\varepsilon\ 4\pi\ r^2 \tag{3.1}$$

where q_1 and q_2 are the magnitude of the charges (normally $+1$ and -1), r is the separation distance between these charges and ε is the permittivity of free space. In other media, such as water this equation is modified to

$$V = q_1\ q_2/\varepsilon_o\ 4\pi\ r^2 \tag{3.2}$$

where ε_o is the permittivity of the medium and is related to the dielectric constant (ε_r) by

$$\varepsilon_r = \varepsilon/\varepsilon_o \tag{3.3}$$

When the medium is water this has a value of approximately ~80 whilst methanol has a value of ~34 and a hydrophobic solvent such as benzene has a value of ~2 (a perfect vacuum has a value of 1 by definition). It has been estimated that each charge–pair interaction located on the surface of a protein may contribute less than 5 kJ mol^{-1} to the overall stability of a protein and this must be compared with the much stronger disulfide bridge (~100–200 kJ mol^{-1}). Occasionally charge interactions exist within non-polar regions in proteins and under these conditions with a low dielectric medium and the absence of water the magnitude of the charge–charge interaction can be significantly greater.

A variation occurring in normal charge–charge interactions are the partial charges arising from hydrogen bonding along helices. Hydrogen bonding between amide and carbonyl groups gives rise to a net dipole moment for helices. In addition the peptide bonds pointing in the same direction contribute to the accumulative polarization of helices. As a result longer helices

[1] Sometimes called electrostatic, ionic or salt-bridge interactions.

have a greater macrodipole. The net result is that a small positive charge is located at the N-terminal end of a helix whilst a negative charge is associated with the C-terminal end. For a helix of average length this is equivalent to +0.5–0.7 unit charge at the N-terminus and −0.5–0.7 units at the C terminus. Consequently, to neutralize the overall effect of the helix macrodipole acidic side chains are more frequently located at the positive end of the helix whilst basic side chains can occur more frequently at the negative pole.

Hydrogen bonding

Hydrogen bonds contribute significantly to the stability of α helices and to the interaction of β strands to form parallel or antiparallel β sheets. As a result such hydrogen bonds contribute significantly to the *overall* stability of the tertiary structure or folded state. These hydrogen bonds are between main chain NH and CO groups but the potential exists with protein folding to form hydrogen bonds between side chain groups and between main chain and side chain. In all cases the hydrogen bond involves a donor and acceptor atom and will vary in length from 0.26 to 0.34 nm (this is the distance between heavy atoms, i.e. between N and O in a hydrogen bond of the type N–H- - - O=C) and may deviate in linearity by ±40°. In proteins other types of hydrogen bonds can occur due to the presence of donor and acceptor atoms within the side chains. This leads to hydrogen bonds between side chains as well as side chain–main chain hydrogen bonds. Particularly important in hydrogen bond formation are the side chains of tyrosine, threonine and serine containing the hydroxyl group and the side chains of glutamine and asparagine with the amide group. Frequently, side chain atoms hydrogen bond to water molecules trapped within the interior of proteins whilst at other times hydrogen bonds appear shared between two donor or acceptor groups. These last hydrogen bonds are termed bifurcated. Table 3.4 shows examples of the different types of hydrogen bond.

Table 3.4 Examples of hydrogen bonds between functional groups found in proteins

Residues involved in hydrogen bonding	Nature of interaction and typical distance between donor and acceptor atoms (nm)
Amide–carbonyl Gln/Asn -backbone	N–H—O=C ~0.29 nm
Amide-hydroxyl. Backbone-Ser Asn/Gln -Ser	N–H—O(H)–CH_2 ~0.3 nm
Amide-imidazole. Asn/Gln -His Backbone -His	N–H—N (imidazole ring, NH, H_2C) ~0.31 nm
Hydroxyl-carbonyl Ser/Thr/Tyr–backbone Ser-Asn/Gln Thr- Asn/Gln Tyr- Asn/Gln	H–O—O=C ~0.28 nm
Hydroxyl-hydroxyl Thr,Ser and Tyr Tyr-Ser	H–O—H–O ~0.28 nm

Van der Waals interactions

There are attractive and repulsive van der Waals forces that control interactions between atoms and are very important in protein folding. These interactions occur between adjacent, uncharged and non-bonded atoms and arise from the induction of dipoles due to fluctuating charge densities within atoms. Since atoms are continually oscillating the induction of dipoles is a constant phenomena. It is unwise to view van der Waals forces as simply the interaction between temporary dipoles; the interaction occurs between uncharged atoms and involves several different types. Unlike the electrostatic interactions described above they do not obey inverse square laws and they vary in their contributions to the overall intermolecular attraction. Three attractive contributions are recognized and include in order of diminishing strength: (i) the orientation effect or interaction between permanent dipoles; (ii) the induction effect or interaction between permanent and temporary dipoles; (iii) the dispersion effect or London force, which is the interaction between temporary, induced, dipoles.

The orientation effect is the interaction energy between two permanent dipoles and depends on their relative orientation. For a freely rotating molecule it might be expected that this effect would average out to zero. However, very few molecules are absolutely free to rotate with the result that preferred orientations exist. The energy of interaction varies as the inverse sixth power of the interatomic separation distance (i.e. $\propto r^{-6}$) and is inversely dependent upon the temperature. The induction effect also varies as r^{-6} but is independent of temperature. The magnitude of this effect depends on the polarizability of molecules. The dispersion effect, sometimes called the London force, involves the interactions between temporary and induced dipoles. It arises from a temporary dipole inducing a complementary dipole in an adjacent molecule. These dipoles are continually shifting, depend on the polarization of the molecule and result in a net attraction that varies as r^{-6}.[1]

When atoms approach very closely repulsion becomes the dominant and unfavourable interaction. The repulsive term is always positive but drops away dramatically as the distance between the two atoms increases. In contrast it becomes very large at short atomic distances and is usually modelled by r^{-12} distance dependency, although there are no experimental grounds for this relationship. The combined repulsive and attractive van der Waals terms are described by plots of the potential energy as a function of interatomic separation distance (Figure 3.24). The energy of the van der Waals reactions is therefore described

[1] The dispersion effect is so called because the movement of electrons underlying the phenomena cause a dispersion of light.

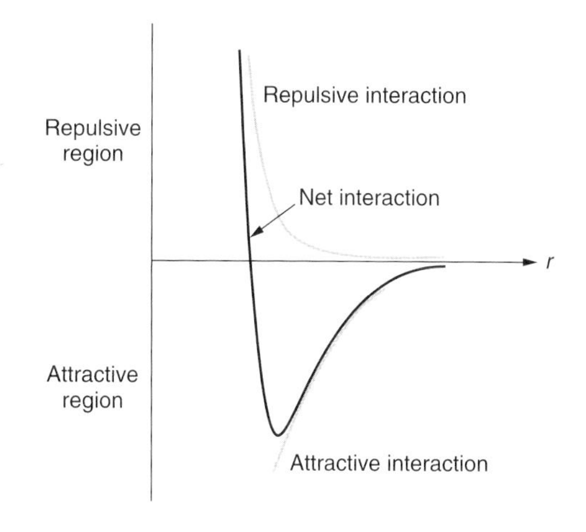

Figure 3.24 The van der Waals interaction as a function of interatomic distance r. The repulsive term increases rapidly as atoms get very close together

by the difference between the attractive (r^{-6}) and repulsive terms (r^{-12})

$$E_{\mathrm{vdw}} \propto -\mathrm{A}/r^{6} + \mathrm{B}/r^{12} \tag{3.4}$$

$$E_{\mathrm{vdw}} = \Sigma\ E\ [(r_{\mathrm{m}}/r)^{12} - 2(r_{\mathrm{m}}/r)^{6}] \tag{3.5}$$

where the van der Waals potential (E_{vdw}) is the sum of all interactions over all atoms, E is the depth of the potential well and r_{m} is the minimum energy interaction distance. This type of interaction is often called a 6–12 interaction, or the Leonard Jones potential. Although van der Waals forces are extremely weak, especially when compared with other forces influencing protein conformation, their large number arranged close together in proteins make these interactions significant to the maintenance of tertiary structure.

A protein's folded state therefore reflects the summation of attractive and repulsive forces embodied by the summation of the electrostatic, hydrogen bonding, disulfide bonding, van der Waals and hydrophobic interactions (see Table 3.5). It is often of considerable surprise to note that proteins show only marginal stability with the folded state being between 20 and 80 kJ mol^{-1} more stable than the unfolded state. The relatively small value of ΔG reflects the differences in non-covalent interactions between the folded and

Table 3.5 Distance dependence and bond energy for interactions between atoms in proteins

Bond	Distance dependence	Approximate bond energy (kJ mol^{-1})
Covalent	No simple dependence	~200
Ionic	$\propto 1/r^2$	<20
Hydrogen bond	No simple expression	<10
van der Waals	$\propto 1/r^6$	<5
Hydrophobic	No simple expression	<10

unfolded states. This arises because an unfolded protein normally has an identical *covalent* structure to the folded state and any differences in protein stability arise from the favourable non-covalent interactions that occur during folding.

The organization of proteins into domains

For proteins larger than 150 residues the tertiary structure may be organized around more than one structural unit. Each structural unit is called a domain, although exactly the same interactions govern its stability and folding. The domains of proteins interact together although with fewer interactions than the secondary structural elements within each domain. These domains can have very different folds or tertiary structures and are frequently linked by extended relatively unstructured regions of polypeptide.

Three major classes of domains can be recognized. These are domains consisting of mainly α helices, domains containing mainly β strands and domains that are mixed by containing α and β elements. In this last class are structures containing both alternating α/β secondary structures as well as proteins made up of collections of helices and strands (α + β). Within each of these three groups there are many variations of the basic themes that lead to further classification of protein architectures (see Figure 3.25).

Protein domains arise by gene duplication and fusion. The result is that a domain is added onto another protein to create new or additional properties. An example of this type of organization is seen in the cytochrome b_5 superfamily. Cytochrome b_5 is a small globular protein containing both α helices and β strands functioning as a soluble reductant to methaemoglobin in the erythrocyte or red blood cell. The cytochrome contains a non-covalently bound heme group in which the iron shuttles between the ferric (Fe^{3+}) and ferrous (Fe^{2+}) states (Figure 3.26). The reversible redox chemistry of the iron is central to the role of this protein as an electron carrier within erythrocytes. When united with a short hydrophobic chain this protein assumes additional roles and participates in the fatty acyl desaturase pathway. In mitochondria the enzyme sulfite oxidase converts sulfite to sulfate as part of the dissimilatory pathway for sulfur within cells. Sulfite oxidase contains cytochrome b_5 linked to a molybdenum-containing domain. Further use of the b_5-like fold occurs in nitrate reductase where a flavin binding domain is linked to the cytochrome. The result is a plethora of new proteins containing different domains that allow the basic redox role of cytochrome b_5 to be exploited and enhanced to facilitate the catalysis of new reactions.

A rigorous definition of a domain does not exist. One acceptable definition is the presence of an autonomously folding unit within a protein. Alternatively a domain may be defined as a region of a protein showing structurally homology to other proteins. However, in all cases domains arise from the folding of a *single* polypeptide chain and are distinguished from quaternary structure (see below) on this basis.

Super-secondary structure

The distinctions between secondary structure and super-secondary structure or between tertiary structure and super-secondary structure are not well defined. However, in some proteins there appears to be an intermediate level of organization that reflects groups of secondary structural elements but does not encompass all of the structural domain or tertiary fold. The β barrel found in enzymes such as triose phosphate isomerase could form an element of super secondary structure since it does not represent *all* of the

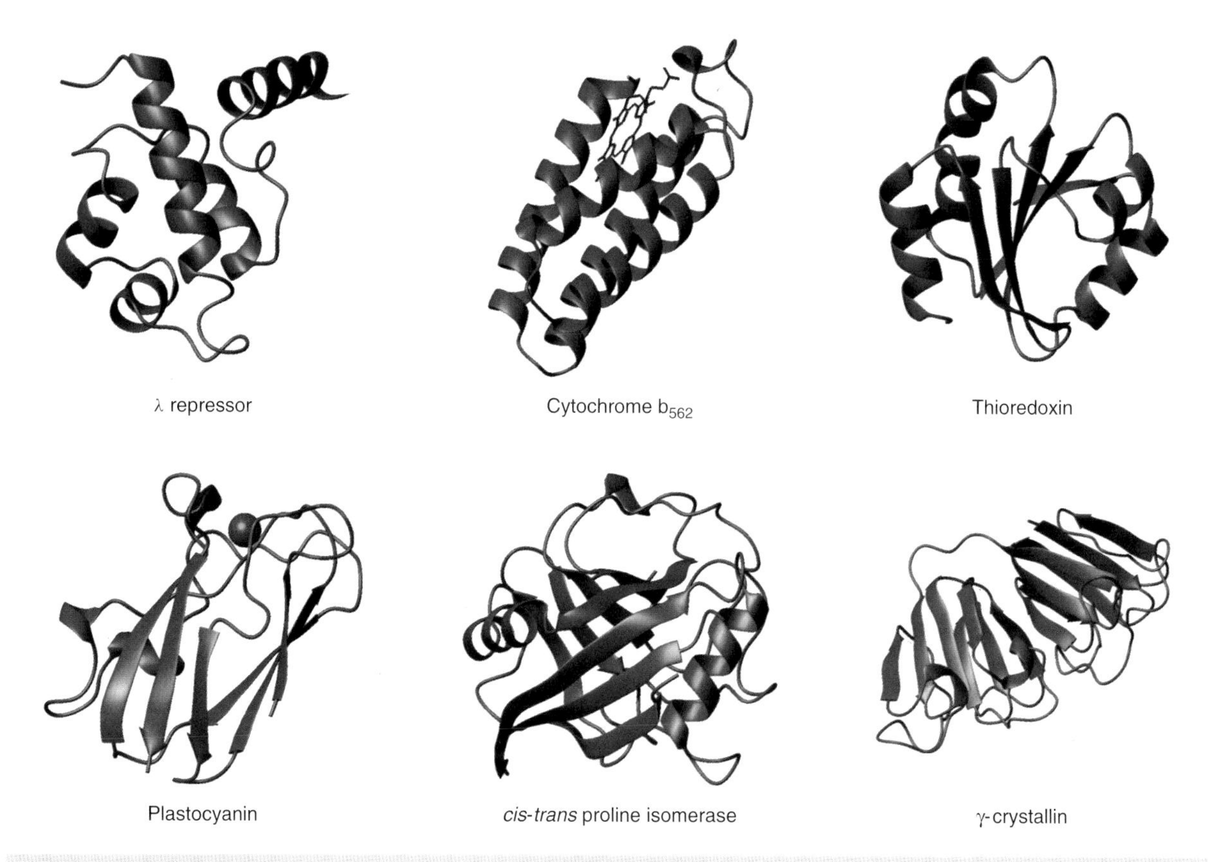

Figure 3.25 The secondary structure elements found in monomeric proteins. The λ repressor protein (PDB: 1LMB) contains the helix turn helix (HTH) motif; cytochrome b-562: (PDB: 256b) is a four-helix bundle heme binding domain; human thioredoxin: (PDB: 1ERU), a mixed α/β protein containing a five-stranded twisted β sheet. spinach plastocyanin: (PDB: 1AG6) a single Greek key motif binds Cu (shown in green); human *cis-trans* proline isomerase (PDB: 1VBS), a small extensive β domain containing a collection of strands that fold to form a 'sandwich'; human γ-crystallin: (PDB: 2GCR), two domains each of which is an eight-stranded β barrel type structure composed of two Greek key motifs

folded domain yet represents a far greater proportion of the structure than a simple β strand. Amongst other elements of super-secondary structure recognized in proteins are the β-α-β motif, Rossmann fold, four-helix bundles, the Greek-key motif and its variants, and the β meander. The cartoon representations in Figure 3.27 demonstrate the arrangement of β strands in some of these motifs and the careful viewer may identify these motifs in some figures in this chapter.

The β meander motif is a series of antiparallel β strands linked by a series of loops or turns. In the β meander the order of strands across the sheet reflects their order of appearance along the polypeptide sequence. A variation of this design is the so-called Greek key motif, and it takes its name from the design found on many ancient forms of pottery or architecture. The Greek key motif shown here links four antiparallel β strands with the third and fourth strands forming the outside of the sheet whilst strands 1 and 2 form on the inside or middle of the sheet. The Greek key motif can contain many more strands ranging from 4 to 13. The Cu binding metalloprotein plastocyanin contains eight β strands arranged in a Greek key motif.

A β sandwich forms normally via the interaction of strands at an angle and connected to each other

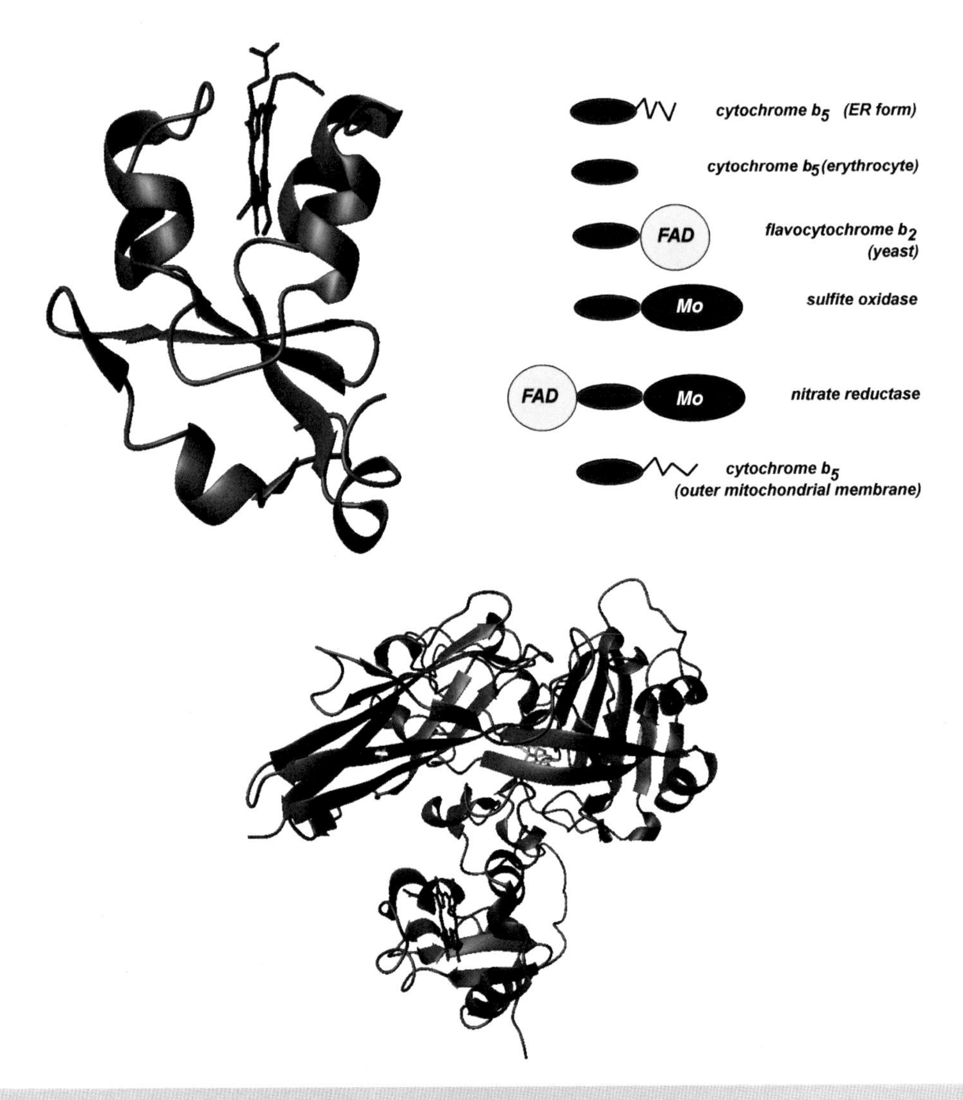

Figure 3.26 The heme binding domain of cytochrome b_5 (PDB:3B5C), a schematic representation of the duplication of this domain (red) to form part of multi-domain proteins by linking it to FAD domains, Mo-containing domains or hydrophobic tails. The arrangement of domains in proteins such as nitrate reductase, yeast flavocytochrome b_2 and sulfite oxidase is shown (left). The structure of sulfite oxidase (bottom) (PDB: 1SOX) shows the cytochrome domain in the foreground linked to a larger molybdenum-pterin binding domain

via short loops. In some cases these strands can originate from a different polypeptide chain but the emphasis is on two layers of β strands interacting together within a globular protein. The layers of the sandwich can be aligned with respect to each other or arranged orthogonally. An example of the second arrangement is shown in human *cis-trans* proline peptidyl isomerase.

The Rossmann fold, named after its discoverer Michael Rossmann, is an important super-secondary structure element and is an extension of the β-α-β domain. The Rossmann fold consists of three parallel β strands with two intervening α helices i.e. β-α-β-α-β. These units are found together as a dimer – so the Rossmann fold contains six β strands and four helices – and this collection of secondary structure

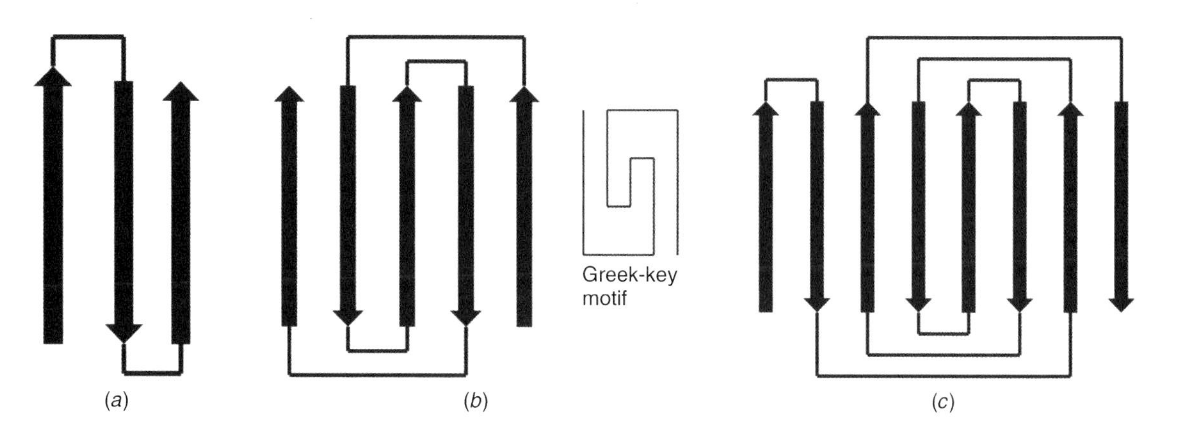

Figure 3.27 Cartoon representations of (a) the β meander, (b) Greek key and (c) Swiss or Jelly roll motifs

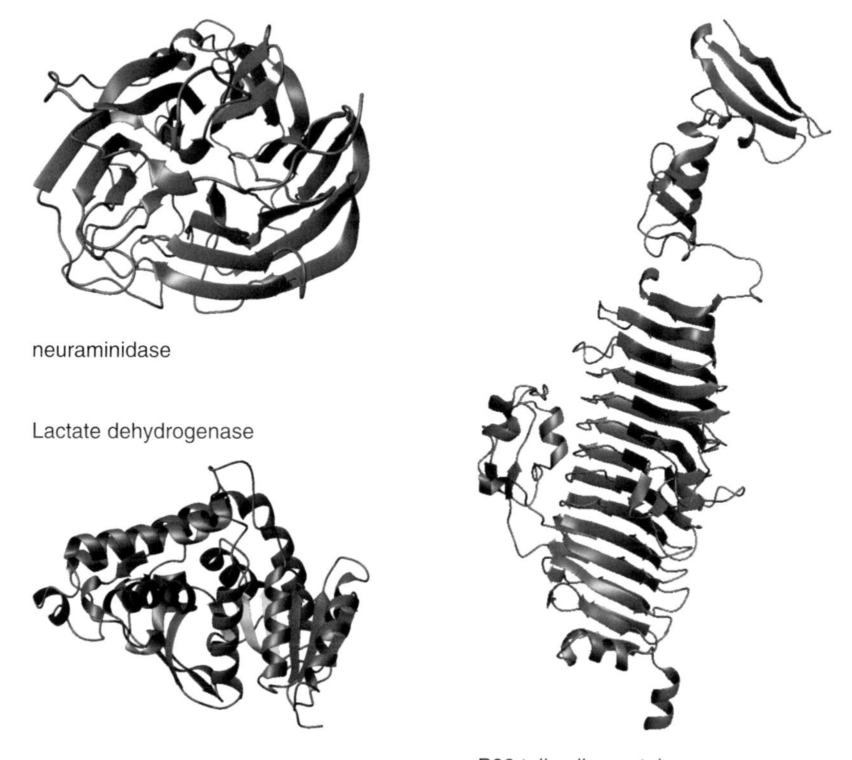

Figure 3.28 The structure of complex motifs involving β strands. Native influenza neuraminidase showing characteristic β propellor (PDB:1F8D). A β helix found in the tailspike protein of bacteriophage P22 (PDB:1TSP). The β-α-β-α-β motif is shown in the Rossmann fold of Lactate dehydrogenase with each set of three strands shown in different colours (PDB:1LDH)

frequently forms a nucleotide-binding site. Nucleotide binding domains are found in many enzymes and in particular, dehydrogenases, where the co-factor nicotinamide adenine dinucleotide is bound at an active site. Examples of proteins or enzymes containing the Rossmann fold are lactate dehydrogenase, glyceraldehyde-3-phosphate dehydrogenase, alcohol dehydrogenase, and malate dehydrogenase. However, it is clear that this fold is found in other nucleotide proteins beside dehydrogenases, including glycogen phosphorylase and glyceroltriphosphate binding proteins.

Elements of super secondary structure are frequently used to allow protein domains to be classified by their structures. Most frequently these domains are identified by the presence of characteristic folds. A fold represents the 'core' of a protein domain formed from a collection of secondary structures. In many cases these folds occur in more than one protein allowing structural relationships to be established. These characteristic folds include four-helix bundles (cytochrome b_{562} in Figure 3.25), helix turn helix motifs (the λ repressor in Figure 3.25), β barrels, and the β sandwich as well as more complicated structures such as the β propellor and β helix (see Figure 3.28). The β helix is an unusual arrangement of secondary structure – β strands align in a parallel manner one above another forming inter-strand hydrogen bonds but collectively twisting as a result of the displacement of successive strands. The strands all run in the same direction and the displacement result in the formation of a helix. Both left-handed and right-handed β helix proteins have been discovered, and a prominent example of a right handed β helix occurs in the tailspike protein of bacteriophage P22. In this protein the tailspike protein is actually a trimer containing three interacting β helices. The arrangement of strands within a β helix gives any subunit with this structural motif a very elongated appearance and leads to the hydrophobic cores being spread out along the long axis as opposed to a typical globular packing arrangement.

Quaternary structure

Many proteins contain more than one polypeptide chain. The interaction between these chains underscores quaternary structure. The interactions are exactly the same as those responsible for tertiary structure, namely disulfide bonds, hydrophobic interactions, charge–pair interactions and hydrogen bonds, with the exception that they occur between one or more polypeptide chains. The term subunit is often used instead of polypeptide chain.

Quaternary structure can be based on proteins with identical subunits or on non-identical subunits (Figure 3.29). Triose phosphate isomerase, HIV protease and many transcription factors function as homodimers. Haemoglobin is a tetramer containing two different subunits denoted by the use of Greek letters, α and β. The protein contains two α and two β subunits in its tetrameric state and for haemoglobin this is normally written as $\alpha_2\beta_2$. Although it might be thought that aggregates of subunits are an artefact resulting from crystal packing it is abundantly clear that correct functional activity requires the formation of quaternary structure and the specific association of subunits. Subunits are held together predominantly by weak non-covalent interactions. Although individually weak these forces are large in number and lead to subunit assembly as well as gains in stability.

Quaternary structure is shown by many proteins and allows the formation of catalytic or binding sites at the interface *between* subunits. Such sites are impossible for monomeric proteins. Further advantages of oligomeric proteins are that ligand or substrate-binding causes conformational changes within the whole assembly and offer the possibility of regulating biological activity. This is the basis for allosteric regulation in enzymes.

In the following sections the quaternary organization of transcription factors, immunoglobulins and oxygen-carrying proteins of the globin family are described as examples of the evolution of biological function in response to multiple subunits. The presence of higher order or quaternary structure allows greater versatility of function, and by examining the structures of some of these proteins insight is gained into the interdependence of structure and function. A common theme linking these proteins is that they all bind other molecules such as nucleic acid in the case of transcription factors, small inorganic molecules such as oxygen and bicarbonate by the globins and larger proteins or peptides in the case of immunoglobulins.

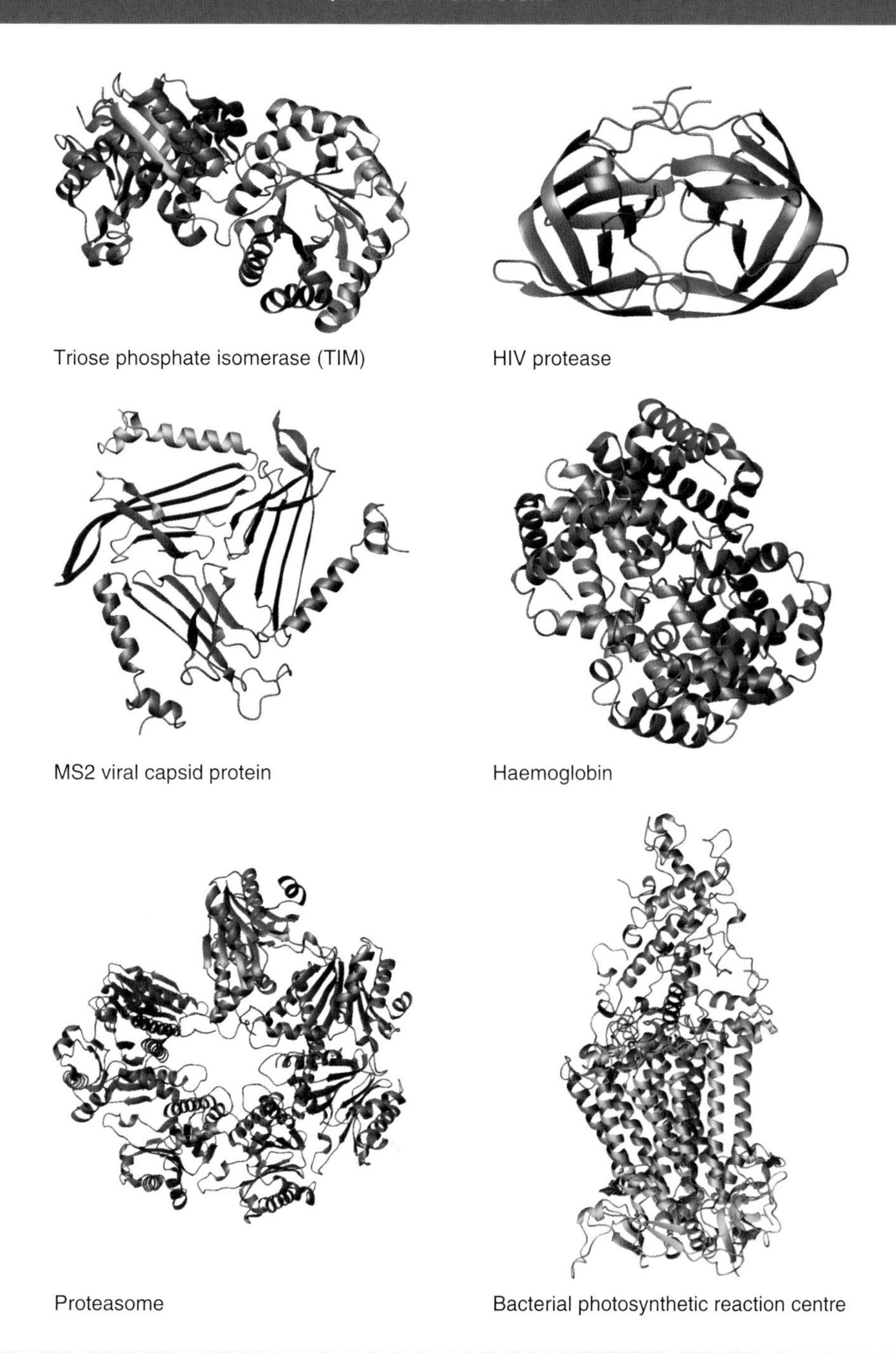

Figure 3.29 The structures of large oligomeric proteins. Dimers are shown with triose phosphate isomerase and HIV protease. Trimers are represented by the MS2 viral capsid protein. Haemoglobin is a tetramer composed of two pairs of identical subunits. The proteasome consists of four concentric rings each made up of seven subunits. Only one ring is shown in the current view. The bacterial photosynthetic reaction centre contains four different subunits, the H,M,L and C subunits

Dimeric DNA-binding proteins

Many dimeric proteins exist in the proteomes of cells and one of the most common occurrences is the use of two subunits and a two-fold axis of symmetry to bind DNA. DNA binding proteins are a very large group of proteins typified by transcription factors. Transcription factors bind to promoter regions of DNA sequences called, in eukaryotic systems, the TATA box, and in prokaryotes a Pribnow box. The TATA box is approximately 25 nucleotides upstream of the transcription start site and as their name suggests these factors mediate the transcription of DNA into RNA by promoting RNA polymerase binding to this region of DNA in a pre-initiation complex (Figure 3.30).

In prokaryotes the mode of operation is simpler with fewer modulating elements. Prokaryotic promoter contains two important zones called the −35 region and −10 region (Pribnow box). The −35 bp region functions in the initial recognition of RNA polymerase and possesses a consensus sequence of TTGACAT. The −10 region has a consensus sequence (TATAAT) and occurs about 10bp before the start of a bacterial gene.

The cI and cro proteins from phage λ were amongst the first studied DNA binding proteins and act as regulators of transcription in bacteriophages such as 434 or λ. The life cycle of λ bacteriophage is controlled by the dual action of the cI and cro proteins in a complicated series of reactions where DNA binding close to initiation sites physically interferes with gene transcription by RNA polymerase (Figure 3.31). For this reason these proteins are also called repressors. When phage DNA enters a bacterial host cell two outcomes are possible: lytic infection results in the production of new viral particles or alternatively the virus integrates into the bacterial genome lying dormant for a period known as the lysogenic phase. Lytic and lysogenic phases are initiated by phage gene expression but competition between the cro and cI repressor proteins determines which pathway is followed. Cro and cI repressor compete for control by binding to an operator region of DNA that contains at least three sites that influence the lytic/lysogenic switch. The bacteriophage remains in the lysogenic state if cI proteins predominate. Conversely the lytic cycle is followed if cro proteins dominate.

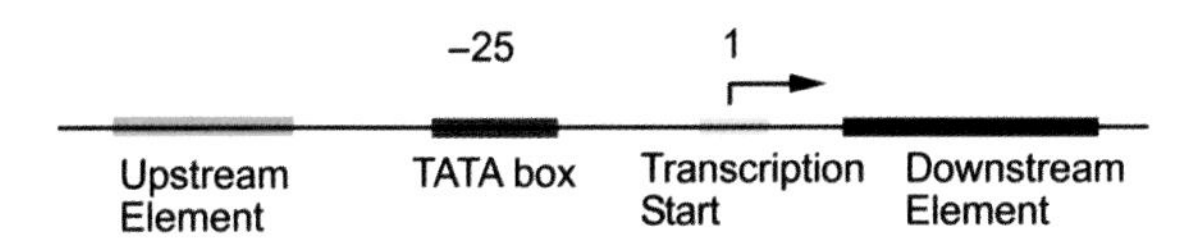

Figure 3.30 Transcription starts at the initiation site (+1) but is promoted by binding to the consensus TATA box sequence (−25) of the RNA polymerase complex that includes transcription factors. Upstream DNA sequences that facilitate transcription have been recognized (CAT and GC boxes ∼−80 and ∼−90 bases upstream of start site) whilst downstream elements lack consensus sequences but exist for some transcription systems

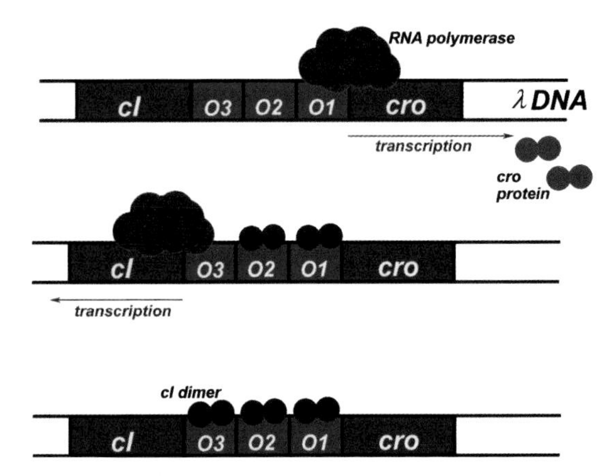

Figure 3.31 The control of transcription in bacteriophage λ by cro and cI repressor proteins. The cI dimer may bind to any of three operators, although the order of affinity is O1 ≈ O2 > O3. In the absence of cI proteins, the cro gene is transcribed. In the presence of cI proteins only the cI gene may be transcribed. High cI concentrations prevents transcription of both genes

Determining the structure of the cI repressor in the presence of a DNA containing a consensus binding sequence uncovered detailed aspects of the mechanism of nucleic acid binding. The mechanism of DNA binding is of considerable interest in view of the widespread occurrence of transcription factors in all cells and the growing evidence of their involvement in many disease states. In the absence of DNA a

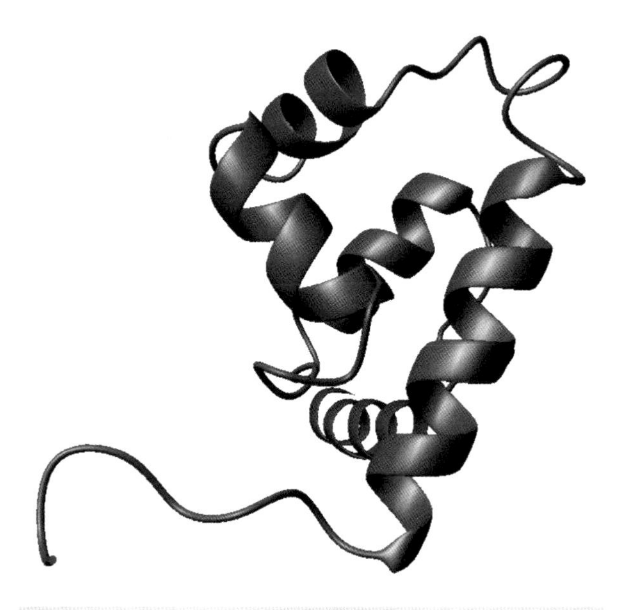

Figure 3.32 The monomer form of the cI repressor (PDB: 1LMB)

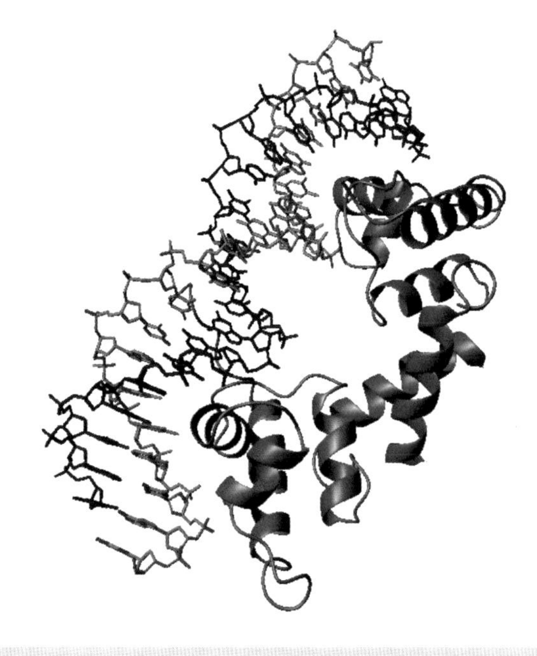

Figure 3.33 The structure of the N terminal domain of the λ repressor in the presence of DNA. The fifth helix forms part of the dimerization domain that allows two monomer proteins to function as a homodimer. In each case helices 2 and 3 bind in the major groove of DNA since the spatial separation of each HTH motif is comparable with the dimensions of the major groove

structure for the cI repressor revealed a polypeptide chain containing five short helices within a domain of ~70 residues (Figure 3.32). Far more revealing was the structure in the presence of DNA and the use of helices 2 and 3 as a helix-turn-helix or HTH motif to fit precisely into the major groove (Figure 3.33).

The N- and C-terminal domains of one subunit are separated by mild proteolysis. Under these conditions two isolated N-terminal domains form a less stable dimer that has lowered affinity for DNA when compared with the complete repressor. The HTH motif is 20 residues in length and is formed by helices 2 and 3 found in the N terminal domain. Although the HTH motif is often described as a 'domain' it should be remembered it is part of a larger protein and does not fold into a separate, stable element of structure. Helix 3 of the HTH motif makes a significant number of interactions with the DNA. It is called the recognition helix with contact points involving Gln33, Gln44, Ser45, Gly46, Gly48, and Asn52. These residues contain a significant proportion of polar side chains and bind directly into the major groove found in DNA. In comparison helix 2 makes fewer contacts and has a role of positioning helix 3 for optimal recognition. Of the remaining structure helices 4 and 5 form part of a dimer interface whilst the C terminal domain is involved directly in protein dimerization. Dimerization is important since it allows HTH motifs to bind to successive major grooves along a sequence of DNA. Determination of the structure of the cro repressor protein showed a very similar two-domain structure to the cI repressor with comparable modes of DNA binding (see Figure 3.34). The overwhelming similarity in structure between these proteins suggested a common mechanism of DNA binding and one that might extend to the large number of transcription factors found in eukaryotes.

Comparison of the sequences of HTH motifs from λ and cro repressors showed sequence conservation (Figure 3.35). The first seven residues of the 20 residue HTH motif form helix 2, a short 4 residue turn extends from positions 8 to 11 and the next (recognition) helix is formed from residues 12 to 20. The sequences of

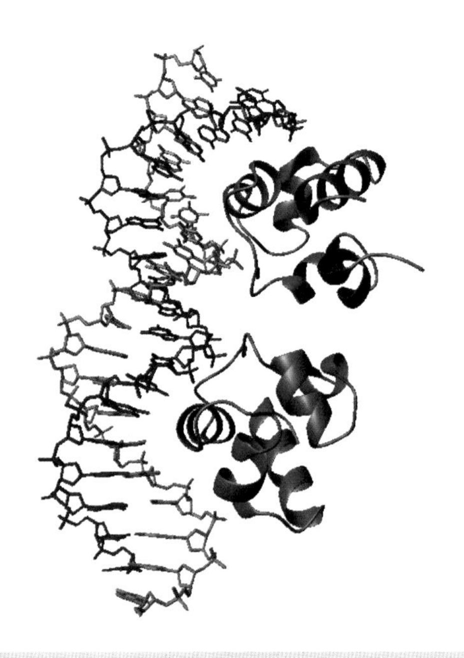

Figure 3.34 The cro repressor protein from bacteriophage 434 bound to a 20-mer DNA helix (PDB:4CRO)

```
Cro    : QTKTAKDLGVYQSAINKAIH
434Cro : QTELATKAGVKQQSIQLIEA
cI     : QESVADKMGMGQSGVGALFN
```

Figure 3.35 The sequence homology in the cro and cI repressors. Residues highlighted in red are invariant whilst those shown in yellow illustrate conservative changes in sequence

the HTH motif in the cro and λ repressors helped define structural constraints within this region. Residue 9 is invariant and is always a glycine residue. It is located in the 'turn' region and larger side chains are not easily accommodated and would disrupt the turn and by implication the positioning of the DNA binding helix. Residues 4 and 15 are completely buried from the solvent whilst residues 8 and 10 are partially buried, non-polar side chains charged side chains unfavourable at these locations. Proline residues are never found in HTH motifs. Further structural constraints were apparent at residue 5 located or wedged between the two helices. Large or branched side chains would cause a different alignment of the helices and destroy the primary function of HTH motif. DNA binding was critically dependent on the identity of side chains for residues 11–13, 16–17 and 20. Residues with polar side chains were common and were important in forming hydrogen bonds to the major groove.

The globin family and the role of quaternary structure in modulating activity

Myoglobin occupies a pivotal position in the history of protein science. It was the first protein structure to be determined in 1958. Myoglobin and haemoglobin, whose structure was determined shortly afterwards, have been extensively studied particularly in relation to the inter-dependence of structure *and* function. Many of the structure–function relationships discovered for myoglobin have proved to be of considerable importance to the activity of other proteins.

The evolutionary development of oxygen carrying proteins was vital to multicellular organisms where large numbers of cells require circulatory systems to deliver oxygen to tissues as well as specialized proteins to carry oxygen around the body. In vertebrates the oxygen-carrying proteins are haemoglobin and myoglobin. Haemoglobin is located within red blood cells and its primary function is to convey oxygen from the lungs to regions deficient in oxygen. This includes particularly the skeletal muscles of the body where oxygen consumption as a result of mechanical work requires continuous supplies of oxygen for optimal activity. In skeletal muscle myoglobin functions as an oxygen storage protein.

For both haemoglobin and myoglobin the oxygen-carrying capacity arises through the presence of a heme group. The heme group is an organic carbon skeleton called protoporphyrin IX made up of four pyrrole groups that chelate iron at the centre of the ring (Figure 3.36). The heme group gives myoglobin and haemoglobin a characteristic colour (red) that is reflected by distinctive absorbance spectra that show intense peaks around 410 nm due to the Fe-protoporphyrin IX group. The iron group is found predominantly in the ferrous state (Fe^{2+}) and only this form binds oxygen. Less frequently the iron group is found in an oxidized state as the ferric iron

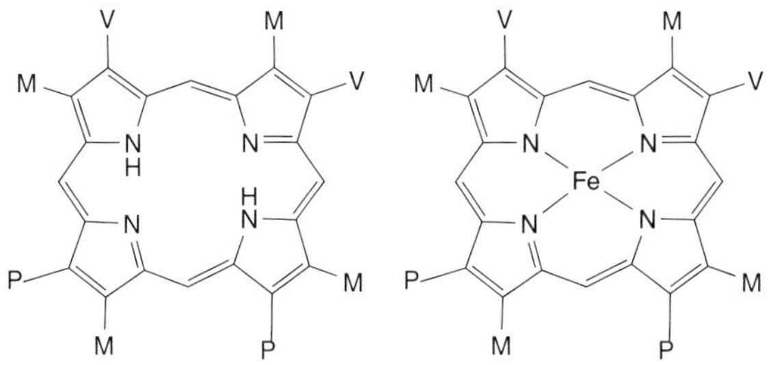

Figure 3.36 The structure of protoporphyrin IX and heme (Fe-protoporphyrin-IX)

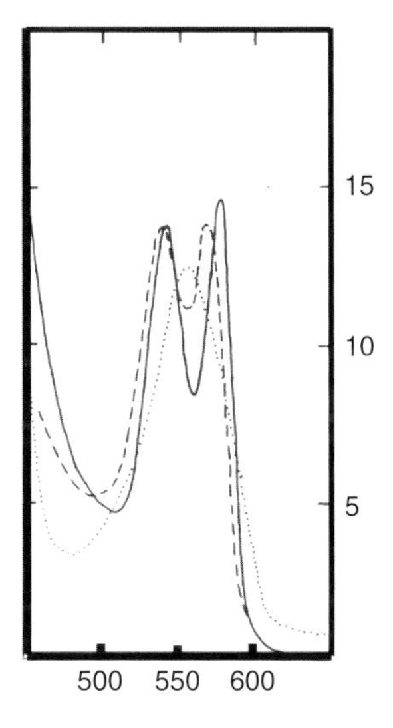

Figure 3.37 The absorbance spectra of oxy, met and deoxyhaemoglobin. The dotted line is the spectrum of deoxyHB, the solid line is metHb whilst the dashed line is oxyHb. DeoxyHb shows a single maximum at 550 nm that shifts upon oxygen binding to give two peaks at 528 and 563 nm. The extinction coefficients associated with these peaks are $\varepsilon_{555} = 12.5$ mM cm^{-1}, $\varepsilon_{541} = 13.5$ mM cm^{-1}, and $\varepsilon_{576} = 14.6$ mM cm^{-1}

(Fe^{3+}), a state not associated with oxygen binding. The binding of oxygen (and other ligands) to the sixth coordination site of iron in the heme ring is accompanied by distinctive changes to absorbance spectra (Figure 3.37).

The structure of myoglobin

Detailed crystallographic studies of myoglobin provided the first three-dimensional picture of a protein and established many of the ground rules governing secondary and tertiary structure described earlier in this chapter. Myoglobin was folded into an extremely compact single polypeptide chain with dimensions of ~4.5 × 3.5 × 2.5 nm. There was no free space on the inside of the molecule with the polypeptide chain folded efficiently by changing directions regularly to create a compact structure. Approximately 80 percent of all residues were found in α helical conformations. The structure of myoglobin (Figure 3.38) provided the first experimental verification of the α helix in proteins and confirmed that the peptide group was planar, found in a *trans* configuration and with the dimensions predicted by Pauling.

Eight helices occur in myoglobin and these helices are labelled as A, B, C etc up to the eighth helix,

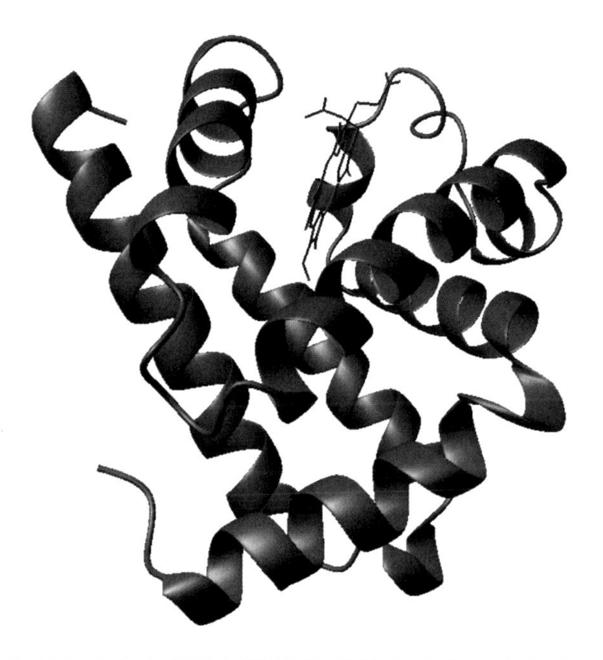

Figure 3.38 The structure of myoglobin showing α helices and the heme group. Helix A is in the foreground along with the N terminal of the protein. The terminology used for myoglobin labels the helices A–H and residues within helices as F8, A2 etc.

helix H. Frequently, although not in every instance, the helices are disrupted by the presence of proline residues as for example occurs between the B and C helices, the E and F helices and the G and H helices. Even in low-resolution structures of myoglobin (the first structure produced had a resolution of ~6 Å) the position of the heme group is easily discerned from the location of the electron-dense iron atom.

The heme group was located in a crevice surrounded almost entirely by non-polar residues with the exception of two heme propionates that extended to the surface of the molecule making contact with solvent and two histidine residues, one of which was in contact with the iron and was termed the proximal (F8) histidine (Figure 3.39). The imidazole side chain of F8 provided the fifth ligand to the heme iron. A second histidine, the distal histidine (E7), was more distant from the Fe centre and did *not* provide the sixth ligand. An obvious result of this arrangement was an asymmetric conformation for the iron where it was drawn out of the heme plane towards the proximal histidine.

The structure of myoglobin confirmed the partitioning of hydrophobic and hydrophilic side chains. The interior of the protein and the region surrounding the heme group consisted almost entirely of non-polar residues. Here leucine, phenylalanine and valine were common whilst hydrophilic side chains were located on the exterior or solvent accessible surface.

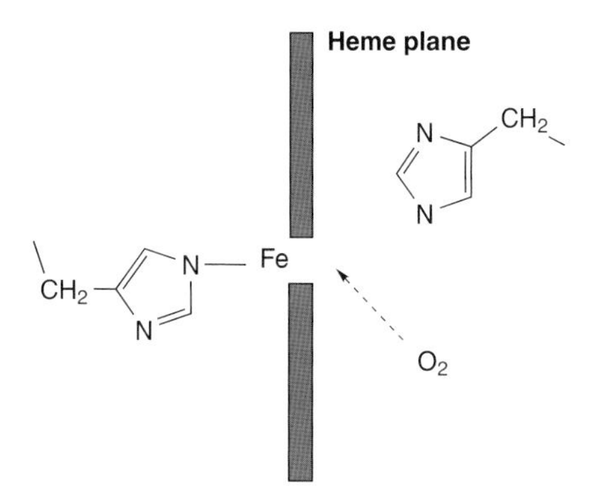

Figure 3.39 Diagram of the proximal and distal histidine side chains forming part of the oxygen binding site of myoglobin with the ferrous iron pulled out of the plane of the heme ring. The vacant sixth coordination position is the site of oxygen binding

There are approximately 200 structures, including mutants, of myoglobin deposited in the PDB but the three most important structures of relevance to myoglobin are those of the oxy and deoxy forms together with ferrimyoglobin (metmyoglobin) where the iron is present as the ferric state. The structures of all three forms turned out to be remarkably similar with one exception located in the vicinity of the sixth coordination site. In oxymyoglobin a single oxygen molecule was found at the sixth coordination site. In the deoxy form this site remained vacant whilst in metmyoglobin water was found in this location. The unique geometry of the iron favours its maintenance in the reduced state but oxygen binding resulted in movement of the iron approximately 0.2 Å towards the plane of the ring (Figure 3.40). An analysis of the heme binding site emphasizes how the properties of the heme group are modulated by the polypeptide. The identical ferrous heme group in cytochromes undergoes oxidation in the presence of oxygen to yield the ferric (Fe^{3+}) state, other enzymes such as catalase or cytochrome oxidase convert the oxygen into hydrogen peroxide and water respectively but in globins the oxygen is bound with the iron remaining in the ferrous state.

Myoglobin has a very high affinity for oxygen and this is revealed by binding curves or profiles that record the fractional saturation versus the concentration of oxygen (expressed as the partial pressure of oxygen, pO_2). The oxygen-binding curve of myoglobin is

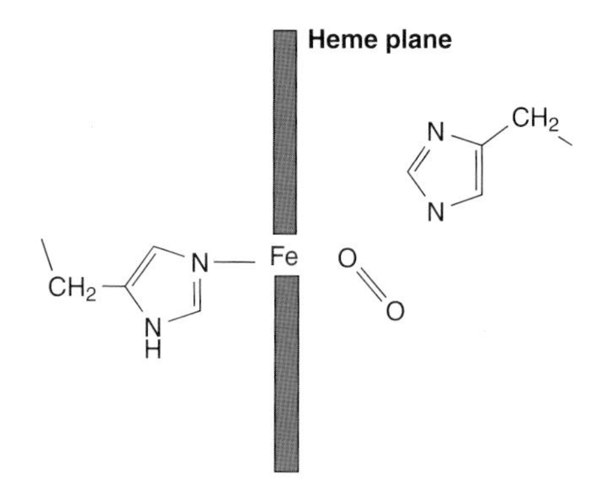

Figure 3.40 Binding of oxygen in oxymyoglobin and the movement of the iron into the approximate heme plane

hyperbolic and rapidly reaches a saturating level seen by the asymptotic line above pO_2 levels of 30 torr. The affinity for oxygen is expressed as an equilibrium

$$MbO_2 \leftrightharpoons Mb + O_2 \tag{3.6}$$

where the equilibrium constant, K, is

$$K = [Mb][O_2]/[MbO_2] \tag{3.7}$$

The fractional saturation of myoglobin (Y) is simply expressed as the number of oxygenated myoglobin (MbO_2) molecules divided by the total number of myoglobin molecules ($MbO_2 + Mb$). Thus the fractional saturation (Y)

$$Y = [MbO_2]/([MbO_2] + [Mb]) \tag{3.8}$$

and substitution of Equation 3.7 into 3.8 yields

$$Y = pO_2/(pO_2 + K) \tag{3.9}$$

where pO_2 reflects the concentration, strictly partial pressure, of oxygen in the atmosphere surrounding the solution. Equation 3.9 may be written by equating the equilibrium constant in terms of the partial pressure of oxygen necessary to achieve 50 percent saturation leading to

$$Y = pO_2/(pO_2 + P_{50}) \tag{3.10}$$

The oxygen-binding curve of haemoglobin revealed a very different profile. The curve is no longer hyperbolic but defines a sigmoidal or S-shaped profile. In comparison with myoglobin the haemoglobin molecule becomes saturated at much higher oxygen concentrations (Figure 3.41). The profile defines a binding curve where initial affinity for oxygen is very low but then increases dramatically before becoming resistant to further oxygenation. This is called cooperativity and these differences are fundamentally dependent on the structures of myoglobin and haemoglobin. In order to understand the basis for these differences it is necessary to compare and contrast the structure of haemoglobin with myoglobin.

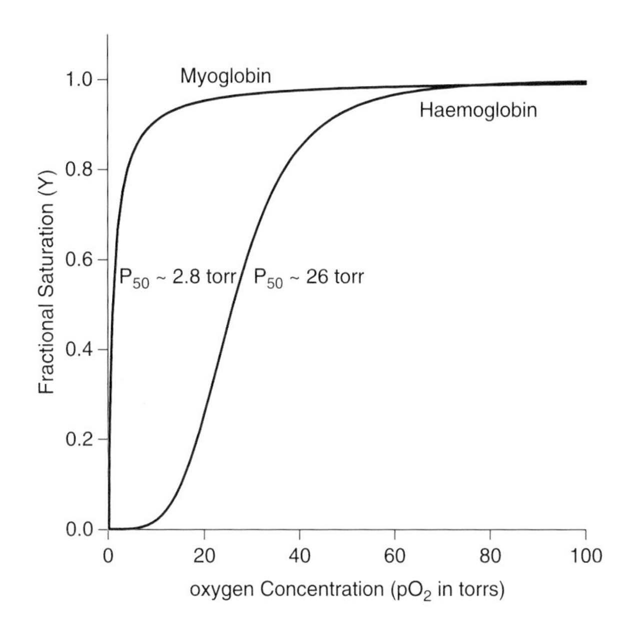

Figure 3.41 The oxygen-binding curves of myoglobin and haemoglobin show different profiles. Whilst the affinity of myoglobin for oxygen is reflected by a hyperbolic curve the affinity curve of haemoglobin is sigmoidal. The P_{50} of myoglobin is 2.8 torr whilst the P_{50} of haemoglobin in red blood cells is 26 torr (760 torr = 1 atm). In the tissues oxygen concentrations of between 20 and 40 torr are typical whilst in the lungs much higher partial pressures of oxygen exist above 100

The structure of haemoglobin

The most obvious difference between haemoglobin and myoglobin is the presence of quaternary structure in the former (Figure 3.42). In mammals adult haemoglobin is composed of four polypeptide chains containing two different primary sequences in the form of 2α chains and 2β chains. The determination of the structure of haemoglobin revealed that each α and β subunit possessed a conformation similar to myoglobin and showed a low level of sequence homology that reflected evolution from a common ancestral protein. The structure of haemoglobin revealed that each globin chain contained a heme prosthetic group buried within a crevice whilst the four subunits packed together forming a compact structure with little free space yet this time with the crucial difference of additional interactions *between* subunits.

Although not immediately apparent the tertiary structures of myoglobin, α globin and β globin chains

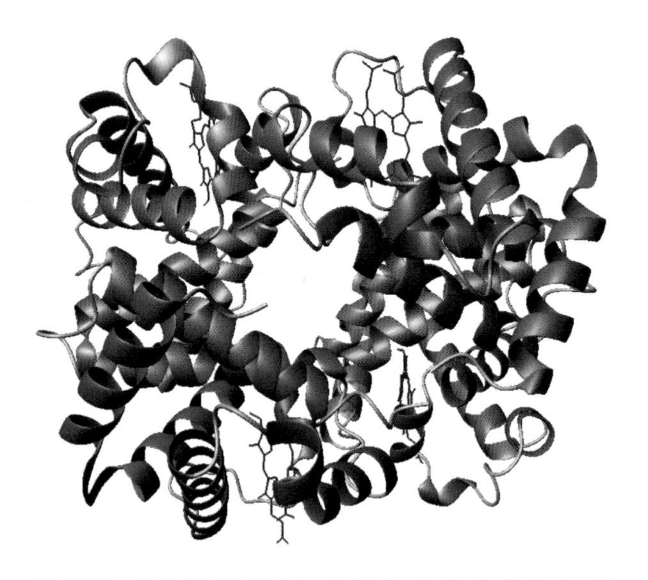

Figure 3.42 The quaternary structure of haemoglobin showing the four subunits each with a heme group packing together. The heme groups are shown in red with the α globin chain in purple and the β globin chain in green

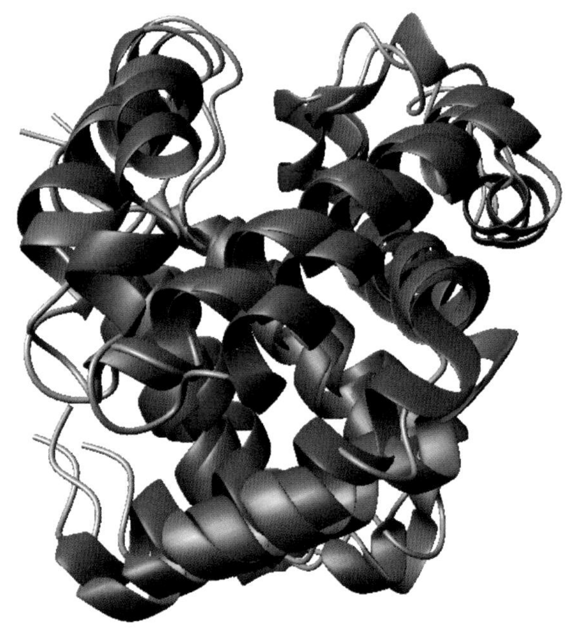

Figure 3.43 Superposition of the polypeptide chains of myoglobin, α globin and β globin. Structural similarity occurs despite only 24 out of 141 residues showing identity. The proximal and distal histidines are conserved along with several residues involved in the heme pocket structure such as leucine (F4) and a phenylalanine between C and D helices

are remarkably similar despite differences in primary sequence (Figure 3.43). With similar tertiary structures for myoglobin and the α and β chains of haemoglobin it is clear that the different patterns of oxygen binding must reflect additional subunit interactions in haemoglobin. In other words cooperativity arises as a result of quaternary structure.

The pattern of oxygenation in haemoglobin is perfectly tailored to its biological function. At high concentrations of oxygen as would occur in the lungs ($pO_2 > 100$ torr or $\sim 0.13 \times$ atmospheric pressure) both myoglobin and haemoglobin become saturated with oxygen. Myoglobin would be completely oxygenated carrying 1 molecule of oxygen per protein molecule. Complete oxygenation of haemoglobin would result in the binding of 4 oxygen molecules. However, as the concentration of oxygen decreases below 50 torr myoglobin and haemoglobin react differently. Myoglobin remains fully saturated with oxygen whilst haemoglobin is no longer optimally oxygenated; sites on haemoglobin are only 50 percent occupied at $\sim$26 torr whereas myoglobin exhibits a much lower P_{50} of $\sim$3 torr.

The physiological implications of these binding properties are profound. In the lungs haemoglobin is saturated with oxygen ready for transfer around the body. However, when red blood cells (containing high concentrations of haemoglobin) reach the peripheral tissues, where the concentration of oxygen is low, unloading of oxygen from haemoglobin occurs. The oxygen released from haemoglobin is immediately bound by myoglobin. The binding properties of myoglobin allow it to bind oxygen at low concentrations and more importantly facilitate the transport of oxygen from lungs to muscles or from regions of high concentrations to tissues with much lower levels.

The cooperative binding curve of haemoglobin arises as a result of structural changes in conformation that occur upon oxygenation. In the deoxy form the sixth coordination site is vacant and the iron is drawn out of the heme plane towards the proximal histidine.

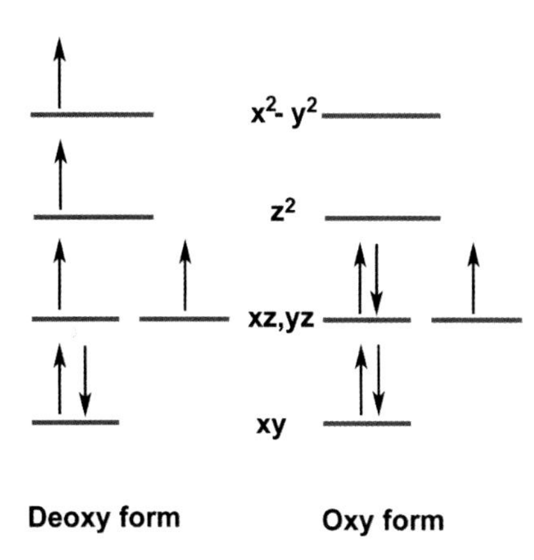

Figure 3.44 The spin state changes that occur in the deoxy and oxy ferrous forms of haemoglobin showing the population of the d orbitals (xy, xz, yz, z^2, x^2-y^2) by electrons. The deoxy form is high spin (S = 2) ferrous state with an ionic radius too large to fit between the four tetra pyrrole ring nitrogens. The low spin (S = 1/2) state of ferrous iron with a different distribution of orbitals has a smaller radius and moves into the plane of the heme

Oxygenation results in conformational change mediated by oxygen binding to the iron. The events are triggered by changes in the electronic structure of the iron as it shifts from a high spin ferrous centre in the deoxy form to a low spin state when oxygenated. Changes in spin state are accompanied by reorganization of the orbital structure and decreases in ionic radius that allows the iron to move closer to the plane of the heme macrocycle (Figure 3.44).

Changes at the heme site are accompanied by reorientation of helix F, the helix containing the proximal histidine, and results in a shift of approximately 1 Å to avoid unfavourable contact with the heme group. The effect of this conformational change is transmitted throughout the protein but particularly to interactions *between* subunits. Oxygen binding causes the disruption of ionic interactions between subunits and triggers a shift in the conformation of the tetramer from the deoxy to oxy state. By examining the structure of haemoglobin in both the oxy and deoxy states many of these interactions have been highlighted and another significant milestone was a structural understanding for the cooperative oxygenation of haemoglobin proposed by Perutz in 1970.

The mechanism of oxygenation

Perutz's model emphasized the link between cooperativity and the structure of the tetramer. The cooperative curve derives from conformational changes exhibited by the protein upon oxygenation and reflects the effect of two 'competing' conformations. These conformations are called the R and T states and the terms derive originally from theories of allosteric transitions developed by Monod, Wyman and Changeaux (MWC). According to this model haemoglobin exists in an equilibrium between the R and T states where the R state signifies a 'relaxed' or active state whilst the T signifies a 'tense' or inactive form.

The deoxy state of haemoglobin is resistant to oxygenation and is equated with the inactive, T state. In contrast the oxy conformation is described as the active, R state. In this model haemoglobin is an equilibrium between R and T states with the observed oxygen-binding curve reflecting a combination of the binding properties of each state. The shift in equilibrium between the R and T states is envisaged as a two state or concerted switch with only weak and strong binding states existing and intermediate states containing a mixture of strong and weak binding subunits specifically forbidden. This form of the MWC model could be described as an 'all or nothing' scheme but is more frequently termed the 'concerted' model. In general detailed structural analysis has retained many of the originally features of the MWC model.

At a molecular level the relative orientation between the $\alpha_1\beta_1$ and $\alpha_2\beta_2$ dimers differed in the oxy and deoxy states. Oxygenation leads to a shift in orientation of $\sim$15° and a translation of $\sim$0.8 Å for one pair of α/β subunits relative to the other. It arises as a result of changes in conformation at the interfaces between these pairs of subunits. One of the most important regions involved in this conformational switch centres around His97 at the boundary between the F and G helices of the β_2 subunit. This region of the β subunit makes contact with the C helix of the α_1 subunit and lies close to residue 41. In addition the folding of the β_2 subunit leads to the C terminal residue, His146, being situated above the same helix (helix C)

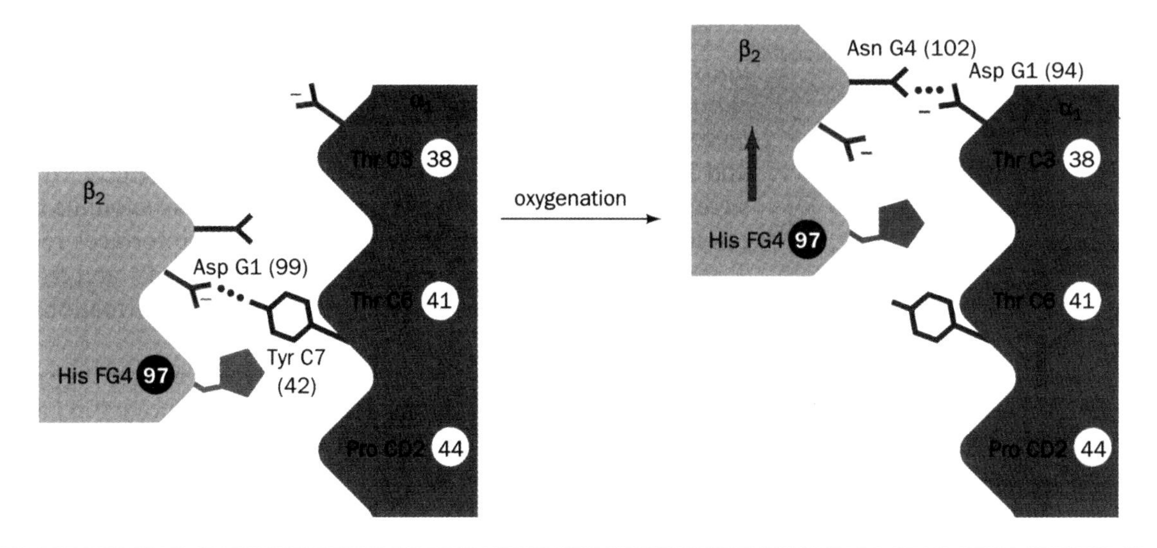

Figure 3.45 Interactions at the $\alpha_1\beta_2$ interface in deoxyhaemoglobin and the changes that occur after oxygenation. In the T state His97 of the β2 subunit fits next to Thr41 of the α1 chain. In the R state conformation changes result in this residue lying alongside Thr38. Although a salt bridge between Asp99 and Tyr42 is broken in the T > R transition a new interaction between Asn102 and Asp94 results. There is no detectable intermediate between the T and R states and the precisely interacting surfaces allow the two subunits to move relative to each other easily. (Reproduced with permission from Voet, D., Voet, J.G & Pratt, C.W., *Fundamentals of Biochemistry*. Chichester, John Wiley & Sons, 1999.)

and it forms a series of hydrogen bonds and salt bridges with residues in this region (38–44). In the deoxy state His 146 is restrained by an interaction with Asp94. As a result of symmetry an entirely analogous set of interactions occur at the $\alpha_2\beta_1$ interface (Figure 3.45).

In the T state the iron is situated approximately 0.6 Å out of the plane of the heme ring as part of a dome directed towards HisF8. Oxygenation causes a shortening of the iron–porphyrin bonds by 0.1 Å due to electron reorganization and causes the iron to move into the plane dragging the proximal His residue with it. Movement of the His residue by 0.6 Å is unfavourable on steric grounds and as a result the helix moves to compensate for changes at the heme centre. Helix movement triggers conformational changes throughout the subunits and more importantly between the subunits. Other important conformational switches include disruption of a network of ion pairs within and between subunits. In particular ionic interactions of the C terminal residue of the α subunit (Arg141) with Asp126 and the amino terminal of the α1 subunit and that of the β subunit (His146) with Lys40 (α) and Asp94(β) are broken upon oxygenation. Since these interactions stabilize the T state their removal drives the transition towards the R state.

Cooperativity arises because structural changes within one subunit are coupled to conformational changes throughout the tetramer. In addition the nature of the switch means that intermediate forms cannot occur and once one molecule of oxygen has bound to a T state subunit converting it to the R state all other subunits are transformed to the R state. As a result the remaining subunits rapidly bind further oxygen molecules, leading to the observed shape of the oxygen binding curve seen in Figure 3.46.

Allosteric regulation and haemoglobin

The new properties of haemoglobin extend further than the simple binding of oxygen to the protein. Haemoglobin's affinity for oxygen is modulated by

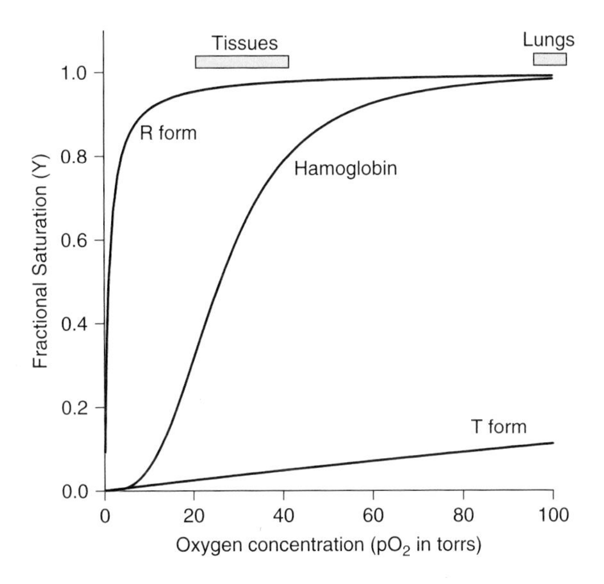

Figure 3.46 Oxygen binding profiles of the R and T states of haemoglobin. The R or active form of haemoglobin binds oxygen readily and shows a hyperbolic curve. The T or inactive state is resistant to oxygenation but also shows a hyperbolic binding curve with a higher P_{50}. The observed binding profile of haemoglobin reflects the sum of affinities of the R and T forms for oxygen and leads to the observed P_{50} of ~26 torr for isolated haemoglobin. Shown in yellow is the approximate range of partial pressures of oxygen in tissues and lungs

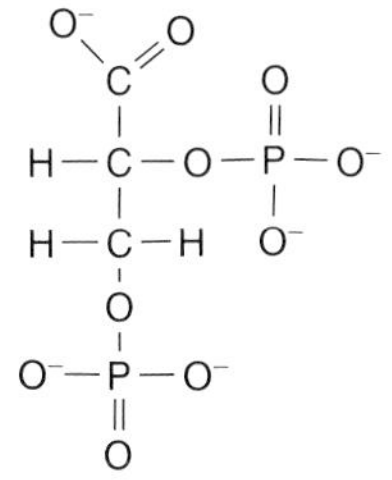

Figure 3.47 2,3-Bisphosphoglycerate is a negatively charged molecule

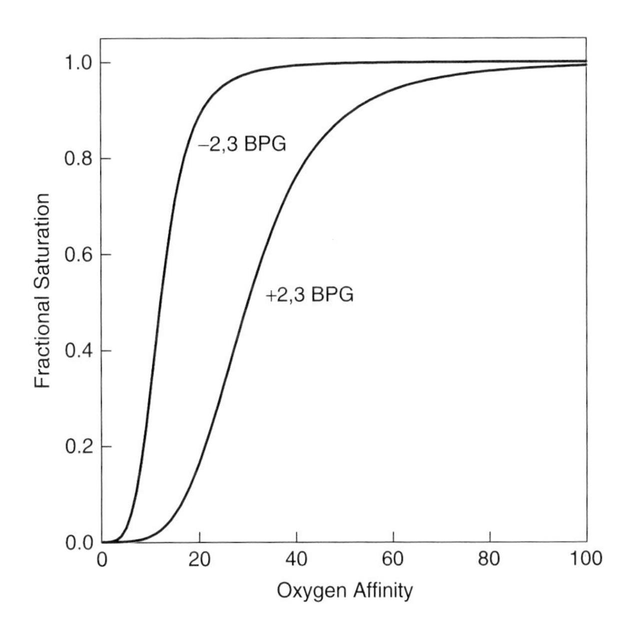

Figure 3.48 Modulation of oxygen binding properties of haemoglobin by 2,3 BPG. The 2,3 BPG modulates oxygen binding by haemoglobin raising the P_{50} from 12 torr in isolated protein to 26 torr observed for haemoglobin in red blood cells. This is shown by the respective oxygen affinity curves

small molecules (effectors) as part of an important physiological control process. This mode of regulation is called allostery. An allosteric modulator binds to a protein altering its activity and in the case of haemoglobin the effector alters oxygen binding. The most important allosteric regulator of haemoglobin is 2,3-bisphosphoglycerate (2,3 BPG) (Figure 3.47).[1]

The allosteric modulator of haemoglobin, 2,3 BPG, raises the P_{50} for oxygen binding in haemoglobin from 12 torr in isolated protein to 26 torr observed for haemoglobin in red blood cells (Figure 3.48). Indeed the existence of an allosteric modulator of haemoglobin was first suspected from careful comparisons of the oxygen-binding properties of isolated protein with that found in red blood cells. In erythrocytes 2,3 BPG is found at high concentrations as a result of highly active glycolytic pathways and is formed from the key intermediates 3-phosphoglycerate or 1,3 bisphosphoglycerate. The result is to create concentrations of 2,3 BPG approximately equimolar with haemoglobin within cells. In the erythrocyte this results in 2,3 BPG concentrations of ~5 mM.

[1] Sometimes called 2,3 DPG = 2,3-diphosphoglycerate.

The effect arises by differential binding – the association constant (K_a) for 2,3 BPG for deoxyhaemoglobin is $\sim 4 \times 10^4$ M^{-1} compared with a value of ~ 300 M^{-1} for the oxy form. In other words it stabilizes the T state, which has low affinity for oxygen whilst binding to the oxy state is effectively inhibited by the presence of bound oxygen. The effect of 2,3 BPG on the oxygen-binding curve of haemoglobin is to shift the profile to much higher P_{50} (Figure 3.48). Detailed analysis of the structures of the oxy and deoxy states showed that 2,3 BPG bound in the central cavity of the deoxy form and allowed a molecular interpretation for its role in allostery. In the centre of haemoglobin between the four subunits is a small cavity lined with positively charged groups formed from the side chains of His2, His143 and Lys82 of the β subunits together with the two amino groups of the first residue of each of these chains (Val1). The result of this charge distribution is to create a strong binding site for 2,3 BPG. In oxyhaemoglobin the binding of oxygen causes conformational changes that lead to a closure of the allosteric binding site. These changes arise as a result of subunit movement upon oxygenation and decrease the cavity size to prevent accommodation of 2,3 BPG. Both oxygen and 2,3 BPG are reversibly bound ligands yet each binds at separate sites and lead to opposite effects on the R→T equilibrium.

2,3 BPG is not the only modulator of oxygen binding to haemoglobin. Oxygenation of haemoglobin causes the disruption of many ion pairs and leads to the release of ~0.6 protons for each oxygen bound. This effect was first noticed in 1904 by Christian Bohr and is seen by increase oxygen binding with increasing pH at a constant oxygen level (Figure 3.49). The Bohr effect is of utmost importance in the physiological delivery of oxygen from the lungs to respiring tissues and also in the removal of CO_2 produced by respiration from these tissues and its transport back to the lungs. Within the erythrocyte CO_2 is carried as bicarbonate as a result of the action of carbonic anhydrase, which rapidly catalyses the slow reaction

$$CO_2 + H_2O \leftrightharpoons H^+ + HCO_3^- \qquad (3.11)$$

In actively respiring tissues where pO_2 is low the protons generated as a result of bicarbonate formation favour the transition from R > T and induce the haemoglobin to unload its oxygen. In contrast in the lungs the oxygen levels are high and binding to haemoglobin occurs readily with the disruption of T state ion pairs and the formation of the R state. A further example of the Bohr effect is seen in very active muscles where an unavoidable consequence of high rates of respiration is the generation of lactic acid. The production of lactic acid will cause a lowering of the pH and an unloading of oxygen to these deficient tissues. Carbon dioxide can also bind directly to haemoglobin to form carbamates and arises as a result of reaction with the N terminal amino groups of subunits. The T form (or deoxy state) binds more carbon dioxide as carbamates than the R form. Again the physiological consequences are clear. In capillaries with high CO_2 concentrations the T state is favoured leading to an unloading of oxygen from haemoglobin.

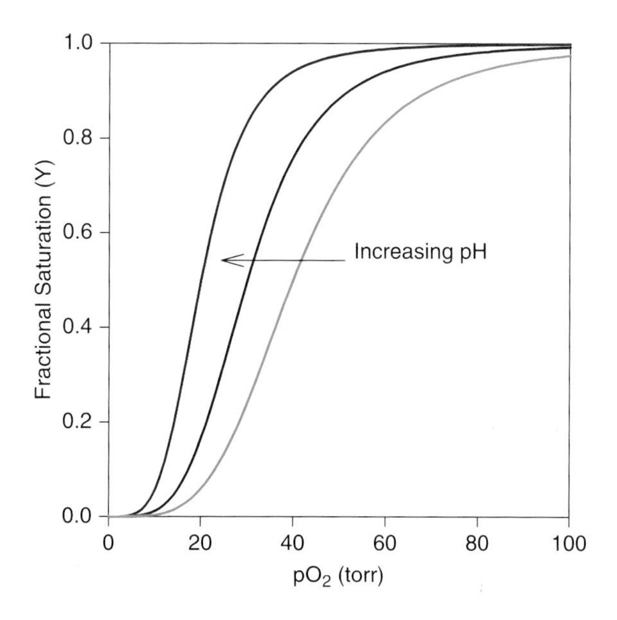

Figure 3.49 The Bohr effect. The oxygen affinity of haemoglobin increases with increasing pH

Immunoglobulins

The immunoglobulins are a large group of proteins found in vertebrates whose unique function is to bind foreign substances invading a host organism. By breaking the physical barrier normally represented by skin or mucous membranes pathogens enter a system, but once this first line of defence is breached an immune response is normally started. In most instances the immune response involves a reaction to bacteria or viruses but it can also include isolated proteins that break the skin barrier. Collectively these foreign substances are termed antigens.

The immune response is based around two systems. A circulating antibody system based on B cells sometimes called by an older name of humoral immune response. The cells were first studied in birds where they matured from a specialized pocket of tissue known as the bursa of Fabricius. However, there is no counterpart in mammals and B cells are derived from the bone marrow where the 'B' serves equally well to identify its origin. B cells have specific proteins on their surfaces and reaction with antigen activates the lymphocyte to differentiate into cells capable of antibody production and secretion. These cells secrete antibodies binding directly to antigens and acting as a marker for macrophages to destroy the unwanted particle.

This system is supported by a second cellular immune response based around T lymphocytes. The T cells are originally derived from the thymus gland and these cells also contain molecules on their membrane surfaces that recognize specific antigens and assist in their destruction.

Early studies of microbial infection identified that all antigens are met by an immune response that involves a collection of heterogeneous proteins known as antibodies. Antibodies are members of a larger immunoglobulin group of proteins and an important feature of the immune response is its versatility in responding to an enormous range of antigens. In humans this response extends from before birth and includes our lifelong ability to fight infection. One feature of the immune response is 'memory'. This property is the basis of childhood vaccination and arises from an initial exposure to antigen priming the system so that a further exposure leads to the rapid production of antibodies and the prevention of disease. From a biological standpoint the immune response represents an enigma. It is a highly specific recognition system capable of identifying millions of diverse antigens yet it retains an ability to 'remember' these antigens over considerable periods of time (years). The operation of the immune system was originally based around the classic observations reported by Edward Jenner at the end of the 18th century. Jenner recognized that milk maids were frequently exposed to a mild disease known as cowpox, yet rarely succumbed to the much more serious smallpox. Jenner demonstrated that cowpox infection, a relatively benign disease with complete recovery taking a few days, conferred immunity against smallpox. Exposure to cowpox triggered an immune response that conferred protection against the more virulent forms of smallpox, at that time relatively common in England. It was left to others to develop further the ideas of vaccination, notably Louis Pasteur, but for the last 100 years vaccination has been a key technique in fighting many diseases. The basis to vaccination lies in the working of the immune system.

The mechanism of specific antibody production remained puzzling but a considerable advance in this area arose with the postulation of the clonal selection theory by Neils Jerne and Macfarlane Burnet in 1955. This theory is now widely supported and envisages stem cells in the bone marrow differentiating to become lymphocytes each capable of producing a single immunoglobulin type. The immunoglobulin is attached to the outer surfaces of B lymphocytes and when an antigen binds to these antibodies replication of the cell is stimulated to produce a clone. The result is that only cells experiencing contact with an antigen are stimulated to replicate. Within the group of cloned B cells two distinct populations are identified. Effector B cells located in the plasma will produce soluble antibodies. These antibodies are comparable to those bound to the surface of B cells but lack membrane-bound sequences that anchor the antibodies to the lipid bilayer. The second group within the cloned B cell populations are called 'memory cells'. These cells persist for a considerable length of time even after the removal of antigen and allow the rapid production of antibodies in the event of a second immune reaction. The clonal selection theory was beautiful in that it explained how individuals distinguish 'self' and 'non-self'. During embryonic development immature B cells encounter 'antigens' on the surfaces of cells. These B cells do not replicate but are destroyed thereby removing antibodies that would react against the host's own proteins. At birth the only B cells present are those that are capable of producing antibodies against non-self antigens.

Immunoglobulin structure

Despite the requirement to recognize enormous numbers of potential antigens all immunoglobulin molecules are based around a basic pattern that was elucidated by the studies of Rodney Porter and Gerald Edelman. Porter showed that immunoglobulin G (IgG) with a mass of ~150 000 could be split into three fragments each retaining biological activity through the action of proteolytic enzymes such as papain. Normal antibody contains two antigen-binding sites (Figure 3.50) but after treatment with papain two fragments each binding a single antigen molecule were formed, along with a fragment that did not bind antigen but was necessary for biological function. The two fragments were called the F_{ab} (F = fragment, ab = antigen binding) and F_c (c = crystallizable) portions. The latter's name arose from the fact that its homogeneous composition allowed the fragment to be crystallized in contrast to the F_{ab} fragments.

The IgG molecule is dissociated into two distinct polypeptide chains of different molecular weight via the action of reducing agents. These chains were called the heavy (H) and light (L) polypeptides (Figure 3.51).

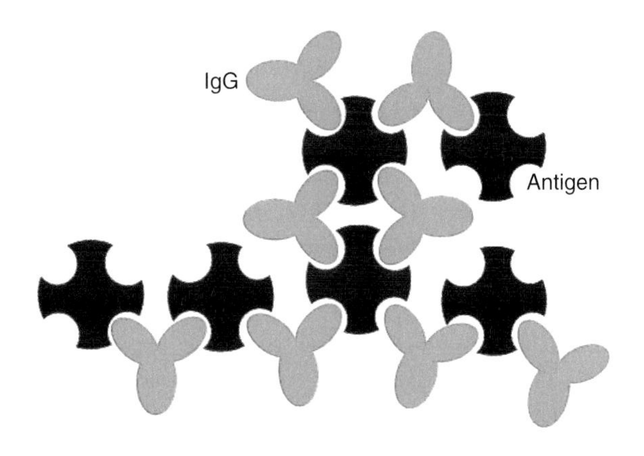

Figure 3.50 Schematic diagram showing bivalent interaction between antigen and antibody (IgG) and extended lattice

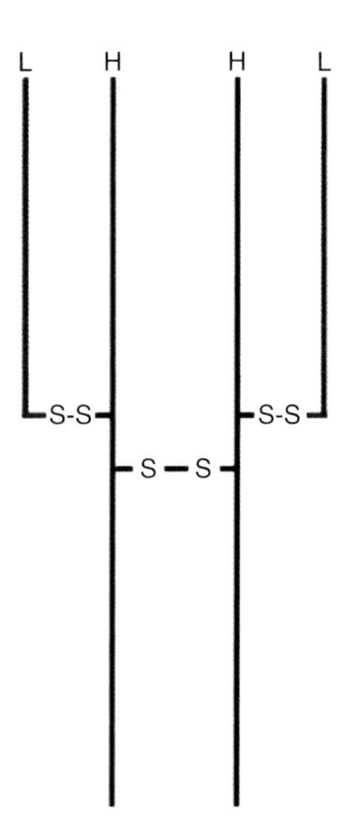

Figure 3.51 The subunit structure of IgG is H_2L_2

Further studies by Porter revealed that IgG was reconstituted by combining 2H and 2L chains, and a model for immunoglobulin structure was proposed envisaging each L chain binding to the H chain via disulfide bridges, whilst the H chains were similarly linked to each other. The results of papain digestion were interpretable by assuming cleavage of the IgG molecule occurred on the C terminus side of the disulfide bridge linking H and L chains. The F_{ab} region therefore contained both L chains and the amino terminal region of 2 H chains whilst the F_c region contained only the C terminal half of each H chain.

The amino acid sequence revealed not only the periodic repetition of intrachain disulfide bonds in the H and L chains but also the existence of regions of sequential homology between the H and L chains and within the H chain itself. Portions of the molecule, now known as the variable regions in the H and L chains, showed considerable sequence diversity, whilst constant regions of the H chain were internally homologous, showing repeating units. In addition the constant regions were homologous to regions on the L chain. This pattern of organization immediately suggested that immunoglobulins were derived from simpler antibody molecules based around a single domain of ~100 residues, and by a process of gene duplication the structure was assimilated to include multiple domains and chains.

Although the organization of IgG molecules is remarkably similar antigen binding is promoted through the use of sequence variability. These differences are not distributed uniformly throughout the H and L chains but are localized in specific regions (Figure 3.53). The regions of greatest sequence variability lie in the N terminal segments of the H and L chains and are called the V_L or V_H regions. In the L chain the region extends for ~110 residues and is accompanied by a more highly conserved region of the same size called the constant region of C_L. In the H chain the V_H region extends for ~110 residues whilst the remainder of the sequence is conserved, the C_H region. The C_H region is subdivided into three homologous domains: C_H1, C_H2 and C_H3.

Figure 3.52 The immunoglobulin fold containing a sandwich formed by two antiparallel sets of β strands. Normally seven β strands make up the two sets. In this domain additional sheet structures occur and small regions of a helix are found shown in blue. The important loop and turn regions are shown in grey. In the IgG light chain this fold would occur twice; once in the variable region and once again in the constant region. In the heavy chain it occurs four times

Fractionation studies identified antigen-binding sites at the N terminal region of the H and L chains in a region composed of the V_H and V_L domains where the ability to recognize antigens is based on the surface properties of these folds. Within the V_H and V_L domains the sequences of highest variability are three short segments known as hypervariable sequences and collectively known as complementarity determining regions (CDRs). They bind antigen and by arranging these CDRs in different combinations cells generate vast numbers of antibodies with different specificity.

The light chain consists of two discrete domains approximately 100–120 residues in length whilst the heavy chain is twice the size and has four such domains. Each domain is characterized by common structural topology known as the immunoglobulin fold and is repeated within the IgG molecule. The immunoglobulin fold is a sandwich of two sheets of antiparallel β strands each strand linked to the next via turns or large loop regions, with the sheets usually linked via a disulfide bridge. The immunoglobulin fold is found in other proteins operating within the immune system but is also noted in proteins with no obvious functional similarity (Figure 3.52).

The hypervariable regions are located in turns or loops of variable length that link together the different β strand elements. Collectively these regions form the antigen-binding site at the end of each arm of the immunoglobulin molecule and insight into the interaction between antibody and antigen has been gained from crystallization of an F_{ab} fragment with hen egg white lysozyme.

Raising antibodies against lysozyme generated several antibody populations – each antibody recognizing a different site on the surface of the protein. These sites are known as epitopes or antigenic determinants. By studying one antibody–antigen complex further binding specificity was shown to reside in the contact of the V_L and V_H domains with lysozyme at the F_{ab} tip (Figure 3.54). Interactions based around hydrogen bonding pairs involved at least five residues in the V_L domain (Tyr32, Tyr50, Thr53, Phe91, Ser93) and five residues in the V_H domain (Gly53, Asp54, Asp100, Tyr101, Arg102). These donor and acceptors hydrogen

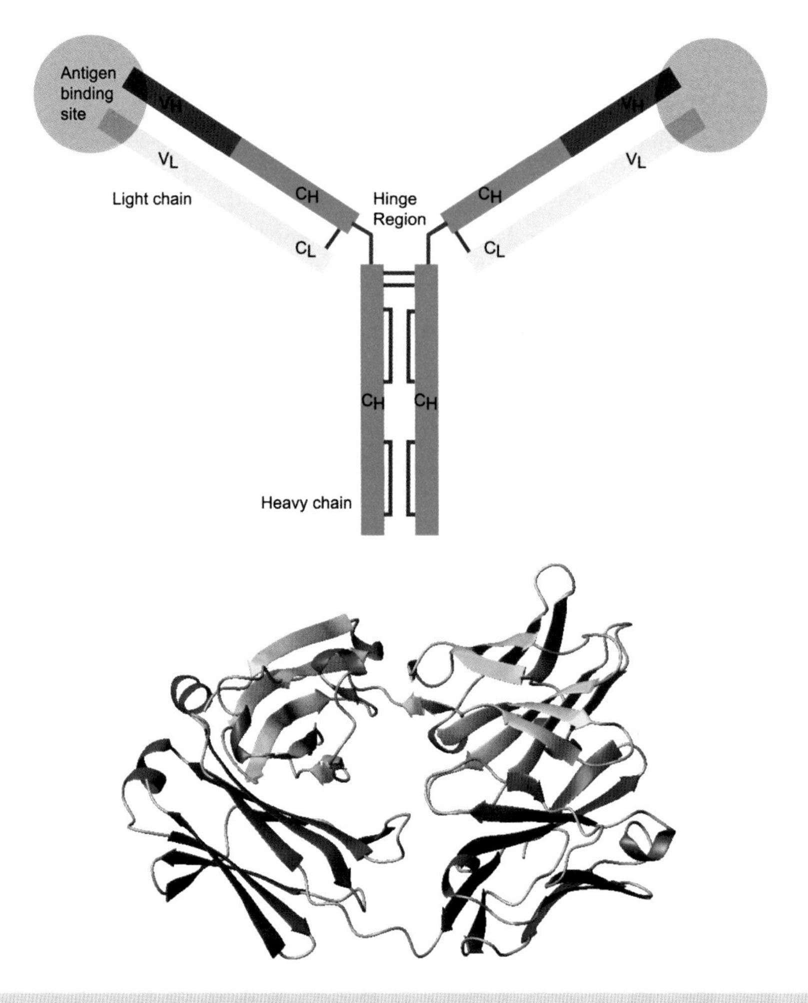

Figure 3.53 The structure and organization of IgG. The location of V_L, V_H and C_H within the H and L chains of an antibody together with a crystal structure determined for the isolated F_{ab} region (PDB: 7FAB). The H chain is shown with yellow strands whilst the L chain has strands shown in green. In each chain two immunoglobulin folds are seen

bonded to two groups of residues on lysozyme with the V_L domain interacting with Asp18, Asn19 and Gln121 whilst those on the V_H domain formed hydrogen bonds with 7 residues on the surface of lysozyme (Gly22, Ser24, Asn27, Gly117, Asp119, Val120, and Gln121).

The epitope on the surface of lysozyme is made up of two non-contiguous groups of residues (18–27 and 117–125) found on the antigen's surface. All six hypervariable regions participated in epitope recognition, but on the surface of lysozyme Gln121 was particularly important protruding away from the surface and forming a critical residue in the formation of a high-affinity antibody–antigen complex. The antibody forms a cleft that surrounds Gln121 with a hydrogen bond formed between the side chain and Tyr101 on the antibody (Figure 3.55). Elsewhere, van der Waals, hydrophobic and electrostatic interactions play a role in binding the other regions of the interacting surfaces between antibody and antigen.

Five different classes of immunoglobulins (Ig) have been recognized called IgA, IgD, IgE, IgG and IgM. These classes of immunoglobulins differ in the composition of their heavy chains whilst the light chain is based in all cases around two sequences identified

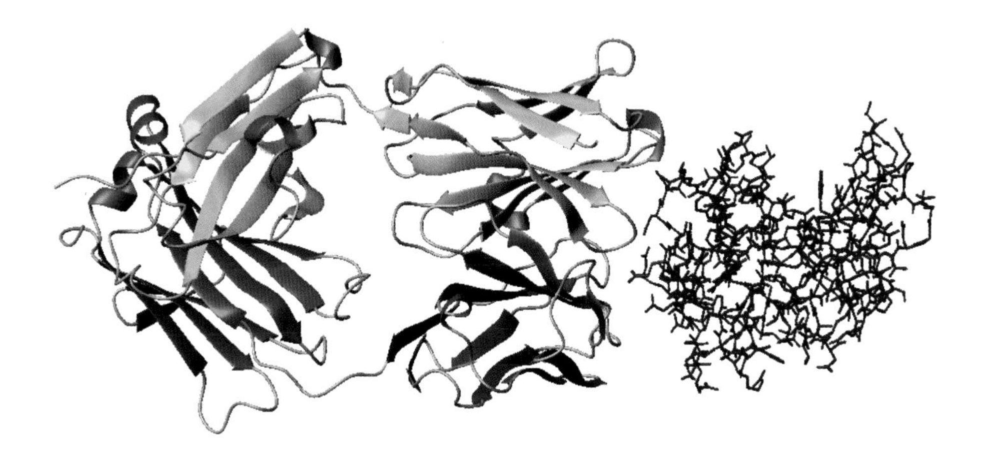

Figure 3.54 Interaction between monoclonal antibody fragment F_{ab} and lysozyme. The lysozyme is shown in blue on right and the prominent side chain of Gln121 is shown in red (PDB:1FDL)

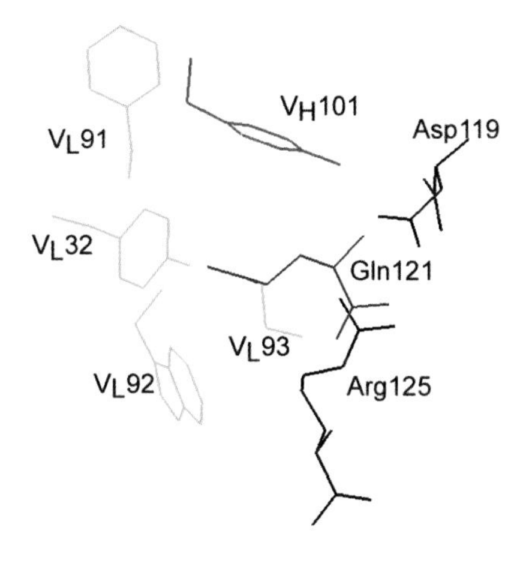

Figure 3.55 Interaction between Gln121 and residues formed by a pocket on the surface of the antibody in a complex formed between lysozyme and a monoclonal antibody

by the symbols κ and λ. The heavy chains are called α, δ, ε, γ and μ by direct analogy to the parent protein (Table 3.6).

As a consequence of remarkable binding affinities exemplified by dissociation constants between 10^4-10^{10} M antibodies have been used extensively as 'probes' to detect antigen. In the area of clinical diagnostics this is immensely valuable in detecting infections and many diseases are routinely identified via cross-reaction between antibodies and serum containing antigen. Antibodies are routinely produced today by injecting purified protein into a subject animal. The animal recognizes a protein as foreign and produces antibodies that are extracted and purified from sera. One extension of the use of antibodies as medical and scientific 'tools' was pioneered by César Milstein and Georges Köhler in the late 1970s and has proved popular and informative. This is the area of monoclonal antibody production, and although not described in detail here, the methods are widely used to identify disease or infection, to specifically identify a single antigenic site and increasingly as therapeutic agents in the battle against cancer and other disease states.

The humoral immune response involves the secretion of antibodies by B cells and the aggregation of antigen. Aggregation signals to macrophages that digestion should occur along with the destruction of pathogen. The cellular immune response to antigen involves different cells and a very different mechanism. On the surface of cytotoxic T lymphocytes sometimes more evocatively called 'killer T cells' are proteins whose structure and organization resembles the F_{ab} fragments of IgG. T cells identify antigenic peptide fragments that are bound to a surface protein known as the major histocompatibility complex

Table 3.6 Chain organization within the different Ig classes together with their approximate mass and biological roles

Class	Heavy chain	Light chain	Organization	Mass (kDa)	Role
IgA	α	κ or λ	$(\alpha_2\kappa_2)_n$ or $(\alpha_2\lambda_2)_n$	360–720	Found mainly in secretions
IgD	δ	κ or λ	$\delta_2\kappa_2$ or $\delta_2\lambda_2$	160	Located on cell surfaces
IgE	ε	κ or λ	$\varepsilon_2\kappa_2$ or $\varepsilon_2\lambda_2$	190	Found mainly in tissues. Stimulates mast cells to release histamines
IgG	γ	κ or λ	$\gamma_2\kappa_2$ or $\gamma_2\lambda_2$	150	Activates complement system. Crosses membranes
IgM	μ	κ or λ	$(\mu_2\kappa_2)_5$ or $(\mu_2\lambda_2)_5$	950	Early appearance in immune reactions. Linked to complement system and activates macrophages

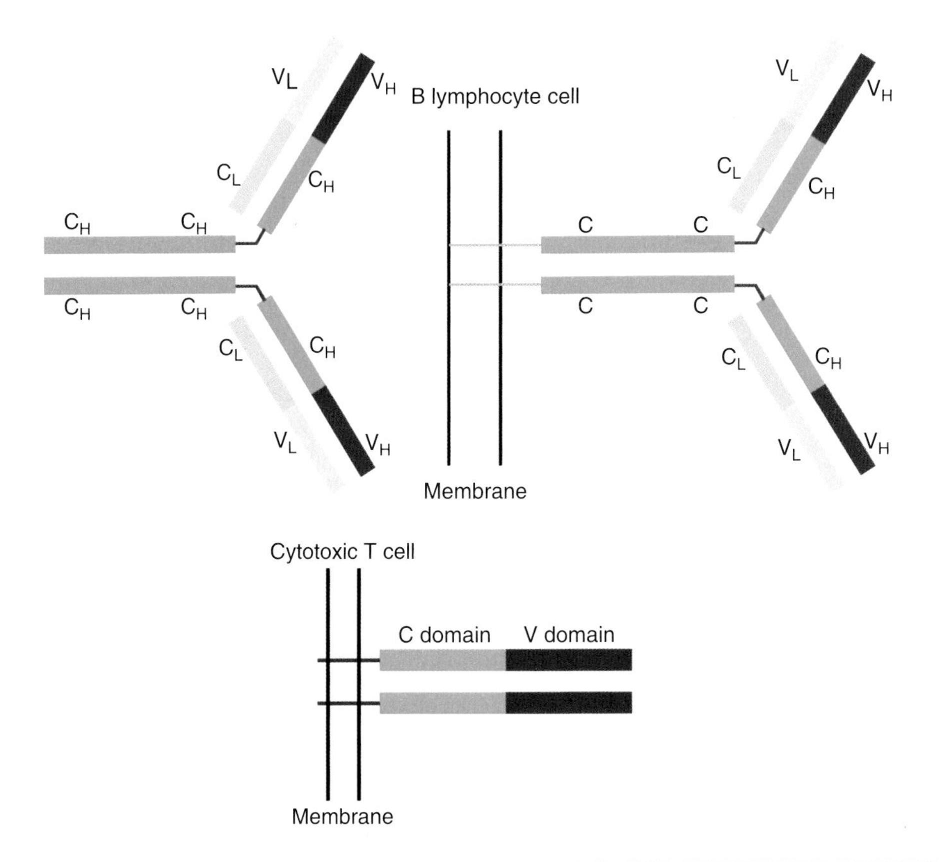

Figure 3.56 The use of the immunoglobulin fold in cell surface receptors by B and T lymphocytes

(MHC). Cytotoxic or killer T cells recognize antigen through receptors on their surface (Figure 3.56) and release proteins that destroy the infected cell. The MHC complex is based around domains carrying the immunoglobulin fold.

Cyclic proteins

Until recently it was thought that cyclic proteins were the result of unusual laboratory synthetic reactions without any counterparts in biological systems. This is now known to be untrue and several systems have been shown to possess cyclic peptides ranging in size from approximately 14 residues to the largest cyclic protein currently known a highly basic 70 residue protein called AS-48 isolated from *Enterococcus faecalis* S-48. Cyclic proteins are a unique example of tertiary structure but employ exactly the same principle as linear proteins with the exception that their ends are linked together (Figure 3.57). In view of the fact that many globular proteins have the N and C terminals located very close together in space there is no conceptual reason why the amino and carboxy terminals should not join together in a further peptide bond.

Cyclic proteins are distinguished from cyclic peptides such as cyclosporin. These small peptides have been known for a long time and are synthesized by microorganisms as a result of multienzyme complex reactions. The latter products contain unusual amino acid residues often extensively modified and are not the result of transcription. In contrast, cyclic proteins are known to be encoded within genomes and appear to have a broad role in host defence mechanisms.

Many cyclic proteins are of plant origin and a common feature of these proteins appears to be their size (~30 residues), a cyclic peptide backbone coupled with a disulfide rich sequence containing six conserved cysteine residues and three disulfide bonds. Unusually the disulfides cross to form a knot like arrangement and the cyclic backbone coupled with the cysteine knot has lead to the recognition of a new structural motif called the CCK motif or cyclic cysteine knot. The term cyclotides has been applied to these proteins. A common feature of all cyclotides is their derivation from longer precursor proteins in steps that involve both cleavage and cyclization. Although the gene sequences for the precursors are known the putative cleaving and cyclizing enzymes have not yet been reported. Many of these proteins have assumed enormous importance with the demonstration that several are natural inhibitors of enzymes such as trypsin whilst others are seen as possible lead compounds in the development of new pharmaceutical products directed against viral pathogens, and in particular anti-HIV activity. Whilst the cyclotide family (see Table 3.7) appear to have a common structural theme based around the CCK motif and the presence of three β strands other cyclic proteins such as bacteriocin AS-48 contain five short helices connected by five short turn regions that enclose a compact hydrophobic core. Despite different tertiary structure, all of the cyclic proteins are characterized by high intrinsic stability (denaturation only at very high temperatures) as well as resistance to proteolytic degradation.

Summary

Proteins fold into precise structures that reflect their biological roles. Within any protein three levels of organization are identified called the primary, secondary and tertiary structures, whilst proteins with more than one polypeptide chain exhibit quaternary levels of organization.

Primary structure is simply the linear order of amino acid residues along the polypeptide chain from the N to C terminals. Long polymers of residues cannot fold into any shape because of restrictions placed on conformational flexibility by the planar peptide bond and interactions between non-bonded atoms.

Conformational flexibility along the polypeptide backbone is dictated by ϕ and ψ torsion angles. Repetitive values for ϕ and ψ lead to regular structures known as the α helix and β strand. These are elements of secondary structure and are defined as the spatial arrangement of residues that are close together in the primary sequence.

The α helix is the most common element of secondary structure found in proteins and is characterized by dimensions such as pitch 5.4 Å, the translation distance 1.5 Å, and the number of residues per turn (3.6). A regular α helix is stabilized by hydrogen bonds

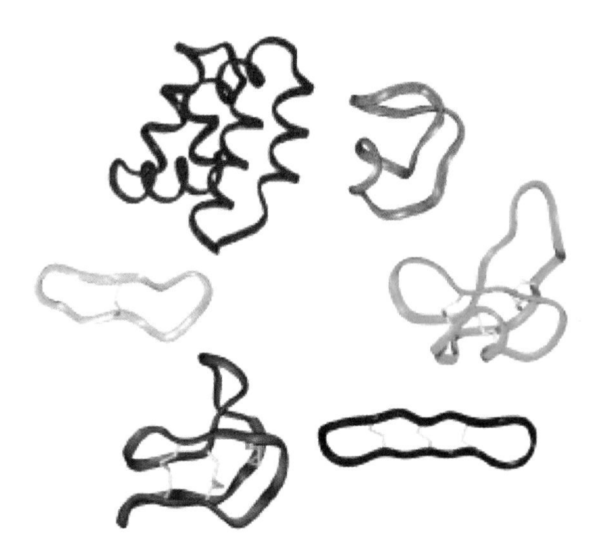

Figure 3.57 The topology of naturally occurring circular proteins. Three-dimensional structures of naturally occurring proteins. Clockwise from the upper left the proteins are: bacteriocin AS-48 (PDB:1E68) from *Enterococcus faecalis*, microcin J25 (PDB:1HG6) from *E. coli*, MCoTI-II (PDB:1HA9, 1IB9) from bitter melon seeds, RTD-1 (PDB:1HVZ) from the leukocytes of Rhesus macaques, kalataB1 (PDB:1KAL) from several plants of the *Rubiaceae* and *Violaceae* plant families, and SFTI-1 (PDB:1SFI, 1JBL) from the seeds of the common sunflower. Disulfide bonds are shown in yellow. (Reproduced with permission from Trabi, M., & Craik, D.J. *Trends Biochem. Sci.* 2002, **27**, 132–138. Elsevier)

orientated parallel to the helix axis and formed between the CO and NH groups of residues separated by four intervening residues.

In contrast the β strand represents an extended structure as indicated by the pitch distance ~7 Å, the translation distance 3.5 Å and fewer residues per turn (2). Strands have the ability to hydrogen bond with other strands to form sheets – collections of β strands stabilized via inter-strand hydrogen bonding. Numerous variations on the basic helical and strand structures are found in proteins and truly regular or ideal conformations for secondary structure are rare.

Tertiary structure is formed by the organization of secondary structure into more complex topology or folds by interaction between residues (side chain and backbone) that are often widely separated in the primary sequence.

Several identifiable folds or motifs exist within proteins and these units are seen as substructures within a protein or represent the whole protein. Examples include the four-helix bundle, the β barrel, the β helix, the HTH motif and the β propeller. Proteins can now be classified according to their tertiary structures and this has led to the description of proteins as all α, α + β and α/β. The recognition that proteins show similar tertiary structures has led to the concept of structural homology and proteins grouped together in related families.

Tertiary structure is maintained by the magnitude of favourable interactions outweighing unfavourable ones. These interactions include covalent and more frequently non-covalent interactions. A covalent bond formed between two thiol side chains results in a disulfide bridge, but more common stabilizing forces include charged interactions, hydrophobic forces, van der Waals interactions and hydrogen bonding. These interactions differ significantly in their strength and number.

Quaternary structure is a property of proteins with more than one polypeptide chain. DNA binding proteins function as dimers with dimerization the result of specific subunit interaction. Haemoglobin is the classic example of a protein with quaternary structure containing 2α,2β subunits.

Proteins with more than one chain may exhibit allostery; a modulation of activity by smaller effector molecules. Haemoglobin exhibits allostery and this is shown by sigmoidal binding curves. This curve is described as cooperative and differs from that shown by myoglobin. Oxygen binding changes the structure of one subunit facilitating the transition from deoxy to oxy states in the remaining subunits. Historically the study of the structure of haemoglobin provided a platform from which to study larger, more complex, protein structures together with their respective functions.

One such group of proteins are the immunoglobulins. These proteins form the body's arsenal of defence mechanisms in response to foreign macromolecules and are collectively called antibodies. All antibodies are based around a Y shaped molecule composed of two heavy and two light chains held together by covalent and non-covalent interactions.

Table 3.7 Source, size and putative roles for cyclotides characterized to date

Protein	Source	Size (residues)	Role
SFTI-1	*Helianthus annuus*	14	Potent trypsin inhibitor
Microcin J25	*Escherichia coli*	21	Antibacterial
Cyclotide family	*Rubiaceae* and *Violaceae* sp	28–37	Wide range of activities ~45 proteins
McoTI-I and II	*Momordica cochinchinensis*	34	Seed-derived trypsin inhibitor
Bacteriocin AS-48	*Enterococcus faecalis*	70	Hydrophobic antibacterial protein
RTD-1	*Macaca mulatta*	18	Antibiotic defensin from primate leukocytes

Antigen binding sites are formed from the hypervariable regions at the end of the heavy and light chains. These hypervariable regions allow the production of a vast array of different antibodies within five major classes and allow the host to combat many different potential antigens.

The basic immunoglobulin fold is widely used within the immune system with antibody-like molecules found as the basis of many cell surface receptors particularly in cells such as helper and killer T cells and parts of the MHC complex.

Problems

1. Draw a diagram of a typical polypeptide backbone for a pentapeptide. Label on your diagram the following; the α carbon, the side chains, use a box to define the atoms making up the peptide bond and finally identify the torsion angles ϕ and ψ and the atoms/bonds defining these angles.

2. Show how the above pentapeptide changes when the third residue is proline.

3. Poly-lysine and poly-glutamate can switch between disordered structures and helical structures. What conditions might promote this switch and how does this drive formation of helical structure?

4. Using the bond lengths given for C–C and C–N bonds in Chapter 2 together with the dimensions found in α helices and β strands calculate the length of (i) a fully extended polypeptide chain of 150 residues, (ii) a chain made up entirely of one long regular α helix and, (iii) a chain composed of one long β strand. Comment on your results.

5. From Figure 3.2 use the primary sequences of myoglobin to define the extent of each element of helical secondary structure. Explain your reasoning in each case.

6. What limits the conformational space sampled by ϕ and ψ?

7. Some proteins when unfolded are described as a random coil. Why is this a misleading term?

8. List the interactions that stabilize the folded structures of proteins. Rank these interactions in terms of their average 'energies' and give an example of

each interaction as it occurs between residues within proteins.

9. How would you identify turn regions within the polypeptide sequence of proteins? Why do turn regions occur more frequently on protein surfaces? What distinguishes a turn from a loop region?

10. What are the advantages of more than one subunit within a protein? How are multimeric proteins stabilized?

4

The structure and function of fibrous proteins

Historically a division into globular and fibrous proteins has been made when describing chemical and physical properties. This division was originally made to account for the very different properties of fibrous and globular proteins as well as the different roles each group occupied within cells. Today it is preferable to avoid this distinction and to treat proteins as belonging to families that exhibit structural or sequential homology (see Chapter 6). However, despite very different properties the structures of fibrous proteins were amongst the first to be studied because of their accumulation in bodies such as hair, nails, tendons and ligaments. These proteins demonstrate new aspects of biological design not shown by globular proteins that requires separate description.

Fibrous proteins were named because they were found to make up many of the 'fibres' found in the body. Here, fibrous proteins had a common role in conferring strength and rigidity to these structures as well as physically holding them together. Subsequent studies have shown that these proteins are more widely distributed than previously supposed, being found in cells as well as making up connective tissues such as tendons or ligaments. More importantly, these proteins occupy important biological roles that arise from their wide range of chemical and physical properties. These properties are distinct from globular proteins and arise from the individual amino acid sequences. A common feature of most fibrous proteins is their long, drawn-out, or filamentous structure. Essentially, these proteins tend to occur as 'rod-like' structures extended more in two out of the three possible dimensions and lacking the compactness of globular proteins. As a result fibrous proteins tend to possess architectures based around regular secondary structure with little or no folding resulting from long-range interactions. In other words they lack true tertiary structure.

The second large class of proteins distinct from globular proteins are the membrane proteins. Members of this group of proteins probably make up the vast majority of all proteins found in cells. For many years the purification of these proteins remained very difficult and limited our knowledge of their structure and function. However, slowly membrane proteins have become amenable to biochemical characterization and Chapter 5 will deal with the properties of this important group to highlight in successive sections globular, fibrous and membrane proteins.

The amino acid composition and organization of fibrous proteins

In fibrous proteins at least three different structural plans or designs have been recognized in construction.

Proteins: Structure and Function by David Whitford

These designs include: (i) structure composed of 'coiled-coils' of α helices and represented by the α keratins; (ii) structures made up of extended antiparallel β sheets and exemplified by silk fibroin a collection of proteins made by spiders or silkworms; and (iii) structures based on a triple helical arrangement of polypeptide chains and shown by the collagen family of proteins. The structures of each class of fibrous proteins will be described in the following sections, highlighting how the structure is suitable for its particular role and also emphasizing how defects in fibrous proteins can lead to serious and life threatening conditions.

An analysis of the amino acid composition of typical fibrous proteins (Table 4.1) reveals considerable differences in their constituent amino acids to that described earlier for globular proteins (see Table 2.2, Chapter 2). More significantly the amino acid compositions of fibrous proteins differ with each group. For example, collagen has a proline content in excess of 20 percent whilst in silk fibroin this value is below 1 percent. Similarly in α keratin the cysteine content is 11.2 percent but in collagen and silk fibroin the levels of cysteine are essentially undetectable. In each case the amino acid composition influences the secondary structure formed by fibrous proteins.

Table 4.1 The amino acid composition of three common classes of fibrous proteins in mole percent

Amino acid	Fibroin (silk)	α-keratin	Collagen
Gly	44.6	8.1	32.7
Ala	29.4	5.0	12.0
Ser	12.2	10.2	3.4
Glx	1.0	12.1	7.7
Cys	0	11.2	0
Pro	0.3	7.5	22.1
Arg	0.5	7.2	5.0
Leu	0.5	6.9	2.1
Thr	0.9	6.5	1.6
Asx	1.3	6.0	4.5
Val	2.2	5.1	1.8
Tyr	5.2	4.2	0.4
Ile	0.7	2.8	0.9
Phe	0.5	2.5	1.2
Lys	0.3	2.3	3.7
Trp	0.2	1.2	0
His	0.2	0.7	0.3
Met	0	0.5	0.7

In collagen considerable amounts of hydroxylated lysine and proline are found. Adapted from *Biochemistry*, 3rd edn. Mathews, van Holde & Ahern (eds). Addison Wesley Longman, London.

Keratins

Keratins are the major class of proteins found in hair, feathers, scales, nails or hooves of animals. In general the keratin class of proteins are mechanically strong, designed to be unreactive and resistant to most forms of stress encountered by animals. At least two major groups of keratins can be identified; the α keratins are typically found in mammals and occur as a large number of variants whilst β keratins are found in birds and reptiles as part of feathers and scales containing a significantly higher proportion of β sheet. The β keratins are analogous to the silk fibroin structures produced by spiders and silkworms and described later in this chapter. The α keratins are a subset of a much larger group of filamentous proteins based on coiled-coils called intermediate filaments (IF). The distribution of intermediate filaments is not restricted to mammals but appears to extend to most animal cells as major components of cytoskeletal structures.

In mammals approximately 30 different variants of keratin have been identified with each appearing to be expressed in cells in a tissue-specific manner. In each keratin the 'core' structure is similar and is based around the α helix so that the following discussion of the conformation applies equally well to all proteins within this group. Although the basic unit of keratin is an α helix this structure is slightly distorted as a result of interactions with a second helix that lead to the formation of a left handed coiled-coil. The most common arrangement for keratin is a coiled-coil of two α helices although three helical stranded arrangements are known for extracellular keratins domains whilst in insects four stranded coiled coils have been found. In 1953 Francis Crick postulated that the stability of α helices would be enhanced if pairs of helices interacted not as straight rods but in a simple coiled-coil arrangement (Figure 4.1). This coiled coil arrangement is sometimes called a super helix with in this instance α keratin found as a left-handed super helix. Detailed structural studies of coiled-coils confirmed this arrangement of helices and diffraction studies showed a periodicity of 1.5 Å and 5.1 Å.

The coiled-coil is formed by each helix interacting with the other and by burying their hydrophobic residues away from the solvent interface. The hydrophobic or non-polar side chains are not randomly located within the primary sequence but occur at regular intervals throughout the chain. This non-random distribution of hydrophobic residues is also accompanied by a preference for residues with charged side chains at positions within helices that are in contact with solvent. As a result of this periodicity a repeating unit of seven residues occurs along each chain or primary sequence. This is called the heptad repeat and the residues within this unit are labelled a, b, c, d, e, f and g. To facilitate identification of each residue within a helix the positions are frequently represented by a helical wheel diagram (Figure 4.2).

Although a regular α helix has 3.6 residues per turn, the coiled-coil arrangement leads to a slight decrease in the number of residues/turn to 3.5. Each helix is inclined at an angle of approximately 18° towards the other helix and this allows for the contacting side chains to make a precise inter-digitating surface. It also leads to a slightly different pitch for the helix of 5.1 Å compared with 5.4 Å in a regular α helix. The hydrophobic residues located at residues a and d of each heptad form a hydrophobic surface interacting with other hydrophobic surfaces. The hydrophobic residues form a seam that twists about each helix. By interacting with neighbouring hydrophobic surfaces helices are forced to coil around each other forming the super helix or coiled-coil. The interleaving of side chains has been known as 'knobs-into-holes' packing and in other proteins as a leucine zipper arrangement, although this last term is slightly misleading. A coiled coil arrangement not only enhances the stability of the intrinsically unstable single α- helix but also confers considerable mechanical strength in a manner analogous to the intertwining of rope or cable. Further aggregation of the coiled-coils occurs and leads to larger aggregates with even greater strength and stability. Besides the high content of hydrophobic residues α keratins also have significant proportions of cysteine (see Table 4.1). The cysteine residues participate in disulfide bridges that cross-link neighbouring coiled-coils to build up a filament or bundle and ultimately the network of protein constituting hair or nail.

The coiled-coil containing many heptad repeats extends on average for approximately 300–330 residues and is flanked by amino and carboxy terminal domains. These amino and carboxy domains vary greatly in size. In some keratins very small domains of approximately 10 residues can exist whilst in other homologous proteins, such as nestin, much larger domains in excess of 500 residues are found. More significantly these regions show much greater sequence variability when compared with the coiled-coil regions suggesting that these domains confer specificity on the individual keratins and are tailored towards increasing functional specificity.

The keratins are often classified as components of the large group of filamentous proteins making up the cytoskeletal system and in particular they are classified as IF (Table 4.2). IF are generally between 8 and 10 nm in diameter and are more common in cells that have to withstand stress or extreme conditions.

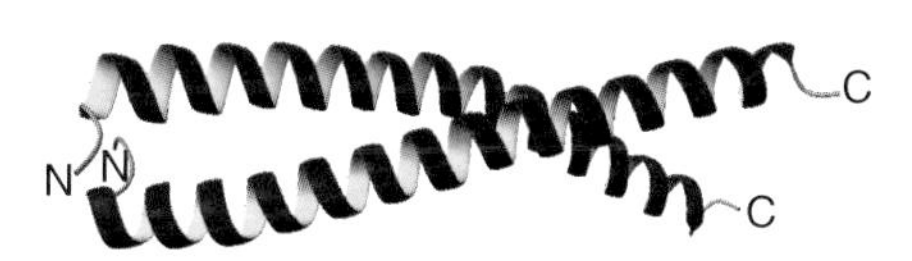

Parallel coiled coil

Anti-parallel coiled coil

Figure 4.1 Two individual α helices distorted along their longitudinal axes as a result of twisting. The two helices can pack together to form coiled-coils

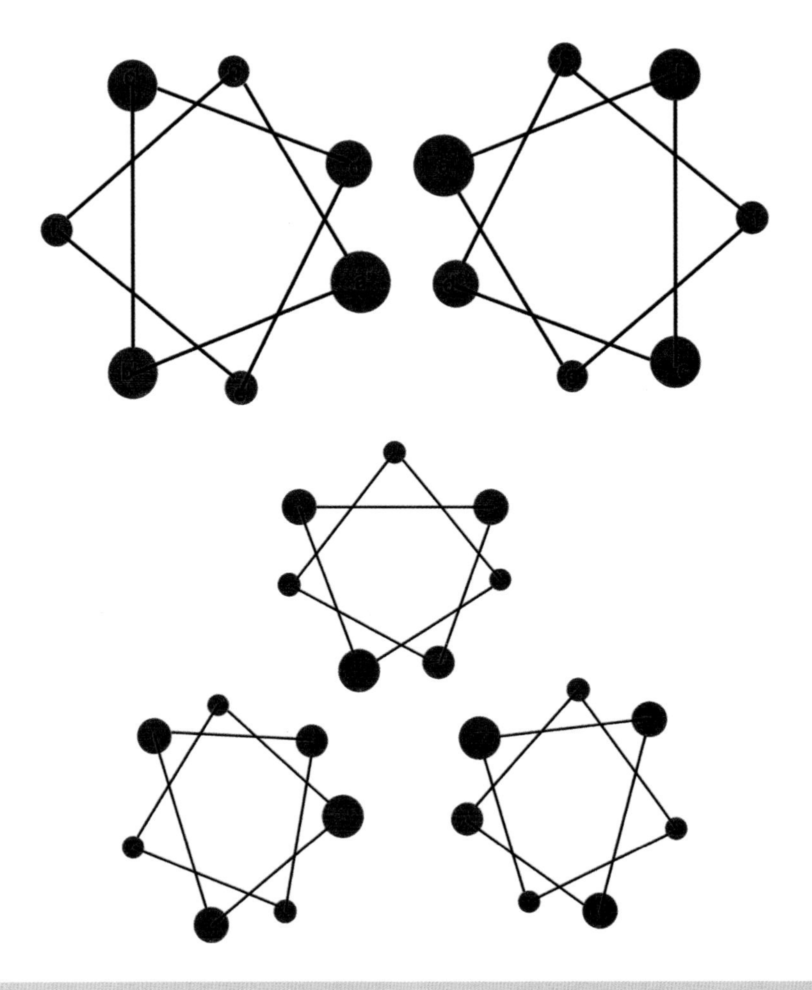

Figure 4.2 A helical wheel representation of the heptad repeat of coiled-coil keratins. The diagram shows a dimer (top) and trimer (bottom) of helices and emphasizes the hydrophobic contact regions between each helix as a result of residues a and d packing together. Leu, Ile and Ala are found frequently at positions a and d. Glu and Gln occur frequently at positions e and g whilst Arg and Lys are found frequently at position g

At least six different IF have been identified with classes I and II represented by acidic and basic keratins. An assignment of the terms 'acidic' and 'basic' to the N and C terminal globular domains of keratin refer to their overall charge. These domains vary in composition from one keratin to another. The acidic domains found at the end of the central α helical regions contain more negatively charged side chains (Glu and Asp) than positively charge ones (Lys, Arg) and consequently have isoelectric points in the pH range from 4 to 6.

Acidic and basic monomers are found within the same cell and the coiled coil or dimer contains one of each type (a heterodimer). Each coiled-coil aligns in a head to tail arrangement and in two staggered rows to form a protofilament. The protofilament dimerizes to form a protofibril with four protofibrils uniting to make a microfibril. The aggregation of protein units is still not finished since further association between microfibrils results in the formation of a macrofibril in reactions that are still poorly understood. The assembly of coiled coils into microfibrils is shown in Figure 4.3.

Table 4.2 The classification of intermediate filaments

Type	Example	Mass (M_r) × 10^3	Location
I	Acidic keratins	40–64	Epithelial
II	Basic keratins	52–68	Epithelial
III	Glial fibrillary acidic protein	51	Astroglia
IV	Vimentin	55	Mesenchymal
VI	Desmin	53	Muscle
	Peripherin	54	Neuronal
	Neurofilaments (L, M, H);	68, 110, 130	Neuronal
	Internexin	66–70	Neuronal
V	Lamins A, B, C	58–70	Most cell types
Unclassified	Septins A, B, C		Some unclassified IFs appear to be found in invertebrates
	Filensin	100	
	Lens	50–60	

At least 6 different classes have been identified from homology profiles although new IF like proteins are causing these schemes to expand. IF are larger than the thin microfilaments (7–9 nm diameter) often made up of actin subunits and smaller than the thick microtubules (~25 nm diameter) made up of tubulin.

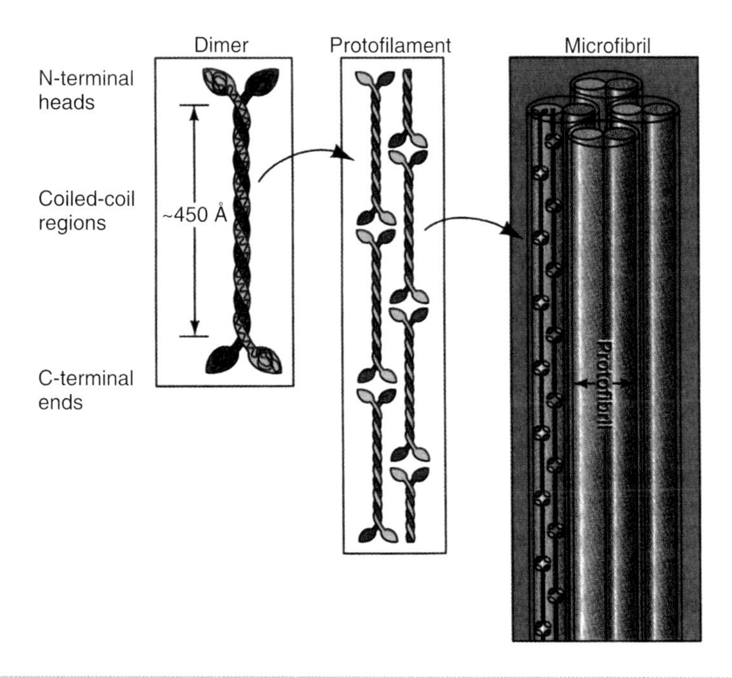

Figure 4.3 Higher order α keratin structure. Left: heterodimeric arrangement of two α helices to form a coiled-coil with both acidic and basic domains. Middle: protofilaments formed by the association of two coiled coils in a head-tail order and in two staggered or offset rows. Right: dimerization of protofilaments to form a protofibril followed by four protofibrils uniting to form a macrofibril (Reproduced from Voet, D, Voet, J.G & Pratt, C.W. *Fundamentals of Biochemistry*, John Wiley & Sons Inc, Chichester, 1999)

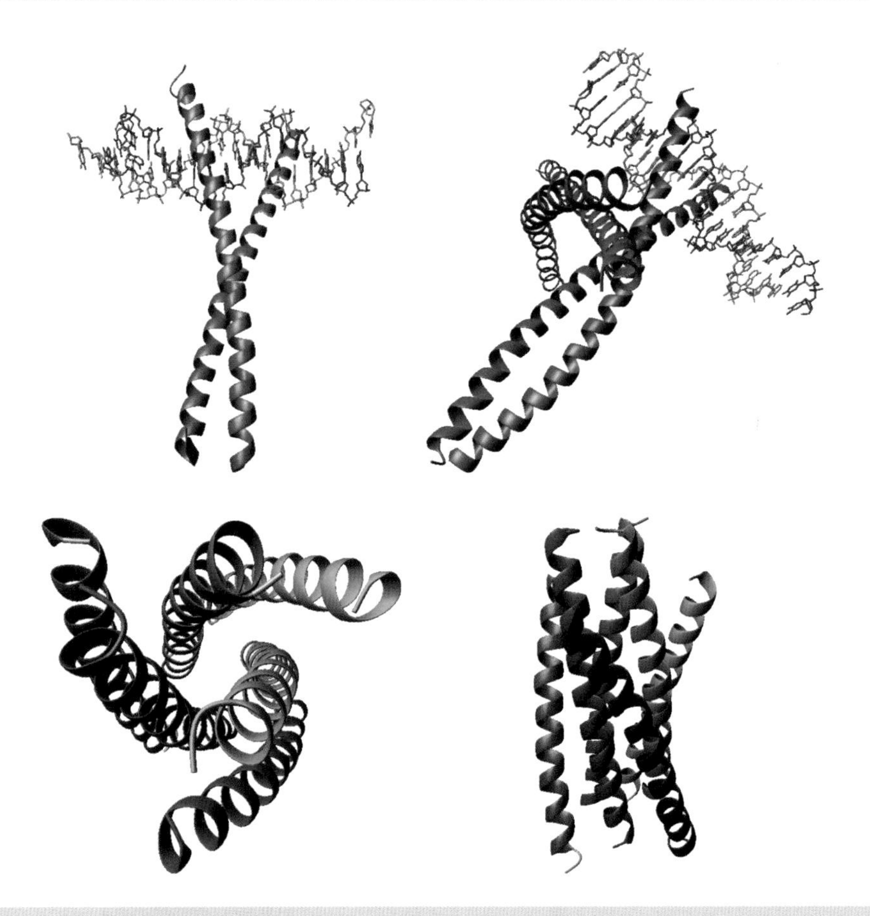

Figure 4.4 Examples of coiled coil motifs occurring in non-keratin based proteins. In a clockwise direction from the top left the diagrams show part of the leucine zipper DNA binding protein GCN4 (PDB:1YSA); a heterodimer of the c-jun proto-oncogene (the transcription factor Ap-1) dimerized with c-fos and complexed with DNA (PDB:1FOS); and two views of the gp41 core domain of the simian immunodeficiency virus showing the presence of six α helices coiled together as two trimeric units – an inner trimer often called N36 and an outer trimer called C34 (PDB:2SIV)

Crystallographic studies have shown that the coiled-coil motif occurs in many other proteins as a recognizable motif. It occurs in viral membrane-fusion proteins including the gp41 domain found as part of the human and simian immunodeficiency virus (HIV/SIV), in the haemagglutinin component of the influenza virus as well as transcription factors such as the leucine zipper protein GCN4. It is also found in muscle proteins such as tropomyosin and is being identified with increasing regularity in proteins of diverse function and cellular location. Although first recognized as part of long fibrous proteins the coiled-coil is now established as a structural element of many other proteins with widely differing folds (Figure 4.4). In the coiled-coil structures the α helices do not have to run in the same direction for hydrophobic interactions to occur. Although a parallel conformation is the most common arrangement antiparallel orientations, where the chains run in opposite directions, occur in dimers but are very rare in higher order aggregates.

Mutations in the genes coding for keratin lead to impaired protein function. In view of the almost ubiquitous distribution of keratin these genetic defects can have severe consequences on individuals. Defects prove particularly deleterious to the integrity of skin and several inherited disorders are known where cell adhesion, motility and proliferation are severely disturbed. Since many human cancers arise in epithelial tissues where keratins are prevalent such defects may predispose individuals to more rapid tumour development.

Keratins, the most abundant proteins in epithelial cells, are encoded by two groups of genes designated type I and type II. There are >20 type I and >15 type II keratin genes occurring in clusters at separate loci in the human genome. A distinction is often made on the type of cell from which the keratin is derived or linked. This leads to type II keratin proteins from soft epithelia labelled as K1–K8 whilst those derived from hard epithelia (such as hair, nail, and parts of the tongue) are designated Hb1–Hb8. Similarly type I keratins are comprised of K9–K20 in soft epithelia and Ha1–Ha10 in hard epithelia. All α keratins are rich in cysteine residues that form disulfide bonds linking adjacent polypeptide chains. The term 'hard' or 'soft' refers to the sulfur content of keratins. A high cysteine (i.e. sulfur) content leads to hard keratins typical of nails and hair and is resistant to deformation whilst a low sulfur content due to a lower number of cysteine residues will be mechanically less resistant to stress.

In vitro the combination of any type I and type II keratins will produce a fibrous polymer when mixed together but *in vivo* the pairwise regulation of type I and II keratin genes in a tissue specific manner gives rise to 'patterns' that are very useful in the study of epithelial growth as well as disease diagnosis. The distribution of some of these keratins is described in Table 4.3 and with over 30 different types of keratin identified there is a clear preferential location for certain pairs of keratins.

In all complex epithelia a common set of keratin genes are transcribed consisting of the type II K5 and the type I K14 genes (along with variable amounts of K15 or K19, two additional type I keratins). Post-mitotic, suprabasal cells in these epithelia transcribe other pairs of keratin genes, the identity of which depends on the differentiation route of these cells. Thus the K1 and K10 pair is characteristic of cornifying epithelia such as the epidermis, whilst the K4 and K13 pair is expressed in epithelia found lining the oral cavity, the tongue and the oesophagus, and the K3 and K12 pair is found in the cornea of the eye.

Table 4.3 Distribution of type I and II keratin in different cells

Type I (acidic)	Type II (basic)	Location
K10	K1	Suprabasal epidermal keratinocytes
K9	K1	Suprabasal epidermal keratinocytes
K10	K2e	Granular layer of epidermis
K12	K3	Cornea of eye
K13	K4	Squamous epithelial layers
K14	K5	Basal layer keratinocytes
K15	K5	Basal layer of non-keratinizing epithelia
K16	K6a	Outer sheath of hair root, oral epithelial cells, hyperproliferative keratinocytes
K17	K6b	Nails
	K7	Seen in transformed cells
K18	K8	Simple epithelia
K19		Follicles, simple epithelia
K21		Intestinal epithelia

In skin diseases such as psoriasis and atopic dermatitis it is noticed that keratin 6 and 16 predominate whilst a congenital blistering disease, epidermolysis bullosa simplex, arises from gene defects altering the structure of keratins 5 and 14 at the basal layer.

A property of disulfide bridges between cysteine residues is their relative ease of reduction with reducing agents such as dithiothreitol, mecaptoethanol or thioglycolate. In general these reagents are called mercaptans, and for hair the use of thioglycolate allows the reduction of disulfide bridges and the relaxing of hair from a curled state to a straightened form. Removal of the reducing agent and the oxidation of the thiol groups allows the formation of new disulfide bridges and in this way hair may be reformed in a new 'curled', 'straightened' or 'permed' conformation. The springiness of hair is a result of the extensive number of coiled-coils and their tendency in common with a conventional spring to regain conformation after initial stretching. The reduction of disulfide bridges in hair allows a keratin fibre to stretch to over twice their original length and in this very extended 'reduced' conformation the structure of the polypeptide chains shift towards the β sheet conformations found in feathers or in the silk-like sheets of fibroin. In the 1930s W.T. Astbury showed that a human hair gave a characteristic X-ray diffraction pattern that changed upon stretching the hair, and it was these two forms that were designated α and β.

Fibroin

A variation in the structure of fibrous proteins is seen in the silk fibroin class made up of an extended array of β strands assembled into a β sheet. Insect and spiders produce a variety of silks to assist in the production of webs, cocoons, and nests. Fibroin is produced by cultivated silkworm larvae of the moth *Bombyx mori* and has been widely characterized. Silk consists of a collection of antiparallel β strands with the direction of the polypeptide backbone extending along the fibre axis. The high content of β strands leads to a microcrystalline array of fibres in a highly ordered structure. The polypeptide backbone has the extended structure typical of β strands with the side chains projecting above and below the plane of the backbone. Of great significance in silk fibroin are the long stretches of repeating composition. A six residue repeat of $(\text{Gly-Ser-Gly-Ala-Gly-Ala})_n$ is observed to occur frequently and it is immediately apparent that this motif lacks large side chains. These three residues appear to represent over 85 percent of the total amino acid composition with approximate values for the individual fractions being 45 percent Gly, 30 percent Ala and 15 percent Ser in silk fibroin (see Table 4.1). The sequence of six residues is part of a larger repeating unit

$$\text{-(Gly-Ala)}_2\text{-Gly-Ser-Gly-Ala-Ala-Gly-}$$

$$\text{(Ser-Gly-Ala-Gly-Ala)}_8\text{-Tyr-}$$

that may be repeated up to 50 times leading to masses for silk polypeptides between 300 000 and 400 000 (see Figure 4.5). Glycine, alanine and serine are the three smallest side chains in terms of their molecular volumes (see Table 2.3) and this is significant in terms of packing antiparallel strands. More importantly the order of residues in this repeating sequence places the glycine side chain (simply a hydrogen) on one side of the strand whilst the Ser and Ala side chains project to the other side. This arrangement of side chains leads to a characteristic spacing between strands that represents the interaction of Gly residues on one surface and the interaction of Ala/Ser side chains on the other. The interaction between Gly surfaces yields an inter-sheet spacing or regular periodicity of 0.35 nm whilst the interaction of the Ser/Ala rich surfaces gives a spacing of 0.57 nm between the strands in silk fibroin. Larger amino acid side chains would tend to disrupt the regular periodicity in the spacing of strands and they tend to be located in regions forming the links between the antiparallel β strands. The structure of these linker regions has not been clearly defined.

Silk has many remarkable properties. Weight-for-weight it is stronger than metal alloys such as steel, it is more resilient than synthetic polymers such as Kevlar©, yet is finer than a human hair. It is no exaggeration to say that silk is nature's high-performance polymer fine-tuned by evolution over several hundred million years. As a result of these desirable properties there have been many attempts to mimic the properties of silk with the development of new materials in the area of biomimetic chemistry. Silk is extremely strong because in the fully extended conformation of β strands any further extension would require the breakage of strong covalent bonds. However, this strength is coupled with surprisingly flexibility that arises as a result of the weaker van der Waals interactions that exist between the antiparallel β strands. These desirable physical properties are very difficult to reproduce in most synthetic polymers.

Collagen

Collagen is a major component of skin, tendons, ligaments, teeth and bones where it performs a wide variety of structural roles. Collagen provides the framework that holds most multicellular animals together and constitutes a major component of connective tissue. Connective tissue performs many functions including

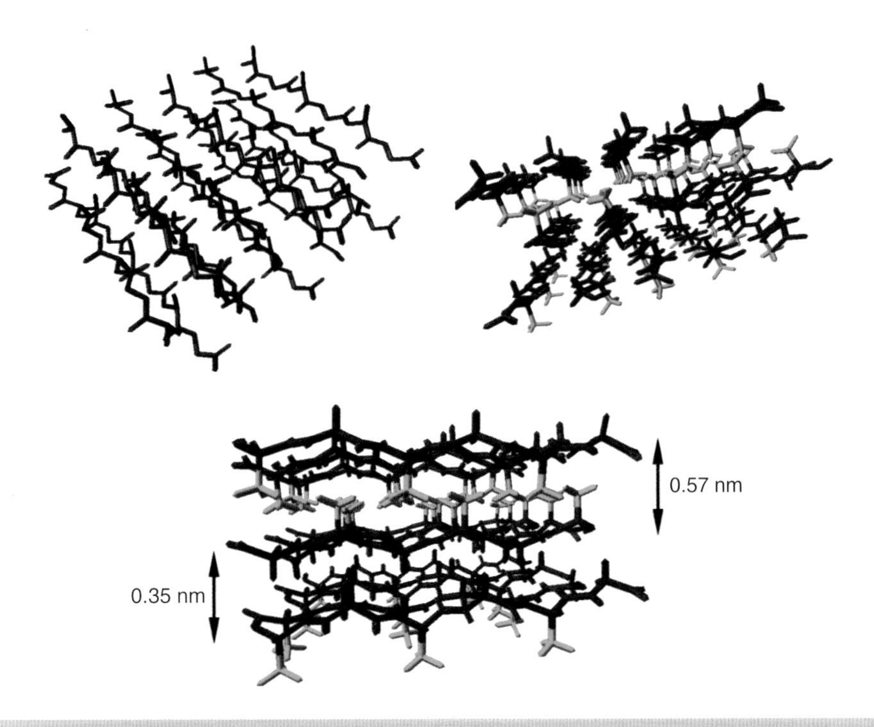

Figure 4.5 The interaction of alternate Gly and Ala/Ser rich surfaces in antiparallel β strands in the silk fibroin structure from *Bombyx mori*. The spacing between strands alternates between 0.35 and 0.57 nm as shown in the bottom figure. The structures of silk were generated for an alternating series of Gly/Ala polymers (PDB: 2SLK). Top left: the arrangement of strands in the antiparallel β sheet structure showing backbone only. Top right: The antiparallel β strands showing the interaction between Ala residue (shown by cyan colour) in an end on view where the backbone is running in a direction proceeding into and out of the page. In the bottom diagram the strands are running laterally and the Ala/Ser rich interface shown by the cyan side chains is wider than the smaller interface formed between glycine residues

binding together body structures and providing support and protection. Connective tissue is the most abundant tissue in vertebrates and depends for its structural integrity primarily on collagen. In vertebrates connective tissue appears to account for approximately 30 percent of the total mass. Although collagen is often described as the single most abundant protein it has many diverse biological roles and to date at least 30 distinct types of collagen have been identified from the respective genes with each showing subtle differences in the amino acid sequence along their polypeptide chains. Collagen is usually thought of as a protein characteristic of vertebrates such as mammals but it is known to occur in all multicellular animals. Sequencing of the nematode (flatworm) genome of *Caenorhabditis elegans* has revealed over 160 collagen genes. In this nematode collagen proteins are the major structural component of the exoskeleton with the collagen genes falling into one of three major gene families.

All collagens have the structure of a triple helix described in detail below and are assembled from three polypeptide chains. Since these chains can be combined in more than one combination a great many distinct collagens can exist and several have been identified as occurring predominantly in one group of vertebrate tissues. In humans at least 19 different collagens are assembled from the gene products of approximately 30 distinct and identified genes. Within these 19 structural types four major classes or groupings are generally identified. These groupings are summarized in Table 4.4. Type I collagen consists of two identical chains called α_1(I) chains and a third chain called α_2.

Table 4.4 The major collagen groups

Type	Function and location
Type I	The chief component of tendons, ligaments, and bones
Type II	Represents over 50 % of the protein in cartilage. It is also used to build the notochord of vertebrate embryos
Type III	Strengthens the walls of hollow structures like arteries, the intestine, and the uterus
Type IV	Forms the basal lamina (sometimes called a basement membrane) of epithelia. For example, a network of type IV collagens provides the filter for the blood capillaries and the glomeruli of the kidneys

In contrast Type II collagen contain three identical α_1 chains.

In a mature adult collagen fibres are extremely robust and insoluble. The insolubility of collagen was for many years a barrier to its chemical characterization until it was realized that the tissues of younger animals contained a greater proportion of collagen with higher solubility. This occurred because the extensive cross-linking of collagen characteristic of adults is lacking in young animals and it was possible to extract the fundamental structural unit called tropocollagen.

The structure and function of collagen

Tropocollagen is a triple helix of three similarly sized polypeptide chains each on average about 1000 amino acid residues in length. This leads to an approximate M_r of 285 000, an average length of ~300 nm and a diameter of ~1.4 nm. A comparison of the dimensions of tropocollagen with an average globular protein is useful and instructive. The length of collagen is approximately 100 times greater than myoglobin yet its diameter is only half that of the globular protein emphasizing its extremely elongated or filamentous nature. The polypeptides of tropocollagen are unusual in their amino acid composition and are defined by high proportions of glycine residues (see Table 4.2) as well as elevated amounts of proline. Collagen has a repetitive primary sequence in which every third residue is glycine. The sequence of the polypeptide chain can therefore be written as

-Gly-Xaa-Yaa–Gly-Xaa-Yaa-Gly-Xaa-Yaa-

where Xaa and Yaa are any other amino acid residue. However, further analysis of collagen sequences reveals that Xaa and Yaa are often found to be the amino acids proline or lysine. Many of the proline and lysine residues are hydroxylated via post-translational enzymatic modification to yield either hydroxyproline (Hyp) or hydroxylysine (Hyl) (see Chapter 8). The sequence Gly-Pro-Hyp occurs frequently in collagen. The existence of repetitive sequences is a feature of collagen, keratin and fibroin proteins and is in marked contrast to globular proteins where repetitive sequences are the exception.

Each polypeptide chain intertwines with the remaining two chains to form a triple helix (Figure 4.6). The helix arrangement is very different to the α helix and shows most similarity with the poly-proline II helices described briefly in Chapter 3. Each chain has the sequence Gly-Xaa-Yaa and forms a left-handed super helix with the other two chains. This leads to the triple helix shown in Figure 4.6. When viewed 'end-on' the super helix can be seen to consist of left handed polypeptide chain supercoiled in a right handed manner about a common axis.

The rise or translation distance per residue for each chain in the triple helix is 0.286 nm whilst the number of residues per turn is 3.3. Combining these two figures yields a value of ~0.95 nm for the helix pitch and reveals a more extended conformation especially when compared with α helices or 3_{10} helices (pitch ~0.5–0.54 nm).

Glycine lacking a chiral centre and possessing considerable conformational flexibility presents a significant contrast to proline. In proline conformational restraint exists as a result of the limited variation in the torsion angle (ϕ) permitted by a cyclic pyrrolidone ring. The presence of large amounts of proline in tropocollagen is also significant because the absence

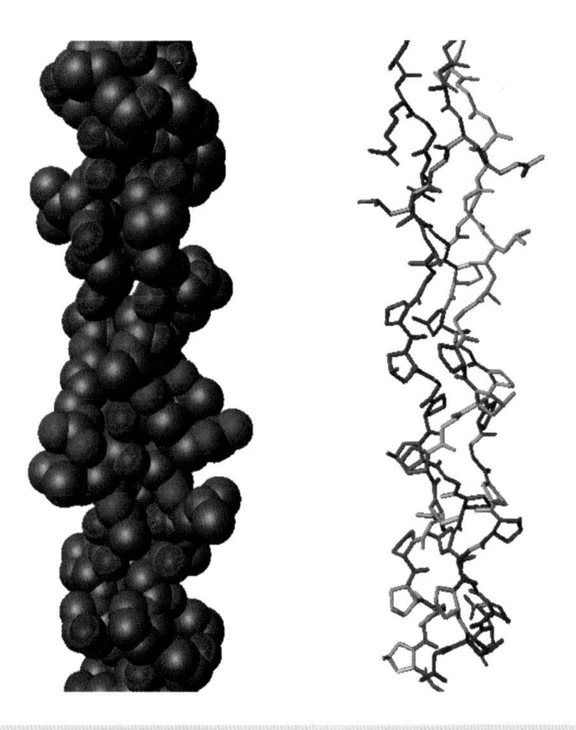

Figure 4.6 The basic structure of the triple helix of collagen. Space filling representation of the triple helix and a simpler wireframe view of the different chains (orange, dark green and blue chains). Careful analysis of the wireframe view reveals regular repeating Pro residues in all three chains

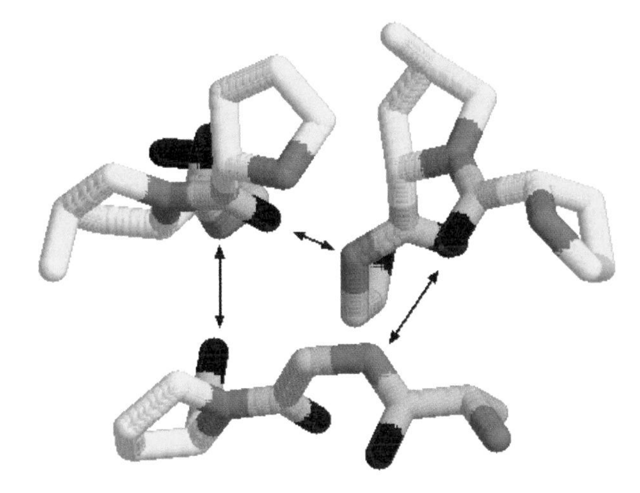

Figure 4.7 The interaction of glycine residues at the centre of the collagen helix. Note the side chains of residues Xaa and Yaa are located towards the outside of the helix on each chain where they remain sterically unhindered. In each of the three chains proline and hydroxyproline side chains are shown in yellow with the remaining atoms shown in their usual CPK colour scheme. Only heavy atoms are shown in this representation. The arrows indicate the hydrogen bonds from the glycine NH to the CO of residue Xaa in the neighbouring chain. Each chain is staggered so that Gly and Xaa and Yaa occur at approximately the same level along the axis of the triple helix (derived from PDB:1BKV)

of the amide hydrogen (HN) eliminates any potential hydrogen bonding with suitable acceptor groups. As a result of the presence of both glycine and proline in high frequency in the collagen sequences the triple helix is forced to adopt a different strategy in packing polypeptide chains. Since the glycine residues are located at every third position and make contact with the two remaining polypeptide chains it is clear that only a very small side chain (i.e. glycine) can be accommodated at this position. Any side chain bigger than hydrogen would disrupt the conformation of the triple helix. As a result there is very little space along the helix axis of collagen and glycine is always the residue closest to the helix axis (Figure 4.7). The side chains of proline residues along with lysine and other residues are on the outside of the helix.

The close packing of chains clearly stabilizes the triple helix through van der Waals interactions, but in addition extensive hydrogen bonding occurs between polypeptide chains. The hydrogen bonds form between the amide (NH) group of one glycine residue and the backbone carbonyl (C=O) group of residue Xaa on adjacent chains. The direction of the hydrogen bonds are transverse or across the long axis of the helix. Interactions within the triple helix are further enhanced by hydrogen bonding between amide groups and the hydroxyl group of Hyp residues. An indication of the importance of hydrogen bonding interactions in collagen helices has been obtained through constructing synthetic peptides and determining the melting temperature of the collagen triple helix. The melting temperature is the temperature

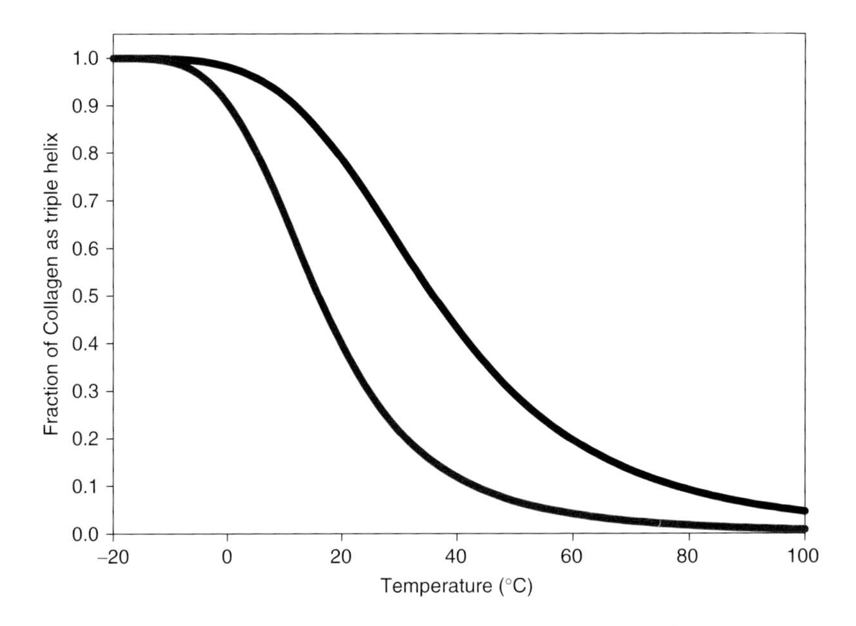

Figure 4.8 Thermal denaturation curve for collagen. In normal collagens the transition midpoint temperature or T_m is related to the normal body temperature of the organism and for mammals is above 40 °C as shown by the blue line in the above graph

at which half the helical structure has been lost and is characterized by a curve showing a sharp transition at a certain temperature reflecting the loss of ordered structure. The loss of structural integrity reflects denaturation of the triple helix and is accompanied by a progressive loss of function.

The importance of hydroxyproline to the transition temperature is shown by synthetic peptides of (Gly-Pro-Pro)$_n$ and (Gly-Pro-Hyp)$_n$. The former has a transition mid point temperature (T_m) of 24 °C whilst the latter exhibits a much higher T_m of ~60 °C (Figure 4.8). This experiment strongly supports the idea of triple helix stabilization through hydroxylation of proline (Hyp) and the formation of hydrogen bonds with neighbouring chains. Heating of collagen forms gelatin a disordered state in which the triple helix has dissociated. Although cooling partially regenerates the triple helix structure much of the collagen remains disordered. The reasons underlying this observation were not immediately apparent until the route of collagen synthesis was studied in more detail. Collagen is synthesized as a precursor termed procollagen in which additional domains at the N and C terminal specifically modulate the folding process. Mature collagen lacks these domains so any unfolding or disordering that occurs remains difficult to reverse.

Although hydrogen bonds and van der Waals interactions impart considerable stability to the tropocollagen triple helix and underpin its use as a structural component of many cells further strength arises from the association of tropocollagen molecules together as part of a collagen fibre. Each tropocollagen molecule is approximately 300 nm in length and packs together with neighbouring molecules to produce a characteristic banded appearance of fibres in electron micrographs. The banded appearance arises from the overlapping of each triple helix by approximately 64 nm thereby producing the striated appearance of collagen fibrils. This pattern of association relies on further cross-linking both within individual helices, known as *intra*molecular cross-links, as well as bonds between helices where they are called *inter*molecular cross-links. Both cross-links are the result of covalent bond formation.

The covalent cross-links among collagen molecules are derived from lysine or hydroxylysine and involve

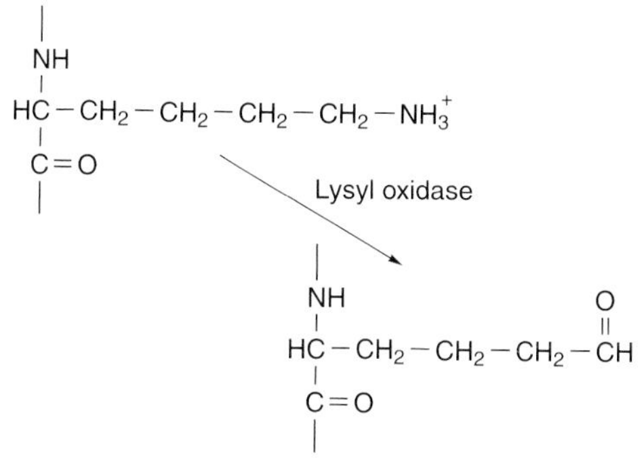

Figure 4.9 Oxidation of ε-amino group of lysine to form aldehyde called allysine

the action of an enzyme called lysyl oxidase. This copper dependent enzyme oxidizes the ε-amino group of a lysine side chain and facilitates the formation of a cross link with neighbouring lysine residues (Figure 4.9). Lysine sidechains are oxidized to an aldehyde called allysine and this promotes a condensation reaction between two chains forming strong covalent cross-links (Figure 4.10). A reaction between lysine residues in the same collagen fibre results in an intramolecular cross-link whilst reaction between different triple helices results in intermolecular bridges. In view of the presence of hydroxylysine and lysine in collagen these reactions can occur between two lysine, two hydroxylysines or between one hydroxylysine and one lysine. The products are called hydroxylysinonorleucine or lysinonorleucine. Further cross-linking can form trifunctional cross-links and a hydroxypyridinoline structure.

Cross-linking of collagen is a progressive process but does not occur in all tissues to the same extent. In general, younger cells have less cross-linking of their collagen than older cells, with a visible manifestation of this process being the increase in the appearance of wrinkled skin in the elderly, especially when compared with that found in a newborn baby. It is also the reason why meat from older animals is tougher than that derived from younger individuals.

Collagen biosynthesis

Collagen is a protein that undergoes significant post-translational modification and serves to introduce a subject that is described in more detail in Chapter 8. The initial translation product synthesized at the ribosome is very different to the final product and without these subsequent modifications it is extremely unlikely that the initial translation product could perform the same biological role as mature collagen. It is also true to say that any process interfering with modification of collagen tends to result in severe forms of disease.

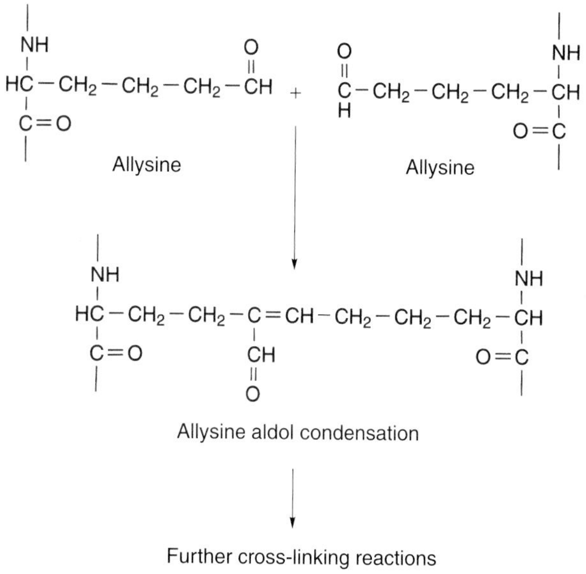

Figure 4.10 Outline of the reaction pathway leading to the formation of lysine crosslinks in collagen fibres. The first oxidative step results in the deamination of the ε-amino group and the formation of allysine (or hydroxyallysine). Two allysine residues condense to form a stable cross-link which can undergo further reactions that heighten the complexity of the cross-link. The allysine route predominates in skin whilst the hydroxyallysine route occurs in bone and cartilage

The biosynthesis of collagen is divided into discrete reactions that differ not only in the nature of the modification but their cellular location. Step 1 is the initial formation of preprocollagen, the initial translation product formed at the ribosome. In this state the collagen precursor contains a signal sequence that directs the protein to the endoplasmic reticulum membrane and

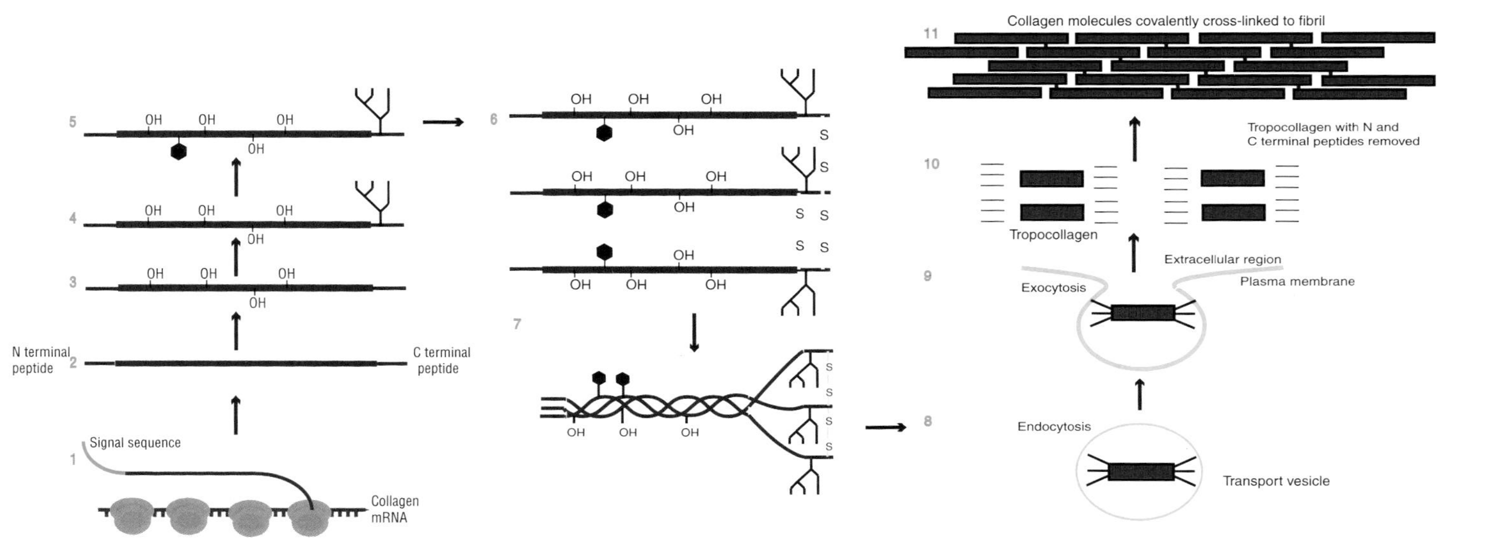

Figure 4.11 Processing of procollagen removes N and C terminal extensions and allows association into staggered, cross-linked fibrils.
1. Synthesis on ribosomes. Entry of chains into lumen of endoplasmic reticulum occurs with first processing reaction removing signal peptide
2. Collagen precursor with N and C terminal extensions.
3. Hydroxylation of selected proline and lysines
4. Addition of Asn-linked oligosaccharides to collagen
5. Initial glycosylation of hydroxylysine residues
6. Alignment of three polypeptide chains and formation of inter-chain disulfide bridges
7. Formation of triple helical procollagen
8. Transfer by endocytosis to transport vesicle
9. Exocytosis transfers triple helix to extracellular phase
10. Removal of N and C terminal propeptides by specific peptidases
11. Lateral association of collagen molecules coupled to covalent cross linking creates fibril

facilitates its passage into the lumen where the signal peptide is cleaved by the action of specific proteases. However, while still associated with the ribosome, this polypeptide is hydroxylated from the action of prolyl hydroxylase and lysyl hydroxylase, resulting in the formation of hydroxyproline and hydroxylysine, and is followed by transfer of the polypeptide into the lumen of the endoplasmic reticulum. Here, a third step involves glycosylation of the collagen precursor and the attachment of sugars, chiefly glucose and galactose, occurs via the hydroxyl group of Hyl. Frequently, the sugars are added as disaccharide units. In this state the pro-α-chains join forming procollagen whilst the N- and C-terminal regions form inter-chain disulfide bonds and the central regions pack into a triple helix. In this state the collagen is termed procollagen and it is transported to the Golgi system prior to secretion from the cell. Procollagen peptidases remove the disulfide-rich N and C terminal extensions leaving the triple helical collagen in the extracellular matrix (Figure 4.11) where it can then associate with other collagen molecules to form staggered, parallel arrays (Figure 4.12). These arrays undergo further modification by the formation of cross-links through the action of lysyl oxidase, as described above.

Collagenases degrade collagen and have been shown to be one member of a large group of enzymes called matrix metalloproteinases (MMPs). These enzymes degrade the extracellular matrix. Abnormal metalloproteinase expression leads to premature degradation of the extracellular matrix and is implicated in diseases such as atherosclerosis, tumour invasion and rheumatoid arthritis.

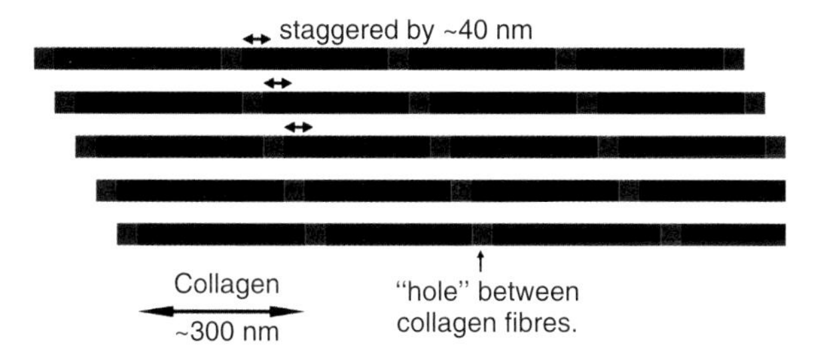

Figure 4.12 Association of procollagen into staggered parallel arrays making up collagen fibres. Each collagen fibre is approximately 300 nm in length and is staggered by ~40 nm from adjacent parallel fibres. This 'hole' can represent the site of further attachment of extracellular proteins and can become filled with calcium phosphate

Disease states associated with collagen defects

The widespread involvement of collagen in not only tendons and ligaments, but also the skin and blood vessels, means that mutations in collagen genes often result in impaired protein function and severe disorders affecting many organ systems. Mutations in the 30 collagen genes discovered to date give rise to a large variety of defects in the protein. In addition defects have also been found in the enzymes responsible for the assembly and maturation of collagen creating a further group of disease states. In humans defects in collagen as a result of gene mutation lead to osteogenesis imperfecta, hereditary osteoporosis and familial aortic aneurysm.

Several hereditary connective tissue diseases have been identified as arising from mutations in genes encoding collagen chains. Most common are single base mutations that result in the substitution of glycine by a different residue thereby destroying the characteristic repeating sequence of Gly-Xaa-Yaa. A further consequence of these mutations is the incorrect assembly or folding of collagen. Two particularly serious diseases attributable to defective collagen are osteogenesis imperfecta and Ehlers–Danlos syndrome. The molecular basis for these diseases in relation to the structure of collagen will be described.

Osteogenesis imperfecta is a genetic disorder characterized by bones that break comparatively easily often without obvious cause. It is sometimes called brittle bone disease. At least four different types of osteogenesis imperfecta are recognized by clinicians (Table 4.5) although all appear to arise from mutation of the collagen genes coding for either the α_1 or α_2 chains of type I collagen.

Osteogenesis imperfecta is caused by a mutation in one allele of either the α_1 or α_2 chains of the major collagen in bone, type I collagen. Type I collagen contains two α_1 chains together with a single

Table 4.5 Classification of osteogenesis imperfecta. Currently, diagnosis of type II form of osteogenesis imperfecta is determined *in utero* whilst diagnosis of other forms of the disease can only be made antenatal

Type	Inheritance	Description
I	Dominant	Mild fragility, slight deformity, short stature. Presentation at young age (2–6)
II	Recessive	Lethal: death at pre- or perinatal stage
III	Dominant	Severe, progressive deformity of limbs and spine
IV	Dominant	Skeletal fragility and osteoporosis, bowing of limbs. Most mild form of disease

α_2 chain and the presence of one mutant collagen allele is sufficient to produce a defective collagen fibre. The incidence of all forms of osteogenesis imperfecta is estimated to be approximately 1 in 20 000; from screening individuals it was found that glycine substitution was a frequently observed event with cysteine, aspartate and arginine being the most frequent replacement residues. The disruption of the Gly-Xaa-Yaa repeat has pathological consequences that vary considerably, as shown by Table 4.5 where four different types of osteogenesis imperfecta ranging in severity from lethal to mild are observed. One consequence of glycine substitution is defective folding of the collagen helix and this is accompanied by increased hydroxylation of lysine residues N-terminal to the mutation site. Since hydroxylation occurs on unfolded chains one possible effect of mutation is to inhibit the rate of triple helix formation rendering the proteins susceptible to further modification or interaction with molecular chaperones or enzymes of the endoplasmic reticulum that alter their processing or secretion. A further sequela of mutant collagen fibres is incorrect processing by N-propeptide peptidases (see section on collagen biosynthesis) and where mutant collagen molecules are incorporated into fibrils there is evidence of poor mineralization. It seems very likely that the exact site of the mutation of α_1 or α_2 chains will influence the overall severity of the disease with some evidence suggesting mutations towards the C-terminal produce more severe phenotypes. However, the precise relationship between disease state, mutation site and perturbation of collagen structure remain to be elucidated.

A second important disease arising from mutations in a single collagen gene is Ehlers–Danlos syndrome. This syndrome results in widely variable phenotypes, some relatively benign whilst others are life threatening. Classically the syndrome has been recognized by physicians from the presentation of patients with joint hypermobility as well as extreme skin extensibility, although many other diagnostic traits are now recognized including vascular fragility. Many different types of the disease are recognized both medically and at the level of the protein. The variability arises from mutations at different sites and in different types of collagens leading to the variety of phenotypes. These mutations can lead to changes in the levels of collagen molecules, changes in the cross-linking of fibres, a decreased hydroxylysine content and a failure to process collagen correctly by removal of the N-terminal regions. The common effect of all mutations is to create a structural weakness in connective tissue as a result of a molecular defect in collagen.

Related disorders characterized at a molecular level

Marfan's syndrome is an inherited disorder of connective tissue affecting multiple organ systems including the skeleton, lungs, eyes, heart and blood vessels. For a long time Marfan's syndrome was believed to be caused by a defect in collagen but this defect is now known to reside in a related protein that forms part of the microfibrils making up the extracellular matrix that includes collagen. The condition affects both men and women of all ethnic groups with an estimated incidence of ~1 per 20 000 individuals throughout the world.

Before identification of Marfan's syndrome became routine and surgical management of this disease was common most patients died of cardiovascular complications at a very early age and usually well before the age of 50. In 1972 the average life expectancy for a Marfan sufferer was ~32 years but today, with increased research facilitating recognition of the disease and surgical intervention alleviating many of the health problems, the expected life span had increased to over 65 years.

Marfan's syndrome is caused by a molecular defect in the gene coding for fibrillin, an extracellular protein found in connective tissue, where it is an integral component of extended fibrils. Microfibrils are particularly abundant in skin, blood vessels, perichondrium, tendons, and the ciliary zonules of the eye. The elastin-based fibres form part of an extracellular matrix structure that provides the elastic properties to tissues. Both morphological and biochemical characterization of fibres reveal an internal core made up primarily of the protein elastin together with a peripheral layer of microfibrils composed primarily of fibrillin. Humans have two highly homologous fibrillins, fibrillin-1 (Figure 4.13) and fibrillin-2, mapping to chromosomes 15 and 5 respectively. In 1991, the first mutation in fibrillin 1 was reported and subsequently over 50 mutations in individuals with Marfan syndrome have been described. A characteristic feature of both fibrillins is their mosaic composition where numerous small modules combine to produce the complete, very large, protein of 350 kDa.

The majority of fibrillin consists of epidermal growth factor-like subunits (47 epidermal growth factor (EGF)-like modules) of which 43 have a consensus sequence for calcium binding. Each of these domains is characterized by six cysteine residues three disulfide bridges and a calcium binding consensus sequence of D/N-x-D/N-E/Q-x_m-D/N*-x_n-Y/F (where m and n are variable and * indicates possible post-translational modification by hydroxylation). Other modules found in fibrillin-1 including motifs containing eight Cys residues, hybrid modules (two) along with sequences unique to fibrillin (three). These domains are interspersed throughout the molecule with the major differences between fibrillins residing in a proline-rich region close to the N-terminus in fibrillin-1 that is replaced by a glycine-rich region in fibrillin-2.

The EGF domain occurs in many other proteins including blood coagulation proteins such as factors X, VII, IX, and the low density lipoprotein receptors.

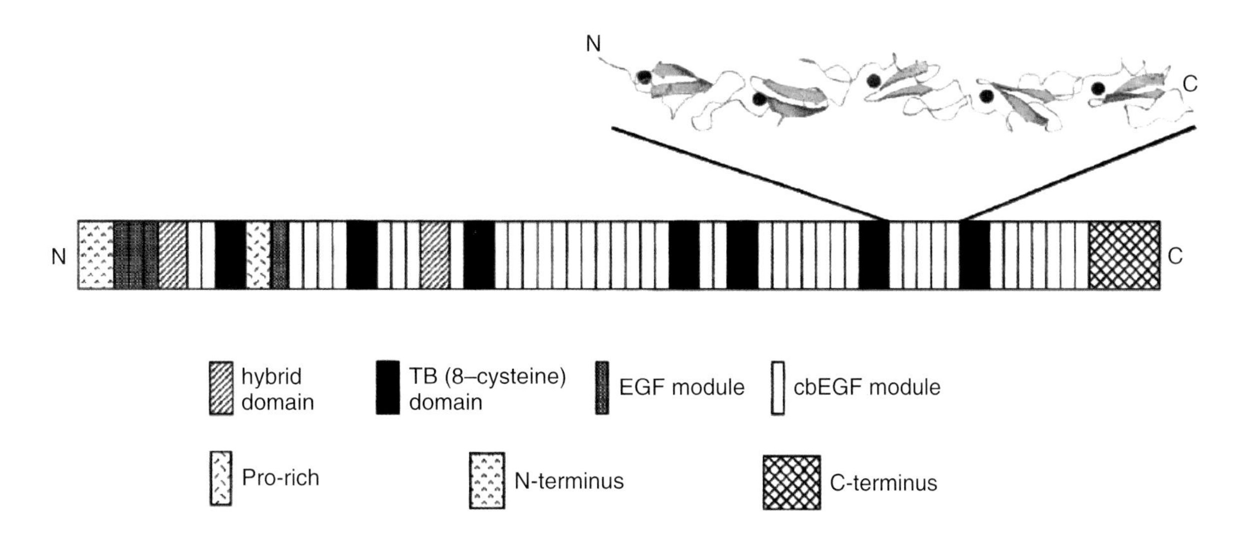

Figure 4.13 The modular organization of fibrillin-1 (reproduced with permission from Handford, P.A. *Biochim. Biophys. Acta* 2000, **1498** 84–90. Elsevier)

Figure 4.14 The structure of a pair of calcium binding EGF domains from fibrillin-1 (PDB:1EMN)

The structure of a single EGF like domain and a pair of calcium-binding domains confirmed a rigid rod-like arrangement stabilized by calcium binding and hydrophobic interactions (Figure 4.14). Mutations known to result in Marfan's syndrome lead to decreased calcium binding to fibrillin, and this seems to play an important physiological role. The importance of the structure determined for the modules of fibrillin was that it offered an immediate explanation of the defect at a molecular level. The structural basis for Marfan's syndrome resides in the disruption of calcium binding within fibrillin as a result of single site mutation.

Marfan's syndrome is an autosomal dominant disorder affecting the cardiovascular, skeletal, and ocular systems. The major clinical manifestations are progressive dilatation of the aorta (aortic dissection), mitral valve disorder, a tall stature frequently associated with long extremities, spinal curvature, myopia and a characteristic thoracic deformity leading to a 'pigeon-chest' appearance. The majority of the mutations known today are unique (i.e. found only in one family) but at a molecular level it results in the substitution of a single amino acid residue that disrupts the structural organization of individual EGF-like motifs.

Flo Hyman, a famous American Olympian volleyball player, was a victim of Marfan's syndrome and a newspaper extract testifies to the sudden onset of the condition in apparently healthy or very fit athletes. "During the third game, Hyman was taken out in a routine substitution. She sat down on the bench. Seconds later she slid silently to the floor and lay there, still. She was dead". Many victims of Marfan's disease are taller than average. As a result basketball and volleyball players are routinely screened for the genetic defect which despite phenotypic characteristics such as pigeon chest, enlarged breastbone, elongated fingers and tall stature often remains undiagnosed until a sudden and early death. It has also been speculated largely on the basis of their physical appearances that Abraham Lincoln (16th president of the United States, 1809–1865) and the virtuoso violinist Niccolò Paganini (1782–1840) were suffering from connective tissue disorders.

Summary

Fibrous proteins represent a contrast to the normal topology of globular domains where compact folded tertiary structures exist as a result of long-range interactions. Fibrous proteins lack true tertiary structure, showing elongated structures and interactions confined to local residues.

The amino acid composition of fibrous proteins departs considerably from globular proteins but also varies widely within this group. This variation reflects the different roles of fibrous proteins.

Three prominent groups of fibrous proteins are the collagens, silk fibroin and keratins and all occupy pivotal roles within cells.

Collagen, in particular, is very abundant in vertebrates and invertebrates where the triple helix provides a platform for a wide range of structural roles in the extracellular matrix delivering strength and rigidity to a wide range of tissues.

The triple helix is a repetitive structure containing the motif (Gly-Xaa-Yaa) in high frequency with Xaa and Yaa often found as proline and lysine residues. Repeating sequences of amino acids are a feature of many fibrous proteins and help to establish the topology of each protein. In collagen the presence of glycine at every third residue is critical because its small side chain allows it to fit precisely into a region that forms from the close contact of three polypeptide chains. Larger side chains would effectively disrupt this region and perturb the triple helical structure.

Although based around a helical design collagen differs considerably in dimensions to the typical α helix. The triple helix of collagen undergoes considerable post-translational modification to increase strength and rigidity.

Keratins make up a considerable proportion of hair and nails and contain polypeptide chains arranged in an α helical conformation. The helices interact via supercoiling to form coiled-coils. Of importance to coiled-coils are specific interactions between residues in different helices via non-polar interactions that confer significant stability. The basis of this interaction is a heptad of repeating residues along the primary sequence.

A heptad repeat possesses leucine or other residues with hydrophobic side chains arranged periodically to favour inter-helix interactions. Helices are usually arranged in pairs. Significantly, this mode of organization is found in other intracellular proteins as well as in viral proteins. Several DNA binding proteins contain coiled-coil regions and the hydrophobic-rich domains are often called leucine zippers.

In view of their widespread distribution in all animal cells mutations in fibrous proteins such as keratins or collagens lead to serious medical conditions. Many disease states are now known to arise from inherited disorders that lead to impaired structural integrity in these groups of proteins.

Problems

1. Describe the different amino acid compositions of collagen, silk and keratins?
2. Explain why glycine and proline are found in high frequency in the triple helix of collagen but are not found frequently within helical regions of globular proteins?
3. Why is silk both strong and flexible?
4. Why is wool easily stretched or shrunk? Why is silk more resistant to these deformations.
5. Poly-L-proline is a synthetic polypeptide that adopts a helical conformation with dimensions comparable to those of a single helix in collagen. Why does poly-L–proline fail to form a triple helix. Will the sequence poly-(Gly-Pro-Pro)$_n$ form a triple helix and how does the stability of this helix compare with native collagen. Does poly-(Gly-Pro-Gly-Pro)$_n$ form a collagen like triple helix?
6. A mutation is detected in mRNA in a region in frame with the start codon (AUG) CCCUAA$\underline{\text{AUG}}$....... GGACCCAAAGGACCUAAG**U**GUCCAUCUGGU CCGAAGGGGUCCAACGGACCCAAGGGU...... Establish the identity of the peptide and describe the possible consequences of this mutation.
7. Describe four post-translational modifications of collagen and list these modifications in their order of occurrence?
8. Gelatin is primarily derived from collagen, the protein responsible for most of the remarkable strength of connective tissues in tendons and other tissues. Gelatin is usually soft and floppy and generally lacking in strength. Explain this observation.
9. Ehlers–Danlos syndromes are a group of collagen based diseases characterized by hyperextensible joints and skin. What is the most probable cause of this disorder in terms of the structure of collagen?
10. Keratin is based on a seven residue or heptad repeat. Describe the properties of this sequence that favour coiled-coil structures.

5

The structure and function of membrane proteins

By 2004 the Protein Data Bank of structures contained over 22 000 submissions. Almost all of these files reflect the atomic coordinates of soluble proteins and less than 1 percent of all deposited coordinates belong to proteins found within membranes. It is clear from this small fraction that membrane proteins present special challenges towards structure determination.

Many of these challenges arise from the difficulty in isolating proteins embedded within lipid bilayers. For many decades membrane proteins have been represented as amorphous objects located in a sea of lipid with little consideration given to their organization. This is an unfortunate state of affairs because the majority of proteins encoded by genomes are located in membranes or associated with the membrane interface. Slowly, however, insight has been gained into the different types of membrane proteins, their domain structure and more recently the three dimensional structures at atomic levels of resolution. This chapter addresses the structural and functional properties of membrane proteins.

The molecular organization of membranes

Although this book is primarily about proteins it is foolish to ignore completely the properties of membranes since the lipid bilayers strongly influence protein structure and function. The term lipid refers to *any* biomolecule that is soluble in an organic solvent, and extends to quinones, cholesterol, steroids, fatty acids, triacylglycerols and non-polar vitamins such as vitamin K_3. All of these components are found in biological membranes. However, for most purposes the term lipid refers to the collection of molecules formed by linking a long-chain fatty acid to a glycerol-3-phosphate backbone; these are the major components of biological membranes.

Two long chain fatty acids are esterified to the C_1 and C_2 positions of the glycerol backbone whilst the phosphate group is frequently linked to another polar group. Generically all of these molecules are termed glycerophospholipids (Figure 5.1). Phosphatidic acid occurs in low concentrations within membranes and the addition of polar head groups such as ethanolamine, choline, serine, glycerol, and inositol forms phosphatidylethanolamine (PE), phosphatidylcholine (PC, also called lecithin), phosphatidylserine (PS), phosphatidylglycerol (PG) and phosphatidylinositol (PI), the key components of biological membranes (Table 5.1). Fatty acyl chains usually containing 16 or 18 carbon atoms occupy the R1 and R2 positions although chain length can range from 12 to 24 carbons.

Proteins: Structure and Function by David Whitford

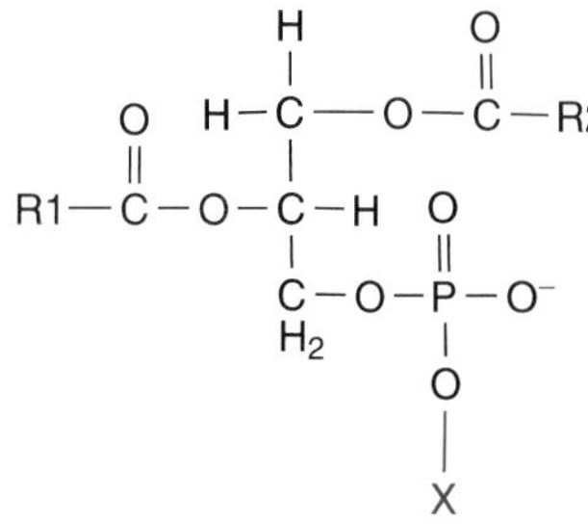

Figure 5.1 The structure of glycerophospholipids involves the esterification of two long-chain fatty acids to a glycerol-3-phosphate backbone with further additions of polar groups occurring at the phosphate group. When X=H the parent molecule phosphatidic acid is formed

In animal membranes the C_1 position is typically a saturated lipid of 16–18 carbons whilst the C_2 position often carries an unsaturated C_{16}–C_{20} fatty acid. Unsaturated acyl chains (Table 5.2) contain one or more double bonds usually found at carbons 5, 6, 8, 9, 11, 12, 14 and 15. Where more than one double bond occurs the sites are separated by three carbons so that, for example in linolenic acid the double bonds are at 9, 12 and 15, whilst in arachidonic acid the four double bonds are located at carbons 5, 8, 11 and 14. The double bonds are found in the *cis* state with *trans* isomers rarely found. With a large number of possible head groups coupled to a diverse number of fatty acyl chains (each potentially differing in their degree of saturation) the number of possible combinations of non-polar chain/polar head group is enormous.

Naming phospholipids follows the identity of the fatty acid residues; so a glycerophospholipid containing two 16 carbon saturated fatty acyl chains and a choline head group is commonly referred to as dipalmitoylphosphatidylcholine (DPPC). Although

Table 5.1 The common head groups attached to the glycerol-3-phosphate backbone of glycerophospholipids

Donor	Head group	Phospholipid
Water	$-H$	Phosphatidic acid
Ethanolamine	$-CH_2-CH_2-NH_3^+$	Phosphatidylethanolamine
Choline	$-CH_2-CH_2-N^+(CH_3)_3$	Phosphatidylcholine
Serine	$-CH_2-CH(NH_3^+)-C(=O)-O^-$	Phosphatidylserine
Inositol	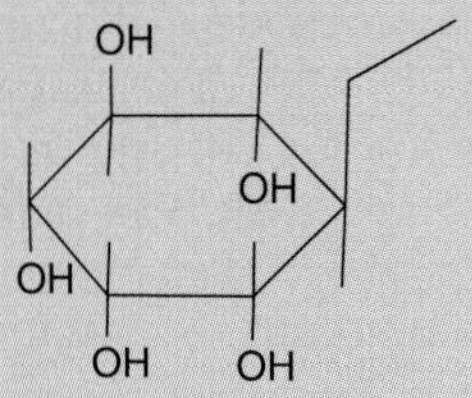 (inositol ring with OH groups)	Phosphatidylinositol
Glycerol	$-CH_2-CH(OH)-CH_2-OH$	Phosphatidylglycerol

Table 5.2 Common acyl chains found in lipids of biological membranes

Name	Numeric representation	Name	Numeric representation
Lauric	12:0	Stearic	18:1
Myristic	14:0	Linoleic	18:2
Palmitic	16:0	Linolenic	18:3
Palmitoleic	16:1	Arachidic	20:0
Oleic	18:0	Arachidonic	20:4

For molecules such as linolenic acid the position of the three double bonds are denoted by the nomenclature $\Delta^{9,12,15}$ with carbon 1 starting from the carboxyl end. The IUPAC name for this fatty acid is 9,12,15 octadecenoic acid. A third alternative identifies the number of carbon atoms in the acyl chain, the number of double bonds and the geometry and carbon number associated with each bond. Thus, linolenic acid is 18:3 9-*cis*, 12-*cis*, 15-*cis*

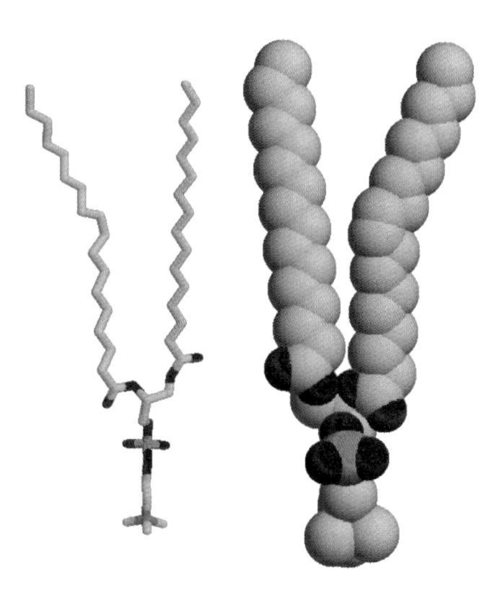

Figure 5.2 The diagram shows a wireframe and space filling model of two acyl chains, one 16 carbons in length, the other 18 carbons in length containing a *cis* double bond. A phosphatidylcholine head group is linked to a glycerol backbone. Protons have been omitted to increase clarity

glycerophospholipids are the most common lipids of biological membranes other variants occur such as sphingolipids, sulfolipids and galactolipids. In plant membranes sulfolipids and galactolipids occur alongside phospholipids, increasing the variety of lipids.

A phospholipid such as DPPC is described as amphipathic, containing polar and non-polar regions. Space filling models (Figure 5.2) show that introducing a single *cis* double bond produces a 'kink' in the acyl chain and an effective volume increase. This has pronounced effects on physical properties of lipids such as transition temperatures, fluidity and structures formed in solution.

When dispersed in aqueous solutions lipid molecules minimize contact between the hydrophobic acyl chain and water by forming structures where the polar head groups interact with aqueous solvents whilst the non-polar 'tails' are buried away from water. The size and shape of the lipid assembly is a complex combination of factors, including polar head group volume, charge, degree of unsaturation and length. Thus lipids with short acyl chains form micelles and pack efficiently into a spherical shape where the large volume of the hydrated polar head group exceeds that of the tail and favours micellar structure. Lipids with a single acyl chain, as well as detergents, adopt this shape. Longer acyl chains, particularly those with one or more double bonds, cannot pack efficiently into a spherical micelle and produce bilayers in most cases although alternative structures, such as inverted micelles where the polar head group is buried on the inside, can be formed by some of the common lipids. Additionally lipid aggregates are modulated by ionic strength or pH, but a detailed discussion of these states is beyond the scope of this text.

The ability of lipids or detergents to aggregate in solution is often expressed in terms of their critical micelle concentration (CMC). At concentrations above their CMC lipids are found as aggregated structures whilst below this concentration monomeric species exist in solution. The CMC of dodecylmaltoside (a single 12 carbon fatty acyl chain linked to a maltose group) is estimated at $\sim$150 μM whilst the corresponding value for DPPC is 0.47 nM and means that this polar lipid is always found as an aggregated species. An isolated bilayer dispersed in a polar solvent is unlikely to be stable since at some point the ends of the bilayer will be exposed leading

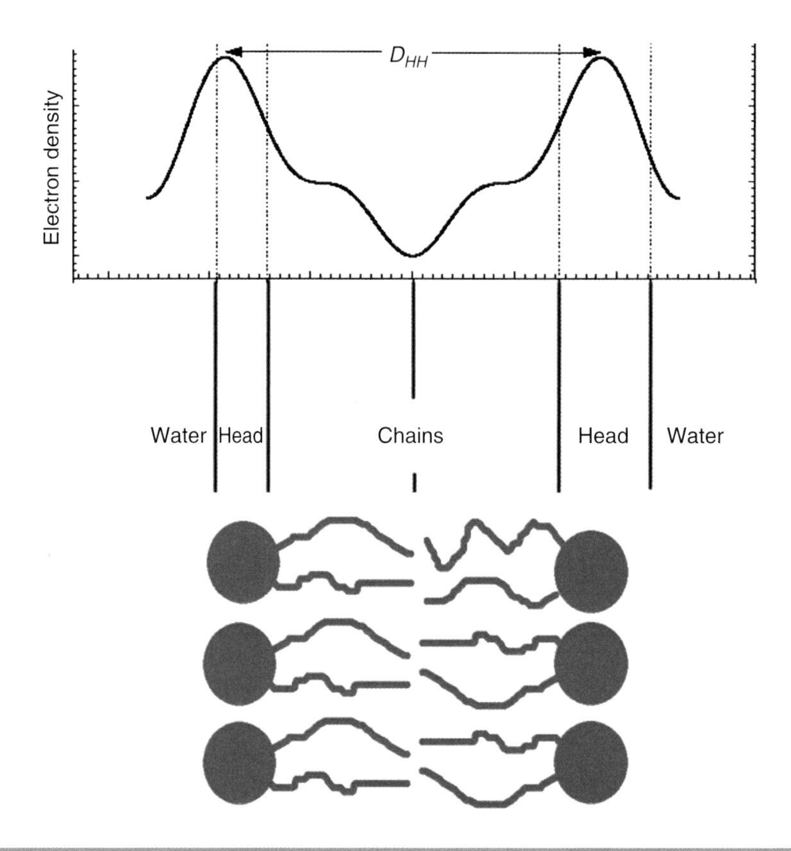

Figure 5.3 Electron density profile determined for a single bilayer. The electron dense regions correspond to the head group region whilst the non-polar domain are less electron dense. Such data established the basic dimensions of a bilayer (D_{HH}) at around 40 nm and the profile is shown alongside a diagram of a bilayer showing relative positions of head groups and acyl chains

to unfavourable interactions between non-polar and polar surfaces. As a result, extended bilayers in aqueous solution often form closed, sealed, solvent-filled structures known as liposomes or single lamellar vesicles that have proved to be good models for biological membranes.

Since integral membrane proteins are located in bilayers the properties and arrangement of lipids dictates the structure of any protein residing in this region. The structure of a lipid bilayer has been examined by neutron and X-ray diffraction to confirm a periodic arrangement for the distribution of electron density. The polar head groups are electron dense whilst the acyl chains represent comparatively 'diffuse' areas of electron density (Figure 5.3).

Fluidity and phase transitions are important properties of lipid bilayers

An important property of lipid bilayers is fluidity. Membrane fluidity depends on lipid composition, temperature, pH and ionic strength. At low temperatures the fluidity of a membrane is reduced with the acyl chains showing lower thermal mobility and packing into a 'frozen' gel state. At higher temperatures increased translational and rotational mobility results in a less ordered fluid state, sometimes called the liquid crystal (Figure 5.4). Transitions from the gel to fluid states are characterized by phase transition temperatures (Table 5.3). Short chains have lower transition temperatures when compared with longer lipids. The transition temperature increases with chain length since hydrophobic interactions increase with the number of non-polar atoms. The transition temperature decreases with increasing unsaturation and is reflected by greater inherent disorder packing.

Biological membranes exhibit broad phase transition temperatures over a wide temperature range. All biological membranes exist in the fluid state and organisms have membranes whose lipid compositions exhibit transition temperatures significantly below their lowest body temperature. Thus, fish living in the depths of the ocean where temperatures may drop below −20 °C have membranes rich in unsaturated long-chain fatty acids such as linolenic and linoleic acid.

Real membranes are not homogeneous dispersions of lipids but contain proteins, often in considerable quantity. The lipid:protein ratio varies dramatically, with membranes such as the myelin sheath possessing comparatively little protein (Table 5.4). In contrast membranes concerned with energy transduction have larger amounts of protein and exhibit lower lipid:protein ratios.

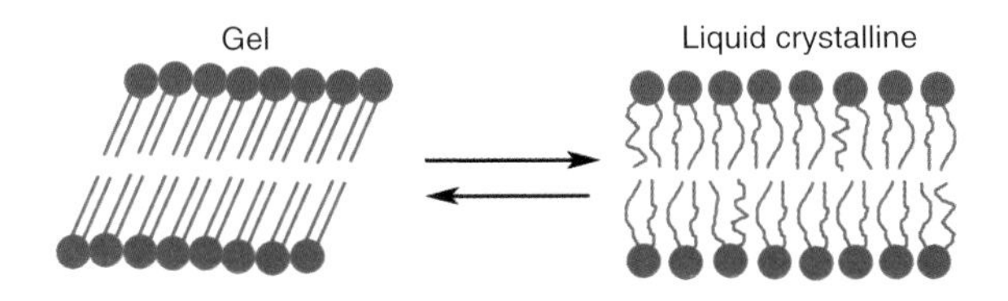

Figure 5.4 The gel to fluid (liquid crystalline) transition in lipid bilayers. In the gel phase the head groups are tightly packed together, the tails have a regular conformation and the bilayer width is thicker than the liquid crystalline state

Table 5.3 Phase transition temperatures of common diacylphospholipids

Diacylphospholipid	Transition temperature °C (T_m)
C22 phosphatidyl choline	75
C18 phosphatidyl choline	55
C16 phosphatidyl choline	41
C14 phosphatidyl choline	24
C18:1 phosphatidyl choline	−22
C16 phosphatidyl serine (low pH)	72
C16 phosphatidyl serine (high pH)	55
C16 phosphatidyl ethanolamine	60
C14 phosphatidyl ethanolamine	50

Phase transition temperatures are measured by differential scanning calorimetry (DSC) where heat is absorbed (ΔH increases) as the lipid goes through a phase transition

The fluid mosaic model

Any model of membrane organization must account for the properties of lipids, such as self-association into stable structures, mobility of lipids within monolayers, and the impermeability of cell membranes to polar molecules. In addition models must account for the presence of proteins both integral and peripheral and at a wide variety of ratios with the lipid component. In 1972, to account for these and many other observations, S.J. Singer and G. Nicolson proposed the fluid mosaic model; a model that has with only minor amendments stood the test of time. In this model proteins were originally viewed as floating like icebergs in a sea of lipid but in view of the known lipid:protein ratios found in some membranes this picture is unrealistic and representations should involve higher proportions of protein.

Table 5.4 Protein and lipid content of different membranes

Membrane	Protein	Lipid
Myelin sheath (Schwann cells)	21	79
Erythrocyte plasma membrane	49	43
Bovine retinal rod	51	49
Mitochondrial outer membrane	52	48
Mitochondrial inner membrane	76	24
Sarcoplasmic reticulum (muscle cells)	67	33
Chloroplast lamellae	70	30

The figures represent the percentage by weight. The figures do not always add up to 100 % reflecting different contents of carbohydrate within each membrane (derived from Guidotti, G. *Ann. Rev. Biochem.* 1972, **41**, 731–752)

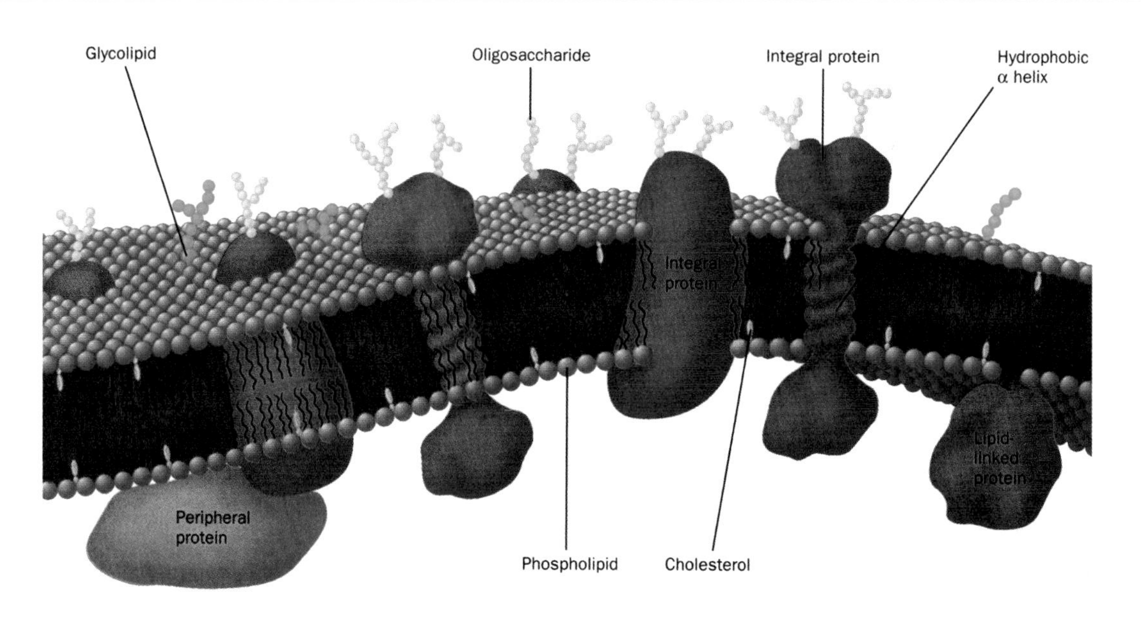

Figure 5.5 The organization of a typical cell membrane. The lipid in blue contains integral and peripheral proteins (red/orange) with cholesterol (yellow) found in animal membranes whilst polysaccharides are attached to the extra-cellular surface of proteins (reproduced with permission from Voet, D., Voet, J.G & Pratt, C.W., *Fundamentals of Biochemistry*, 1999 John Wiley & Sons, Ltd, Chichester)

In 1972 nothing was known about the structure of these proteins but the intervening period has elucidated details of organization within membranes. Interactions between protein and lipid alter the properties of both and in some systems there exists convincing data on the arrangement of proteins with respect to exterior and cytosolic compartments. However, a major problem is a comparative paucity of data on the secondary and tertiary structure of integral proteins that restricts understanding of their function in transport, cell–cell recognition, catalysis, etc. To address these problems it is necessary to examine the composition of membranes in more detail and one of the best model systems is the erythrocyte or red blood cell membrane. Figure 5.5 shows a typical cell membrane.

Membrane protein topology and function seen through organization of the erythrocyte membrane

The erythrocyte membrane is a popular system to study because blood can be obtained in significant quantities containing large amounts of cells (i.e membranes). The cell membrane is readily partitioned from haemoglobin and other cytosolic components by centrifugation. The resulting membranes are called 'ghosts' because they reseal to form colourless particles devoid of haemoglobin. Using ghost membranes identified two major groups of proteins. The first group of proteins shows weak association with membranes and is removed after mild washing treatments such as incubation with elevated concentrations of salt. The second class involved proteins that were only removed from the membrane by extreme procedures that included the use of detergents, organic solvents, chaotropic agents, lipases or mechanical fractionation. This second procedure involved the complete disruption of membrane environment with the result that many protein's lost structure and function. The terms *extrinsic* and *intrinsic* are sometimes applied to these two groups of membrane proteins. Other schemes of classification use the terms *non-integral* or peripheral and *integral* to describe each group. Unsurprisingly the properties of integral and non-integral proteins differ but further characterization of the erythrocyte membrane revealed a third type of

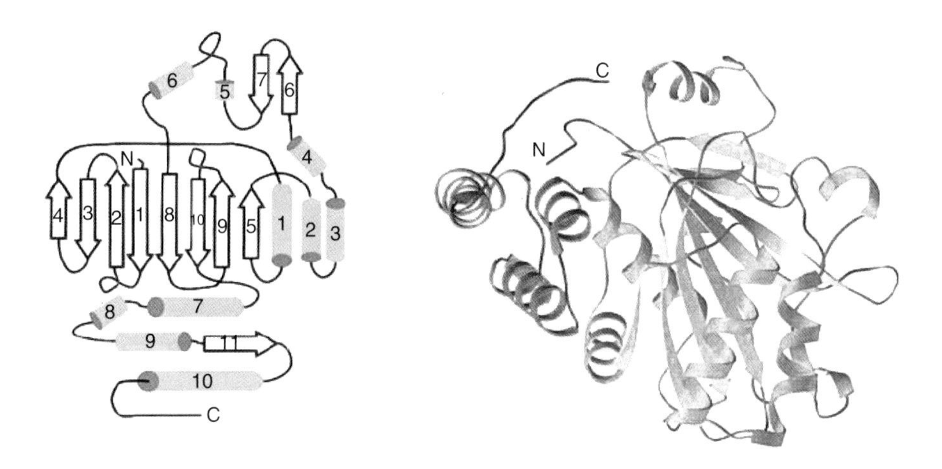

Figure 5.6 Structure of band 3 monomer. The secondary structure distribution (left) and a ribbon diagram of band 3 monomer (right). Regions shown in blue correspond to the peripheral protein binding domain and the red regions to the dimerization arm (reproduced with permission from Zhang, D., Kiyatkin, A., Bolin, J.T. & Low, P.S. *Blood* 2000, **96**, 2925–2933. The American Society of Hematology)

protein. This integral protein was attached firmly to the bilayer yet had substantial proportions of its sequence accessible to the aqueous phase.

An intrinsic protein of the erythrocyte membrane: band 3 protein

The name 'band 3' originally arose from sodium dodecyl sulfate–polyacrylamide gel electrophoresis studies of the protein composition of erythrocyte membranes where it was the subunit with the third heaviest molecular mass. It is the most abundant protein found in the erythrocyte membrane facilitating anion exchange and the transfer of bicarbonate (HCO_3^-) and chloride (Cl^-) ions. Band 3 protein is hydrophobic with many non-polar side chains and is released from the membrane only after extraction with detergents or organic solvents. As such it is typical of many membrane proteins and until recently there was little detail on the overall tertiary structure.

Band 3 consists of two independent domains (Figure 5.6). The N-terminal cytoplasmic domain links the membrane to the underlying spectrin–actin-based cytoskeleton using ankyrin and protein 4.1 as bridging proteins. The C-terminal domain mediates anion exchange across the erythrocyte membrane. As a result of its cytoplasmic location the structure of the N-terminal domain has been established by crystallography to a resolution of 2.6 Å but the membrane C-terminal domain remains structurally problematic. The cytoplasmic domain contains 11 β strands and 10 helical segments and belongs to the $\alpha + \beta$ fold class. Eight β-strands assemble into a central β-sheet of parallel and antiparallel strands whilst two of the remaining strands (β6 and β7, residues 176–185) form a β-hairpin that constitutes the core of the ankyrin binding site.

The approach of studying large extracellular regions belonging to membrane proteins is widely used to provide information since it represents in many instances the only avenue of structural analysis possible for these proteins.

Membrane proteins with globular domains

Integral membrane proteins are distinguished from soluble domains by their high content of hydrophobic residues. Residues such as leucine, isoleucine and valine often occur in blocks or segments of approximately 20–30 residues in length. Consequently, it is possible to identify membrane proteins by the distribution of residues with hydrophobic side chains throughout a primary sequence. Hydropathy plots reflect the preference of amino acid side chains for polar and non-polar environments and as such numerical parameters are assigned to each residue (Table 5.5). The hydropathy values reflect measurements of the free energy of transfer of an amino acid from non-polar to polar solvents. By assigning numerical values to each residue along a sequence and then averaging these values over a pre-defined window size a hydropathy profile is established. For a small window size the precision is not great but for a window size of between 18 and 22 accurate prediction of transmembrane regions is possible. Increasingly sophisticated algorithms allow transmembrane region recognition and this approach is routinely performed on new, unknown, proteins.

Transmembrane regions occur as peaks in the hydropathy profile. An analysis of band 3 protein reveals a primary sequence containing many hydrophobic residues arranged in at least 13 transmembrane domains. This was therefore consistent with its role as an integral membrane protein. In contrast the 'integral' protein glycophorin also found in the erythrocyte has a low content of non-polar residues compatible with a single transmembrane helix (Figure 5.7). This was unexpected in view of its location and shows the dramatic variation of integral protein organization and secondary structure.

Table 5.5 Kyte and Doolittle scheme of ranking hydrophobicity of side chains

Amino acid	Parameter	Amino acid	Parameter
Ala	1.80	Leu	3.80
Arg	−4.50	Lys	−3.90
Asn	−3.50	Met	1.90
Asp	−3.50	Phe	2.80
Cys	2.50	Pro	−1.60
Gln	−3.50	Ser	−0.80
Glu	−3.50	Thr	−0.70
Gly	−0.40	Tyr	−0.90
His	−3.20	Trp	−1.30
Ile	4.50	Val	4.20

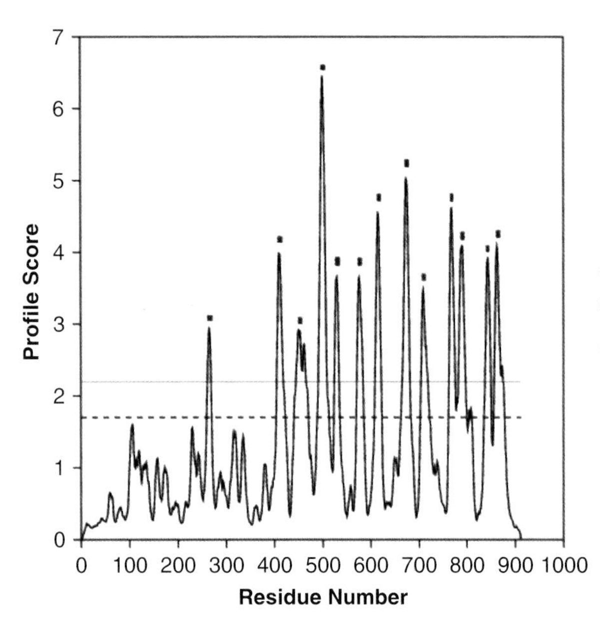

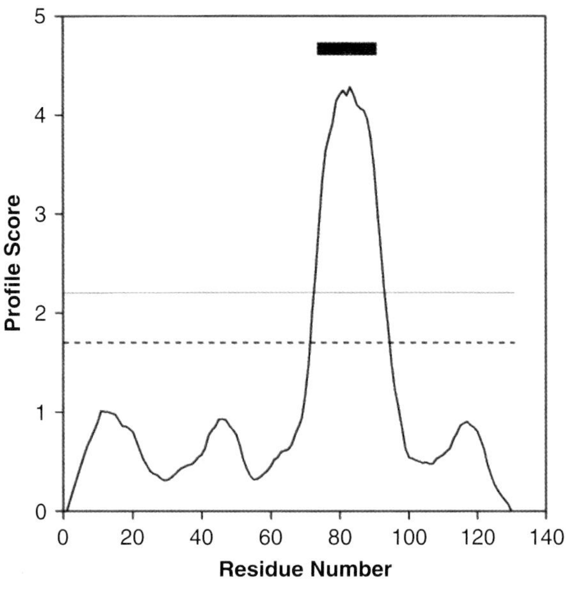

Figure 5.7 The putative transmembrane domains of band 3 and glycophorin. The red bars indicate the positions of the transmembrane domains

Figure 5.8 Sequence and postulated structure of glycophorin A. The external N terminal region has many polar residues and is extensively glycosylated. There is no glycosylation of the hydrophilic domain on the cytoplasmic side of the membrane. The non-polar region stretches for approximately 20 residues. A common genetic variant of glycophorin occurs at residues 1 and 5 with Ser and Gly replaced by Leu and Glu

Careful inspection of the hydropathy plot for glycophorin reveals that a single hydrophobic domain extends for approximately 20–25 residues and that less than 20 % of the protein's 131 residues are compatible with membrane locations. This observation was supported by amino acid sequencing of glycophorin, the first membrane protein to be sequenced, showing significant numbers of polar residues at the N and C terminals (Figure 5.8).

A combination of biochemical techniques established the arrangement of glycophorin – large hydrophilic domains protruded from either side of the erythrocyte membrane. One approach was to use membrane-impermeant agents that reacted with specific residues to yield fluorescent or radioactive labels that facilitated subsequent detection. Using this approach only the N-terminal domain was labelled and it was deduced that this region protruded out from the cell's surface. Rupturing the erythrocyte membrane labelled C-terminal domains but the transmembrane region was never labelled. Detailed studies established an N-terminal region of approximately 70 residues (1–72) with the hydrophobic domain extending from 73–92 and the C-terminal region occupying the remaining fraction of the polypeptide from residues 92–131. *In vivo* the hydrophobic domain promotes dimerization and studies have shown that fusion of this protein sequence to 'foreign' proteins not only promotes membrane localization but also favours association into dimers. Dimerization is sequence dependent with mutagenesis of critical residues completely disrupting protein association. However, although considerable detail has been acquired on the organization of proteins within the erythrocyte membrane a description at an atomic level was largely lacking. In part this stemmed from the difficulty in isolating proteins in forms suitable for analysis by structural methods.

What was required was a different approach. A new approach came with two very different integral membrane proteins. The first was the purple protein, bacteriorhodopsin, found in the bacterium *Halobacterium halobium*. This membrane protein functions as a light-driven proton pump. The second system was a macromolecular photosynthetic complex from purple non-sulfur bacteria that ultimately converted light into chemical energy. The results from studies of these proteins would revolutionize our understanding of membrane protein structure.

Bacteriorhodopsin and the discovery of seven transmembrane helices

The study of the protein bacteriorhodopsin represents a landmark in membrane protein structure determination because for the first time it allowed an insight into the organization of secondary and tertiary structure, albeit at low resolution. Bacteriorhodopsin, a membrane protein of 247 residues ($M_r \sim 26$ kDa), accumulates within the membranes of *H. halobium* – a halophilic (salt loving) bacterium that lives in extreme conditions and is restricted to unusual ecosystems. The cell membrane of *H. halobium* contains patches approximately 0.5 μm wide consisting predominantly of bacteriorhodopsin along with a smaller quantity of lipid. These patches contribute to a purple appearance of membranes and contain protein arranged in an ordered two-dimensional lattice whose function is to harness light energy for the production of ATP via light driven proton pumps. This process is assisted by the presence of retinal co-factor linked covalently to lysine 216 in bacteriorhodopsin (Figure 5.9).

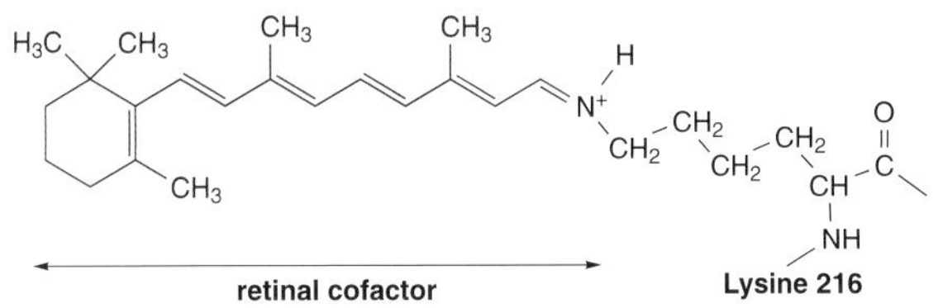

Figure 5.9 The retinal chromophore of bacteriorhodopsin linked to lysine 216 of bacteriorhodopsin. The retinal is responsible for the purple colour observed in the membranes of *H. halobium*

Bacteriorhodopsin is the simplest known proton pump and differs from those found in other systems in not requiring associated electron transfer systems. By dispersing purified purple membranes containing bacteriorhodopsin in lipid vesicles along with bovine heart ATP synthetase (the enzyme responsible for synthesizing ATP) Efraim Racker and Walter Stoeckenius were able to demonstrate that light-driven proton fluxes resulted in ATP synthesis. Not only was this a convincing demonstration of the light-driven pump of bacteriorhodopsin but it was also an important milestone in emphasizing the role of proton gradients in ATP formation.

Richard Henderson and Nigel Unwin used the two-dimensional patches of protein to determine the structure of bacteriorhodopsin using electron crystallography a technique (see Chapter 10) analogous to X-ray crystallography. By a process of reconstruction Henderson and Unwin were able to produce low-resolution structural images of bacteriorhodopsin in 1975 (Figure 5.10). This was the first time the structure of a membrane protein had been experimentally determined at anything approaching an atomic level of resolution.

The structure of bacteriorhodopsin revealed seven transmembrane helices and it was assumed that these regions were linked by short turns that projected into the aqueous phase on either side of the purple membrane. Shortly after initial structure determination the primary sequence of this hydrophobic protein was determined. This advance allowed a correlation

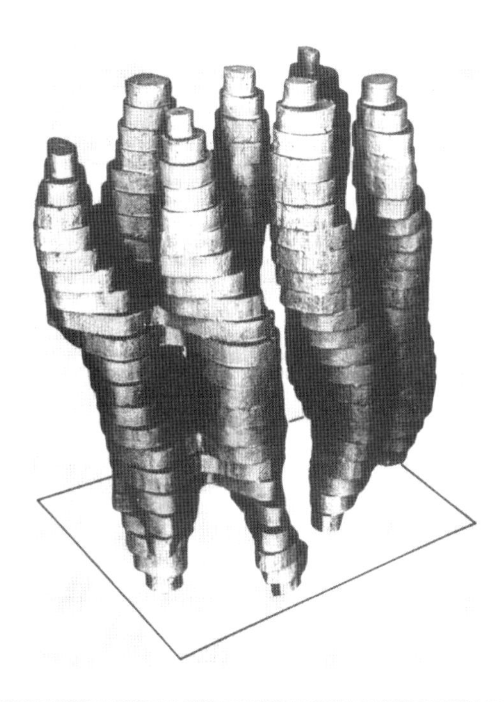

Figure 5.10 A model proposed for the structure of bacteriorhodopsin derived using electron microscopy (after Henderson, R. & Unwin, P.N.T *Nature* 1975, **257**, 28–32). The view shown lies parallel to the membrane with the helices running across the membrane. Although clearly 'primitive' and of low resolution this representation of seven transmembrane helices had a dramatic effect on perceptions of membrane protein structure

between structural and sequence data. Hydrophobicity plots of the amino acid sequence identified seven hydrophobic segments within the polypeptide chain and this agreed with the transmembrane helices seen in electron density maps. The primary sequence of bacteriorhodopsin was pictured as a block of seven transmembrane helices (Figure 5.11).

Subsequently, determination of the structure of bacteriorhodospin by X-ray crystallography to a resolution of 1.55 Å confirmed the organization seen by electron diffraction and showed that the seven transmembrane helices were linked by extra- and intracellular loops as originally proposed by Henderson and Unwin (Figure 5.12). Although not apparent from the early low-resolution structures a short antiparallel β strand is found in a loop between helices B and C. More significantly this structure allowed insight into the mechanism of signal transduction and ion pumping, the major biological roles of bacteriorhodopsin.

The light-driven pump is energized by light absorption by retinal and leads to the start of a photocycle where the conversion of retinal from an all-*trans* isomer to a 13-*cis* isomer is the trigger for subsequent reactions. Photobleaching of pigment leads to the vectorial release of a proton on the extracellular side of the membrane and a series of reactions characterized by spectral intermediates absorbing in different regions of the UV–visible spectrum.

The discovery that bacteriorhodopsin contained seven transmembrane helices arranged transversely across the membrane has assisted directly in understanding the structure and function of a diverse range of proteins. Of direct relevance is the presence of rhodopsin in visual cycle of the eye in vertebrates and invertebrates as well as a large family of proteins based on seven transmembrane helices.

Seven transmembrane helices and the discovery of G-protein coupled receptors

These receptors are embedded within membranes and are not easily isolated let alone studied by structural methods. However, cloning cDNA sequences for many putative receptors allowed their amino acid sequences to be defined with the recognition of similarity within this large class of receptors. The proteins were generally 400–480 residues in length with sequences containing seven conserved segments or blocks of hydrophobic residues. These seven regions of hydrophobic residues could only mean, in the light of the work on bacteriorhodopsin, transmembrane helices linked by cytoplasmic or extracellular loops. Amongst the first sequences to be recognized with this topology was human β_2 adrenergic receptor (Figure 5.13). Seven transmembrane helices approximately 24 residues in length were identified together with a glycosylated N-terminal domain projecting out from the extracellular surface and on the other side of the membrane a much larger C-terminal domain.

Prior to sequence analysis receptors such as the β_2 adrenergic receptors had been subjected to enormous experimental study. It was shown that receptors bind

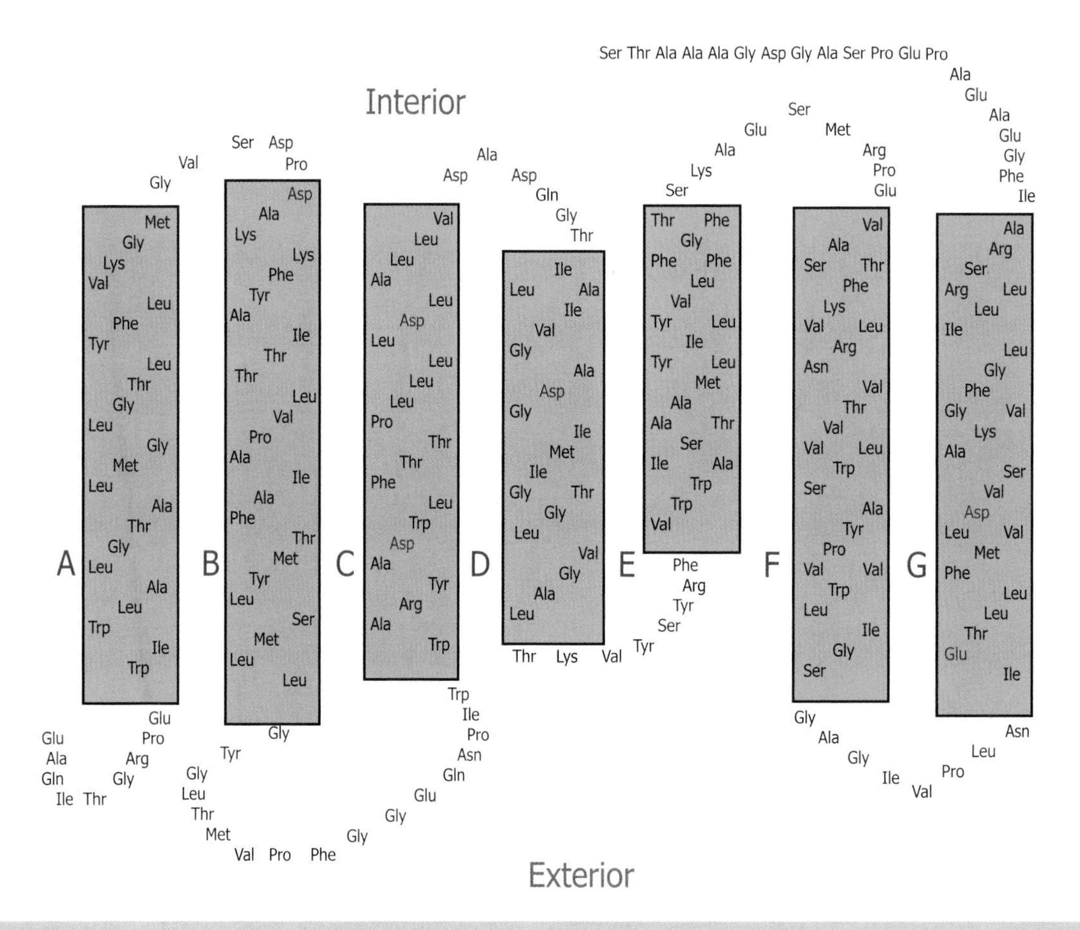

Figure 5.11 The primary sequence of bacteriorhodopsin superimposed on a seven transmembrane helix model. Charged side chains are shown in either red (negative) or blue (positive). Charged residues are usually found either in the loop regions or at the loop–helix interface. The presence of charged residues within a helix is rare and normally indicates a catalytic role for these side chains. This is seen in helix C and G of bacteriorhodopsin

ligands or hormones with kinetics resembling those exhibited by enzymes where the hormone (substrate) bound tightly with dissociation constants measured in the picomolar range (10^{-12} M). Binding studies also helped to establish the concept of agonist and antagonist. The use of structural analogues or agonists to the native hormone yielded a binding reaction and elicited a normal receptor response. In contrast antagonists compete for hormone binding site on the receptor inhibiting the normal biological reaction. One important hormone antagonist is propranolol – a member of a group of chemicals widely called β blockers – that competes with adrenaline (epinephrine) for binding sites on adrenergic receptors. These compounds have been developed as drugs to control pulse rate and blood pressure. Not only have useful drugs emerged from agonist and antagonist binding studies of adrenergic receptors but four major groups of adrenergic receptor have been recognized in the membranes of vertebrates (α_1, α_2 β_1 and β_2).

These receptors play a role in signal transduction and a major discovery in 1977 was the distinction between β adrenergic receptors and adenylate cyclase. Adenylate cyclase was shown in classic research on hormonal control of glycogen levels to be a membrane bound receptor for adrenaline. This led to the incorrect

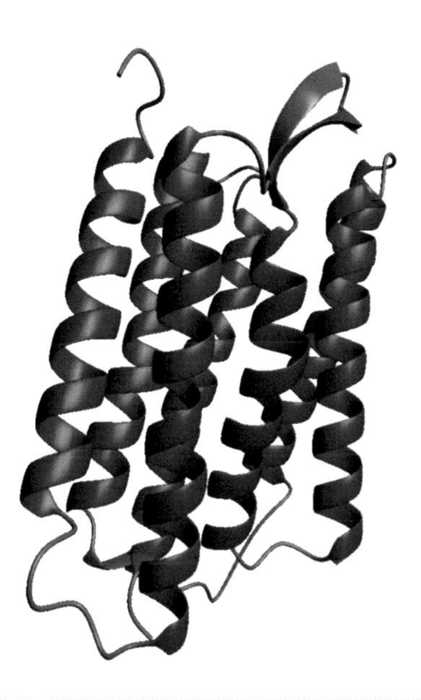

Figure 5.12 High resolution structure of bacteriorhodopsin (PDB:1C3W) showing seven transmembrane helices linked by extra and intracellular loops

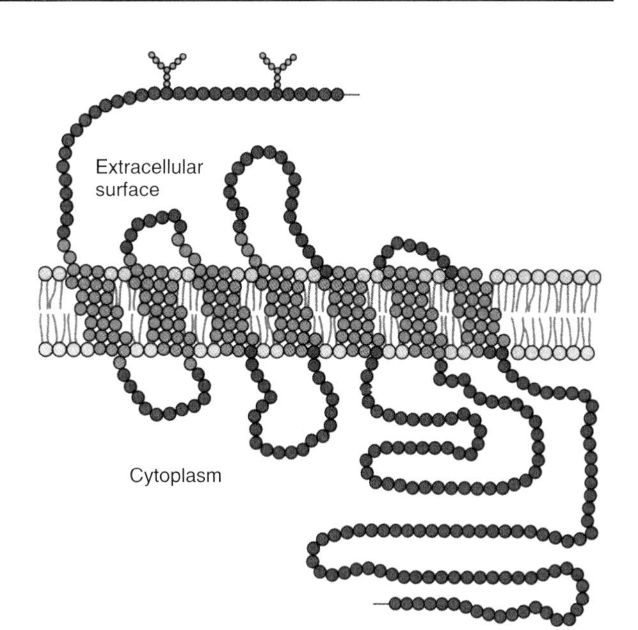

Figure 5.13 The amino acid sequence of the β_2 adrenergic receptor defines the organization of the protein into seven transmembrane helices. The adrenergic receptors are so called because they interact with the hormone adrenaline (after Dohlman *et al. Biochemistry* 1987, **26**, 2660–2666). The receptor interacts with cellular proteins in the C-terminal region and interaction is controlled by the reversible phosphorylation of Ser/Thr residues

idea that receptors were adenylate cyclase – a role shown later to be performed by G protein coupled receptors. Both adenylate cyclase and membrane bound receptors catalyse the activation or inhibition of cyclic AMP synthesis. A crucial discovery was the observation that transduction of the hormonal signal to adenylate cyclase involved not only a membrane bound receptor but a separate group of proteins called G proteins. G proteins are so called because they bind guanine nucleotides such as GTP/GDP. As a result the overall picture of signal transduction mechanisms changed to involve a hormone receptor, adenylate cyclase *and* a G protein. All of these proteins are associated with the membrane.

Studies of bacteriorhodopsin played a key role in understanding rhodopsin, the photoreceptive pigment at the heart of the visual cycle and responsible for the transduction of light energy in the eyes of vertebrates and invertebrates. Rhodopsin is the best characterized receptor and much of our current knowledge of signal transduction arose from studies of G proteins and G protein-coupled receptors in the visual cycle. The model of signal transduction receptors as seven transmembrane helices with extra and intracellular loops was in part possible because of the initial structural studies of bacteriorhodopsin – at first glance a completely unrelated protein.

Within the retina of the eye rhodopsin is the major protein of the disc membrane of rod cells and by analogy to the bacterial protein it contains a chromophore 11-*cis* retinal that binds to opsin (the protein). Rhodopsin absorbs strongly between wavelengths of 400 and 600 nm and absorption of light triggers the conversion of the bound chromophore to the all-*trans* form. Several conformational changes later metarhodopsin II is formed as an important intermediate in the multiple step process of visual transduction. The release of *trans*-retinal from the protein and the conversion of retinal back to the initial 11-*cis* state by the action of retinal isomerase signals the end of the cycle (Figure 5.14).

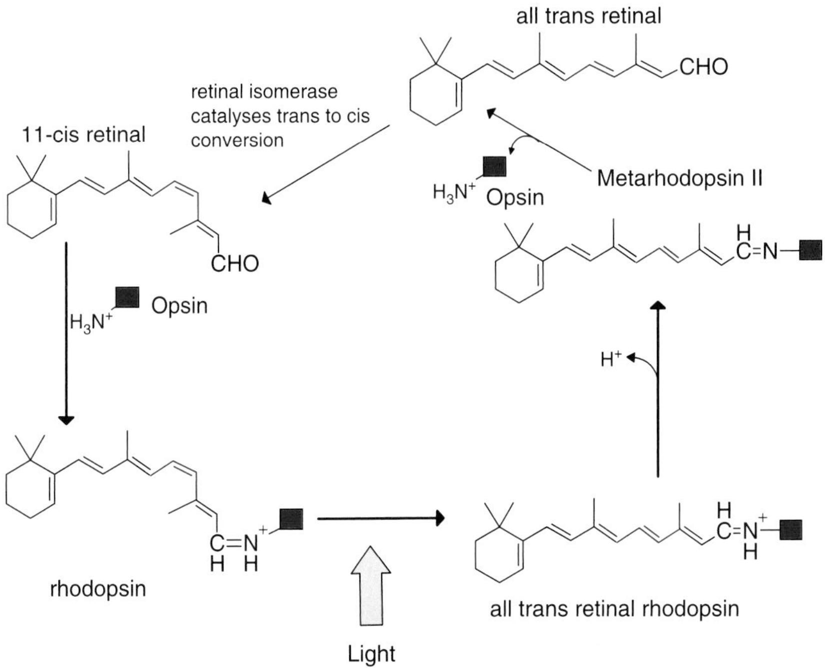

Figure 5.14 The basic visual cycle involving chemical changes in rhodopsin. The critical 11-*cis* bond is shown in red with retinal combining with opsin to form rhodopsin. Light converts the rhodopsin into the *trans*-retinal followed by the formation of metarhodopsin II. This intermediate reacts with transducin before dissociating after ~1 s into *trans*-retinal and opsin

Photochemically induced conformational changes in metarhodopsin II lead to interactions with a second protein transducin found in the disc membrane of retinal rods and cones. Transducin is a heterotrimeric guanine nucleotide binding protein (G protein) containing three subunits designated as α, β and γ of molecular weight 39 kDa, 36 kDa and 8 kDa. Photoexcited rhodopsin interacts with the α subunit of transducin promoting GTP binding in place of GDP and a switch from inactive to active states. The transducin α subunit is composed of two domains. One domain contains six β strands surrounded by six helices whilst the other domain is predominantly helical. Although in the visual cycle the extracellular stimulus and biochemical endpoint are different to hormone action the transmembrane events are remarkably similar.

Interaction between rhodopsin and G protein causes dissociation with the G_α subunit separating from the $G_{\beta\gamma}$ dimer (Figure 5.15). The transducin G_α–GTP complex activates a specific phosphodiesterase enzyme by displacing its inhibitory domain and thus triggering subsequent events in the signalling pathway. The phosphodiesterase cleaves the cyclic nucleotide guanosine 3′, 5′-monophosphate (cyclic GMP) in a process analogous to the production of cyclic AMP in hormone action and this in turn stimulates visual signals to the brain.

As a signalling pathway in humans the rhodopsin–transducin system contributed significantly to our understanding of other G proteins and G-protein coupled receptors. Subsequent work has suggested that all G proteins share a common structure of three subunits. The identification of human cDNA encoding 24 α proteins, 5 β proteins and 6 γ proteins testifies to the large number of G protein catalysed reactions and their importance within cell signalling pathways. The α subunits of G proteins are one group of an

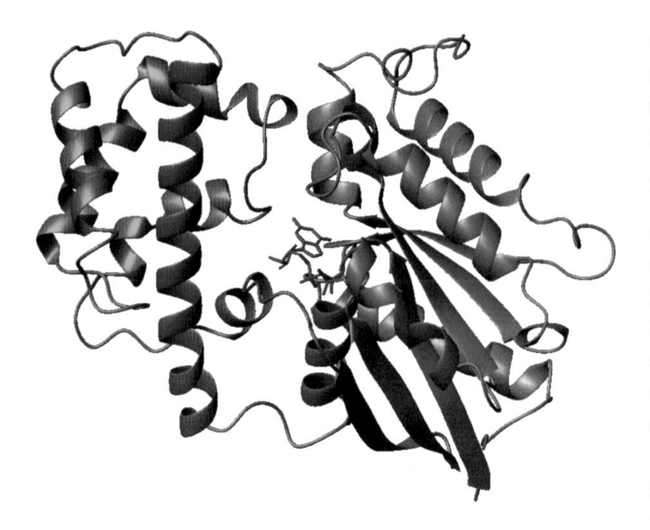

Figure 5.15 The structure of the G_α subunit of transducin. Two domains are shown in dark green (helical) and blue ($\alpha + \beta$) together with a GTP analogue (magenta) bound at the active site (PDB: 1TND)

important class of nucleotide binding proteins whose activity is enhanced by GTP binding and diminished by GDP binding. Other members with this group of G proteins include the elongation factors involved in protein synthesis and the products of oncogenes such as Ras.

Addressing hormone action is an entirely analogous process and is best visualized through inspection of the structure of the β_2 adrenergic receptor. Hormone binding at the extracellular surface promotes conformational change and the interaction of the C-terminal region with the G_α subunit. This stimulates GDP/GTP exchange and in its active state the G_α subunit binds to adenylate cyclase causing the formation of cAMP which functions downstream in protein phosphorylation reactions that stimulate or inhibit metabolic processes (Figure 5.16). G proteins are further sub-divided into two major classes: G_s and G_i. The former stimulate adenylate cyclase whilst the latter inhibit although both interact with a wide range of receptors and interact with other proteins besides adenylate cyclase.

Rhodopsin and β_2 adrenergic receptors are seven transmembrane receptor proteins that activate heterotrimeric GTP binding proteins and are grouped under the term G-protein coupled receptors (GPCRs). GPCRs are part of an even larger group of seven transmembrane receptors and represent the fourth largest superfamily found in the human genome. It is estimated that within the human genome there are over 600 genes encoding seven transmembrane helices receptors. Within this family there are three major structural subdivisions with the rhodopsin-like family representing the largest class. In almost all cases there is a common mechanism involving G protein activation and two types of intracellular or secondary messengers, calcium (Ca^{2+}) and cyclic AMP (cAMP). Approximately 70 % of GPCRs signal through the cAMP pathway and 30 % signal through the Ca^{2+} pathway.

Detailed structural data on GPCRs remains limited and this is unfortunate because many proteins within this family are important and lucrative pharmaceutical targets. Many drugs in clinical use in humans are directed at seven transmembrane receptors. Despite overwhelming sequence data and biophysical data the structure of a non-microbial rhodopsin has only recently been achieved. The crystal structure confirmed, perhaps belatedly, the eponymous seven transmembrane helical model but more importantly this structure provided a model for other GPCRs that allows insight into receptor–G protein complex formation and an understanding of the mechanism by which agonist-receptor binding leads to G protein activation.

The photosynthetic reaction centre of Rhodobacter viridis

Studies of bacteriorhodopsin defined a basic topology of helices traversing the membrane but did not reveal sufficient detail of helical arrangement with respect to each other and the membrane. Detailed analysis of membrane proteins required the use of X-ray crystallography, a technique whose resolution is often superior to electron crystallography. However, this experimental approach requires the formation of three-dimensional crystals of suitable size and this had proved extremely difficult for non-globular proteins. Bacteriorhodopsin forms a two-dimensional array of molecules naturally within the membrane but for almost all other membrane proteins the formation of crystals is a major hurdle to be overcome before acquisition of structural information.

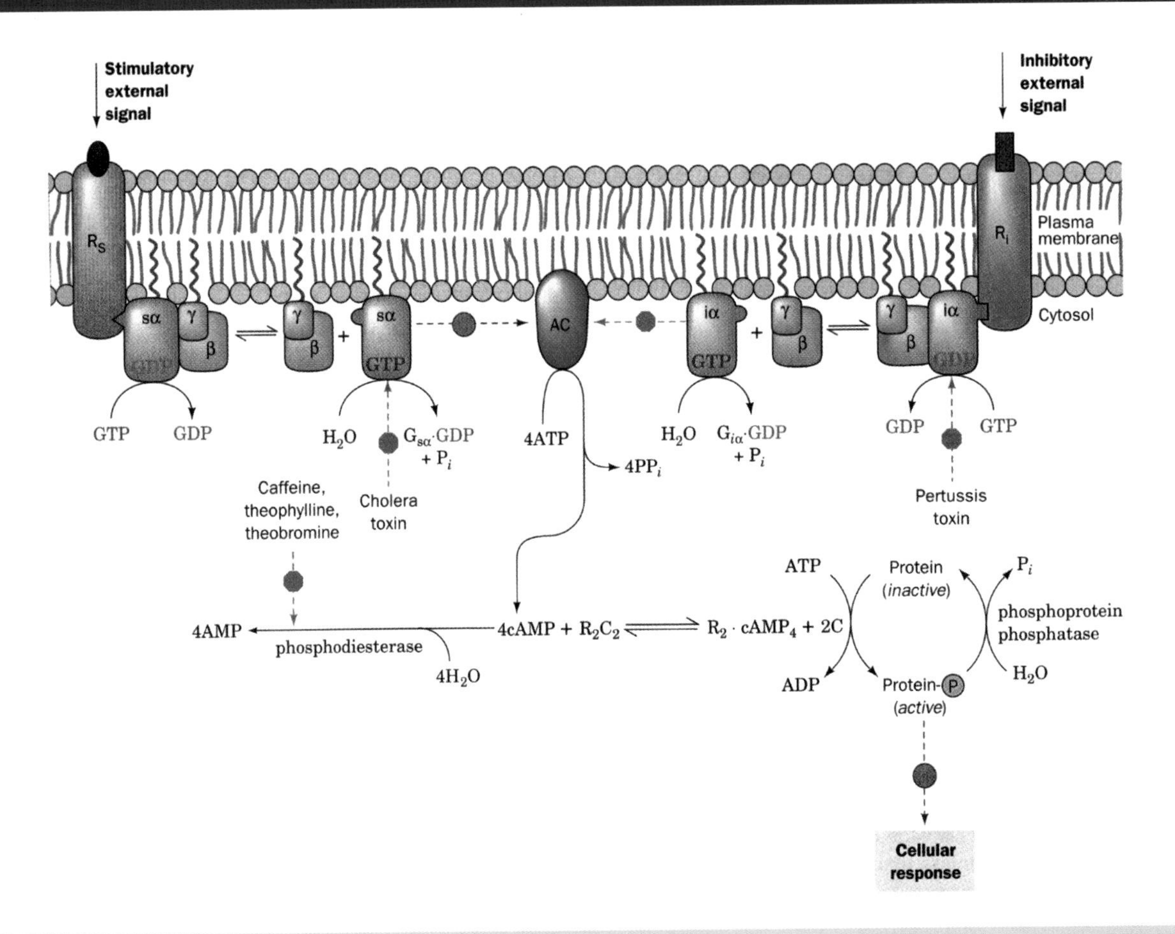

Figure 5.16 The adenylate cyclase signalling system. Binding of hormone to either receptor caused a stimulatory (R_S) or inhibitory (R_i) activity of adenylate cyclase. These receptors undergo conformational changes leading to binding of the G protein and the exchange of GDP/GTP in the Gα subunit. The action of the Gα subunit in stimulating or inhibiting adenylate cyclase continues until hydrolysis of GTP occurs. Adenylate cyclase forms cAMP which activates a tetrameric protein kinase by promoting the dissociation of regulatory and catalytic subunits (reproduced with permission from Voet, C, Voet, J.G & Pratt, C.W. *Fundamentals of Biochemistry* John Wiley & Sons, Ltd, Chichester, 1999).

Removing membrane proteins from their environment as a prelude to crystallization was frequently unsuccessful since the lipid bilayer plays a major role maintaining conformation. Moreover, exposure of hydrophobic residues that were normally in contact with non-polar regions of the bilayer to aqueous solutions resulted in many proteins unfolding with a concomitant loss of activity. One approach adopted in an attempt to improve methods of membrane protein crystallization was to substitute detergents that mimicked the structure and function of lipids. In many instances these detergents failed to preserve structure or function satisfactorily. The situation changed drastically in 1982 when Hartmut Michel prepared highly ordered crystals of the photosynthetic reaction centre from *Rhodobacter viridis* (originally *Rhodopseudomonas viridis*).

The reaction centre is a macromolecular complex located in heavily pigmented membranes of photosynthetic organisms (Figure 5.17). Within most

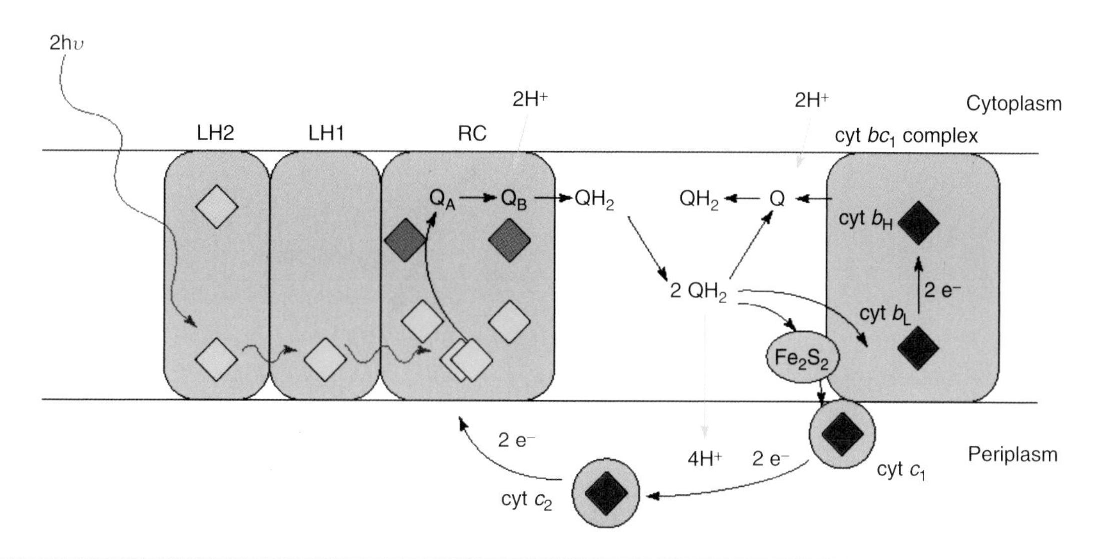

Figure 5.17 The organization of reaction centres and the photosynthetic electron transfer system in purple bacteria such as *R. sphaeroides*. In *R. viridis* a bound cytochrome is linked to the reaction centres but is absent in most species. Bacteriochlorophyll molecules are represented by blue diamonds, bacteriopheophytin molecules by purple diamonds and hemes by red diamonds. The red arrows represent excitation transfer from light harvesting chlorophyll complexes and the black and blue arrows correspond to electron and proton transfer, respectively (reproduced with permission Verméglio, A & Joliot, P. *Trends Microbiol.* 1999, **7**, 435–440. Elsevier).

bacterial membranes the centre is part of a cyclic electron transfer pathway using light to generate ATP production.

Absorption of a photon by light-harvesting complexes (LH1 and LH2) funnels energy to the reaction centre initiating primary charge separation where an electron is transferred from the excited primary donor, a bacteriochlorophyll dimer, to a quinone acceptor (Q_B) via other pigments. After a second turnover, the reduced quinone picks up two protons from the cytoplasmic space to form a quinol (QH_2) and is oxidized by cytochrome c_2, a reaction catalysed by the cytochrome bc_1 complex. This reaction releases protons to the periplasmic space for use in ATP synthesis and the cyclic electron transfer pathway is completed by the reduction of the photo-oxidized primary electron donor.

Fractionation studies of *R. viridis* membranes using detergents such as lauryl dimethylamine oxide (LDAO) yield a 'core' reaction centre complex containing four polypeptides lacking the light-harvesting chlorophyll and cytochrome bc_1 complexes normally associated with these centres. The subunits (H, M, and L) possessed all of the components necessary for primary photochemistry and were associated with a membrane bound cytochrome.

Analysis of the photosynthetic complex showed that its ability to perform primary photochemical reactions was due to the presence of chromophores that included four bacteriochlorophyll molecules, two molecules of a second related pigment called bacteriopheophytin, carotenoids, a non-heme iron centre, and two quinone molecules (ubiquinone and menaquinone). Using spectroscopic techniques such as circular dichroism, laser flash photolysis and electron spin resonance the order and time constants associated with electron transfer processes was established (Figure 5.18). In all cases electron transfer was characterized by extremely rapid reactions. The light harvesting complexes act as antennae and transfer energy to a pair of chlorophyll molecules (called P870, and indicating the wavelength associated with maximum absorbance) in the reaction centre located close to the periplasmic membrane surface. Absorption of a photon causes an excited state and at room temperature the photo-excited pair reduces

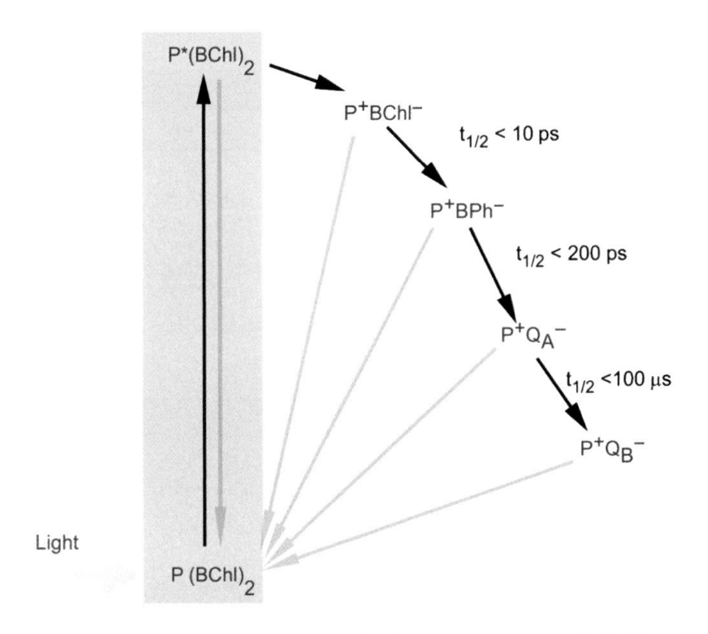

Figure 5.18 The sequence of donors and acceptors in the reaction centres of *R. viridis*. Q_A is menaquinone and Q_B is ubiquinone. Rapid back reactions occur at all stages and the primary photochemical reactions occur very rapidly (ps timescale) presenting numerous technical difficulties in accurate measurement

a pheophytin molecule within 3 ps. The initial primary charge separation generates a strong reductant and reduces the primary acceptor bacteriophaeophytin. Within a further 200 ps the electron is transferred from bacteriopheophytin to a quinone molecule. These reactions are characterized by high quantum efficiency and the formation of a stable charge separated state. Normally the quinone transfers an electron to a second quinone within 25 μs before further electron transfer to other acceptors located within membrane protein complexes. In isolated reaction centres electrons reaching the quinone recombine with a special pair of chlorophyll by reverse electron transfer.

The electron transfer pathway was defined by spectroscopic techniques but the arrangement of chromophores together with the topology of protein subunits remained unknown until Michel obtained ordered crystals of reaction centres.

Crystallization requires the ordered precipitation of proteins. Membrane proteins have both hydrophilic and hydrophobic segments due to their topology, with the result that the proteins are rarely soluble in either aqueous buffers or organic media. By partitioning the membrane protein in detergent (LDAO) micelles it was possible to purify subunits but it proved very difficult to crystallize reaction centres from LDAO. As a general rule macromolecular protein complexes will only aggregate into ordered crystalline arrays if their exposed polar surfaces can approach each other closely. The use of detergents, particularly if they formed micelles of large size, tended to limit the close approach of polar surfaces preventing crystallization. The major advance of Michel was to introduce small amphiphilic molecules such as heptane 1,2,3 triol into the detergent mixture. Although the precise mechanism of action of these amphiphiles is unclear they are thought to displace large detergent molecules from the crystal lattice, thereby assisting aggregation. In addition they do not form micelles themselves but lower the effective concentration of detergent by becoming incorporated within micelles and limiting protein denaturation. Lastly, amphiphiles interact with protein but do not prevent close proximity of solvent-exposed polar regions.

The structure of the reaction centre revealed the topology of structural elements within the H,M,L

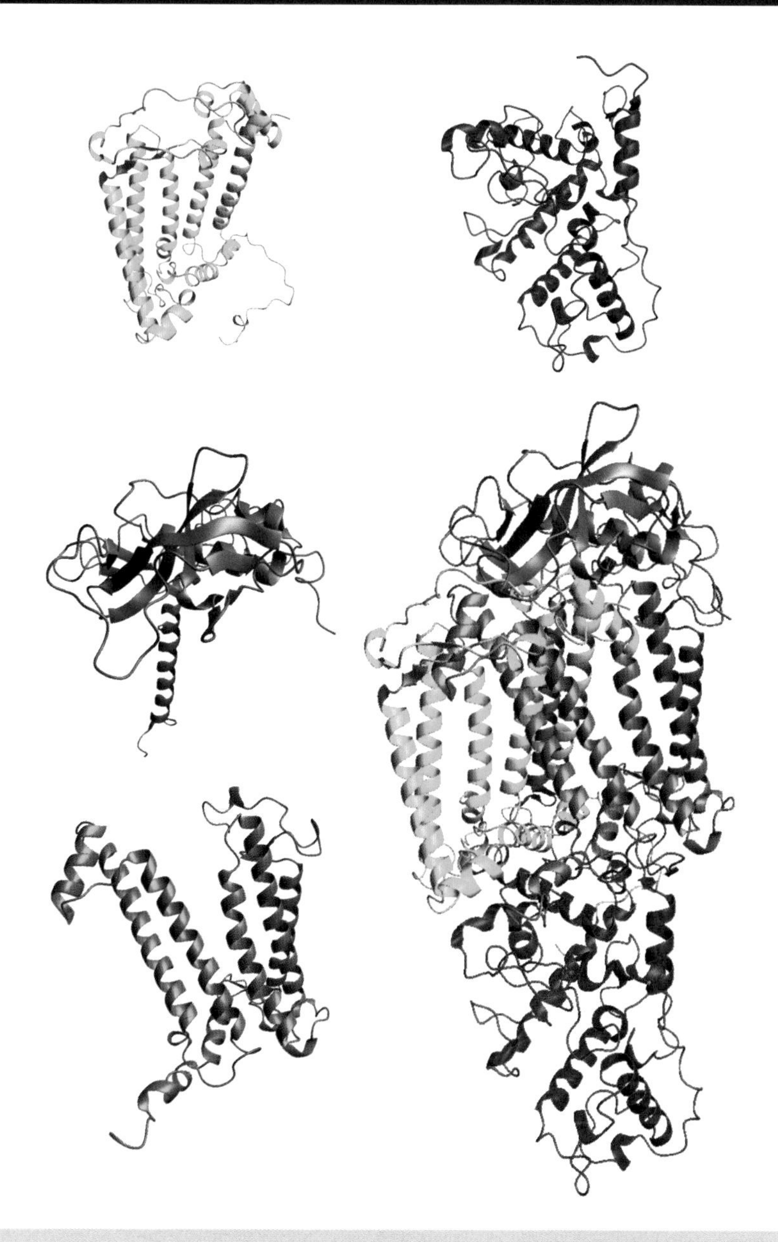

Figure 5.19 Distribution of secondary structure in each subunit of the photosynthetic reaction centre together with the whole complex. The cytochrome subunit is shown in orange, the L subunit in light green, the M subunit in dark green and the H subunit in blue (PDB:1PRC)

and cytochrome subunits and importantly the position of the pigments (Figure 5.19). The structure offered the possibility of understanding how reaction centres achieved picosecond charge separation following light absorption.

The structure of the bacterial reaction centre

One surprise from the crystal structure of the four-subunit complex was the H subunit. Sequencing the gene for subunit H revealed that it was the smallest subunit in terms of number of residues. Its migration on SDS gels was anomalous when compared to the L and M subunits leading to slightly inaccurate terminology. Careful inspection of Figure 5.19 reveals that the H subunit is confined to the membrane by a single transmembrane α helix (shown in blue) whilst most of the protein is found as a cytoplasmic domain. The physiological role of the H subunit is unclear since it lacked all pigments. The structure of subunit H is divided into three distinct segments; a membrane-spanning region, a surface region from residues 41 to 106, and a large globular domain formed by the remaining residues. The membrane-spanning region of the H subunit is a single regular α helix from residues 12 to 35 with some distortion near the cytoplasmic membrane surface. Following the single transmembrane helix are ~65 residues forming a surface region interacting with the L and M subunits and based around a single α helix and 2 β strands. In the cytoplasmic domain of subunit H an extended system of antiparallel and parallel β strands are found together with a single α helix from residues 232 to 248.

The L and M subunits each contain five transmembrane helices (Table 5.6) but more significantly each subunit is arranged with an approximate two-fold symmetry that allows the superposition of atoms from one subunit with those on the other. The superposition is surprising and occurs with low levels of sequence identity (~26 percent). In each subunit the five helices are designated A–E; three (A, B and D) are extremely regular and traverse the membrane as straight helices whilst C and E are distorted. Helix E in both subunits shows a kink caused by a proline residue that disrupts regular patterns of hydrogen bonding. Although the helices show remarkably regular torsion angles and hydrogen bonding patterns they are inclined at angles of ~11° with respect to the membrane. Elements of β strand were absent and the results confirmed those obtained with bacteriorhodopsin. In the reaction centre between 22 and 28 residues were required to cross the membrane each helix linked by a short loop.

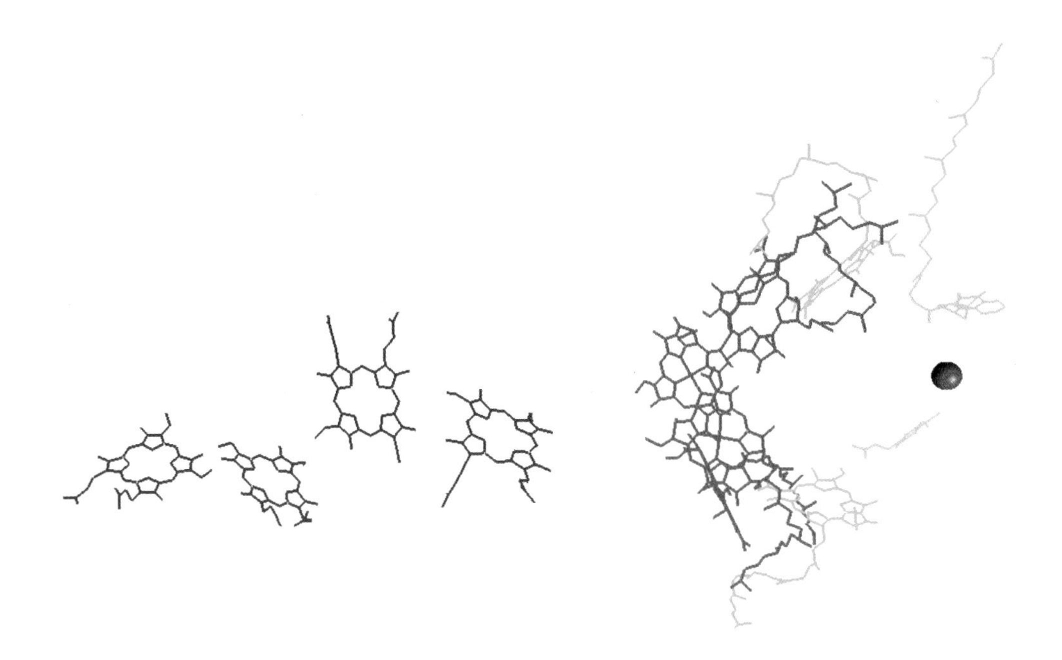

Figure 5.20 The arrangement of heme, bacteriochlorophyll, bacteriopheophytin, quinone and non heme iron in the hydrophobic region between L and M subunits. Electron transfer proceeds from the hemes shown on the left through to the quinones shown on the right

Table 5.6 The transmembrane helices found in the L/M subunits of the reaction centre

Helical segments in subunits L and M		
Region and number of residues in brackets		
Helix	Subunit L	Subunit M
A	L33–L53 (21)	M52–M76 (25)
B	L84–L111 (28)	M111–M137 (27)
C	L116–L139 (24)	M143–M166 (24)
D	L171–L198 (28)	M198–M223 (26)
E	L226–L249 (24)	M260–M284 (25)

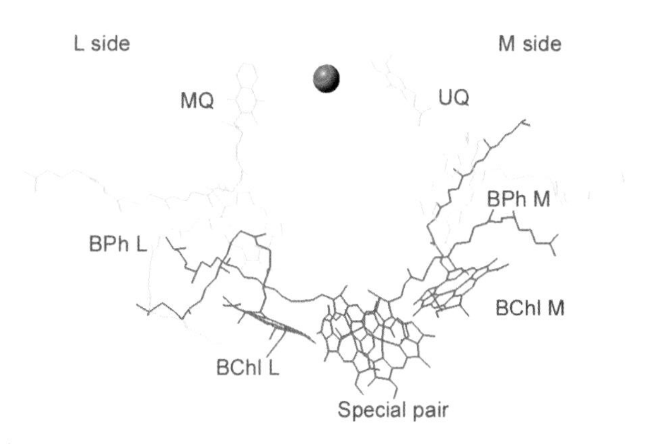

Figure 5.21 Arrangement of chromophores in the bacterial reaction centre. The cytochrome redox co-factors have been removed

For the L and M subunits the five transmembrane helices defined the bilayer region of the membrane. Surprisingly, direct interactions between subunits are limited by the presence of chromophores between the subunits and most of the subunit contacts were restricted to periplasmic surfaces close to the H subunit.

The cytochrome subunit contained 336 residues with a total of nine helices and four heme groups covalently linked to the polypeptide via two thioether bridges formed with cysteine residues. The fifth and six ligands to the hemes were methionine and histidine for three of the centres whilst the fourth showed bis-histidine ligation. The heme groups have different redox potentials and serve to shuttle electrons to the special pair chlorophyll after excitation. Although predominantly in the aqueous phase the cytochrome remains membrane bound by a covalently linked diglyceride attached to the N-terminal Cys residue.

It is, however, the arrangement of pigments that provides the most intriguing picture and sheds most light on biological function. The active components (4 × BChl, 2 × BPh, 2 × Q + Fe) are confined to a region between the L and M subunits in a hydrophobic core with each chromophore bound via specific side chain ligands. The special pair of chlorophyll molecules is located on the cytoplasmic side of the membrane close to one of the heme groups of the cytochrome. Stripping the L and M polypeptides reveals the two-fold symmetry of the chromophores (see Figures 5.20 & 5.21). The structural picture agrees well with the order of electron transfer established earlier using kinetic techniques as

$$\mathrm{B(Chl)_2} \longrightarrow \mathrm{BPh} \longrightarrow \mathrm{Q_A} \longrightarrow \mathrm{Q_B} \longrightarrow \text{other acceptors}$$

The additional chlorophylls found in the reaction centre do not participate directly in the electron transfer process and were called 'voyeur' chlorophylls.

The co-factors between the L and M subunits are arranged as two arms diverging from the special pair of chlorophylls. A single 'voyeur' chlorophyll together with a pheophytin molecule lies on each side and they lead separately to menaquinone and ubiquinone molecules with a single non-heme iron centre located between these acceptors. It was therefore a considerable surprise to observe that despite the high degree of inherent symmetry in the arrangement of co-factors the left hand branch (L side) was the more favoured electron transfer route by a ratio of 10:1 over the right hand or M route.

The underlying reasons for asymmetry in electron transfer are not clear but rates of electron transfer are very dependent on free energy differences between excited and charge separated states as well as the distance between co-factors. Both of these parameters are influenced by the immediate protein surroundings of each co-factor and studies suggest that rates of electron transfer are critically coupled to the properties of the bacteriochlorophyll monomer (the voyeur) that lies between the donor special pair and bacteriopheophytin acceptor.

The X-ray crystal structure of the *R. viridis* reaction centre was determined in 1985 and was followed by the structure of the *R. sphaeroides* reaction centre that confirmed the structural design and pointed to fold conservation. Research started with crystallization of the bacterial photosynthetic resulted in one of the outstanding achievements of structural biology and Hartmut Michel, Johan Diesenhofer, and Robert Huber were rewarded with the Nobel Prize for Chemistry in 1988. Over the last two decades this membrane protein has been extensively studied as a model system for understanding the energy transduction, the structure and assembly of integral membrane proteins, and the factors that govern the rate and efficiency of biological electron transfer.

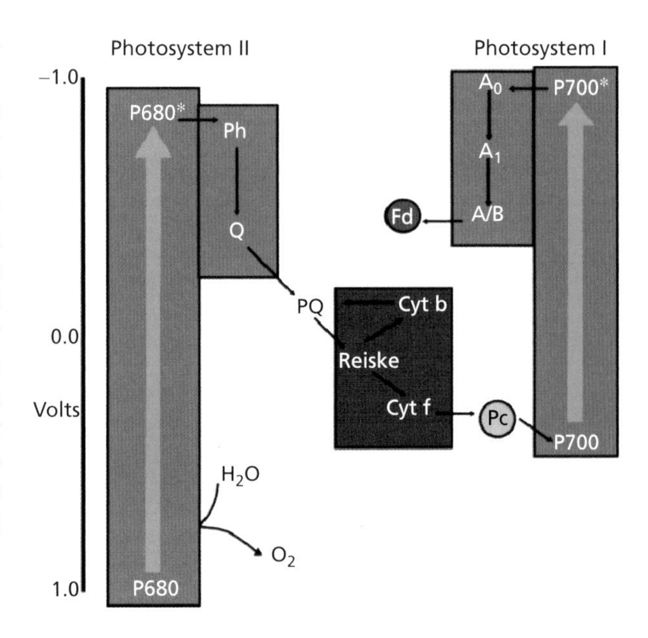

Figure 5.22 The oxygenic photosynthetic system found in cyanobacteria and eukaryotic algae and higher plants. Two photosystems acting in series generate reduced NADPH and ATP via the absorption of light. Photosystem II precedes photosystem I

Oxygenic photosynthesis

The photosynthetic apparatus of most bacteria is simpler than that found in algae and higher plants. Here two membrane-bound photosystems (PSI and PSII) act in series splitting water into molecular oxygen and generating ATP/NADPH via photosynthetic electron transfer (Figure 5.22). The enzymatic splitting of water is a unique reaction in nature since it is the only enzyme to use water as a principal substrate. Oxygen is one of the products and the evolution of this reaction represents one of the most critical events in the development of life. It generated an atmosphere rich in oxygen that supported aerobic organisms. The geological record suggests the appearance of oxygen in the atmosphere fundamentally changed the biosphere.

The water splitting reaction is carried out by PSII and involves Mn ions along with specific proteins. Within PSII are two polypeptides called D1 and D2 that show a low level of homology with the L and M subunits of anoxygenic bacteria. The structure of PSII from the thermophilic cyanobacterium (sometimes called blue–green algae) *Synechococcus elongatus* confirmed structural homology with bacterial reaction centres. PSII is a large membrane-bound complex comprising of at least 17 protein subunits, 13 redox-active co-factors and possibly as many as 30 accessory chlorophyll pigments. Of these subunits 14 are located within the photosynthetic membrane and include most significantly the reaction centre proteins D1 (*PsbA*) and D2 (*PsbD*). The observation that D1 and D2 each contained five transmembrane helices arranged in two interlocked semicircles and related by the pseudo twofold symmetry axis emphasized the staggering similarity to the organization of the L and M subunits (Figure 5.23). Moreover the level of structural homology between D1/D2 and L/M subunits suggested that all photosystems evolved from common ancestral complexes.

Photosystem I

Oxygenic photosynthesis requires two photosystems acting in series and the structure of the cyanobacterial PSI reaction centre further confirmed the design pattern of heterodimers (Figure 5.24). Two subunits, PsaA

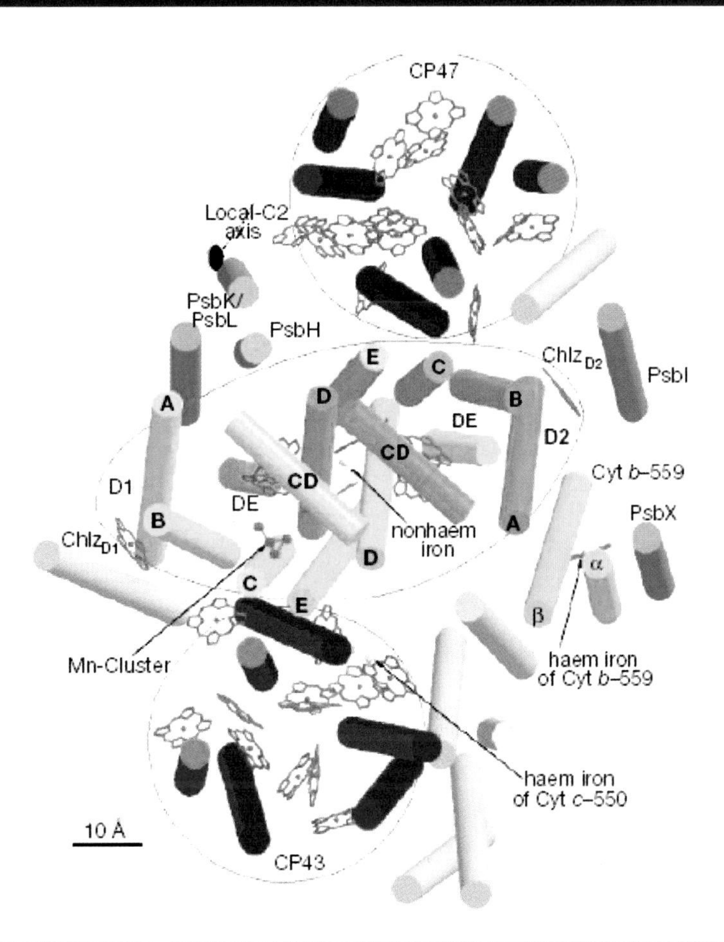

Figure 5.23 Structure of PSII with assignment of protein subunits and co-factors. Arrangement of transmembrane α helices and co-factors in PSII. Chl a and hemes are indicated by grey wireframe representations. The direction of view is from the lumenal side, perpendicular to the membrane plane. The α helices of D1, D2 are shown in yellow and enclosed by an ellipse. Antennae chlorophylls in subunits CP43 and CP47 are enclosed by circles. Unassigned α helices are shown in grey (reproduced with permission from Zouni, A, *et al. Nature* 2001 **409** 739–743. Macmillan)

and PsaB make up a reaction centre that contains eleven transmembrane helices. By now this sounds very familiar, and five helices from each subunit traverse the membrane and together with a third protein, PsaC, contain all of the co-factors participating in the primary photochemistry. A further six subunits contribute transmembrane helices to the reaction centre complex whilst two subunits (PsaD and E) bind closely to the complex on the stromal side of the membrane.

Support for the hypothesis that reaction centres evolved from a common ancestor were strengthened further by the observation that the five C-terminal helices of PsaA and PsaB in PSI showed homology to the D1/D2 and L/M heterodimers described previously. Reaction centres are based around a special pair of chlorophyll molecules that transfer electrons via intermediary monomeric states of chlorophyll and quinone molecules to secondary acceptors that are quinones in type II reaction centres and iron–sulfur proteins in type I reaction centres. Figure 5.25 summarizes the organization of reaction centre complexes.

Oxygenic photosynthesis is the principal energy converter on Earth converting light energy into biomass and generating molecular oxygen. As a result of the

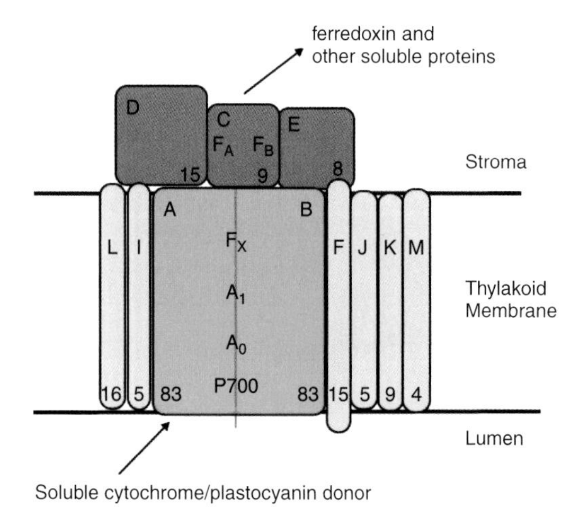

Figure 5.24 The organization of PSI showing subunit composition and co-factors involved in electron transfer

success of membrane protein structure determination we are beginning to understand the molecular mechanisms underlying catalysis in complexes vital for continued life on this planet.

Membrane proteins based on transmembrane β barrels

So far membrane proteins have been characterized by the exclusive use of the α helix to traverse membranes. However, some proteins form stable membrane protein structure based on β strands. Porins are channel-forming proteins found in the outer membrane of Gram-negative bacteria such as *Escherichia coli* where they facilitate the entry of small polar molecules into the periplasmic space. Porins are abundant proteins within this membrane with estimates of ∼100 000

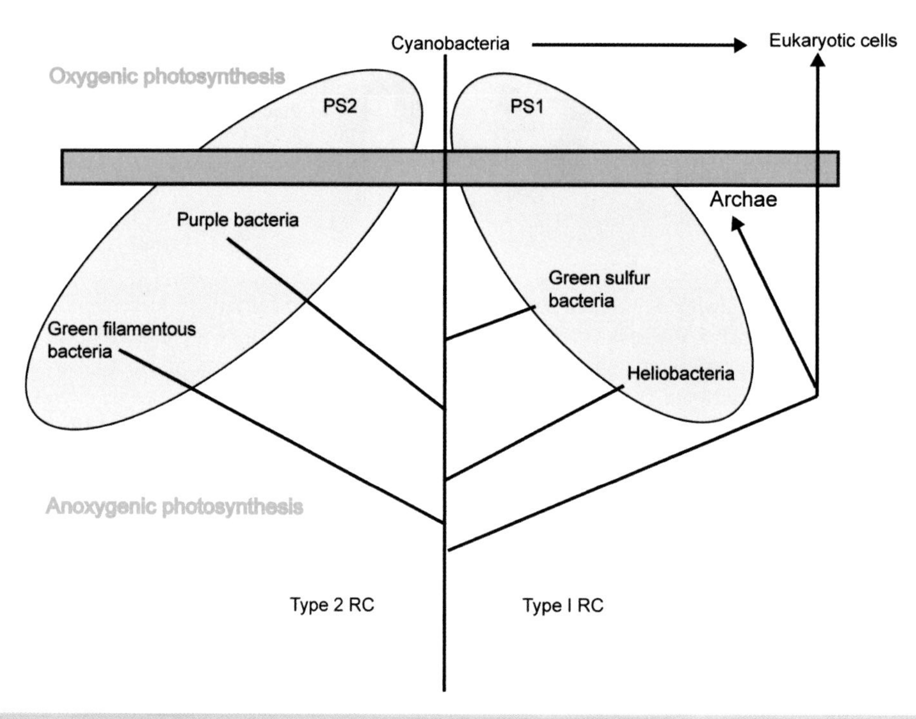

Figure 5.25 A summary of the distribution of photosynthetic reaction centre complexes in pro- and eukaryotes. Oxygenic photosynthesis is based around two reaction centre complexes, PSI and PSII, while anoxygenic photosynthesis uses only one reaction centre. This reaction centre can be either type II or type I. Type-II reaction centres are found in purple non-sulfur bacteria, green filamentous bacteria and oxygenic PSII and type-I reaction centres are typical of green sulfur bacteria, heliobacteria and oxygenic PSI

copies of this protein per cell. Porins have a size limit of less than 600 for transport by passive diffusion through pores. Although most porins form non-specific channels a few display substrate-specificity such as maltoporin.

The OmpF porin from the outer membrane *of E. coli* contains 16 β strands arranged in a barrel-like fold (Figure 5.26). The significance of this topology is that it overcomes the inherent instability of individual strands by allowing inter-strand hydrogen bonding. OmpF appears to be a typical *E. coli* porin and will serve to demonstrate structural properties related to biological function. Although shown as a monomer the functional unit for porins is more frequently a trimeric complex of three polypeptide barrels packed closely together within the bilayer.

The individual OmpF monomer contains 362 residues within a subunit of 39.3 kDa where the length of the monomer (5.5 nm) is sufficient to span a lipid bilayer and to extrude on each side into the aqueous phase. In the barrel hydrophobic groups, particularly aromatic side chains, occupy regions directly in contact with the lipid bilayer (Figure 5.27). As a consequence porins have a 'banded' appearance when the distribution of hydrophobic residues is plotted over the surface of the molecule. The barrel fold creates a cavity on the inside of this protein that is lined with polar groups. These groups located away from the hydrophobic bilayer form hydrogen bonds with water found in this cavity.

The cavity itself is not of uniform diameter – the pore range is 1.1 nm at its widest but it is constricted near the centre with the diameter narrowing to ~0.7 nm. In this cavity are located side chains that influence the overall pore properties. For example, OmpF shows a weak cation selectivity whilst highly homologous porins such as PhoE phosphoporin (63 percent identity) shows a specificity towards anions. The molecular basis for this charge discrimination has become clearer since the structures of both porins have been determined. A region inside the barrel contributes to a constriction of the channel and in most porins this region is a short α helix. In OmpF the short helix extends from residues 104 to 112 and is followed by a loop region that, instead of forming the hairpin

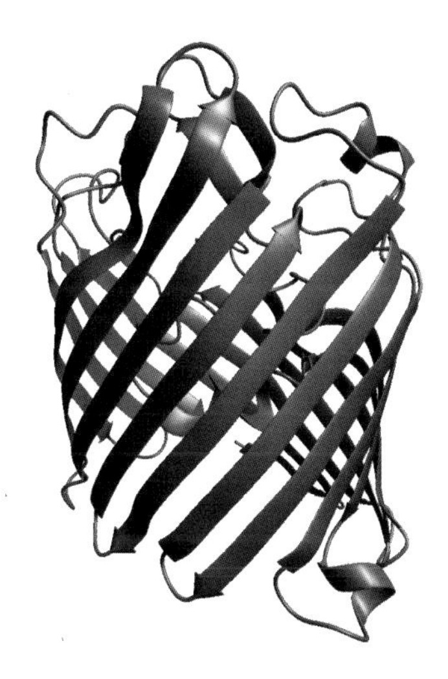

Figure 5.26 The arrangement of strands in OmpF from the outer membrane of *E. coli*. OmpF stands for outer membrane protein f. The strands make an angle of between 35 and 50° with the barrel axis with the barrel often described as having a rough end, marked by the presence of extended loop structures and a smooth end where shorter loop regions define the channel

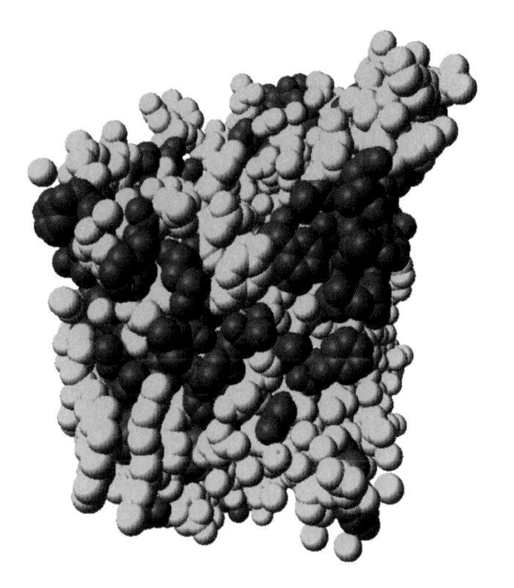

Figure 5.27 Distribution of hydrophobic residues around the barrel of OmpF porin. The space filling model of porin shows the asymmetric distribution of hydrophobic side chains (Gly, Ala, Val, Ile, Leu, Phe, Met, Trp and Pro) in a relatively narrow band. This band delineates the contact region with the lipid bilayer

Figure 5.28 Two views of the OmpF monomer showing the L3 loop region (yellow) located approximately half way through the barrel and forming a constriction site

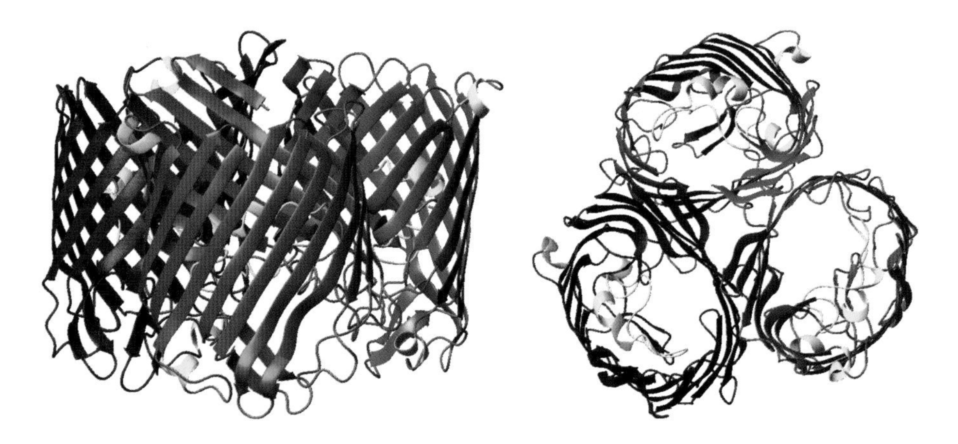

Figure 5.29 The biological unit of OmpF (and other porins) is a trimer of barrels with interaction occurring via the hydrophobic surfaces of each monomer. The three monomers are shown in a side view (viewed from a position within the membrane) and a top view

structure between two strands, projects into the channel (Figure 5.28). In the vicinity of this constriction but on the opposite side of the barrel are charge side chains that influence ion selectivity. The organization and properties of OmpF is shared by the PhoE porin but also by evolutionary more distant porins from *Rhodobacter* and *Pseudomonas* species.

Porins are stable proteins because the 16-stranded antiparallel β-barrel is closed by a salt bridge formed between the N- and C-terminal residues (Ala and Phe, respectively) whilst the biological unit packs together as a trimeric structure enhanced in stability by hydrophobic interactions between the sides of the barrel (Figure 5.29).

Another important example of a membrane protein based on a β barrel is α-haemolysin, a channel-forming toxin, exported by the bacterium *Staphylococcus aureus*. Unusually this protein is initially found as a water soluble monomer (M_r~33 kDa) but assembles in a multistep process to form a functional heptamer on

Figure 5.30 The structure of α-haemolysin showing the complex arrangement of β strands. The heptamer is mushroom shaped as seen from the above figure with the transmembrane domain residing in the 'stalk' portion of the mushroom and each protomer contributing two strands of the 14-stranded β barrel. This arrangement is not obvious from the above structure but becomes clearer when each monomer is shown in a different colour and from an altered perspective (Figure 5.31)

the membranes of red blood cells. Once assembled and fastened to the red blood cell membrane α-haemolysin causes haemolysis by creating a pore or channel in the membrane of diameter ~2 nm. If left untreated infection will result in severe illness.

The monomeric subunit associates with membranes creating a homoheptameric structure of ~230 kDa that forms an active transmembrane pore that allows the bilayer to become permeable to ions, water and small solutes. The three-dimensional structure of α-haemolysin (Figures 5.30 and 5.31) shows a structure containing a solvent-filled channel of length 10 nm that runs along the seven-fold symmetry axis of the protein. As with porins the diameter of this channel varies between 1.4 and 4.6 nm.

Porins and the α-haemolysins make up a large class of membrane proteins based on β barrels. Previously thought to be just an obscure avenue of folding followed by a few membrane proteins it is now clear that this class is increasing in their number and distribution. Porins are found in a range of bacterial membranes as well as the mitochondrial and chloroplast membranes. The OmpF structure is based on 16 β strands and porins from other membranes

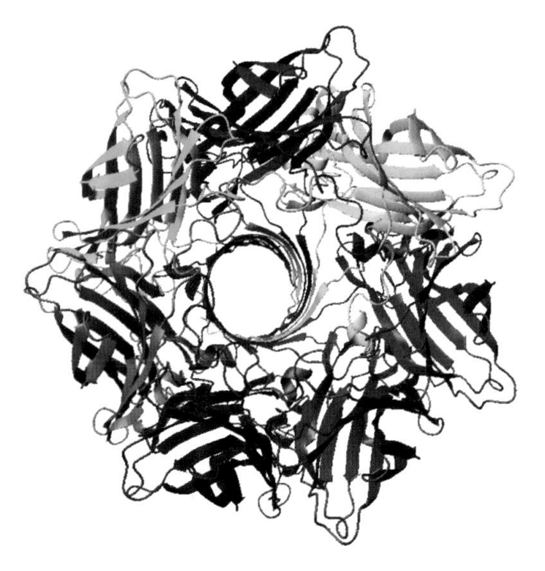

Figure 5.31 The structure of α-haemolysin showing the contribution of two strands from each protomer to the formation of a 14-stranded β barrel representing the major transmembrane channel. The channel or pore is clearly visible from a top view looking down on the α-haemolysin (PDB: 7AHL)

contain between 8 and 22 strands (in all cases the number of strands is even) sharing many structural similarities.

Respiratory complexes

Structure determination of bacterial reaction centres provided enormous impetus to solving new membrane protein structures. In the next decade success was repeated for membrane complexes involved in respiration.

Respiration is the major process by which aerobic organisms derive energy and involves the transfer of reducing equivalents through series of electron carriers resulting in the reduction of dioxygen to water. In eukaryotes this process is confined to the mitochondrion, an organelle with its own genome, found in large numbers in cells of metabolically active tissues such as flight muscle or cardiac muscle. The mitochondrion is approximately 1–2 μm in length although it may sometimes exceed 10 μm, with a diameter of ~0.5–1 μm and contains a double membrane. The outer membrane, a semipermeable membrane, contains porins analogous to those described in previous sections. However, the inner mitochondrial membrane is involved in energy transduction with protein complexes transferring electrons in steps coupled to the generation of a proton gradient. The enzyme ATP synthetase uses the proton gradient to make ATP from ADP in oxidative phosphorylation.

Purification of mitochondria coupled with fractionation of the inner membrane and the use of specific inhibitors of respiration established an order of electron transfer proteins from the oxidation of reduced substrates to the formation of water. The inner mitochondrial membranes called cristae are highly invaginated with characteristic appearances in electron micrographs and are the location for almost all energy transduction. This process is dominated by four macromolecular complexes that catalyse the oxidation of substrates such as reduced nicotinamide adenine dinucleotide (NADH) or reduced flavin adenine dinucleotide ($FADH_2$) through the action of metalloproteins such as cytochromes and iron sulfur proteins (Figure 5.32).

Some complexes of mitochondrial respiratory chains are amenable to structural analysis and although structures of complex I (NADH-ubiquinone oxidoreductase; see Figure 5.33) and complex II (succinate dehydrogenase) are not available, advances in defining complex III (ubiquinol-cytochrome c oxidoreductase) and the final complex of the respiratory chain, complex IV or cytochrome oxidase, have assisted understanding of catalytic mechanisms.

Complex III, the ubiquinol-cytochrome c oxidoreductase

The cytochrome bc family developed early in evolution and universally participates in energy transduction. All but the most primitive members of this family of protein complexes contain a b type cytochrome, a c type cytochrome[1] and an iron sulfur protein (ISP). In addition there are often proteins lacking redox groups with non-catalytic roles controlling complex assembly. In all cases the complex oxidizes quinols and transfers electrons to soluble acceptors such as cytochrome c.

Two-dimensional crystals of low resolution (~20 Å) defined the complex as a dimer with 'core' proteins lacking co-factors projecting out into solution. Much later the structure of the extrinsic domains of the iron–sulfur protein and the bound cytochrome were derived by crystallography but the main body of the complex, retained within the confines of the lipid bilayer, remained structurally 'absent'. At the beginning of the 1990s several groups obtained diffraction quality crystals of the cytochrome bc_1 complex from bovine and chicken mitochondria. An extensive structure for the cytochrome bc_1 complex from bovine heart mitochondria was produced by S. Iwata and co-workers in 1998 at a resolution of 3 Å after several groups had produced structures for the majority of the complex. The structures showed similar organization although small structural differences were noticed and appeared to be of functional importance (see below).

Isolated cytochrome bc_1 complexes from eukaryotic organisms contain 10 (occasionally 11) subunits

[1] In plants and algae the c type cytochrome is sometimes called cytochrome f but in all other respects it has analogous structure and function. The f refers to the latin for leaf, *frons*

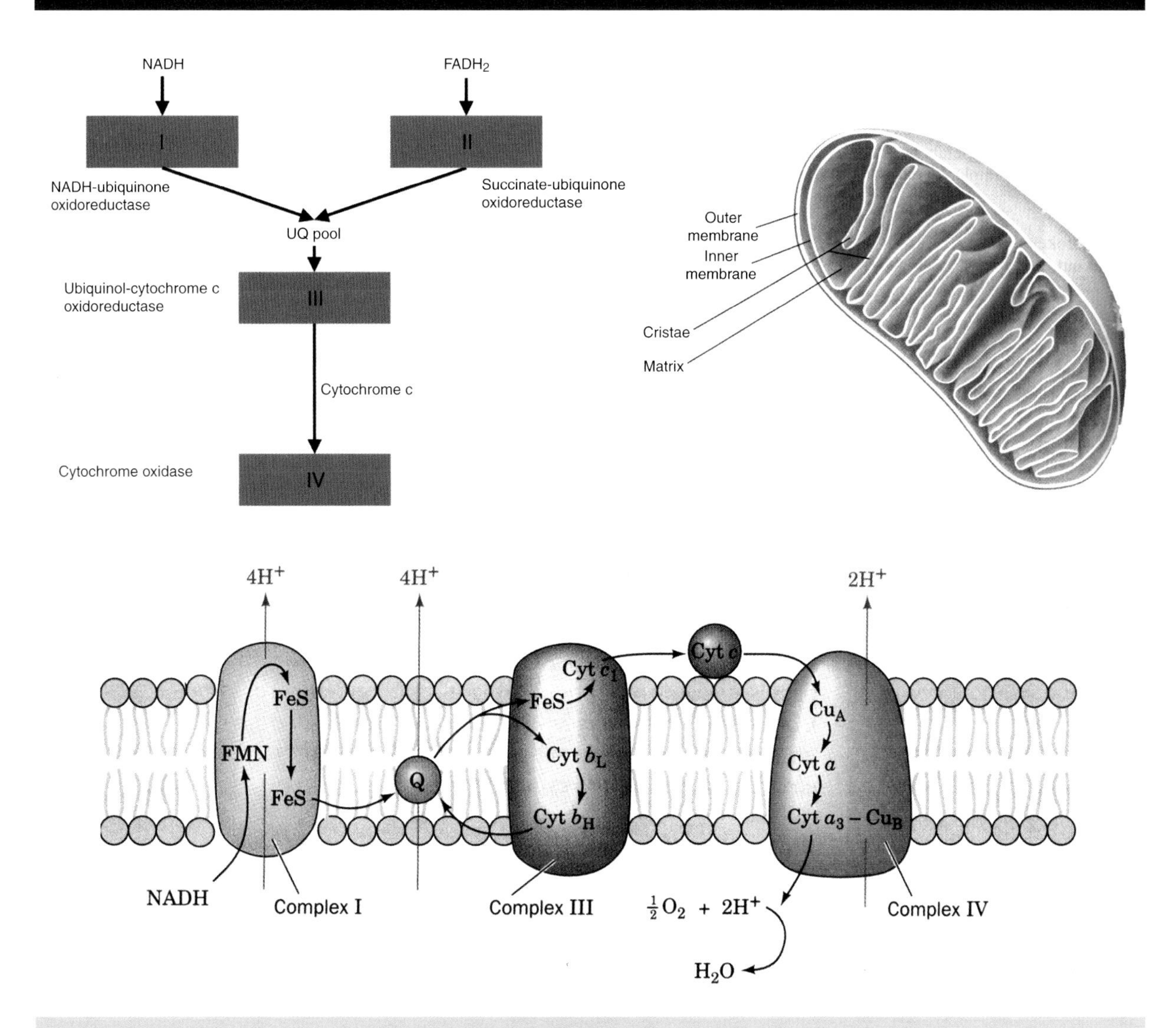

Figure 5.32 The organization of the respiratory chain complexes within the inner mitochondrial membrane. The traditional view of mitochondria (centre) was derived from electron microscopy of fixed tissues

including a b type cytochrome with two heme centres, an iron–sulfur protein (called the Reiske protein after its discoverer) and a mono heme c type cytochrome. DNA sequences for all subunits are known and considerable homology exists between the three metazoan branches represented by bovine, yeast and potato. Detailed UV-visible absorbance studies combined with potentiometric titrations established that the two hemes of cytochrome b had slightly different absorbance maxima in the reduced state (α bands at 566 and 562 nm) and different equilibrium redox potentials (E_m~120 and −20 mV respectively). This led to each heme of cytochrome b being identified as b_H or b_L for high and low potential.

Further understanding of electron transfer in the bc_1 complex came from detailed spectroscopic investigations coupled with the use of inhibitors that block electron (and proton) transfer at specific sites by binding to different protein subunits. One commonly used inhibitor is an anti-fungal compound called

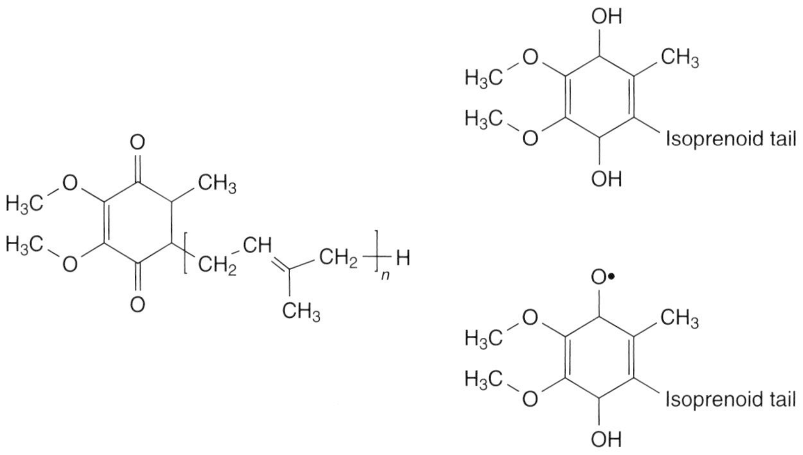

Figure 5.33 The structure of ubiquinone, sometimes called coenzyme Q_{10}, contains an isoprenoid tail linked to a substituted benzoquinone ring. The isoprenoid tail contains a series of isoprenyl units (CH_2-CH=C(CH_3)-CH_2). Various numbers of these units can occur but 10 is the most common leading to the designation UQ_{10} or Q_{10}. The molecules form semiqinone and quinol states with the pattern of reduction complicated by protonation leading to species such as anionic semiquinones

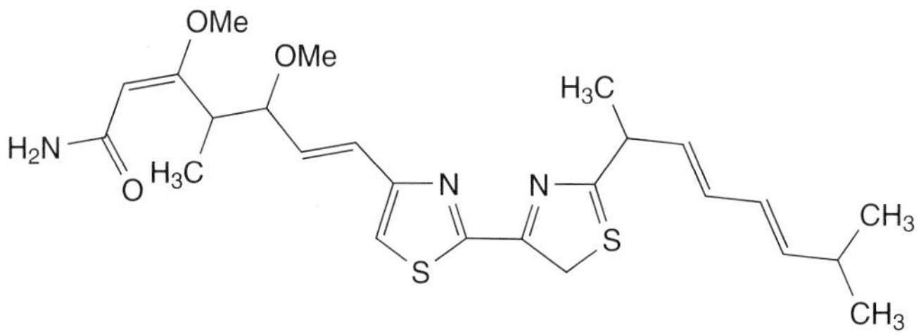

Figure 5.34 The mitochondrial cytochrome bc_1 complex inhibitor myxathiazol

myxathiazol (Figure 5.34) that binds to the Reiske iron–sulfur protein blocking quinol oxidation. It was observed to block reduction of the low potential b heme. Other inhibitors such as stigmatellin, 5-*n*-undecyl-6-hydroxy-4,7-dioxobenzothiazol (UHDBT) and β-methoxyacrylate (MOA)–stilbene showed similar inhibitory profiles. Another group of inhibitors bound differently to the bc_1 complex allowing reduction of the iron-sulfur centre and cytochrome b but prevented their oxidation. These inhibitors included antimycin A and led to the idea of two distinct inhibitor-binding sites within the cytochrome bc_1 complex.

Inhibition of electron transfer by antimycin A yielded unexpected patterns of reduction and oxidation; the normal reduction of cytochrome c_1 by ubiquinol was prevented but was accompanied by an increased reduction of cytochrome b. This observation indicated a branch point in the electron transfer chain where electrons were shuttled to cytochromes c_1 *and* b. The electron transfer pathway (Figure 5.35) from ubiquinol to cytochrome c was rationalized by the Q cycle. Whilst cytochrome c_1 is oxidized by the soluble electron carrier (cytochrome c) the only possible acceptor of electrons from cytochrome b (b_H) is ubiquinone.

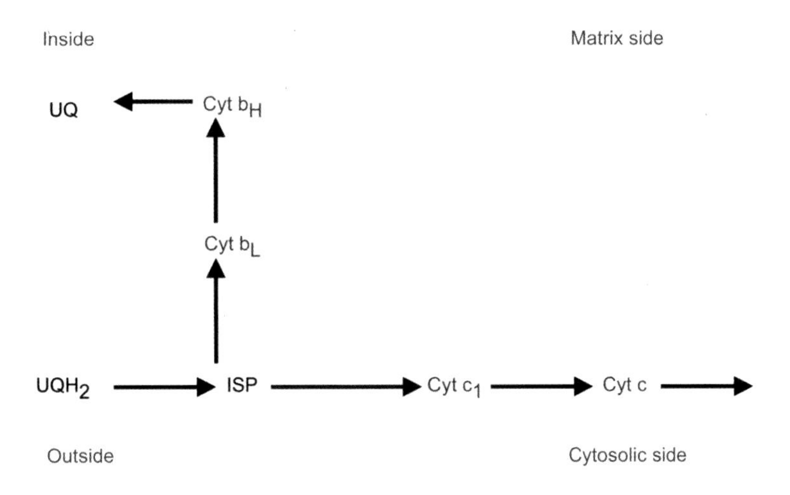

Figure 5.35 Outline of electron transfer reactions in cytochrome bc_1 complex. UQ/UQH_2 is ubiquinone/ubiquinol; ISP, Reiske iron sulfur protein; Cyt c_1, cytochrome c_1; b_L and b_H low and high potential hemes of cytochrome b

The Q cycle accounts for the fact that ubiquinol is a $2e/2H^+$ carrier whilst the bc_1 complex transfers electrons as single entities. Quinones have roles as oxidants *and* reductants. Quinol is oxidized at a Q_o site within the complex and one electron is transferred to the high-potential portion of the bc_1 complex whilst the other electron is diverted to the low-potential part (cytochrome b) of the pathway. The high-potential chain, consisting of the Reiske iron–sulfur protein, cytochrome c_1 and the natural acceptor of the inter-membrane space (cytochrome c) transfers the first electron from quinol through to the final complex of the respiratory chain, cytochrome oxidase. The low-potential portion consists of two cytochrome b hemes and forms a separate pathway for transferring electrons across the membrane from the Q_o site to a Q_i site. The rate-limiting step is quinol (QH_2) oxidation at the Q_o site and it is generally supposed that an intermediate semiquinone is formed at this site. In order to provide two electrons at the Q_i site for reduction of quinone, the Q_o site oxidizes two molecules of quinol in successive reactions. The first electron at the Q_i site generates a semiquinone that is further reduced to quinol by the second electron from heme b_H. The overall reaction generates four protons in the inter-membrane space, sometimes called the P-side, and the formation of quinol uses up two protons from the mitochondrial matrix. The whole reaction is summarized by the scheme

$$QH_2 + 2\text{cyt c}(Fe^{3+}) + 2H^+ \longrightarrow Q + 2\text{cyt c}(Fe^{2+}) + 4H^+$$

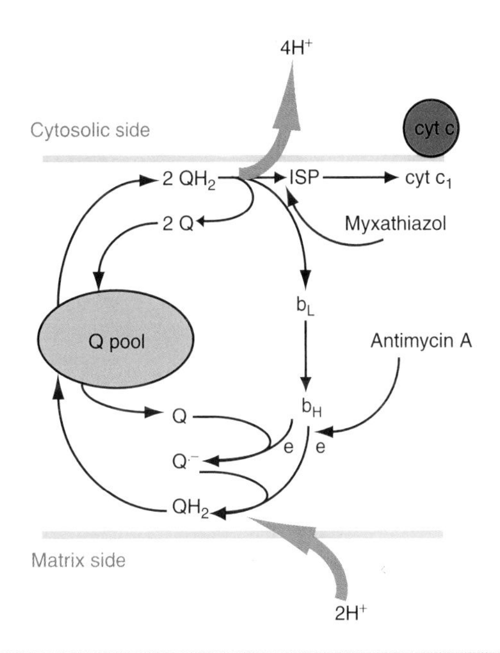

Figure 5.36 The Q cycle and the generation of a vectorial proton gradient

but does not reflect the subtlety of this reaction pathway that allows the bc_1 complex to generate a proton gradient via electron flow through the complex.

The Q cycle (Figure 5.36) became an established concept long before the structure of the cytochrome bc_1 complex was determined. However, this scheme places stringent constraints on the topology of redox co-factors. With the determination of the structure of the cytochrome bc_1 complex came the opportunity to critically assess the Q cycle and to explore the potential mechanism of charge transfer.

One of the first impressions gained from the structure of the bc_1 complex is the large number of extrinsic domains. The Reiske protein is anchored to the membrane by two N terminal helices. The ligands are the side chains of cysteines 139 and 158 and, unlike most iron–sulfur proteins, two histidines (His 141 and 161). The nitrogen containing ligands lead to a higher redox potential for the iron ($\sim$100 mV) than is normal in this class of proteins. The iron sulfur centre located in a layer of β sheets is solvent accessible and probably accounts for interaction with quinones and inhibitors such as stigmatellin.

Cytochrome c_1 is also anchored to the bilayer by a single transmembrane helix although in this case it is located at the C terminus. The transmembrane helix is short (20 residues) and the majority of the protein projects $\sim$4 nm into the inter-membrane space. This domain contains a single heme centre and presents a binding surface for cytochrome c containing large numbers of negatively charged residues.

On the matrix side of the complex are two unfortunately named 'core' proteins. These proteins surprisingly project outwards into the aqueous phase and are certainly not located at the core of the complex. Sequence analysis has ascertained that these core subunits show significant homology to a family of heterodimeric Zn proteases and current views of their function envisages a role for these domains in processing matrix proteins by specific peptidase activity. It remains unclear why these proteins remain affiliated to the cytochrome bc_1 complex (Figure 5.37).

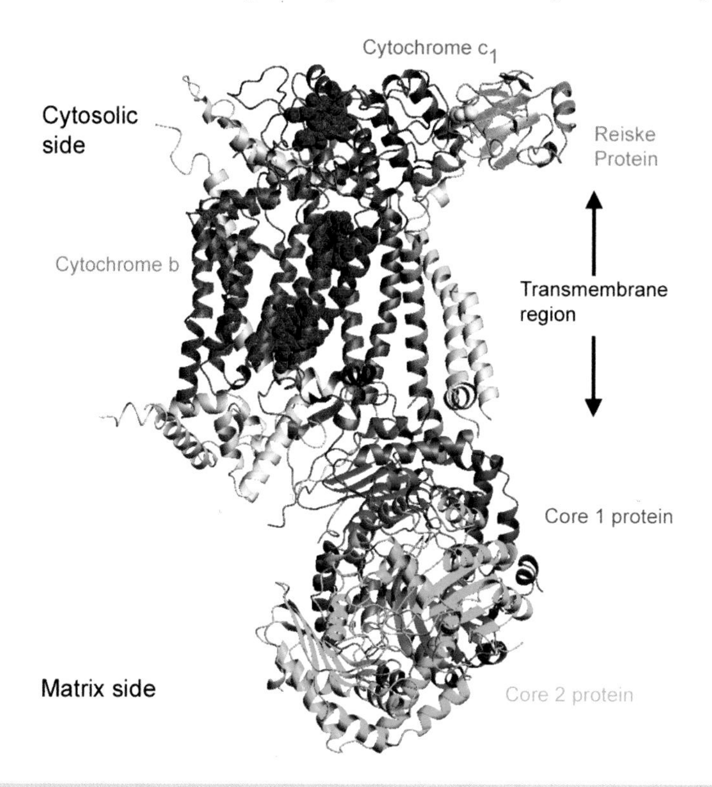

Figure 5.37 The structure of monomeric bovine cytochrome bc_1 complex (PDB:1BE3)

The major transmembrane protein is cytochrome b and the protein has two hemes arranged to facilitate transmembrane electron transfer exactly as predicted by Q cycle schemes. Each heme is ligated by two histidine residues and these side chains arise from just two (out of a total of eight) transmembrane helices found in cytochrome b. The two hemes are separated by 20.7 Å (Fe–Fe centre distances) with the closest approach reflected by an edge–edge distance of 7.9 Å – a distance compatible with rapid inter-heme electron transfer. Cytochrome b is exclusively α helical and lacks secondary structure based on β strands. The helices of cytochrome b delineate the membrane-spanning region and are tilted with respect to the plane of the membrane. The remaining subunits of the bc_1 complex (see Table 5.7) are small with masses below 10 kDa with undefined roles.

Although all of the structures produced from different sources are in broad agreement it was noticed that significant differences existed in the position of the Reiske iron–sulfur protein with respect to the other redox co-factors. This different position reflected crystallization conditions such as the presence of inhibitors and led to a position for the iron–sulfur protein that varied from a location close to the surface of cytochrome b to a site close to cytochrome c_1. In the presence of the inhibitor stigmatellin the iron–sulfur protein shifted towards a position proximal to cytochrome b and suggested intrinsic mobility. The concept of motion suggested that the Reiske protein might act as a 'gate' for electron transfer modulating the flow of electrons through the complex (Figure 5.38).

Table 5.7 The conserved subunits of bovine mitochondrial cytochrome bc_1

Subunit	Identity	No of residues	Mr
1	Core 1	446	53604
2	Core 2	439	46524
3	Cyt b	379	42734
4	Cyt c_1	241	27287
5	Reiske	196	21609
6	Su6	111	13477
7	QP-Cb	82	9720
8	Hinge	78	9125
9	Su10	62	7197
10	Su 11	56	6520

The colour shown in the first column identifies the relevant subunit in the structure shown in Figure 5.37.

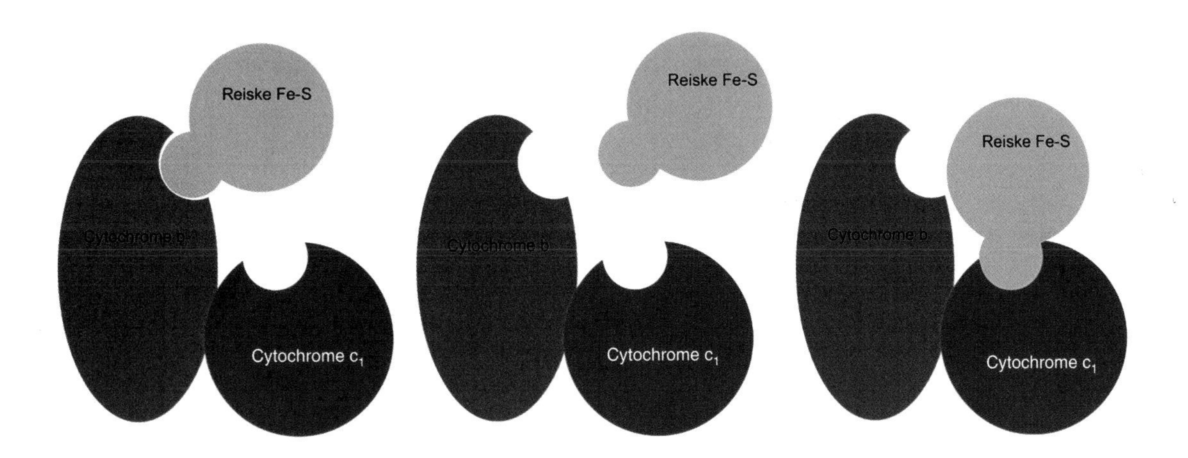

Figure 5.38 The Reiske iron–sulfur protein acts as a redox gate. The schematic diagrams show the Reiske Fe-S centre associated with the cytochrome b, in an intermediate state and associated with cytochrome c_1. Motion at the Reiske Fe-S centre is believed to control electron and proton transfer at the cytochrome bc_1 complex

Complex IV or cytochrome oxidase

Cytochrome c oxidase is the final complex of the respiratory chain catalyzing dioxygen reduction to water. For many years it has occupied biochemists thoughts as they attempted to unravel the secrets of its mechanism. Isolation of cytochrome oxidase has two heme groups (designated a and a_3) together with 2 Cu centres (called Cu_A and Cu_B). Techniques such as electron spin resonance and magnetic circular dichroism alongside conventional spectrophotometric methods showed that heme *a* is a low spin iron centre, whilst heme a_3 was a high spin state compatible with a five-coordinate centre and a sixth ligand provided by molecular oxygen. The catalytic function of cytochrome oxidase is summarized as

$$4\ \text{Cyt c}(Fe^{2+}) + O_2 + 8H^+ \longrightarrow 4\ \text{Cyt c}(Fe^{3+}) + 2H_2O + 4H^+$$

where oxidation of ferrous cytochrome c leads to four-electron reduction of dioxygen and the generation of a proton electrochemical gradient across the inner membrane. Armed with knowledge of the overall reaction spectroscopic studies established an order for electron transfer of

$$\text{Cyt c} \longrightarrow Cu_A \longrightarrow \text{heme a} \longrightarrow \text{heme } a_3\text{–}Cu_B$$

Cu_B and heme a_3 were located close together modulating their respective magnetic properties and this cluster participated directly in the reduction of oxygen with inhibitors such as cyanide binding tightly to heme a_3–Cu_B and abolishing respiration. Transient intermediates are detected in the reduction of oxygen and included the formation of dioxygen adjunct (Fe^{2+}–O_2), the ferryl oxide (Fe^{4+}–O) and the hydroxide (Fe^{3+}–OH). Other techniques established histidine side chains as the ligands to hemes a and a_3–Cu_B on subunit I and Cu_A as a binuclear centre.

The catalytic cycle occurs as discrete steps involving electron transfer from cytochrome c via Cu_A, heme a and finally to the heme a_3–Cu_B site binuclear complex (Figure 5.39). The initial starting point for catalysis involves a fully oxidized complex with ferric (Fe^{3+}) and cupric centres (Cu^{2+}).

Complex IV, with oxidized heme a_3–Cu_B centres, receives two electrons in discrete (1e) steps from the preceding donors. Both centres are reduced and bind two protons from the mitochondrial matrix. In this state oxygen binds to the binuclear centre and is believed to bridge the Fe and Cu centres. Reduction of dioxygen forms a stable peroxy intermediate that remains bound (Fe^{3+}– O^-=O^-–Cu^{2+}) and further single electron (from cytochrome c) and proton transfers lead to the formation of hydroxyl and then ferryl (Fe^{4+}) states. Rearrangement at the catalytic centre during the last two stages completes the cycle forming two molecules of water for every oxygen molecule reduced.

In the absence of a structure for cytochrome oxidase the scheme proposed would need to consider the route and mechanism of proton uptake for the reduction of oxygen from the matrix side of the enzyme to the catalytic site, the mechanism and route of substrate (O_2) binding to the active site and finally the route for product removal from the catalytic site. A further difficulty in understanding the catalytic mechanism of cytochrome oxidase lies with the observation that in addition to the protons used directly in the reduction of water it appears from measurements of H^+/e ratios that up to four further protons are pumped across the membrane to the inter-membrane space. It seems likely that conformational changes facilitating electron and proton transfer lie at the heart of oxidase functional activity.

Cytochrome oxidase from denitrifying bacteria such as *Paracoccus denitrificans* has a simpler subunit composition with only three subunits: all of the redox components are found on the two heaviest polypeptides (subunits I and II). In eukaryotic cytochrome oxidase at least 13 subunits are present with subunits I–III, comparable to those observed in prokaryotic oxidases, encoded by the mitochondrial genome. The crystallization of the complete 13 subunit complex of bovine cytochrome oxidase by Yoshikawa and colleagues in 1995 resolved or suggested answers to several outstanding questions concerning the possible electron transfer route through the complex, a mechanism and pathway of proton pumping and the organization of the different subunits.

The mitochondrially coded subunits are the largest and most hydrophobic subunits and contain four redox centres. Surprisingly the oxidase binds two additional

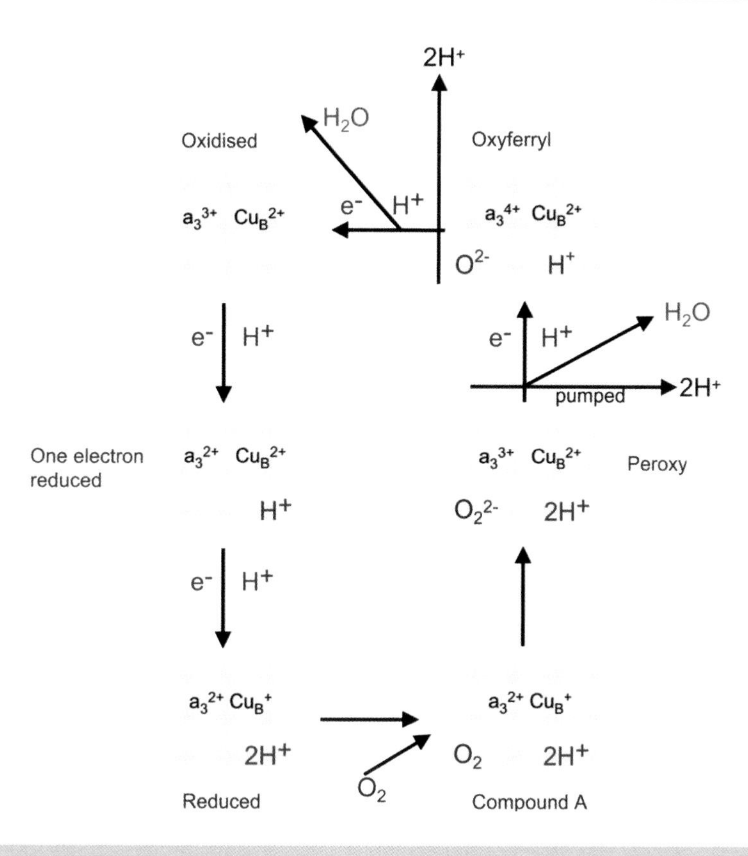

Figure 5.39 A possible scheme for catalytic cycles of cytochrome oxidase. The redox state of the binuclear heme a_3–Cu_B site together with reduction of oxygen, proton binding and protons pumped across membrane is highlighted

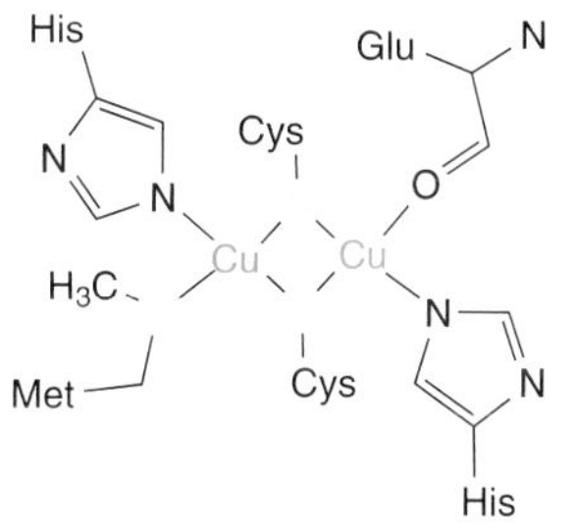

Figure 5.40 The ligands of the Cu_A centre of subunit II of bovine mitochondrial cytochrome oxidase

divalent metal ions namely magnesium and zinc. The zinc is ligated to a nuclear coded subunit (called subunit Vb) whilst the magnesium is held by subunit II. The Cu_A centre as expected was a binuclear cluster of Cu atoms and was contained within subunit II on a cytosolic projecting domain. The Cu_A centre lies ~0.8 nm above the surface of the membrane and is separated from the Fe atoms of hemes a and a_3 by distances of ~1.9 and 2.2 nm respectively. The two Cu atoms are joined in a bridge formed by two sulfur atoms from the side chains of Cys residues in a geometry reminiscent of that found in [2Fe–2S] clusters of iron–sulfur proteins (Figure 5.40). Besides the two sulfur ligands provided by cysteine side chains at positions 196 and 200 of subunit II additional ligands originated from side chains of His161 and His204, Met207 together with the backbone oxygen from Glu198. Glu 198 is also involved in binding the divalent Mg^{2+} ion. In this case the ligand originates from the side chain oxygen atom as opposed to backbone groups with the cation located on a pathway between the Cu_A and heme a_3 centres.

Heme a is coordinated by the side chains of His61 and His378 of subunit I. The iron is a low spin ferric centre lying in the plane of the heme with two imidazole ligands – a geometry entirely consistent with the spin properties. The heme plane is positioned perpendicularly to the membrane surface with the hydroxyethylfarnesyl sidechain of the heme pointing towards the membrane surface of the matrix side in a fully extended conformation. The remaining two redox centres heme a_3 and Cu_B make up the oxygen reduction site and lie very close to heme a. Hemes a and a_3 are within 0.4 nm of each other and this short distance is believed to facilitate rapid electron transfer.

Of some considerable surprise to most observers was the location in the oxidase structure of the fifth ligand to heme a_3. The side chain of histidine 376 provides the fifth ligand to this heme group and lies extremely close to one of the ligands to heme a (Figure 5.41). In the oxidase structure the Cu_B site is close to heme a_3 ($\sim$0.47 nm) and has three imidazole ligands (His240, 290 and 291) together with a strong possibility that a fourth unidentified ligand to the metal ion exists. It is generally believed that trigonal coordination is unfavourable and a more likely scenario will involve a tetrahedral geometry around the copper.

A hydrogen-bond network exists between the Cu_A and heme a and includes His204, one of the ligands to Cu_A, a peptide bond between Arg438 and Arg439 and the propionate group of heme a. Although direct electron transfer between these centres has not been observed and they are widely separated the arrangement of bonds appears conducive for facile electron transfer. In addition, a network connects Cu_A and heme a_3 via the magnesium centre and may represent an effective electron transfer path directly between these centres. Direct electron transfer between Cu_A and heme a_3 has never been detected with kinetic measurements, suggesting that electron transfer from Cu_A to heme a is at least two orders of magnitude greater than the rate from Cu_A to heme a_3. From the arrangement of the intervening residues in subunit I the structural reasons underlying this quicker electron transfer from $Cu_A \longrightarrow$ heme a than $Cu_A \longrightarrow$ heme a_3 are not obvious.

Before crystallization the composition and arrangement of subunits within eukaryotic oxidases was unclear. This has changed dramatically with the crystallization of oxidases from bovine heart mitochondria and *P. denitrificans* so that there is now precise information on the arrangement of subunits (Table 5.8). In the crystal structure of bovine cytochrome oxidase the enzyme is a homodimer with each monomer containing 13 polypeptides and six co-factors. The assembly of the 13 subunits into a complex confirmed that these polypeptides are intrinsic constituents of the active enzyme and not co-purified contaminants. The crystal structure confirmed the presence of Zn and Mg atoms as integral components albeit of unknown function.

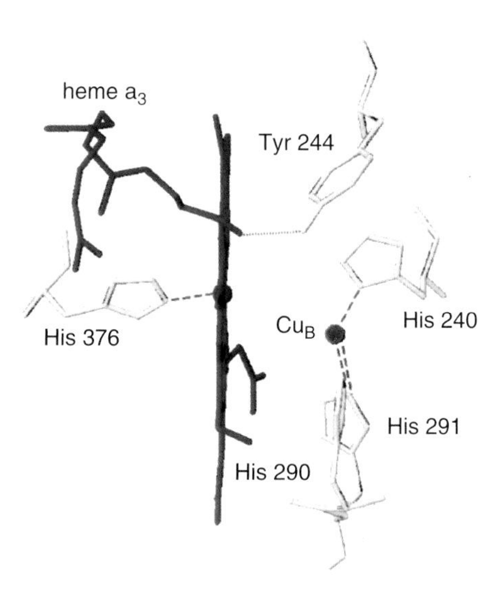

Figure 5.41 The heme a_3–Cu_B redox centre of cytochrome oxidase. The dotted line denotes a hydrogen bond and broken lines denote coordination bonds. Heme a_3 is shown in blue, Cu_B in pink and amino acid residues in green. (Reproduced with permission from Yoshikawa, S. *Curr. Opin. Struct. Biol.* 1997, **7**, 574–579)

In the mammalian enzyme the 13 subunits of cytochrome oxidase contain 10 transmembrane proteins that contribute a total of 28 membrane-spanning helices. The three subunits encoded by mitochondrial genes form a core to the monomeric complex with the 10 nuclear-coded subunits surrounding these subunits and seven contributing a single transmembrane helix. The remaining three nuclear coded subunits (subunits Va, Vb, VIb) do not contain transmembrane regions.

Table 5.8 The polypeptide composition of eukaryotic and prokaryotic oxidases

Subunit	Prokaryotic	Eukaryotic
I	554, 620112	514, 57032
II	279, 29667	227, 26021
III	273, 30655	261, 29919
IV	*49, 5370	147, 19577
Va	Absent	109, 12436
Vb		98, 10670
VIa		97, 10800
VIb		85, 10025
VIc		73, 8479
VIIa		80, 9062
VIIb		88, 10026
VIIc		63, 7331
VIII		70, 7639

**Paracoccus denitrificans* has a fourth subunit that is unique to prokaryotes with no obvious homology to that found in eukaryotes. Three of the subunits of the eukaryotic enzyme, VIa, VIIa, and VIII, have two isoforms. Of the two isoforms one is expressed in heart and skeletal muscle and the other is expressed in the remaining tissues. The masses and length of the peptides includes signal sequences for nuclear-coded proteins.

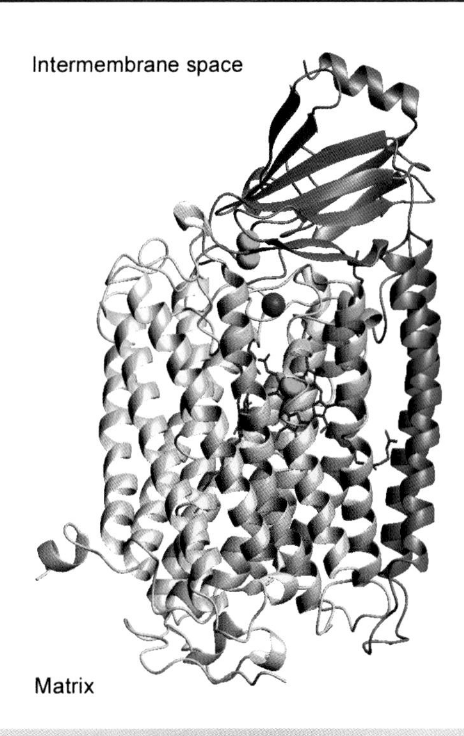

Figure 5.42 The monomeric 'core' of bovine cytochrome oxidase showing subunits I and II. Subunit I is shown in gold whilst subunit II is shown in blue. The limited number of transmembrane helices in subunit II is clear from the diagram in contrast to subunit I. The Cu atoms are shown in green, Mg in purple and the hemes in red (PDB: 1OCC)

In addition to the three metal centre (heme a, heme a_3 and Cu_B) subunit I has 12 transmembrane α helices and this was consistent with structural predictions based on the amino acid sequence. Subunit I is confined to the bilayer region with few loops extending out into the soluble phase. The subunit is the largest found in cytochrome oxidase containing just over 510 residues (M_r ~57 kDa) (Figure 5.42).

Subunits II and III both associate with transmembrane regions of subunit I without forming any direct contact with each other. From the distribution of charged residues on exposed surfaces of subunit II an 'acidic' patch faces the inter-membrane space and defines a binding site for cytochrome c. Unsurprisingly in view of its role as the point of entry of electrons into the oxidase subunit II has just two transmembrane helices interacting with subunit I, but has a much larger polar domain projecting out into the inter-membrane space. This domain consists of a 10-stranded β barrel with some structural similarities to class I copper proteins, although it is a binuclear Cu centre. The magnesium ion is located at the interface between subunits I and II.

Subunit III lacks redox co-factors but remains a significant part of the whole enzyme complex. It has seven α helices and is devoid of large extra-membrane projections. The seven helices are arranged to form a large V-shaped crevice that opens towards the cytosolic side where it interacts with subunit VIb (Figure 5.43).

Subunits Va and Vb (Figure 5.44) are found on the matrix side and both are predominantly located in the soluble phase. Subunit Va is extensively helical with subunit Vb possessing both helical- and strand-rich domains and binding a single Zn ion within a series

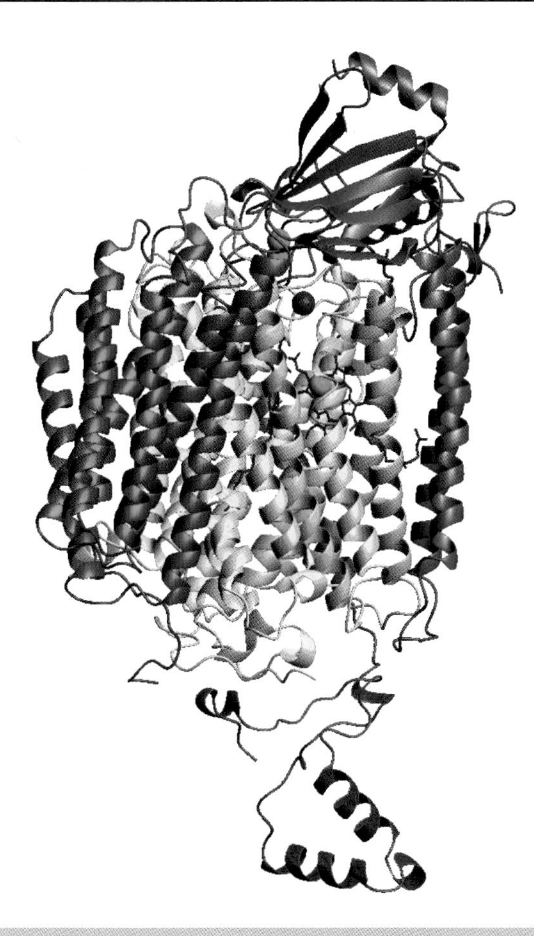

Figure 5.43 To the monomeric core of subunits I and II the remaining mt DNA coded subunit (III) has been added along with subunit IV. Subunit IV (red) has a single helix crossing the membrane with a helix rich domain projecting into the matrix phase. Subunit III (green) does not make contact with subunit II and consists almost exclusively of transmembrane helices

of strands linked by turns that form a small barrel-like structure.

Subunit VIa contributes a single transmembrane helix and is located primarily towards the cytosolic side of the inner mitochondrial membrane. Subunit VIb is located on the same side of the membrane but does not contribute transmembrane helices to the oxidase structure. It is affiliated primarily with subunit III. Subunit VIc contributes a single transmembrane helix and interacts predominantly with subunit II (Figure 5.45). Subunits VIIa, VIIb and VIIc, together with subunit VIII (Figure 5.46), are small subunits with less than 90 residues in each of the polypeptide chains. Their biological function remains unclear. Unlike complex III a far greater proportion of cytochrome oxidase is retained with the boundaries imposed by the lipid bilayer.

The structural model for cytochrome oxidase provides insight into the mechanism of proton pumping, a vital part of the overall catalytic cycle of cytochrome oxidase. Proton pumping results in the vectorial movement of protons across membranes and

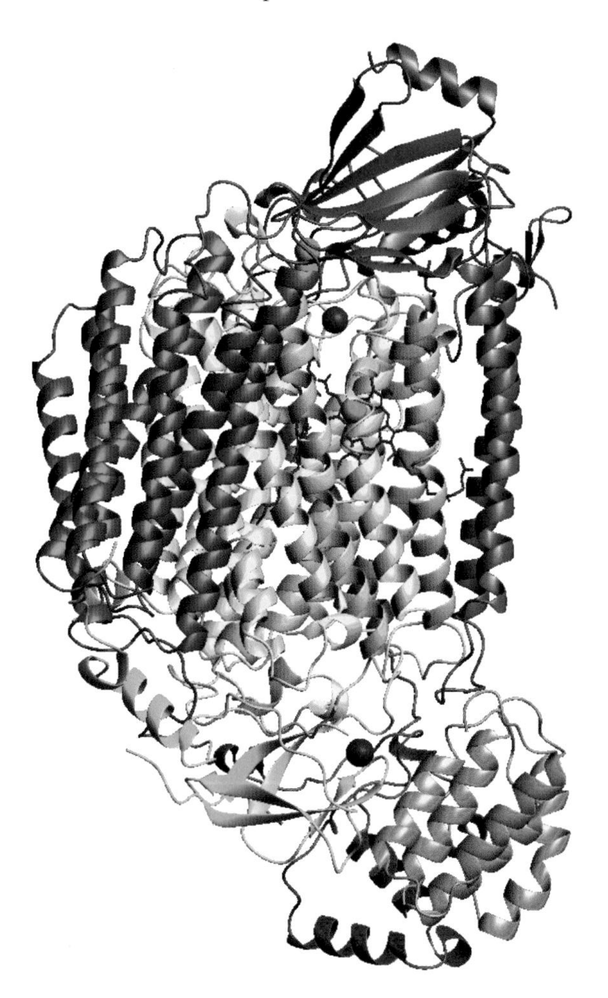

Figure 5.44 The structure of cytochrome oxidase monomer (PDB: 1OCC). Subunits Va (orange) and Vb (aquamarine) have been added. Subunit Vb binds Zn (shown in purple) found in the crystal structure

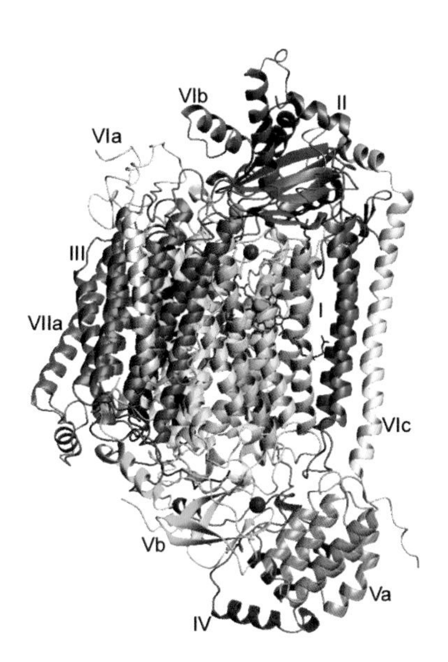

Figure 5.45 The three-dimensional structure of cytochrome oxidase monomer (PDB: 2OCC). The seven smaller subunits VIa (grey) VIb (purple), VIc (light blue), VIIa (pink), VIIb (beige), VIIc (brown) and VIII (black) have been added to the structure shown in Figure 5.44. Subunits VIIb, VIIc, and VIII are largely hidden by the larger subunits (I and III) in this view

in bacteriorhodopsin occurred via strategically placed Asp and Glu residues facilitating proton movement from one side of the membrane to another. This picture, together with that provided by ion channels such as porins, suggests a similar mechanism in cytochrome oxidase. The exact mechanism of proton pumping is hotly debated but it was of considerable interest to observe that the bovine crystal structure contained two possible proton pathways that were shared by simpler bacterial enzymes (Figure 5.47).

The oxygen binding site is located in the hydrophobic region of the protein approximately (35 Å) from the M side. Pathways were identified from hydrogen bonds between side chains, the presence of internal cavities containing water molecules, and structures that could form hydrogen bonds with small possible conformational changes to side chains. Two pathways called the D- and K-channels and named after residues Asp91 and Lys319 of bovine subunit I were identified. (In Paracoccus subunit I these residues are Asp124 and Lys354.) Despite significant advances the proton pumping mechanism of cytochrome oxidase remains largely unresolved. This arises mainly from difficulties associated with resolving proton pumping events during dioxygen

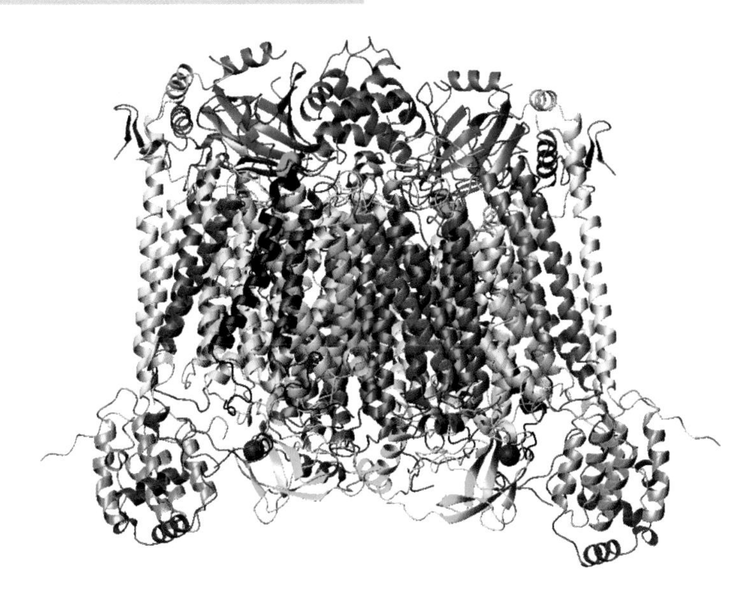

Figure 5.46 The homodimer of bovine cytochrome oxidase using previous colour schemes. This view shows subunits VIIb, VIIc, and VIII in more detail and is from the opposite side to that of the preceding figures

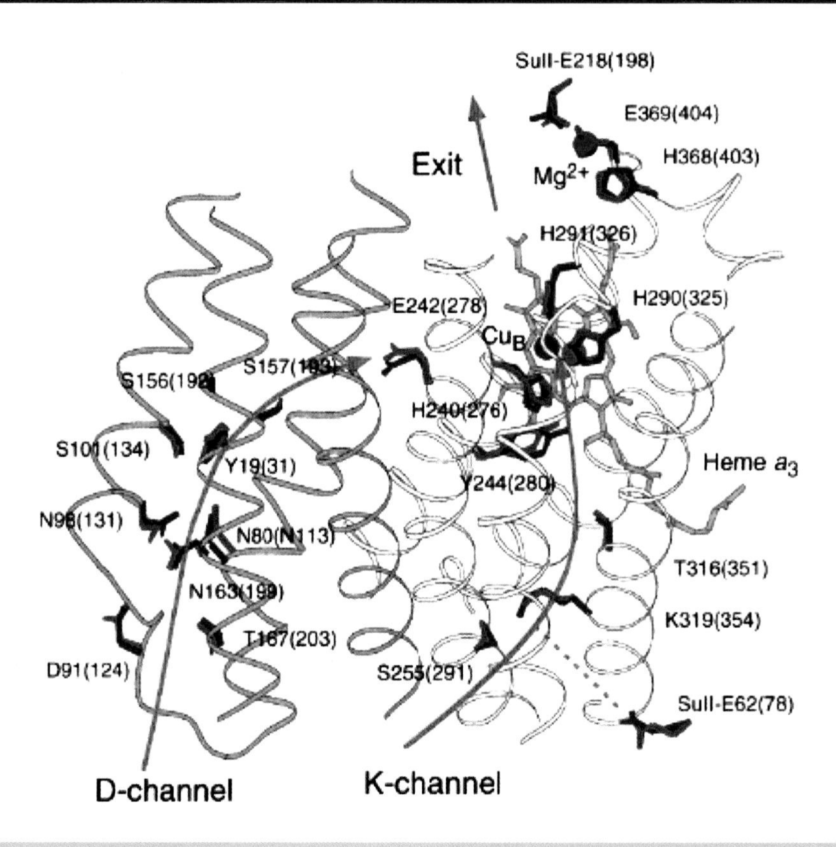

Figure 5.47 The proton pathways in bovine and *Paracoccus denitrificans* cytochrome oxidase. Residues implicated in proton translocation in bovine cytochrome c oxidase are shown in red and in blue for P. denitrificans oxidase. The structures of helices, heme a_3 and metal centres such as the Mg^{2+} binding site are based on PDB: 1ARI (reproduced with permission from Abramson, S *et al. Biochim. Biophys. Acta* 2001, **1544**, 1–9. Elsevier)

reduction and it is currently unclear whether pumping precedes oxygen reduction and what structural rearrangements in the reaction cycle occur. In bacteriorhodopsin intermediate structures were detected and associated with proton movements by combining genetic, spectroscopic and structural studies, and it is likely that such methods will resolve the mechanism of proton transfer in cytochrome oxidase in the near future.

From the study of different membrane proteins a common mode of proton (or cation) transfer involves specifically located aspartate and glutamate residues to shuttle protons across membranes. Both bacteriorhodopsin and cytochrome oxidase provide compelling evidence in their structures for this method of proton conduction across membranes.

The structure of ATP synthetase

It has been estimated that even the most inactive of humans metabolizes kg quantities of ATP in a normal day. This results from repeated phosphorylation reactions with on average each molecule of ADP/ATP being 'turned over' about 1000 times per day. The enzyme responsible for the synthesis of ATP is the F_1F_o ATPase, also called the ATP synthase or synthetase. Dissection of the enzyme complex and genetic studies show the synthase to contain eight different subunits in *E. coli* whilst a greater number of subunits are found in mammalian enzymes. Despite variations in subunit number similarity exists in their ratios and primary sequences with most enzymes having a combined mass between 550 and 650 kDa. With a central role in energy conservation and a presence in bacterial membranes, the thylakoid membranes of plant and algae chloroplasts and the mitochondrial inner membrane of plants and animals the enzyme is one of the most abundant complexes found in nature.

The link between electron transfer and proton transfer was the basis of the chemiosmotic theory proposed in its most accessible form by Peter Mitchell at the beginning of the 1960s. One strong line of evidence linking electron and proton transfer was the demonstration that addition of chemicals known as 'uncouplers' dissipates proton gradients and halts ATP synthesis. Many uncouplers were weak, membrane permeable, acids and a major objective involved elucidating the link between transmembrane proton gradients and the synthesis of ATP from ADP and inorganic phosphate.

Fractionation and reconstitution studies showed that ATP synthases contained two distinct functional units each with different biological and chemical properties. In electron micrographs this arrangement was characterized by the appearance of the enzyme as a large globular projection extending on stalks from the surface of cristae into the mitochondrial matrix (Figure 5.48). A hydrophobic, membrane-binding, domain called F_o interacted with a larger globular domain peripherally associated with the membrane. F_o stands for 'factor oligomycin' referring to the binding of this antibiotic to the hydrophobic portion whilst F_1 is simply "factor one". F_o is pronounced 'ef-oh'. The peripheral protein, F_1, was removed from membranes by washing with solutions of low ionic strength or those containing chelating agents. This offered a convenient method of purification. Although not strictly a membrane protein it is described in this section because its physiological function relies intimately on its association with the F_o portion of ATP synthetase, a typical membrane protein. In an isolated state the F_1 unit hydrolysed ATP but could not catalyse ATP synthesis. For this reason the F_1 domain is often described as an ATPase reflecting this discovery. A picture of ATP synthesis evolved where F_o formed a 'channel' that allowed a flux of protons and their use by F_1 in ATP synthesis.

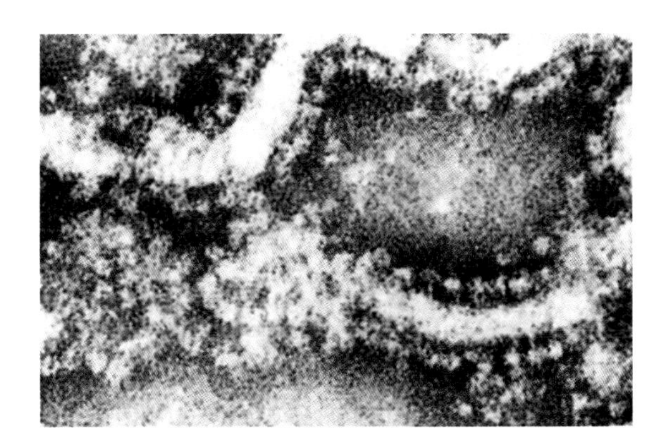

Figure 5.48 An electron micrograph of cristae from isolated mitochondria showing the F_1 units projecting like 'lollipops' from the membrane and into the matrix (reproduced with permission from Voet, D., Voet, J.G & Pratt, C.W. *Fundamentals of Biochemistry*. John Wiley & Sons, Chichester, 1999)

In *E. coli* F_o is extremely hydrophobic and consists of three subunits designated a, b and c (Table 5.9). In *E. coli* a ratio of a_1, b_2, c_{9-14} exists with the a and c subunits making contact and appearing to form a proton channel. Within both subunits are residues essential for proton translocation and amongst those to have been identified from genetic and chemical modification studies are Arg210, His245, Glu196, and Glu219 of the *a* subunit and Asp61 of the *c* subunit. Since all of these residues possess side chains that bind protons the concept of a proton channel formed by these groups is not unrealistic.

Table 5.9 The composition of ATP synthetases

Subunits in F_1 region	Eukaryotic (bovine) $\alpha_3\beta_3\gamma\delta\varepsilon$
α	509 residues, M_r~55,164
β	480, 51,595
γ	272, 30,141
δ	190, 20,967
ε	146, 15,652
Subunits in F_o region	Prokaryotic (*E. coli*) a, b_2, c_{9-14}
a	271, 30,285
b	156, 17,202
c	79, 8,264

The much larger F_1 unit contains five different subunits ($\alpha-\varepsilon$) occurring with a stoichiometry $\alpha_3\beta_3\gamma_1\delta_1\varepsilon_1$. The total mass of the F_1F_o ATPase from bovine mitochondria is ~450 kDa with the F_1 unit having a mass of ~370 kDa and arranged as a spheroid of dimensions 8×10 nm wide supported on a stem or stalk 3 nm in length. Chemical modification studies identified catalytic sites on the β subunits with affinity labelling and mutagenesis studies highlighting the importance of Lys155 and Thr156 in the *E. coli* β subunit sequence ^{149}Gly-Gly-Ala-Gly-Val-Gly-Lys-Thr-Ala^{157}.

A scheme for ATP synthesis envisaged three discrete stages:

1. Translocation of protons by the F_o
2. Catalysis of ATP synthesis *via* the formation of a phosphoanhydride bond between ADP and Pi by F_1
3. A coupling between synthesis of ATP and the controlled dissipation of the proton gradient *via* the coordinated action and interaction of F_1 and F_o.

The mechanism of ATP synthesis

A mechanism for ATP synthesis was proposed by Paul Boyer before structural data was available and has like all good theories stood the test of time and proved to be compatible with structural models derived subsequently for F_1/F_o ATPases. Boyer envisaged the F_1 ATPase as three connected, interacting, sites each binding ADP and inorganic phosphate. The model embodied known profiles for the kinetics of catalysed reactions and the binding of ATP and ADP by isolated F_1. Each catalytic site was assumed to exist in a different conformational state although inhibition of ATP synthesis by modification of residues in the β subunit suggested that these sites interacted cooperatively.

Rate-limiting steps were substrate binding or product release (i.e ATP release or ADP binding) but not the formation of ATP. Careful examination of the ATP hydrolysis reaction suggested three binding sites for ATP each exhibiting a different K_m. When ATP is added in sub-stoichiometric amounts so that only one of the three catalytic sites is occupied substrate binding is very tight (low K_d) and ATP hydrolysis occurs slowly. Addition of excess ATP leads to binding at all three sites but with lowered affinity at the second and third sites. The K_d for binding ATP at the first site is <1 nM whereas values for sites 2 and 3 are ~1 μM and ~30 μM. Upon occupancy of the third site, the rate of overall ATP hydrolysis increases by a factor of $\sim 10^4-10^5$.

As long as the results from hydrolysis of ATP can be related directly to that of the synthetic reaction and isolated F_1 behaves comparably to the complete F_1F_o complex then these results point towards key features of the catalytic mechanism. Isolated F_1 exhibits negative cooperativity with respect to substrate binding where ATP binding at the site with the lowest K_m makes further binding of ATP molecules difficult. This occurs because the K_m for other sites is effectively increased. In contrast the enzyme displays positive cooperativity with respect to catalytic activity with the rate increasing massively in a fully bound state. To explain these unusual properties, Boyer proposed the 'binding change' mechanism (Figure 5.49).

A central feature of this hypothesis are three catalytic sites formed by the three α/β subunit pairs each with a different conformation at any one time. One site is open and ready for substrate binding (ATP or ADP + P_i) whilst the second and third sites are partly open and closed, respectively. Substrate binding results in the closure of the open site and produces a cooperative conformational change in the other two sites; the closed one becomes partly open and the partly open one becomes fully open. In this manner the sites alternate between conformational states. Obvious questions from this model arise. How are conformational changes required in the binding change mechanism propagated? How are these events linked to proton translocation and the F_o part of the ATPase?

From a consideration of the organization of F_o G.B. Cox independently suggested a rotary motor involved the c subunit ring turning relative to the a and b subunits. This idea was adopted and incorporated as a major refinement of the binding change mechanism with the rotary switch causing a sequential change in binding. At the time (1984) the concept of enzymes using rotary motion to catalyse reactions was unusual. The cooperative kinetics of the enzyme (negative for ATP binding and positive for catalysis) supported the

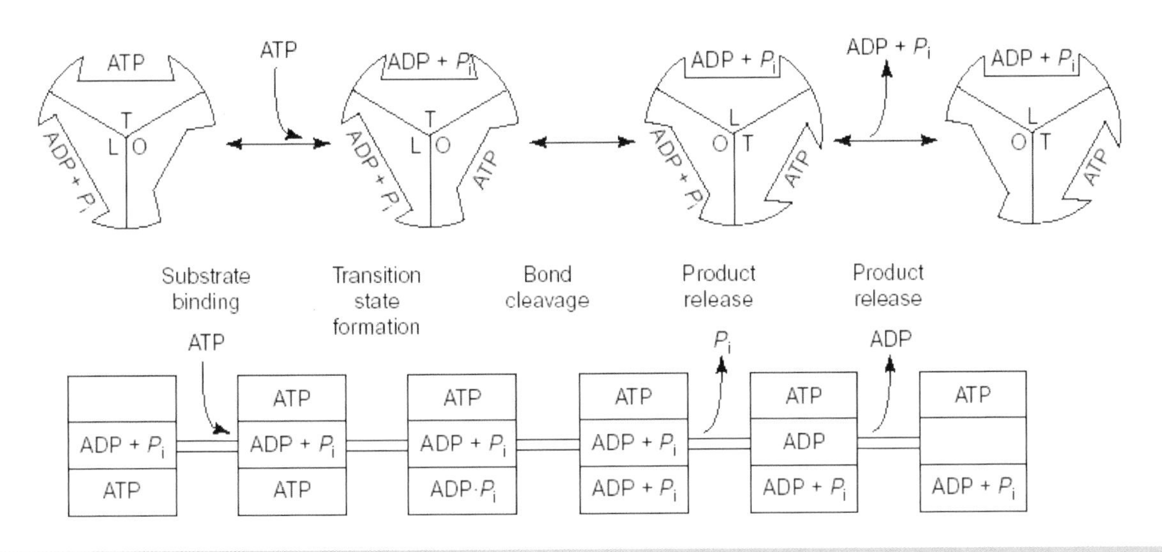

Figure 5.49 The binding change hypothesis and catalytic site activity at the β subunit. Each catalytic site cycles through three states: T, L and O. ATP binds to the O (open and empty) site converting it into a T (tight and ATP-occupied) site. After bond cleavage at this site the T site is converted into the L (loose and ADP-occupied) site where the products escape to recover the O state. The concerted switching of states in each of the sites results in the hydrolysis of one ATP molecule. Sub-steps in the hydrolysis of one ATP molecule based on kinetic and inhibitor studies are shown for a three-site mechanism. Thus, ATP binding in the open site (top) leads to transition state formation and then bond cleavage in a closed site (bottom), followed by release from a partly open site as it opens fully and releases ADP (middle) (reproduced with permission from *Trends Biochem. Sci* 2002, **27**, 154–160. Elsevier)

binding-change mechanism. However, the idea of a rotary switch was so novel that there were few serious attempts to test it until over a decade later when in 1995 the first structures for bovine F_1 ATPase produced by John Walker and co-workers appeared to support Boyer's model.

The structure of the F_1 ATPase

As a peripheral membrane protein the water-soluble F_1 unit is easily purified and the structure showed an alternate arrangement of three α and three β subunits based around a central stalk formed by two α helices running in opposite directions derived from the γ subunit. These two helices form a coiled-coil at the centre of F_1 and the remaining part of the γ subunit protrudes from the $(\alpha\beta)_3$ assembly and interacts with polar regions of the c subunit of F_o (Figure 5.50).

The α and β subunits exhibit sequence identity of ~20 percent and show similar overall conformations

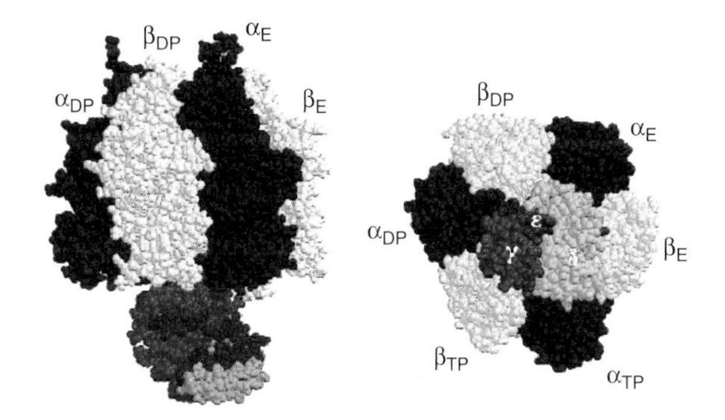

Figure 5.50 A space filling model of bovine F1-ATPase from a side view (left) and from the membrane (right) showing α subunits (red) and β subunits (yellow) and the attached central stalk containing the γ, δ and ε subunits (blue, green and magenta, respectively). The β_{TP}, β_{DP} and β_E subunits had AMP-PNP, ADP and no nucleotide bound, respectively. The α subunits are named according to the catalytic interface to which they contribute

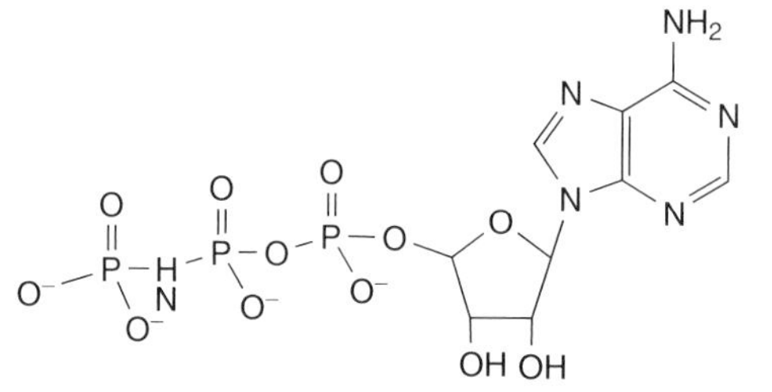

Figure 5.51 The structure of AMP-PNP – a non-hydrolysable derivative of ATP

arranged around the central stalk made up of the γ, δ and ε subunits. F_1 was co-crystallized with ADP and AMP-PNP (5′-adenylyl-imidophosphate, a non-hydrolysed analogue of ATP; Figure 5.51). Only five of the six nucleotide binding sites were occupied by substrate. Each of the α subunits bound AMP-PNP and exhibited identical conformation.

In contrast one β subunit bound AMP-PNP (β_{TP}), another bound ADP (β_{DP}) and the third subunit remained empty (β_E). The structure provided compelling evidence for three different modes of nucleotide binding as expected from the Boyer model. Although both α and β subunits have nucleotide binding sites only those on the β subunits are catalytically important.

Each α and β subunit contains three distinct domains (Figure 5.52). A β barrel located at the N-terminal is followed by a central nucleotide-binding domain containing a Rossmann fold. The β barrels form a crown that sits on top of the central nucleotide-binding domain. At the C terminal is a helix rich domain. The conformation of the β subunit changes with nucleotide-binding forming states associated with high, medium and low nucleotide affinities. These three states are most probably related to release of product (ATP), binding of substrates (ADP and inorganic phosphate), and formation of ATP, respectively. In an isolated state the structure of the β subunit has the open form with nucleotide binding causing transition to closed forms. On the β subunit the principal residues involved in ADP binding have been determined by examining the crystal structure of this subunit in the presence of ADP-fluoroaluminate (ADP-AlF_3) (Figure 5.53). Positively charged side chains of arginine and lysine have a major role in binding ADP-fluoroaluminate whilst the side chain of Glu188 binds a water molecule and the amide of Gly159 binds the second phosphate group of ADP. Arginine 373 is part of the α subunit and contributes to the binding site for ADP. ADP-fluoroaluminate mimics the transition state complex with fluorine atoms occupying similar conformations to the extra phosphate group of ATP but acting as a potent inhibitor of ATP synthesis. Significantly, the structure adopted by the β subunit in the presence of ADP-AlF_3 is very similar to the catalytically active conformation for this subunit (β_{TP}). These results further suggest that once the substrate is bound only small conformational changes are required to promote synthesis of ATP.

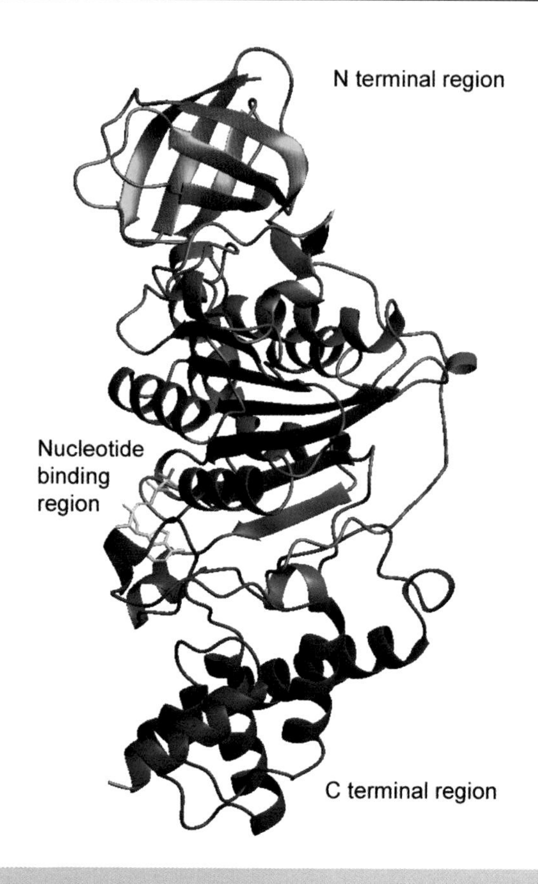

Figure 5.52 An individual β subunit showing the three distinct domains together with the bound nucleotide – in this case ADP. The ADP is shown in yellow, the barrel structure in dark green, the nucleotide binding region in blue and the helix rich region in violet (PDB:1BMF)

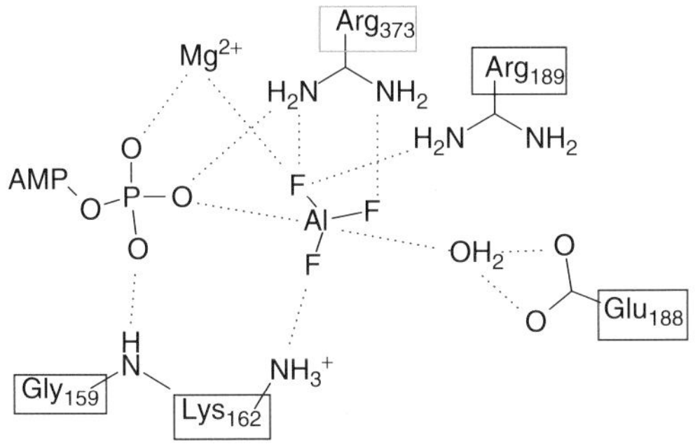

Figure 5.53 Schematic representation of the nucleotide-binding site of the β subunit of the AlF_3-F_1 complex, showing the coordination of the aluminofluoride group. Possible hydrogen-bond interactions are shown by dotted lines

The stoichiometry of the bovine mitochondrial F_1 ATPase means that the single γ, δ and ε subunits lack symmetry, and in the case of the γ subunit the intrinsic asymmetry causes a different interaction with each of the three catalytic β subunits of F_1. This interaction endows each β subunit with different nucleotide affinity and arises from rotation of the central stalk. Rotation of the γ subunit driven by proton flux through the F_o region drives ATP synthesis by cycling through structural states corresponding to the tight, open and loose conformations.

Initial structures of F_1 ATPase showed disorder for the protruding end of the γ subunit from the $(\alpha\beta)_3$ hexamer. Examining enzyme inhibited with dicyclohexylcarbodiimide (DCCD), a known stoichiometric inhibitor of ATP synthesis that reacts specifically with Glu 199 of the β subunit, resulted in improved order and confirmed the orientation of the central pair of helices of the γ subunit, the structures of the δ and ε subunits and their interaction with the F_o region at the end of the central stalk. The protruding part of the central stalk is composed of approximately half of the γ subunit and the entire δ and ε subunits. The overall length of the central stalk from the C-terminus of the γ subunit to the foot of the protruding region is ~11.4 nm. It extends ~4.7 nm from the $(\alpha\beta)_3$ domain thereby accounting for 'lollipop' appearance in electron micrographs (Figure 5.48). The central stalk is the key rotary element in the catalytic mechanism. The γ subunit interacts directly with the ring of c subunits located in the membrane of F_o and in the absence of covalent bonding between c subunits and the stalk region this interaction must be sufficiently robust to permit rotation. In the γ subunit, three carboxyl groups (γAsp 194, γAsp 195 and γAsp 197) are exposed on the lower face of the foot suggesting that they may interact with basic residues found in loop regions of the c-ring.

It is established that all species contain a, b and c subunits within the F_o part of the enzyme although there is currently no clear agreement on the total number of subunits. In yeast ATP synthetase at least thirteen subunits are found in the purified enzyme. As well as the familiar α–ε subunits of F_1 and the a, b and c subunits of F_o additional subunits called f and d are found in F_o whilst OSCP, ATP8 and h occur in F_1. Using the yeast enzyme the F_1 region was co-crystallized with a ring of 10 c subunits and although the loss of c subunits during preparation cannot be eliminated definitively the lack of symmetry relations between the hexameric collection of α/β subunits and the 10 c subunits was surprising. The structure of the c subunit monomer was established independently by NMR spectroscopy (Figure 5.55) and showed two helical regions that were assumed to be transmembrane segments linked by an extramembranous loop region. These loop regions make contact with γ and δ subunits and represent sites of interaction (Figure 5.56). Only the a and c subunits are essential for proton translocation (Figure 5.57) with Arg210, His245, Glu196, and Glu219 (a subunit) and Asp61 of the c subunit implicated as essential residues.

Further refinement of the ATP synthetase structure has shown a second, peripheral, stalk connecting the F_1 and F_o domains in the enzyme (Figure 5.58). The second stalk visible using cryoelectron microscopy of detergent solubilized F_1F_o from *E. coli*, chloroplast and mitochondrial ATPases is located around the outside of the globular F_1 domain and may act as a stator to counter the tendency of the $(\alpha\beta)_3$ domain to follow the rotation of the central stalk.

With the structural model derived from X-ray crystallography in place for the F_1 subunit a number of elegant experiments established that rotation *is* the link between proton flux and ATP synthesis. The

Figure 5.54 The structures of the central stalk assembly with the individual subunits shown in isolation alongside. The γ subunit contains the long α helical coiled coil region that extends through the $(\alpha\beta)_3$ assembly (top right). The bovine ε subunit is very small (~50 residues in length) with one major helix together with a poorly organized helix (bottom right). The δ subunit contains an extensive β sheet formed from seven β strands together with a distinct helix turn helix region (bottom left)

idea of rotary switches is now no longer heretical but is accepted as another solution by proteins to the mechanistic problem of harnessing energy from proton gradients.

Covalently linking the γ subunit of F_1 to the C terminal domain of the β subunit blocked activity whilst cross-linking to the α subunit was without effect. Complementary studies introducing cysteine residues into the structure of β and γ subunits with the formation of thiol cross links inhibited activity but reducing these bonds followed by further cross-linking resulted in the modification of a different β subunit. This could only occur if rotation occurred within ATPases. Finally, covalently joining the γ, ε and c subunit ring did not inhibit ATP synthesis and suggested that the subunits act in unison as a rotating unit. Direct observation of rotation in the synthetase was obtained using actin filaments attached to the γ subunit. By following the fluorescence of covalently linked fluorophores in a confocal microscope it was shown that ATP hydrolysis led to a 360° rotation of the filament in three

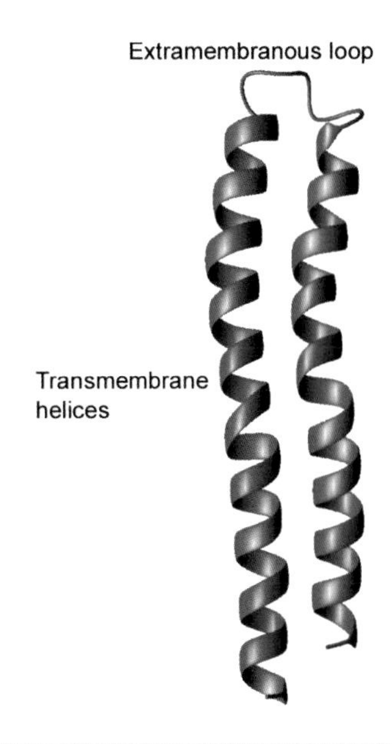

Figure 5.55 The structure of the c ring monomer determined using NMR spectroscopy is a simple HTH structure (PDB:1A91)

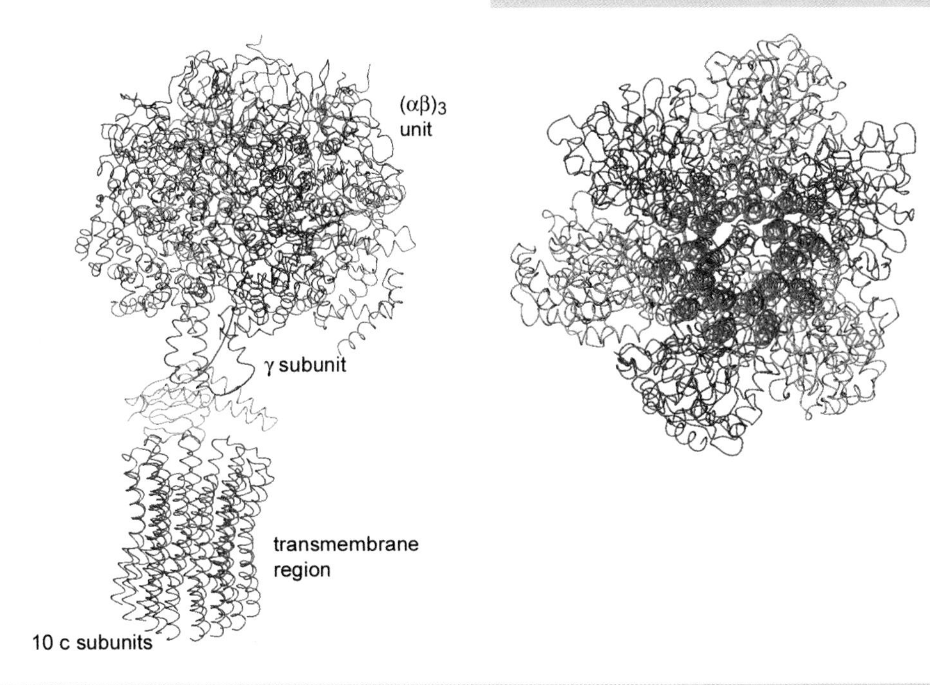

Figure 5.56 The arrangement of the c ring with the F_1 catalytic domain in yeast ATP synthetase. The c ring of 10 monomers is shown in blue, the $(\alpha\beta)_3$ unit is shown in red and orange and the γ subunit in dark green. A view of the F_1 F_o ATPase from the inter-membrane space looking through the membrane region is shown on the right

uni-directional steps of 120°. The complex showed rotation rates of ~130 Hz entirely consistent with a maximal rate of hydrolysis of ATP of ~300 s^{-1}.

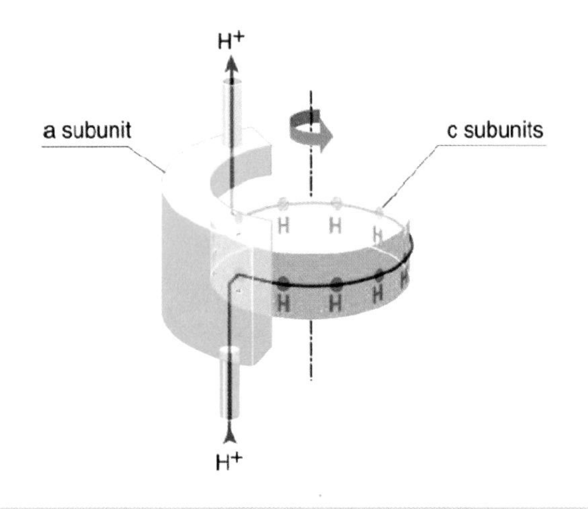

Figure 5.57 Hypothetical model for the generation of rotation by proton transfer through the F_o domain of ATP synthase. The central cylinder (blue) consists of c subunits; the external part (green) corresponds to a single a subunit. The red line indicates the proton path

ATP, the universal carrier of cell energy, is manufactured from ADP and phosphate by the enzyme ATP synthase (or F_1F_o ATPase) using the free energy of an electrochemical gradient of protons. This proton gradient is generated by the combined efforts of numerous respiratory or photosynthetic protein complexes whose catalytic properties include rapid electron transfer coupled to movement of protons across membranes.

ATPase family

The F_1F_o ATPases comprise a huge family of enzymes with members found in the bacterial cytoplasmic membrane, the inner membrane of mitochondria and the thylakoid membrane of chloroplasts. These enzymes are usually classified as proton-transporting F type ATPases to distinguish them from enzymes that hydrolyse ATP as part of their normal catalytic function. Further classes can be recognized in these ATPases through the use of alternative ions to the proton such as Na^+ or Li^+ ions. Minor changes to the a and c subunit sequences alter the coupling ion specificity although the primary mechanism remains very similar. In all cases the ATPases are based around a large globular catalytic domain that projects from the membrane surface but

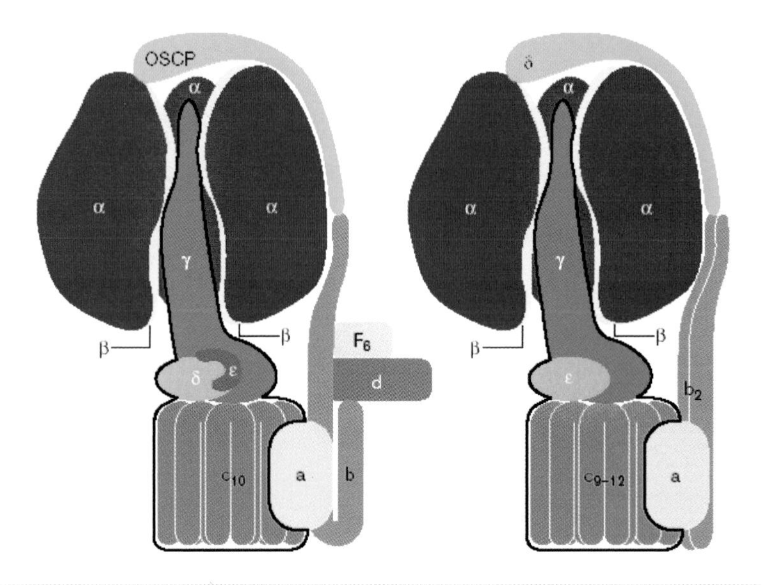

Figure 5.58 Organization of ATP synthases. The bovine F_1 – ATPase (left) and the ATP synthase of *E. coli* (right). (Reproduced with permission from Stock, D. *et al. Curr. Opin. Struct. Biol.* 2000, **10**, 672–679. Elsevier)

interacts with a hydrophobic complex that controls ion translocation via rotary motion.

In all cells there are membrane proteins whose function relies on the hydrolysis of ATP and the coupling of this energy to the movement of ions such as Na^+, K^+ or Ca^{2+} across membranes. Transport proteins represent a major category of membrane proteins and a number of different mechanisms exist for transport. Transport across membranes is divided into mediated and non-mediated processes. Non-mediated transport involves the diffusion of ions or small molecules down a concentration gradient. Transport is closely correlated with solute solubility in the lipid phase with steroids, anaesthetics, narcotics and oxygen diffusing down a concentration gradient and across membranes. Mediated transport is divided into two categories. Passive-mediated transport also called facilitated diffusion relies on the transfer of a specific molecule from high concentration to low concentration. Transport proteins involved in this mechanism of transport have already been described in this chapter and are exemplified by the β barrel porin proteins. Unfortunately these proteins cannot transport substrates against concentration gradients; this requires active transport and the hydrolysis of ATP.

Na^+–K^+ ATPase is one member of a class of membrane bound enzymes that transfer cations with ATP hydrolysis providing the driving force. The Na^+–K^+ ATPases, a highly conserved group of integral proteins, are widely expressed in all eukaryotic cells and are very active with a measure of their importance seen in the estimate that ~25 percent of all cytoplasmic ATP is hydrolysed by sodium pumps in resting human cells. In nerve cells, a much higher percentage of ATP is consumed (>70 percent) to fuel sodium pumps critical to neuronal function.

The Na^+–K^+ ATPase is a transmembrane protein containing two types of subunit. An α subunit of ~110 kDa contains the ion-binding site and is responsible for catalytic function whilst a β subunit, glycosylated on its exterior surface, is of less clear function. Sequence analysis suggests that the α subunit contains eight transmembrane helices together with two cytoplasmic domains. In contrast the β subunit (M_r~55 kDa) has a single transmembrane helix and a much larger domain projecting out into the extracellular environment where glycosylation accounts for ~20 kDa of the total mass of the protein. The β subunit is required for activity and it is thought to play a role in membrane localization and activation of the α subunit. *In vivo* the Na^+–K^+ ATPase functions as a tetramer of composition $\alpha_2\beta_2$.

The Na^+–K^+ ATPase pumps sodium out of the cell and imports potassium with the hydrolysis of ATP at the intracellular surface of the complex. The overall stoichiometry of the reaction results in the movement of 3 Na^+ ions outward and 2 K^+ ions inwards and is represented by the equation

$$3Na^+_{(in)} + 2K^+_{(out)} + ATP + H_2O \rightleftharpoons 3Na^+_{(out)} + 2K^+_{(in)} + ADP + P_I$$

The ATPase is described as an antiport where there is the simultaneous movement of two different molecules in opposite directions, and leads to the net movement of positive charge to the outside of the cell. A uniport involves the transport of a single molecule whilst a symport simultaneously transports two different molecules in the same direction (Figure 5.59).

In the presence of Na^+ ions ATP phosphorylates a specific aspartate residue on the α subunit triggering conformational change. In the presence of high levels of K^+ the dephosphorylation reaction is enhanced and suggests two distinct conformational states for the Na^+–K^+ ATPase. These states, designated E_1 and E_2, differing in conformation show altered affinity and specificity. On the inside of the cell the Na^+–K^+ ATPase (in the E_1 state) binds three sodium ions at specific sites followed by ATP association to yield an E_1–ATP–$3Na^+$ complex. ATP hydrolysis occurs to form an aspartyl phosphate intermediate $E_1 \sim P$–$3Na^+$ that undergoes a conformational change to form the E_2 state ($E_2 \sim P$–$3Na^+$) and the release of bound Na outside the cell. The presence of a 'high energy' phosphoryl intermediate (denoted by ~P) has led to this class of membrane proteins being called P type ATPases. As an E_2–P state the enzyme binds two K^+ ions from the outside to form the E_2–P–$2K^+$ complex. Hydrolysis of the phosphate group to reform the original aspartate side chain causes conformation change with the result that $2K^+$ ions are released on the inside of the cell. The ATPase ion pump, upon release of the potassium ions, forms the E_1 state capable

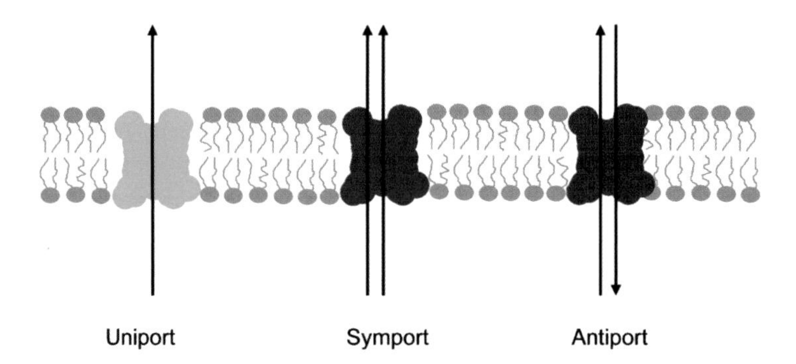

Figure 5.59 Diagrammatic representation of uniport, symport and antiport systems used in processes of active transport across membranes

of binding three Na^+ ions and restarting the cycle again.

Study of the function of Na^+–K^+ ATPases has been aided by the use of cardiac glycosides, natural products that increase the intensity of muscular contraction in impaired heart muscle. Abnormalities have been identified in the function of Na^+–K^+ ATPases and are involved in several pathologic states including heart disease and hypertension. Several types of heart failure are associated with significant reductions in the myocardial concentration of Na^+–K^+ ATPase or impaired activity. Similarly excessive renal re-absorption of sodium due to oversecretion of the hormone aldosterone is associated with forms of hypertension. Oubain is a naturally occurring steroid and is still prescribed as a cardiac drug. Its action is to block the efflux of Na^+ ions leading to an increase in their intracellular concentration (Figure 5.60). The resulting increase in Na^+ concentration causes a stimulation of secondary active transport systems in cardiac muscle such as the Na^+–Ca^{2+} antiport system. This pump removes Na^+ ions from the cell and transports Ca^{2+} to the inside. Transient increases in cytosolic Ca^{2+} ion concentration often triggers many intracellular responses one of which is muscular contraction. The Ca^{2+} ions stimulate cardiac muscle contraction producing a larger than normal response to aid cardiovascular output. Although of major

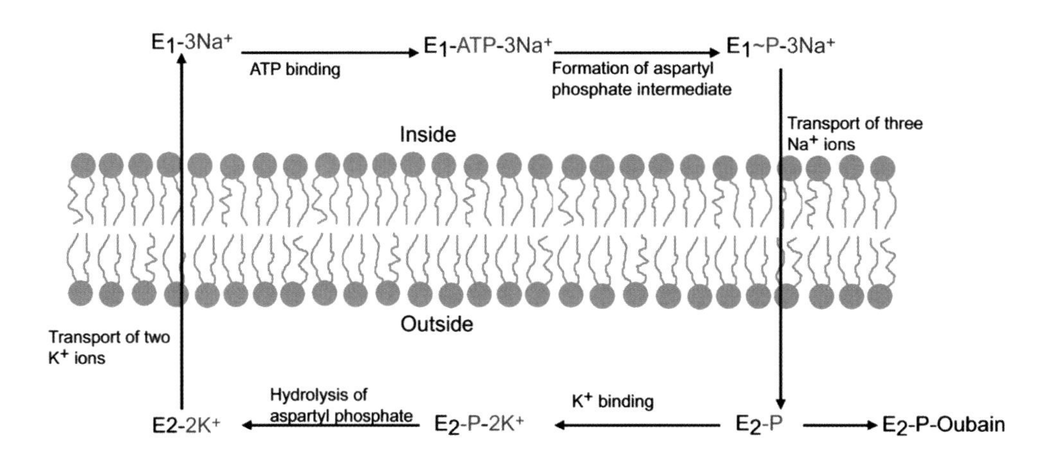

Figure 5.60 The active transport of Na^+ and K^+ ions by the Na^+–K^+ ATPase. Oubain (pronounced 'wabane') is a cardiac glycoside promoting increased intracellular Na^+ concentration

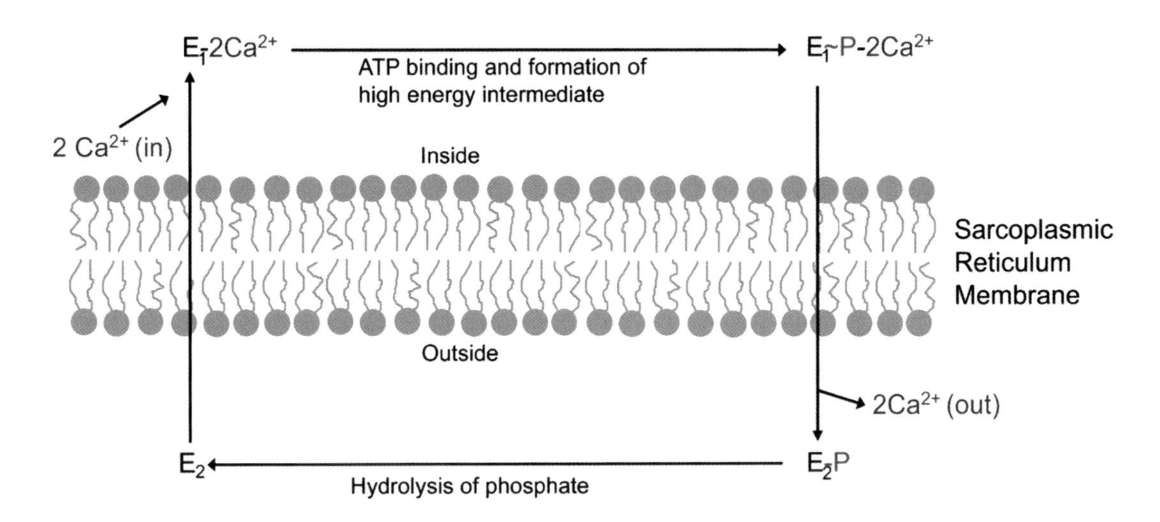

Figure 5.61 Scheme for active transport of Ca^{2+} by SR Ca ATPase. The inside refers to the cytosol and the outside represents the lumen of the sarcoplasmic reticulum of skeletal muscle cells. Similar Ca pumps exist in the plasma membranes and ER

importance to the physiology of all cells detailed studies of the Na^+–K^+ ATPase protein and ultimately the development of improved drugs is hampered by the absence of a well-defined structure.

Fortunately, another prominent member of the P type ATPases, the Ca^{2+}–ATPase from the sarcoplasmic reticulum (SR) of muscle cells, has been crystallized to reveal some aspects of the organization of transport proteins. SR Ca^{2+} ATPase is amongst the simplest proteins in this large family consisting of a single polypeptide chain of ~110 kDa. Its biological function is to transport $2Ca^{2+}$ ions against a concentration gradient for every molecule of ATP hydrolysed (Figure 5.61). Within the cytosol the concentration of Ca^{2+} is ~0.1 μM – about 10000 times lower than the extracellular or lumenal concentration (~1.0–1.5 mM). When muscle contracts large amounts of Ca^{2+} stored in the SR are released into the cytosol of muscle cells. In order to relax again after contraction the Ca concentration must be reduced by pumping ions back into the SR against a concentration gradient.

As an ion pump the Ca^{2+} ATPase works comparatively slowly with a turnover number of ~60 Ca^{2+} ions per second. Pumps such as the Na^+–K^+ ATPase transfer ions at rates of $10^6\ s^{-1}$ and in order to overcome this rate deficiency for Ca^{2+} transfer the cell compensates by accumulating large amounts of enzyme within the SR membrane. The accumulation of enzyme in this membrane leads to ~60 percent of the total protein being Ca^{2+} ATPase.

The structure of the SR Ca^{2+} ATPase reveals two Ca^{2+} ions bound in a transmembrane domain of 10 α helical segments. In addition to the transmembrane region three cytosolic regions were identified as phosphorylation (P), nucleotide binding (N) and actuator (A) domains. This last domain transmits conformational change through the protein leading to movement of Ca across membranes. The organization of these domains can be seen in the beautiful structure of Ca bound SR-ATPase where ten transmembrane α helices (M1–M10) segregate into two blocks (M1–M6 and M7–M10), a structural organization consistent with the lack of M7–M10 in simpler bacterial P-type ATPases (Figure 5.62).

The lengths and inclination of the ten helices differ substantially; helices M2 and M5 are long and straight whilst others are unwound (M4 and M6) or kinked (M10) in the middle of the membrane. The Ca ions are located side by side surrounded by four transmembrane helices (M4, M5, M6 and M8), with the ligands to Ca^{2+} provided at site I by side-chain oxygen atoms

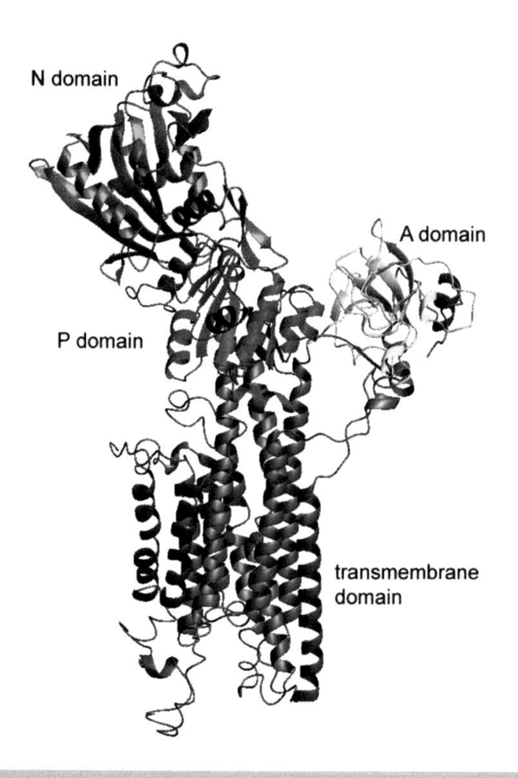

Figure 5.62 The structure of SR Ca-ATPase determined to a resolution of A. (PDB: 1EUL). The four distinct domains are shown in different colours; transmembrane domain (blue), Actuator or A domain (yellow), phosphorylation or P domain (magenta) and the nucleotide binding or N domain (green). Two bound Ca ions are shown in red. In this view the lumen of the SR is at the bottom whilst the cytosol is found at the top containing the A, P and N domains

derived from Asn768, Glu771 (M5), Thr799, Asp800 (M6) and Glu908 (M8). Site II is formed almost entirely by helix M4 with backbone oxygen atoms of Val304, Ala305 and Ile307 together with the side-chain oxygens of Asn796, Asp800 (M6) and Glu309 (M4) providing the key ligands. The coordinating ligands provided by residues 304, 305, 307 and 309 occur in residues located either sequentially or very close together in the primary sequence. Such an arrangement is not compatible with regular helical conformations and is only possible as a result of local unfolding of the M4 helix in this region. In this context a proline residue at position 308 almost certainly contributes to unfolding whilst the sequence PEGL is a key motif in all ATPases. Although the position of the Ca^{2+} is clearly seen within the transmembrane region the structure of the SR-CaATPase does not offer any clues as to the route of ion conduction. There were no obvious vestibules in the bound state, as seen in porins, and conformational changes facilitate transport of Ca ions across the membrane by coupling events in the transmembrane domain with those in the cytoplasmic domains. This involves Ca binding and release, nucleotide binding as well as Asp351 phosphorylation and dephosphorylation.

Summary

Following the determination of the structure of the bacterial photosynthetic reaction centre there has been a progressive increase in the number of refined structures for membrane bound.

Compared with the large number of structures existing for globular domains in the protein databank it is clear that the number of highly resolved (<5 Å) structures available for membrane proteins is much smaller. This will change rapidly as more membrane proteins become amenable to methods of structure determination.

Peripheral or extrinsic proteins are associated with the membrane but are generally released by mild disruptive treatment. In contrast, integral membrane proteins remain firmly embedded within the hydrophobic bilayer and removal from this environment frequently results in a loss of structure and function.

Crystallization of membrane proteins requires controlled and ordered association of subunits via the interaction of exposed polar groups on the surfaces of proteins. The process is assisted by the introduction of amphiphiles that balance conflicting solvent requirements of membrane proteins.

The vast majority of proteins rely on the folding of the primary sequence into transmembrane helical structures. These helices have comparable geometry to those occurring in soluble proteins exhibiting similar bond lengths and torsion angles.

The major differences with soluble proteins lie in the relative distribution of hydrophobic amino acid residues.

Table 5.10 Some of the membrane protein structures deposited in PDB; Most structures were determined using X-ray crystallography

Protein	PDB	Resolution (Å)	Function
Rhodopsin family and G-protein coupled receptors			
Bacteriorhodopsin (EM)	2BRD	3.5	Light driven proton pump
(X-ray)	1C3W	1.55	
Halorhodopsin	1E12	1.8	
Rhodopsin	1F88	2.8	Visual cycle transducing light energy into chemical signals
Sensory rhodopsin II	1H68	2.1	
Photosynthetic and light harvesting complexes			
R. viridis reaction centre	1PRC	2.3	Primary charge separation initiated by a photon of light
R. sphaeroides reaction centre	1PSS	3.0	
Light harvesting chlorophyll complexes from *R. acidophila*	1 KZU	2.5	Gathering light energy and focussing photons to the photosynthetic reaction centre
Photosystem I from *S. elongatus*	1JB0	2.5	Primary photochemistry
Photosystem II from *S. elongatus*	1FE1	3.8	Oxygenic photosynthesis
Bacterial and mitochondrial respiratory complexes			
Cytochrome oxidase (aa_3) from *P. denitrificans*	1AR1	2.8	Catalyses terminal step of aerobic respiration
Cytochrome oxidase (ba_3) from *T. thermophilus*	1EHK	2.4	
Cytochrome oxidase from bovine heart mitochondria	1OCC	2.8	
Bovine heart mitochondria bc_1 complex	1BGY	2.8	Electron transfer complex that oxidizes ubiquinonols and reduces cytochrome c
Chicken heart cytochrome bc_1 complex	1BCC	3.2	
S. cerevisiae cytochrome bc_1 complex	1EZV	2.3	

(*continued*)

Table 5.10 (*continued*)

Protein	PDB	Resolution (Å)	Function
Other energy transducing membrane proteins			
ATP synthase (F_1 c10) *S. cerevisiae*	1E79	3.9	Membrane portion forms proton channel
Fumarate reductase complex from *Wolinella succinogenes*	*1QLA*	2.2	
β barrel membrane proteins of porin family			
Porin from *R. capsulatus*	2POR	1.8	Semi-selective pore proteins
Porin from *R. blastica*	1PRN	2.0	
OmpF from *E. coli.*	2OMF	2.4	
PhoE from *E. coli.*	1PHO	3.0	
Maltoporin from *S. typhimurium*	2MPR	2.4	
Maltoporin from *E. coli*	1MAL	3.1	
OmpA from *E. coli.* (NMR)	1G90		
(X-ray)	1BXW	2.5	
Ion and other channel proteins			
Calcium ATPase from SR of rabbit	1EUL	2.6	Controls Ca flux in sarcoplasmic reticulum
Toxins			
α-haemolysin from *S. aureus*	7AHL	1.9	Channel forming toxin
LukF from *S. aureus*	3LKF	1.9	Homologous in structure and function to α-haemolysin

Membrane proteins such as toxins or pore proteins are based on β strands assembling into compact, barrel structures that assume great stability via inter-strand hydrogen bonding.

Sequencing proteins assists in the definition of transmembrane domains *via* identification of hydrophobic side chains. Proteins containing high proportions of residues with non-polar side chains such as Leu, Val, Ile, Phe, Met and Trp in blocks of ~20–30 residues are probably membrane bound.

The structures of bacteriorhodopsin revealed seven transmembrane helices of ~23 residues in length correlating with seven 'blocks' in the primary sequence composed predominantly of residues with non-polar side chains.

The seven transmembrane helix structure is a common architecture in membrane proteins with G-protein coupled receptors sharing these motifs.

Methods developed for crystallization of the bacterial reaction centre from *R. viridis* proved applicable to other classes of membrane proteins including respiratory complexes found in mitochondrial membranes and in aerobic bacteria. These complexes generate proton gradients vital for the production of ATP and represent the 'power stations' providing energy for all cellular processes.

The structure of Ca ATPase reveals the coordinated action of cytosolic and membrane located domains in Ca transport across the membrane and points the way to understanding the movement of charged ions across impermeant lipid bilayers.

ATP is produced by ATP synthase a large enzyme composed of a peripheral F_1 complex together with a smaller membrane bound region (F_o). These domains form a rotary motor where the flux of protons through F_o drives rotation of c subunits via transmission to a central stalk (γ, δ and ε subunits) that mediates conformational changes at the catalytic centres in $(\alpha\beta)_3$ subunits promoting ATP formation.

Problems

1. Having determined the nucleotide sequence of a putative membrane protein describe how you would use this data to obtain further information on this protein.
2. List the post-translational modifications added to proteins to retain them within or close to the lipid bilayer. Give an example protein for each modification.
3. Find the membrane-spanning region in the following sequence: .........GELHPDDRSKITKPSESIITTIDS NPSWWTNWLIPAISALFVALIYHLYTSEN.......
4. How many residues are required to span a lipid bilayer if they are all form within a regular α helix? How many turns of a regular α helix are required to cross the membrane once?
5. Using the crystal structure for the bacterial reaction centre estimate the average length of the transmembrane helices. Are there any differences and if so how do you account for them?
6. Describe the advantages and disadvantages of using the erythrocyte membrane as a model for studying membrane protein structure and organization.
7. Use protein databases to find membrane protein structures determined at resolution below 0.3 nm and those between 0.3 and 0.5 nm.
8. Gramicidin is a small polypeptide of 15 residues that adopts a helical conformation. This molecule has been observed to act as an ionophore allowing K^+ ions to cross the membrane. How might this occur?

6

The diversity of proteins

The incredible diversity shown by the living world ranges from bacteria and viruses to unicellular organisms, eventually culminating with complex multicellular systems of higher plants, including gymnosperms and angiosperms, and animals such as those of the vertebrate kingdom. However, the living world is based around proteins made up of the *same* 20 amino acids. There is no fundamental difference between the amino acids and proteins making up a bacterium such as *Escherichia coli* to those found in higher vertebrates. As a result the principles governing protein structure and function are equally applicable to all living systems. This pattern of similarity is not surprising if one considers that all living systems are related by their evolutionary origin to primitive ancestors that had acquired the basic 20 amino acids to use in the synthesis of proteins. Higher levels of complexity were acquired by evolutionary divergence that led to a subtle alteration in primary sequence and the generation of new or altered functional properties. Although the exact nature of the 'first' ancestral cell is unclear along with details of self-replicating systems advances have been made in understanding the origin of proteins.

Prebiotic synthesis and the origins of proteins

The origin of life represents one of the greatest puzzles facing scientists today. What originally seemed to be an impossible problem has gradually become better understood via experimentation. In order to synthesize proteins it is first necessary to make amino acids containing carbon, hydrogen, oxygen and nitrogen, and occasionally sulfur. The source of all carbon, hydrogen, oxygen and nitrogen would have been the original atmospheric gases of carbon dioxide, nitrogen, water vapour, ammonia, and methane, but not atmospheric oxygen since this was almost certainly lacking in early evolutionary periods. When this series of events is placed in a time span we are describing reactions that occurred more than 3.6 billion years ago.

In a carefully designed experiment Stanley Miller and Harold Urey showed that simulating the primitive conditions present on the Earth around 4 billion years ago could result in the production of biomolecules. This study involved adding inorganic molecules to a closed system under a reducing atmosphere (lacking oxygen). The gaseous mixture of mainly ammonia, hydrogen and methane simulated the early Earth's atmosphere. The whole mixture was refluxed in a closed evacuated system with the water phase representative of the 'oceans' of the early earth analysed at the end of an experiment lasting several days (Table 6.1). Subjecting the system to electrical discharge (lightning) and high amounts of ultraviolet light (Sun) formed biomolecules, including the amino acids glycine, alanine and aspartate. The formation of hydrogen cyanide, aldehydes and other cyano

Proteins: Structure and Function by David Whitford

$$RCHO + HCN + H_2O \longrightarrow H_2N\text{-}CHR\text{-}COOH$$

Aldehyde Hydrogen Cyanide Amino acid

Figure 6.1 Prebiotic synthesis of amino acids from simple organic molecules

Table 6.1 Yields of biomolecules from simulating prebiotic conditions using a mixture of methane, ammonia, water and hydrogen

Biomolecule	Approximate yield (%)
Formic acid	4.0
Glycine	2.1
Glycolic acid	1.9
Alanine	1.7
Lactic acid	1.6
β-Alanine	0.76
Propionic acid	0.66
Acetic acid	0.51
Iminodiacetic acid	0.37
α-Hydroxybutyric acid	0.34
Succinic acid	0.27
Sarcosine	0.25
Iminoaceticpropionic acid	0.13
N-Methylalanine	0.07
Glutamic acid	0.051
N-Methylurea	0.051
Urea	0.034
Aspartic acid	0.024
α-Aminoisobutyric acid	0.007

Shown in red are constituents of proteins (after Miller, S.J. & Orgel, L.E. *The Origins of Life on Earth*. Prentice-Hall, 1975).

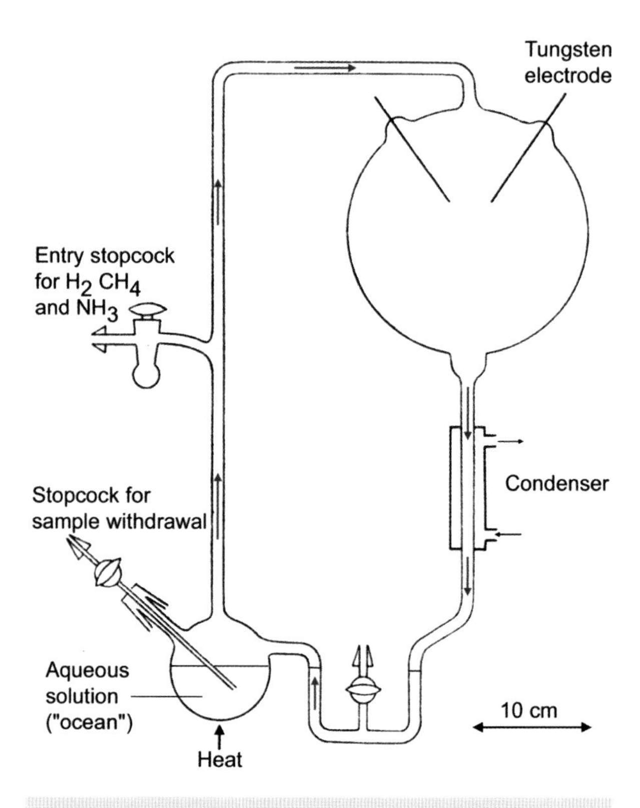

Figure 6.2 An example of the apparatus used by Urey and Miller to demonstrate prebiotic synthesis of organic molecules

compounds was also important since these simple compounds undergo a wide range of further reactions. The general reaction is summarized by a simple addition reaction between aldehydes and hydrogen cyanides in the presence of water (Figure 6.1), although in practice the reaction may form a nitrile derivative in the atmosphere followed by hydrolysis in the 'ocean' to yield simple amino acids.

Performing this type of experiment with different mixtures of starting materials yielded additional biomolecules, including adenine. Variations on this basic theme have suggested that although the Earth's early atmosphere lacked oxygen the presence of gases such as CO, CO_2 and H_2S were vital for prebiotic synthesis. The presence of sulfur enhances the number and type of reactions that could occur. More recently, examination of ocean floors has revealed the presence of deep sea vents sometimes called 'smokers' or fumaroles. These vents release hot gases and minerals from the Earth's crust into the ocean and are also prime sites for organic synthesis. This suggests that many potential sites and sources of energy were available for prebiotic synthesis.

The next barrier to the evolution of life involved the formation of polymers from precursors. The genetic systems found in cells today are specialized polymers. They are able to direct the synthesis of proteins from messenger ribonucleic acid (mRNA), the latter representing the information present in deoxyribonucleic acid (DNA). In addition these polymers are capable of directing their own synthesis in a macromolecular world of DNA/RNA and protein.

$$\text{DNA} \longrightarrow \text{RNA} \longrightarrow \text{protein}$$

These systems are self-replicating and their evolution represents one of the greatest hurdles to be overcome in the development of living systems. Replicating DNA requires proteins to assist in the overall process whilst the scheme above demonstrates that protein synthesis requires DNA and RNA. This creates a paradox often called the 'chicken and egg' puzzle of molecular biology of which came first.

The recent demonstration that RNA molecules have catalytic function analogous to conventional enzymes has revolutionized views of prebiotic synthesis. Catalytic forms of RNA called ribozymes mean that RNA molecules, in theory, at least have the means to direct their own synthesis and to catalyse a limited number of chemical reactions. For this reason a prevalent view of molecular evolution involves a world dominated by RNA molecules that gradually evolved into a system in which proteins carried out catalysis, whilst nucleic acid performed a role of information storage, transfer and control. The details of this transition remain far from complete but supportive lines of evidence include: (i) the existence of different forms of RNA such as rRNA, mRNA, tRNA and genomic RNA; (ii) molecules such as nicotinamide adenine dinucleotide (NAD), adenosine tri- and diphosphate (ATP/ADP), and flavin adenine dinucleotide (FAD) found universally throughout cells are composed of adenine units analogous to those occurring in RNA; (iii) tRNAs have a tertiary structure; (iv) ribosomes represent hybrid RNA–protein systems where catalysis is RNA based; and (v) the enzyme RNaseP from *E. coli* catalyses the degradation of polymeric RNA into smaller nucleotide units in a reaction where RNA is the active component.

A major objection to the view of an 'RNA world' has been the comparative instability of RNA (especially when compared to DNA). RNA is easily degraded and it is difficult to see how stable systems capable of replication and catalysis evolved. Additionally the synthesis of polymers of RNA under conditions similar to those found early in the earth's history has proved remarkably difficult. Despite these problems most researchers view RNA as a likely intermediate between the 'primordial soup' and the systems of replication and catalysis found in modern cell types.

The fossil record evidence shows that bacteria-like organisms were present on earth 3.6 billion years ago. This implies that the systems of replication present today in living cells had already evolved. The 'RNA-directed world' was therefore a comparatively short time interval of ~0.5 billion years!

Having evolved a primitive replication and catalysis system based on RNA the simple amino acids could be used in protein biosynthesis. It is very unlikely that all amino acids were present in the primordial soup since some of the amino acids are relatively unstable especially under acidic conditions, and this includes the side chains of asparagine, glutamine, and histidine. In addition the amino acids found in proteins represent a very small subset of the total number of amino acids known to exist. This might suggest that the present class of amino acids evolved over millions of years to reflect a blend of chemical and physical properties required by proteins although it was essentially complete and intact 3.6 billion years ago.

Evolutionary divergence of organisms and its relationship to protein structure and function

The precise details of the origin of life involving the generation of a self-replicating system and its evolution into the complex multicellular structures found in higher plants and animals are unclear but the fossil record shows that primitive bacteria were present on earth in the pre-Cambrian period nearly 3.6 billion years ago. These fossilized cells resemble a class of bacteria found on present day Earth called cyanobacteria. Although it is surprising that bacteria leave fossil records cyanobacteria often form a significant cell wall together with layered structures called stromatolites (Figure 6.3). These structures form a mat as cyanobacteria grow trapping sediment and helping the fossilization process. Cyanobacteria are prokaryotic cells, lacking a nucleus and internal membranes, capable of both photosynthetic and respiratory growth and are represented today by genera such as *Nostoc*, *Anacystis* and *Synechococcus*.

Less ambiguously and more importantly the fossil record demonstrates progressive increases in complexity from the simple prokaryotic cell lacking a nucleus to more complicated structures similar to modern eukaryotic cells. Eukaryotic cells became multicellular and evolved by specialization towards specific cellular functions (Figure 6.4). These cells increased in structural complexity by internal compartmentalization with genome organization becoming more complex along with the variety of biochemical reactions catalysed within these cells. Whilst the fossil record aided our understanding of evolution one of the best methods of deciphering evolutionary pathways has come from comparing protein and DNA sequences.

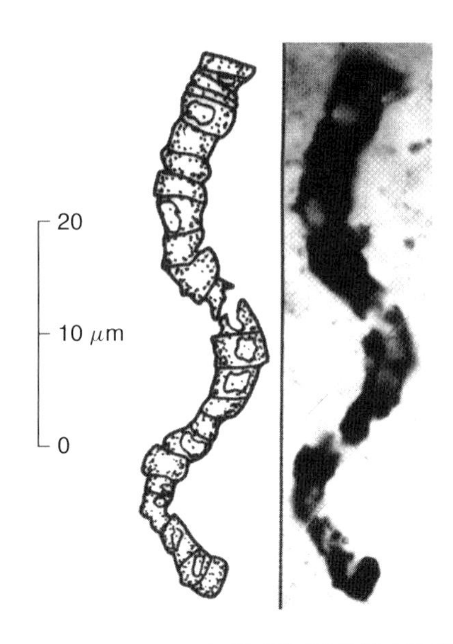

Figure 6.3 Microfossil of filamentous bacterial cells. The fossil shown alongside an interpretive drawing is from Western Australia and rocks dated at ~3.4×10^9 years (Reproduced with permission from Voet, D., Voet, J.G. and Pratt, C.W. *Fundamentals of Biochemistry*. John Wiley & Sons, Ltd, Chichester, 1999)

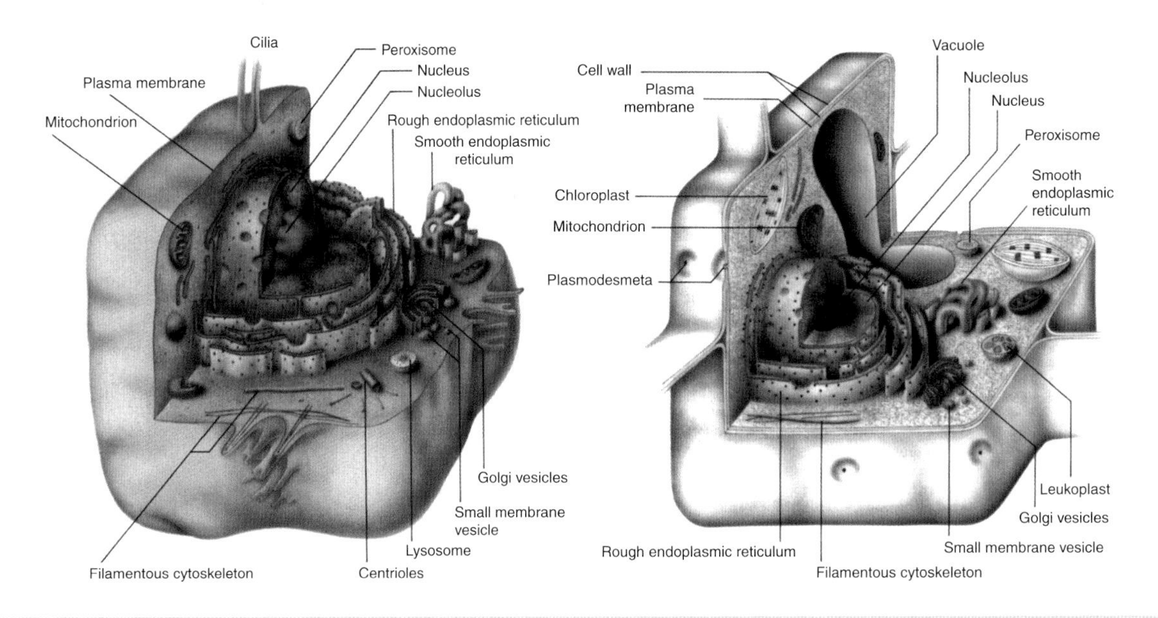

Figure 6.4 Generalized eukaryotic cells (plant and animal) (reproduced courtesy of Darnell, J.E. *et al. Molecular Cell Biology*. Scientific American, 1990)

In a protein of 100 amino acid residues there are 20^{100} unique or possible sequences. It is clear that biological sequences represent only a small fraction of the total number of permutations. Protein sequences often show similarities with these relationships governed by evolutionary lineage. Over millions of years a sequence can change but it cannot cause complete loss of function unless gene duplication has occurred. If mutation provides new enhanced functional activities any selective advantage conferred on the host organism possessing this protein will lead to improved survival and gene perpetuation.

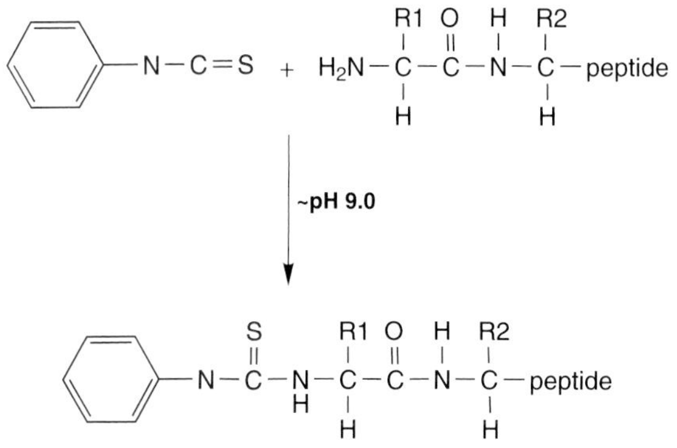

Protein sequence analysis

Protein sequencing

The sequencing of DNA has advanced so rapidly that this method is now by far the most common and effective way of determining the sequence of a protein. By translating the order of nucleotide bases along a DNA sequence one can simply derive the sequence of amino acid residues. However, there are occasions when it becomes important to sequence a protein directly and this might include determining the extent of post-translational processing or arranging peptide fragments in a linear order.

Protein sequencing is an automated technique carried out using sophisticated instruments (sequenators) and based on methods devised by Pehr Edman (it is often called Edman degradation). The unknown polypeptide is reacted under alkaline conditions (pH ~9) with phenylisothiocyanate (PTC) where the free amino group at the N-terminal forms a phenylthiocarbamoyl derivative, which is hydrolysed from the remaining peptide using anhydrous trifluoroacetic acid (Figures 6.5 and 6.6). PTC makes the first peptide bond less stable and easily hydrolysed. Residue rearrangement in aqueous acidic solution yield a phenylthiohydantoin (PTH) derivative of the N-terminal amino acid that is identified using chromatography or mass spectrometry (Figure 6.7). The significance of this series of reactions is that the N-terminal amino acid is 'tagged' by attaching PTC but the remaining polypeptide chain (now containing $n - 1$ residues) remains intact and can undergo further reactions with PTC at its new N-terminal residue. The Edman degradation is a repetitive, cyclical, series of reactions, although as with most repetitive procedures, errors accumulate and progressively degrade the accuracy of the whole process. Errors include: random breakage of the polypeptide chain producing a second free amino terminal residue; incomplete reaction between PTC and the N-terminal amino acid leading to its appearance in the next reactive cycle; and side reactions that compete with the reaction between PTC and the polypeptide chain. Sequenators are very sensitive

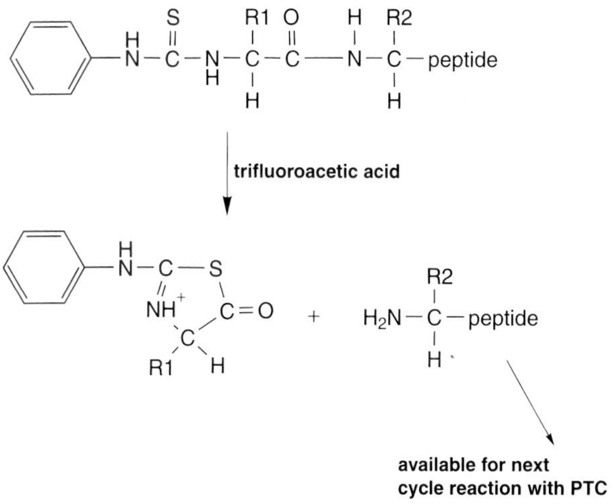

Figure 6.5 The reaction of the N-terminal amino acid residue with phenylisothiocyanate in the first step of the Edman degradation

Figure 6.6 Hydrolysis of the phenylthiocarbamoyl derivative of the peptide to yield a protein of $n - 1$ residues and a free 'labelled' amino acid

~1M HCl

PTH-amino acid

Figure 6.7 Re-arrangement of the PTC-derivative to form a phenylthiohydantoin (PTH) derivative of the N-amino acid

instruments capable of sequencing picomole amounts of polypeptide. Usually an upper limit on the length of the polypeptide chain that can be sequenced directly is about 70 residues. Since most proteins contain far more than 70 residues the sequencing procedures relies on 'chopping' the polypeptide chain into a series of smaller fragments that are each sequenced independently.

Generation of smaller peptide fragments involves using hydrolytic enzymes that cleave the polypeptide at specific sequences or by the use of cyanogen bromide that splits polypeptide chains after methionine residues (Table 6.2).

After purification of the individual fragments the shorter peptides are sequenced, although the major problem is now to deduce the respective order of each

Table 6.2 Enzymes or reagents for generating peptide fragments suitable for sequencing

Enzyme/reagent	Cleavage site
Trypsin	-Arg -↑-Yaa or Lys -↑-Yaa-
Endoprotease Arg-C	-Arg -↑-Yaa
Chymotrypsin	-Phe -↑-Yaa, -Tyr -↑-Yaa, -Trp -↑-Yaa
Clostripain	-Arg -↑-Yaa
Asp-N	-Xaa-↑-Asp
Thermolysin	-Xaa-↑-Leu, -Xaa-↑-Ile, -Xaa-↑-Val, -Xaa-↑-Met,
V-8 protease	-Asp -↑-Yaa, -Glu -↑-Yaa
Cyanogen bromide (CNBr)	-Met -↑-Yaa

In many cases the above enzymes show wider specificity. For example chymotrypsin will cleave other large side chains particularly Leu and care needs to be exercised in interpreting the results of proteolytic cleavage. In other instances the identity of Xaa/Yaa can influence whether cleavage occurs. For example Lys-Pro is not cleaved using trypsin.

Table 6.3 Fragments derived by digestion of unknown protein with Asp-N and trypsin

Digestion with trypsin	
Mass	Peptide sequence
4905.539	ITKPSESIITTIDSNPSWW TNWLIPAISALFVALIYHLYTSEN
2205.928	EQAGGDATENFEDVGHSTDAR
1511.749	FLEEHPGGEEVLR
1412.717	TFIIGELHPDDR
1186.599	YYTLEEIQK
1160.646	STWLILHYK
738.403	VYDLTK
650.299	AEESSK
599.290	HNNSK
476.271	ELSK
317.218	AVK
234.145	SK
Digestion with Asp-N	
Mass	Peptide Sequence
4100.113	AEESSKAVKYYTLEEIQK HNNSKSTWLILHYKVY
3621.805	DSNPSWWTNWLIPAISA LFVALIYHLYTSEN
2411.184	DLTKFLEEHPGGEEVLREQAGG
1825.981	DARELSKTFIIGELHP
1789.007	DRSKITKPSESIITTI
825.326	DATENFE
615.273	DVGHST
134.045	D

peptide. This problem is resolved by repeating the digestion of the intact protein with a second enzyme that reacts at different sites producing fragments whose relationship to the first set is established by sequencing to determine an unambiguous order of residues along a polypeptide chain. The principle of this method is demonstrated with an 'unknown' protein of 133 residues and its digestion with two enzymes; trypsin and Asp-N (Table 6.3 and Figure 6.8). Trypsin shows substrate specificity for lysine and arginine residues cleaving peptide bonds on the C terminal side of these residues whilst Asp-N is a protease that cleaves before aspartate residues.

To verify the primary sequence a total amino acid analysis is usually performed on the unknown protein by complete hydrolysis of the protein into individual amino acid residues. Quite clearly the total amino acid composition must equate with the combined number of amino acids derived from the primary sequence.

Amino acid analysis consists of three steps: (i) hydrolysis of the protein into individual amino acids; (ii) separation of the amino acids in this mixture; and (iii) identification of amino acid type and its quantification. Hydrolysis of the protein is normally complete after dissolving a small amount of the sample in 6 M HCl and heating the sample in a vacuum at 110 °C for 24 hours. The peptide bonds are broken leaving a mixture of individual amino acids. This approach destroys the amino acid tryptophan completely whilst cysteine residues may be oxidized and partially destroyed by these conditions. Similarly, acid hydrolysis of glutamine and asparagine side chains can form aspartate and glutamate and it is not usually possible to distinguish Asn/Asp and Glu/Gln in protein hydrolysates. For this reason protein sequences may be written as Glx or Asx representing the combined number of glutamine/glutamate and asparagine/aspartate residues.

Separation is achieved by cation exchange chromatography (Figure 6.9) using resins, supported within stainless steel or glass columns, containing negatively charge groups such as sulfonated polystyrenes. The negatively charge amino acids such as aspartate and glutamate elute rapidly whilst the flow of positively charged amino acids through the column is retarded due to interaction with the resin. By altering the polarity of the eluting solvent the interaction of hydrophobic amino acids with the column is enhanced. Amino acids with large hydrophobic side chains, for example phenylalanine and isoleucine, elute more slowly than smaller amino acids such as alanine and glycine.

To enhance detection the amino acids are reacted with a colored reagent such as ninhydrin, fluorescein, dansyl chloride or PTC. If this procedure is performed prior to column separation the absorbance of derivatized amino acids as they elute from the column is readily recorded at ~540 nm or from their fluorescence. Due to the high reproducibility of these

```
                    *        20         *        40         *
Unknown  :  AEESSKAVKYYTLEEIQKHNNSKSTWLILHYKVYDLTKFLEEHPGGEEVL  :   50
Unknown  :  AEESSKAVKYYTLEEIQKHNNSKSTWLILHYKVYDLTKFLEEHPGGEEVL  :   50

                   60         *        80         *       100
Unknown  :  REQAGGDATENFEDVGHSTDARELSKTFIIGELHPDDRSKITKPSESIIT  :  100
Unknown  :  REQAGGDATENFEDVGHSTDARELSKTFIIGELHPDDRSKITKPSESIIT  :  100

                    *       120         *
Unknown  :  TIDSNPSWWTNWLIPAISALFVALIYHLYTSEN  :  133
Unknown  :  TIDSNPSWWTNWLIPAISALFVALIYHLYTSEN  :  133
```

Figure 6.8 The figure shows the unknown protein whose primary sequence can be deduced from sequencing smaller fragments derived by digestion with trypsin (top line) and Asp-N (bottom line). For clarity alternate fragments derived using trypsin are shown in red and purple whilst the Asp-N fragments are shown in magenta and green

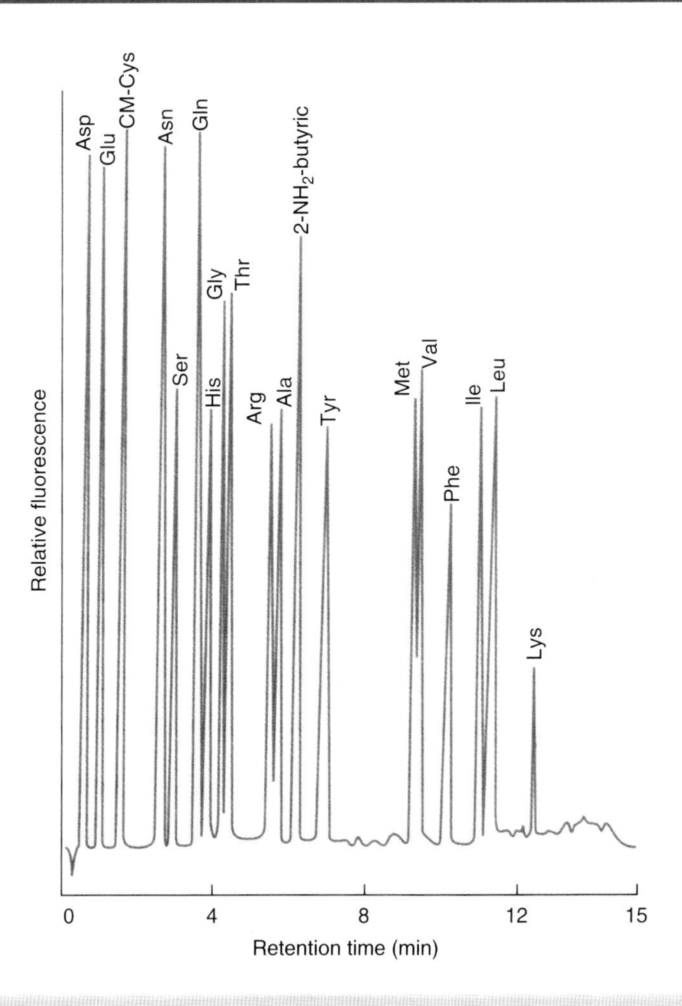

Figure 6.9 Elution of protein hydrolysate from a cation exchange column. The amino acids were derivatized first with a fluorescent tag to aid detection (after Hunkapiller, M.W. *et al. Science* 1984, **226**, 339–344)

profiles the amino acids of each type can be identified and their relative numbers quantified.

DNA sequencing

DNA sequencing methods are now routine and the recent completion of genomic sequencing projects testifies to the efficiency and accuracy of these techniques. The first genome sequencing projects were completed in the early 1980s for viruses and bacteriophages such as φX174 as well as organelles such as mitochondria that contain small circular DNA genomes. However in the last decade massive DNA sequencing projects were initiated and have resulted in vast numbers of primary sequences. In 2001 the human genome project was completed as a 'first draft' and the genomes of many other prokaryotic and eukaryotic organisms have been sequenced. This list includes the completion of the genome of the fruit fly *Drosophila melanogaster* as well as the genomes of many bacteria including pathogenic strains such as *Haemophilus influenzae*, *Helicobacter pylori*, *Yersinia pestis*, *Pseudomonas aeruginosa*, *Campylobacter jejuni*, as well as *E. coli* strain K-12. This has been complemented by completion of the genomes of *Saccharomyces pombe* and *cerevisiae* (the fission and baker's yeast, respectively), the nematode worm *Caenorhabditis elegans* and the plant genome of *Arabidopsis thaliana*. Unfortunately it remains largely true that we have no idea about the structure or function of many of the proteins encoded within these sequenced genomes.

Nowadays it is common to determine the order of bases along a gene and in so doing deduce the primary sequence of a protein. After locating the relevant start codon (ATG) it is straightforward to use the genetic code to translate the remaining triplet of bases into amino acids and thereby determine the primary sequence of the protein. There are many computer programs that will perform these tasks using the primary sequence data to derive additional properties about the protein. These properties can include overall charge, isoelectric point (pI), hydrophobicity and secondary structure elements and are part of the bioinformatics revolution that has accompanied DNA and protein sequencing.

For eukaryotes translation of the order of bases along a gene is complicated by the presence of introns or non coding sequences. The recognition of these sites or the use of cDNA derived from processed mRNA allows protein sequences to be routinely translated. In databases it is normal for DNA sequences of eukaryotic cells to reflect the coding sequence although occasionally sequences containing introns are deposited.

Gene sequencing is an automated technique that involves determining the order of only four components (adenine, cytosine, guanine, and thymine, i.e. A, C, G and T) compared with 20 different amino acids. With fewer components it becomes critical to correctly establish the exact order of bases since a mistake, such as an insertion or deletion, will result in a completely different translated sequence. DNA is amplified to produce many complementary copies of the template using the polymerase chain reaction (PCR) a process that exploits the activity of thermostable DNA polymerases. The PCR technique is divided into three steps: denaturation, annealing and extension (Figure 6.10). Each step is optimized with respect to time and temperature. However, the process is generally performed at three different temperatures of ~96, ~55 and ~72 °C with each phase lasting for about 30, 30 and 120 s, respectively.

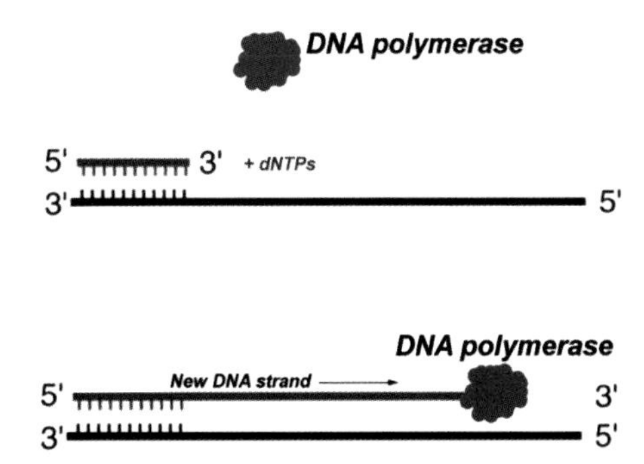

Figure 6.10 General principle of primer-directed DNA synthesis by polymerases. The PCR extends this process in a cyclical series of reactions since the polymerase is thermostable and withstands repeated cycles of high temperature

DNA polymerase activity extends new strands from primers (15–25 bases in length) that are complementary to the two template strands (Figure 6.11). A polymerase commonly used in these studies is that isolated from *Thermus aquaticus* and often referred to as *Taq* polymerase. The double helix is dissociated at high temperatures (~96 °C) and on cooling the suspension primers anneal to each DNA strand (~55 °C). With a suitable supply of nucleotide triphosphates (dNTPs, Figure 6.12) new DNA strand synthesis by the polymerase occurs rapidly at 72 °C forming two complementary strands. After one cycle the PCR doubles the number of copies and the beauty of the process is that it can be repetitively cycled to provide large amounts of identical DNA. It can be seen that starting with one copy of DNA leads after thirty rounds of replication to 2^{30} copies of DNA – all identical to the initial starting material.[1] This procedure is used to amplify DNA fragments as part of a cloning, mutagenesis or forensic study but it can also be used to sequence DNA.

If PCR techniques are carried out with dNTPs and a small amount of dideoxy (ddNTPs) nucleotides then chain termination will result randomly along the new DNA strand. Dideoxynucleotides were first introduced by Frederick Sanger as part of a sequencing strategy based on the absence of an oxygen atom at the

[1] Only with 'proofreading' DNA polymerases. Taq polymerase is not proof reading and exhibits an error rate of ~1 in 400 bases.

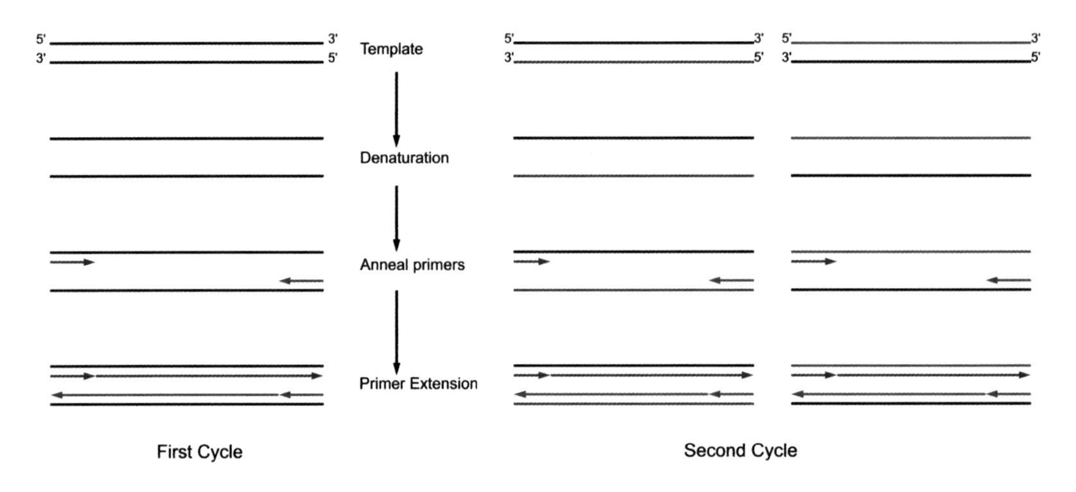

Figure 6.11 The use of PCR-based methods for amplification and replication of template DNA

Figure 6.12 The base and sugar components of NTPs. In this case the deoxyadenosine and dideoxyadenosine (note absence of hydroxyl at C3′ position)

C3′ position of the ribose ring. The effect of the missing oxygen is to prevent further elongation of the nucleotide chain in the 5′–3′ direction via an inhibition of phosphodiester bond formation.

When small amounts of the four nucleotides are present as ddNTPs random incorporation will result in chain termination (Figure 6.13). On average the PCR will result in a series of DNA fragments truncated at every nucleotide, each fragment differing from the previous one by just one base in length. Quite clearly if we can establish the identity of the last base we can gradually establish the DNA sequence. The second step of DNA sequencing separates these DNA fragments. To aid identification each ddNTP is also labelled with a fluorescent probe based on the chromophores fluoroscein and rhodamine 6 G. Each of the four dideoxy nucleotides is tagged with a different fluorescent dye that has an emission maxima (λ_{max}) that allows the final base to be discriminated. Dye-labelled DNA fragments are separated according to mass by running through polyacrylamide gels or capillaries. Today the most sophisticated DNA sequencing systems use capillary electrophoresis with the advantages of high separation efficiency, fast separations at high voltages, ease of use with small sample volumes ($\sim$1 μ l), and high reproducibility. 'Labelled' DNA exits the capillaries and laser-induced fluorescence detected by a charge coupled device (CCD) is interpreted as a fluorescence profile by a computer leading to routine sequencing of over 1000 nucleotides with very low error rates (Figure 6.14).

Sequence homology

Advances in both DNA and protein sequencing have generated enormous amounts of data. This data represents either the order of amino acids along a polypeptide chain or the order of bases along a nucleotide chain and contains within this 'code' significant supplementary information on proteins. With the introduction of powerful computers sequences can be analysed and compared with each other. In particular computational analysis has

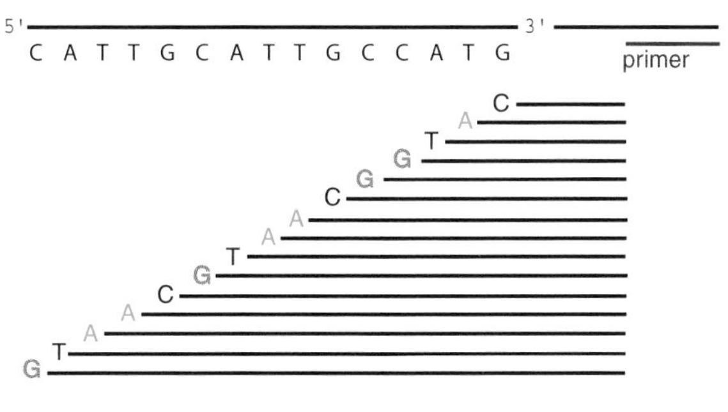

Figure 6.13 Chain termination of an extending DNA sequence by incorporation of dideoxynucleotide triphosphates (ddNTPs)

allowed the recognition of sequence similarity. For any sequence there is a massive number of permutations and similarity does not arise by chance. Instead sequence similarity may indicate evolutionary links and in this context the term homology is used reservedly. Protein sequences can be similar without needing to invoke an evolutionary link but the term homology implies evolutionary lineage from a common ancestor. Both DNA and protein sequences can show homology. In order to establish that protein sequences are homologous we have to establish rules governing this potential similarity. Consider the partial sequences

-D-E-A-L-V-S-V-A-F-T-S-I-V-G-G-

-D-E-A-F-T-S-I-V-G-G-M-D-D-P-G-

This represents a small section of the polypeptide chain (15 residues) in each of two sequences. Are they

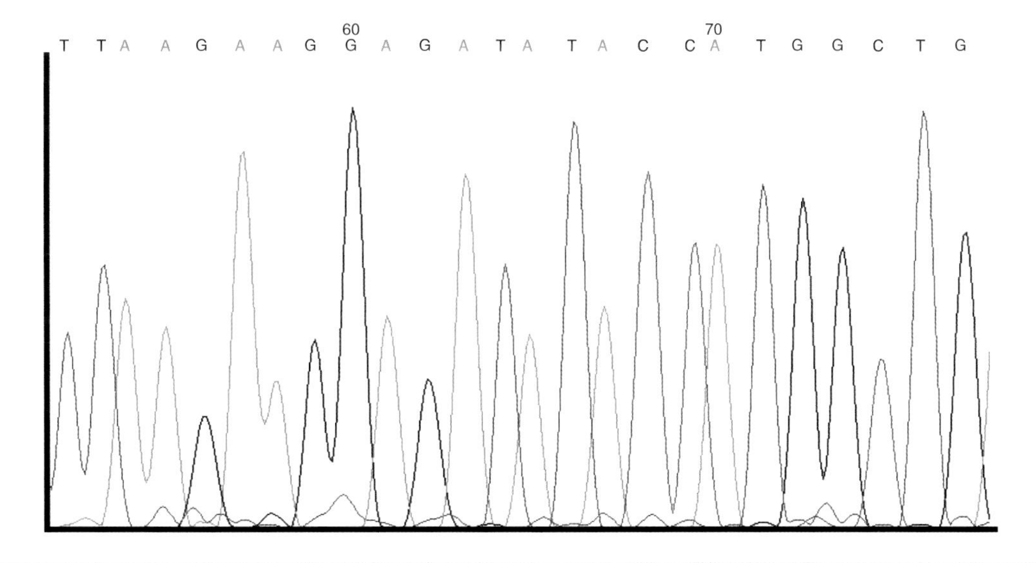

Figure 6.14 A small section of the computer-interpreted fluorescence profile of a DNA sequence. A is denoted by the green trace, C in blue, G in black and T by the red trace. Sequence anomalies arise when 'clumps' of bases are found together, for example, a block of T residues and may require user intervention

similar and if so how similar or are they different? An initial inspection 'by eye' might fail to establish any relationship. However, closer examination would reveal that the two sequences could be 'aligned' by introducing a gap of seven residues into the second sequence. When the sequences are re-written as

D-E-A-L-V-S-V-A-F-T-S-I-V-G-G-
D-E-*-*-*-*-*-*-*-T-S-I-V-G-G-M-D-D-P-G

a good alignment exists for some of the residues and significantly these residues are identical. Clearly to some observers this is significant similarity (8 out of 15 residues are identical) whilst to others it could be viewed as a major difference (7 residues are missing and information is lacking about the similarity of the last 5 residues in the second sequence!). What is needed is a way of quantifying this type of problem.

Alignment of protein sequences is the first step towards quantifying similarity between one or more sequences. As a result of point mutations or larger mutational events sequences change giving proteins containing different residues. This obscures relationships between proteins and one reason for comparing and aligning sequences is to deduce these relationships. For newly determined sequences this allows identification to previously characterized proteins and highlights a shared common origin.

In Chapter 3 the tertiary structures of haemoglobin α and β chains were superimposed to reveal little difference in respective fold (Figure 3.43). This suggested an evolutionary relationship as a result of ancestral gene duplication. Structural homology, supported by significant sequence homology between α and β chains, reveals a level of identity between the two chains of over 40 percent (62/146) (Figure 6.15).

Domains are key features of modular or mosaic proteins. Sequence alignments reveal that gene duplication leads to a proliferation of related domains in different proteins. As a result, proteins are related by the presence of similar domains – one example is the occurrence of the SH3 domain in proteins that share little else in common except the presence of this motif. SH3 (or Src Homology 3) domains are small, non-catalytic, modules of 50–70 residues that mediate protein–protein interactions by binding to proline-rich peptide sequences. The domain was discovered in tyrosine kinase as one module together with SH1 (tyrosine kinase) and SH2 (phosphotyrosine binding). The domain is found in kinases, lipases, GTPases, structural proteins and viral regulatory proteins.

Alignment methods offer a way of pictorially representing similarity between one or more 'test' sequences and a library of 'known' sequences derived from databases. In addition alignment methods offer a route towards quantifying the extent of this similarity by incorporating 'scoring' schemes. A number of different approaches exist for aligning sequences. A prevalent approach is called a 'pairwise similarity' and involves comparing each sequence in the database (library) with a 'test' sequence (Figure 6.16). The observation of 'matches' indicates sequence similarity. A second level of comparison involves comparing families of sequences with libraries to establish relationships (Figure 6.17). This approach establishes 'profiles' for the initial family and then attempts to fit this profile to other members of the database.

A third approach is to use known motifs found within proteins. These motifs are invariant or highly conserved blocks of residues characteristic of a protein family. These motifs are used to search databases for other sequences bearing the corresponding motif.

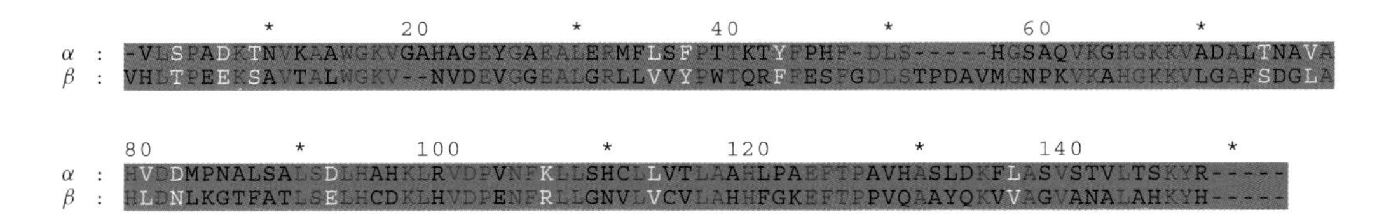

Figure 6.15 The sequence of the α and β chains of haemoglobin. Identical residues are shown in red whilst conserved residues are shown in yellow

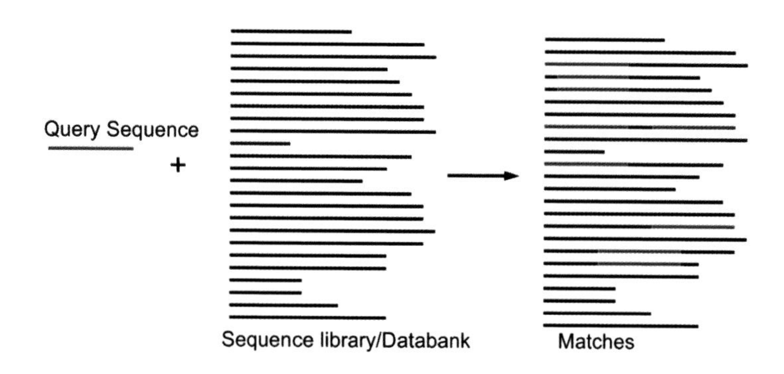

Figure 6.16 Pairwise similarity search of the databank using single 'query' sequence

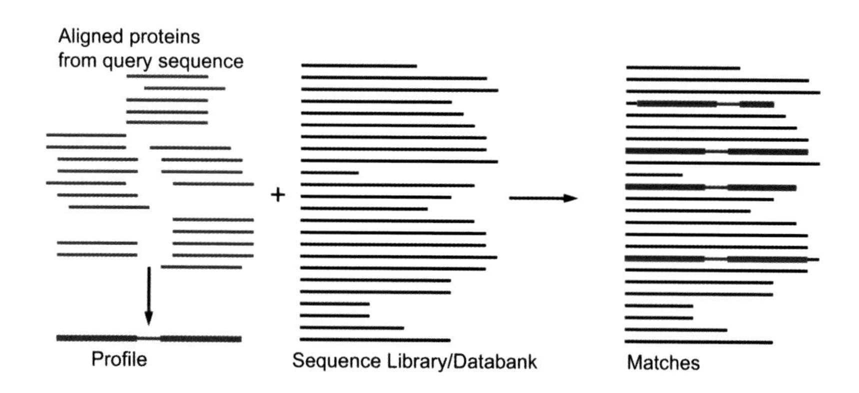

Figure 6.17 The use of a profile established from aligned proteins to improve quality of matches

Table 6.4 A selection of the characteristic motifs used to identify proteins

Motif sequence	Protein family	Example
CX_2CH	Class I soluble c type cytochromes	Bacterial cytochrome c551
$F(Y)L(IVMK)X_2HPG(A)G$	Cytochrome b_5 family	Nitrate reductase
$CX_7L(FY)X_6F(YW)XR(K)X_8CXCX_6C$	Ribosomal proteins	L3 protein
$A(G)X_4GKS(T)$.	ATPases	ATP synthetase
$CX_2CX_3L(IVMFYWC)X_8HX_3H$	Zn finger-DNA binding proteins	TFIIIA
LKEAExRAE	Tropomyosin family	Tropomyosin

Motifs are simply short sequences of amino acid residues within a polypeptide chain that facilitate the identification of related proteins. The residues in parenthesis are alternative residues that make up the consensus motif. The X indicates the number of intervening residues lacking any form of consensus. The number of intervening residues can show minor variations in length.

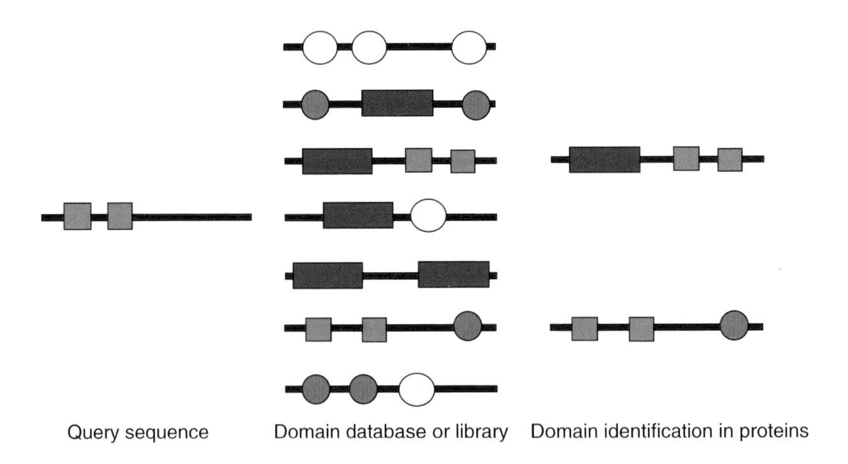

Figure 6.18 A query sequence is used with a domain-based database to identify similar proteins containing recognized domains

Diagnostic motifs occur regularly in proteins with whole databases devoted to their identification (Table 6.4). Many protein families have conserved sequence motifs and it is often sufficient to 'query' the database only with this motif to identify related proteins (Figure 6.18).

The best forms of sequence alignment are derived via computational methods known as dynamic programming that detect optimal pairwise alignment between two or more proteins. The details of computational programming are beyond the scope of this book but the procedure compares two or more sequences by looking for individual characters, or a series of character patterns, that are in the same order in each sequence. Identical or similar characters are placed in the same column, and non-identical characters can be placed in the same column either as a mismatch or opposite a 'gap' in the other sequences. Irrespective of the approach used to derive alignments a query sequence is compared with all sequences within a selected database to yield a 'score' that indicates a level of similarity. An optimal alignment will result in the arrangement of non-identical characters and gaps minimized to yield identical or similar characters as a vertical register. The art of these methods is to create a feasible scoring scheme, and the identity matrix or matrices based on physicochemical properties often fail to establish relationships, even for proteins known to be related.

The most important improvement in scoring schemes came from methods based on the observed changes in amino acid sequence in homologous proteins. This allowed mutation rates to be derived from their evolutionary distances and was pioneered by Margaret Dayhoff in the 1970s. The study measured the frequencies with which residues are changed as a result of mutation during evolution and involved carefully aligning (by eye) all proteins recognized to be within a single family. The process was repeated for different families and used to construct phylogenetic trees for groups of proteins. This approach yielded a table of relative frequencies describing the rate of residue replacement by each of the nineteen other residues over an evolutionary period. By combining this table with the relative frequency of occurrence of residues in proteins a family of scoring matrices were computed known as point accepted mutation (PAM) matrices. The PAM matrices are based on estimated mutation rates from closely related proteins and are effective at 'scoring' similarities between sequences that diverged with evolutionary time. In the PAM 250 matrix the data reflects aligned protein sequences extrapolated to a level of 250 amino acid replacements per 100 residues per 100 million years. A score above 0 indicates that amino acids replace each other more frequently than expected from their distribution in proteins and this usually means that such residues are functionally equivalent. As expected mutations involving the substitution of D > E or D > N are relatively common whilst transitions such as D > R or D > L are rare.

An alternative approach is the BLOCKS database based on ungapped multiple sequence alignments that correspond to conserved regions of proteins. These 'blocks' were constructed from databases of families of related proteins (such as Pfam, ProDom, InterPro or Prosite). This has yielded approximately 9000 blocks representing nearly 2000 protein families. The BLOCKS database is the basis for the BLOSUM substitution matrices that form a key component of common alignment programs such as BLAST, FASTA, etc. (BLOSUM is an acronym of BLOcks SUbstitution Matrices.) These substitution matrices are widely used for scoring protein sequence alignments and are based on the observed amino acid substitutions in a large set of approximately 2000 conserved amino acid patterns, called blocks. The blocks act as identifiers of protein families and are based around a greater dataset than that used in the PAM matrices. The BLOSUM matrices detect distant relationships and produce alignment that agrees well with subsequent determination of tertiary structure. In general if a test sequence shares 25–30 percent identity with sequences in the database it is likely to represent a homologous protein. Unfortunately, when the level of similarity falls below this level of identity it proves difficult to draw firm conclusions about homology.

Structural homology arising from sequence similarity

With large numbers of potential sequences for proteins the number of different folded conformations might also be expected to be large. However, the tertiary or folded conformations of proteins are less diverse than would be expected from the total number of sequences. Protein folds have been conserved during evolution with conservation of structure occurring despite changes in primary sequence.

In most cases structural similarities arise as a result of sequence homology. However, in a few instances structural homology has been observed where there is no obvious evolutionary link. One family of proteins that clearly indicates both sequence and structural homology is the cytochrome c family. Vertebrate cytochrome c isolated from mitochondria, such as those from horse and tuna, share very similar primary sequences (only 17 out of 104 residues are different) and this is emphasized by comparable tertiary structures (Figure 6.20). The conservation of structure is expected as both horse and tuna proteins occupy similar functional roles. Yeast cytochrome c also exhibits homology and a similar tertiary structure but a reduced level of sequence identity to either horse or tuna cytochrome c (~59 out of 104 residues are identical) reflects a more distant evolutionary lineage (hundreds of millions of years). In bacteria a wide range of *c* type cytochromes are known all containing the heme group covalently ligated to the polypeptide via two thioether bridges derived from cysteine residues. The proteins are soluble, contain between 80 and 130 residues, and function as redox carriers.

If the sequences of cytochrome c_2 from *Rhodobacter rubrum*, cytochrome c-550 from *Paracoccus denitrificans* and the mitochondrial cytochromes c from yeast, tuna and horse are compared only 18 residues remain invariant (Figure 6.19). These residues include His

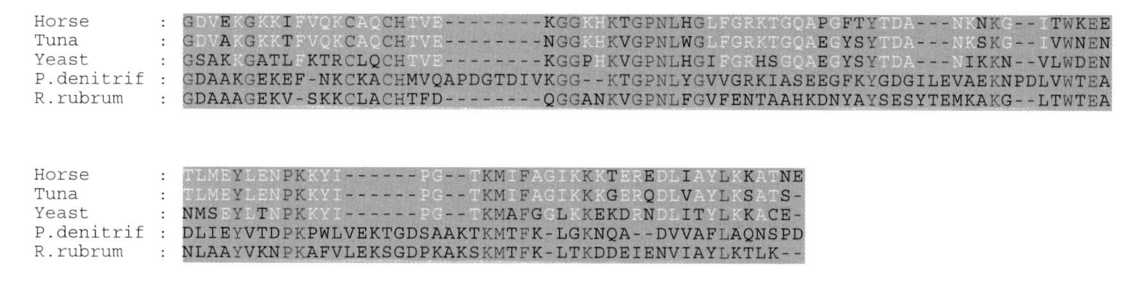

Figure 6.19 Sequence homology between the cytochromes c of horse, tuna, yeast, *R. rubrum* and *P. denitrificans*. Shown in red are identical residues between the sequences whilst yellow residues highlight those residues that are conserved within the horse, tuna and yeast eukaryotic sequences. Only 18 out of 104 residues show identity

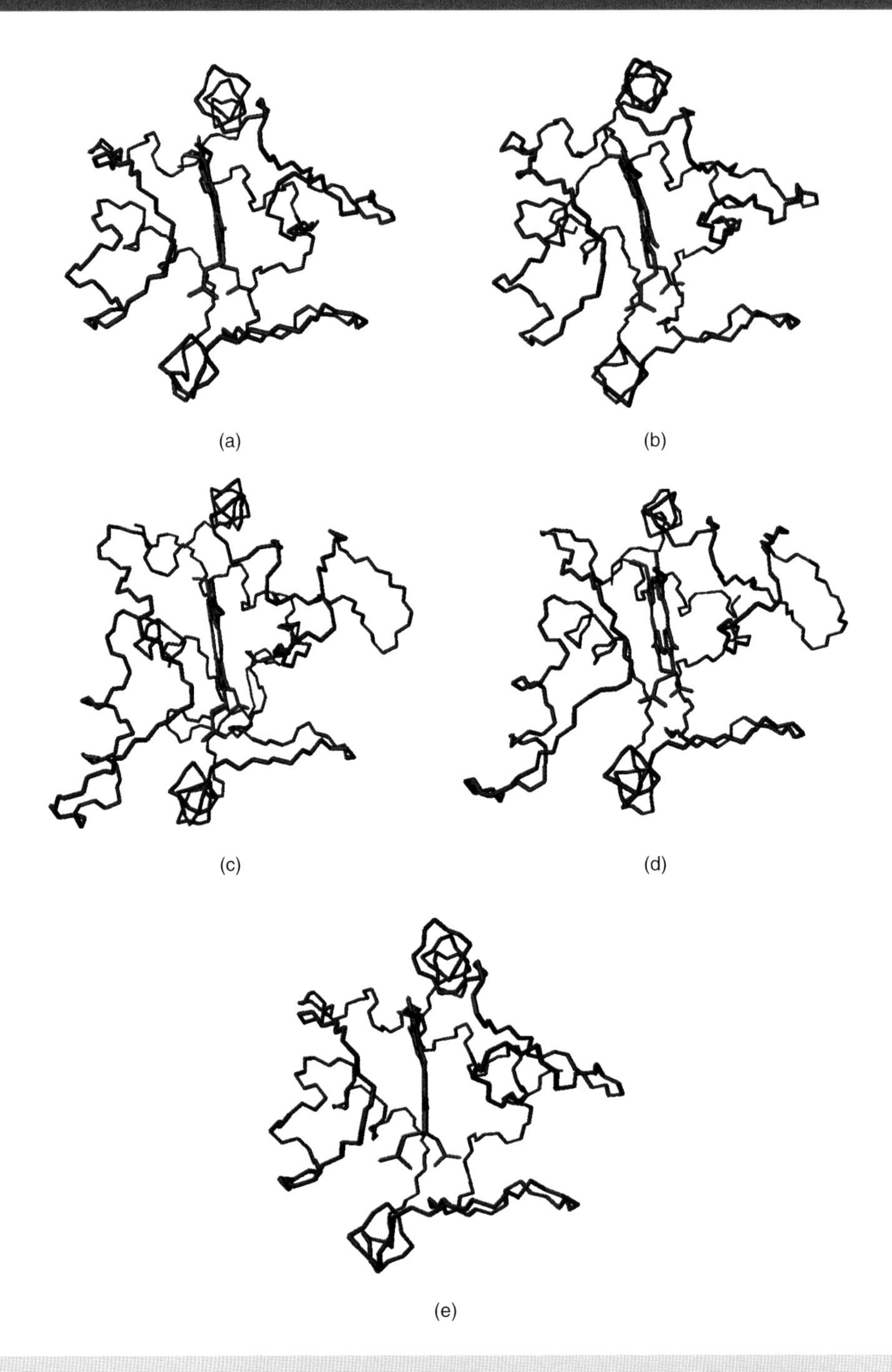

Figure 6.20 The structures of cytochrome c from different source organisms. The structures shown for five cytochrome c were obtained from (a) tuna (PDB: 3CYT), (b) yeast – the iso-1 form (PDB:1YCC), (c) *P. denitrificans* (PDB:155C), (d) *R. rubrum* cytochrome c_2 (PDB:1C2R), and (e) horse (PDB:1CRC). The insertions of residues in the bacterial sequences can be seen in the additional backbone structure shown in the bottom left region of the molecule

18, Met80, Cys14 and Cys17 and are critical to the functional role of cytochrome c in electron transfer. Consequently their conservation is expected. There is very little sequence similarity between cytochrome c_2 and horse cytochrome c but an evolutionary lineage is defined by tracking progressive changes in sequence through micro-organisms, animal and plants. In this manner it is clear that cytochrome c represents the systematic evolution of a protein designed for biological electron transfer from a common ancestral protein. A more thorough analysis of the sequences reveals that although changes in residue occur at many positions the majority of transitions involve closely related amino acid residues. For example, near the C terminus of each cytochrome c a highly conserved Phe residue is usually found but in the sequence of *P. denitrificans* this residue is substituted with Tyr.

Irrespective of the changes in primary sequence for these cytochromes c considerable structural homology exists between all of these proteins. This homology extends from the protein found in the lowliest prokaryote to that found in man. It is only by comparing intermediate sequences between cytochrome c_2 and horse cytochrome c that evolutionary links are established but structural homology is a strong indicator that the proteins are related. As a caveat although structural homology is normally a good indicator of relationships it is not *always* true (see below) and must be supported by sequence analysis.

The serine proteases are another example of structural homology within an evolutionarily related group of proteins. This family of enzymes includes familiar proteins from higher organisms such as trypsin, chymotrypsin, elastase and thrombin and they have the common function of proteolysis. If the sequences of trypsin, chymotrypsin and elastase are compared (Figure 6.21) they exhibit an identity of ~40 percent and this is comparable to the identity shown between the haemoglobin α and β chains. The three-dimensional structure of all of these enzymes are known and they share similar folded conformations with invariant His and Ser residues equivalent to positions 57 and 195 in chymotrypsin located in the active sites along with the third important residue of Asp 102. Together these residues make up the catalytic triad and from their relative positions other members of the family of serine proteases have been identified. The similarity in

Figure 6.21 The structures of chymotrypsin, trypsin and elastase shown with elements of secondary structure superimposed. Chymotrypsin (blue, PDB: 2CGA), elastase (magenta, PDB: 1QNJ) and trypsin (green, PDB: 1TGN). Shown in yellow in a cleft or active site are the catalytic triad of Ser, His and Asp (from left to right) that are a feature of all serine protease enzymes

sequence and structure between chymotrypsin, elastase and trypsin indicates that these proteins arose from gene duplication of an ancestral protease gene with subsequent evolution accounting for individual differences.

By following changes in sequence for different proteins in a family an average mutation rate is calculated and reveals that 'house-keeping' proteins such as histones, enzymes catalysing essential metabolic pathways, and proteins of the cytoskeleton evolve at very slow rates. This generally means the sequences incorporate between 1 and 10 mutations per 100 residues per 100 million years. Consequently to obscure all evolutionary information, which generally requires ~250–350 substitutions per 100 residues, takes a considerable length of time. As a result of this slow

Table 6.5 Rate of evolution for different proteins (adapted from Wilson, A.C. *Ann. Rev. Biochem.* 1977, 46, 573–639)

Protein	Accepted point mutations/100 residues/10^8 years	Protein	Accepted point mutations/100 residues/10^8 years
Histone H2	0.25	Insulin	7
Collagen α_1	2.8	Glucagon	2.3
Cytochrome c	6.7	Triose phosphate isomerase	5.3
Cytochrome b_5	9.1	Lactate dehydrogenase M chain	7.7
Lysozyme	40	α-lactalbumin	43
Ribonuclease A	43	Immunoglobulin V region	125
Myoglobin	17	Haemoglobin α	27
Histone H1	12	Haemoglobin β	30

rate of evolution 'house-keeping' proteins are excellent tools with which to trace evolutionary relationships over hundreds of millions of years. Higher rates of evolution are seen in proteins occupying less critical roles.

Mutations arising in DNA are not always converted into changes in protein primary sequence. Some mutations are silent due to the degeneracy inherent in the genetic code. A nonsense mutation results from the insertion of a stop codon within the open reading frame of mRNA and gives a truncated polypeptide chains. The generation of a stop codon close to the start codon will invariably lead to a loss of protein activity whilst a stop codon located relatively near to the original 'stop' sight may well be tolerated by the protein. Missense mutations change the identity of a residue by altering bases within the triplet coding for each amino acid. From the standpoint of evolutionary analysis it is these mutations that are detected *via* changes in primary sequences.

Proteins with different structures and functions evolve at significantly different rates. This is seen most clearly by comparing proteins found in *Homo sapiens* and *Rhesus* monkeys. For cytochrome c the respective primary sequences differ by less than 1 percent of their residues but for the α and β chains of haemoglobin these differences are at a level of 3–5 percent whilst for fibrinopeptides involved in blood clotting the differences are ~30 percent of residues. This emphasizes the need to use families of proteins to extrapolate rates of protein evolution. Systematic studies have determined rates of evolution for a wide range of proteins of different structure and function (Table 6.5) with mutation rates expressed as the number of point mutations per 100 residues per 10^8 years.

Conotoxins are a family of small peptides derived from the *Conus* genus of predatory snails. These snails have a proboscis containing a harpoon-like organ that is capable of injecting venom into fish, molluscs and other invertebrates causing rapid paralysis and death. The venom contains over 75 small toxic peptides that are between 13 and 35 residues in length and disulfide rich. The toxins have been purified and exhibit varying degrees of toxicity on the acetylcholine receptor of vertebrate neurones. The peptides represent a molecular arms race whereby rapid evolution has allowed *Conus* to develop 'weapons' in the form of toxins of different sequence and functional activity. As potential prey adapt to the toxins *Conus* species evolve new toxins based around the same pattern. Whereas proteins such as histones show evolutionary rates of change of ~0.25 point mutations/100 residues/10^8 years the conotoxins have much higher rates of change estimated at ~60–180 point mutations/100 residues/10^8 years. This pattern is supported by analysis of other toxins such as those from snake venom where typical mutation rates are ~100 point mutations/100 residues/10^8 years.

The rapid development of molecular cloning techniques, DNA sequencing methods, sequence comparison algorithms and powerful yet affordable computer workstations has revolutionized the importance of protein sequence data. Thirty years ago protein sequence determination was often one of the final steps in the characterization of a protein, whilst today one premise of the human genome mapping project is that sequencing all of the genes found in man will uncover their function via data analysis. There is no doubt that this premise is beginning to yield rich rewards with the identification of new homologues as well as new open reading frames (ORFs), but in many cases sequence data has remained impervious to analysis.

The above examples highlight structural homology that has persisted from a common ancestral protein despite the slow and gradual divergence of protein sequences by mutation. Occasionally structural homology is detected where there is no discernable relationship between proteins. This is called convergent evolution and arises from the use of similar structural motifs in the absence of sequence homology. In subtilisin and serine carboxypeptidase II a catalytic triad of Ser-Asp-His is observed (Figure 6.22) that might imply a serine protease but the arrangement of residues within their primary sequences are different to chymotrypsin and these proteins differ in overall structure. It is therefore very unlikely that these three proteins could have arisen from a common ancestor of chymotrypsin or a related serine protease since the order

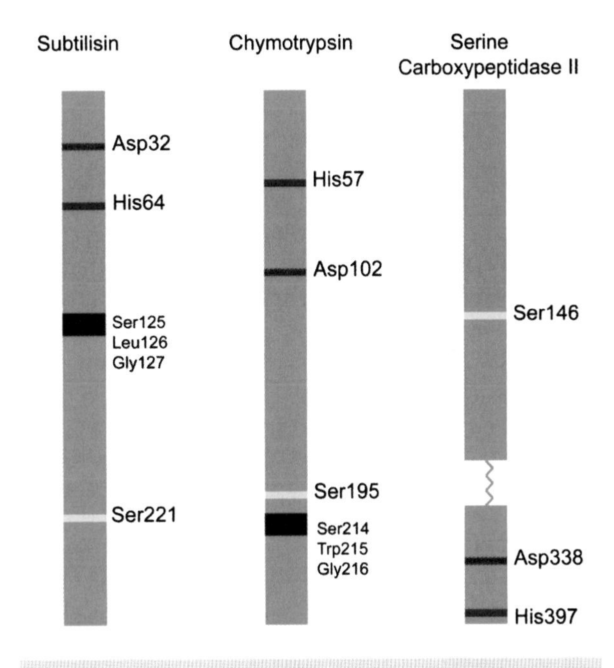

Figure 6.22 The positions of the catalytic triad together with hydrophobic residues involved in substrate binding in unrelated serine proteases. Subtilisin, chymotrypsin and serine carboxypeptidase II

Table 6.6 Potential examples of convergent evolution in proteins

Protein 1	Protein 2	Functional motif
Chymotrypsin	Subtilisin Carboxypeptidase II	Catalytic triad of Ser/Asp/His residues in the proteolysis of peptides
Triose phosphate isomerase	As many as 17 different groups of enzymes possess β barrel. Luciferase, pyruvate kinase and ribulose 1,5 bisphosphate carboxylase-oxygenase (rubisco)	TIM barrel fold of eight parallel β strands forming a cylinder connected by eight helices arranged in an outer layer of the protein
Carbonic anhydrase (α)	Carbonic anhydrase (β) form plants.	Zn ion ligated to protein and involved in catalytic conversion of CO_2 to HCO_3^-
Thermolysin	Carboxypeptidase A	Zn ion ligated to two imidazole side chains and the carboxyl side chain of Glu. Also coordinated by water molecule playing a crucial role in proteolytic activity.

of residues in the catalytic triad differs as do other structural features present in each of the enzymes. This phenomenon is generally viewed as an example of convergent evolution where nature has discovered the same catalytic mechanism on more than one occasion (see Table 6.6).

The term convergent evolution has been applied to the Rossmann fold found in nucleotide binding domains consisting of three β strands interspersed with two α helices with a common role of binding ligands such as NAD^+ or ADP. In many proteins these structural elements are identified, but little sequence homology exists between nucleotide-binding proteins. The sequences reflect the production of comparable topological structures for nucleotide-binding domains by different permutations of residues. Are these domains diverged from a common ancestor or do they result from convergent evolution? The large number of domains found in proteins capable of binding ATP/GTP or NAD/NADP suggests that it is unlikely that proteins have frequently and independently evolved a nucleotide-binding motif domain. It is now thought more likely that the Rossmann fold evolved with numerous variations arising as a result of divergent evolution from a common and very distant ancestral protein. The β barrel exemplified by triose phosphate isomerase is also widely found in proteins and may also represent a similar phenomenon.

Protein databases

A number of important databases are based on the sequence and structure information deposited and archived in the Protein Data Bank. These databases attempt to order the available structural information in a hierarchical arrangement that is valuable for an analysis of evolutionary relationships as well as enabling functional comparisons. In the SCOP (Structural Classification of Proteins) database all of the deposited protein structures are sorted according to their pattern of folding. The folds are evaluated on the basis of their arrangement and in particular the composition and distribution of secondary structural elements such as helices and strands. Frequently construction of this hierarchical system is based on manual classification of protein folds and as such may be viewed as subjective. Currently, non-subjective methods are being actively pursued in other databases in an attempt to remove any bias in classification of protein folds. The levels of organization are hierarchical and involve the classes of folds, superfamilies, families and domains (the individual proteins).

Domains within a family are homologous and have a common ancestor from which they have diverged. The homology is established from either sequence and/or functional similarity. Proteins within a superfamily have the same fold and a related function and therefore also probably have a common ancestor. However, within a superfamily the protein sequence composition or function may be substantially different leading to difficulty in reaching a conclusive decision about evolutionary relationships. At the next level, the fold, the proteins have the same topology, but there is no evidence for an evolutionary relationship except a limited structural similarity.

The CATH database attempts to classify protein folds according to four major hierarchical divisions; Class, Architecture, Topology and Homology (see Figure 6.23 for an example). It also utilizes algorithms to establish definitions of each hierarchical division. Class is determined according to the secondary structure composition and packing within a protein structure. It is assigned automatically for most structure (>90 percent) with manual inspection used for 'difficult' proteins. Four major classes are recognized; mainly α, mainly β and α–β protein domains together with domains that have a very low secondary structure content. The mixed α–β class can be further divided into domains with alternating α/β structures and domains with distinct α rich and β rich regions (α + β). The architecture of proteins (A-level hierarchy) describes the overall shape of the domain and is determined by the orientations of the individual secondary structural elements. It ignores the connectivity between these secondary structures and is assigned manually using descriptions of secondary structure such as β barrel or β–α–β sandwich. Several well-known architectures have been described in this book and include the β propellor, the four-helix bundle, and the helix-turn-helix motif. Structures are grouped into fold families or topologies at the next

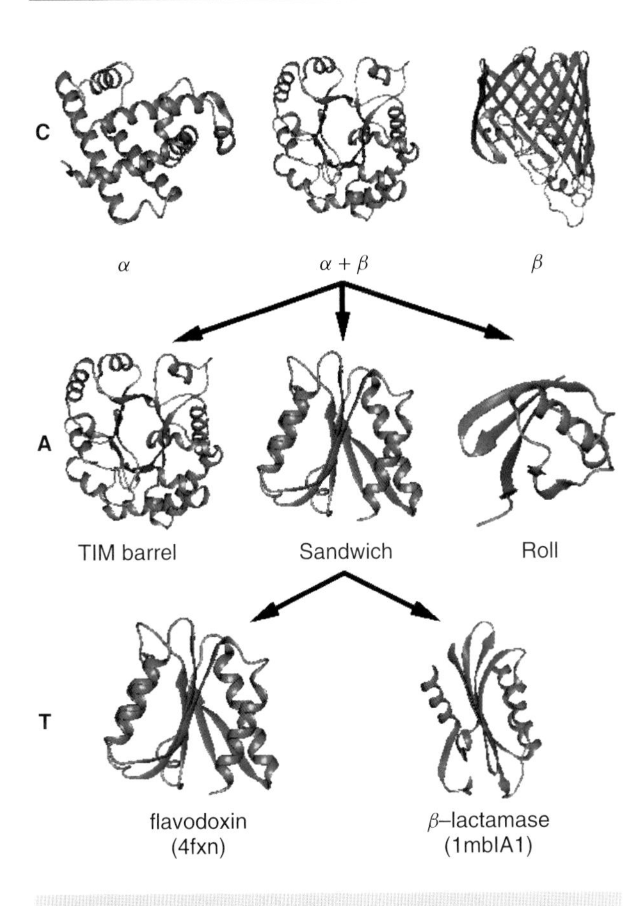

Figure 6.23 The distinguishing of flavodoxin and β lactamase; two α/β proteins based around a β sandwich structure

level of organization. Assignment of topology depends on the overall shape and connectivity of the secondary structures and has started to be automated via algorithm development. A number of topologies or fold families have been identified, with some, such as the β two-layer sandwich architecture and the α–β three-layer sandwich structures, being relatively common. The H level of classification represents the homologous superfamily and brings together protein domains that share a common ancestor. Similarities are identified first by sequence comparisons and subsequently by structural comparisons. Hierarchical levels of organization are shown at the C, A and T states for α/β containing proteins.

Gene fusion and duplication

The preceding sections highlight that some proteins share similar sequences reflected in domains of similar structure. The α + β fold of cytochrome b_5 first evolved as an electron transfer protein and then became integrated as a module found in larger multidomain proteins. The result was that ligated to a hydrophobic tail the protein became a membrane-bound component of the endoplasmic reticulum fatty acid desaturase pathway. When this domain was joined to a Mo-containing domain an enzyme capable of converting sulfite into sulfate was formed, sulfite oxidase. Joining the cytochrome to a flavin-containing domain formed the enzyme, nitrate reductase. Proteins sharing homologous domains arose as a result of gene duplication and a second copy of the gene. This event is advantageous to an organism since large genetic variation can occur in this second copy without impairing the original gene.

The globin family of proteins has arisen by gene duplication. Haemoglobin contains 2α and 2β chains and each α and β chain shows homology and is similar to myoglobin. This sequence homology reflects evolutionary origin with a primitive globin functioning simply as an oxygen storage protein like myoglobin. Subsequent duplication and evolution allowed the subtle properties of allostery as well as the differences between the α and β subunits. During embryogenesis other globin chains are observed (ξ and ε chains) and fetal haemoglobin contains a tetramer made up of $\alpha_2\gamma_2$ subunits that persists in adult primates at a level of ~1 percent of the total haemoglobin. The δ chain is homologous to the β subunit and leads to the evolution of the globin chains in higher mammals as a genealogical tree, with each branch point representing gene duplication that gives rise to further globin chains (Figure 6.24).

Secondary structure prediction

One of the earliest applications of using sequence information involved predicting elements of secondary structure. If sufficient numbers of residues can be placed in secondary structure then, in theory at least, it is possible to generate a folded structure based on

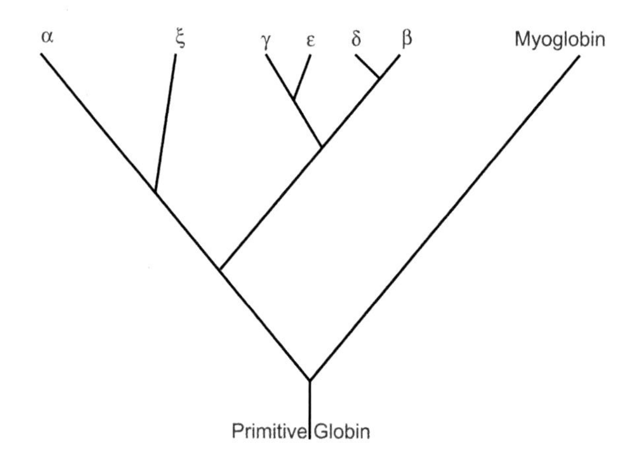

Figure 6.24 Evolution of the globin chains in primates. Each branch point represents a gene duplication event with myoglobin evolving early in the evolutionary history of globin subunit

Table 6.7 Propensity for a given amino acid residue to be found in helix, strand or turn

Residue	P_α	P_β	P_t	Residue	P_α	P_β	P_t
Ala	1.41	0.72	0.82	Leu	1.34	1.22	0.57
Arg	1.21	0.84	0.90	Lys	1.23	0.84	0.90
Asn	0.76	0.48	1.34	Met	1.30	1.14	0.52
Asp	0.99	0.39	1.24	Phe	1.16	1.33	0.65
Cys	0.66	1.40	0.54	Pro	0.34	0.31	1.34
Gln	1.27	0.98	0.84	Ser	0.57	0.96	1.22
Glu	1.59	0.52	1.01	Thr	0.76	1.17	0.90
Gly	0.43	0.58	1.77	Trp	1.02	1.35	0.65
His	1.05	0.80	0.81	Tyr	0.74	1.45	0.76
Ile	1.09	1.67	0.47	Val	0.90	1.87	0.41

(After Chou, P.Y. & Fasman, G.D. *Ann. Rev. Biochem* 1978, **47**, 251–276; and Wilmot, C.M. & Thornton, J.M. *J. Mol. Biol.* 1988, **203**, 221–232).
A value of 1.41 observed for alanine in helices implies that alanine occurs 41 percent more frequently than expected for a random distribution

the packing together of helices, turns and strands. Secondary structure prediction methods are often based on the preference of amino acid residues for certain conformational states.

One algorithm for secondary structure prediction was devised by P.Y. Chou and G.D. Fasman, and although superior methods and refinements exist today it proved useful in defining helices, strands and turns in the absence of structural data. It was derived from databases of known protein structures by estimating the frequency with which a particular amino acid was found in a secondary structural element divided by the frequency for all other residues. A value of 1 indicates a random distribution for a particular amino acid whilst a value greater than unity suggests a propensity for finding this residue in a particular element of secondary structure. A series of amino acid preferences were established (Table 6.7).

The numbers in the first three columns, P_α, P_β P_t, are equivalent to preference parameters for the 20 amino acids for helices, strands and reverse turns respectively. From this list one can assign the residues on the basis of the preferences; Ala, Arg, Glu, Gln, His, Leu, Lys, and Met are more likely to be found in helices; Cys, Ile, Phe, Thr, Trp, Tyr and Val are likely to be found in strands while Asn, Asp, Gly, Pro and Ser are more frequently located in turns. The algorithm of Chou and Fasman pre-dated the introduction of inexpensive computers but contained a few simple steps that could easily be calculated (Table 6.8). In practice the Chou–Fasman algorithm has a success rate of ~50–60 percent, although later developments have succeeded in reaching success rates above 70 percent. A major failing of the Chou–Fasman algorithm is that it considers only local interactions and neglects long-range order known to influence the stability of secondary structure. In addition the algorithm makes no distinction between types of helices, types of turns or orientation of β strands.

The availability of large families of homologous sequences (constructed using the algorithms described above) has revolutionized methods of secondary structure prediction. Traditional methods, when applied to a family of proteins rather than a single sequence have proved much more accurate in identifying secondary structure elements. Today the combination of sequence data with sophisticated computing techniques such as neural networks has given accuracy levels in excess of 70 percent. Besides the use of advanced computational techniques recent approaches to the problem of secondary structure prediction have focussed on the inclusion of additional constraints and parameters to assist precision. For example, the regular periodicity of

Table 6.8 Algorithm of Chou and Fasman

Step	Procedure
1	Assign all of the residues in the peptide the appropriate set of parameters
2	Scan through the peptide and identify regions where four out of 6 consecutive residues have $P_{\alpha} > 1.0$ This region is declared to be an α helix
3	Extend the helix in both directions until a set of four residues yield an average $P_{\alpha} < 1.00$. This represents the end of the helix
4	If the segment defined in step 3 is longer than 5 residues and the average value of $P_{\alpha} > P_{\beta}$ for this segment it can also be assigned as a helix
5	Repeat this procedure to locate all helical regions
6	Identify a region of the sequence where 3 out of 5 residues have a value of $P_{\beta} > 1.00$. These residues are in a β strand
7	Extend the sheet in both directions until a set of four contiguous residues yielding an average $P_{\beta} < 1.00$ is reached. This represents the end of the β strands
8	Any segment of the region located by this procedure is assigned as a β-strand if the average $P_{\beta} > 1.05$ and the average value of $P_{\beta} > P_{\alpha}$ for the same region
9	Any region containing overlapping helical and strand assignments are taken to be helical if the average $P_{\alpha} > P_{\beta}$ for that region. If the average $P_{\beta} > P_{\alpha}$ for that region then it is declared a β strand

the α helix of 3.6 residues per turn means that in proteins many regular helices are amphipathic with polar residues on the solvent side and non-polar residues facing the inside of the protein. Recognition of a periodicity of i, i + 3, i + 4, i + 7, etc. in hydrophobic residues has proved particularly effective in predicting helical regions in membrane proteins.

Genomics and proteomics

The completion of the human genome sequencing project has provided a large amount of data concerning the number and distribution of proteins. It seems likely that the human genome contains over 25 000 different polypeptide chains, and most of these are currently of unknown structure and function. Genomics reflects the wish to understand more about the complexity of living systems through an understanding of gene organization and function. Advances in genomics have provided information on the number, size and composition of proteins encoded by the genome. It has highlighted the organization of genes within chromosomes, their homology to other genes, the presence of introns together with the mechanism and sites of gene splicing and the involvement of specific genes in known human disease states. The latter discovery is heralding major advances in understanding the molecular mechanisms of disease particularly the complex interplay of genetic and environmental factors. Genomics has stimulated the discovery and development of healthcare products by revealing thousands of potential biological targets for new drugs or therapeutic agents. It has also initiated the design of new drugs, vaccines and diagnostic DNA kits. So, although genomic based therapeutic agents include traditional 'chemical' drugs, we are now seeing the introduction of protein-based drugs as well as the very exciting and potentially beneficial approach of gene therapy.

However, as the era of genomics reaches maturity it has expanded from a simple definition referring to the mapping, sequencing, and analysis of genomes to include an emphasis on genome function. To reflect this shift, genome analysis can now be divided into 'structural genomics' and 'functional genomics'. Structural genomics is the initial phase of genome analysis with the end point represented by the high-resolution genetic maps of an organism embodied by a complete DNA sequence. Functional genomics refers to the analysis of gene expression and the use of information provided by genome mapping projects to study the products of gene expression. In many instances this involves the study of proteins and a major branch of functional genomics is the new and expanding area of proteomics.

Proteomics is literally the study of the proteome via the systematic and global analysis of *all* proteins encoded within the genome. The global analysis of proteins includes specifically an understanding of structure and function. In view of the potentially large number of polypeptides within genomes this requires the development and application of large-scale, high throughput, experimental methodologies to examine not only vast numbers of proteins but also their interaction with other proteins and nucleic acids. Proteomics is still in its infancy and current scientific research is struggling with developing methods to deal with the vast amounts of information provided by the genomic revolution. One approach that will be used in both genomic and proteomic study is statistical and computational analysis usually called bioinformatics.

Bioinformatics

Bioinformatics involves the fusion of biology with computational sciences. Almost all of the areas described in the previous sections represent part of the expanding field of bioinformatics. This new and rapidly advancing subject allows the study of biological information at a gene or protein level. In general, bioinformatics deals with methods for storing, retrieving and analysing biological data. Most frequently this involves DNA, RNA or protein sequences, but bioinformatics is also applied to structure, functional properties, metabolic pathways and biological interactions. With a wide range of application and the huge amount of 'raw' data derived from sequencing projects and deposited in databanks (Table 6.9) it is clear that bioinformatics as a major field of study will become increasingly important over the next few decades.

Bioinformatics uses computer software tools for database creation, data management, data storage, data mining and data transfer or communication. Advances in information technology, particularly the use of the Internet allows rapid access to increasing amounts of biological information. For example, the sequenced genomes of many organisms are widely available at multiple web sites throughout the world. This allows researchers to download sequences, to manipulate them or to compare sequences in a large number of different ways.

Central to the development of bioinformatics has been the explosion in the size, content and popularity

Table 6.9 A selection of databases related to protein structure and function; some have been used frequently throughout this book to source data

Database/repository/resource	URL (web address)
The Protein DataBank.	http://www.rcsb.org/pdb
Expert Protein Analysis System (EXPASY)	http://www.expasy.ch/
European Bioinformatics Institute	http://www2.ebi.ac.uk
Protein structure classification (CATH)	http://www.biochem.ucl.ac.uk/bsm/cath
Structural classification of proteins (SCOP)	http://scop.mrc-lmb.cam.ac.uk/scop
Atlas of proteins side chain interactions.	http://www.biochem.ucl.ac.uk/bsm/sidechains
Human genome project (a tour!)	http://www.ncbi.nlm.nih.gov/Tour
An online database of inherited diseases	http://www.ncbi.nlm.nih.gov/Omim
Restriction enzyme database	http://rebase.neb.com
American Chemical Society	http://pubs.acs.org

of the world wide web, the most recognized component of the internet.

Initially expected to be of use only to scientists the world wide web is a vast resource allowing data transfer in the form of pages containing text, images, audio and video content. Pages, linked by pointers, allow a computer on one side of the world to access information anywhere in the world via series of connected networks. The pointers refer to URL's or 'uniform resource locators' and are the basic 'sites' of information. So, for example, the protein databank widely referred to in this book has a URL of http://www.rcsb.org. The 'http' part of a web address simply refers to the method of transferring data and stands for hypertext transfer protocol. Hypertext is the language of web pages and all web pages are written in a specially coded set of instructions that governs the appearance and delivery of pages known as hypertext markup language (HTML). Finally, the web pages are made comprehensible (interpreted) by 'browsers' – software that reads HTML and displays the content. Browsers include Internet Explorer, Netscape and Opera and all can be used to view 'online content' connected with this book. In 10 years the web has become a familiar resource.

The development of computers has also been rapid and has seen the introduction of faster 'chips', increased memory and expanded disk sizes to provide a level of computational power that has enabled the development of bioinformatics. In a trite formulation of a law first commented upon by Gordon Moore, and hence often called Moore's law, the clock speed (in MHz) available on a computer will double every 18 months. So in 2003 I am typing this book on a 1.8 GHz-based computer. Although computer speeds will undoubtedly increase there are grounds for believing rates of increase will slow. Searching and manipulating large databases, coupled with the use of complex software tools to analyse or model data, places huge demands on computer resources. It is likely that in the future single powerful computer workstations with a very high clock speed, enormous amounts of memory and plentiful storage devices will become insufficient to perform these tasks. Instead bioinformatics will have to harness the power of many computers working either in parallel or in grid-like arrays. These computers cumulatively will allow the solution to larger and more complex problems. However, today bioinformatics is an important subject in its own right that directs biological studies in directions likely to result in favourable outcomes.

One of the results of genome sequencing projects is that software tools can be developed to compare and contrast genetic information. One particular avenue that is being pursued actively at the moment is the prediction of protein structure from only sequence information. The determination of three dimensional protein structures is an expensive, time consuming and formidable task usually involving X-ray crystallography or NMR spectroscopy. Consequently any method that allows a 'by-pass' of this stage is extremely attractive especially to pharmaceutical companies where the prime objective is often product development. One approach is to compare a protein sequence to other proteins since sequential homology (identity >25 percent) will always be accompanied by similar topology. If the structure of a sequentially homologous protein is known then the topology of the new protein can be deduced with considerable certainty. A more likely scenario, however, is that the sequence may show relatively low levels of identity and one would like to know how similar, or different, the three dimensional structures might be to each other. This is a far more difficult problem. Comparative modelling will work when sequentially homologous proteins are compared but may fail when the levels of identity fall below a benchmark of ~25 percent. Additionally some proteins show very similar structures with very low levels of sequence homology.

A second approach is the technique of 'threading'. This approach attempts to compare 'target sequences' against libraries of known structural templates. A comparison produces a series of scores that are ranked and the fold with the best score is assumed to be the one adopted by the unknown sequence. This approach will fail when a new 'fold' or tertiary structure is discovered and it relies on a representative database of structures and sequences.

The *ab initio* approach (Figure 6.25) ignores sequence homology and attempts to predict the folded state from fundamental energetics or physicochemical properties associated with the constituent residues. This involves modelling physicochemical parameters in terms of force fields that direct the folding of the

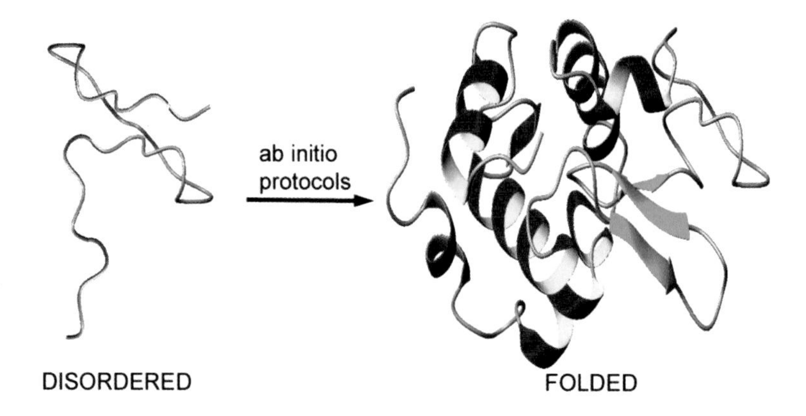

Figure 6.25 The *ab initio* approach to fold prediction

primary sequence from an initial randomized structure to one satisfying all constraints. These constraints will reflect the energetics associated with charge, hydrophobicity and polarity with the aim being to find a single structure of low energy. The resulting structures should have very few violations with respect to bond angles and length and can be checked for consistency against the Ramachandran plot. *Ab initio* protocols do *not* utilize experimental constraints but depend on the generation of structure from fundamental parameters *in silico*.

This approach is based on the thermodynamic argument that the native structure of a protein is the global minimum in the free energy profile. *Ab initio* methods place great demands on computational power but have the advantage of not using peripheral information. Despite considerable technical difficulties success is being achieved in this area as a result of regular 'contests' held to judge the success of *ab initio* protocols. These proceedings go under the more formal name of critical assessment of techniques for protein structure prediction (CASP). As well as involving the comparative homology based methods CASP involves the use of *ab initio* methods to predict tertiary structures for 'test' sequences (Figure 6.26). The emphasis is on *prediction* as opposed to "postdiction" and involves a community wide attempt to determine structures for proteins that have been assessed independently (but are not available in public domain databases). Independent structural knowledge allows an accurate assessment of the eventual success of the different *ab initio* approaches. Generally the results are expressed as rmsd (root mean square deviations), reflecting the difference in positions between corresponding atoms in the experimental and calculated (predicted) structures. In successful predictions rmsd values below 0.5 nm were seen for small proteins (<100 residues) using *ab initio* approaches. Although the agreement between predicted and experimentally determined structures is still relatively poor the resolution allows in favourable

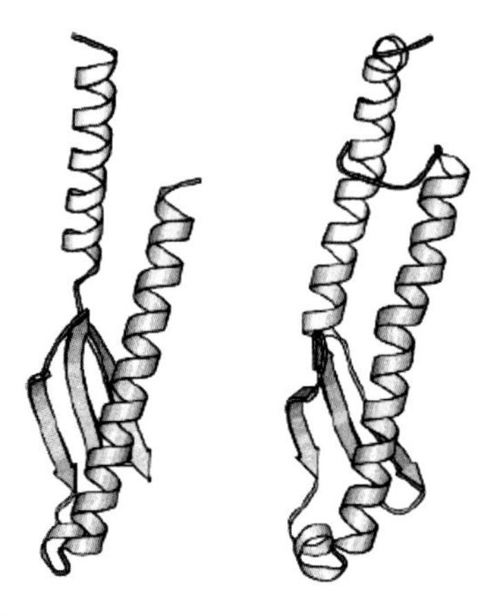

Figure 6.26 One good model predicted for target protein (T0091) in CASP4. The native structure of T0091 is shown on the left and the predicted model is shown on the right. Structurally equivalent residues are marked in yellow (reproduced with permission Sippl, M.J. *et al. Proteins: Structure, Function, and Genetics*, Suppl. 5, 55–67. John Wiley & Sons, Chichester, 2001)

circumstances the backbone of 'target proteins' to be defined with reasonable accuracy and for overall fold to be identified. For larger proteins the refinement of structure is worse but *ab initio* methods are becoming steadily better and offer the possibility of true protein structure prediction in the future.

Summary

Despite the incredible diversity of living cells all organisms are made up of the same 20 amino acid residues linked together to form proteins. This arises from the origin of the amino acid alphabet very early in evolution before the first true cells. All subsequent forms of life evolved using this basic alphabet.

Prebiotic synthesis is the term applied to non-cellular based methods of amino acid synthesis that existed over 3.6×10^9 years ago. The famous experiment of Urey and Miller demonstrated formation of organic molecules such as adenine, alanine and glycine that are the precursors today of nucleic acids and protein systems.

A major development in molecular evolution was the origin of self-replicating systems. Today this role is reserved for DNA but the first replicating systems were based on RNA a molecule now known to have catalytic function. An early prebiotic system involving RNA molecules closely associated with amino acids is thought to be most likely.

The fossil record shows primitive prokaryotic cells resembling blue-green cyanobacteria evolved 3.6 billion years ago. Evolution of these cells through compartmentalization, symbiosis and specialization yielded single cell and metazoan eukaryotes.

Protein sequencing is performed using the Edman procedure and involves the labelling and identification of the N-terminal residue with phenylisothiocyanate in a cyclic process. The procedure can be repeated ~50–80 times before cumulative errors restrict the accuracy of sequencing.

Nucleic acid sequencing is based on dideoxy chain termination procedures. The result of efficient DNA sequencing methods is the completion of genome sequencing projects and the prevalence of enormous amounts of bio-information within databases.

Databases represent the 'core' of the new area of bioinformatics – a subject merging the disciplines of biochemistry, computer science and information technology together to allow the interpretation of protein and nucleic acid sequence data.

One of the first uses of sequence data was to establish homology between proteins. Sequence homology arises from a link between proteins as a result of evolution from a common ancestor. Serine proteases show extensive sequence homology and this is accompanied by structural homology. Chymotrypsin, trypsin and elastase share homologous sequences and structure.

Structural homology will also result when sequences show low levels of sequence identity. The c type cytochromes from bacteria and mitochondria exhibit remarkably similar folds achieved with low overall sequence identity. The results emphasize that proteins evolve with the retention of the folded structure and the preservation of functional activity.

The bioinformatics revolution allows analysis of protein sequences at many different levels. Common applications include secondary structure prediction, conserved motif recognition, identification of signal sequences and transmembrane regions, determination of sequence homology, and structural prediction *ab initio*.

In the future bioinformatics is likely to guide the directions pursued by biochemical research by allowing the formation of new hypotheses to be tested *via* experimental methods.

Problems

1. Use the internet to locate some or all of the web pages/databases listed in Table 6.9. Find any web page(s) describing this book.
2. A peptide containing 41 residues is treated with cyanogen bromide to liberate three smaller peptides whose sequences are

(i) Phe-Leu-Asn-Ser-Val-Thr-Val-Ala-Ala-Tyr-Gly-Gly-Pro-Ala-Lys-Pro-Ala-Val-Glu-Asp-Gly-Ala-Met

(ii) Ala-Ser-Ser-Glu-Glu-Lys-Gly-Met and

(iii) Val-Ser-Thr-Asn-Glu-Lys-Ala-Ala-Val-Phe

Trypsin digestion of the same 41 residue peptide yields the sequences:

(i) Ala-Ala-Val-Phe

(ii) Gly-Met-Phe-Leu-Asn-Ser-Val-Thr-Val-Ala-Ala-Tyr-Gly-Gly-Pro-Ala-Lys-Pro-Ala-Val-Glu-Asp-Gly-Ala-Met-Val-Ser-Thr-Asn-Glu-Lys and

(iii) Ala-Ser-Ser-Glu-Glu-Lys

Can the sequence be established unambiguously?

3. Obtain the amino acid sequences of human α-lactalbumin and lysozyme from any suitable database. Use the BLAST or FASTA tool to compare the amino acid sequences of human α-lactalbumin and lysozyme. Are the sequences sufficiently similar to suggest homology?
4. Identify the unknown protein of Table 6.3.
5. Download the following coordinate files from the protein databank 1CYO and 1NU4. In each case use your molecular graphic software to highlight the number of Phe residues in each protein, any co-factors present in these proteins and the number of charged residues present in each protein.
6. What is the EC number associated with the enzyme β-galactosidase. Locate the enzyme derived from *E. coli* in a protein or sequence database. How many domains does this protein possess? Are these domains related? Describe the structure of each domain. Estimate the residues encompassing each domain. Identify the active site and any conserved residues.
7. Find a program on the internet that allows the prediction of secondary structure. Use this program to predict the secondary structure content of myoglobin. Does this value agree with the known secondary structure content from X-ray crystallography? Repeat this trial with your 'favourite' protein? Now repeat the analysis with a different secondary structure prediction algorithm. Do you get the same results?
8. Identify proteins that are sequentially homologous to (a) α chain of human haemoglobin, (b) subtilisin E from *Bacillus subtilis*, (c) HIV protease. Comment on your results and some of the implications?
9. The β barrel structure is a common topology found in many proteins. An example is triose phosphate isomerase. Use databases to find other proteins with this structural motif. Do these sequences exhibit homology? Comment on the evolutionary relationship of β barrels.
10. Assuming a point accepted DNA mutation rate of 20 (20 PAMs/100 residues/10^8 years) establish the degree of DNA homology for two proteins each of 200 residues that diverged 500 million years ago. What is the maximal level of amino acid residue homology between proteins and what is the lowest level?

7

Enzyme kinetics, structure, function, and catalysis

One important function performed by proteins is the ability to catalyse chemical reactions. Catalytic function was amongst the first biological roles recognized in proteins through the work of Eduard Buchner and Emil Fischer. They identified and characterized the ability of some proteins to convert reactants into products. These proteins, analogous to chemical catalysts, increased rates of reaction but did not shift the equilibrium formed between products and reactants. The biological catalysts were named enzymes – the name derived from the Greek for 'in yeast' – 'en' 'zyme'.

Within all cells every reaction is regulated by the activity of enzymes. Enzymes catalyse metabolic reactions and in their absence reactions proceed at kinetically insignificant rates incompatible with living, dynamic, systems. The presence of enzymes results in reactions whose rates may be enhanced (catalysed) by factors of 10^{15} although enhancements in the range 10^3-10^9 are more typical. Enzymes participate in the catalysis of many cellular processes ranging from carbohydrate, amino acid and lipid synthesis, their breakdown or catabolic reactions (see Figure 7.1), DNA repair and replication, transmission of stimuli through neurones, programmed cell death or apoptosis, the blood clotting cascade reactions, the degradation of proteins, and the export and import of proteins across membranes. The list is extensive and for each reaction the cell employs a unique enzyme tailored via millions of years of evolution to catalyse the reaction with unequalled specificity.

This chapter explores how enzyme structure allows reactions to be catalysed with high specificity and rapidity. In order to appreciate enzyme-catalysed reactions the initial sections deal with simple chemical kinetics introducing the terms applicable to the study of reaction rates. Analysis of chemical kinetics proved to be very important in unravelling mechanisms of enzyme catalysis. Kinetic studies uncovered the properties of enzymes, products and reactants, as well as the steps leading to their formation. *In vivo* enzyme activity differs from that observed *in vitro* by its modulation by other proteins or ligands. The modulation of enzyme activity occurs by several mechanisms but is vital to normal cell function and homeostasis. In principle, regulation allows cells to respond rapidly to environmental conditions, turning on or off enzyme activity to achieve metabolic control under fluctuating conditions. When conditions change and the enzyme is no longer required activity is reduced without wasting valuable cell resources. Feedback mechanisms control enzyme activity and are a feature of allostery. Allosteric enzymes exist within the biosynthetic pathways of all cells.

Proteins: Structure and Function by David Whitford

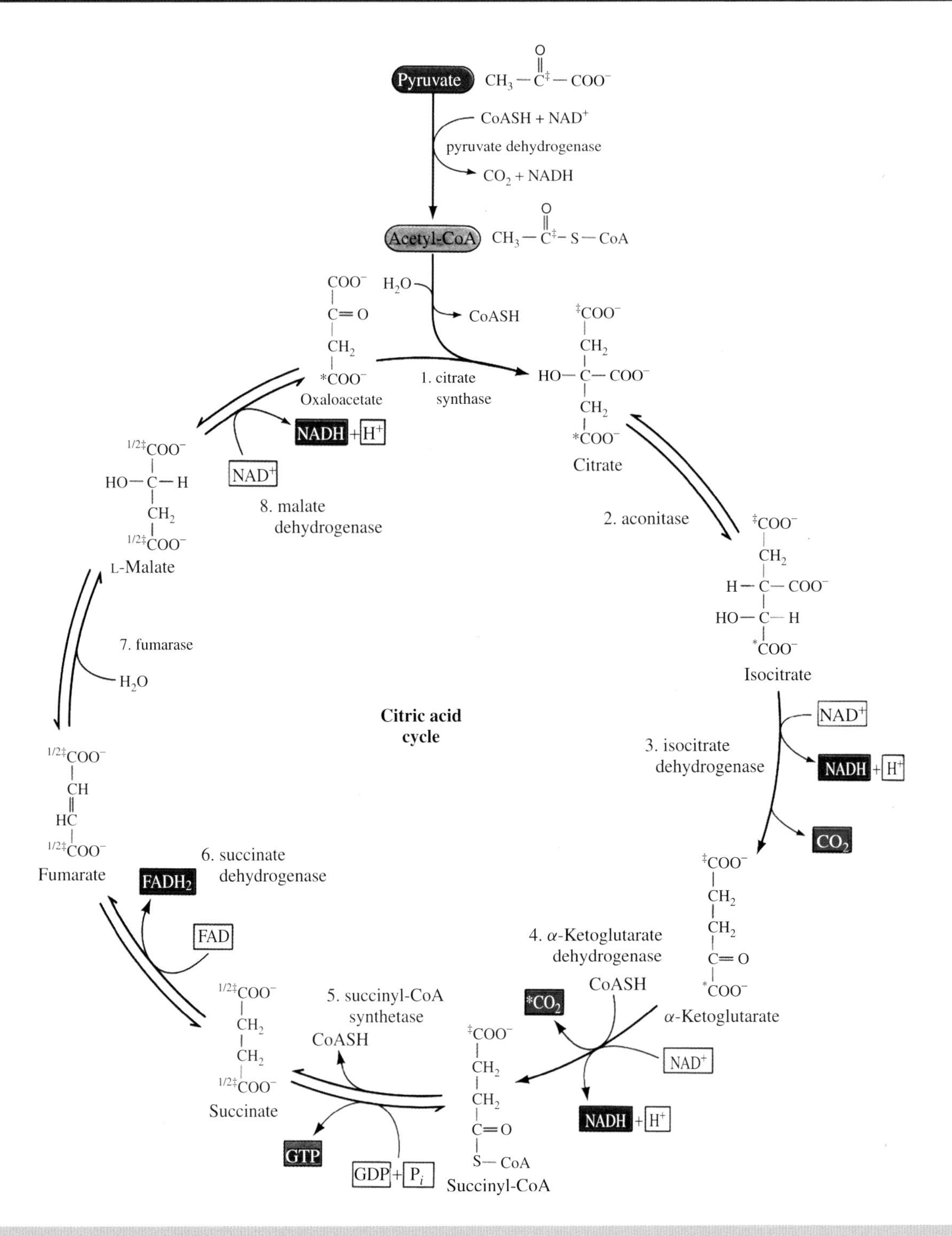

Figure 7.1 The conversion of pyruvate to acetate from the reactions of glycolysis lead to the Krebs citric acid cycle. Reproduced from Voet *et al.* (1999) by permission of John Wiley & Sons, Ltd. Chichester

Table 7.1 Examples of enzymes belonging to the six major classes

Enzyme class	Typical example	Reaction catalysed
Oxidoreductases	Lactate dehydrogenase (EC1.1.1.27)	Reduction of pyruvate to lactate with the corresponding formation of NAD
Transferases	Hexokinase (EC 2.7.1.1)	Transfer of phosphate group to glucose from ATP
Hydrolases	Acetylcholinesterase (EC 3.1.1.7)	Hydrolysis of acetylcholine to choline and acetate
Lyases	Phenylalanine Ammonia Lyase (EC 4.3.1.5)	Splits phenylalanine into ammonia and *trans*-cinnamate
Isomerases	Triose phosphate isomerase (EC 5.3.1.1)	Reversible conversion between dihydroxyacetone phosphate and glyceraldehyde-3-phosphate
Ligases	T4 DNA ligase (EC 6.5.1.1)	Phosphodiester bond linkage and conversion of ATP → AMP

Before reviewing these areas it is necessary to establish the basis of enzyme nomenclature and to establish the broad groups of enzyme-catalysed reactions occurring within all cells. For each catalysed reaction a unique enzyme exists; in glycolysis the conversion of 3-phosphoglycerate phosphate to 3-phosphoglycerate is catalysed by phosphoglycerokinase (PGK).

$$\text{3-phosphoglyceroyl phosphate} + \text{ADP} \rightleftharpoons \text{3-phosphoglycerate} + \text{ATP}$$

No other enzyme catalyses this reaction within cells.

Enzyme nomenclature

The naming of enzymes gives rise to much confusion. Some enzymes have 'trivial' names such as trypsin where the name provides no useful clue to biological role. Many enzymes are named after their coding genes and by convention the gene is written in *italic* script whilst the protein is written in normal type. Examples are the *Escherichia coli* gene *polA* and its product DNA polymerase I; the *lacZ* gene and β-galactosidase. To systematically 'label' enzymes and to help with identification of new enzymes the International Union of Biochemistry and Molecular Biology (IUBMB) devised a system of nomenclature that divides enzymes into six broad classes (Table 7.1 gives examples of these). Further division into subclasses groups enzymes sharing functional properties. The major classes of enzyme are:

1. Oxidoreductases: catalyse oxidation–reduction (redox) reactions.
2. Transferases: catalyse transfer of functional groups from one molecule to another.
3. Hydrolases: perform hydrolytic cleavage of bonds.
4. Lyases: remove groups from (or add a group to) a double bond or catalyse bond scission involving electron re-arrangement.
5. Isomerases: catalyse intramolecular rearrangement of atoms.
6. Ligases: joins (ligates) two molecules together.

Each enzyme is given a four-digit number written, for example, as EC 5.3.1.1. The first number indicates an isomerase whilst the second and third numbers identify further sub-classes. In this case the second digit

(3) indicates an isomerase that acts as an intramolecular isomerase whilst the third digit (1) indicates that the substrates for this enzyme are aldose or ketose carbohydrates. The fourth number is a serial number that frequently indicates the order in which the enzyme was recognized. The enzyme EC 5.3.1.1 is triose phosphate isomerase found in the glycolytic pathway catalysing inter-conversion of glyceraldehyde 3 phosphate and dihydroxyacetone phosphate. A complete list of enzyme classes is given in the Appendix. At the end of 2003 the structures of over 8000 enzymes were deposited in the PDB database whilst other databases recognized over 3700 different enzymes (http://www.expasy.ch/enzyme/). These lists are continuing to be updated and grow with the completion of further genome sequencing projects.

Until recently it was thought that all biological catalysts (enzymes) were proteins. This view has been revised with the observation that RNA molecules (ribozymes) catalyse RNA processing and protein synthesis. The observation of catalytic RNA raises an interesting and perplexing scenario where RNA synthesis is catalysed by proteins (e.g. RNA polymerase) but protein synthesis is catalysed by RNA (the ribosome). This chapter will focus on the properties and mechanisms of protein based enzymes, in other words those molecules synonymous with the term enzyme.

Enzyme co-factors

Many enzymes require additional co-factors to catalyse reactions effectively. These co-factors, also called coenzymes, are small organic molecules or metal ions (Figure 7.2). The nature of the interaction between cofactor and enzyme is variable, with both covalent and non-covalent binding observed in biological systems. A general scheme of classifying co-factors recognizes the strength of their interaction with enzymes. For example, some metal ions are tightly bound at active sites and participate directly in the reaction mechanism whilst others bind weakly to the protein surface and have remote roles (Table 7.2). Co-factors are vital to effective enzyme function and many were recognized originally as vitamins. Alongside vitamins metal ions modulate enzyme activity; in the apo (metal-free) state metalloenzymes often exhibit impaired catalytic activity or decreased stability. The metal ions, normally divalent cations, exhibit a wide range of biological roles (Table 7.3). In carboxypeptidase Zn^{2+} stabilizes intermediates formed during hydrolysis of peptide bonds whilst in heme and non-heme enzymes iron participates in redox reactions.

Chemical kinetics

In order to understand the details of enzyme action it is worth reviewing the basics of chemical kinetics since these concepts underpin biological catalysis. For a simple unimolecular reaction where A is converted irreversibly into B

$$\mathrm{A} \rightarrow \mathrm{B}$$

the rate of reaction is given by the rate of disappearance of reactant or the rate of appearance of product

$$-\mathrm{d}[\mathrm{A}]/\mathrm{d}t \quad \text{or} \quad \mathrm{d}[\mathrm{B}]/\mathrm{d}t \qquad (7.1)$$

The disappearance of reactants is proportional to the concentration of A

$$-\mathrm{d}[\mathrm{A}]/\mathrm{d}t = k[\mathrm{A}] \qquad (7.2)$$

where k is a first order rate constant (units s^{-1}). Rearranging this equation and integrating between

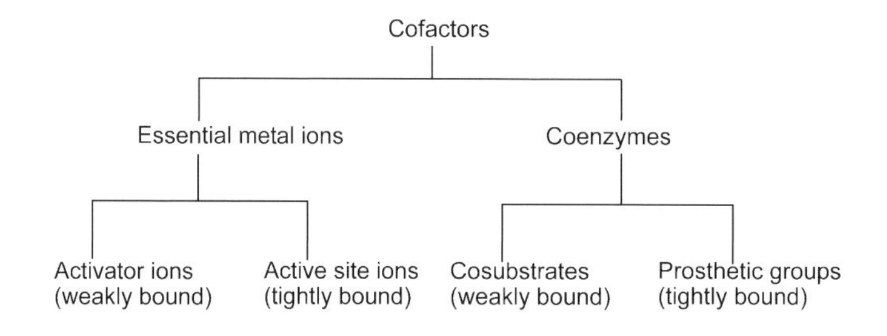

Figure 7.2 Different co-factors and their interactions with proteins

Table 7.2 Co-factors found in enzymes together with their biological roles

Component	Co-factor	Metabolic role
Vitamin A	Retinal	Visual cycle.
Thiamine (B_1)	Thiamine pyrophosphate	Carbohydrate metabolism, transient shuttle of aldehyde groups
Riboflavin (B_2)	Flavin adenine dinucleotide (FAD) Flavin mononucleotide (FMN).	Redox reactions in flavoenzymes
Niacin (Nicotinic acid)	Nicotinamide adenine dinucleotide (NAD)	Redox reactions involving NAD linked dehydrogenases
Panthothenic acid (B_3)	Coenzyme A	Acyl group activation and transfer
Pyridoxal phosphate (B_6)	Pyridoxine	Transaminase function
Vitamin B_{12}	Cyanocobalamin	Methyl group transfer or intramolecular rearrangement
Folic acid	Tetrahydrofolate	Transfer of formyl or hydroxymethyl groups
Biotin	Biotin	ATP dependent carboxylation of substrates
Vitamin K	Phylloquinone	Carboxylation of Glu residues.
Coenzyme Q	Ubiquinone	Electron and proton transfer

Table 7.3 Role of metal ions as co-factors in enzyme catalysis

Metal	Metalloenzyme	General reaction catalysed
Fe	Cytochrome oxidase	Reduction of O_2 to H_2O
Co	Vitamin B_{12}	Transfer of methyl groups
Mo	Sulfite oxidase	Reduction of sulfite to sulfate.
Mn	Water splitting enzyme	Photosynthetic splitting of water to oxygen
Ni	Urease	Hydrolysis of urea to ammonia and carbamate
Cu	Superoxide dismutase	Dismutation of superoxide into O_2 and H_2O_2
Zn	Carboxypeptidase A	Hydrolysis of peptide bonds

$t = 0$ and $t = t$ leads to the familiar first order equation

$$-\frac{d[A]}{[A]} = k\,dt \quad (7.3)$$

$$\int_{[A]_0}^{[A]} \frac{d[A]}{[A]} = -\int_0^t k dt \quad (7.4)$$

$$[A] = [A]_o e^{-kt} \quad (7.5)$$

A plot of ln ($[A]/[A]_o$) versus time (t) gives a straight line of slope $-k$ and provides a graphical method of estimating first order rate constants (Figure 7.3).

An important quantity for any first order reaction is the half-life of the reaction; the time taken for

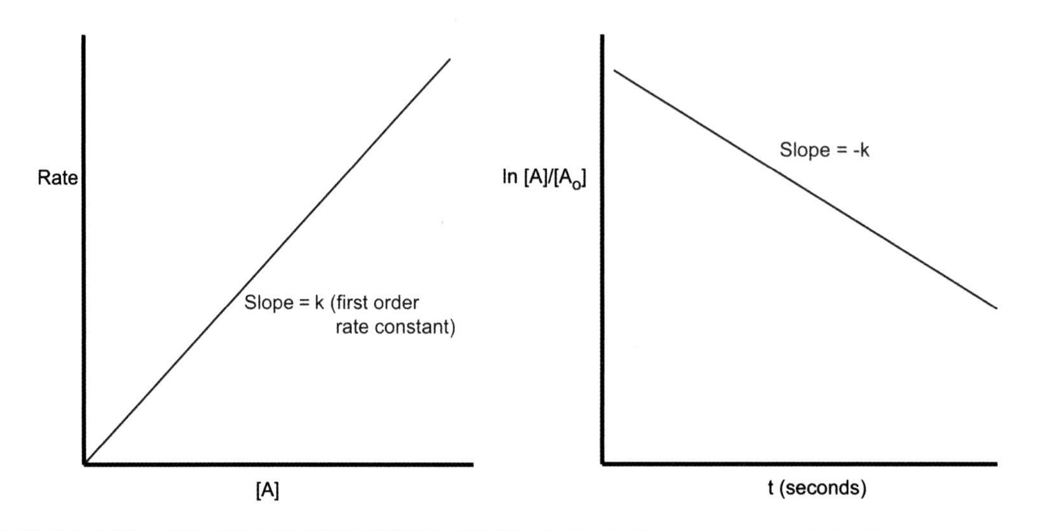

Figure 7.3 The rate of reaction as function of the concentration of A in the first order reaction of A→B and a graphical plot of ln ([A]/[A]$_o$) versus time (t) for the reaction that allows the first order rate constant (k) to be estimated

the concentration of the reactant to drop to half its initial value. Rearranging Equation 7.5 yields upon substitution of [A] = [A]$_o$/2

$$\ln[A]_o/2/[A]_0 = -kt_{1/2} \quad (7.6)$$

where

$$t_{1/2} = \ln 2/k = 0.693/k \quad (7.7)$$

Measuring the time for a reactant to fall to half its original concentration allows an estimation of the first order rate constant. In general for a reaction of the nth order the half life is proportional to $1/[A]_o^{n-1}$. These equations occur commonly in reaction kinetics involving biomolecules and extend to related areas such as folding, stability and complex formation. A simple bimolecular reaction is described as an irreversible process where

$$A + B \longrightarrow \text{product}$$

In contrast the reaction between enzyme and substrate is usually considered as an equilibrium involving forward and backward reactions and it is useful to examine kinetics applicable to (i) a reversible first order reaction and (ii) consecutive reactions, since these occur frequently in biological studies. Examples include

$$A \rightleftharpoons B$$

$$A + B \underset{k_{-1}}{\overset{k_1}{\rightleftharpoons}} AB \overset{k_2}{\longrightarrow} P$$

If we consider the reaction scheme for a reversible reaction defined by forward and backward rate constants of k_1 and k_{-1} the change in the concentration of A with time is described by the equation

$$-d[A]/dt = k_1[A] - k_{-1}[B] \quad (7.8)$$

with the reverse reaction written similarly as

$$-d[B]/dt = k_{-1}[B] - k_1[A] \quad (7.9)$$

This yields

$$[B]/[A] = k_1/k_{-1} = K_{eq} \quad (7.10)$$

where K_{eq} is the equilibrium constant for the reaction. A variation occurs when a reversible reaction for conversion of A into B is followed by a consecutive reaction converting B into C. The formation of AB is a bimolecular reaction between A and B whilst the breakdown of AB to yield P is unimolecular. The easiest way to deal algebraically with this reaction is to describe the formation of AB. This leads to the equation

$$d[AB]/dt = k_1[A][B] - k_{-1}[AB] - k_2[AB] \quad (7.11)$$

Under conditions where either $k_2 \gg k_1$ or $k_{-1} \gg k_1$ the concentration of the encounter complex [AB] remains low at all times. This assumption leads to the equality $d[AB]/dt \approx 0$ and is known as the steady state approximation where

$$k_1[A][B] = k_{-1}[AB] + k_2[AB] \quad (7.12)$$

$$= (k_{-1} + k_2)[AB] \quad (7.13)$$

$$[AB] = k_1[A][B]/(k_{-1} + k_2) \quad (7.14)$$

The rate of formation of product P is described by the rate of breakdown of AB

$$d[P]/dt = k_2[AB] \quad (7.15)$$

$$= k_2k_1[A][B]/(k_{-1} + k_2) \quad (7.16)$$

By assimilating the terms $k_2k_1/(k_{-1} + k_2)$ into a constant called k_{obs} the bimolecular reaction $A + B \rightarrow AB \rightarrow P$ can be described as a one step reaction governed by a single rate constant (k_{obs}).

In the above equation two interesting boundary conditions apply. When the dissociation of AB reflected by the constant k_{-1} is much smaller than the rate k_2 ($k_{-1} \ll k_2$) then the overall reaction k_{obs} approximates to k_1 and is the diffusion-controlled rate. The second condition applies when $k_2 \ll k_{-1}$ and is known as the reaction-controlled rate leading to an observed rate constant of

$$k_{obs} = k_2k_1/k_{-1} \quad (7.17)$$

or

$$k_{obs} = K_{eq}k_2 \quad (7.18)$$

The transition state and the action of enzymes

Understanding enzyme action was assisted by transition state theory developed by Henry Eyring from the 1930s onwards. For a bimolecular reaction involving three atoms there exists an intermediate state of high energy that reflects the breaking of one covalent bond and the formation of a new one. For the reaction of a hydrogen *atom* with a *molecule* of hydrogen this state equates to the breakage of the $H_A–H_B$ bond and the formation of a new $H_B–H_C$ bond.

$$H_A–H_B + H_C \longrightarrow H_A + H_B–H_C$$

An intermediate called the transition state is denoted by the symbol "‡". The free energy associated with the transition state, $\Delta G^{\ddagger}$, is higher than the free energy associated with either reactant or products (Figure 7.4).

Eyring formulated the transition state in quantitative terms where reactants undergo chemical reactions by their respective close approach along a path of minimum free energy called the reaction coordinate. Reactions are expressed by transition state diagrams that reflect changes in free energy as a function of reaction coordinate. Favourable reactions have negative values for $\Delta G (\Delta G < 0)$ and are called exergonic. Other reactions require energy to proceed and are described as endergonic ($\Delta G > 0$). The free energy changes as the reaction progresses from reactants to product. For the reaction $R \rightarrow P$ we identify average free energies associated with R and P, as G_R and G_P, respectively, with the free energy associated with the products being lower than that of the reactants; the change in free energy is negative and the reaction is favourable. However, although the reaction is favourable the profile does not reveal anything about the rate of reaction nor does it offer any clues about why some reactions occur faster than others.

The concept of activation energy recognized by Arrhenius in 1889 from the temperature dependence of reactions follows the relationship

$$k = A \exp^{-E_a/RT} \quad (7.19)$$

where k is a rate constant, A is a frequency factor and E_a is the activation energy. Eyring realized that a new view of chemical reactions could be obtained by postulating an intermediate, the transition state, occurring between products and reactants. The

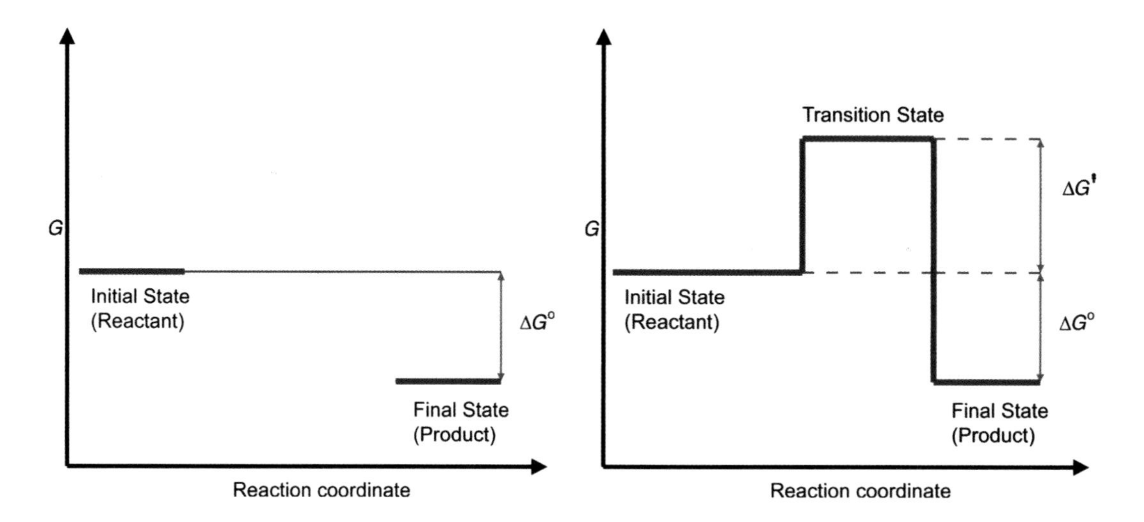

Figure 7.4 Transition state diagrams. In the absence of a transition state the pathway from reactants to products is obscure but one can recognize that the products have a lower free energy than the reactants ($G_P < G_R$)

transition state exists for approximately 10^{-13}–10^{-14} s leading to a small percentage of reactants being found in this conformation at any one time. The formation of product required crossing the activation barrier otherwise the transition state decomposed back to the reactants. With this view of a chemical reaction the rate-limiting step became the decomposition of the transition state to products or reactants involving a non-conventional equilibrium established between reactants and the transition state. For the reaction

$$A + B \rightleftharpoons X^{\ddagger} \longrightarrow \text{products}$$

the equilibrium constant ($K^{\ddagger}$) is expressed in terms of the activated complex ($X^{\ddagger}$) and the reactants (A and B).

$$K^{\ddagger} = X^{\ddagger}/[A][B] \tag{7.20}$$

The rate of the reaction is simply the product of the concentration of activated complex multiplied by the frequency of crossing the activation barrier (ν).

$$\text{Rate} = \nu[X^{\ddagger}] = \nu[A][B]K^{\ddagger} \tag{7.21}$$

Since the rate of reaction is also

$$= k[A][B] \tag{7.22}$$

this leads to

$$k = \nu K^{\ddagger} \tag{7.23}$$

Although evaluation of k depends on our ability to deduce ν and $K^{\ddagger}$ this can be achieved via statistical mechanics to leave the result

$$k = k_B T/h\ K^{\ddagger} \tag{7.24}$$

where k_B on the right-hand side of the equation is the Boltzmann constant, T is temperature and h is Planck's constant. Equation 7.24 embodies thermodynamic quantities (ΔG, ΔH and ΔS) since

$$\Delta G^{\ddagger} = -RT \ln K^{\ddagger} \tag{7.25}$$

and the rate constant can be expressed as

$$k = k_B T/h\ e^{-\Delta G^{\ddagger}/RT} \tag{7.26}$$

where $\Delta G^{\ddagger}$ is the free energy of activation. Since the parameter $k_B T/h$ is independent of A and B the rate of reaction is determined solely by $\Delta G^{\ddagger}$, the free energy of activation. Since

$$\Delta G^{\ddagger} = \Delta H^{\ddagger} - T\Delta S^{\ddagger} \tag{7.27}$$

a thermodynamic formulation of transition state theory is often written as

$$k = k_B T/h \; e^{\Delta S^{\ddagger}/R} \; e^{-\Delta H^{\ddagger}/RT} \qquad (7.28)$$

where $\Delta S^{\ddagger}$ and $\Delta H^{\ddagger}$ are the entropy and enthalpy of activation. For uncatalysed reactions reaching a transition state represents an enormous energy barrier to the formation of product. A large activation energy barrier is reflected by a slow reaction that arises because few reactants acquire sufficient thermal energy to cross this threshold.

Many reactions, particularly those in biological systems, consist of several steps. For reactions such as

$$A \longrightarrow I \longrightarrow P$$

the shape of the transition state diagram (Figure 7.5) reflects the overall reaction. If the activation energy of the first step ($A \rightarrow I$) is greater than that of $I \rightarrow P$ then the first step is slower than that of the second. The reaction with the highest activation energy represents the rate-determining step for the overall process.

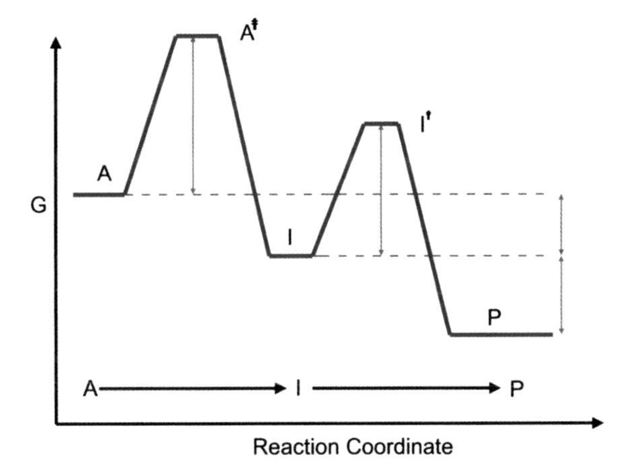

Figure 7.5 Transition state diagram for a two-stage reaction where the step A → I is the rate determining event. The shape of the transition barrier is described by 'box like' profiles (Figure 7.3), curves or gradients but these 2D representations are inaccurate since the pathway to the transition state is likely to be convoluted and reached by several 'routes'

Catalysts lower the energy associated with a transition state and increase the rate of a chemical reaction (termed velocity) without undergoing a net change over the entire process. The efficiency of a catalyst is reflected by lowering the activation energy with the difference in free energy ($\Delta\Delta G^{\ddagger}$) allowing catalytic rate enhancements to be calculated ($e^{-\Delta\Delta G^{\ddagger}/RT}$).

Enzymes are biological catalysts and increase the rate of a reaction of energetically favourable processes. As with all catalysts they are unchanged in chemical composition after a reaction. Enzymes produce dramatic rate enhancement as high as 10^{15}-fold when catalysed and uncatalysed reactions are compared although normal ranges are from $10^3 - 10^9$. Enzymes reduce the free energy of the transition state by promoting binding reactions that enhance intermediate stability. Enzymes employ a number of different mechanisms to achieve efficient catalysis. These mechanisms are influenced by the structure and dynamics of active site regions where environments differ significantly to those found in bulk solvent. Catalytic mechanisms identified within enzymes include acid–base catalysis, covalent catalysis, metal-ion mediated catalysis, electrostatic catalysis, proximity and orientation effects, and preferential binding of transition state complexes.

The kinetics of enzyme action

In enzyme-catalysed reactions the concentration of catalyst is usually much lower than that of the substrate. As a result enzymes catalyse reactions several times, leading to relatively slow changes in the concentration of product and substrate whilst the concentration of enzyme can be viewed as constant. Early studies established kinetic principles through observation of the changes in concentration of substrate (or product) with time with the enzyme remaining an unobservable quantity. Often these changes were followed by spectrophotometric methods. For a simple enzyme-catalysed reaction (i.e. a reaction not involving allosteric modulators) it was recognized that as the concentration of substrate increased so the rate of appearance of product increased when the concentration of enzyme was held at a fixed level. However, at certain concentrations of substrate the rate of reaction no longer increased but reached a plateau (Figure 7.6).

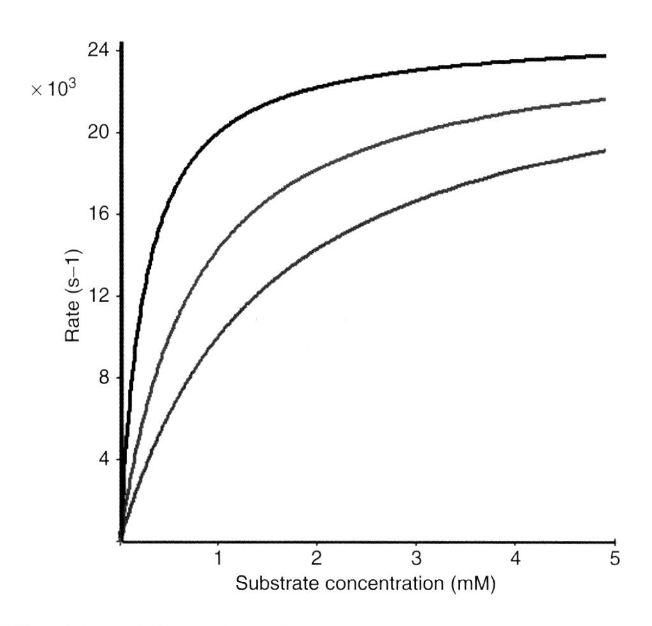

Figure 7.6 Three different enzyme-catalysed reactions showing different K_m values but similar V_{max} values. See text for discussion of V_{max} and K_m. Each V_{max} is estimated from the asymptote approximating to the curve's maximum value and despite appearances these curves actually reach the same value at very high concentrations of substrate

The rate of reaction (v) is proportional to the rate of removal of substrate or alternatively the rate of formation of product.

$$v = -\mathrm{d}[\mathrm{S}]/\mathrm{d}t = \mathrm{d}[\mathrm{P}]/\mathrm{d}t \qquad (7.29)$$

At low substrate concentrations a linear or first order dependence on [S] exists whilst at higher substrate concentrations this shifts to zero order or substrate independent. The profile suggests that the enzyme-catalysed conversion of substrate into product involved the formation of an intermediate called the enzyme–substrate complex (Figure 7.7).

From a minimal reaction scheme, together with two assumptions originally formulated by Leonor Michaelis and Maud Menten, we can derive important enzyme kinetic relationships. The two assumptions made by Michaelis and Menten were that the enzyme exists either as a free form [E] or as enzyme substrate complex [ES] and secondly that the enzyme substrate complex [ES] is part of an equilibrium formed between E and S during the time scale of the experiment or observation. These assumptions require that k_2, the rate of breakdown of the enzyme substrate complex, is much smaller than the remaining two rate constants. The total concentration of enzyme active sites ($\mathrm{E_T}$) is expressed as the sum of the free enzyme concentration (E) plus that found in the enzyme substrate complex. The initial velocity of an enzyme-catalysed reaction is

$$v = -\mathrm{d}[\mathrm{S}]/\mathrm{d}t\ \mathrm{d} = [\mathrm{P}]/\mathrm{d}t = k_2[\mathrm{ES}] \qquad (7.30)$$

with a maximal velocity achieved at high substrate concentrations when the enzyme is saturated with substrate. Defining this maximal velocity as V_{max} leads to the following equalities

$$[\mathrm{ES}] = [\mathrm{E_T}] \Rightarrow V_{max} = k_2[\mathrm{E_T}] = k_{cat}[\mathrm{E_T}] \qquad (7.31)$$

$$\mathrm{E} + \mathrm{S} \underset{k_{-1}}{\overset{k_1}{\rightleftharpoons}} \mathrm{ES} \xrightarrow{k_2} \mathrm{E} + \mathrm{P}$$

Figure 7.7 The basic enzyme-catalysed reaction converts substrate into product

The enzyme substrate complex is called the Michaelis complex and the equilibrium formed with E and S is described by a constant called the Michaelis constant (K_m). From the scheme above the Michaelis constant K_m is equal to

$$K_m = (k_2 + k_{-1})/k_1 = ([E_T] - [ES])[S]/[ES] \quad (7.32)$$

and since $[E_T] - [ES] = [E]$ reorganization of this equation leads to

$$K_m[ES] = [E_T][S] - [ES][S] \quad (7.33)$$

$$K_m v/k_2 = V_{max}/k_2[S] - v/k_2[S] \quad (7.34)$$

A final re-arrangement of the above equation leads to the velocity of an enzyme-catalysed reaction, v, expressed as a function of substrate concentration–the famous Michaelis–Menten equation

$$v = V_{max}[S]/K_m + [S] \quad (7.35)$$

The velocity is related to substrate concentration through two parameters (V_{max} and K_m) that are specific and characteristic for a given enzyme under a defined set of conditions (pH, temperature, ionic strength, etc.). From the above equation it can be seen that when

$$[S] \ll K_m \qquad v = V_{max}[S]/K_m$$

$$[S] \gg K_m \qquad v = V_{max}$$

$$[S] = K_m \qquad v = V_{max}/2 \quad (7.36)$$

The Michaelis–Menten equation is usually analysed by linear plots to derive the parameters K_m and V_{max}. Most frequently used in enzyme kinetics is the Lineweaver–Burk plot of $1/v$ versus $1/[S]$ (Figure 7.8).

Although widely used the Lineweaver–Burk plot can be unsatisfactory because it fails to adequately define deviations away from linearity at high and low substrate concentrations. This problem can be neglected with the use of line fitting or linear regression analysis methods. More satisfactory in terms of data analysis is the Eadie–Hofstee plot where rearranging the Michaelis–Menten equation yields

$$v = -K_m\, v/[S] + V_{max} \quad (7.37)$$

Experimental data are usually derived by measuring the initial velocity of the enzyme-catalysed reaction at a range of different substrate concentrations before substantial substrate depletion or product accumulation.

The steady state approximation and the kinetics of enzyme action

An alternative approach to enzyme kinetic analysis proposed by G.E. Briggs and J.B.S. Haldane showed

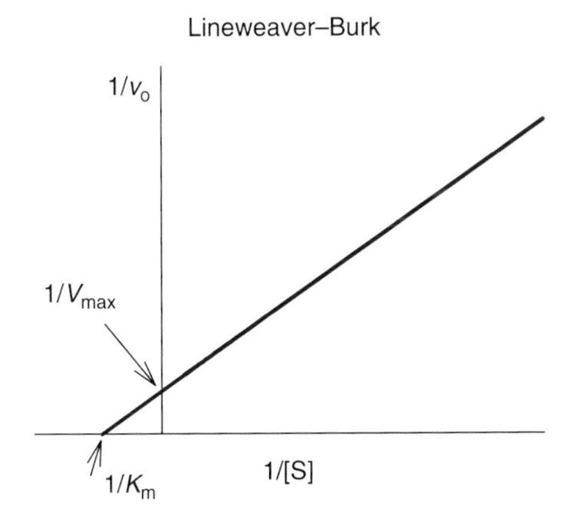

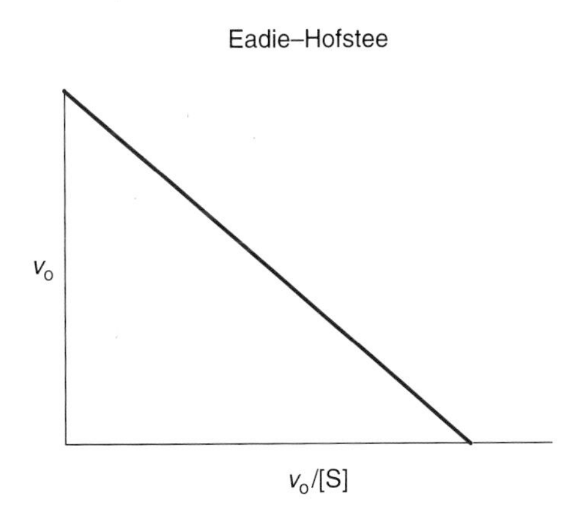

Figure 7.8 Lineweaver–Burk and Eadie–Hofstee plots derived from the Michaelis–Menten equation. The slopes associated with each plot are K_m/V_{max} and $-K_m$ respectively

$$E + S \underset{k_{-1}}{\overset{k_1}{\rightleftharpoons}} ES \xrightarrow{k_2} E + P$$

Figure 7.9 The basic scheme for an enzyme-catalysed reaction

$$E + S \underset{k_{-1}}{\overset{k_1}{\rightleftharpoons}} ES \underset{k_{-2}}{\overset{k_2}{\rightleftharpoons}} EP \underset{k_{-3}}{\overset{k_3}{\rightleftharpoons}} E + P$$

Figure 7.10 A more detailed reaction scheme that describes an enzyme catalysed reaction

that it was unnecessary to assume that enzyme and substrate are in thermodynamic equilibrium with the ES complex (Figure 7.9).

After mixing together enzyme and substrate the concentration of ES rises rapidly reaching a constant level called the 'steady state' that persists until substrate depletion. Most rate measurements are made during the steady state period after the initial build up of ES called the pre-steady state and before substrate depletion. The 'steady state approximation' leads to the following relationships

$$d[ES]/dt = 0 = k_1[E][S] - (k_2 + k_{-1})[ES] \quad (7.38)$$

$$d[E]/dt = 0 = k_2 + k_{-1}[ES] - k_1[E][S] \quad (7.39)$$

The rate of formation of product is proportional to the concentration of ES

$$d[P]/dt = k_2[ES] \quad (7.40)$$

and from the steady state relationships defined above we have

$$k_1[E][S] = (k_2 + k_{-1})[ES] \quad (7.41)$$

This leads with the substitution of $[E] = [E_T] - [ES]$ to

$$[E_T] = (k_2 + k_{-1})[ES]/k_1[S] + [ES] \quad (7.42)$$

$$= [ES](1 + (k_2 + k_{-1})/k_1[S]) \quad (7.43)$$

The Michaelis constant K_m is defined as $(k_2 + k_{-1})/k_1$ and allows the above equation to be further simplified

$$[ES] = [E_T]/(1 + K_m/[S]) \quad (7.44)$$

Having defined the concentration of the ES complex we can now relate the formation of product ($d[P]/dt$) to substrate concentration since

$$d[P]/dt = k_2[ES] = k_2[E]_T/(1 + K_m/[S]) \quad (7.45)$$

$$= k_2[E]_T[S]/([S] + K_m) \quad (7.46)$$

A second assumption made in the Michaelis–Menten theory was that the enzyme exists in two forms: the enzyme and the enzyme–substrate complex. Observations derived from kinetic studies indicate several forms of enzyme–substrate complex and enzyme–product complex co-exist with the free enzyme. This allows the simple kinetic scheme to be redefined including reversible reactions and enzyme–product complex formed before product release (Figure 7.10).

K_m is the initial substrate concentration that allows the reaction to proceed at the half maximal velocity. K_m is sometimes defined as the substrate dissociation constant ($K_s = k_{-1}/k_1$) although this applies *only* under a restricted set of conditions when the catalytic constant k_{cat} is much smaller than either k_1 or k_{-1}. Under these conditions, and only these conditions, the parameter K_m describes the affinity of an enzyme for its substrate.

K_m discriminates between hexokinase and glucokinase; a rare example of two enzymes that catalyse the phosphorylation of glucose. The K_m for glucokinase is ~10 mM whilst the K_m for hexokinase is ~100 μM. The K_m of each enzyme differs by a factor of ~100 and although catalysing the same reaction these enzymes have different biological functions. Hexokinase is distributed throughout all cells and is the normal route for glucose conversion into glucose-6-phosphate. In contrast glucokinase is found in the liver and is used when blood glucose levels are high, for example after a carbohydrate-rich meal. It seems likely that enzymes possess K_m values reflecting intracellular levels of substrate to ensure turnover at rates close to maximal velocity.

More rigorous descriptions of enzyme activity involve the use of k_{cat} and K_m. K_m is a measure of the stability of the enzyme–substrate [ES] complex whilst k_{cat} is a first order rate constant describing the breakdown of the ES complex into

Table 7.4 The efficiency of enzymes; k_{cat}/K_m ratios or specificity constants show a wide range of values

Enzyme	Substrate	K_m (mM)	k_{cat} (s^{-1})	k_{cat}/K_m (M^{-1} s^{-1})
Carbonic anhydrase	CO_2	12.0	1.0×10^6	8.3×10^7
Catalase	H_2O_2	25.0	4.0×10^5	1.5×10^7
Pepsin	Phe-Gly of peptide chain	0.3	0.5	1.7×10^3
Ribonuclease	$(RNA)_n$	7.9	7.9×10^2	1.0×10^5
Tyrosl-tRNA synthetase	tRNA	0.9	7.6	8.4×10^3
Glucose isomerase	Glucose	140.0	1.04×10^3	7.4
Fumarase	Fumarate	0.005	8.0×10^2	1.6×10^8
Lysozyme	$(NAG\text{-}NAM)_3$	6	0.5	83
DNA polymerase I	dNTPs	0.025	1	4.0×10^4
Acetylcholinesterase	Acetylcholine	0.095	1.4×10^4	1.5×10^8

products. The magnitude of k_{cat} varies enormously with enzymes such as superoxide dismutase, carbonic anhydrase and catalase exhibiting high k_{cat} values ($>10^5$ s^{-1}) although values between 10^2–10^3 are more typical of enzymes such as trypsin, pepsin or ribonuclease. By comparing the ratio of the turnover number, k_{cat}, to the K_m a measure of enzyme efficiency is obtained (Table 7.4). Some enzymes achieve high k_{cat}/K_m values by maximizing catalytic rates but others enzymes achieve high efficiency by binding substrate at low concentrations ($K_m <$ 10 μM).

Further insight into the significance of k_{cat}/K_m values is obtained by considering the Michaelis–Menten equation at low substrate concentration. Under conditions where $[S] \ll K_m$ most of the enzyme is free and does not contain bound substrate, i.e. $[E]_t \approx [E]$. The equation becomes

$$\nu \approx k_{cat}/K_m[E][S] \tag{7.49}$$

The term k_{cat}/K_m is a second order rate constant for the reaction between substrate and free enzyme and is a true estimate of enzyme efficiency since it measures binding and catalysis under conditions when substrate and enzyme are not limiting. These values are particularly useful for comparing the efficiency of enzymes with different substrates. Proteolytic enzymes such as chymotrypsin show a wide range of substrate specificity, but measuring k_{cat}/K_m values for different substrates demonstrates a clear preference for cleavage of polypeptide chains after aromatic residues such as phenylalanine (Table 7.5). Using the amino acid methyl esters (Figure 7.11) as substrates the second order rate constant (k_{cat}/K_m) is greater by a factor of $\sim 10^7$ when the side chain is phenylalanine as opposed to glycine.

Table 7.5 The side chain preference of chymotrypsin in the hydrolysis of different amino acid methyl esters

Amino acid analogue	Side chain	k_{cat}/K_m
Glycine	$-H$	0.13
Norvaline	$-CH_2-CH_2-CH_3$	360
Norleucine	$-CH_2-CH_2-CH_2-CH_3$	3000
Phenylalanine	$-CH_2-$(phenyl ring)	100 000

The k_{cat}/K_m ratio for enzymes such as fumarase, catalase, carbonic anhydrase and superoxide dismutase range from 10^7 to 10^9 $M^{-1}\,s^{-1}$ and are close to the diffusion-controlled limit where rates are determined

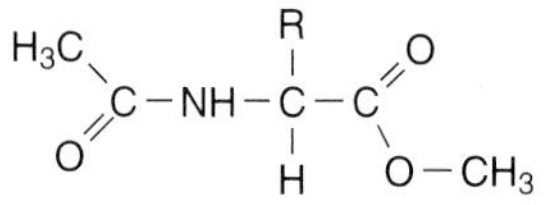

$$H_3C-C(=O)-NH-CH(R)-C(=O)-O-CH_3$$

Figure 7.11 A general structure for *N*-acetyl amino acid methyl esters

by diffusion through solution and all collisions result in productive encounters. For an average-sized enzyme this value is $\sim 10^8 - 10^9\ \text{M}^{-1}\,\text{s}^{-1}$ and such reactions are said to be 'diffusion controlled'. Many enzymes approach diffusion-controlled rates and molecular evolution has succeeded in generating catalysts of maximum efficiency. With knowledge about the kinetics and efficiency of enzymes it is relevant to look at the structural basis for enzyme activity yielding near perfect biological catalysts.

Catalytic mechanisms

Why do enzymes catalyse reactions more rapidly and with higher specificity than normal chemical processes? Answering this question is the basis of understanding catalysis although it is interesting to note that biology employs exact the same atoms and molecules as chemistry. Enzymes have 'fine tuned' catalysis by binding substrates specifically and with sufficient affinity in an active site that contains all functional groups. Enzymes are potent catalysts with 'perfection' resulting from a wide range of mechanisms that include acid–base catalysis, covalent catalysis, metal ion catalysis, electrostatic mechanisms, proximity and orientation effects and preferential binding of the transition state complex. To reach catalytic perfection many enzymes combine one or more of these mechanisms.

Acid–base catalysis

Acid or base catalysis is a common method used to enhance reactivity. In acid-catalysed reactions partial proton transfer from a donor (acid) lowers the free energy of a transition state. One example of this reaction is the keto-enol tautomerization reactions (Figure 7.12) that occur frequently in biology. The uncatalysed reaction is slow and involves the transfer of a proton from a methyl group to the oxygen via a high energy carbanion transition state.

Proton donation to the electronegative oxygen atom decreases the carbanion character of the transition state and increases rates of reaction whilst a similar effect is achieved by the proximity of a base and the abstraction of a proton. In biological reactions both processes can occur as concerted acid–base catalysis. Many reactions are susceptible to acid or base catalysis and in the side chains of amino acid residues enzymes have a wide choice of functional group. The side chains of Arg, Asp, Cys, Glu, His, Lys and Tyr are known to function as acid–base catalysts with active sites containing several of these side chains arranged around the substrate to facilitate catalysis.

Many enzymes exhibit acid–base catalysis although RNase A (Figure 7.13) from bovine pancreas is used here as an illustrative example. The digestive enzyme secreted by the pancreas into the small intestine degrades polymeric RNA into component nucleotides. Careful analysis of the pH dependency of catalysis of RNA hydrolysis observed two ionizable groups with pKs of 5.4 and 6.4 that governed overall rates of reaction. Chemical modification studies suggested these groups were imidazole side chains of histidine residues. The active site of RNase A contains two histidines (12 and 119) with the side chains acting in a concerted fashion to catalyse RNA hydrolysis. A key event towards understanding mechanism was the observation that 2′–3′ cyclic nucleotides could be isolated from reaction mixtures indicating their presence as intermediates.

Hydrolysis is a two-step process; in the first stage His12 acts as a general base abstracting a proton from the 2′–OH group of the ribose ring of RNA. This event promotes nucleophilic attack by the oxygen on the phosphorus centre. His119 acts as an acid catalyst enhancing bond scission by donating a proton to the leaving group (the remaining portion of the polynucleotide chain) (Figure 7.14). The resulting 2′–3′ cyclic nucleotide is the identified intermediate. The cyclic intermediate is hydrolysed in a reaction where His119 acts as a base abstracting a proton from water whilst His12 acts as an acid. The completion of step 2 leaves the enzyme in its original state and ready to repeat the cycle for the next catalytic reaction.

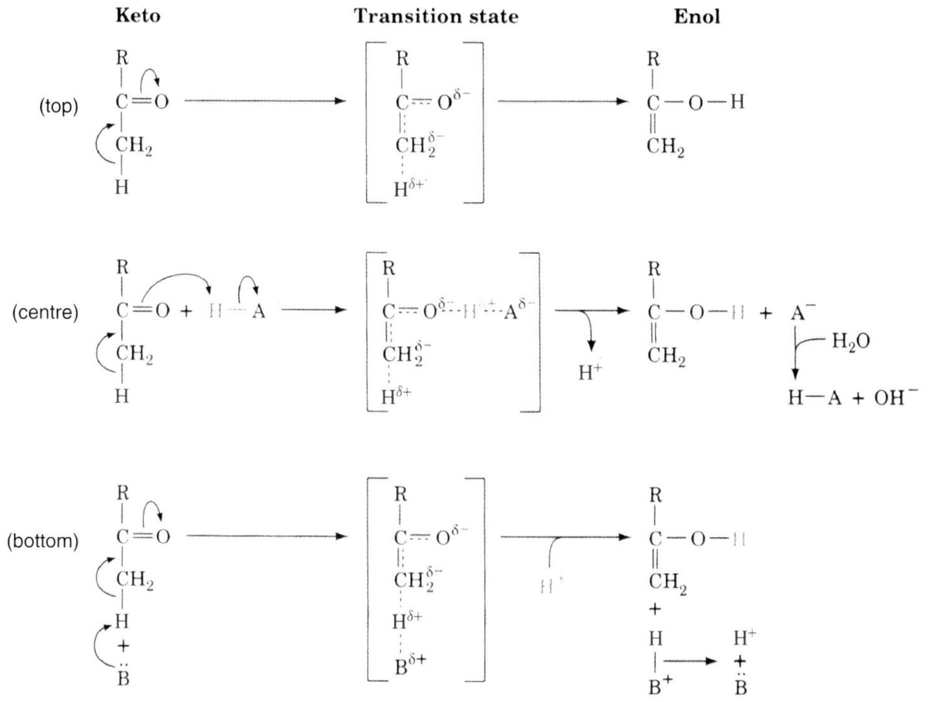

Figure 7.12 Mechanism of keto-enol tautomerization. Top: uncatalysed reaction; centre: general acid-catalysed reaction; bottom: base-catalysed reaction. HA represents an acid B: represents a base. The transition state is enclosed in brackets to indicate 'instability'

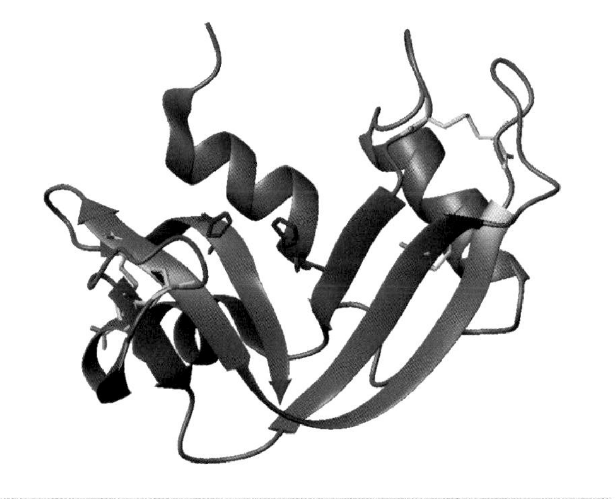

Figure 7.13 The active site of RNase A within the protein structure. The enzyme is based predominantly around two collections of β sheet and four disulfide bridges (yellow). The active site histidines side are shown in red (PDB: 1FS3)

Covalent catalysis

Transient formation of covalent bonds between enzyme and substrate accelerates catalysis. One method of achieving covalent catalysis is to utilize nucleophilic groups on an enzyme and their reaction with centres on the substrate. In enzymes the common nucleophiles are the oxygen of hydroxyl groups, the sulfur atom of sulfhydryl groups, unprotonated amines as well as unprotonated imidazole groups (Figure 7.15). The common link is that all groups have at least one set of unpaired electrons available to participate in catalysis.

Electrophiles include groups with unfilled shells frequently bonded to electronegative atoms such as oxygen. In substrates the most common electrophiles (Figure 7.16) are the carbon atoms of carbonyl groups although the carbon in cationic imine groups (Schiff base) are also relevant to catalysis.

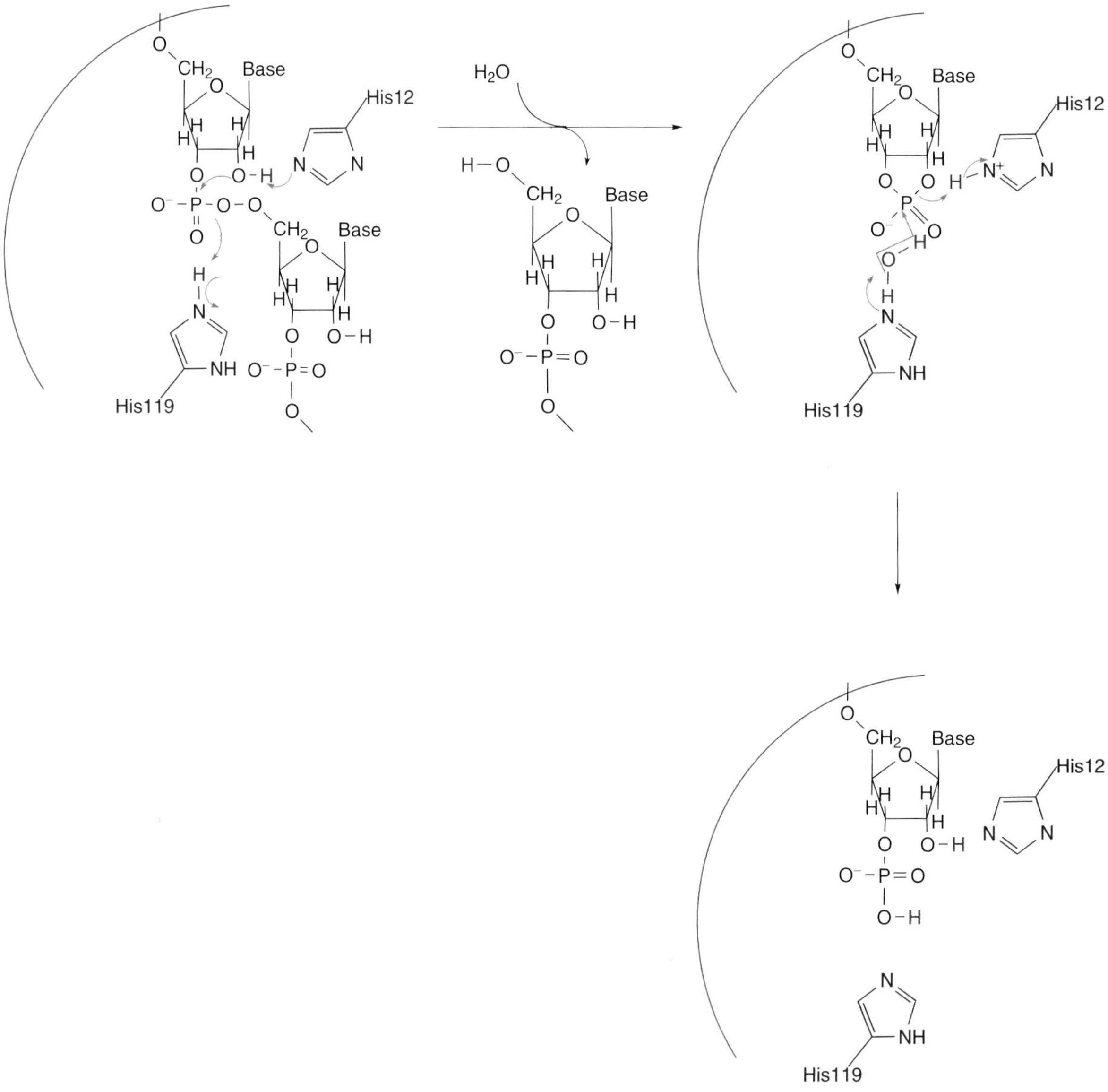

Figure 7.14 The acid–base catalysis by His12 and His119 in bovine pancreatic RNaseA occurs in two discrete steps involving the alternate action of His12 first as a base and second as an acid. His119 acts in the opposite order

Metal ion catalysis

Metalloenzymes use ions such as Fe^{2+}/Fe^{3+}, Cu^{+}/Cu^{2+}, Zn^{2+}, Mn^{2+} and Co^{2+} for catalysis. Frequently, the metal ions have a redox role but additional important catalytic functions include directing substrate binding and orientation via electrostatic interactions. Carbonic anhydrase, a zinc metalloenzyme found widely in cells (Figure 7.17), catalyses the reversible hydration of carbon dioxide

$$CO_2 + H_2O \rightleftharpoons HCO_3^- + H^+$$

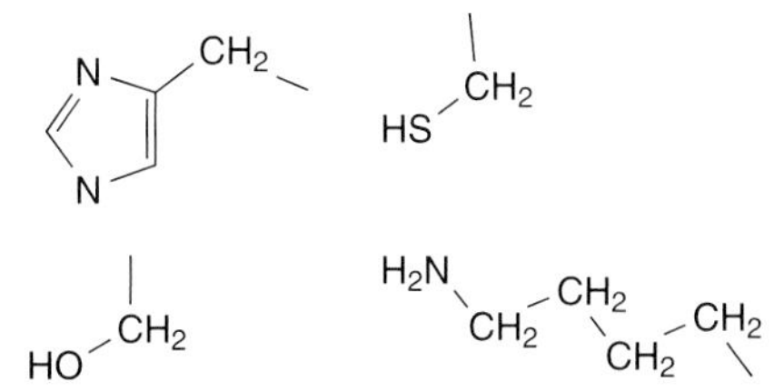

Figure 7.15 Important nucleophiles found in proteins include the imidazole, thiol, amino and hydroxyl groups

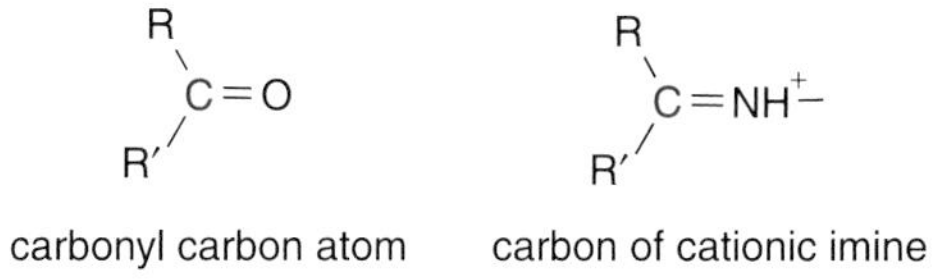

carbonyl carbon atom　　carbon of cationic imine

Figure 7.16 Biologically important electrophilic groups

In red blood cells this reaction is of major importance to gaseous exchange involving transfer of CO_2 from haemoglobin. At the heart of carbonic anhydrase is a divalent Zn cation essential for catalytic activity lying at the bottom of a cleft, ~1.5 nm deep, coordinated in a tetrahedral environment to three invariant histidine residues and a bound water molecule. Enzyme kinetics show pH-dependent catalysis increasing at high pH and modulated by a group with a pK_a of ~7.0. Water molecules bound close to a metal ion tend to be stronger proton donors than bulk water with the result that in carbonic anhydrase the Zn not only orients the water at the active site but assists in base catalysis by removing a proton. This generates a bound hydroxide ion, a potent nucleophile, even at pH 7.0. This model is supported by the observation that anion addition inhibits enzyme activity by competing with OH^- anions at the Zn site.

At least 10 isozymes of carbonic anhydrase are identified in human cells with carbonic anhydrase II of particular pharmaceutical interest because its activity is linked to glaucoma via increased intra-ocular pressure. Uncontrolled increases in pressure in the eye damages the optic nerve leading to a loss of peripheral vision and 'blind spots' in the field of sight. Drugs (enzyme inhibitors) combat glaucoma by binding to the active site. Acetazolamide, a member of the sulfonamide group of antibacterial agents, is the prototypal inhibitor and displaces the hydroxide ion at the Zn active site by ligation through the ionized nitrogen of the primary sulfonamide group (Figure 7.18).

Electrostatic catalysis

Substrate binding to active sites is often mediated via charged groups located in pockets lined with hydrophobic side chains. Any charged side chain located in this region will exhibit a pK_a shifted away from its normal 'aqueous' value and exhibits in theory at least, a massive potential for electrostatic binding and catalysis. Direct involvement of charged residues in catalysis has proved harder to demonstrate, in contrast to many studies showing that substrate binding is enhanced by charge distribution. The aim of electrostatic catalysis is to use the distribution of charged groups at the active site of enzymes to enhance the stability of the transition state. This aspect of electrostatic enhancement of catalysis has been termed the 'circe' effect by W.P. Jencks and encompasses the use of attractive forces to 'lure' the substrate into the active site destabilizing the reactive group undergoing chemical transformation.

Orotidine 5′-monophosphate decarboxylase (ODCase) catalyses the conversion of orotidine 5′-monophosphate to uridine 5′-monophosphate, the last step in the biosynthesis of pyrimidine nucleotides (Figure 7.19). ODCase catalyses exchange of CO_2 for a proton at the C6 position of uridine 5′-monophosphate in a reaction that is exceptionally slow in aqueous solution but is accelerated 17 orders of magnitude by enzyme. The mechanism underlying enhancement has fascinated chemists and biochemists alike and in the active site of the β barrel are a series of charged side chains (Lys42-Asp70-Lys72-Asp75) found close together. Although definitive demonstration of the role of these residues in the activity of ODCase requires further study the frequent presence of charge groups in non-polar active sites points to the involvement of electrostatics in catalysis.

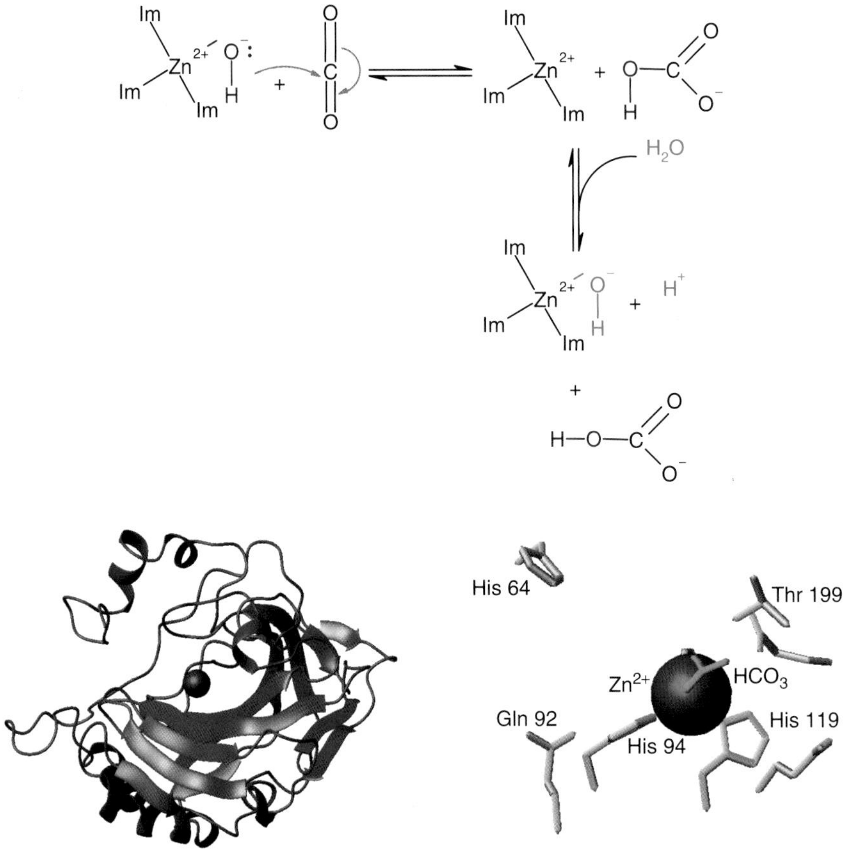

Figure 7.17 The arrangement of imidazole ligands (Im) around the central Zn cation in carbonic anhydrase favours abstraction of a proton from water and the nucleophilic attack by the hydroxide ion on carbon dioxide bound nearby. The imidazole ligands arise from His 94, His96 (hidden by Zn), and His119. A zinc bound hydroxide ion is hydrogen bonded to the hydroxyl side chain of Thr199, which in turn hydrogen bonds with Glu106. Catalytic turnover requires the transfer of the product proton to bulk solvent via shuttle groups such as His64. Shown in the diagram of the active site is the bicarbonate ion bound to the zinc

Catalysis through proximity and orientation effects

Biological catalysis is typified by high specificity coupled with remarkable speed. The gains in efficiency for enzyme-catalysed reactions arise from specific environments found at active sites that promote chemical reactions to levels of efficiency the organic chemist can only dream about. Two parameters that are important in modulating catalysis are proximity and orientation. These superficially nebulous terms include the effects of binding substrates close to specific groups thereby facilitating chemical reactions coupled with the binding of substrate in specific, restrictive, orientations.

The potential effects of proximity on rates of reaction have been demonstrated through experiments involving the non-enzymatic hydrolysis of *p*-bromophenylacetate. The reaction is first performed as a simple bimolecular reaction and compared with a series of intramolecular reactions in which the

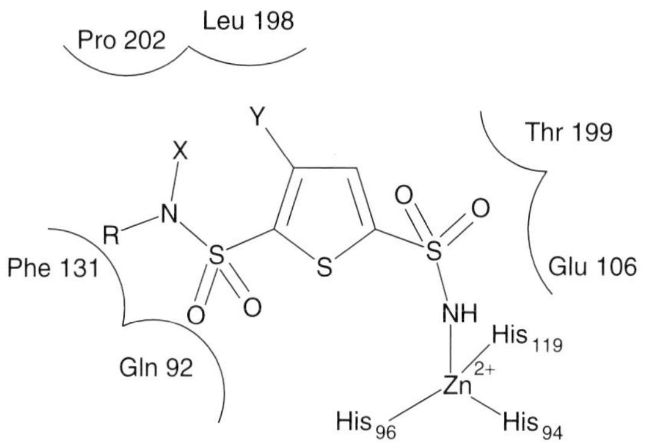

Figure 7.18 General scheme of sulfonamide inhibitor binding to carbonic anhydrase. The 'primary sulfonamide' coordinates to the active site zinc ion and hydrogen bonds with Thr199

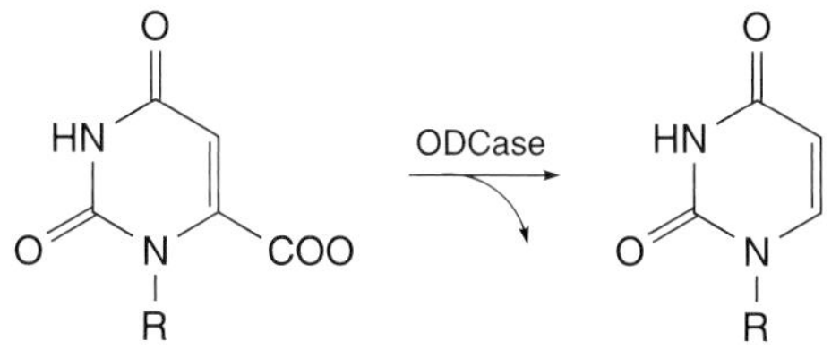

Figure 7.19 The reaction catalysed by orotidine 5′-monophosphate decarboxylase converts orotidine 5′-monophosphate into uridine monophosphate in the final step in pyrimidine biosynthesis. R is a ribose 5′-phosphate group in both reactant and product

reacting groups are connected by a bridge that exerts progressively more conformational restriction on the molecule. In the simple bimolecular reaction *p*-bromophenylacetate is hydrolysed to yield bromophenolate anion with the formation of acetic anhydride (Figure 7.20). The rate of hydrolysis of bromophenol esters increases dramatically as the reactants are held in closer proximity and in restricted conformations. The relative rate constants increase from 1 (on an arbitrary scale) to 10^3, to 10^5 and finally in the most restricted form to a value approaching 10^8 times greater than that exhibited by the bimolecular reaction. Although the unimolecular reactions described here do not involve true catalysis they emphasize the significant rate enhancement that results from controlling proximity and orientation. The intramolecular reactions model the binding and positioning of two substrates within an enzyme active site and although sizeable rate enhancements occur they do not account for all of the observed increase in rates seen in enzyme versus uncatalysed reactions (for example orotidine 5′-monophosphate decarboxylase enhances rates $\sim 10^{17}$-fold). Enzymes have yet further mechanisms to enhance catalysis.

Preferential binding of the transition state causes large rate enhancements

Acid–base catalysis and covalent catalytic mechanisms can perhaps enhance rates by factors of 10–10^2 whilst proximity and orientation can achieve enhancements of up to 10^8, although frequently it is much less. To achieve the magnitude of catalytic enhancement seen for some enzymes additional mechanisms are required to achieve the ratios of 10^{14}–10^{17} seen for urease, alkaline phosphatase and ODCase. One of the most important catalytic mechanisms exhibited by enzymes is the ability to bind transition states with greater affinity than either products or reactants.

For many organic reactions the concept of strain is well known and it would not be surprising if enzymes adopted similar mechanisms. Enzymes are envisaged as 'straining' or 'distorting' their substrates towards the transition state geometry by the formation of binding sites that do not allow productive binding of 'native' substrates. Thus, the rate of a catalysed reaction compared with an uncatalysed one is related to the magnitude of transition state binding. Catalysis results from preferential transition state binding or the stabilization of the transition state relative to the substrate. In this manner a rate enhancement of 10^6 would require a 10^6 enhancement in transition state binding relative to that of the substrate. This corresponds at room temperature (298 K) to an increased binding affinity of $\sim 34\ kJ\,mol^{-1}$. In terms of bond energies this is the

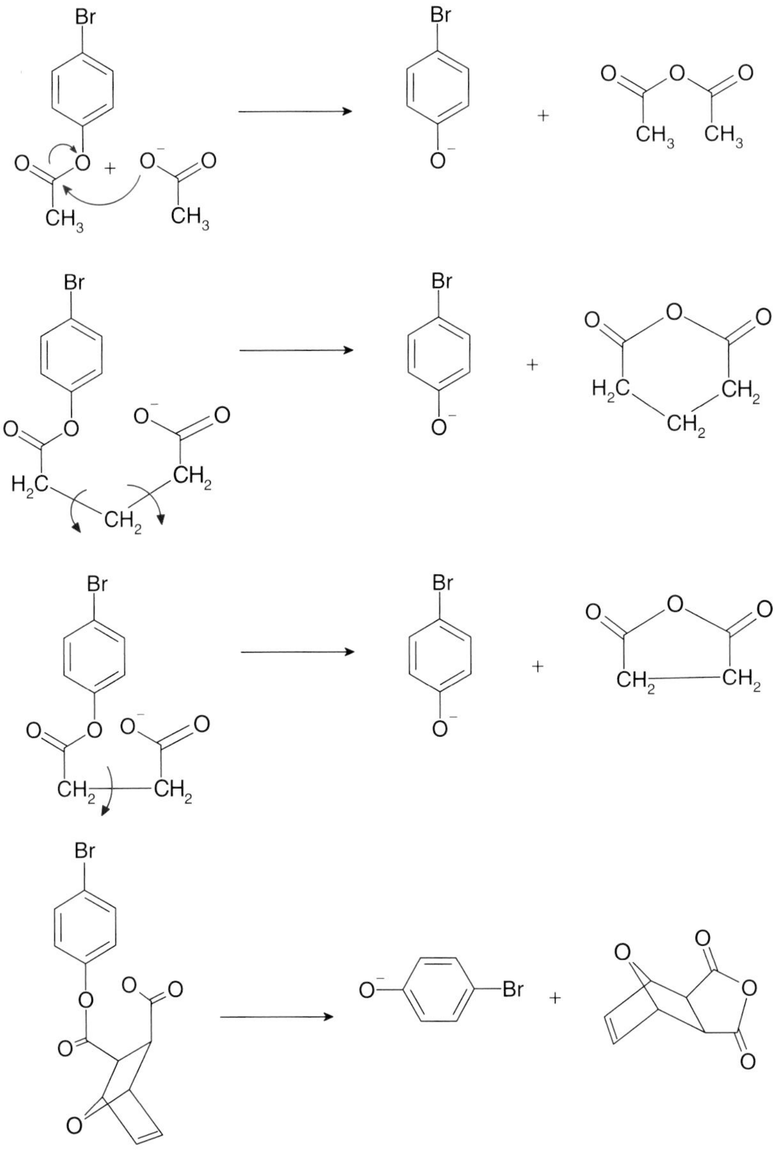

Figure 7.20 Bimolecular and unimolecular reactions describing the hydrolysis of bromophenol esters. The reactions illustrate how proximity effects enhance catalysis. A progressive increase in rate constant occurs from a relative value of 1 in the top reaction to 10^3, 10^5 and finally 10^8

equivalent of a handful of hydrogen bonds. Very large rate enhancements will result from the formation of new bonds between transition state and enzyme and this mechanism underpins rate enhancement by enzymes. Additional bonds can involve non-reacting regions of the substrate and many proteolytic enzymes show substrate specificity that depends on residues bordering the scissile bond at the P_4,P_3 subsites, for example. The concept of transition state binding also explains why some enzymes bind substrates with acceptable

E + DHAP ⇌ E-DHAP ⇌ E-intermediate ⇌ E-G3P ⇌ E + G3P

Figure 7.21 Conversion of dihydroxyacetone phosphate (DHAP) into glyceraldehyde-3-phosphate (G3P) by triose phosphate isomerase

affinity but fail to show significant catalytic enhancement. These 'poor' substrates do not form transition states and hence do not exhibit high turnover rates. As a corollary it is not accurate to say that enzymes bind substrates with the highest affinity – this is reserved for transition state complexes.

The concept of transition state binding receives further support from the observation that transition state analogues, stable molecules that resemble the transition state, are potent catalytic inhibitors. Many transition state analogues are known and used as enzyme inhibitors. 2-Phosphoglycolate is an inhibitor of triose phosphate isomerase. The reaction (Figure 7.21) involves general acid–base catalysis and the formation of enediol/enediolate intermediates (Figure 7.22). 2-Phosphoglycolate resembles the intermediate and was observed to bind to the enzyme with greater affinity than either dihydroxyacetone phosphate or glyceraldehyde-3-phosphate. Increased binding arises from the partially charged oxygen atom found on 2-phosphoglycolate and the transition state but not on either substrate.

Enzyme structure

One of the prime reasons for wanting to determine the structure of enzymes is to verify different catalytic mechanisms. In many instances structural description of enzymes refines or enhances particular catalytic schemes, whilst in a few cases it has dictated a fresh look into enzyme mechanism.

Lysozyme

No discussion of enzyme structure would be complete without a description of lysozyme, a hydrolytic enzyme or hydrolase. It destroys bacterial cell walls by hydrolysing the β(1→4) glycosidic bonds between *N*-acetylglucosamine (NAG) and *N*-acetylmuramic acid (NAM) that form the bulk of peptidoglycan cell walls (Figure 7.23). It will also hydrolyse poly-NAG chains

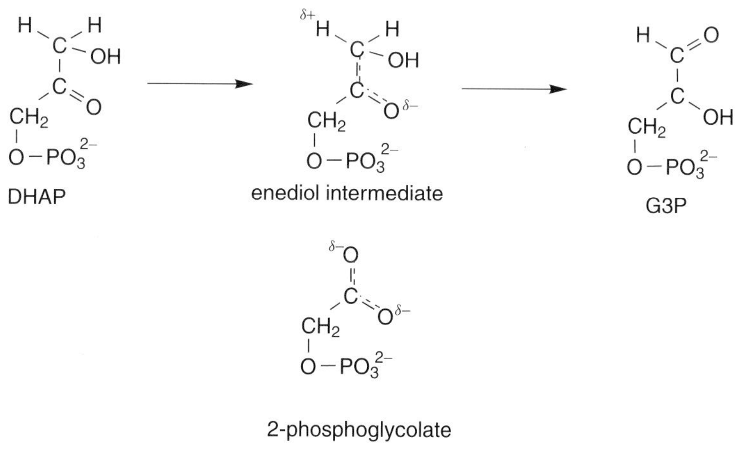

DHAP → enediol intermediate → G3P

2-phosphoglycolate

Figure 7.22 The first enediol intermediate in the conversion of dihydroxyacetone phosphate (DHAP) into glyceraldehyde-3-phosphate (G3P) and the structure of 2-phosphoglycolate, a transition state analogue and an inhibitor of triose phosphate isomerase

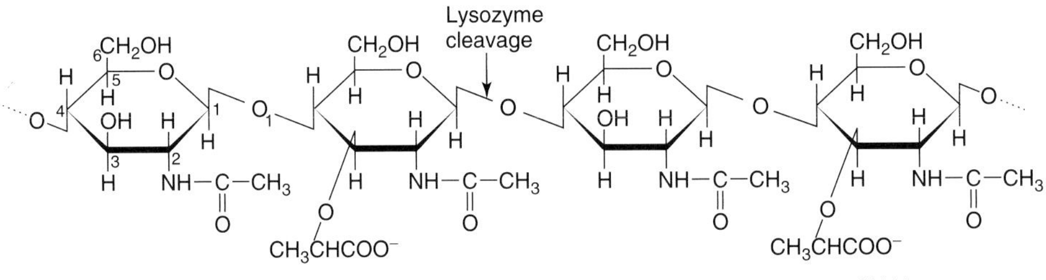

Figure 7.23 The lysozyme cleavage site after the β(1→4) linkage in alternating NAG–NAM units of the polysaccharide components of cell walls. Reproduced from Voet *et al.* (1998) by permission of John Wiley & Sons, Ltd. Chichester

Figure 7.24 The three-dimensional structure of lysozyme (PDB: 1HEW). The distribution of secondary structure is shown together with the binding of a poly-NAG trimer. A space filling model from the same view emphasizes the cleft into which the polysaccharide, shown in magenta, fits precisely. The cleft can accommodate six sugar units

such as those in chitin a key extracellular component of fungal cell walls and the exoskeletons of insects and crustaceans.

The structure of lysozyme determined by David Phillips in 1965 represented the first three-dimensional structure for an enzyme (Figure 7.24). The protein was elliptical in shape (3.0 × 3.0 × 4.5 nm), with the most striking feature being a prominent cleft that traversed one face of the molecule. Although a picture of substrate binding sites had been established from kinetic and chemical modification studies this feature provided a potent image of 'active sites'. Lysozyme contains five helical segments together with a triple stranded anti-parallel β sheet that forms one wall of the binding site for NAG–NAM polymers and a smaller β strand region. Models of lysozyme complexed with saccharides reveal a cleft accommodating six units at individual sites arranged linearly along the binding surface. The sites are designated A–F; NAM binds to sites B, D and F because there is not enough space for the bulkier side chains at positions A, C and E. The remaining sites are occupied by NAG.

Sites B and F accommodate NAM without distortion but at site D the sugar molecule will only fit with distortion from its normal chair conformation into a half-chair arrangement (Figure 7.25). Moreover, trimers of NAG bound at sites A–C are not hydrolysed at significant rates, and by performing lysozyme-mediated catalysis in the presence of H_2O^{18} it was demonstrated that hydrolysis occurred between the C_1 and O atoms of the glycosidic bond located at subsites D and E (Figure 7.26). At this site two side chains from Glu35 and Asp52 are located close to the C_1 position of the distorted sugar. These two side chains are the

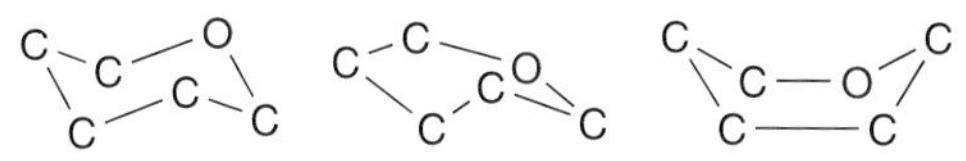

Figure 7.25 Chair, half chair and boat conformations for hexose sugars

only potential catalytic groups in the vicinity of the target C–O bond.

Glu35 is located in a non-polar environment and has an elevated pK_a of ~6.5. Asp52 in contrast has a more typical value around 3.5. The pH optimum for lysozyme is 5.0, midway between the pK_as associated with each of these side chains. Glu35, as a result of its unusual pK, is protonated at pH 5.0 and acts as an acid catalyst donating a proton to the oxygen atom involved in the glycosidic bond between the D and E sites (Figure 7.27). Cleavage occurs and a portion of

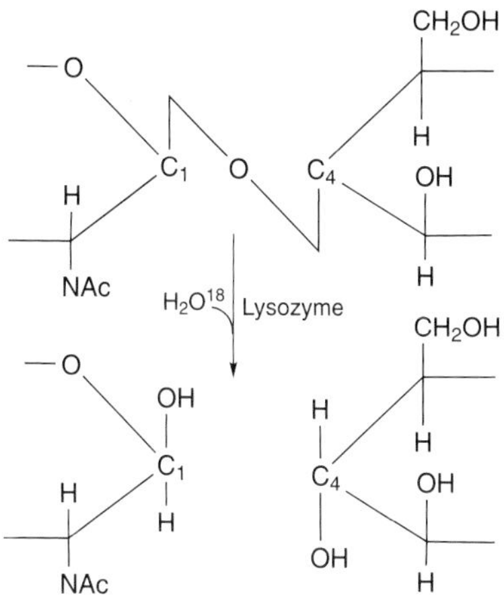

Figure 7.26 Hydrolysis of NAM–NAG polymers by lysozyme in the presence of ^{18}O-labelled H_2O

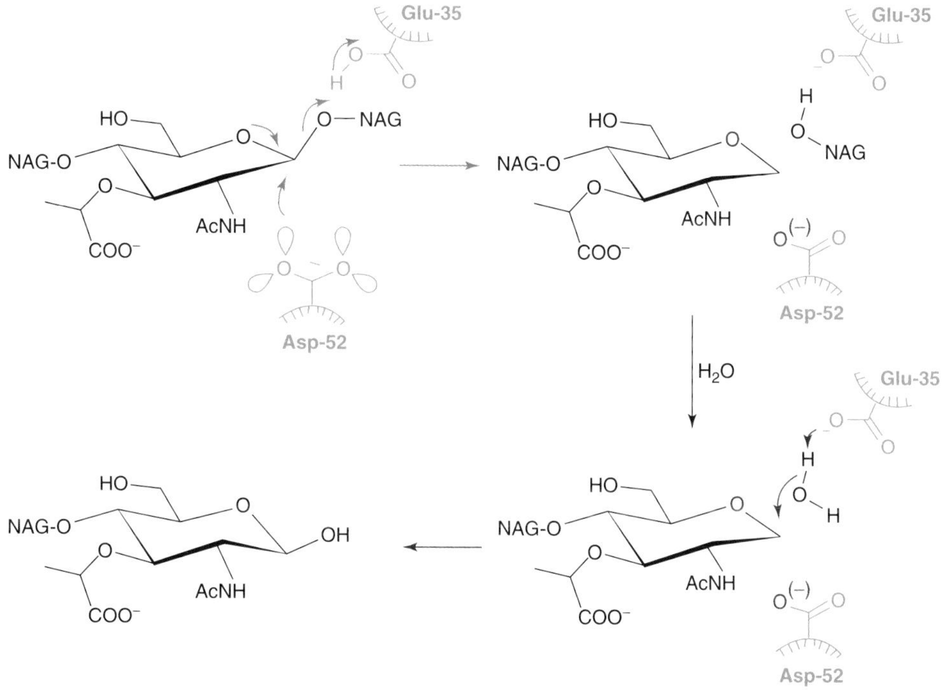

Figure 7.27 The hydrolysis of NAG–NAM polysaccharides catalysed by lysozyme. The substrate is bound so that the leaving group oxygen, the 4-OH group of an *N*-acetylglucosamine (NAG) residue, is protonated as it leaves via the COOH group of Glu35. High pK favours protonation. Enzyme groups are coloured green (reproduced with permission from Kirby, A.J. *Nature Struct. Biol.* 2001, **8**, 737–739. Macmillan)

the substrate bound at sites E and F can diffuse out of the cleft to be replaced by water. The resulting cleavage leads to the formation of a carbocation which interacts with the carboxylate group of Asp 52 (nucleophile) to form a glycosyl–enzyme intermediate in a concerted S_N2-type reaction. The carboxylate group is then displaced from the glycosyl–enzyme intermediate by a water molecule that restores the original configuration at the C_1 centre.

The serine proteases

The serine proteases include the common proteolytic enzymes trypsin and chymotrypsin, as well as acetylcholinesterase, an enzyme central to the transmission of nerve impulses across synapses, thrombin (an enzyme with a pivotal role in blood coagulation) and elastase (a protease important in the degradation of connective fibres). As proteases these enzymes hydrolyse peptide bonds in different target substrates but are linked by the presence of an invariant serine residue at the enzyme active site that plays a pivotal role in catalysis.

For trypsin, chymotrypsin, elastase and thrombin the enzymes cut polypeptides on the C-terminal side of specific residues. These residues are not random but show a substrate specificity that differs widely from one enzyme to another. Trypsin cuts preferentially after positively charged side chains such as arginine or lysine whilst chymotrypsin shows a different specificity cutting after bulking hydrophobic residues with a clear preference for phenylalanine, tyrosine, tryptophan and leucine. In contrast, elastase cuts only after small and neutral side chains whilst thrombin shows a restricted specificity for arginine side chains (Figure 7.28). The substrate specificity arises from the properties of the active site pocket and in particular the identity of residues at subsites making up the substrate binding region. The subsites that bind substrate residues preceding the scissile bond are termed S1, S2, S3, whilst those after this bond are given the names S1′, S2′, S3′, etc. A similar terminology is applied to the peptide substrate where P3, P2 and P1 sites precede the scissile bond and P1′, P2′, and P3′ continue after it.

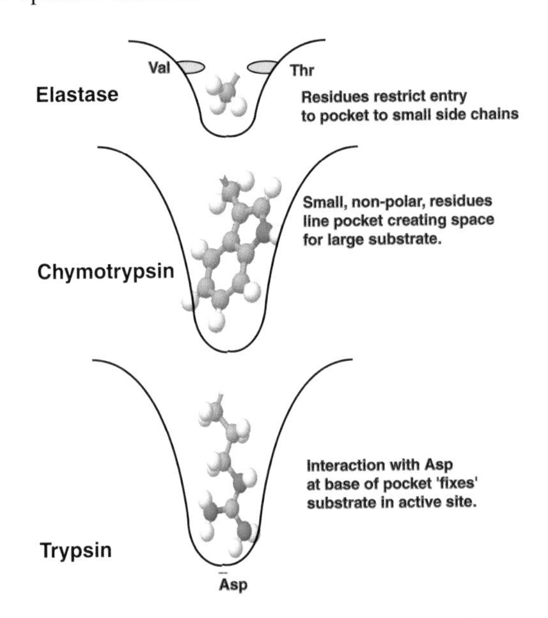

Figure 7.28 Representation of substrate specificity pockets for elastase, chymotrypsin and trypsin

Serine proteases show similar structures with the active site containing three invariant residues arranged in close proximity. These residues are His57, Asp102 and Ser195 and together they form a catalytic triad linked by a network of hydrogen bonds. In addition to these three invariant residues the serine proteases possess a pocket, located close to the active site serine that strongly influences substrate specificity. In trypsin the pocket is relatively long and narrow with a strategically located carboxylate side chain found at the bottom of the pocket. In view of the preferred specificity of trypsin for cleavage after lysine or arginine side chains this pocket appears to have been tailored to 'capture' the substrate very effectively. In chymotrypsin the pocket is larger, wider, lacks charged residues but contains many hydrophobic groups. This is consistent with specificity towards bulky aromatic groups. Elastase, in contrast, has a shallower pocket formed by valine and threonine side chains. Its specificity for cleavage on the C-terminal side of small, uncharged, side chains again reflects organization of the pocket.

Despite tailoring substrate specificity through the properties of the pocket all serine proteases show similar catalytic mechanisms based around a highly reactive serine residue identified from labelling and chemical inactivation studies. Agents based around organophosphorus compounds such as di-isopropylphosphofluoridate (DIPF) were identified via their action as potent neurotoxins (Figure 7.29). Neurotoxicity arises

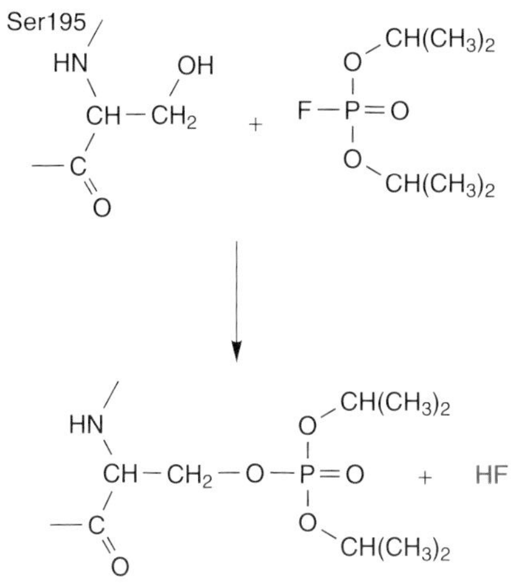

Figure 7.29 Inactivation of serine proteases by di-isopropylphosphofluoridate with the formation of DIP-enzyme

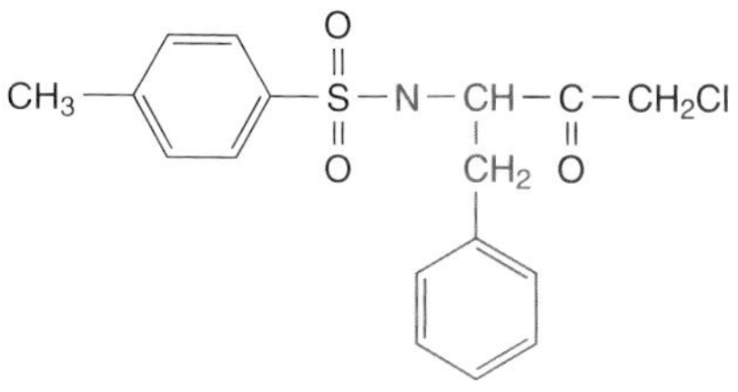

Figure 7.30 Tosyl-L-phenylalanine chloromethylketone (TPCK)

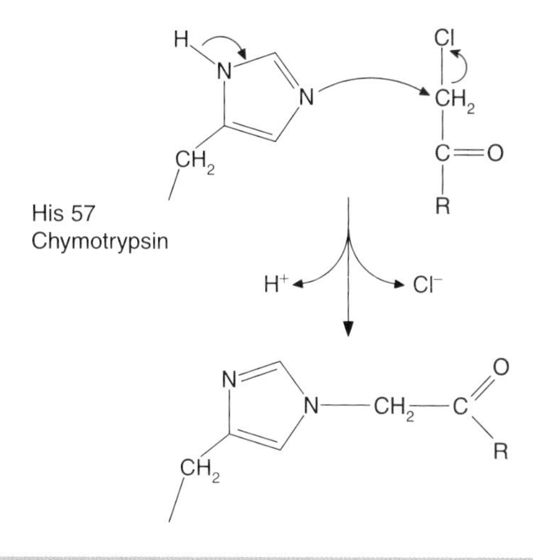

Figure 7.31 The reaction of TPCK with His57 of chymotrypsin. Only the active part of TPCK is shown (red)- alkylation causes enzyme inactivation

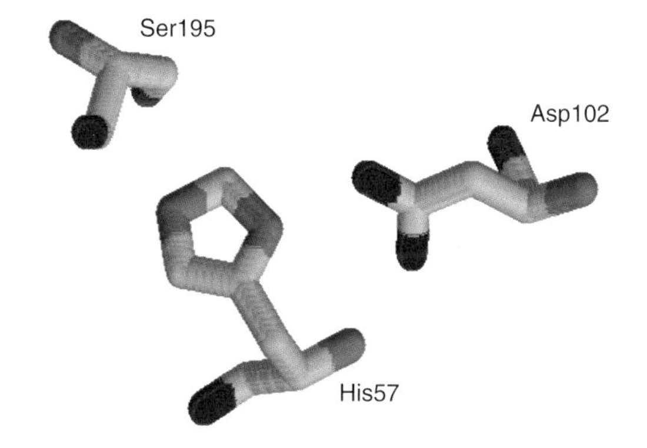

Figure 7.32 The spatial arrangement of residues in the catalytic triad (His57, Asp102 and Ser195) in the active site of chymotrypsin (PDB: 2CGA)

from modification of an active site serine residue in acetylcholinesterase and indicated the similarity of this enzyme to other serine proteases such as trypsin and chymotrypsin where DIPF covalently reacts with Ser195 to form an adduct lacking catalytic activity. Labelling studies also identified the importance of a second residue of the catalytic triad, His57. Chymotrypsin binds tosyl-L-phenylalanine chloromethylketone (TPCK, Figure 7.30) as a result of the presence of a phenyl group, a preferred substrate, within the analogue. The reactive group forms a stable covalent bond with the imidazole side chain (Figure 7.31).

The structure of chymotrypsin revealed the close proximity of His57 and Ser195 together with Asp102 in the active site despite their separation in the primary sequence (Figure 7.32). The first step of catalysis is substrate binding to the active site region and specificity is largely dictated by residues at the bottom of a well-defined cleft. Interactions at subsites either side of the scissile bond are limited. Specific binding leads to close proximity between Ser195 and the carbonyl group of the bond to be cleaved. Under normal conditions the pK_a of the –OH group of serine is very high (>10) but in chymotrypsin the proximity of His57 leads to proton transfer and formation of a charged imidazole group. Protonation of His57 is further stabilized by the negative charge associated with Asp102 – the third member of the catalytic triad – that lies in a hydrophobic environment. The formation of a strong nucleophile in the activated serine attacks the carbonyl group as the first step in peptide bond cleavage. Nucleophile generation is triggered by small conformational changes upon substrate binding that cause Asp102 and His57 to form a hydrogen bond. The influence of this hydrogen bond raises the pK of the imidazole side chain from $\sim$7 to $\sim$11 increasing its basicity. Electrons driven towards the second nitrogen atom form a stronger base that abstracts a proton from the normally

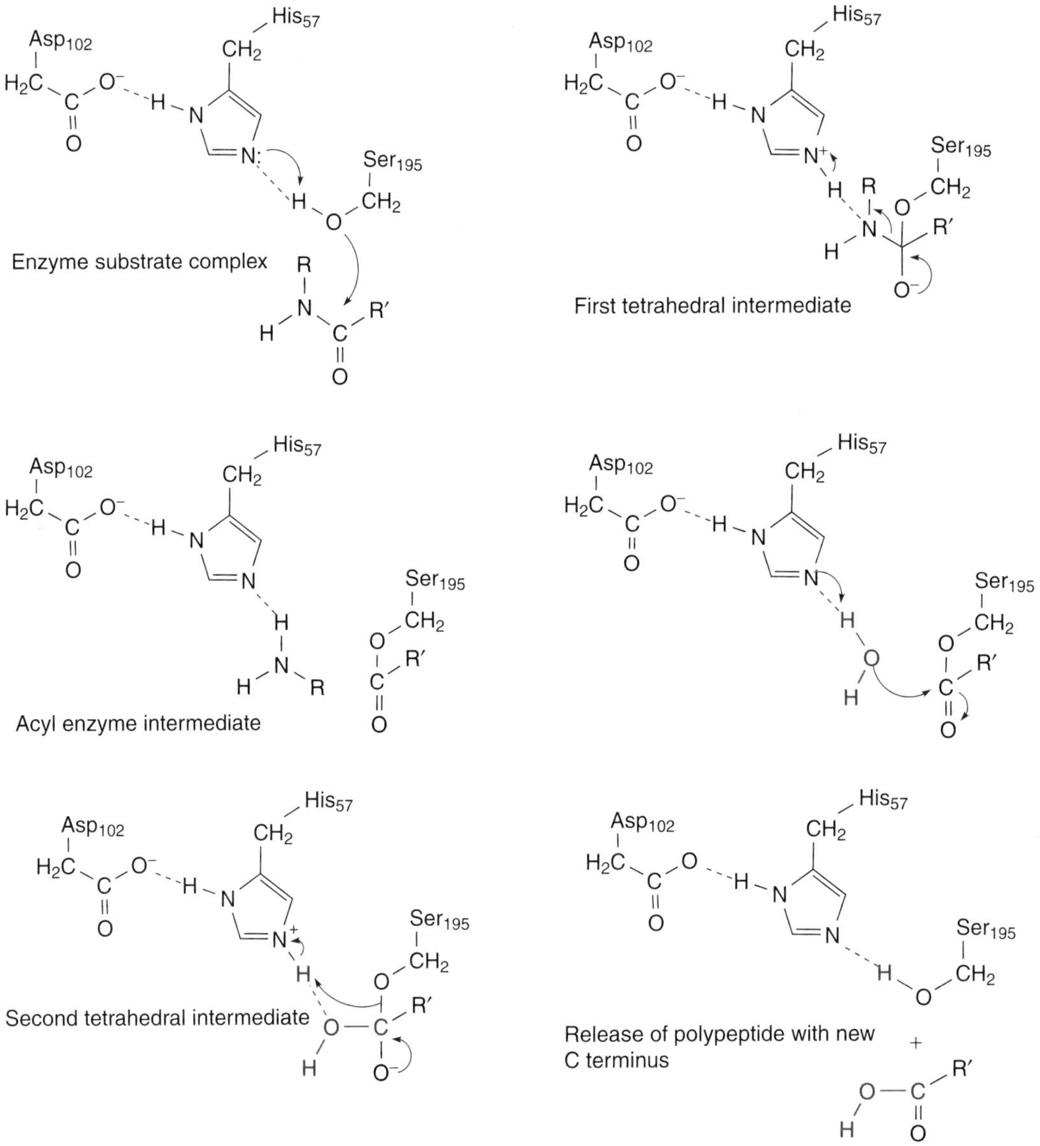

Figure 7.33 The catalytic cycle of serine proteases such as chymotrypsin

unreactive side chain of Ser195 creating the potent nucleophile.

Nucleophilic attack results in a tetrahedral transition state followed by cleavage of the peptide bond. The N-terminal region of the peptide substrate remains covalently linked to the enzyme forming an acyl enzyme intermediate. The proton originally attached to the serine residue and located on the histidine is abstracted by the C-terminal peptide to form a new terminal amino group and this region of the peptide does not bind strongly to the enzyme. As a hydrolytic enzyme, water is involved in the overall reaction summarized in Figure 7.33.

In chymotrypsin water displaces C-terminal peptide from the active site and hydrolyses the acyl-enzyme intermediate. The water molecule donates a proton to the imidazole side chain and forms a second tetrahedral transition state with the acyl intermediate. This intermediate is hydrolysed as a result of proton transfer from His57 back to Ser195 with restoration of the active site to its original state with the release of the N-terminal peptide. Serine proteases follow a ping-pong mechanism (Figure 7.34); the substrate is first bound to the enzyme, the C-terminal peptide is released followed by binding a water molecule and the formation of an acyl–enzyme intermediate. This is completed by release of the N-terminal peptide from hydrolysis of the acyl–enzyme intermediate. The type of group transfer where one or more products are released before all substrates are added is also known as a double-displacement reaction.

It is interesting to note that serine proteases utilize the whole gamut of catalytic mechanisms. Proximity effects play a role in generation of the bound substrates whilst transition state stabilization ensures effective binding prior to catalysis. This is complemented by the use of covalent catalysis (the serine nucleophile) together with acid–base catalysis by histidine to yield efficient enzyme action.

Triose phosphate isomerase

Triose phosphate isomerase has been widely studied and catalyses the conversion between DHAP and G3P (Figure 7.35). The equilibrium for the reaction lies in favour of DHAP, but the rapid removal of G3P by successive reactions in glycolysis drives the reaction towards the formation of G3P.

Triose phosphate isomerase merits further discussion as an example of enzyme catalysis because of two interesting phenomena: the use of the imidazolate form of histidine in catalysis and the evolution of the enzyme towards a near perfect biological catalyst. Influencing both of these phenomena is the structure of triose phosphate isomerase with crystallographic structures

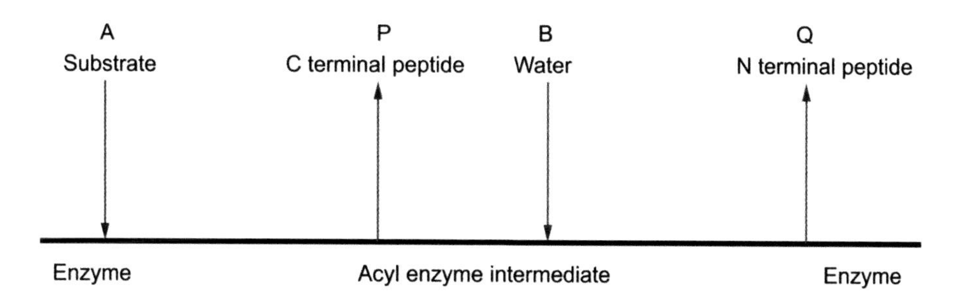

Figure 7.34 The ping-pong mechanism for serine proteases. Using the normal convention A and B represent substrates in the order that they are bound by the enzyme whilst P and Q represent products in the order that they leave the enzyme. The activity of trypsin and other serine proteases are therefore summarized by a scheme

$$E + A \rightleftharpoons E.A \rightleftharpoons E.Q + P$$

$$EQ + B \rightleftharpoons E\text{-}Q.B \rightleftharpoons E + Q\text{-}B$$

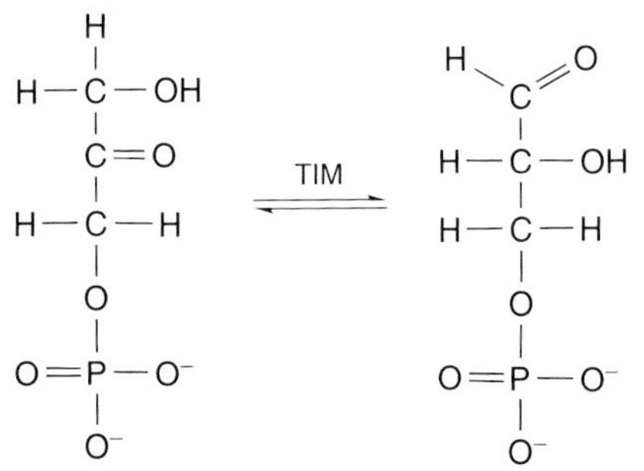

$$H_2C(OH){-}C(=O){-}CH_2{-}O{-}PO_3^{2-} \underset{}{\overset{TIM}{\rightleftharpoons}} H{-}C(=O){-}CH(OH){-}CH_2{-}O{-}PO_3^{2-}$$

Figure 7.35 The interconversion of dihydroxyacetone phosphate (left) and glyceraldehyde-3-phosphate catalysed by triose phosphate isomerase

of the enzyme available in a free state, bound with DHAP and plus 2-phosphoglycolate. Kinetic studies on the pH dependence of the catalytic process revealed two inflection points in the profile corresponding to pH values of ~6.0 and 9.0. When considered alongside the structure of triose phosphate isomerase (Figure 7.36) these values implicated the side chains of glutamate and histidine. Affinity labelling studies using bromohydroxyacetone phosphate modified a glutamate residue within a hexapeptide (Ala-Tyr-Glu-Pro-Val-Trp) and this residue was identified as Glu165. The active site lies near the top of the β barrel with Glu165 close to His 95 as well as Lys12, Glu97 and Tyr208. Both Glu165 and His 95 are essential for catalytic activity and the enzyme provides a cage or reaction vessel holding substrate and protecting intermediates through a series of 11 residues immediately after Glu165 (residues 166–176) that form a loop and acts as a lid for the active site.

Dihydroxyacetone phosphate binds to the active site with the carbonyl oxygen forming a hydrogen bond with the neutral imidazole side chain of His95. The glutamate side chain is charged and facilitates abstraction of a proton from the C1 carbon of dihydroxyacetone phosphate. In this state the His95 forms a strong hydrogen bond with the carbonyl oxygen (C2) of the enediolate intermediate that leads to protonation of this atom. Protonation of the oxygen atom forms an imidazolate side chain on His95 that has a negative charge and is a strong base. Proton abstraction from the hydroxyl group attached to the C1 carbon results in the formation of a second enediolate intermediate with this unstable intermediate converted to the final product by donation of a proton from Glu165 to the C2 carbon. These reactions are shown in Figure 7.37.

Kinetic and thermodynamic characterization of triose phosphate isomerase show measurable binding

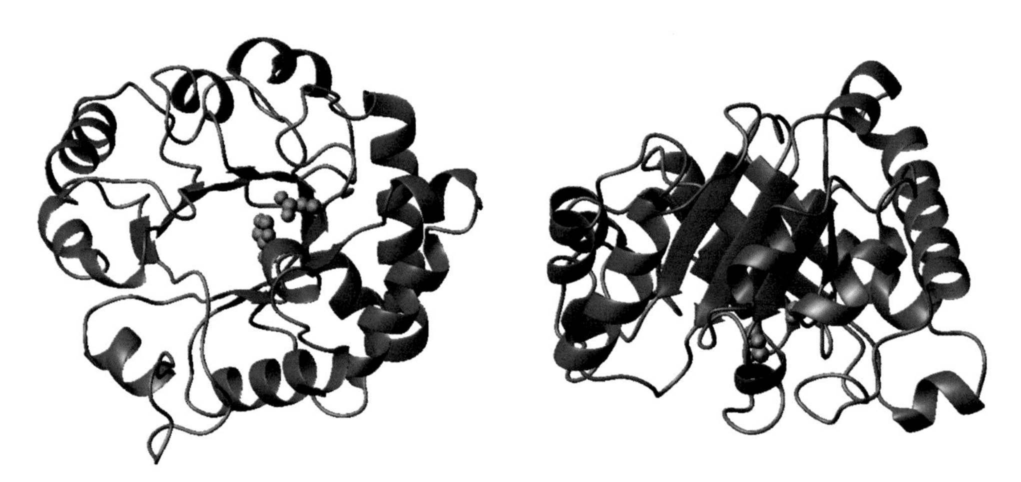

Figure 7.36 Structure of triose phosphate isomerase monomer together with the active site residues surrounding substrate. Strands of the β barrel are shown in purple, helices in blue. The two critical residues of His95 and Glu165 are shown as solid spheres. The enzyme is shown from two orientations

Enediolate intermediate

Enediolate intermediate

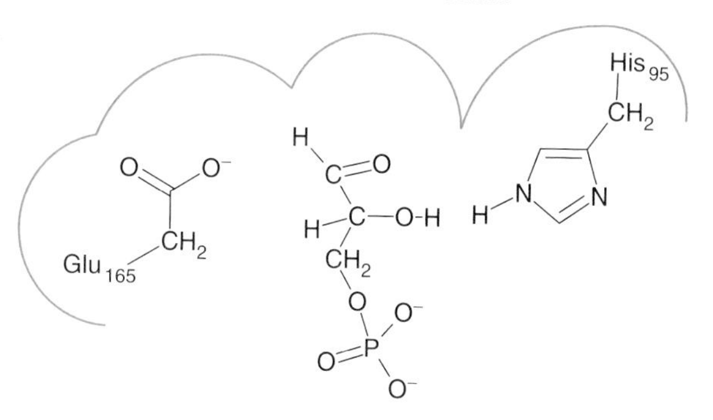

Glyceraldehyde-3-phosphate

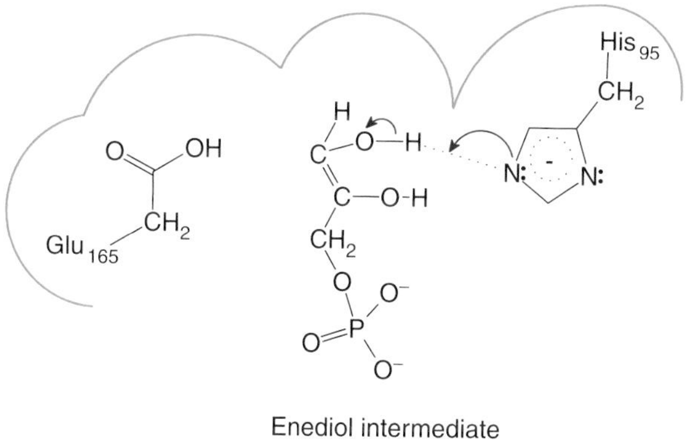

Enediol intermediate

Figure 7.37 The mechanism for conversion of DHAP to G3P in reactions catalysed by triose phosphate isomerase. The enzyme employs general acid–base catalysis with the uncharged histidine side chain acting as a proton donor whilst the charged imidazolate state acts as an acceptor. Alternative mechanisms have been proposed for this reaction including the removal of a proton from the OH group attached to the C1 and donation to the C2 oxygen by Glu165

of DHAP, conversion of DHAP to enediol intermediate, formation of G3P and the release of product (Figure 7.38). The equilibrium lies in favour of DHAP by a factor of 100 with the enzyme showing a high specificity constant of 4.8×10^8 $\mathrm{M}^{-1}\,\mathrm{s}^{-1}$. The slowest reaction will have the greatest activation energy ($\Delta G^{\ddagger}$) but in the four reactions encompassing conversion of DHAP to G3P it was observed that each reaction has approximately the same activation energy. One reaction could not be clearly defined as the rate-determining

$$\mathrm{E + DHAP} \rightleftharpoons \mathrm{E\text{-}DHAP}$$

$$\mathrm{E\text{-}DHAP} \rightleftharpoons \mathrm{E\text{-}enediol}$$

$$\mathrm{E\text{-}enediol} \rightleftharpoons \mathrm{E\text{-}G3P}$$

$$\mathrm{E\text{-}G3P} \rightleftharpoons \mathrm{E + G3P}$$

Figure 7.38 The reactions of triose phosphate isomerase in a minimal kinetic scheme

step. This is unlikely to have arisen by accident and reflects catalytic 'fine tuning' to allow maximal efficiency. A k_{cat} of 4×10^3 s^{-1} is not very high but the superb enzyme efficiency arises from a combination of fast turnover, effective tight binding of substrate ($K_m \sim 10$ μM) *and* transition state stabilization.

Tyrosyl tRNA synthetase

The attachment of amino acid residues to tRNA molecules is a major part of the overall reaction of protein synthesis. The reaction involves the addition of activated amino acids by a group of enzymes known as amino acyl tRNA synthetases, and is summarized as

$$\text{Amino acid} + \text{tRNA} + \text{ATP} \rightleftharpoons \text{Aminoacyl-tRNA} + \text{AMP} + \text{PP}_i$$

Any error at this stage is disastrous to protein synthesis and would lead to incorrect amino acid incorporation into a growing polypeptide chain. The error rate for most tRNA synthetases is extremely low and part of this efficiency comes from high discrimination during binding. Tyrosyl tRNA always binds tyrosine but never the structurally similar phenylalanine and the binding constant for tyrosine of 2×10^{-6} M is at least five orders of magnitude smaller than that of phenylalanine.

Amino acids attach to tRNA molecules at the acceptor arm containing a conserved base sequence of -CCA-3′. Tyrosyl tRNA synthetase represents one of the simplest catalytic mechanisms known in biology with all activity deriving from binding interactions at the active site. The three-dimensional structure of tyrosyl tRNA synthetase from the *Bacillus stearothermophilus* is a dimer of subunits each containing 418 residues. (A dimer is an exception for most Class I tRNA synthetases which are generally monomeric, see Chapter 8.) The structure of the protein (Figure 7.39) has been determined with tyrosine and tyrosine-adenylate and is predominantly α helical with a central six-stranded β sheet forming the core of the enzyme. The enzyme contains the Rossmann fold with tyrosyl adenylate bound at the bottom of the sheet with the phosphate groups reaching the surface of the protein. Although functionally a dimer the structure shown

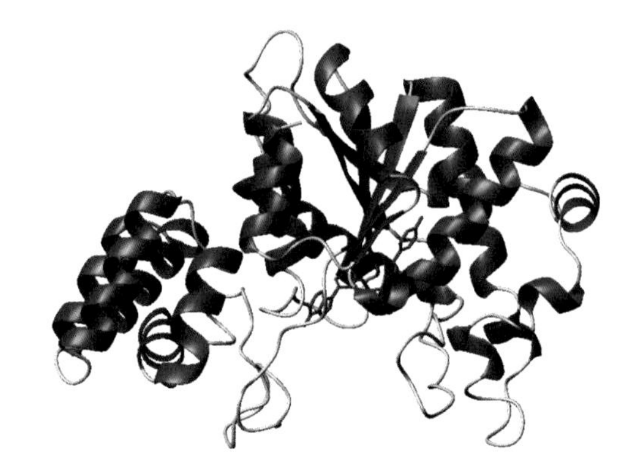

Figure 7.39 The structure of tyrosyl tRNA from *B. stearothermophilus* in a complex with tyrosyl adenylate (PDB: 3TSI). Helices are shown in blue and the six stranded sheet in green. The bound tyrosyl adenylate is shown in red, with the adenylate portion situated on the left and the tyrosine pointing away from the viewer

is that of the monomer and represents the first 320 residues of the protein with remaining 99 residues at the C terminal found disordered. The monomer contains 220 residues in an amino terminal domain with α/β structure together with a helix rich domain extending from residues 248–318 (shown on left of Figure 7.39).

The reaction catalysed by tyrosyl tRNA synthetase is divided into the formation of tyrosine adenylate (Tyr-AMP) and pyrophosphate from the initial reactants of tyrosine and ATP and then in a second reaction the Tyr-AMP is transferred to a specific tRNA. The structures of the enzyme bound with tyrosine, ATP and the tyrosyl adenylate intermediate are known and interactions between polypeptide chain and substrates or activated intermediate have been identified. These structures allowed individual residues to be identified as important for the formation of tyrosyl-adenylate – an intermediate stabilized by the enzyme and seen in crystal complexes.

Binding of substrate to tyrosyl tRNA synthetase is a random kinetic process with either tyrosine or ATP bound first. However, an examination of binding constants for each substrate suggests that ATP binding is weaker leading to an ordered process in which tyrosine binds first followed by ATP (Figure 7.40).

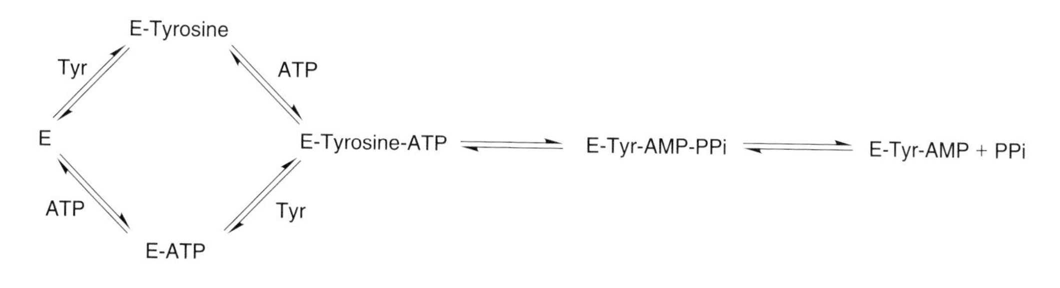

Figure 7.40 Kinetic schemes for the formation of tyrosyl adenylate from tyrosine and ATP. This reaction represents the first part of the tyrosyl tRNA synthetase catalysed process. The products of the first reaction are tyrosyl adenylate (enzyme bound) and pyrophosphate (dissociates from ternary complex)

From the crystal structures in the presence of tyrosine and ATP it is apparent that residues Asp78, Tyr169 and Glu173 form a binding site for the α-amino group of the tyrosine substrate. Not unexpectedly, the side chain of tyrosine forms hydrogen bonds via the phenolate group with residues Tyr34 and Asp176. The crystal structure of the complex with tyrosyl adenylate indicated that Cys35, Thr51 and His48 interact with the ribose ring (Figure 7.41).

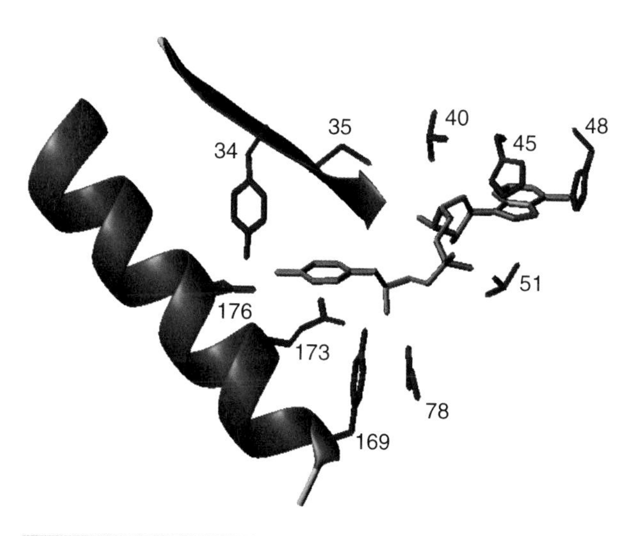

Figure 7.41 Tyrosyl-adenylate binding to tyrosyl tRNA synthetase (PDB: 3TSI). Shown in orange is Tyr-AMP together with the side chains of Tyr34, Cys35, Thr40, His45, His48, Thr51, Asp78, Tyr169, Glu173 and Asp176

The role of residues in substrate binding was defined using mutagenesis to perturb the overall efficiency of tyrosyl tRNA synthetase. Cys35 was the first mutated residue and although in contact with the ribose ring of ATP and having a role in substrate binding it is remote from the critical active site chemistry based around the phosphate groups. Mutation of Cys35 to Gly estimated the effect of deleting the side chain hydrogen bonding interaction and revealed an enzyme operating with a lower efficiency by $\sim$6 kJ mol^{-1}. Modifying Tyr34 to Phe34 decreased the affinity of the enzyme for tyrosine, lowered the stability of the Tyr-AMP enzyme complex and decreased the ability of the enzyme to discriminate between tyrosine and phenylalanine. Mutagenesis combined with kinetic studies showed that changing Thr40 and His45 produced dramatic decreases in rates of Tyr-AMP formation. Changing Thr40 to Ala decreased Tyr-AMP formation by approximately 7000 fold whilst the mutation His45 to Gly caused a 200-fold decrease. A double mutant, bearing both substitutions, showed a decrease of $\sim 10^5$. In each case the affinity of the enzyme for substrate was unaltered but the affinity of the enzyme for the intermediate (Tyr-AMP..PP$_i$) was lowered suggesting that Thr40 and His45 played key roles in stabilizing the transition state complex via interactions with inorganic pyrophosphate groups.

The single most important property of tRNA synthetases is substrate binding in pockets in restrained and extended conformation. The geometry lowers the activation energy barrier for the reaction and allows formation of the amino acyl adenylate complex. Two characteristic motifs with sequences His-Ile-Gly-His (HIGH) and Met-Ser-Lys (MSK) are important to formation of the transition state. The close approach of negatively charge carboxyl groups (from the amino acid substrate) and the α phosphate group of the ATP promotes formation of a transition state. The α phosphate group is stabilized by interaction with His43 and Lys270 of the HIGH and MSK motifs. The pyrophosphate acts as a leaving group and the amino acyl adenylate transition state complex is stabilized by interactions with specific residues, as described in the preceding section.

The second half of the reaction, the transfer of the amino acid to the correct tRNA, has been examined through crystallization of other class I and class II tRNA synthetases (Table 7.6) with tRNA bound in a tight complex. The acceptor arm of tRNA and in particular the $C_{74}C_{75}A_{76}$ base sequence binds alongside ATP in the active site. The arrangement of substrate in the active site of glutaminyl tRNA synthetase (Figure 7.42) has been derived by using glutaminyl-adenylate analogues (Figure 7.43) that co-crystallize with the enzyme defining interactions between residues involved in tRNA and amino acyl binding (Figure 7.44).

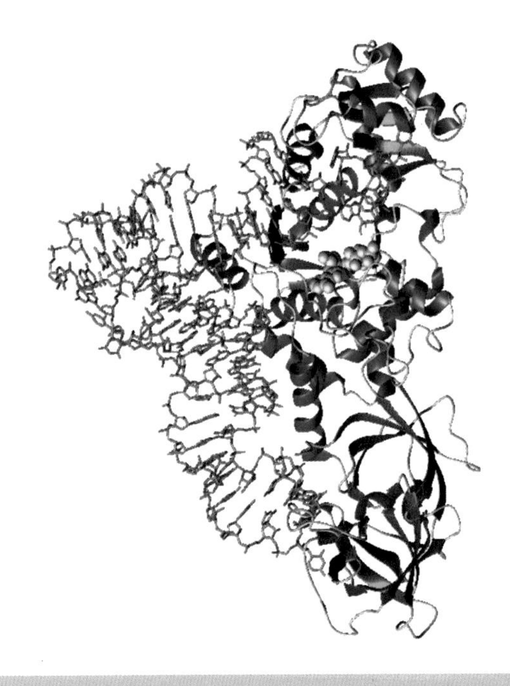

Figure 7.42 Gln-tRNA synthetase showing complex of tRNA and ATP (PDB: 1GTG). The tRNA molecule is shown in orange with ATP in yellow. The secondary structure elements of enzyme are shown in blue (helices), green (strands) and grey (turns)

Gln tRNA synthetase must avoid selecting either glutamate or asparagine – two closely related amino acids. Discrimination between amino acids is favoured by recognition of both hydrogen atoms of the nitrogen in the glutamine side chain. Interactions involve the hydroxyl group of Tyr211 and a water molecule as obligatory hydrogen-bond acceptors in a network of side chains and water molecules. In molecular recognition discrimination against glutamate and glutamic acid occurs because they lack one or both hydrogen atoms whilst the side chain of asparagine is too short.

Table 7.6 Principal features of class I and class II aminoacyl tRNA synthetases

Property	Class 1	Class II
Conserved substrate binding motifs	HIGH KMSKS	Motif 2 (FRxE) Motif 3 (GxGxGXER)
Dimerization motifs	None	Motif 1 (little consensus)
Active site fold	Parallel β sheet of Rossmann fold	Antiparallel β sheet
ATP conformation	Extended	Bent
Amino acylation site	2′ OH	3′OH
tRNA binding	Variable loop faces solvent Binds to minor groove of acceptor stem	Variable loop faces protein Binds to major groove of acceptor stem

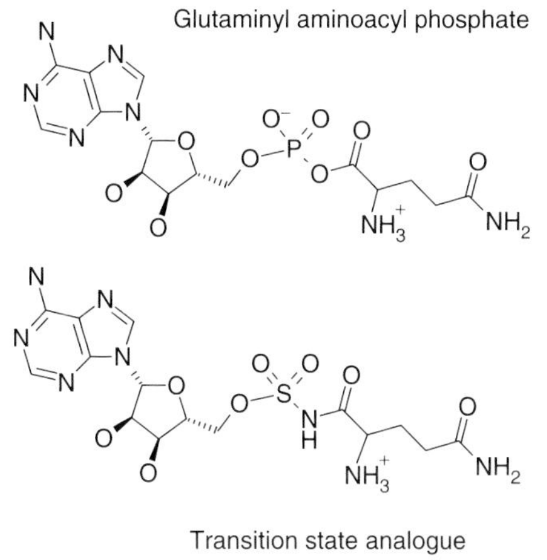

Glutaminyl aminoacyl phosphate

Transition state analogue

Figure 7.43 Glutaminyl amino acyl-adenylate and its transition state analogue

EcoRI restriction endonuclease

Restriction enzymes were characterized through the 1960s and early 1970s, largely through the efforts of Werner Arber, Hamilton O. Smith and Daniel Nathans. The discovery of type II restriction endonucleases represented a key event at the beginning of a new chapter of modern biochemistry. Subsequently, the use of type II restriction enzymes would allow manipulation of DNA molecules to create new sequences that would form the basis of molecular biology.

In addition to type II restriction endonucleases there are also type I and type III enzymes. Type I restriction enzymes have a combined restriction/methylation function within the same protein and cleave DNA randomly often at regions remote from recognition sites and often into a large number of fragments. They are not used in molecular cloning protocols. Type III restriction enzymes have a combined restriction/methylation function within the same protein but bind to DNA and at specific recognition sites. Again this class of enzymes is not routinely used in cloning procedures. In type II enzymes the restriction function is not accompanied by a methylase activity. Type II enzymes do not require ATP for activity although all enzymes appear to require Mg^{2+} ions.

Currently more that 3000 type II restriction endonucleases have been discovered with more than 200 different substrate specificities. These enzymes are widely used in 'tailoring' DNA for molecular cloning applications. The substrate is double stranded DNA

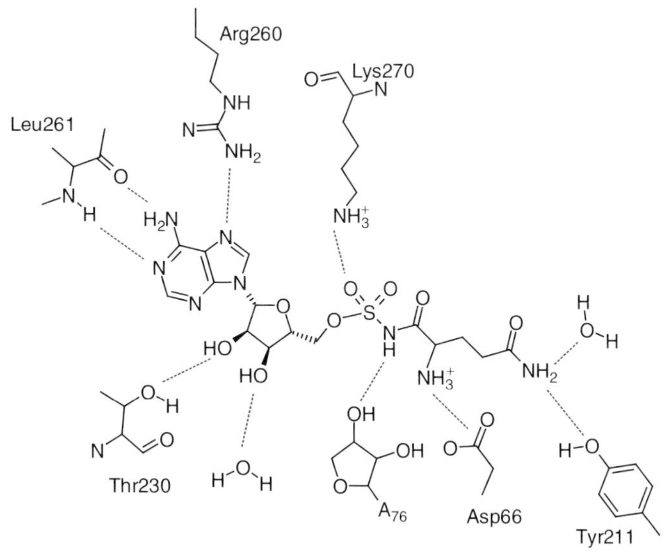

Figure 7.44 The interaction between glutaminyl adenylate analogue and the active site pocket of Gln-tRNA synthetase

Table 7.7 A selection of restriction endonucleases showing cleavage sequences and sources. The arrows between bases indicate the site of cleavage within palindromic sequences

Enzyme	Recognition site	Source
AvaI	C↓(T/C)GG (A/G)G	*Anabaena variablis*
BamHI	G↓GATCC	*Bacillus amyloliquefaciens* H
EcoRI	G↓AATTC	*Escherichia coli* RY13
HindIII	A↓AGCTT	*Haemophilus influenzae* Rd
NcoI	C↓CATGG	*Nocardia corallina*
PvuI	CGAT↓CG	*Proteus vulgaris*
SmaI	CCC↓GGG	*Serratia marcescens*
XhoI	C↓TCGAG	*Xanthomonas holcicola*

and in the majority of cases restriction endonucleases recognize short DNA sequences between 4 and 8 base pairs in length although in some instances longer sequences are identified. The names of restriction enzymes derive from the genus, species and strain designations of host bacteria (Table 7.7). So for example *Eco*RI is produced by *E. coli* strain RY13 whilst *Nco*I is obtained from *Nocardia corallina* and *Hin*dIII from *Haemophilus influenzae*.

Restriction enzymes recognize a symmetrical sequence of DNA where the top strand is the same as the bottom strand read backwards (palindromic). When *Eco*RI cuts the motif GAATTC it leaves overhanging chains termed 'sticky ends'. The uncut regions of DNA remain base paired together. Sticky ends are an essential part of genetic engineering since they allow similarly cut fragments of DNA to be joined together.

*Eco*RI is one of the most widely used restriction enzymes in molecular biology and its structure has been determined by X-ray crystallography (Figure 7.45). The protein is a homodimer with each subunit containing a five-stranded β sheet flanked by helices. Subsequent determination of the structure of other restriction enzymes such as *Bam*HI, *Bgl*II, *Cfr*10I, *Mun*I, and *Ngo*MIV revealed significant fold similarity despite relatively low levels of sequence identity. This further suggested a common mechanism of catalytic action.[1] The *Eco*RI family share a substructure based around five β strands and two α helices called the common core motif (CCM). The CCM is the result of a β meander followed by a binding pocket that interacts with recognition site DNA and is formed predominantly by two helices called the outer and inner helices. The core is readily seen by comparing structures of *Eco*RI complexed with DNA and devoid of its recognition partner.

The above view is, however, slightly misleading since whilst it portrays the secondary structure elements of the CCM it does not accurately reflect the action of the homodimer of subunits in surrounding DNA. The subunits bind to specific bases (GAATTC) in the major groove of the DNA helix with the minor groove exposed to the solvent. The dimers form a

[1]The other major structural family of restriction endonucleases is based around the structure of EcoRV.

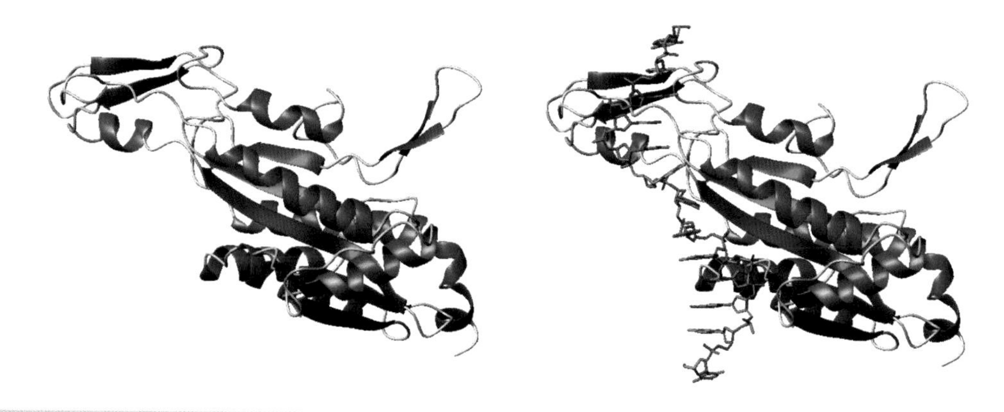

Figure 7.45 The structure of *Eco*RI is shown with and without a recognition sequence of DNA. The five-stranded β sheet making up most of the CCM is shown in the centre of the protein in green

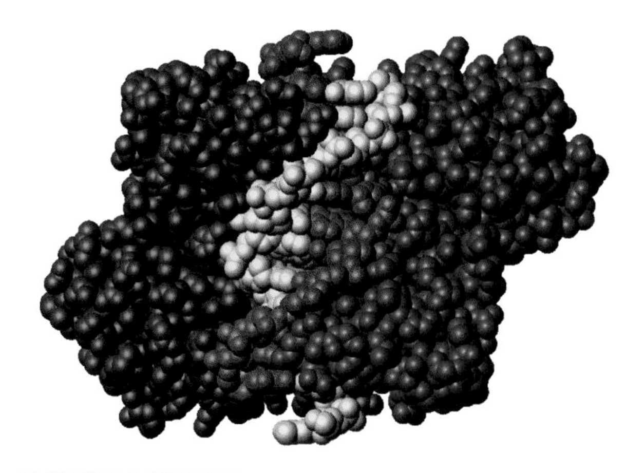

Figure 7.46 Space filling representation of homodimer of *Eco*RI bound to DNA. The subunits of the enzyme are shown in blue and green whilst the DNA is shown in orange/yellow. The minor groove is exposed to the solvent

cleft lined with basic residues that associate with sugar-phosphate chains via charge interactions (Figure 7.46). Of more importance are arginine side chains at residues 145 and 200, together with a glutamate side chain (Glu144) in each monomer that form hydrogen bonds with base pairs of the recognition motif. This confers specificity to the binding reaction with DNA and avoids interactions with incorrect, but often closely related, sequences. Sequence specificity is mediated by 16 hydrogen bonds originating from α helical recognition regions. Twelve of the hydrogen bonds occur between protein and purine whilst four exist between protein and pyrimidine. *Eco*RI 'reads' all six bases making up the recognition sequence by forming interactions with each of them. Arg200 forms two hydrogen bonds with guanine while Glu144 and Arg145 form four hydrogen bonds to adjacent adenine residues. In addition to the hydrogen bonds an elaborate network of van der Waals contacts exist between bases and recognition helices. It is these interactions that discriminate the *Eco*RI hexanucleotide GAATTC from all other hexanucleotide sequences. Changes in base identity lose hydrogen bonds and effectively discriminate between different sequences. Hydrogen bond formation distorts the DNA helix introducing strain and a torsional kink in the middle of its recognition sequence leading to local DNA unwinding. The unwinding involves movement of DNA by $\sim 28^\circ$ and groove widening by ~ 0.35 nm where 'strain' may enhance catalytic rates.

The presence of divalent cations (Mn^{2+}, Co^{2+}, Zn^{2+} and Cd^{2+}) results in bond scission but in their absence enzymes shows low activity. The use of electron dense ions pinpoints the metal centre location as close to Glu111 in co-crystals of protein–DNA–divalent ion and emphasizes the role of a metal centre in catalysis. Metal ions are coordinated in the vicinity of the active site acidic side chains. The observation of mono and binuclear metal centres within active sites of restriction endonucleases suggests that although the structure of the active site-DNA binding region of many enzymes is similar the catalytic mechanisms employed may differ from one enzyme to another.

Restriction endonucleases contain a signature sequence D-(X)$_n$-D/E-Z-K, in which three residues are *weakly* conserved (D, D/E and K). The two acidic residues are usually Asp, although Glu occurs in *Eco*RI and several other restriction endonucleases, including *Bam*HI. These residues are separated by 9–20 residues and are followed by a basic lysine side chain preceded by Z, a residue with a hydrophobic side chain. In *Eco*RI the catalytic residues are Asp91, Glu111 and Lys113, whilst Asp59, preceding the conserved cluster, also appears to be important to catalysis. These residues make up the active site, whilst in *Bam*HI the corresponding residues are Asp94, Glu111, and Glu113 with Glu77 preceding the cluster of charged residues. *Bam*HI cuts the sequence GGATTC – a hexanucleotide sequence almost identical to that recognized by *Eco*RI (GAATTC) – yet each enzyme shows absolute specificity for its own sequence. Both enzymes possess homodimeric structures with a catalytic core region of ~200–250 residues per monomer consisting of five strands and two helices. However, the method of sequence recognition and potential mechanism of catalysis in two clearly related enzymes appears to differ. The catalytic mechanism for *Eco*RI involves a single metal ion whilst that of *Bam*HI involves two cations (Figure 7.47). Similarly, DNA binding through base contacts in the major groove uses a set of loops between strands together with β strands (*Eco*RI) whilst *Bam*HI lacks these regions and uses the N-terminal end of a helix bundle. Despite a lack of agreement conclusions can be drawn about the requirements for phosphodiester bond cleavage in any restriction enzyme. At least three components are required; a general base is needed to activate water, a Lewis acid is necessary to stabilize negative charge formed in the transition state complex and a general acid is required to protonate the leaving group oxygen.

Enzyme inhibition and regulation

Enzymes are inhibited and regulated. Studying each process sheds light on the control of catalysis, catalytic pathways, substrate binding and specificity, as well as the role of specific functional groups in active site chemistry. Inhibition and regulation are inter-related with cells possessing proteins that inhibit enzymes as part of a normal cellular function.

Reversible inhibition involves non-covalent binding of inhibitor to enzyme. This mode of inhibition is shown by serpins, proteins that act as natural trypsin inhibitors preventing protease activity. The complex formed between bovine pancreatic trypsin inhibitor (BPTI) and trypsin has extraordinary affinity with a large association constant (K_a) of $\sim 10^{13}$ M^{-1}. Tight association between proteins causes effective enzyme

Figure 7.47 A general mechanism of catalysis for type II restriction endonucleases based on *Bam*HI. The reaction occurs through an SN2 mechanism, with an in-line displacement of the 3′-OH group and an inversion of configuration of the 5′-phosphate group

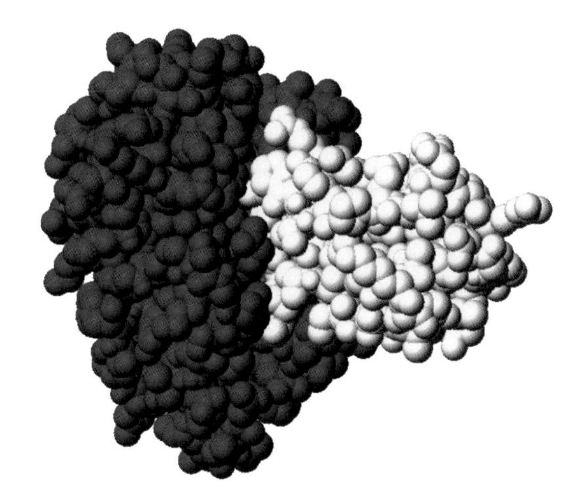

Figure 7.48 Binary complex formed between trypsin and BPTI showing insertion of Lys15 (red) into the specificity pocket. Trypsin is shown in blue, BPTI in yellow (PDB: 2PTC)

inhibition and crystallography of the protein complex reveals a tightly packed interface between BPTI and trypsin formed by numerous intermolecular hydrogen bonds (Figure 7.48). On the surface of BPTI Lys15 occupies the specificity pocket of trypsin with a Lys–Ala peptide bond representing the scissile region. However, bond cleavage does not occur despite the proximity of Ser195 and the formation of a tetrahedral intermediate. Absence of bond cleavage results from tight binding between BPTI and enzyme that prevents diffusion of the leaving group (peptide) away from the active site and restricts access of water to this region.

By measuring reaction kinetics at different conditions activity profiles (k_{cat}/K_m) are determined for individual enzymes. Extremes of pH lead to a loss of activity as a result of enzyme denaturation and solutions of pH 2.0 do not in general support folded structure and lead to a loss of activity that is restored when the pH returns to normal values. Reversibility is observed because covalent modification of the enzyme does not occur.

Investigating kinetic profiles for enzyme-catalysed reactions in the presence of an inhibitor may identify mechanisms of inhibition and provides important clues in the absence of structural data of the active site geometry and composition. Armed with structural knowledge the pharmaceutical industry has produced new inhibitors with improved binding or inhibitory action that play crucial roles in the battle against disease. The starting point for enzyme inhibitor studies is often identification of a natural inhibitor. When inhibitor data is available with detailed structural knowledge of an enzyme's active site it becomes possible to enter the area of rational drug design. This combines structural and kinetic data to design new improved compounds and is often performed on computers. With a family of 'lead in' compounds chemical synthesis of a restricted number of favourable targets, followed by assays of binding and inhibition, can save years of experimental trials and allows drugs to enter clinical trials and reach the marketplace. Such approaches combining classical enzymology with bioinformatics and molecular modelling allows development of new pharmaceutical arsenals and a new approach to molecular medicine. Using these techniques enzyme inhibitors have been introduced to counteract emphysema, HIV infection, alcohol abuse, inflammation and arthritis.

The most important type of reversible inhibition is that caused by molecules that compete for the active sites of enzymes (Figure 7.49). The competitor is almost always structurally related to the natural substrate and can combine with the enzyme to form an enzyme–inhibitor complex analogous to that formed between the enzyme and the substrate. Occasionally the 'inhibitor' is converted into product but more frequently inhibitors bind to active sites without undergoing further catalytic reactions.

When the enzyme binds inhibitor it is unable to catalyse reactions leading to products and this is observed by a decrease in enzyme activity at a given substrate concentration. A usual presentation of competitive inhibition involves Eadie–Hofstee or Lineweaver–Burk plots, and is marked by a decrease in K_m although the overall V_{max} remains similar between inhibited and uninhibited reactions (Figures 7.50 and 7.51). V_{max} is unchanged because addition of substrate in high concentrations will effectively displace inhibitor from the active site of an enzyme leading to unimpeded rates of reaction.

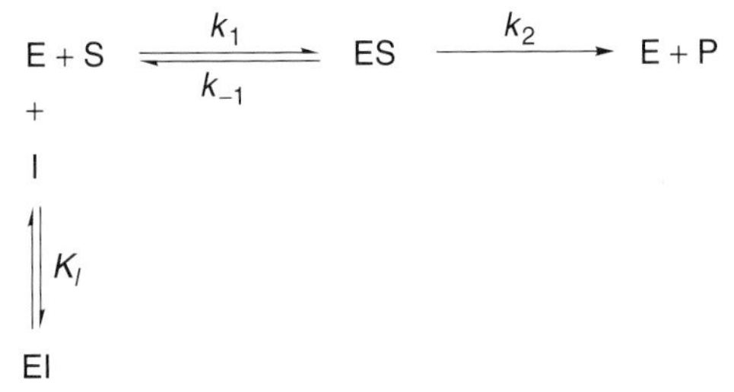

$$\begin{array}{ccccc} E + S & \underset{k_{-1}}{\overset{k_1}{\rightleftharpoons}} & ES & \xrightarrow{k_2} & E + P \\ + & & & & \\ I & & & & \\ \updownarrow K_I & & & & \\ EI & & & & \end{array}$$

Figure 7.49 Competitive inhibition leads to additional reactions in Michaelis–Menten schemes

The rate equations are solved exactly as before, except that the total enzyme concentration is equivalent to

$$[E]_T = [E] + [ES] + [EI] \quad (7.48)$$

and leads to an expression for the initial velocity v

$$v = k_{cat}[E]_T[S]/K_m(1 + [I]/K_I) + [S] \quad (7.49)$$

The K_m is increased by a factor of $(1+[I]/K_I)$ with the second 'apparent' K_m shown in the following equation as K_{APP}

$$v = V_{max}[S]/K_{APP} + [S] \quad (7.50)$$

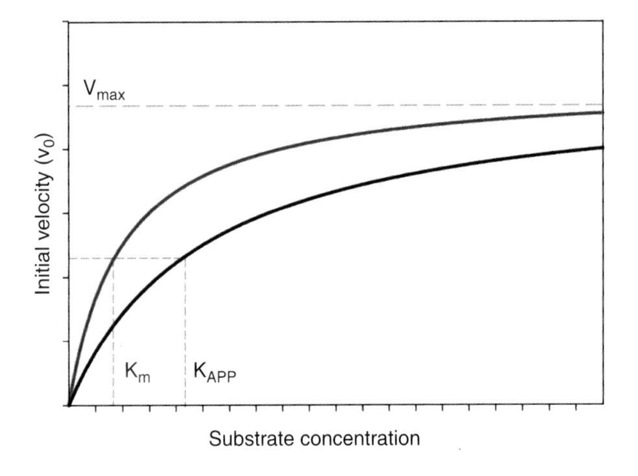

Figure 7.50 An enzyme-catalysed reaction showing a plot of initial velocity against substrate concentration in the presence (black trace) and absence (red trace) of inhibitor

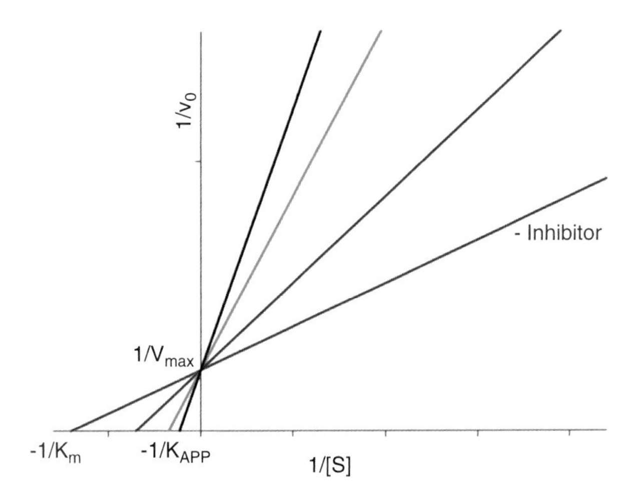

Figure 7.51 Lineweaver–Burk plot of an enzyme-catalysed reaction with competitive inhibition. With K_{APP} determined at several different inhibitor concentrations a plot of K_{APP} versus I will yield a straight line whose gradient is K_m/K_I, where K_m can be estimated from the line's intercept where $[I] = 0$. Alternatively K_m can be estimated from the Lineweaver–Burk plot

One example of competitive inhibition is seen in succinate dehydrogenase and the conversion of succinate to fumarate as part of the citric acid cycle. Malonate is a competitive inhibitor of succinate dehydrogenase and with structural similarity to succinate, binds at the active site but does not undergo dehydrogenation (Figure 7.52). The observation of competitive inhibition suggests binding involves the two carboxylate groups. A further example of competitive inhibition is benzamidine, a structural analogue of arginine, that binds to the active site of trypsin (Figure 7.53).

The remaining patterns of reversible inhibition are called uncompetitive and non-competitive (or mixed) inhibition. Uncompetitive inhibition involves binding directly to enzyme–substrate complexes but not to free enzyme (Figure 7.54). As such there is no requirement for an uncompetitive inhibitor to resemble the natural substrate and it is generally believed that inhibition arises from a distortion of the active site region preventing further turnover.

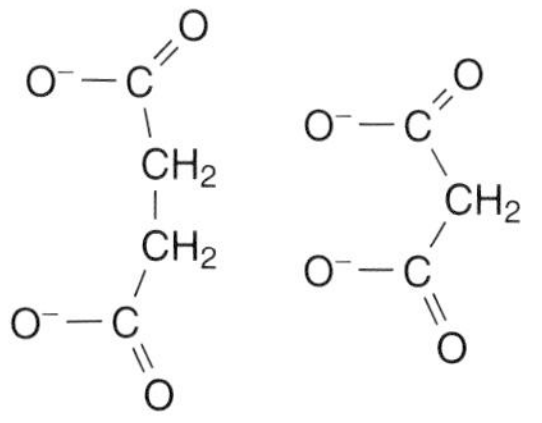

Figure 7.52 The structures of succinate and malonate

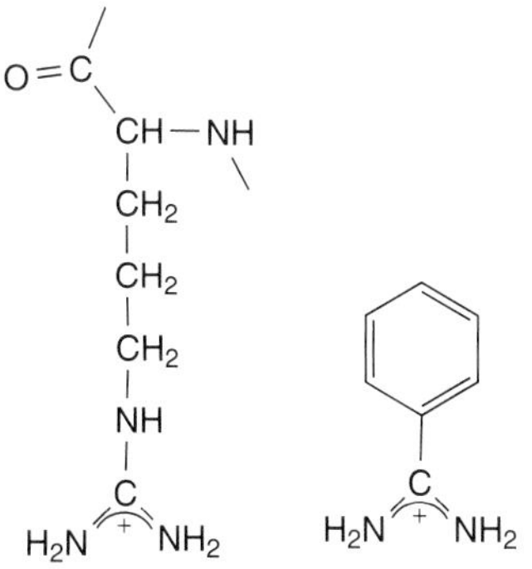

Figure 7.53 Benzamidine and arginine. Competitive inhibitor and substrate in trypsin

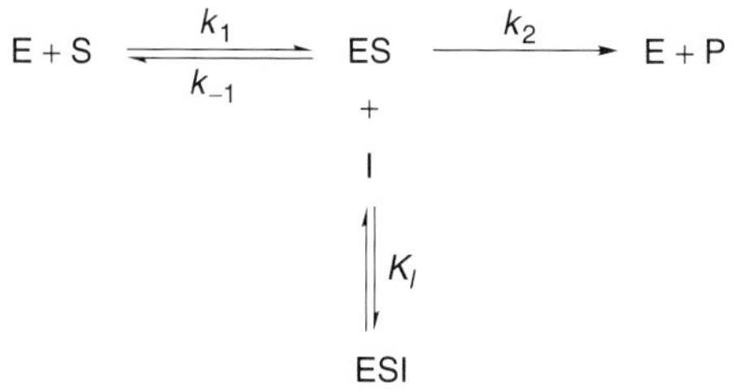

$$E + S \underset{k_{-1}}{\overset{k_1}{\rightleftharpoons}} ES \xrightarrow{k_2} E + P$$
$$ES + I \overset{K_I}{\rightleftharpoons} ESI$$

Figure 7.54 Kinetic scheme illustrating uncompetitive inhibition of enzyme-catalysed reaction

Non-competitive inhibition (Figure 7.55) involves inhibitor binding to free enzyme and enzyme substrate complexes to form inactive EI or ESI states. It is comparatively rare but is exhibited by some allosteric enzymes. In these cases it is thought that inhibitor binds to the enzyme causing a change in conformation that permits substrate binding but prevents further catalysis.

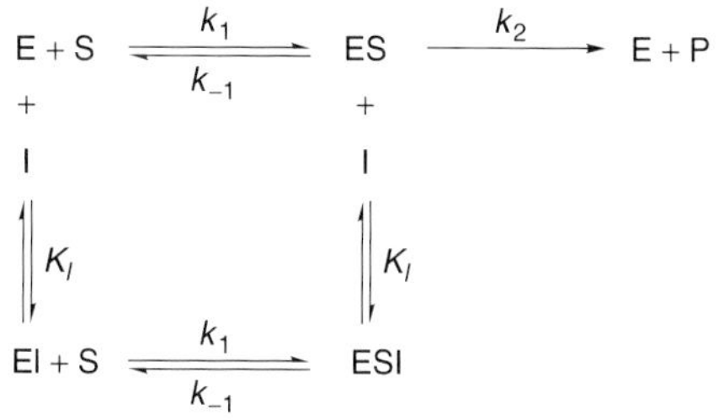

$$E + S \underset{k_{-1}}{\overset{k_1}{\rightleftharpoons}} ES \xrightarrow{k_2} E + P$$
$$E + I \overset{K_I}{\rightleftharpoons} EI; \quad ES + I \overset{K_I}{\rightleftharpoons} ESI$$
$$EI + S \underset{k_{-1}}{\overset{k_1}{\rightleftharpoons}} ESI$$

Figure 7.55 Kinetic scheme describing non-competitive inhibition. In the above example numerous kinetic simplifications have been introduced. The complex ESI can sometimes exhibit catalytic activity at reduced rates, or the binding of inhibitor may alter substrate binding and/or k_{cat}. As a result of these differential effects the term mixed inhibition is used

Since a non-competitive inhibitor binds to both E and ES removing catalytic competence in the latter the effect of this is to lower the number of effective enzyme molecules. The result is a decrease in V_{max} as a result of changes in k_{cat}. Since non-competitive inhibitors bind at sites distinct from the substrate there is not normally any effect on K_m.

Irreversible inhibition of enzyme activity

Irreversible inhibition can involve the inactivation of enzymes by extremes of pH, temperature or chemical reagents. Most frequently, however, it involves the covalent modification of the enzyme in a manner that destroys activity. Most substances that cause irreversible inhibition are toxic to cells, although in the case of aspirin the 'toxic' effect is beneficial. Many irreversible inhibitors have been identified; some are naturally occurring metabolites whilst others result from laboratory synthesis (Table 7.8).

The inhibition of enzymes involved in bacterial cell wall synthesis is exploited through the use of penicillin as an antibiotic that binds to bacterial enzymes involved in cell wall synthesis causing inactivation. The inability to extend bacterial cell walls prevents division and leads to the cessation of bacterial growth and limits the spread of the infection.

Table 7.8 Irreversible enzyme inhibitors and sites of action

Compound	Common source and site of action
Diisopropylfluorophosphate (DIFP), Sarin	Synthetic; inactivate serine residues of serine protease enzymes
Ricin/abrin	Castor bean; ribosomal inactivating proteins; inhibit ribosome function via depurination reactions
N-tosyl-L-phenylalanine chloromethylketone (TPCK)	Synthetic inhibitor; binds to active site histidine in chymotrypsin
Penicillin	Naturally occurring antibiotic plus many synthetic variants; inhibits cell wall biosynthesis enzymes particularly in Gram-positive bacteria
Cyanide	Released during breakdown of cyanogenic glycosides found in many plants; inhibits respiration
Aspirin	Natural product – methyl salicylate plus synthetic variants. All bind to active site of cyclo-oxygenase inhibiting substrate binding

Aspirin: an inhibitor of cyclo-oxygenases

Aspirin's pain-killing capacity was known to Hippocrates in the 5th century BC but provides a marvelous example of how studies of enzyme inhibition combined with structural analysis leads to advances in health care.

Prostaglandins are hormones created by cells that act only in the local or surrounding area before they are broken down. Unlike most hormones they are not transported systemically around the body but control 'local' cellular processes such as the constriction of blood vessels, platelet aggregation during blood clotting, uterine constriction during labour, pain transmission and the induction of inflammation. Although these diverse processes are controlled by different prostaglandins they are all created from a common precursor molecule called arachidonic acid. Cyclo-oxygenase catalyses the addition of two oxygen molecules to arachidonic acid in a pathway that leads to the formation of prostaglandins.

Aspirin-like drugs, sometimes called non-steroidal anti-inflammatory drugs or NSAIDs (Figure 7.56), prevent prostaglandin biosynthesis through inhibition of cyclo-oxygenases such as COX-1 (Figure 7.57).

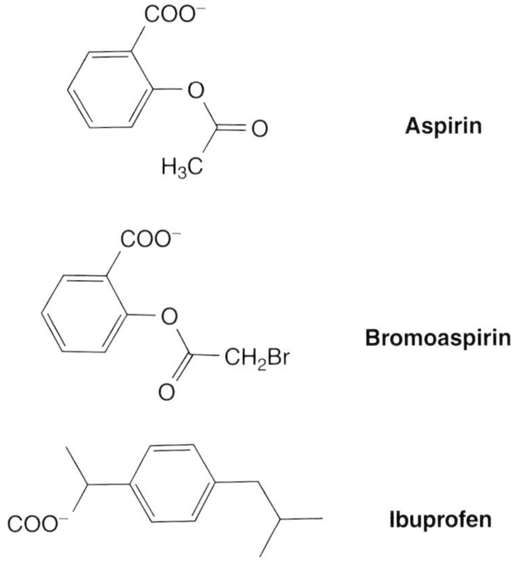

Figure 7.56 Three common non-steroidal anti-inflammatory drugs (NSAIDs)

Crystallization of cyclo-oxygenase by R. Garavito showed a membrane protein comprised of three

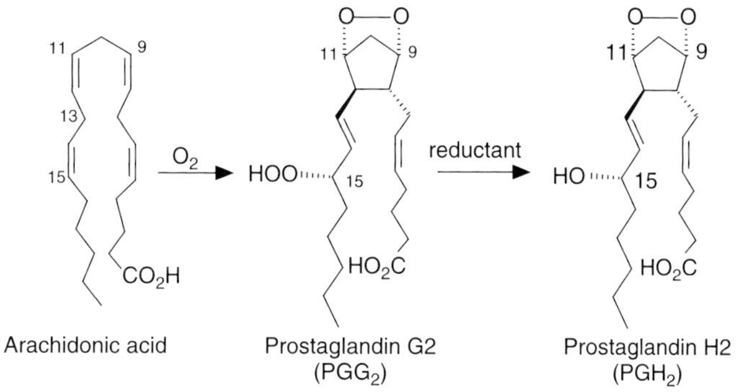

Arachidonic acid $\xrightarrow{O_2}$ Prostaglandin G2 (PGG_2) $\xrightarrow{\text{reductant}}$ Prostaglandin H2 (PGH_2)

Figure 7.57 Formation of prostaglandin H_2 from arachidonic acid by cyclo-oxygenase

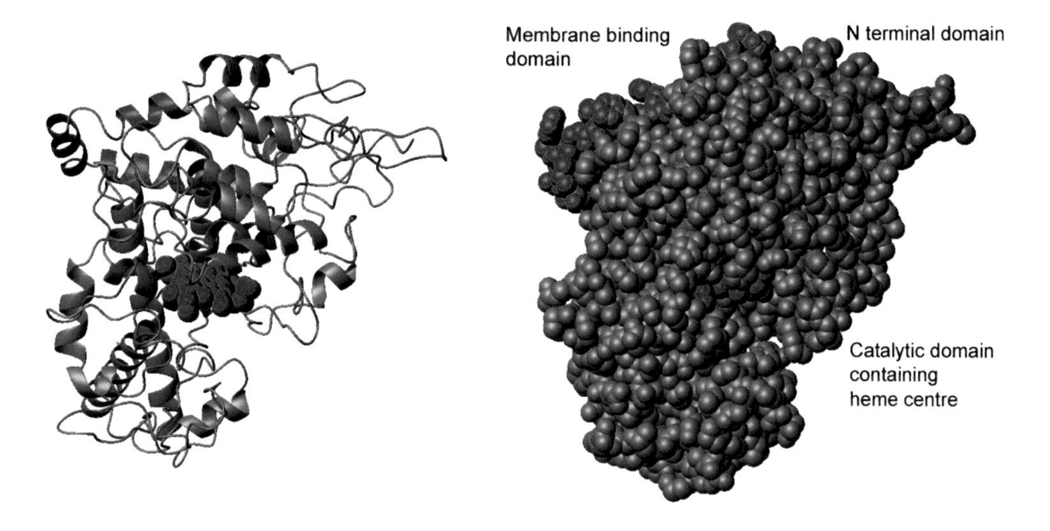

Figure 7.58 The three-dimensional structure of cyclo-oxygenase (PDB:1PRH). On the left the distribution of secondary structure is shown for one subunit of the dimer found in the crystal lattice. The space filling model highlights the different domains and the buried heme group. (blue, residues 33–72; orange, membrane-binding region formed from residues 73–116; and purple the globular domain containing the catalytic heme group)

domains in a chain of 551 residues (Figure 7.58). Residues 33–72 form a small compact module whilst a second domain (residues 73–116) adopts a right-handed spiral of four helical segments along one side of the protein. This domain binds to the membrane via amphipathic helices. The third domain is the catalytic heart of the bifunctional enzyme performing cyclo-oxygenation and peroxidation. Cyclo-oxygenase is a homodimer, and should be more correctly called prostaglandin H synthase (or PGHS-1) because the cyclo-oxygenase reaction is only the first of two functional reactivities.

The cyclo-oxygenase reaction occurs in a hydrophobic channel that extends from the membrane-binding domain to the catalytic core and a non-covalently bound heme centre. The arachidonic acid is positioned in this channel in an extended L-shaped conformation and catalysis begins with abstraction of the 13-pro-*S*-hydrogen to generate a substrate bound radical. At this point, a combination of active site residue interactions and conformational changes control the regio- and stereospecific additions of molecular oxygen and ring closures resulting in the product.

Insight into the mechanism of action of aspirin stemmed from crystallization of this protein in the presence of inhibitors. In the bromoaspirin-inactivated structure Ser530 is bromoacetylated (as opposed to simple acetylation with aspirin) with the salicylate leaving-group bound in the tunnel as part of the active site (Figure 7.59). The bound salicylate group blocks substrate access to the heme centre whilst other NSAIDs do not covalently modify the enzyme but prevent conversion of arachidonic acid in the first reaction of prostaglandin synthesis. The active site is essentially a hydrophobic channel of highly conserved residues such as Arg120, Val349, Tyr348, Tyr355, Leu352, Phe381, Tyr385, and Ser530. At the mouth of the channel lie Arg120 and Tyr 355 and these residues interact with the carboxylate groups of arachidonic acid with the remainder of the eicosanoid chain extending into the channel and binding to hydrophobic residues close to Ser530. Although the details of catalysis and inhibition remain to be clarified the heme group lies at one end of the channel and it is thought that heme-catalysed peroxidase activity involves the formation of a tyrosyl radical to initiate the process.

Aspirin is an extremely beneficial drug but concern exists over the damage caused to the stomach and duodenum in significant numbers of patients. About one in three people who take NSAIDs for long periods develop gastrointestinal bleeding. The underlying reasons for the unwanted side effects became clear with the realization that two forms of cyclo-oxygenase (COX-1 and COX-2) exist *in vivo*. COX-1 is the constitutive isoform whilst COX-2 is induced by inflammatory stimuli. The beneficial anti-inflammatory actions of NSAIDs are due to inhibition of COX-2 whereas unwanted side effects such as irritation of the stomach lining arise from inhibition of COX-1. It is clearly desirable to obtain enzyme inhibitors (drugs) that selectively target (inhibit) COX-2.

Despite different physiological roles COX-1 and COX-2 show similar structures and catalytic functions, with a small number of amino acid substitutions giving rise to subtle differences in ligand interaction between each isoform. These differences form the basis of developing structure–function relationships in cyclo-oxygenases and allow rational drug design aimed at improving the desirable properties of aspirin without

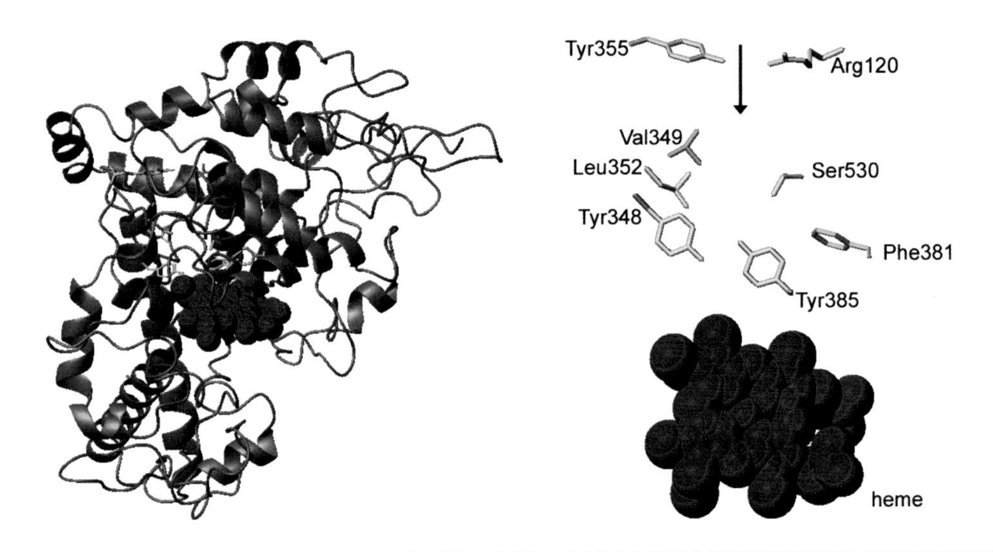

Figure 7.59 The active site of cyclo-oxygenase (COX-1) shown by the yellow residues superimposed on the secondary structure elements. Removing the bulk of the protein shows the approximate position of the channel (arrow) and the lining of this pocket by conserved residues. Ser530, acetylated by aspirin, lies approximately midway along the channel

any of the unwanted side effects. As a result new and improved products are reaching the marketplace, with one product, Celebrex®, introduced in 1999 as a potent selective inhibitor of COX-2 designed to relieve pain associated with osteoarthritis and adult rheumatoid arthritis.

Allosteric regulation

Although inhibition is one form of regulation it tends to be 'all or nothing' and the cell has of necessity devised better methods of control. The concept of regulation of protein activity by molecules known as effectors was demonstrated in haemoglobin. Binding to a site between individual subunits by 2,3 DPG caused conformational changes in haemoglobin that modulated oxygen affinity. The principle of regulation by molecules unrelated to substrate is the basis of allostery with studies on haemoglobin pioneering the development of this area.

Allosteric enzymes have at least two subunits (exhibit quaternary organization) with multiple catalytic and binding sites. In general the active and inactive forms undergo conformational changes arising from effector binding at sites distinct from the catalytic centres. Allosteric regulation is shown by large numbers of enzymes but particularly those involved in long metabolic pathways. These pathways include biosynthetic reactions, glycolysis, glycogenesis and β oxidation. Here, the effectors are often components of the metabolic pathway and their presence serves to inhibit or stimulate activity in a manner that allows exquisite control over a reaction pathway. Allosteric modulators bind non-covalently to the enzyme and can precipitate changes in either K_m for the substrate or V_{max} of the enzyme.

Allosteric enzymes are multimeric proteins, but can be composed of identical or non-identical subunits. Allosteric enzymes made up of identical subunits include phosphofructokinase (PFK) glycogen phosphorylase, and glycogen synthase. For enzymes composed of identical subunits each polypeptide chain contains at least one catalytic and one regulatory site whilst for enzymes composed of non-identical subunits these sites can be on different subunits, but this is not obligatory. Allosteric enzymes composed of non-identical subunits

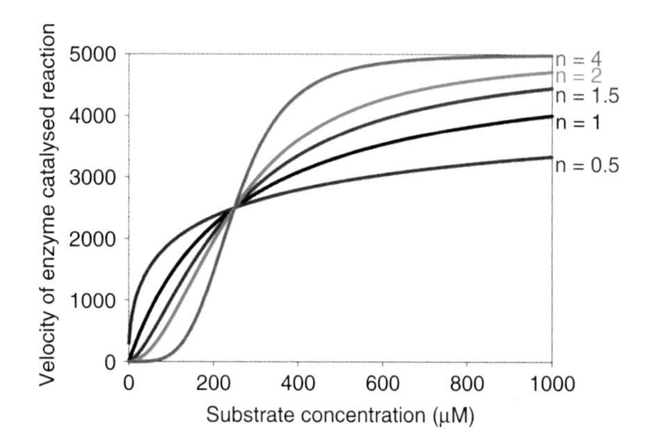

Figure 7.60 A plot of v_0 versus [S] for an allosteric enzyme. In the different curves the V_{max} and $K_{0.5}$ are constant at 1000 and 250 whilst the Hill coefficient ranges from 0.5, 1, 1.5, 2, and 4, respectively

include aspartate transcarbamoylase, lactate dehydrogenase, pyruvate kinase, and protein kinase A.

Allosteric enzymes are identified from plots of initial velocity versus substrate concentration. For at least one substrate this profile will not exhibit the normal hyberbolic variation of v_0 with [S] but shows a sigmoidal profile (Figure 7.60). This is an example of positive substrate cooperativity and a plot of v_0 against [S] is defined by the Hill equation. The Hill equation is expressed as

$$v = V_{max}[S]^n / K_{0.5}{}^n + [S]^n \qquad (7.53)$$

where $K_{0.5}$ is the substrate concentration at half maximal velocity. Since this description is not the Michaelis–Menten equation it should not be called the K_m, although within the literature this term is frequently found. The parameter n, the Hill coefficient, can be greater or less than 1. A value of 1 gives hyberbolic profiles whilst other values yield curves showing cooperativity.

In PFK the profile of v against [S] with ATP as a substrate exhibits Michaelis–Menten kinetics but with fructose-6-phosphate the profile is sigmoidal (Figure 7.61). PFK is found universally in prokaryotes and eukaryotes as one enzyme in the pathway of glycolysis and catalyses the conversion of fructose 6-phosphate into fructose 1,6 diphosphate. In eukaryotic cells PFK activity is inhibited by ATP and citrate

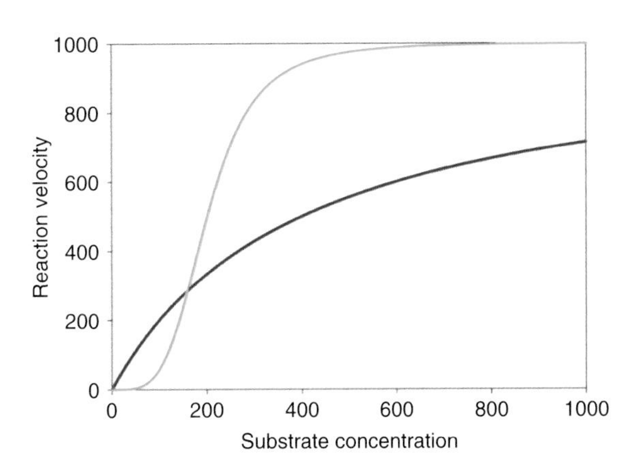

Figure 7.61 The sigmoidal and hyperbolic kinetics for PFK in the presence of effector (fructose-6-phosphate) and normal substrate (ATP) Fructose-6-phosphate + ATP $\rightleftharpoons$ Fructose 1,6diphosphate + ADP

whilst activation occurs with ADP, AMP and cyclic AMP. In bacteria the PFK enzymes are structurally simpler with a diverse array of allosteric modulators. The most studied PFKs are isolated from *B. stearothermophilus* and *E. coli*. These enzymes are inhibited by phosphoenolpyruvate (an end product of glycolysis) and stimulated by ADP as well as GDP.

PFK is a homotetramer containing polypeptide chains of 320 residues each folded into two domains. These domains are of different structure with domain 1 formed by residues 1–141 and residues 256–300 containing α and β secondary structure (α/β) and domain 2 formed by residues 142–255 and the C-terminal sequence forming an α-β-α layered structure (Figure 7.62). Each subunit has 12 α helices and 11β strands and forms a large contact area with an identical subunit creating a dimer. Dimers associate through smaller interfacial contact zones forming the tetramer. Initial studies of the enzyme showed that one of the products of the reaction (ADP) together with Mg^{2+} ions bind to the larger domain (blue) whilst an allosteric regulator (ADP again) is also found attached to the smaller second domain (orange). In this instance the effector molecule (ADP) resembles the structure of one of the substrates (ATP) and represents an example of homotypic regulation (Figures 7.63 and 7.64).

The active site for ADP and fructose-6-phosphate binding is formed between two subunits in a cleft. The T state of PFK has a low affinity for fructose-6-phosphate (substrate) whilst the R form has a much higher affinity.

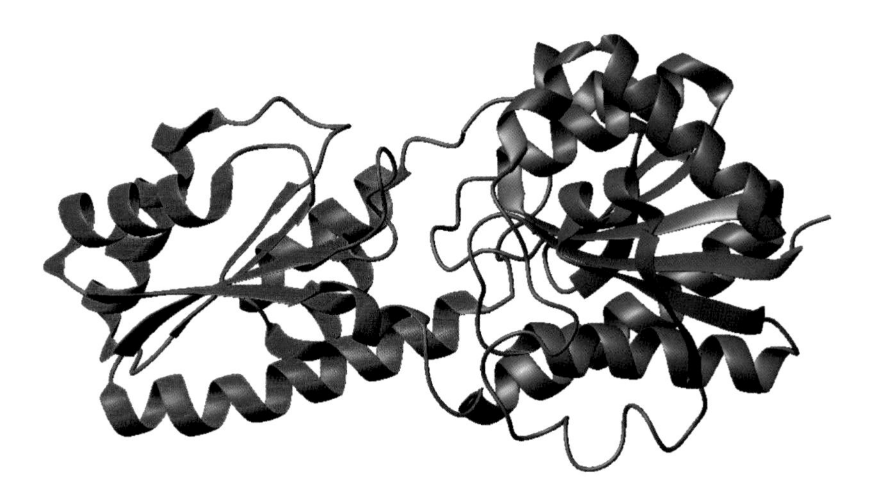

Figure 7.62 The arrangement of domains 1 and 2 within a single subunit of PFK. Shown in blue is the domain encompassing residues 1–138 and 256–305 whilst in orange is the secondary structure of residues 139–255 and 306–319

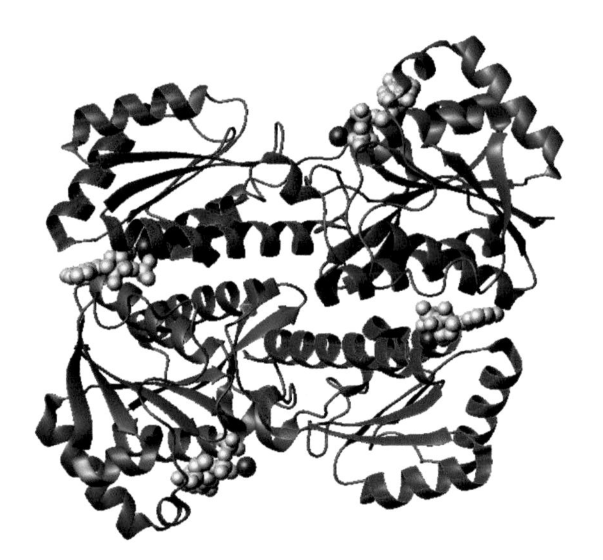

Figure 7.63 The arrangement of domains 1 and 2 within the PFK dimer showing ADP and Mg^{2+} ions bound to the different parts of the enzyme. ADP is shown in yellow, Mg^{2+} ions in green with the large and small domains of each subunit shown using the colour scheme of Figure 7.62. The dimers associate forming the binding site for the effector ADP molecule

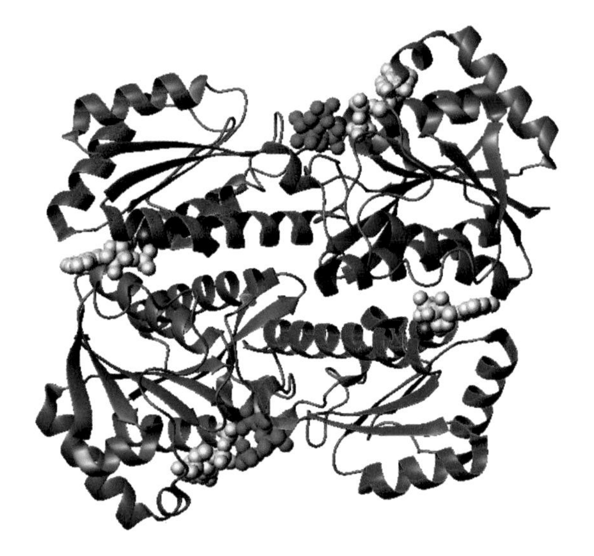

Figure 7.64 The arrangement of domains 1 and 2 within a PFK dimer showing fructose-6-bisphosphate (magenta) and ADP/Mg^{2+} bound to the active site of the large domain. These are the products of the reaction and the structure is the R state enzyme. ADP/Mg^{2+} is also bound at the effector site located on the smaller domain (PDB:1PFK). Only the dimer is shown and it should be remembered an additional dimer of subunits is present in the PFK homotetramer

Alongside fructose-6-phosphate ATP is also bound at the active site whilst at the allosteric site ATP (inhibitor) and ADP (activator) bind in addition to phosphoenolpyruvate (PEP), a second allosteric inhibitor. The allosteric sites are located principally in the smaller effector domain (orange section in Figures 7.63 and 7.64) but form contacts with the active site domain of neighbouring dimers. As a result of this arrangement conformational changes occurring at allosteric sites are easily transmitted to the active site domains of other subunits.

An analogue of PEP, 2-phosphoglycolate, binds tightly to the allosteric site yielding an inhibited form of the enzyme equivalent to the T state. By comparing the structure of T state enzyme with that formed in the presence of products (R state) the extent of conformational changes occurring in PFK were estimated. The crystal structures of PFK determined in the R and T states suggest the structural differences reside in rotation of dimers relative to one another. Comparing the inhibited state with the active state showed that the tetramer twists about its long axis so that one pair of subunits rotates relative to the other pair by about 8° around one of the molecular dyad axes. This rotation partly closes the binding site for the co-operative substrate fructose-6-phosphate and therefore explains its weaker binding to this conformational state. A single subunit contains two domains and the R > T transition also involves rotation of these regions relative to each other by ~4.5° that further closes the active (fructose-6-phosphate) site between domains. Another consequence of these movements is to alter the orientation of residues lining the catalytic pocket. In the R state two arginine side chains form ionic interactions with fructose-6-phosphate whilst a glutamate side chain is displaced out of the pocket. In the T state movement of dimers results in Arg162 pointing away from the binding site and Glu161 facing into the binding site. The effect of replacing a positive charge with a negative charge at the active site is to decrease fructose 6-phosphate binding by repulsion between negatively charged groups. PFK appears to represent a simple allosteric kinetic scheme in which the active and the inhibited conformations differ in their affinities for fructose-6-phosphate but do not differ in their catalytic competence. The presence of two conformational states is very similar to that previously described for haemoglobin.

Aspartyl transcarbamoylase (ATCase) is a large complex composed of 12 subunits of two different types. The enzyme catalyses formation of *N*-carbamoyl aspartate from aspartate and *N*-carbamoyl phosphate, and is an early step in a unique and critical biosynthetic pathway leading to the synthesis of pyrimidines (such as cytosine, uracil and thymine; Figure 7.65).

Both aspartate and carbamoyl phosphate bind cooperatively to the enzyme giving sigmoidal profiles (Figure 7.66) that are altered in the presence of cytidine triphosphate (CTP), a pyrimidine nucleotide, and adenosine triphosphate (ATP), a purine nucleotide. Allosteric inhibition is observed in the presence of CTP whilst ATP induces allosteric activation. Experimentally, a plot of v_o against [S] is observed to be sigmoidal and to shift to the left in the presence of ATP

Aspartate transcarbamoylase

$H_2PO_4^-$

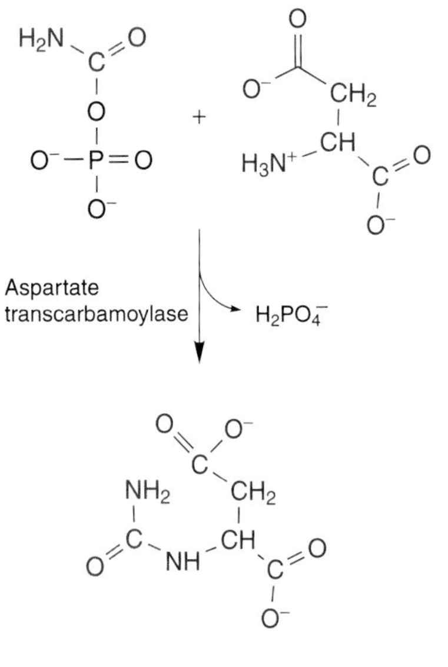

Figure 7.65 Initial reaction of pyrimidine biosynthesis involving condensation of aspartate and carbamoyl phosphate

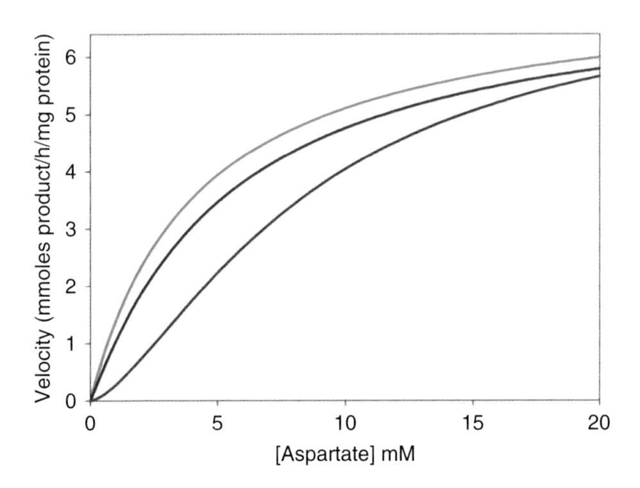

Figure 7.66 The steady-state kinetic behaviour of ATCase. Carbamoyl aspartate formation at a fixed concentration of substrate (blue) is increased in the presence of ATP (green) whilst CTP has the opposite effect acting as an inhibitor

and to the right in the presence of CTP. At a given substrate concentration CTP diminishes the reaction velocity whilst ATP increases it. ATP is therefore a positive heterotropic effector, CTP a negative heterotropic effector, and aspartate is a positive homotropic effector.

The effects of ATP and CTP are not unexpected; CTP is one product of pyrimidine biosynthesis and inhibition of ATCase activity is an example of feedback inhibition where end products modulate biosynthetic reactions to provide effective means of control (Figure 7.67). Consequently, high intracellular levels of CTP decrease pyrimidine biosynthesis whilst high intracellular levels of ATP indicate an imbalance between purine and pyrimidine concentrations and lead to increased rates of pyrimidine biosynthesis.

Structural insight into allosteric regulation of ATCase is provided by crystallographic analysis of the *E. coli* enzyme by William Lipscomb (Figure 7.68). The enzyme has a composition of c_6r_6 where c and r represent catalytic and regulatory subunits respectively arranged as two sets of catalytic trimers complexed with three sets of regulatory dimers. Each regulatory dimer links together two subunits present in different trimers. For both catalytic and regulatory subunits two domains exist within each polypeptide chain.

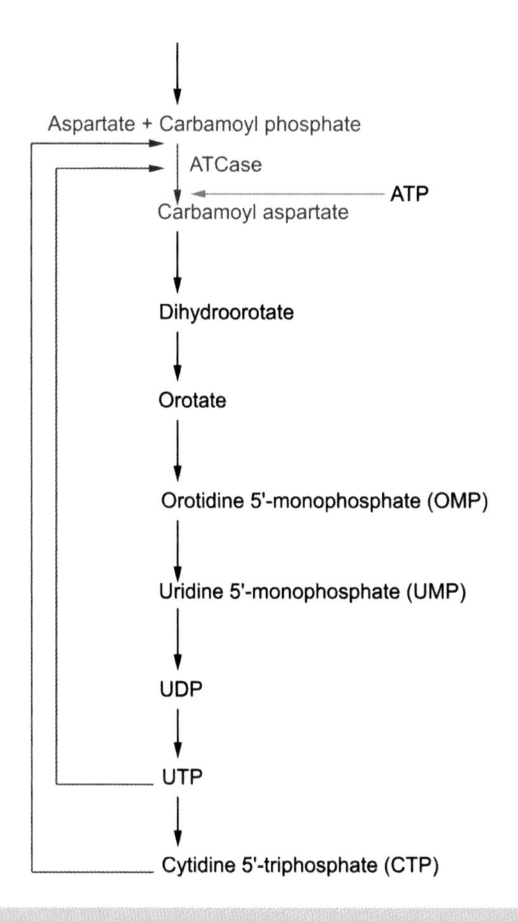

Figure 7.67 The metabolic pathway leading to UTP and CTP formation – feedback inhibition of ATCase is part of the normal control process

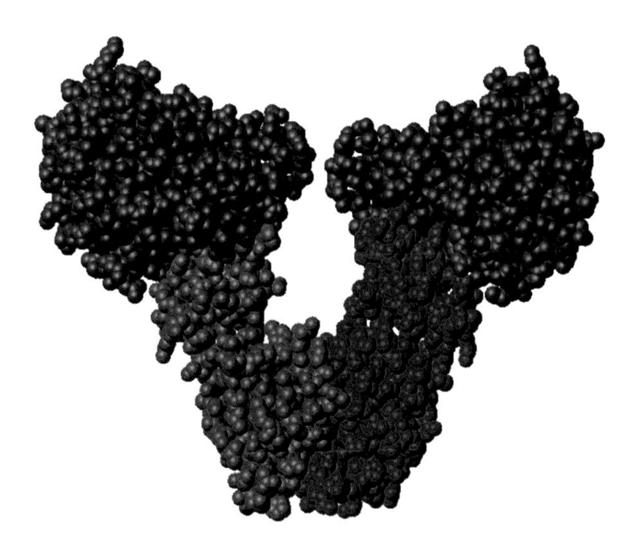

Figure 7.68 The quaternary structure of ATCase from *E. coli* showing the r_2c_2 structure (PDB: 1ATI). Two regulatory units (orange) link to catalytic subunits (green) belonging to *different* trimers

The catalytic subunit (Figure 7.69) contains one domain involved in aspartate binding whilst a second domain binds carbamoyl phosphate with the active site lying between domains. In view of their separation the c subunit must undergo large conformational change to bring the sites closer together for product formation. A bifunctional analogue of carbamoyl phosphate and aspartate called *N*-(phosphonacetyl)-*L*-aspartate (PALA) binds to ATCase, but is unreactive and has facilitated structural characterization of the enzyme–substrate complex (Figure 7.70). PALA binds to the R state, and by comparing the structure of this form of the enzyme with that obtained in the T state (ATCase plus CTP) the conformational changes occurring in transitions from one allosteric form to another were estimated.

The observed kinetic profiles of enzyme activity with ATP and CTP provide further insight into the basis of allosteric regulation. ATP enhances activity and binds to the R state of the enzyme whilst CTP an allosteric inhibitor binds preferentially to the T or low affinity state. Both ATP and CTP bind to the regulatory subunit and compete for the same site. The allosteric binding domain is the larger part of the r subunit containing a Zn^{2+} binding region that forms the interface between the regulatory and catalytic subunits. Displacement of the non-covalently bound Zn using mercurial reagents leads to the absence of binding and the dissociation of the c_6r_6 complex.

The rate of catalysis in isolated trimer is not modulated by ATP or CTP but exceeds the rate observed in the complete enzyme. This kinetic behaviour indicates that r subunits reduce the activity of the c subunits in the complete enzyme. In view of the different kinetic properties exhibited by ATCase in the presence of ATP or CTP it is somewhat surprising to find that both nucleotides bind in similar manner to the r subunit. This makes a structural explanation of

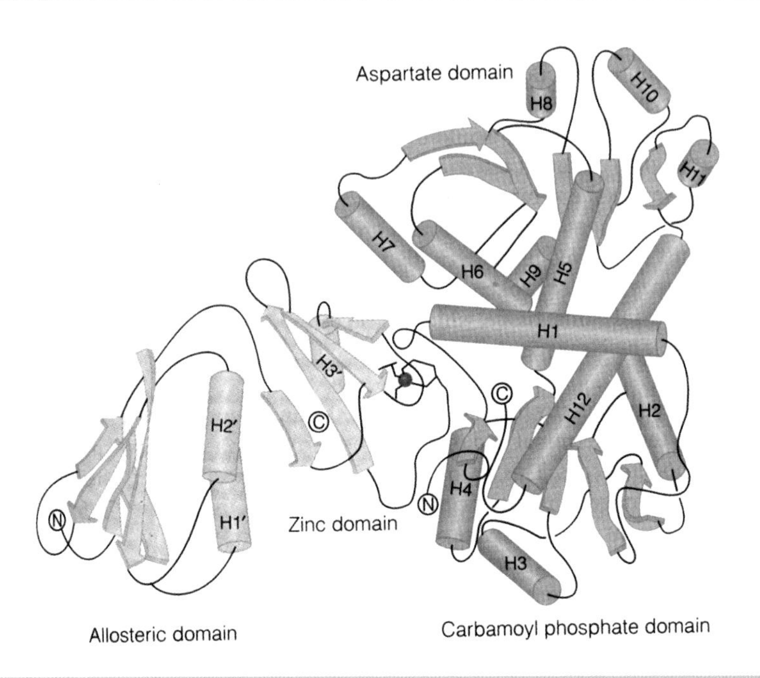

Figure 7.69 The structure of a single catalytic subunit and its adjacent regulatory subunit. The catalytic subunit is shown in green and the regulatory subunit in yellow (reproduced with permission from Lipscomb, W.N. *Adv. Enzymol.* 1994, **73**, 677–751. John Wiley & Sons)

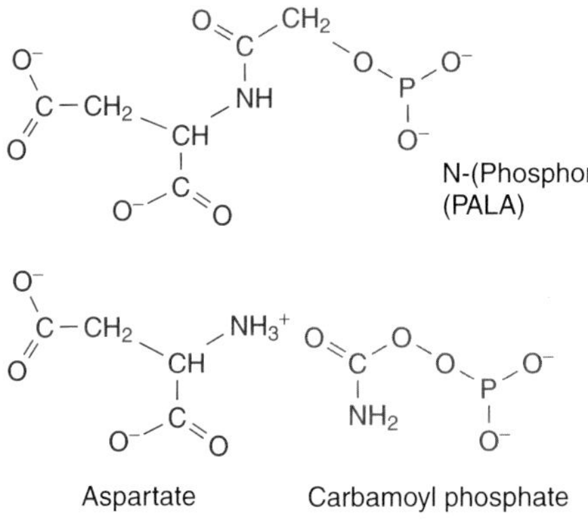

N-(Phosphonacetyl)-L-aspartate (PALA)

Aspartate

Carbamoyl phosphate

Figure 7.70 The structure of the bifunctional substrate analogue PALA and its relationship to carbamoyl phosphate and aspartate

their effects difficult to understand and the structural basis of allostery in ATCase has proved very elusive despite structures for the enzyme in the ATCase–CTP complex (T state) and the ATCase–PALA (R-state) complex (Figure 7.71). A comparison of these two structures shows that in the T>R transition the enzyme's catalytic trimers separate by ~1.1 nm and reorient their axes relative to each other by ~5°. The trimers assume an elongated or eclipse-like configuration along one axis accompanied by rotation of the r

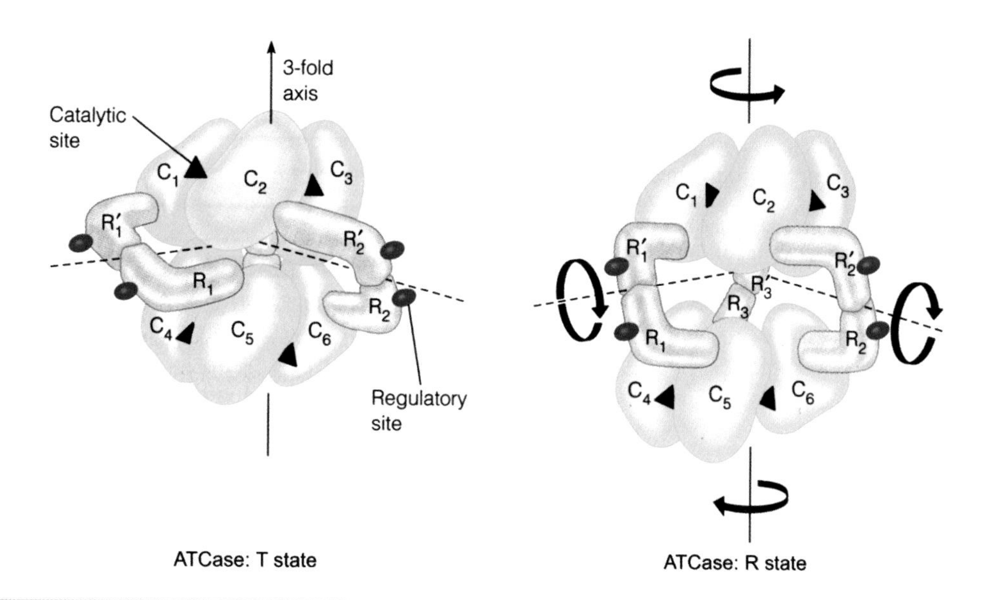

Figure 7.71 A diagram of the quaternary structure of aspartate transcarbamoylase (ATCase). The quaternary structure in the T state (left) shows six catalytic subunits and six regulatory subunits. Between each catalytic subunit lies the catalytic site whilst on each regulatory subunit is a single effector site on its outer surface (reproduced with permission from Lipscomb, W.N. *Adv. Enzymol.* 1994, **73**, 677–751. John Wiley & Sons)

subunits by ~15° about their two-fold rotational symmetry axis.

Covalent modification

Covalent modification regulates enzyme activity by the synthesis of *inactive* precursor enzymes containing precursor sequences at the N terminal. Removal of these regions by proteases leads to enzyme activation. Trypsin, chymotrypsin and pepsin are initially synthesized as longer, inactive, precursors called trypsinogen, chymotrypsinogen and pepsinogen or collectively zymogens. Zymogens are common for proteolytic enzymes such as those found in the digestive tracts of mammals where the production of active enzyme could lead to cell degradation immediately after translation. Acute pancreatitis is precipitated by damage to the pancreas resulting from premature activation of digestive enzymes.

Trypsinogen is the zymogen of trypsin and is modified when it enters the duodenum from the pancreas. In the duodenum a second enzyme enteropeptidase secreted from the mucosal membrane excises a hexapeptide from the N terminus of trypsinogen to form trypsin (Figure 7.72). Enteropeptidase is itself under hormonal control whilst the site of cleavage after Lys6 is important. The product is trypsin with a natural specificity for cleavage after Lys or Arg residues and

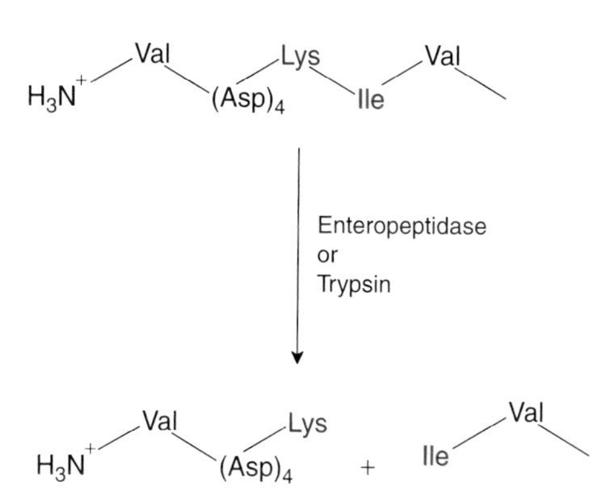

Figure 7.72 The activation of trypsinogen to trypsin. Limited proteolysis of trypsinogen by enteropeptidase and subsequently trypsin removes a short hexapeptide from the N-terminal

the small amount of trypsin produced is able to catalyse the further production of trypsin from the zymogen. As a result the process is often described as auto-catalytic.

Chymotrypsin is closely related to trypsin and is activated by cleavage of chymotrypsinogen between Lys15 and Val16 to form π-chymotrypsin (Figure 7.73). Further autolysis (self-digestion) to remove residues Ser14-Arg15 and Thr147-Asn148 yields α-chymotrypsin, the normal form of the enzyme. The three segments of polypeptide chain produced by zymogen processing remain linked by disulfide bridges.

Trypsin plays a major role in the activation of other proteolytic enzymes. Proelastase, a zymogen of elastase, is activated by trypsin by removal of a short N terminal region along with procarboxypeptidases A and B and prophospholipase A_2. The pivotal role of trypsin in zymogen processing requires tight control of trypsinogen activation. At least two mechanisms exist for tight control. Trypsin catalysed activation occurs slowly possibly as a result of the large amount of negative charge in the vicinity of the target lysine residue (four sequential aspartates). This charge repels substrate from the catalytic pocket that also contains aspartate residues. Zymogens are stored in the pancreas in intracellular vesicles known as zymogen granules that are resistant to proteolytic digestion. As a final measure cells possess serpins that prevent further activation.

A second important group of zymogens are the serine proteases of the blood-clotting cascade (Figure 7.74) that circulate as inactive forms until a stimulus normally in the form of injury to a blood vessel. This rapidly leads to the formation of a clot that prevents further bleeding via the aggregation of platelets within an insoluble network of fibrin. Fibrin is produced from fibrinogen, a soluble blood protein, through the action of thrombin, a serine protease. Thrombin is the last of a long series of enzymes involved in the coagulation process that are sequentially activated by proteolysis of their zymogens. Sequential enzyme activation leads to a coagulation cascade and the overall process allows rapid generation of large quantities of active enzymes in response to biological stimuli.

The concept of programmed cell death or apoptosis is vital to normal development of organisms. For example, the growing human embryo has fingers that are joined by web-like segments of skin. The removal

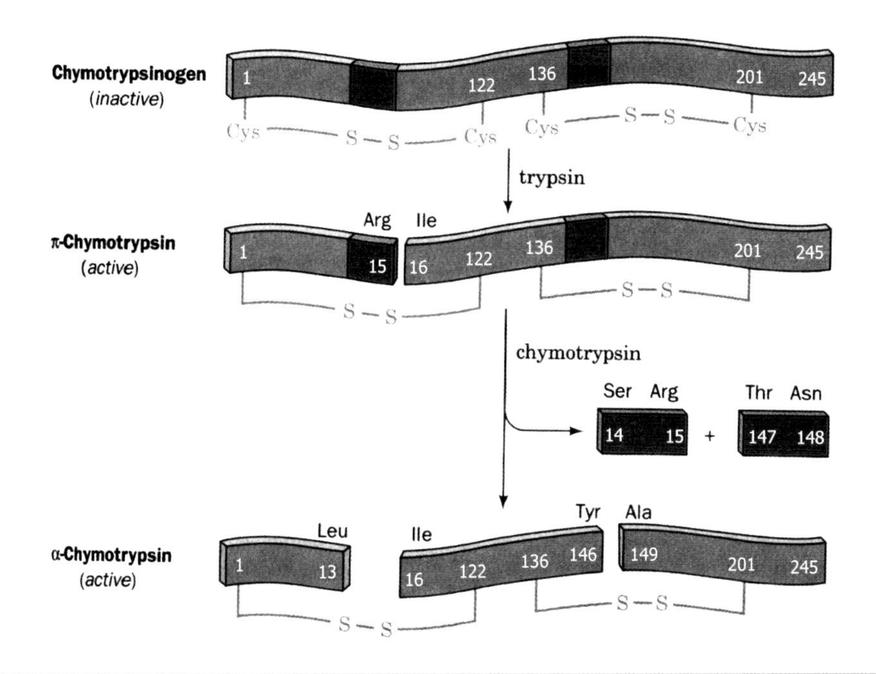

Figure 7.73 Activation of chymotrypsinogen proceeds through proteolytic processing and the removal of short peptide sequences.

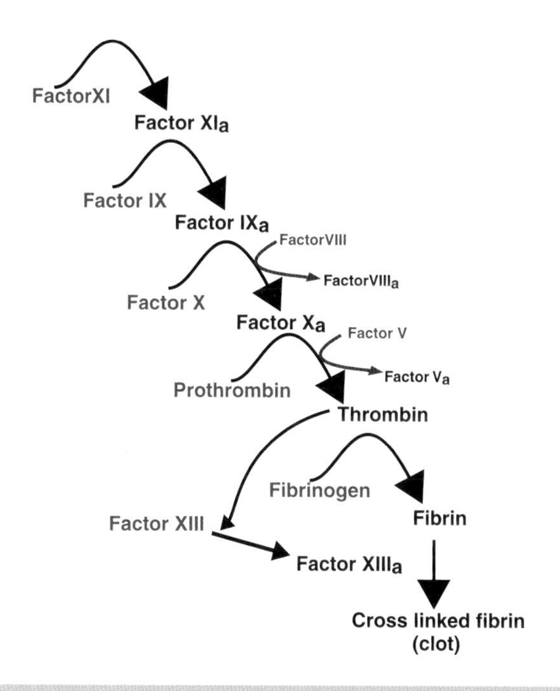

Figure 7.74 The cascade of enzyme activation during the intrinsic pathway of blood clotting involves processing of serine proteases

of these skin folds and the generation of correct pattern formation (i.e. fingers) relies on a family of intracellular enzymes called caspases. The term caspase refers to the action of these proteins as cysteine-dependent aspartate-specific proteases. Their enzymatic properties are governed by specificity for substrates containing Asp and the use of a Cys285 sidechain for catalysing peptide-bond cleavage located within a conserved motif of five residues (QACRG).

The first level of caspase regulation involves the conversion of zymogens to active forms in response to inflammatory or apoptotic stimuli although second levels of regulation involve specific inhibition of active caspases by natural protein inhibitors. Caspases are synthesized as inactive precursors and their activity is fully inhibited by covalent modification (Figure 7.75). It is important to achieve complete inhibition of activity since the cell is easily destroyed by caspase activity. Caspases are synthesized joined to additional domains known as death effector domains because their removal initiates the start of apoptosis where a significant step is the formation of further active caspases.

Yet another form of enzyme modification is the reversible phosphorylation/dephosphorylation of serine, threonine or tyrosine side chains. The phosphorylation of serine, threonine or tyrosine is particularly important in the area of cell signalling. Phosphorylation of tyrosine is an important covalent modification catalysed by enzymes called protein tyrosine kinases (PTKs). These enzymes transfer the γ phosphate of ATP to specific tyrosine residues on protein substrates according to the reaction

$$\text{Protein} + \text{ATP} \rightleftharpoons \text{Protein-P} + \text{ADP}$$

This reaction modulates enzyme activity and leads to the creation of new binding sites for downstream signalling proteins. At least two classes of PTKs are present in cells: the transmembrane receptor PTKs and enzymes that are non-receptor PTKs.

Receptor PTKs are transmembrane glycoproteins activated by ligand binding to an extracellular surface. A conformational change in the receptor results in phosphorylation of specific tyrosine residues on the intracellular domains of the receptor (Figure 7.76) – a process known as autophosphorylation – and initiates activation of many signalling pathways controlling cell proliferation, differentiation, migration

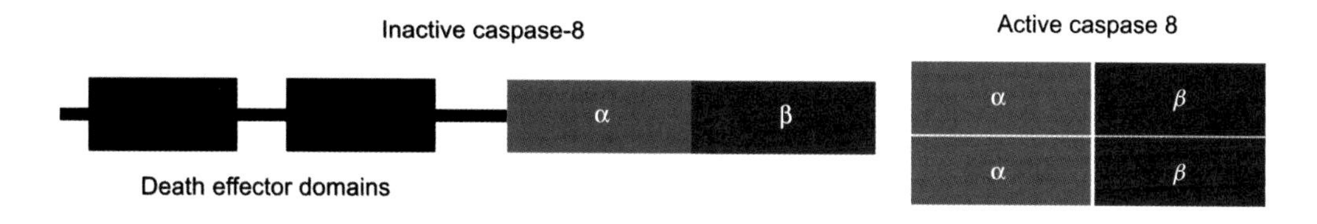

Figure 7.75 The covalent modification by addition of large protein domains to caspases leads to an inactive enzyme. Their removal allows the association of the α and β domains in an active heterotetramer

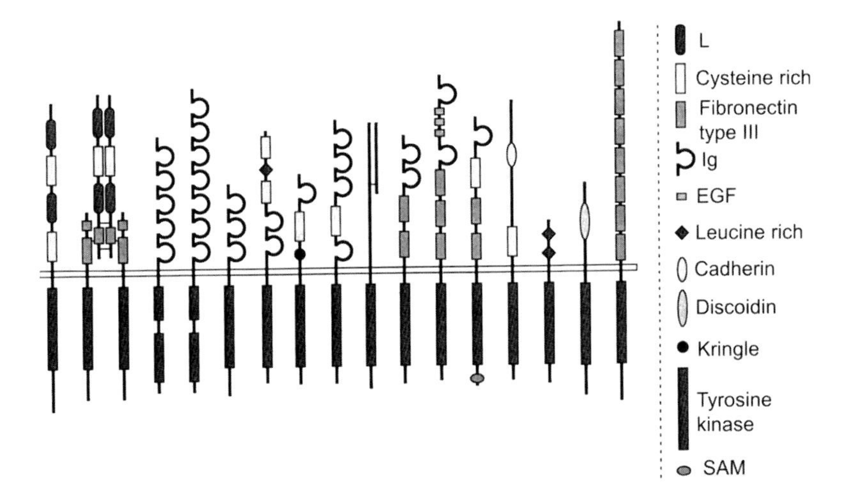

Figure 7.76 Domain organization of receptor tyrosine kinases. The extracellular domain at the top contains a variety of globular domains arranged with modular architecture. In contrast the cytoplasmic segment is almost exclusively a tyrosine kinase domain (reproduced with permission from Hubbard, S.R. & Till, J.R. *Ann. Rev. Biochem.* 2000, **69**, 373–398. Annual Reviews Inc)

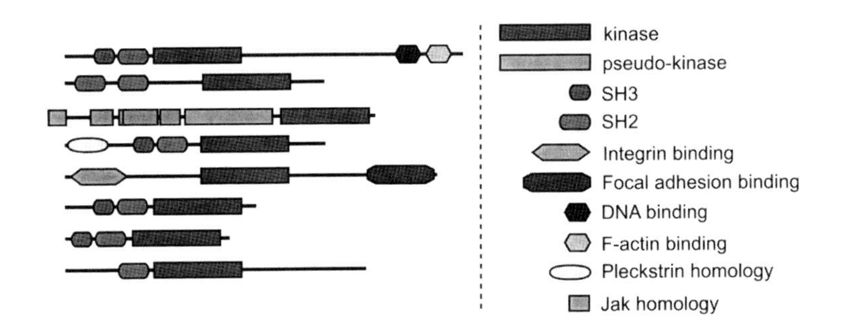

Figure 7.77 Domain organization found within non-receptor protein kinases (reproduced with permission from Hubbard, S.R. & Till, J.R. *Ann. Rev. Biochem.* 2000, **69**, 373–398. Annual Reviews Inc)

and metabolism. The transmembrane receptor tyrosine kinase family includes the receptors for insulin and many growth factors such as epidermal growth factor, fibroblast growth factor and nerve growth factor. Receptors consist of large extracellular regions constructed from modules of globular domains linked together. A single transmembrane helix links to a cytoplasmic domain possessing tyrosine kinase activity.

In contrast non-receptor PTKs (Figure 7.77) are a large family of proteins that are integral components of the signal cascades stimulated by RTKs as well as G-protein coupled receptors. These kinases phosphorylate protein substrates causing a switch between active and inactive states. These tyrosine kinases lack large extracellular ligand-binding regions together with transmembrane regions and most are located in the cytoplasm although some bear N-terminal modifications such as myristoylation or palmitoylation that lead to membrane association. In addition to kinase activity this large group of proteins possesses domains facilitating protein–protein, protein–lipid or protein–DNA interactions and a common set of domains are the Src homology 2 (SH2) and Src homology 3 (SH3) domains.

Phosphotyrosine binding has been studied in SH2 domains where the binding site consists of a deep pocket and centres around interaction of the phosphate with an arginine side chain at the base. The side chains of phosphoserine and phosphothreonine are both too short to form this interaction thereby accounting for the specificity of binding between SH2 and target sequences.

One of the first forms of phosphorylation to be recognized was that occurring during the activation of enzymes involved in glycogen metabolism. Glycogen phosphorylase plays a pivotal role in controlling the metabolism of the storage polysaccharide glycogen. Glycogen is primarily a series of α 1$\rightarrow$4 glycosidic bonds formed between glucose under conditions where the supply of glucose is plentiful. Glycogen phosphorylase catalyses the first step in the degradation of glycogen by releasing glucose-1-phosphate from a long chain of glucose residues for use by muscle tissue during contraction.

$$\underset{(n\text{ residues})}{\text{Glycogen}} + P_i \rightleftharpoons \text{Glucose-1-phosphate} + \underset{(n-1)\text{ residues}}{\text{Glycogen}}$$

As an allosteric enzyme the inactive T state is switched to the active R state by a change in conformation controlled by phosphorylation of a single serine residue (Ser14) in a polypeptide chain of ~850 residues. Reversible phosphorylation leads to transition between the two forms of the enzyme called phosphorylase a and phosphorylase b. The two forms differ markedly in their allosteric properties with the 'a' form generally uncontrolled by metabolites with the exception of glucose whilst the 'b' form is activated by AMP and inhibited by glucose-6-phosphate, glucose, ATP and ADP. Conversion is stimulated by hormone action, particularly the release of adrenaline and glucagon, where these signals indicate a metabolic demand for glucose that is mediated by intracellular production of cAMP by adenylate cyclase. Adenylate cyclase activates cAMP dependent kinases – enzymes causing the phosphorylation of a specific kinase – that subsequently modifies phosphorylase b at Ser14. This cascade is also regulated by phosphatases.

The structural basis of phosphorylation of Ser14 has been established from crystallographic studies of the a and b forms. Glycogen phosphorylase is a dimer of two identical chains. In the absence of Ser14 phosphorylation the N-terminal residues are poorly ordered but upon modification a region of ~16 residues assembles into a distorted 3_{10} helix that interacts with the remaining protein causing an increase in disorder at the C terminal. Phosphorylation orders the N-terminal residues and creates major changes in the position of Ser14 largely from new interactions formed with Arg43 on an opposing subunit and with Arg69 within the same chain. These interactions trigger changes in the tertiary and quaternary structure that lead to activation of a catalytic site located ~4.5 nm from the phosphoserine. The subunits move closer together in the presence of AMP or phosphoserine 14 and are accompanied by opening of active site regions to allow substrate binding and catalysis. The structure of glycogen phosphorylase suggested that inhibition of the enzyme by glucose-6-phosphate and ADP occur by excluding AMP from the binding site and from preventing conformational movements of the N-terminal region of the protein containing the phosphorylated serine residue. It has become clear that phosphorylation is a major mechanism for allosteric control of enzyme activity not only in metabolic systems but also in 'housekeeping' proteins such as those involved in cell division and the cell cycle.

Isoenzymes or isozymes

The term 'isoenzyme' is poorly defined. It commonly refers to multiple forms of enzymes arising from genetically determined differences in primary structure. In other words isoenzymes are the results of different gene products. A second form of isoenzyme involves multimeric proteins where combining subunits in different ratios alters the intrinsic catalytic properties. Cytochromes P450 are a large group of proteins that catalyse the oxidation of exogenous and endogenous organic substrates via a common reaction.

$$RH + 2H^+ + O_2 + 2e^- \rightleftharpoons ROH + H_2O$$

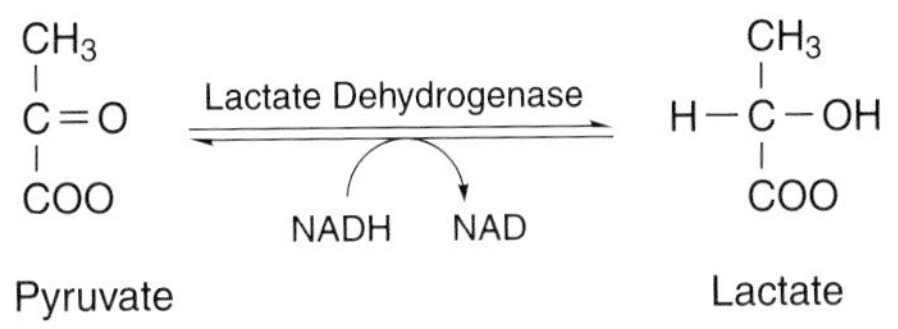

$$\underset{\text{Pyruvate}}{CH_3-C(=O)-COO} \xrightarrow[\text{NADH} \rightarrow \text{NAD}]{\text{Lactate Dehydrogenase}} \underset{\text{Lactate}}{CH_3-CH(OH)-COO}$$

Figure 7.78 Conversion of pyruvate to lactate by lactate dehydrogenase

Located in the endoplasmic reticulum of cells these enzymes are responsible for the detoxification of drugs and xenobiotics as well as the transformation of substrates such as steroids and prostaglandins. They are described as isoenzymes but arise from different genes containing different primary sequences with active site homology.

An excellent example of an isoenzyme in the second group is lactate dehydrogenase. In the absence of oxygen the final reaction of glycolysis is the conversion of pyruvate to lactic acid catalysed by the enzyme lactate dehydrogenase (LDH) with the conversion of NADH to NAD (Figure 7.78).

In skeletal muscle, where oxygen deprivation is common during vigorous exercise, the reaction is efficient and large amounts of lactate are generated. Under aerobic conditions the reaction catalysed by lactate dehydrogenase is not efficient and pyruvate is preferentially converted to acetyl-CoA. The subunit composition of lactate dehydrogenase is not uniform but varies in different tissues of the body. Five major LDH isoenzymes are found in different vertebrate tissues and each molecule is composed of four polypeptide chains or subunits. LDH is a tetramer and it has been shown that three different primary sequences are detected in human tissues. The three types of LDH chains are M (LDH-A) found predominantly in muscle tissues, H (LDH-B) found in heart muscle together with a more limited form called LDH-C present in the testes of mammals. LDH-C is clearly of restricted distribution within the body and is not considered further. The M and H subunits of LDH are encoded by different genes and as a result of differential gene expression in tissues the tetramer is formed by several different combinations of subunits. Significantly, isoenzymes of LDH were originally demonstrated by their different electrophoretic mobilities under non-denaturing conditions. Five different forms of LDH were detected; LDH-1, LDH-2, LDH-3, LDH-4, and LDH-5 with LDH-1 showing the lowest mobility. The reasons for these differences in mobility are clear when one compares the primary sequences of the M and H subunits (Figure 7.79).

The H subunit contains a greater number of acidic residues (Asp,Glu) than the M polypeptide. In heart muscle the gene for the H subunit is more active than the gene for the M subunit leading to an LDH enzyme containing more H subunit. Thus, LDH isoenzyme 1 is the predominant form of the enzyme in cardiac muscle. The reverse situation occurs in skeletal muscle where there is more M than H polypeptide produced and LDH-5 is the major form of the enzyme. More importantly, the conversion of pyruvate to lactate increases with the number of M chains and LDH 5 in skeletal muscle contains 4 M subunits and is ideally suited to the conversion of pyruvate to lactate. In contrast LDH-1 has 4 H subunits and is less effective at converting pyruvate into lactate. Lactate dehydrogenase, like many other enzymes, is also found in serum where it is the result of normal cell death. Dying or dead cells liberate cellular enzymes into the bloodstream. The liberation of enzymes into the circulatory system is accelerated during tissue injury and the measurement of LDH isoenzymes in serum has been used to determining the site and nature of tissue injury in humans. For example, when the blood supply to the heart muscle is severely reduced, as occurs in a heart attack, muscle cells die and liberate the enzyme LDH-1 into the bloodstream. An increase in LDH-1 in serum is indicative of a heart attack and may be used diagnostically. In contrast, muscular dystrophy is accompanied by an increase in the levels of LDH-5 derived from dying skeletal muscle cells.

Summary

Enzymes, with the exception of ribozymes, are catalytic proteins that accelerate reactions by factors up to 10^{17} when compared with the corresponding rate in the uncatalysed reaction. All enzymes can be

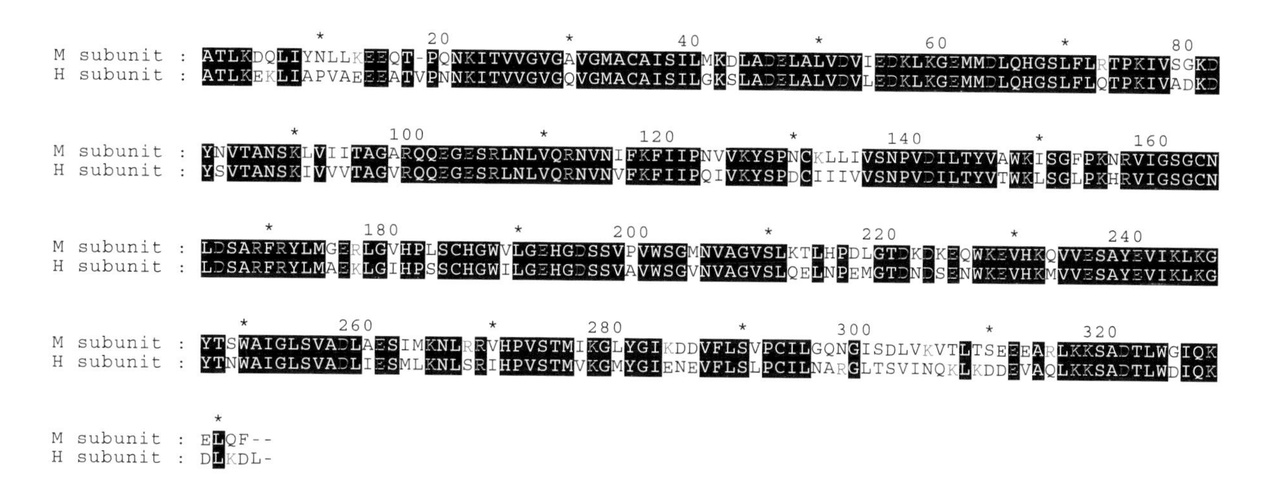

Figure 7.79 The primary sequences of the M and H subunits of lactate dehydrogenase showing sequence identity (blocks) and the distribution of positively charged side chains (red) and negatively charged side chains (blue) in each sequence. Careful summation of the charged residues reveals that the H subunit has more acidic residues

grouped into one of six functional classes each representing the generic catalytic reaction. These classes are oxidoreductases, transferases, hydrolases, lyases, isomerase and lyases.

Many enzymes require co-factors to perform effective catalysis. These co-factors can be tightly bound to the enzyme including covalent linkage or more loosely associated with the enzyme. Co-factors include metal ions as well as organic components such as pyridoxal phosphate or nicotinamide adenine dinucleotide (NAD). Many co-factors or co-enzymes are derived from vitamins.

All elementary chemical reactions can be described by a rate equation that describes the progress (utilization of initial starting material or formation of product) with time. Reactions are generally described by first or second order rate equations.

Enzymes bind substrate forming an enzyme–substrate complex which then decomposes to yield product. A plot of the velocity of an enzyme-catalysed reaction as a function of substrate concentration exhibits a hyperbolic profile rising steeply at low substrate concentrations before reaching a plateau above which further increases in substrate have little effect on overall rates.

The kinetics of enzyme activity are usually described via the Michaelis–Menten equation that relates the initial velocity to the substrate concentration. This analysis yields several important parameters such as V_{max} the maximal velocity that occurs when all the enzyme is found as ES complex, K_m the substrate concentration at which the reaction velocity is half maximal and k_{cat}/K_m, a second order rate constant that represents the catalytic efficiency of enzymes.

Enzymes catalyse reactions by decreasing the activation free energy ($\Delta G^{\ddagger}$), the energy associated with the transition state. Enzymes use a wide range of catalytic mechanisms to convert substrate into product. These mechanisms include acid–base catalysis, covalent catalysis and metal ion catalysis. Within active sites the arrangement of functional groups allows not only the above catalytic mechanisms but further enhancements via proximity and orientation effects together with preferential binding of the transition state complex. The latter effect is responsible for the greatest enhancement of catalytic activity in enzymes.

The catalytic mechanisms of numerous enzymes have been determined from the application of a combination of structural, chemical modification and kinetic analysis. Enzymes such as lysozyme, the serine protease family including trypsin and chymotrypsin, triose phosphate isomerase and tyrosinyl tRNA synthetase are now understood at a molecular level and with this level of understanding comes an appreciation of the catalytic mechanisms employed by proteins. Virtually all enzymes employ a combination of mechanisms to achieve effective (rapid and highly specific) catalysis.

Enzyme inhibition represents a vital mechanism for controlling catalytic function. *In vivo* protein inhibitors bind tightly to enzymes causing a loss of activity and one form of regulation. A common form of enzyme inhibition involves the competition between substrate and inhibitor for an active site. Such inhibition is classically recognized from an increase in K_m whilst V_{max} remains unaltered. Other forms of inhibition include uncompetitive inhibition where inhibitor binds to the ES complex and mixed (non-competitive) inhibition were binding to both E and ES occurs. Different modes of inhibition are identified from Lineweaver–Burk or Eadie–Hofstee plots.

Irreversible inhibition involves the inactivation of an enzyme by an inhibitor through covalent modification frequently at the active site. Irreversible inactivation is the basis of many forms of poisoning but is also used beneficially in therapeutic interventions with, for example, the modification of prostaglandin H_2 synthase by aspirin alleviating inflammatory responses.

Allosteric enzymes are widely distributed throughout many different cell types and play important roles as regulatory units within major metabolic pathways. Allosteric enzymes contain at least two subunits and often possess many more.

Changes in quaternary structure arise as a result of ligand (effector) binding causing changes in enzyme activity. The conformational changes frequently involve rotation of subunits relative to one another and may produce large overall movement. Such movements are the basis for the transitions between high (R state) and low (T state) affinity states of the enzyme with conformational changes leading to large differences in substrate binding. The ability to modify enzymes via allosteric effectors allows 'fine-tuning' of catalytic activity to match fluctuating or dynamic conditions.

Alternative mechanisms of modifying enzyme activity exist within all cells. Most important are covalent modifications that result in phosphorylation or remove parts of the protein, normally the N-terminal region, that limit functional activity. Covalent modification is used in all cells to secrete proteolytic enzymes as inactive precursors with limited proteolysis revealing the fully active protein. Caspases are one group of enzymes using this mechanism of modification and play a vital role in apoptosis. Apoptosis is the programmed destruction of cell and is vital to normal growth and development. Premature apoptosis is serious and caspases must be inactivated via covalent modification to prevented unwanted cell death.

Problems

1. Consult an enzyme database and find three examples of each major class of enzyme. (i.e. three from each of the six groupings).
2. Describe the effect of increasing temperature and pH on enzyme-catalysed reactions.
3. How can some enzymes work at pH 2.0 or at 90 °C. Illustrate your answer with selective example enzymes.
4. Draw a transition state diagram of energy versus reaction coordinate for uncatalysed and an enzyme catalysed reactions.
5. How is the activity of many dehydrogenases most conveniently measured. How can this method be exploited to measure activity in other enzymes?
6. Identify nucleophiles used by enzymes in catalysis?
7. Why does uncompetitive inhibition lead to a decrease in K_m? What happens to the V_{max}?
8. An enzyme is found to have an active site cysteine residue that plays a critical role in catalysis. At what pH does this cysteine operate efficiently in nucleophilic catalysis? Describe how proteins alter the pK of this side chain from its normal value in

solution. What would you expect the effect to be on the catalysed reaction?

9. Construct a plot of initial velocity against substrate concentration from the following data. Estimate values of K_m and V_{max}. Repeat the calculation of K_m and V_{max} to yield improved estimates.

Substrate concentration (μM)	Initial velocity (μmol s^{-1})
25.0	80.0
50.0	133.3
75.0	171.4
100.0	200.0
200.0	266.6
300.0	300.0
400.0	320.0
600.0	342.8
800.0	355.5

10. In a second enzyme kinetic experiment the initial velocity of an enzyme catalysed reaction was followed in the presence of native substrate and at five different concentrations of an inhibitory substrate. The following data were obtained. Interpret the results.

Substrate concentration (μM)	Initial velocity (μmol s^{-1})	Inhibitor 50 μM	Inhibitor 150 μM	Inhibitor 250 μM	Inhibitor 350 μM	Inhibitor 450 μM
20	90.91	62.50	38.46	27.78	21.74	17.86
40	166.67	117.65	74.07	54.05	42.55	35.09
60	230.77	166.67	107.14	78.95	62.50	51.72
80	285.71	210.53	137.93	102.56	81.63	67.80
100	333.33	250.00	166.67	125.00	100.00	83.33
125	384.62	294.12	200.00	151.52	121.95	102.04
150	428.57	333.33	230.77	176.47	142.86	120.00
200	500.00	400.00	285.71	222.22	181.82	153.85
250	555.56	454.55	333.33	263.16	217.39	185.19
300	600.80	500.83	375.78	300.70	250.62	214.85
400	666.67	571.43	444.44	363.64	307.69	266.67
500	714.29	625.00	500.00	416.67	357.14	312.50
600	750.00	666.67	545.45	461.54	400.00	352.94
800	800.00	727.27	615.38	533.33	470.59	421.05
1000	833.33	769.23	666.67	588.24	526.32	476.19

8 Protein synthesis, processing and turnover

Uncovering the cellular mechanisms resulting in sequential transfer of information from DNA (our genes) to RNA and then to protein represents one of major achievements of biochemistry in the 20th century. Beginning with the discovery of the structure of DNA in 1953 (Figure 8.1) by James Watson, Francis Crick, Maurice Wilkins and Rosalind Franklin this area of biodiversity has expanded dramatically in importance. The information required for synthesis of proteins resides in the genetic material of cells. In most cases this genetic material is double-stranded DNA and the underlying principles of replication, transcription and translation remain the same throughout all living systems.

The information content of DNA resides in the order of the four nucleotides (adenine, cytosine, guanine and thymine) along the strands making up the famous double helix. The structure of DNA is one of the most famous images of biochemistry and although this chapter will focus on the reactions occurring during or after protein synthesis it is worth remembering that replication and transcription involve the concerted action of many diverse proteins. The transfer of information from DNA to protein can be divided conveniently into three major steps: replication, transcription and translation, with secondary steps involving the processing of RNA and the modification and degradation of proteins (Figure 8.2).

Cell cycle

Cell division is one part of an integrated series of reactions called appropriately the cell cycle. The cell cycle involves replication, transcription and translation and is recognized as a universal property of cells. Bacteria such as *Escherichia coli* replicate and divide within 45 minutes whilst eukaryotic cell replication is slower and certainly more complicated, with rates of division varying dramatically. Typically, eukaryotic cells pass through distinct phases that combine to make the cell cycle and lead to mitosis. The first phase is G_1, the first gap phase, and occurs after cell division when a normal diploid chromosome content is present. G_1 is followed by a synthetic S phase when DNA is replicated, and at the end of this synthetic period the cell has twice the normal DNA content and enters a second gap or G_2 phase. The combined G_1, S and G_2 phase make up the interphase, a period first recognized by cytogeneticists studying cell division. After the G_2 phase cells enter mitosis, with division giving two daughter cells each with a normal diploid level of DNA and the whole process can begin again (Figure 8.3).

Mitosis is accompanied by the condensation of chromosomes into dense opaque bodies, the disintegration of the nuclear membrane and the formation of a specialized mitotic spindle apparatus containing

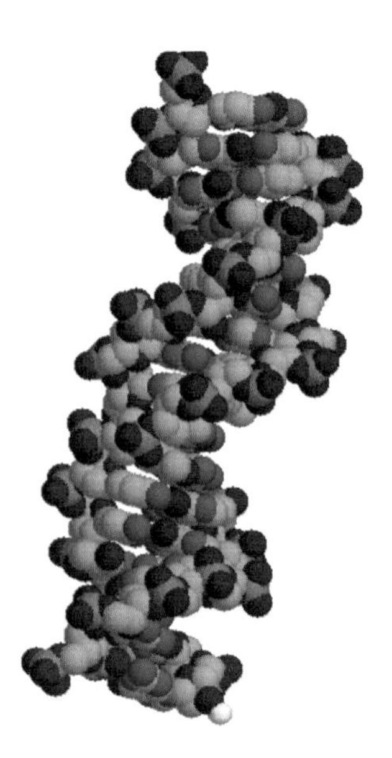

Figure 8.1 The structure of DNA. The bases lie horizontally between the two sugar-phosphate chains. The positions of the major and minor grooves are defined by the outline of phosphate/oxygen atoms (orange/red) arranged along the sugar-phosphate backbone. The pitch of the helix is ~3.4 nm and represents the distance taken for each chain to complete a turn of 360°

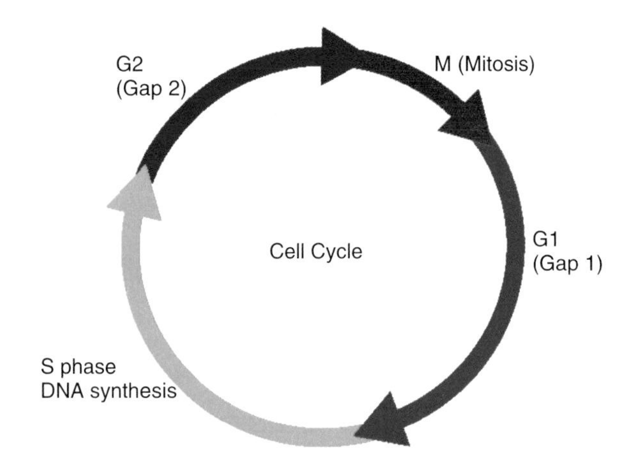

Figure 8.3 Cell cycle diagram showing the G_1, S, G_2 phases in a normal mitotic cell cycle

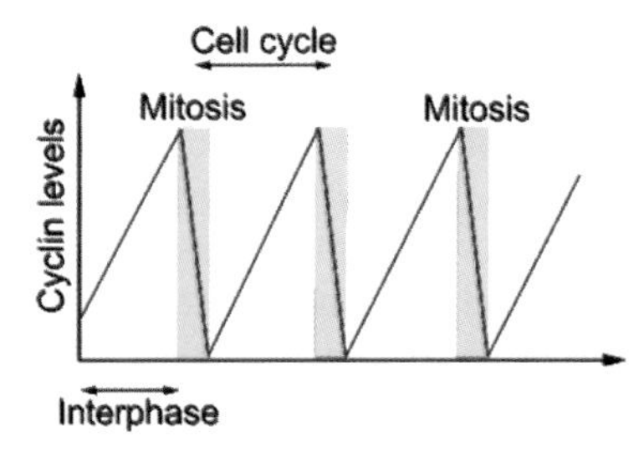

Figure 8.4 The levels of protein vary according to the stage of the cell cycle and gave rise to the name of this protein. Cyclins are abundant during the stage of the cell cycle in which they act and are then degraded

microtubule elements that directs chromosome movements so that daughter cells receive one copy of each chromosome pair. After mitotic division the cell cycle is complete and normally a new G_1 phase starts, although in some cells G_1 arrest is observed and a state called G_0 is recognized.

Proteins with significant roles in the cell cycle were discovered from cyclical variations in their concentration during cell division and called cyclins (Figure 8.4). The first cyclin molecule was discovered in 1982 from sea urchin (*Arbacia*) eggs, but subsequently at least eight cyclins have been identified in human cells.

Cyclins have diverse sequences, but functionally important regions called cyclin boxes show high levels of homology. This homology has facilitated domain identification in other proteins such as transcription factors. Cyclin boxes confer binding specificity towards protein partners. They are found with other

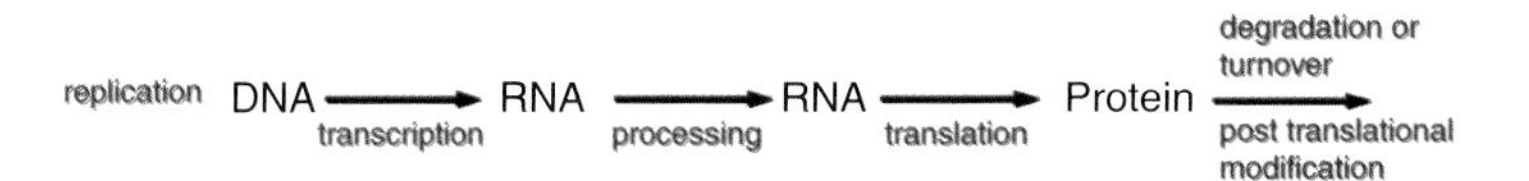

Figure 8.2 From DNA to protein

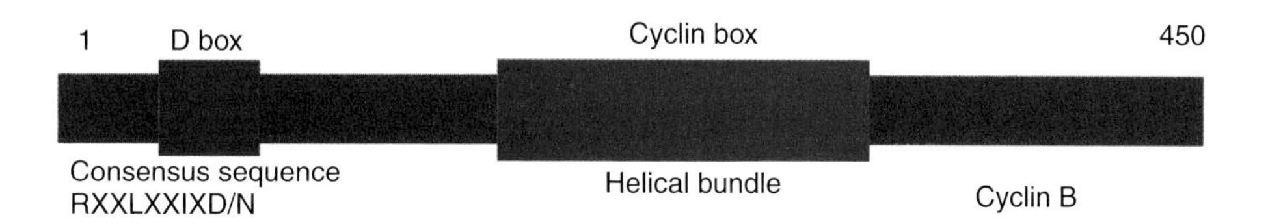

Figure 8.5 Organization of cyclins showing D box and cyclin box segments

characteristic motifs such as the cyclin destruction box with a minimal consensus sequence of Arg-Xaa-Xaa-Leu-Xaa-Xaa-Ile-Xaa-Asn/Asp (Figure 8.5).

The first 'box' structure was determined for a truncated construct of cyclin A representing the final 200 residues of a 450 residue protein. The structure has two domains each with a central core of five regular α helices that is preserved between cyclins despite low levels of sequence identity. Five charged, invariant, residues (Lys266, Glu295, Leu255, Asp240, and Arg21) form a binding site for protein partners such as cyclin-dependent kinases.

Cyclins regulate the activity of cyclin-dependent kinases (Cdks) by promoting phosphorylation of a single Thr side chain. Cdks were originally identified via genetic analysis of yeast cell cycles and purification of extracts stimulating the mitotic phase of frog and marine invertebrate oocytes. However, Cdks are found in all eukaryotic cell cycles where they control the major transitions of the cell cycle. Studies of the cell cycle were aided by the identification of mutant strains of fission yeast (*Schizosaccharomyces pombe*) and budding yeast (*Saccharomyces cerevisiae*) containing genetic defects (called lesions) within key controlling genes. In *S. cerevisiae* Leland Hartwell identified many genes regulating the cell cycle and coined the term 'cell division cycle' or *cdc* gene (Figure 8.6). One gene, *cdc28*, controlled the important first step of the cell cycle – the 'start' gene. Cells normally go through a number of 'checkpoints' to ensure cell division proceeded correctly; at each of these points mutants were identified in cell cycle genes. These studies allowed a picture of the cell cycle in terms of genes and cells would not advance further in the cycle until the 'block' was overcome by repairing defects. Cancerous cells turn out to be cells that manage to evade these 'checkpoints' and divide uncontrollably.

Using the fission yeast, *S. pombe*, Paul Nurse identified a gene (*cdc2*) whose product controlled many reactions in the cell cycle and was identified by homology to cyclin-dependent kinase 1 (Cdk1) found in human cells. The human gene could substitute for the yeast gene in Cdk^- mutants. The genes *cdc2* and *cdc28* from *S. pombe* and *S. cerevisiae* encode homologous

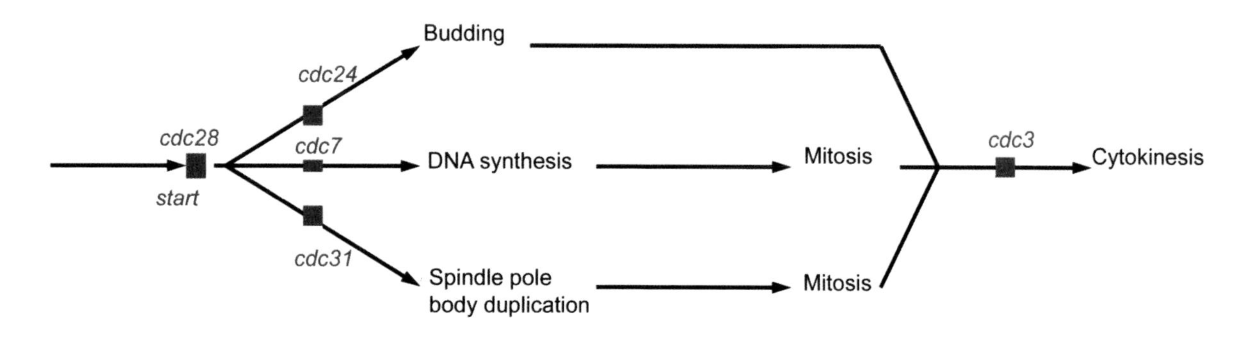

Figure 8.6 The role of mutants and the identification of key genes in the cell cycle of yeast. The genes *cdc28, cdc24, cdc31* and *cdc7* were subsequently shown to code for cdk2, a guanine nucleotide exchange factor, centrin – a Ca-binding protein, and another cyclin-dependent kinase

cyclin-dependent protein kinases occupying the same pivotal role within the cell cycle.

Cdks are identified in all cells and retain activity within a single polypeptide chain ($M_r \sim 35$ kDa). This subunit contains a catalytic core with sequence similarity to other protein kinases but requires cyclin binding for activation together with the phosphorylation of a specific threonine residue. Molecular details of this activity were uncovered by crystallographic studies of Cdks, cyclins and the complex formed between the two protein partners. Cdks catalyse phosphorylation (and hence activation) of many proteins at different stages in the cell cycle: including histones, as part of chromosomal DNA unpackaging; lamins, proteins forming the nuclear envelope that must disintegrate prior to mitosis; oncoproteins that are often transcription factors; and proteins involved in mitotic spindle formation. Cdks control reactions occurring within the cell by the combined action of a succession of phosphorylated Cdk–cyclin complexes that elicit activation and deactivation of enzymes. It is desirable that these systems represent an 'all or nothing' signal in which the cell is compelled to proceed in one, irreversible, direction. The reasons for this are fairly obvious in that once cell division has commenced it would be extremely hazardous to attempt to reverse the procedure. Cdks trigger cell cycle events but also enhance activity of the next cyclin–Cdk complex with a cascade of cyclin–Cdk complexes operating during the G_1/S/G_2 phases in yeast and human cells (Figure 8.7).

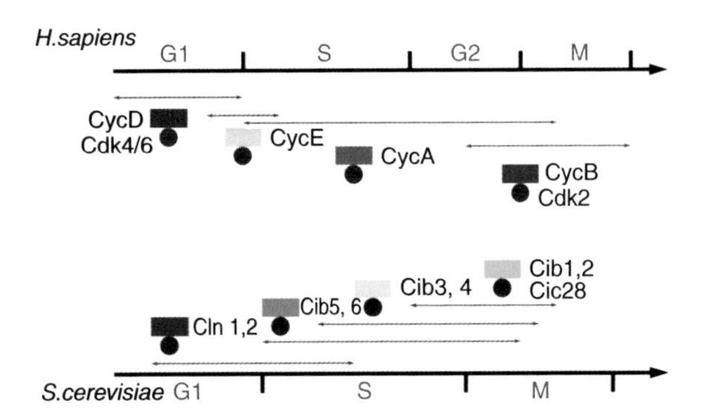

Figure 8.7 Major Cdk-cyclin complexes involved in cell cycle control in humans and budding yeast. Purple arrows indicate approximate timing of activation and duration of the Cdk–cyclin complexes. Cyclins are shown as coloured rectangles with the cognate Cdk shown by circles. The S and M phases overlap in *S. cerevisiae*

The structure of Cdk and its role in the cell cycle

Understanding the structure of Cdks uncovered the mechanism of action of these enzymes with the first structure obtained in 1993 for Cdk2 in a complex with ATP (Figure 8.8). A bilobal protein comprising a small lobe at the N terminal together with a much larger C-domain resembled other protein kinases, such as cAMP-dependent protein kinase. The comparatively small size reflected a 'minimal' enzyme. Within this small structure two regions of Cdk2 stood out as 'different' from other kinases. A single helical segment, called the PSTAIRE helix from the sequence of conserved residues Pro-Ser-Thr-Ala-Ile-Arg-Glu-, was located in the N-terminal lobe and was unique to Cdks. The second distinctive region contained the site of phosphorylation (Thr160) in Cdk and was comparable to other regulatory elements seen in protein kinases. This region was called the T or activation loop.

The smaller N-terminal lobe has the PSTAIRE helix and a five stranded β sheet whilst the larger lobe has six α helices and a small section of two-stranded β sheet. ATP binding occurs in the cleft between lobes with the adenine ring located between the β sheet of the small lobe and the L7 loop between strands 2 and 5. A glycine-rich region interacts with the adenine ring, whilst Lys33, Asp145, and amides in the backbone of the glycine-rich loop interact with phosphate groups (Figure 8.9).

The basic fold of Cdk2 contains the PSTAIRE helix and T loop positioned close to the catalytic cleft, but is an inactive form of the enzyme. Activation by a factor of ~10 000 occurs upon cyclin binding and is generally assessed via the ability of Cdk2 to phosphorylate protein substrates such as histone. Further enhancement of activity occurs with phosphorylation of Thr160 in Cdk2 and suggested that conformational changes might modulate biological activity. The origin and extent of these conformational changes were determined by comparing the structures of Cdk2–cyclin A complexes in phosphorylated (fully active) and unphosphorylated

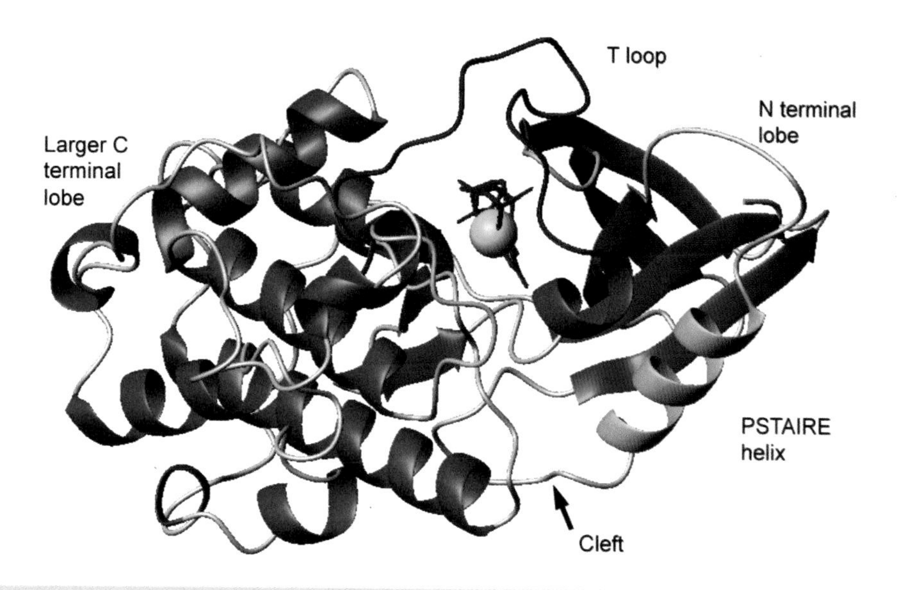

Figure 8.8 Structure of isolated Cdk2 and its interaction with Mg^{2+}-ATP. The ATP is shown in magenta, the Mg^{2+} ion in light green. The secondary structure elements are shown in blue and green with the PSTAIRE helix shown in yellow and the T loop in red. The L12 helix can be seen adjacent to the PSTAIRE helix in the active site of Cdk2

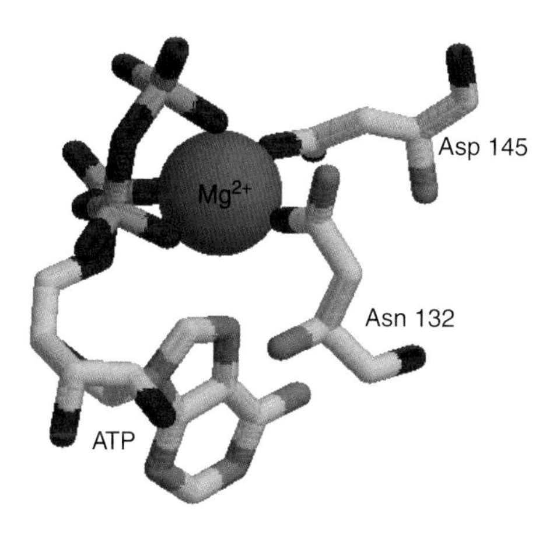

Figure 8.9 Binding of Mg^{2+}-ATP to monomeric Cdk2. Magnesium bound in an octahedral environment with six ligands provided by phosphates of ATP, Asp145, Asn132 and a bound water molecule

(partially active) states together with the isolated (inactive) Cdk2 subunit.

In the Cdk2–cyclin A complex conformational changes were absent in cyclin A with structural perturbations confined to Cdk2. The binding site between cyclin A and Cdk2 involved the cyclin A box containing Lys266, Glu295, Leu255, Asp240, and Arg211 and a binding site located close to the catalytic cleft. In proximity to the cleft hydrogen bonds with Glu8, Lys9, Asp36, Asp38, Glu40 and Glu42 were important to complex formation as well as hydrophobic interactions that underpinned movement of the 'PSTAIRE' helix and T loop. Large conformational changes upon cyclin binding occur for the PSTAIRE helix and the T loop with the helix rotated by ~90° and the loop region displaced. In the absence of cyclin the T loop is positioned above the catalytic cleft blocking the active site ATP to polypeptide substrates. Cyclin binding moves the T loop opening the catalytic cleft and exposing the phosphorylation site at Thr160. Figures 8.10 and 8.11 show the structure of the Cdk2–cyclin A–ATP complex.

Threonine phosphorylation in the T loop is performed by a second kinase called CAK (Cdk-activating kinase), and in an activated state the complex phosphorylates downstream proteins modifying serine or threonine residues located within target (S/T-P-X-K/R) sequences.

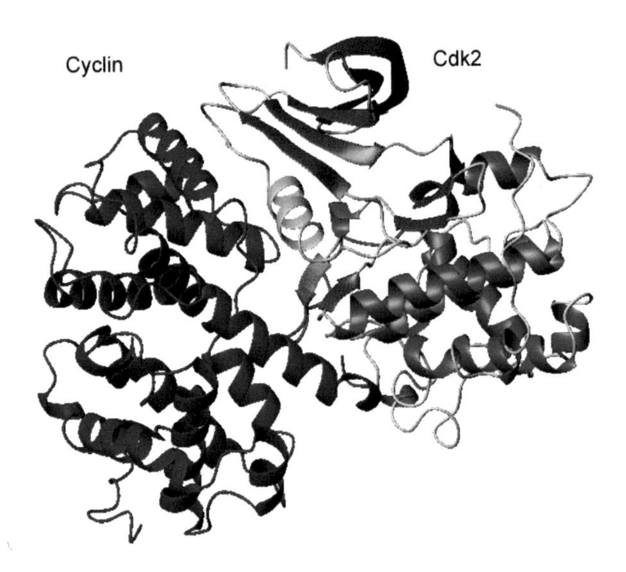

Figure 8.10 Structure of the Cdk2–cyclin A–ATP complex. The structure shows the binding of cyclin A (purple, left) to the Cdk2 protein (green, right). The colour scheme of Figure 8.8 is used with the exception that the β-9 strand of Cdk (formerly the L12 helix) is shown in orange

Figure 8.11 Space-filling representation of the cyclinA (purple) and Cdk2 (green) interaction. Phosphorylated Thr160 (red) of Cdk2 in the T loop together with ATP (magenta) are shown

Cdk–cyclin complex regulation

Cyclin binding offers one level of regulation of Cdk activity but other layers of regulatory control exist within the cell. Phosphorylation of Cdk at Thr160 by CAK is a major route of activation but further phosphorylation sites found on the surface of Cdk lead to enzyme inhibition. In Cdk1 the most important sites are Thr14 and Tyr15. Further complexity arises with the observation of specialized Cdk inhibitors that regulate most kinase activity in the form of two families of inhibitory proteins: the INK4 family bind to free Cdk preventing association with cyclins whilst the CIP family act on Cdk–cyclin complexes (Figure 8.12). The INK4 inhibitors are specific for the G_1 phase Cdks whereas CIP inhibitors show a broader preference. Regulation has become an important issue with the observation that certain tumour lines show altered patterns of Cdk regulation as a result of inhibitor mutation. One inhibitor of CDK4 and CDK6 is p16INK4a, and in about one third of all cancer cases this tumour suppressor is mutated. Another CDK inhibitor, p27Cip2, is present at low levels in cancers that show poor clinical prognosis whilst mutation of Cdk4 in melanoma cells leads to a loss of inhibition by Ink4 proteins.

Cdks are protein kinases transferring the γ-phosphate of ATP onto Ser/Thr side chains in target proteins. This type of phosphorylation is a common regulatory event

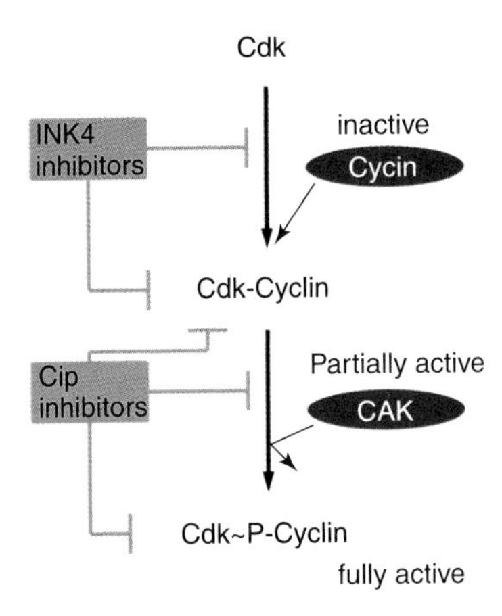

Figure 8.12 The interaction of cyclin–Cdk complexes with INK4 and CIP inhibitors

within cells and underpins signal transduction mediated by cytokines, kinases and growth factors as well as the events within the cell cycle.

In yeast the cell cycle is controlled by one essential kinase known as Cdk1 whilst multicellular eukaryotes have Cdk1 and Cdk2 operating in the M phase and S phase, respectively, along with two additional Cdks (Cdk4 and Cdk6) regulating G_1 entry and exit. Despite the presence of additional Cdks in higher eukaryotes the molecular mechanisms controlling cell cycle events show evolutionary conservation operating similarly in yeasts, insects, plants, and vertebrates, including humans. Research aimed at understanding the cell cycle underpins the development of cancer since the failure to control these processes accurately is a major molecular event during carcinogenesis. It is likely that discoveries in this area will have an enormous impact on molecular medicine. Knowledge of the structure of Cdks together with the structures of Cdk–cyclin–inhibitor complexes will be translated into the design of drugs that modulate Cdk activity and eliminate unwanted proliferative reactions.

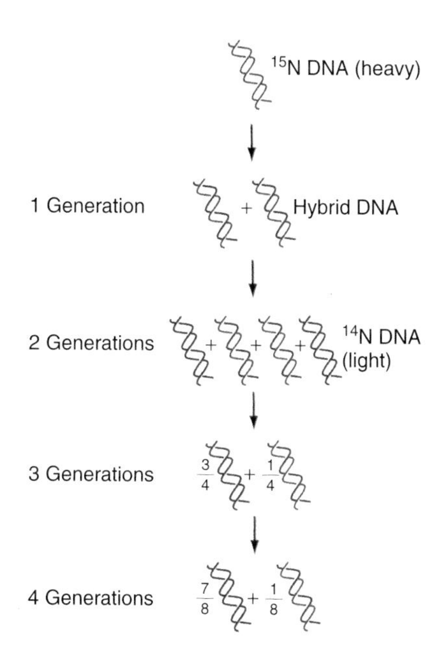

Figure 8.13 Semi-conservative model of DNA replication (reproduced with permission from Voet, D Voet, J.G & Pratt, C.W. *Fundamentals of Biochemistry*. John Wiley & Sons Inc, Chichester, 1999)

DNA replication

A major function of the cell cycle is to integrate DNA synthesis within cell division. The double helix structure of DNA suggested conceptually simple methods of replication; a conservative model where a new double-stranded helix is synthesized whilst the original duplex is preserved, or a semi-conservative scheme in which the duplex unfolds and each strand forms half of a new double-stranded helix.

The semi-conservative' model (Figure 8.13) was demonstrated conclusively by Matthew Meselson and Franklin Stahl in 1958 by growing bacteria (*E. coli*) on media containing the heavy isotope of nitrogen (^{15}N) for several generations to show that the DNA was measurably heavier than that from cells grown on the normal ^{14}N isotope. Small differences in density between these two forms of DNA were demonstrated experimentally by density gradient centrifugation. Isolated DNA from bacteria grown on ^{15}N for many generations gives a single band corresponding to a density of 1.724 g ml^{-1} whilst DNA from bacteria grown on normal (^{14}N) nitrogenous sources yielded a lighter density single band of 1.710 g ml^{-1}. Comparing DNA isolated from bacteria grown on heavy (^{15}N) nitrogenous media and then transferred to ^{14}N-containing media for a single generation led to the observation of a single DNA species of intermediate density (1.717 g ml^{-1}). This was only consistent with the semi-conservative model.

The semi-conservative mechanism suggested that each strand of DNA acted as a template for a new chain. A surprisingly large number of enzymes are involved in these reactions including:

1. *Helicases.* These proteins bind double-stranded DNA catalysing strand separation prior to synthesis of new daughter strands.
2. *Single-stranded binding proteins.* Tetrameric proteins bind DNA stabilizing single-stranded structure and enhancing rates of replication.
3. *Topoisomerases.* This class, sometimes called DNA gyrases, assist unwinding before new DNA synthesis.
4. *Polymerases.* DNA Polymerase I was the first enzyme discovered with polymerase activity. It

is not the primary enzyme involved in bacterial DNA replication, a reaction catalysed by DNA polymerase III. DNA polymerase I has exonuclease activities useful in correcting mistakes or repairing defective DNA. DNA polymerase can be isolated as a fragment exhibiting the 5′–3′ polymerase and the 3′–5′ exonuclease activity but lacking the 5′–3′ exonuclease of the parent molecule. This fragment, known as the Klenow fragment, was widely used in molecular biology prior to the discovery of thermostable polymerases.

5. *Primase*. DNA synthesis proceeds by the formation of short RNA primers that lead to sections of DNA known as Okazaki fragments. The initial priming reaction requires a free 3′ hydroxyl group to allow continued synthesis from these primers. Specific enzymes known as RNA primases catalyse this process.
6. *Ligase*. Nicks or breaks in DNA strands occur continuously during replication and these gaps are joined by the action of DNA ligases.

The synthesis of RNA primers in *E. coli* requires the concerted action of helicases (*DnaB*) and primases (*DnaG*) in the primosome complex (see Table 8.1; $M_r \sim 600\,000$) whilst a second complex, the replisome, contains two DNA polymerase III enzymes engaged in DNA synthesis in a 5′–3′ direction. Figure 8.14 summarizes DNA replication in *E. coli*.

Table 8.1 Primosome proteins

Protein	Organization	Subunit mass
PriA	Monomer	76
PriB	Dimer	11.5
PriC	Monomer	23
DnaT	Trimer	22
DnaB	Hexamer	50
DnaC	Monomer	29
DnaG (primase)	Monomer	60

The complex lacking DnaG is called a pre-primosome. Derived from Kornberg, A. & Baker, T.A. *DNA Replication*, 2nd edn. Freeman, New York, 1992.

Transcription

The first step in the flow of information from DNA to protein is transcription. Transcription forms complementary messenger RNA (mRNA) from DNA through catalysis by RNA polymerase. RNA polymerase copies the DNA coding strand adding the base uracil (U) in place of thymine (T) into mRNA. Most studies of transcription focused on prokaryotes where the absence of a nucleus allows RNA and protein synthesis to occur rapidly, in quick succession and catalysed by a single RNA polymerase. In eukaryotes multiple RNA polymerases exist and transcription is more complex involving larger numbers of accessory proteins.

Structure of RNA polymerase

Most prokaryotic RNA polymerases have five subunits (α, β, β', σ and ω) with two copies of the α subunit found together with single copies of the remaining subunits (Table 8.2). From reconstitution studies the ω subunit is not required for holoenzyme assembly or function. Similarly the σ subunit is easily dissociated to leave a 'core' enzyme retaining catalytic activity. A decrease in DNA binding identified a role for the σ subunit in promoter recognition where it reduces non-specific binding by a factor of $\sim 10^4$.

Table 8.2 Subunit composition of RNA polymerase of *E. coli*; the σ subunit is one member of a group of alternative subunits that have interchangeable roles

Subunit	M_r	Stoichiometry	Function
α	36 500	2	Chain initiation. Interaction with transcription factors and upstream promoter elements
β	151 000	1	Chain initiation and elongation
β'	155 000	1	DNA binding
σ	70 000	1	Promoter recognition
ω	11 000	1	Unknown

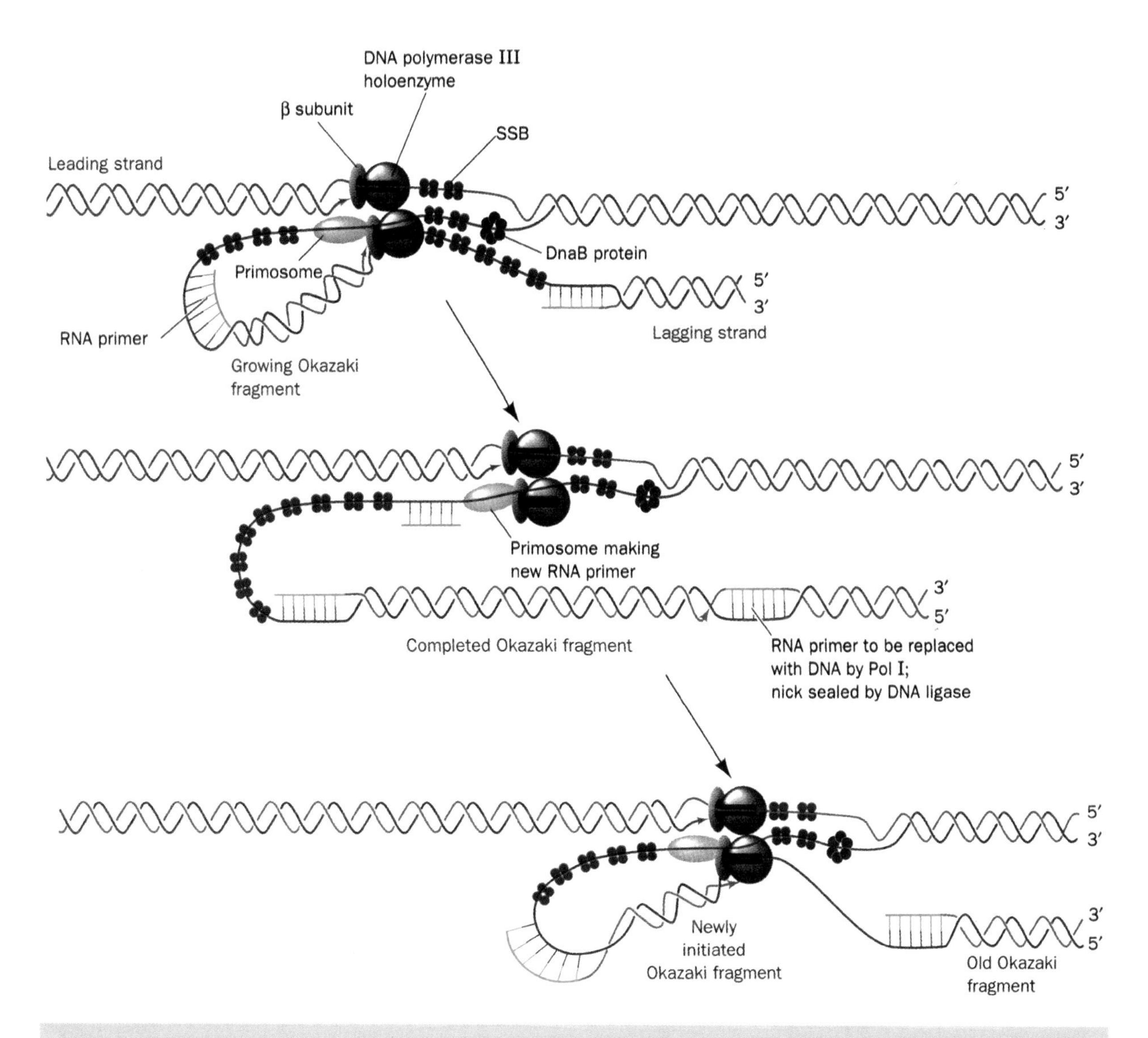

Figure 8.14 The integrated and complex nature of DNA replication in *E. coli*. The replisome contains two DNA polymerase III molecules and synthesizes leading and lagging strands. The lagging strand template must loop around to permit holoenzyme catalysed extension of primed lagging strands. DNA polymerase III releases the lagging strand template when it encounters previously synthesized Okazaki fragments. This signals the initiation of synthesis of a lagging strand RNA fragment. DNA polymerase rebinds the lagging strand extending the primer forming new Okazaki fragments. In this model leading strand synthesis is always ahead of the lagging strand synthesis (reproduced with permission from Voet, D., Voet, J.G & Pratt, C.W. *Fundamentals of Biochemistry*. John Wiley & Sons Inc, Chichester, 1999)

One of the best-characterized polymerases is from bacteriophage T7 (Figure 8.15) and departs from this organization by containing a single polypeptide chain (883 residues) arranged as several domains. It is organized around a cleft that is sufficient to accommodate the template of double-stranded DNA and has been likened to the open right hand divided into palm, thumb and finger regions. The palm, fingers and thumb regions define the substrate (DNA) binding and catalytic sites.

The thumb domain is a helical extension from the palm on one side of the binding cleft and stabilizes the ternary complex formed during transcription by wrapping around template DNA. Mutant T7 RNA polymerases with 'shortened thumbs' have less activity as a result of lower template affinity. The palm domain consisting of residues 412–565 and 785–883 is located at the base of the deep cleft, an integral part of T7 RNA polymerase, and is surrounded by the fingers and the thumb. It contains three β strands, a structurally conserved motif in all polymerases with conserved Asp residues (537 and 812) participating directly in catalysis. The finger domain extends from residues 566–784 and contains a specificity loop involved in direct interactions with specific bases located in the major groove of the promoter DNA sequence. This region also has residues that form part of the active site such as Tyr639 and Lys631. Finally the N-terminal domain forms the front wall of the catalytic cleft and leads to the enzyme's characteristic concave appearance. This domain interacts with upstream regions of promoter DNA and the nascent RNA transcript.

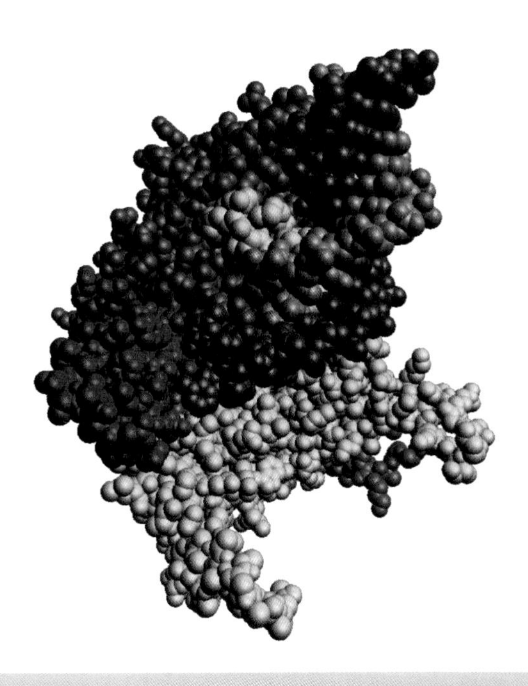

Figure 8.15 The structure of T7 RNA polymerase. A space-filling representation with synthetic DNA (shown in orange and green) has a thumb region represented by residues 325–411 (red), a palm region underneath the DNA and largely obscured by other domains (purple) whilst fingers surround the DNA and are shown in cyan to the right. The N-terminal domain is shown in dark blue (PDB: 1CEZ)

Consensus sequences exist for DNA promoter regions (Figure 8.16) and David Pribnow identified a region rich in the bases A and T approximately −10 bp upstream of the transcriptional start site. A second region approximately 35 bp upstream of the transcriptional start site was subsequently identified with the space between these two sites having an optimal size of 16 or 17 bases in length. Although consensus sequences for the −35 and −10 regions have been identified no natural promoter possesses exactly these bases although all show strong sequence conservation.

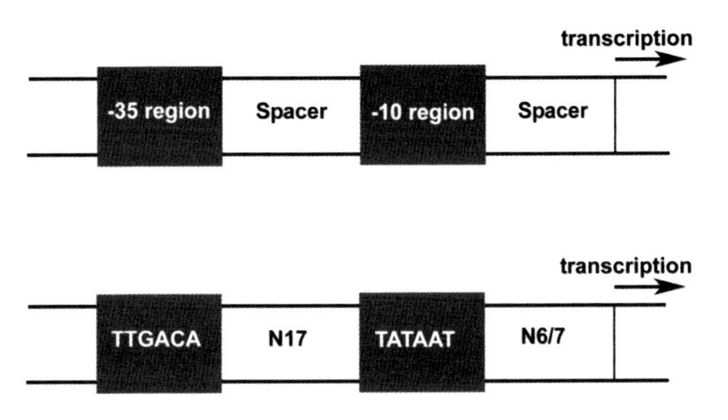

Figure 8.16 The organization of the promoter region of genes in *E. coli* for recognition by RNA polymerase. The consensus sequence TATAAT at a position ~10 bp upstream of the transcription start site is the Pribnow box

Eukaryotic RNA polymerases

In eukaryotes polymerase binding occurs at the TATA box and is more complex. Three nuclear RNA polymerases designated as RNA polymerase I, II and III, together with polymerases located in the mitochondria and chloroplasts are known. The nuclear polymerases are multi-subunit proteins whose activities frequently require the presence of accessory proteins called transcription factors (Table 8.3).

In eukaryotes RNA polymerases I and III transcribe genes coding for ribosomal (rRNA) and transfer RNA (tRNA) precursors. RNA polymerase I produces a large transcript – the 45S pre-rRNA – consisting of tandemly arranged genes that after synthesis in the nucleolus are processed to give 18S, 5.8S and 28S rRNA molecules (Figure 8.17). The 18S rRNA contributes to the structure of the small ribosomal subunit with the 28S and 5.8S rRNAs associating with proteins to form the 60S subunit. RNA polymerase III

Table 8.3 The role of eukaryotic RNA polymerases and their location

RNA Polymerase	Location	Role
I	Nuclear (nucleolus)	Pre rRNA except 5S rRNA
II	Nuclear	Pre mRNA and some small nuclear RNAs
III	Nuclear	Pre tRNA, 5S RNA and other small RNAs
Mitochondrial	Organelle (matrix)	Mitochondrial RNA
Chloroplast	Organelle (stroma)	Chloroplast RNA

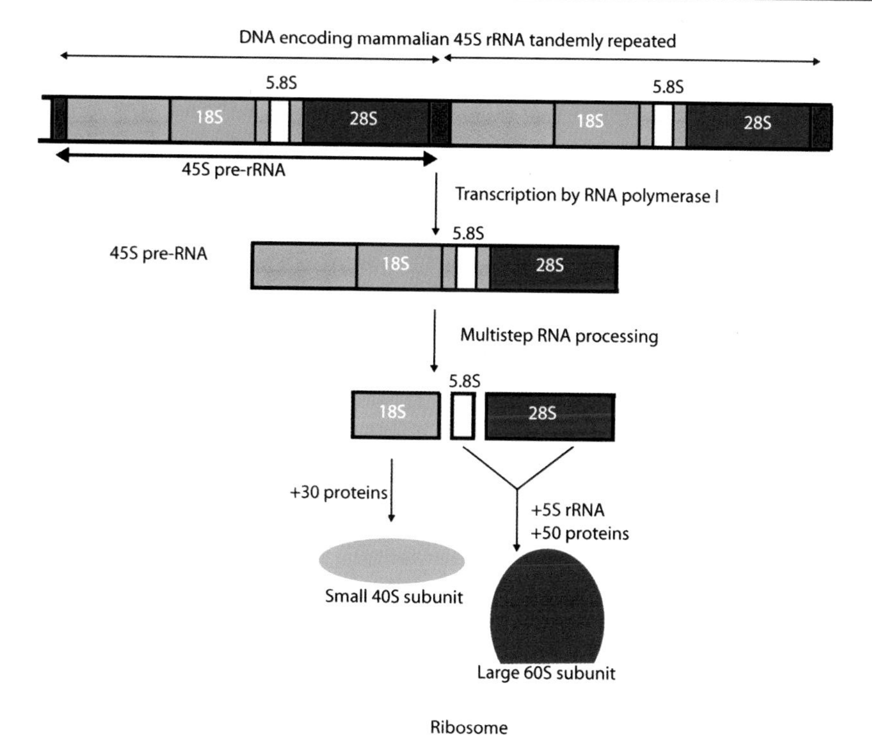

Figure 8.17 The genes for rRNAs exist as tandemly repeated copies separated by non-transcribed regions. The 45S transcript is processed to remove the regions shown in tan to give the three rRNAs

Table 8.4 Basal transcription factors required by eukaryotic RNA polymerase II

Factor	Subunits	Mass (kDa)	Function
TFIID - TBP	1	38	TATA box recognition and TFIIB recruitment.
TAFs	12	15–250	Core promoter recognition
TFIIA	3	12,19,35	Stabilization of TBP binding
TFIIB	1	35	Start site selection.
TFIIE	2	34,57	Recruitment and modulation of TFIIH activity
TFIIF	2	30,74	Promotor targeting of polymerase and destabilization of non specific RNA polymerase-DNA interactions
TFIIH	9	35–90	Promoter melting via helicase activity

Adapted from Roeder, R.G. *Trends Biochem. Sci.* 1996, **21**, 327–335. Subunit composition and mass are those of human cells but homologues are known for rat, *Drosophila* and yeast. TBP, TATA binding protein; TAFs, TATA binding associated factors.

synthesizes the precursor of the 5S rRNA, the tRNAs as well as a variety of other small nuclear and cytosolic RNAs.

RNA polymerase II transcribes DNA into mRNA and is the enzyme responsible for structural gene transcription. However, effective catalysis requires additional proteins with at least six basal factors involved in the formation of a 'pre-initiation' complex with RNA polymerase II (Table 8.4). These transcription factors are called TFIIA, TFIIB, TFIID, TFIIE, TFIIF and TFIIH. The Roman numeral 'II' indicates their involvement in reactions catalysed by RNA polymerase II.

The assembly of the pre-initiation complex begins with TFIID (TBP) identifying the TATA consensus sequence (T-A-T-A-A/T-A-A/T) and is followed by the coordinated accretion of TFIIB, the non-phosphorylated form of RNA polymerase II, TFIIF, TFIIE, and TFIIH (Figure 8.18). Before elongation commences RNA polymerase is phosphorylated and remains in this state until termination when a phosphatase recycles the polymerase back to its initial form. The enzyme can then rebind TATA sequences allowing further transcriptional initiation.

Basal levels of transcription are achieved with TBP, TFIIB, TFIIF, TFIIE, TFIIH, RNA polymerase and the core promoter sequence, and this system has been used to demonstrate minimal requirements for initiation and complex assembly (Figure 8.19). A cycle of efficient re-initiation of transcription is achieved when RNA polymerase II re-enters the pre-initiation complex before TFIID dissociates from the core promoter. Non-basal rates of transcription require proximal and distal enhancer regions of the promoter and proteins that regulate polymerase efficiency (TAFs).

TBP (and TFIID) binding to the TATA box is a slow step but yields a stable protein–DNA complex that has been structurally characterized from several systems. The plant, yeast, and human TBP in complexes with TATA element DNA share similar structures (Figure 8.20) suggesting conserved mechanisms of molecular recognition during transcription. The three-dimensional structure of the conserved portion of TBP is strikingly similar to a saddle; an observation that correlates precisely with function where protein 'sits' on DNA creating a stable binding platform for other transcription factors. DNA binding occurs on the concave underside of the saddle with the upper surface (seat of the saddle) binding transcriptional components.

DNA binding is mediated by the curved, antiparallel β sheet whose distinctive topology provides a large concave surface for interaction with the minor DNA

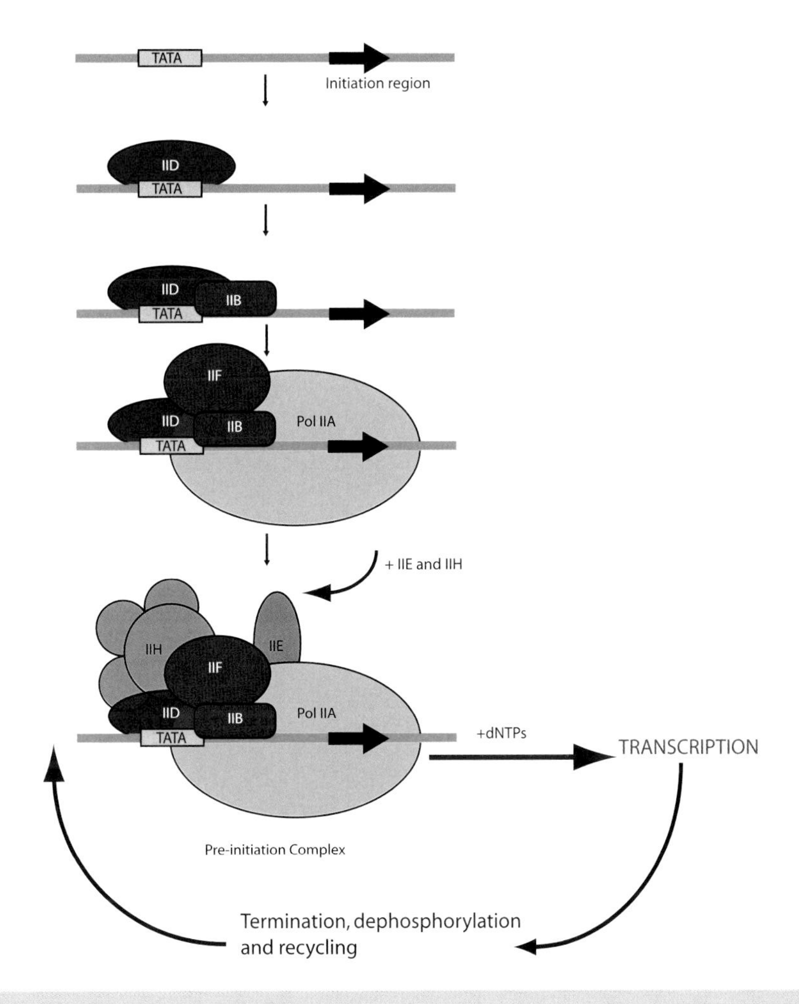

Figure 8.18 The formation of pre-initiation complexes between RNA polymerase II plus TATA sequence DNA

groove of the 8-bp TATA element. The 5′ end of DNA enters the underside of the molecular saddle where TBP causes an abrupt transition from the normal B form configuration to a partially unwound form by the insertion of two Phe residues between the first T:A base step. The widened minor groove adopts a conformation comparable to the underside of the molecular saddle allowing direct interactions between protein side chains and the minor groove edges of the central 6 bp. A second distortion induced by insertion of another two Phe residues into a region between the last two base pairs of the TATA element mediates an equally abrupt return back to the normal B-form DNA.

TFIIB is the next protein to enter the pre-initiation complex and its interaction with TBP was sufficiently strong to allow the structure of TFIIB-TBP-TATA complexes to be determined by Ronald Roeder and Stephen Burley in 1995 (Figure 8.21). A core region based around the C-terminal part of the molecule (cTFIIB) shows homology to cyclin A and binds to an acidic C-terminal 'stirrup' of TBP using a basic cleft on its own surface. The protein also interacts with the DNA sugar-phosphate backbone upstream and downstream of the TATA element. The first domain of cTFIIB forms the downstream surface of the cTFIIB–TBP–DNA ternary complex and in conjunction with the N-terminal domain of TFIIB this

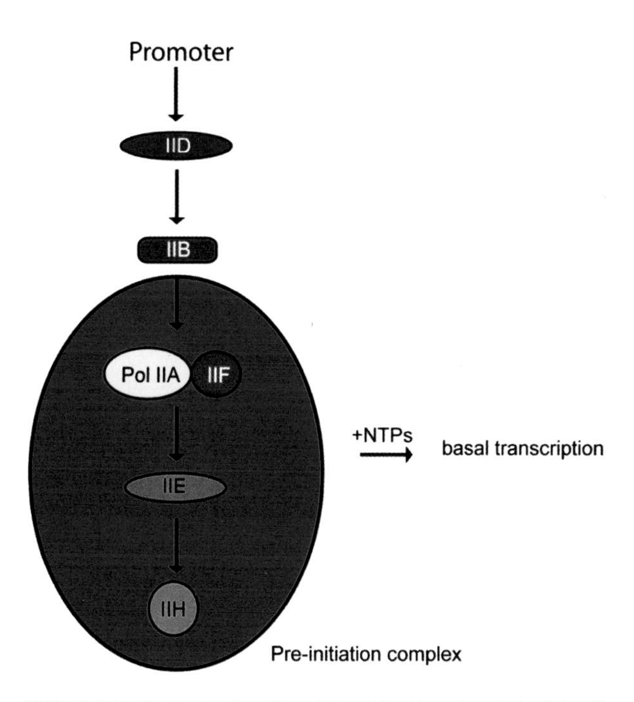

Figure 8.19 Representation of functional interactions modulating basal transcription. The basal factors TBP, TFIIB, TFIIF, TFIIE, and TFIIH together with RNA polymerase II are shown

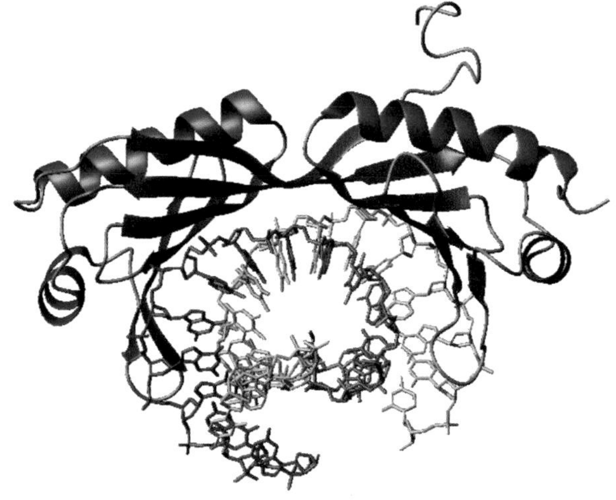

Figure 8.20 The co-complex of TATA element DNA plus TBP showing the remarkable molecular saddle shape. The DNA distorted helix lies on the concave surface of TBP

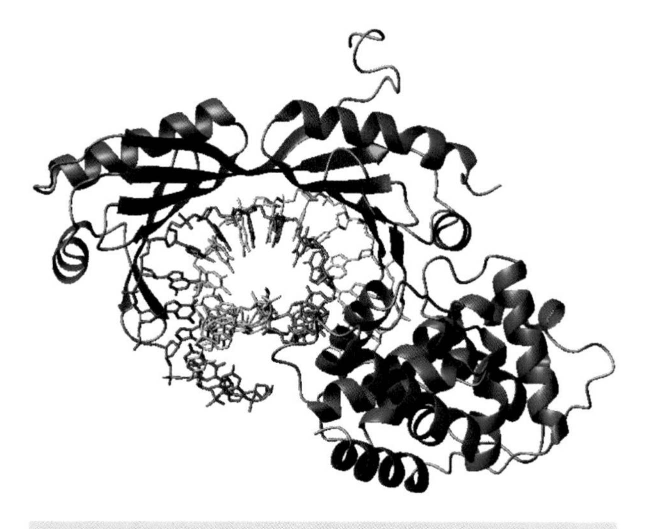

Figure 8.21 The TBP–TFIIB–TATA element complex. TFIIB, shown lower right, is an α helical protein

complex could act between TBP and RNA polymerase II to fix the transcription start site. The rest of the complex reveals solvent accessible surfaces that allow further binding of other transcription factors. The TBP–TATA complex is unaltered in conformation as a result of TFIIB binding and it is likely that a major role of TFIIB is stabilization of the initial protein–DNA complex. After formation of the TFIIB–TFIID–DNA complex three other general transcription factors and RNA polymerase II complete the assembly of the pre-initiation complex.

TFIIA is a co-activator that supports regulation of RNA polymerase II directed transcription. At an early stage in the assembly of the pre-initiation complex TFIIA associates with the TFIID–DNA or the TFIIB–TFIID–DNA complexes recognizing an N-terminal stirrup of TBP and the phosphoribose backbone upstream of the TATA element. This occurs on the opposite face of the double helix to cTFIIB binding and TFIIA preferentially binds to the preformed TBP–DNA complex enhancing its stability.

By comparing structures of the complexes formed by cTFIIB–TBP–DNA with those of TFIIA–TBP–DNA it is possible to create a model of the

TFIIA–TFIIB–TBP–DNA quaternary complex where the mechanism of synergistic action between TFIIB and TFIIA is rationalized by the observation that the two proteins do not interact directly but instead use their basic surfaces to make contact with the phosphoribose backbone on opposite surfaces of the double helix upstream of the TATA element.

Eukaryotic RNA polymerases interact with a number of transcription factors before binding to the promoter and transcription commences. Thus although the fundamental principles of transcription are similar in prokaryotes and eukaryotes the latter process is characterized by the presence of multiple RNA polymerases and additional control processes modulated by a variety of transcription factors.

Structurally characterized transcription factors

The prokaryotic transcription factors cro and λ are examples of proteins containing helix-turn-helix (HTH) motifs that function as DNA binding proteins. The basis for highly specific interactions relies on precise DNA–protein interactions that involve side chains found in the recognition helix of HTH motifs and the bases and sugar-phosphate regions of DNA. A substantial departure in the use of the HTH motif to bind DNA is seen in the *met* repressor of *E. coli.* A homodimer binds to the major groove via a pair of symmetrically related β strands, one from each monomer, that form an antiparallel sheet composed of just two strands. Other repressor proteins such as the arc repressor from a Salmonella bacteriophage (P22) were subsequently shown to use this DNA recognition motif.

Eukaryotic transcription factors: variation on a 'basic' theme

Eukaryotic cells require a greater range of transcription factors, and because genes are often selectively expressed in a tissue-specific manner these processes require an efficient regulatory mechanism. This is achieved in part through a diverse class of proteins. Included in this large and structurally diverse group of proteins are Zn finger DNA binding motifs, leucine zippers, helix loop helix (HLH) motifs (reminiscent of the HTH domains of prokaryotes) and proteins containing a β scaffold for DNA binding. Eukaryotic transcription factors contain distinct domains each with a specific function; an activation domain binds RNA polymerase or other transcription factors whilst a DNA binding domain recognizes specific bases near the transcription start site. There will also be nuclear localization domains that direct the protein towards the nucleus after synthesis in the cytosol. The following sections describe briefly the structural properties of some of the important DNA binding motifs found in eukaryotes.

Zn finger DNA binding motifs

The first Zn finger domain characterized was transcription factor IIIA (TFIIIA) from *Xenopus laevis* oocytes (Figure 8.22). Complexed with 5S RNA was a protein of M_r 40 000 required for transcription *in vitro* that yielded a series of similar sized fragments ($M_r \sim 3000$) containing repetitive primary sequence after limited proteolysis. Sequence analysis confirmed nine repeat units of ~25 residues each containing two cysteine and two histidine residues. Each domain contained a single Zn^{2+} ion.

The proteolytic fragments were folded in the presence of Zn which was ligated in a tetrahedral geometry to the side chains of two invariant Cys and two invariant His residues. In subsequently characterized domains it was found that the Cys_2His_2 ligation pattern varied with sometimes four Cys residues used to bind a single Zn^{2+} ion (Cys_4) or 6 Cys residues used to bind two Zn^{2+} ions. The Zn-binding domains formed small, compact, autonomously folding structures lacking extensive hydrophobic cores, and were christened Zn fingers.

The structures of simple Zn finger domains (Figure 8.23) reveal two Cys ligands located in a short strand or turn region followed by a regular α helix containing two His ligands. From the large number of repeating Zn finger domains and their intrinsic structures it is clear that the ends of the polypeptide chain are widely separated and that complexes with DNA might involve multiple Zn finger domains wrapping around the double helix. This arrangement was confirmed

```
                 *        20         *        40         *        60         *
TFIIIA : MAAKVASTSSEEAEGSLVTEGEMGEKALPVVYKRYICSFADCGAAYNKNWKLQAHLCKHTGEKPFPCKEEGCEKG

            80         *       100         *       120         *       140         *
TFIIIA : FTSLHHLTRHSLTHTGEKNFTCDSDGCDLRFTTKANMKKHFNRFHNIKICVYVCHFENCGKAFKKHNQLKVHQFS

               160         *       180         *       200         *       220
TFIIIA : HTQQLPYECPHEGCDKRFSLPSRLKRHEKVHAGYPCKKDDSCSFVGKTWTLYLKHVAECHQDLAVCDVCNRKFRH

           *       240         *       260         *       280         *       300
TFIIIA : KDYLRDHQKTHEKERTVYLCPRDGCDRSYTTAFNLRSHIQSFHEEQRPFVCEHAGCGKCFAMKKSLERHSVVHDP

                 *       320         *       340         *       360
TFIIIA : EKRKLKEKCPRPKRSLASRLTGYIPPKSKEKNASVSGTEKTDSLVKNKPSGTETNGSLVLDKLTIQ
```

Figure 8.22 The primary sequence of transcription factor TFIIIA from *Xenopus laevi* oocytes. The regions highlighted in yellow are units of 25 residues containing invariant Cys and His residues. Zn binding domains have a characteristic motif $CX_{(2-3)}CX_{(12)}HX_{(3-4)}H$

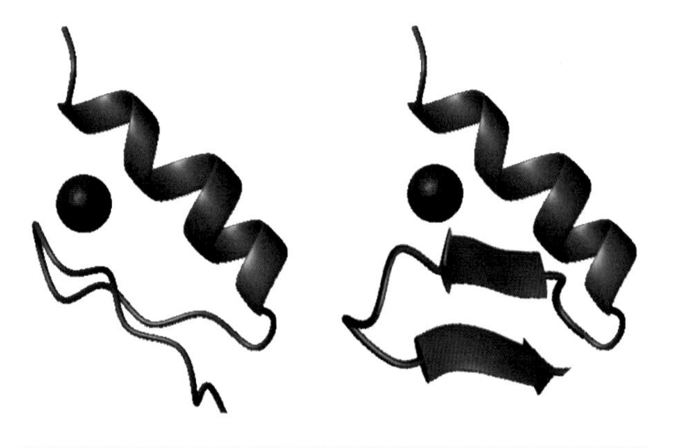

Figure 8.23 A Zn finger domain with helix and either an antiparallel β strand or a loop region

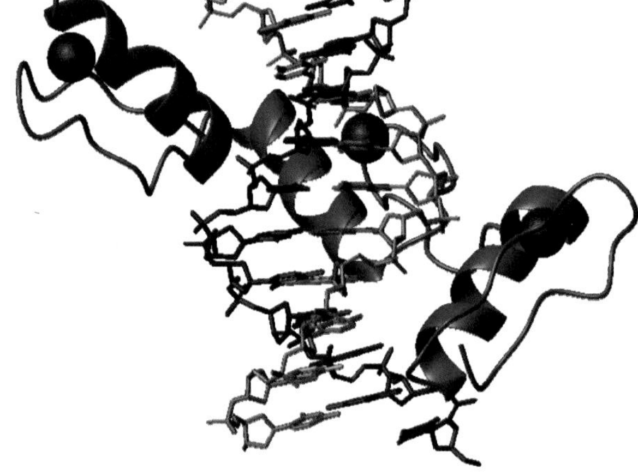

Figure 8.24 The structure of the Zn finger domains of mouse transcription factor Zif268 in a complex with DNA (PDB:1AIH). Three base pairs form hydrogen bonds with the N terminal region of the helix in one strand of the double helix. The periodicity of positively charged side chains allows interactions with phosphate groups

by crystallographic studies of the mouse transcription factor Zif268 where three repeated Zn finger domains complex with DNA by wrapping around the double helix with the β strands of each finger positioning the α helix in the major groove.

Zn finger domains are widely distributed throughout all eukaryotic cells and in the human genome there may be over 500 genes encoding Zn finger domains. Zif268, shown in Figure 8.24, has three such repeats while the homologous *Drosophila* developmental control proteins known as Hunchback and Kruppel have four and five Zn finger domains, respectively. TFIIIA has a comparatively large number of domains (nine) although increasingly proteins are being found with much greater numbers of ‘fingers’. Binuclear Zn clusters exist as Cys_6 Zn fingers. In GAL4, a yeast transcriptional activator of galactose (Figure 8.25), an N terminal

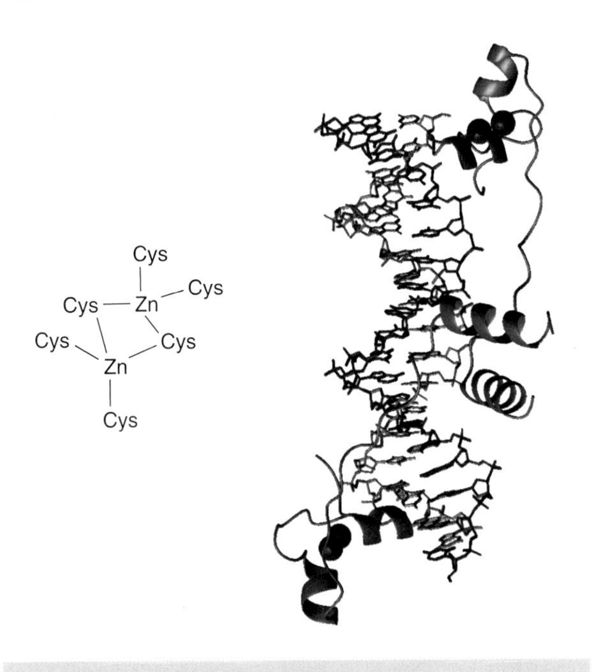

Figure 8.25 The GAL4 DNA binding domain. Tetrahedral geometry formed by six cysteine side chains binds two zinc ions

region containing six Cys residues coordinates two Zn ions. Each Zn cation binds to four Cys side chains in a tetrahedral environment with the central two cysteines ligating both Zn ions.

Leucine zippers

Many transcription factors contain sequences with leucine residues occurring every seventh position. This should strike a note of familiarity because coiled coils such as keratin have similar arrangements called heptad repeats. The heptad repeat is a unit of seven residues represented as $(a\text{-}b\text{-}c\text{-}d\text{-}e\text{-}f\text{-}g)_n$ in which residues *a* and *d* are frequently hydrophobic. The sequences promote the formation of 'coiled coils' by the interaction of residues '*a*' and '*d*' in neighbouring helices. The observation that many DNA binding proteins contained heptad repeats suggested that these regions assist in dimerization – a view supported by the structure of the 'leucine zipper' region of the yeast transcription factor GCN4 (Figure 8.26). The hydrophobic regions do not bind DNA but instead promote the association of subunits containing DNA binding motifs into suitable dimeric structures.

Close inspection of the interaction between helices reveals that the term zipper is inappropriate. The side chains do not interdigitate but show a simpler 'rungs along a ladder' arrangement with stability enhanced by hydrophobic interactions between residues at positions '*a*' and '*d*'. Although misleading in terms of organizational structure the name 'leucine zipper' remains widely used in the literature. Dimerization of the leucine zipper domains allows the N-terminal regions of GCN4, rich in basic residues, to lie in the major groove of DNA. The combination of basic and zipper regions yields a class of proteins with bZIP domains.

Helix-loop-helix motifs

The HLH motif is similar in both name and function to the HTH motif seen in prokaryotic repressors such as cro. The HLH motif is frequently found in transcription factors along with leucine zipper and basic (DNA binding) motifs (see Figure 8.27). As the name implies the structure consists of two α helices often arranged as segments of different lengths – one short and the other long – linked by a flexible loop of between 12 and 28 residues that allows the helices to fold back and pack against each other. The effect of folding is that each helix lies in parallel plane to the other and is amphipathic, presenting a hydrophobic surface of residues on one side with a series of charged residues on the other.

Eukaryotic HTH motifs

The fruit fly *Drosophila melanogaster* is a popular experimental tool for studying gene expression and regulation via changes in developmental patterns. Developmental genes control the pattern of structural gene expression and many *Drosophila* genes share common sequences known as homeodomains or homeoboxes (Figure 8.28). These sequences occur in other animal genomes with an increase in the number of *Hox* genes from one cluster in nematodes, two clusters in *Drosophila*, with the human genome having 39 homeotic genes organized into four clusters. The *Drosophila engrailed* gene encodes a polypeptide that binds to DNA sequences upstream of the transcription start site and imposes its pattern of regulation on other genes as a transcription factor. Crystallization of a 61-residue homeodomain in a complex with DNA showed a HTH motif comparable to the HTH domains of prokaryote repressors such as λ.

```
                       *        20         *        40         *        60
Yeast_GCN4  :  KPNSVVKKSHHVGKDDESRLDHLGVVAYNRKQRSIPLSPIVPESSDPAALKRARNTEAAR  : 60

                       *        80         *       100
Yeast_GCN4  :  RSRARKLQRMKQLEDKVEELLSKNYHLENEVARLKKLVGER  : 101
```

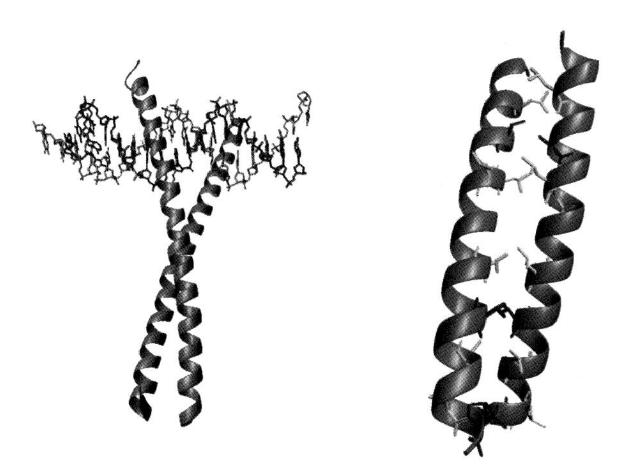

Figure 8.26 A leucine zipper motif such as that found in GCN4 shows a characteristic sequence known as the heptad repeat where every seventh side chain is hydrophobic and is frequently a leucine residue. Interaction between two helices forms a coiled coil and facilitates dimerization of the proteins. The structure of the leucine zipper region of GCN4 complexed with DNA and a detailed view of the interaction between helices (derived from PDB:1YSA and 2TZA)

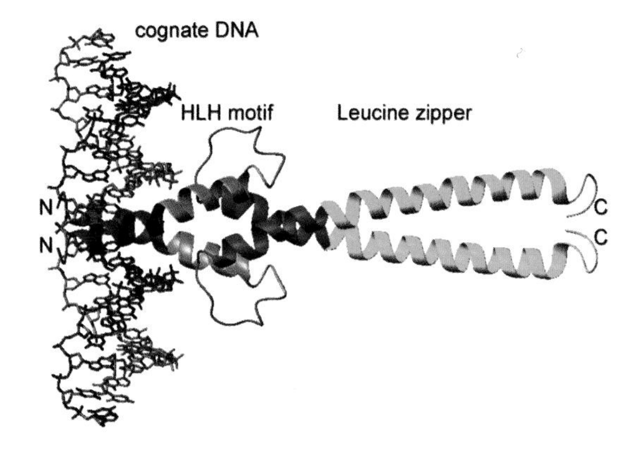

Figure 8.27 The structure determined for the Max–DNA complex. A basic domain (blue) at the N terminal interacts with DNA. The first helix of the HLH motif (orange) is followed by the loop region (red) and second helix (purple). The second helix flows into the leucine zipper (cyan) or dimerization region. This class of transcription factor are often called bHLHZ protein reflecting the three different types of domains namely basic, HLH and zipper (PDB: 1AN2)

Homeodomains are variations of HTH motifs containing three α helices, where the last two constitute a DNA binding domain. Two distinct regions make contact with TAAT sequences in complexes with cognate DNA. An N-terminal arm fits into the minor groove of the double helix with the side chains of Arg3 and Arg5 making contacts near the 5′ end of this 'core consensus' binding site. The second contact site involves an α helix that fits into the major groove with the side chains of Ile47 and Asn51 interacting with the base pairs near the 3′ end of the TAAT site. The 'recognition helix' is part of a structurally conserved HTH unit, but when compared with the structure of the λ repressor the helices are longer and the relationship between the HTH unit and the DNA is significantly different.

It has been estimated that mammals contain over 200 different cell types. These cell lines arise from highly regulated and specific mechanisms of gene expression involving the concerted action of many transcription factors. The human genome project has identified thousands of transcription factors and many carry exotic family names such as Fos and Jun, nuclear factor-κB,

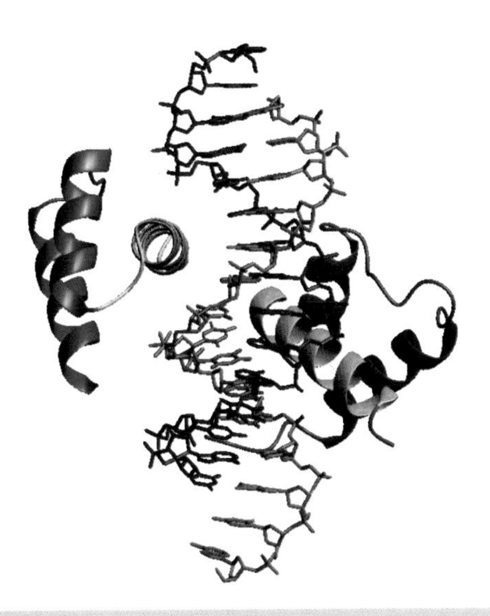

Figure 8.28 A typical homeobox domain showing the HTH motif in contact with its cognate DNA. The structure shown is that of a fragment of the *ubx* or *ultrabithorax* gene product showing the HTH motif. The recognition helix is shown in yellow lying in the major groove (PDB: 1B8I). Similar structures exist in homeodomain proteins such as *antennapedia* (*antp*) or *engrailed* (*eng*). All play a role in the developmental programme of *Drosophila* and each of these proteins acts an a transcriptional activator, binding to upstream regulatory sites and interacting with the RNA polymerase II/TFII complex

Pax and Hox. Growing importance is attached to the study of transcription factors with the recognition that mutations occur in many cancer cells. Wilm's tumour is one of the most common kidney tumours and is prevalent in children as a result of mutation in the WT-1 gene, a gene coding for a transcription factor containing Zn finger motifs. p53 is another transcription factor and is found in a mutated state in over 50 percent of all human cancers. These examples emphasize the link between transcription factors and human disease, where structural characterization has elucidated the basis of DNA–protein interaction.

The spliceosome and its role in transcription

Eukaryotic DNA contains introns or non-coding regions whose function remains enigmatic despite widespread occurrence within chromosomal DNA. Initial mRNA transcripts are processed to remove introns in a series of reactions that occur on the spliceosome. Spliceosomes are ribonucleoprotein complexes with sedimentation coefficients between 50 and 60S and a size and complexity comparable to the ribosome.

Before splicing many eukaryotic RNAs are modified at the 5′ and 3′ ends by capping with GTP and adding polyA tails. GTP is added in a reversed orientation compared to the rest of the polynucleotide chain and is often accompanied by methylation at the N_7 position of the G base. Together with the first two nucleotides the added G base forms a 5′-cap structure (Figure 8.29). Further modifications remove nucleotides from the 3′ end adding 'tails' of 50 to 200 adenine nucleotides. The mechanistic reasons for modification are unclear but involve mRNA stability, mRNA export from the nucleus to the cytoplasm and initiation of translation.

The spliceosome associates with the primary RNA transcript in the nucleus and consists of four RNA–protein complexes called the U1, U2, U4/U6 and U5 or small nuclear ribonucleoproteins (snRNPs, read as 'snurps'). The snRNPs are named after their RNA components so that U1 snRNP contains U1 small nuclear RNA. The U signifies uridine-rich RNAs, a feature found in all small nuclear RNA molecules. The U4 and U6 snRNAs are found extensively base paired and are therefore referred to collectively as the U4/U6 snRNP. The snRNPs pre-assemble in an ordered pathway to form a complex that processes mRNA transcripts.

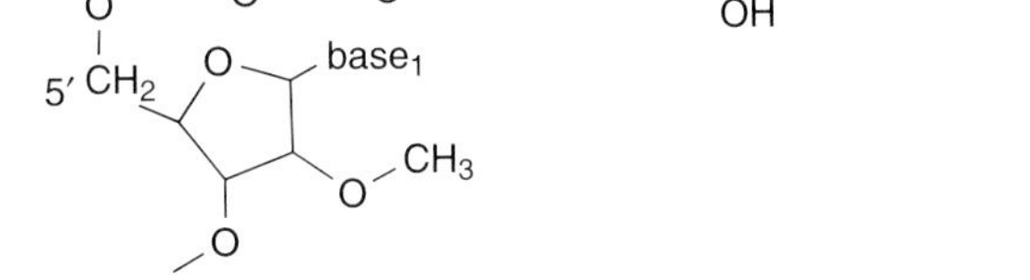

7-methylguanylate

Figure 8.29 Methylation and G-capping of mRNA was discovered with the purification of mRNA

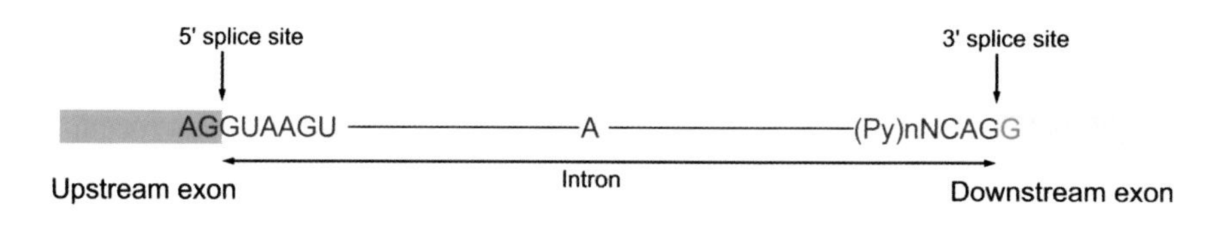

Figure 8.30 Splice sites are specific mRNA sequences and assist intron removal

The first requirement for effective mRNA splicing is to distinguish exons from introns. This is achieved by specific base sequences that signal exon–intron boundaries to the spliceosomes and define precisely the 5′ and 3′ regions to be removed from mRNA. Many introns begin with −5′ GU... and end with −3′ AG and in vertebrates the consensus sequence at the 5′ splice site of introns is AGGUAAGU (Figure 8.30). This sequence base pairs with complementary spliceosomal RNA aligning pre-mRNA and assisting removal of the intron. A second important signal called the branch point is found in the intron sequence as a block of pyrimidine bases together with a special adenine base approximately 50 bases upstream from the 3′ splice site.

Using this sequence information the spliceosome performs complex catalytic reactions to excise the introns in two major reactions. The first step involves the 2′ OH at the branch site and a nucleophilic attack on the phosphate group at the 5′ end of the intron to be removed. For this reaction to occur the mRNA is rearranged to bring two distant sites into close proximity. This movement breaks the pre mRNA at the 5′ end of the intron to leave, at this stage, a branch site with three phosphodiester bonds. The reaction creates a free OH group at the 3′ end with exon 1 no longer attached to the intron although it is retained by the spliceosome. The free OH group at the 3′ end of exon 1 attacks the phosphodiester bond between exon 2 and the intron, with the result that both exons are joined together in the correct reading frame (Figure 8.31). The resulting lasso-like entity is called a lariat structure and is removed from the mRNA (Figure 8.31). The processed mRNA is then exported from the nucleus and into the cytoplasm for translation.

Translation

Conversion of genetic information into protein sequences by ribosomes is called translation and involves the use of a genetic code residing in the order of nucleotide bases. In theory, four nucleotide bases could be used in a variety of different coding systems, but theoretical considerations proposed by George Gamow in the 1950s led to the idea that a triplet of bases is the minimum requirement to code for 20 different amino acids. With the realization that a triplet of bases was involved two experiments critically defined the nature of the genetic code.

The first showed that the genetic code was not 'punctuated' or 'overlapping' but involved a continuous message. Using the bacteriophage T4 Francis Crick and Sydney Brenner showed that insertion or deletion of one or two bases resulted in the formation of nonsense proteins. However, when three bases were added or removed a protein was formed with one residue deleted or added. This demonstrated a code that was an unpunctuated series of triplets arising from a fixed starting point. The second experiment identified codons corresponding to individual amino acids through the use of cell-free extracts, capable of protein synthesis, containing ribosomes, tRNA, amino acids and amino acyl synthetases. When synthetic polynucleotide templates such as polyU were added a polymer of phenylalanine residues was produced. It was deduced that UUU coded for Phe. Similar experiments involving polyC and polyA sequences pointed to the respective codons for proline and lysine. Refinement of this experimental approach allowed the synthesis of polyribonucleotide chains containing repeating sequences that varied according to the frame in which the sequence is read. For example AAGAAGAAGAAGAAG, i.e. $(AAG)_n$ can be read as three different codons (Figure 8.32).

In a variation of this last experiment Philip Leder and Marshall Nirenberg showed that synthetic triplets would bind to ribosomes triggering the binding of specific tRNAs. UUU and UUC lead to Phe-tRNAs

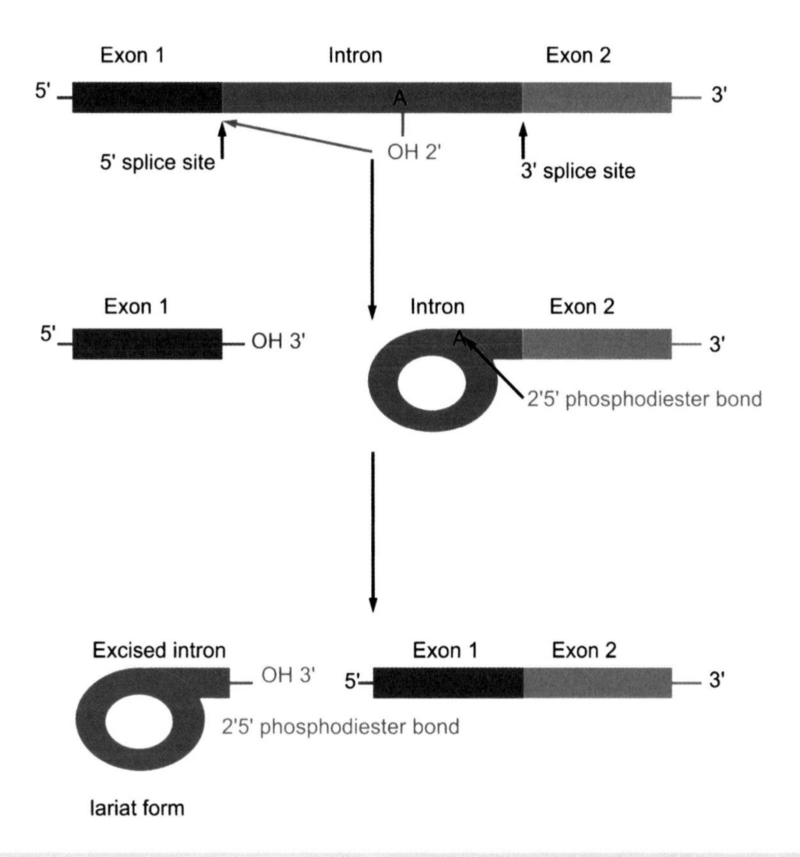

Figure 8.31 Splicing of two exons by intron removal. Exons are drawn in different shades of blue, intron in orange

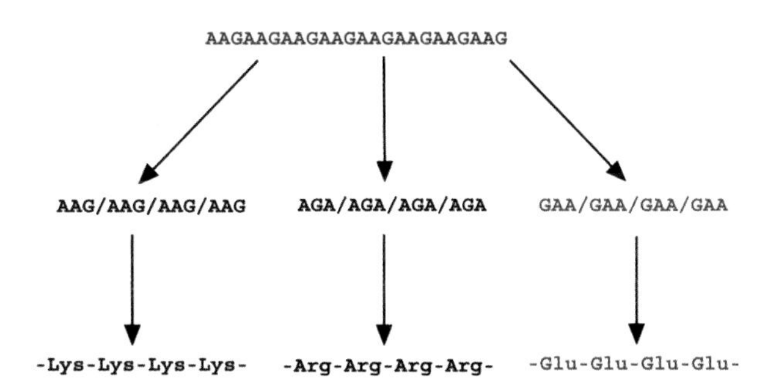

Figure 8.32 Translation of the synthetic polyribonucleotide $(AAG)_n$ leads to the production of three different amino acids in cell free translation systems. The translation of $(AAG)_n$ leads to three possible sequences depending on the 'frame' used to read the message and can yield polylysine, polyarginine or polyglutamate

binding to ribosomes and this approach deciphered the remaining codons of the genetic code confirming degeneracy in the 64 possible triplets.

The genetic code (Figure 8.33) is usually described as universal with the same codons used for the same amino acid in every organism. The ability of bacteria to produce eukaryotic proteins derived from the translation of eukaryotic mRNAs testifies to this universality. However, a few rare exceptions to this universality are documented (Table 8.5) and include alternative use of codons by the mitochondria, organelles with their own transcription and translation systems.

Transfer RNA (tRNA)

In the 1950s Francis Crick postulated that adaptor molecules should contain enzymatically appended amino acids and recognize mRNA codons. The

First position (5' end)	Second base position: U		C		A		G		Thrid position (3' end)
U	UUU	Phe	UCU	Ser	UAU	Tyr	UGU	Cys	U
	UUC		UCC		UAC		UGC		C
	UUA	Leu	UCA		UAA	Stop	UGA	Stop	A
	UUG		UCG		UAG	Stop	UGG	Trp	G
C	CUU	Leu	CCU	Pro	CAU	His	CGU	Arg	U
	CUC		CCC		CAC		CGC		C
	CUA		CCA		CAA	Gln	CGA		A
	CUG		CCG		CAG		CGG		G
A	AUU	Ile	ACU	Thr	AAU	Asn	AGU	Ser	U
	AUC		ACC		AAC		AGC		C
	AUA		ACA		AAA	Lys	AGA	Arg	A
	AUG	Met	ACG		AAG		AGG		G
G	GUU	Val	GCU	Ala	GAU	Asp	GGU	Gly	U
	GUC		GCC		GAC		GGC		C
	GUA		GCA		GAA	Glu	GGA		A
	GUG		GCG		GAG		GGG		G

Figure 8.33 The genetic code. The series of triplets found in mRNA coding for the twenty amino acids are shown with the chain termination signals highlighted in red and the start codon in green

Table 8.5 Rare modifications to the genetic code

Codon	Normal use	Alternative use	Organism
AGA, AGG	Arg	Stop, Ser	Some animal mitochondria; some protozoans
AUA	Ile	Met	Mitochondria
CGG	Arg	Trp	Plant mitochondria
CUX	Leu	Thr	Yeast mitochondria
AUU	Ile	Start codon	Prokaryotic cells
GUG	Val		
UUG	Leu		

'adaptor' molecule was small soluble RNA now called transfer RNA (tRNA). The structure of yeast alanyl-tRNA was reported in 1965 by Robert Holley and contained 76 nucleotides. Base pairing of nucleotides led to a cloverleaf pattern and all tRNAs have lengths between 60 and 95 nucleotides showing comparable structure and function. The cloverleaf structure has distinct regions represented by arms, stems and loops (Figures 8.34 and 8.35).

At the 5′ end the short acceptor stem 7bp in length forms non-Watson–Crick base pairs with bases at the 3′ end of the chain and links via a short stem of 3–4 nucleotides to a region known as the D arm because it contains a modified base, dihydrouridine.[1] A 5bp stem from the D arm leads to the anticodon arm containing a sequence of three bases complementary to the mRNA codon. A variable loop from 3 to 21 nucleotides in length leads to a final arm containing a sequence of bases TψC where ψ represents a modified base called pseudouridine. This region is the T or TψC arm and rejoins an acceptor stem terminated by the sequence -CCA to which amino acids are added by amino acyl tRNA synthetases.

Covalent coupling of amino acids to the tRNA acceptor stem involves activation by reaction with ATP forming an amino adenylate complex, followed by coupling to tRNA in a reaction catalysed by amino acyl tRNA synthetases.

$$\text{Aminoacyl-AMP} + \text{tRNA} \rightleftharpoons \text{aminoacyl-tRNA} + \text{AMP}$$

[1]Non-Watson–Crick base pairs involve, for example, the pairing of G with U.

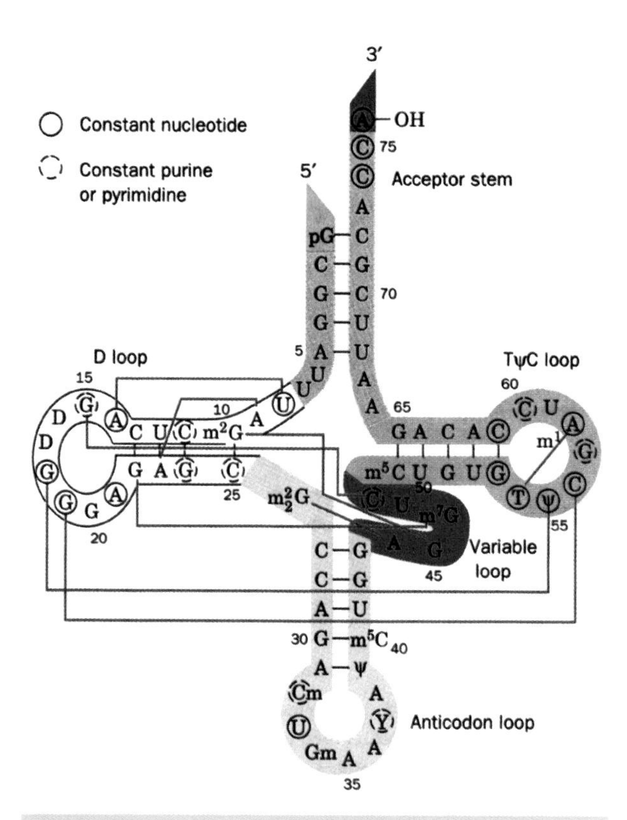

Figure 8.34 The structure of yeast tRNA with the base sequence drawn as a cloverleaf structure highlighting the acceptor stem and four arms; the TψC arm, the variable arm, the anticodon arm and the D arm. Red lines indicate base pairings in the tertiary structure with conserved and semi-conserved nucleotides indicated by solid and dashed circles respectively. The 3′ end is shown in red, the 5′ end in green with the acceptor stem in yellow. The anticodon stem is shown in light green, the D loop is white with the variable loop shown in orange and the TψC arm in cyan

Two classes of synthetases exist: class I enzymes are usually monomeric and attach the carboxyl group of their target amino acid to the 2′ OH of A76 in the tRNA molecule; class II enzymes are dimeric or tetrameric and attach amino acids to the 3′ OH of cognate tRNA (Figure 8.36) with the exception of Phe-tRNA synthetase which uses the 2′ OH. Enzymes within the same class (Table 8.6) show considerable variations in structure and subunit composition.

Table 8.6 Classification of amino acyl tRNA synthetases

Class 1			Class 2		
Amino Acid	Subunit structure and number of residues		Amino Acid	Subunit structure and number of residues	
Arg	α	577	Ala	α_4	875
Cys	α	461	Asn	α_2	467
Gln	α	551	Asp	α_2	590
Gln	α	471	Gly	$\alpha_2\beta_2$	303/689
Ile	α	939	His	α_2	424
Leu	α	860	Lys	α_2	505
Met	α_2	676	Pro	α_2	572
Trp	α_2	325	Phe	$\alpha_2\beta_2$	327/795
Tyr	α_2	424	Ser	α_2	430
Val	α	951	Thr	α_2	642

The composition of prokaryotic and eukaryotic ribosomes

In *E. coli* ribosomes may account for 25 percent of the dry mass and are relatively easily isolated from cells to reveal large structures 25 nm in diameter containing RNA and protein in a 60:40 ratio. The ribosome is always found as two subunits, one with approximately twice the mass of the other.

Prokaryotic ribosomes are described by sedimentation coefficients of 70S with two unequal subunits having individual S values of 50S (large subunit) and 30S (small subunit). Three ribosomal RNA components are identified on the basis of sedimentation coefficients – the 5S, 16S and 23S rRNAs. The small subunit has a single 16S rRNA containing ~1500 nucleotides and the large subunit contains two rRNA molecules – a 23S rRNA of about 2900 nucleotides and a much smaller 5S rRNA 120 nucleotides length. When combined with over 50 different proteins the prokaryotic ribosome has a mass of ~2.5 MDa.

Eukaryotic ribosomes are larger than their prokaryotic counterparts containing more RNA and protein. They have greater sedimentation coefficients (80S vs 70S) with subunits of 60S and 40S. The large subunit

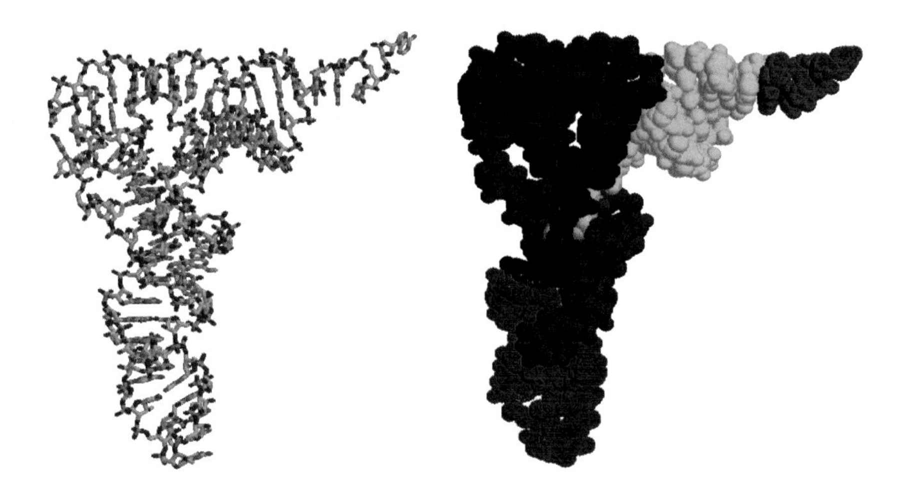

Figure 8.35 The structure of tRNA showing wireframe and spacefilling models for tRNA (PDB:6TNA). The arms are shown in different colours: acceptor stem (yellow), CCA region including A76 (cerise), variable loop (dark green), D arm (blue), the TψC arm (purple) and the anti-codon arm (red)

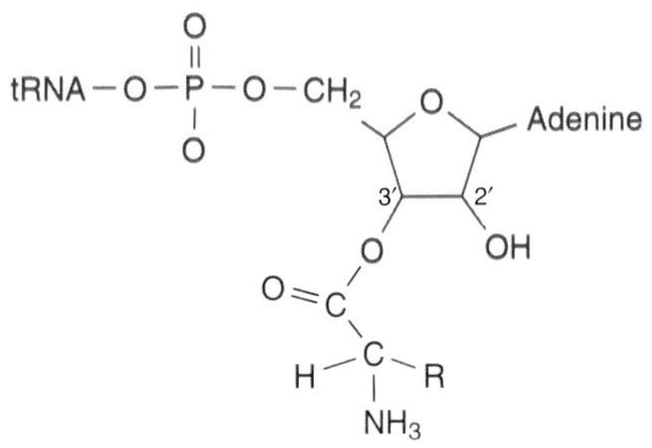

tRNA—O—P(=O)(O)—O—CH$_2$ … Adenine … 3′ … 2′ … OH … O—C(=O)—C(H)(R)(NH$_3$)

Figure 8.36 Esterification of an amino acid to the 3′ OH group of A76 in tRNA

contains 5S, 5.8S and 28S rRNA composed of 120,156 and 4700 nucleotides respectively and about 50 proteins. In contrast the small subunit has a single 18S rRNA about 1900 nucleotides in length and 32 proteins. Despite increased complexity their architecture is fundamentally the same.

By 1984 the sequences of 52 different polypeptide chains were identified in *E. coli* ribosomes with those from the large subunit designated as L1, L2... etc. and those from the small subunit proteins S1...S21 (Figure 8.37). The number of proteins found in ribosomes from different species varies and their homology is much lower than the rRNA components. The *E. coli* large subunit contains 31 different polypeptides each occurring once, with the exception of L7 where four copies are found. Subunit nomenclature refers to migration patterns in two-dimensional electrophoresis and as a result of partial acetylation L7 can have different mobility and was originally thought to be a distinct subunit (L12). The L7/L12 subunits associate with the L10 subunit forming a stable complex that was also mistakenly 'identified' as a separate protein (L8).

The small subunit contains 21 polypeptides, leading to a total of 52 different ribosomal proteins varying in size from small fragments with <100 residues to polypeptide chains containing in excess of 550 residues. Isolated ribosomal proteins were shown to have a high percentage of lysine and arginine residues, a low aromatic content and a topology based around antiparallel β sheets where the strands are connected by regions of α helix. This structure known as the RNA recognition motif (RRM) is seen in L7 and L30 but was also observed in other RNA binding proteins such as spliceosomal U1-snRNP (Figure 8.38).

Recognition motifs are widely found in RNA binding proteins in prokaryotes and eukaryotes suggesting

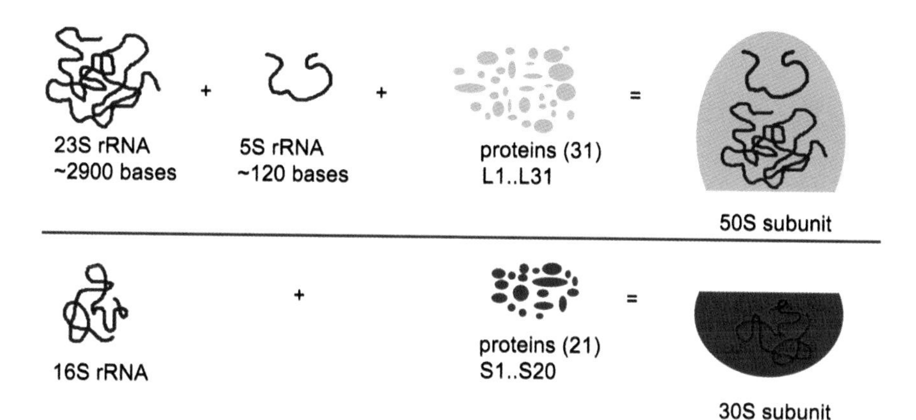

Figure 8.37 Assembly of the prokaryotic ribosome. In prokaryotes the 50 and 30S subunits combine to form a functional 70S ribosome. Association is favoured by increasing concentrations of Mg^{2+} $30S + 50S \rightleftharpoons 70S$. *In vivo* a substantial proportion of the ribosomes are dissociated and this is important during initiation of protein synthesis

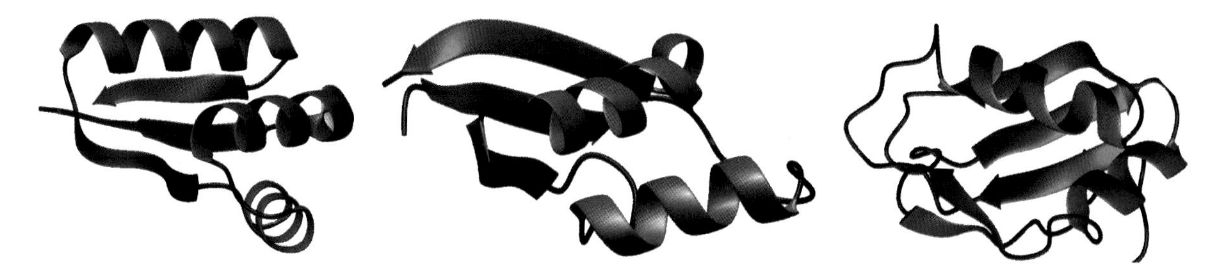

Figure 8.38 The structure of two ribosomal proteins and the U1-snRNP protein showing the homology based around three β strands and two α helices. The structures show the C terminal domain of *E. coli* ribosomal subunit L7/L12 (PDB: 1CTF), the L30 domain of the large ribosomal subunit of *T. thermophilus* (PDB:1BXY) and the N-terminal domain of the spliceosome RNA binding protein U1A (PDB:1FHT)

evolution from an ancestral RNA binding protein. All RRMs consist of a domain of 80–90 residues where the four-stranded antiparallel β sheet and helices occur in the order β-α-β-β-α-β. RNA binds on the flat surface presented by the β sheet a region carrying many conserved Arg/Lys residues on its edge to facilitate interactions with the RNA sugar-phosphate backbone. Groups of non-polar, often aromatic, residues exposed in the sheet region interact directly with purine and pyrimidine bases and in strand 1 a conserved sequence of K/R-G-F/Y-G/A-F-V-x-F/Y is found with alternate residues exposed (blue). Similarly in strand 3 a consensus hexameric sequence L/I-F/Y-V/I-G/K-N/G-L/M is observed.

Low-resolution studies of the ribosome

Despite differences in RNA and protein composition the architecture of all ribosomes is similar with a shape revealed by electron microscopy to contain lobes, ridges, protuberances and even the suggestion of channels as 'anatomical' features (Figure 8.39). The ribosome contains single copies of each protein and allowed antibodies to be raised against exposed epitopes. In combination with electron microscopy the distribution of proteins through the large and small subunits was mapped to enhance the picture of the ribosome (Figure 8.40).

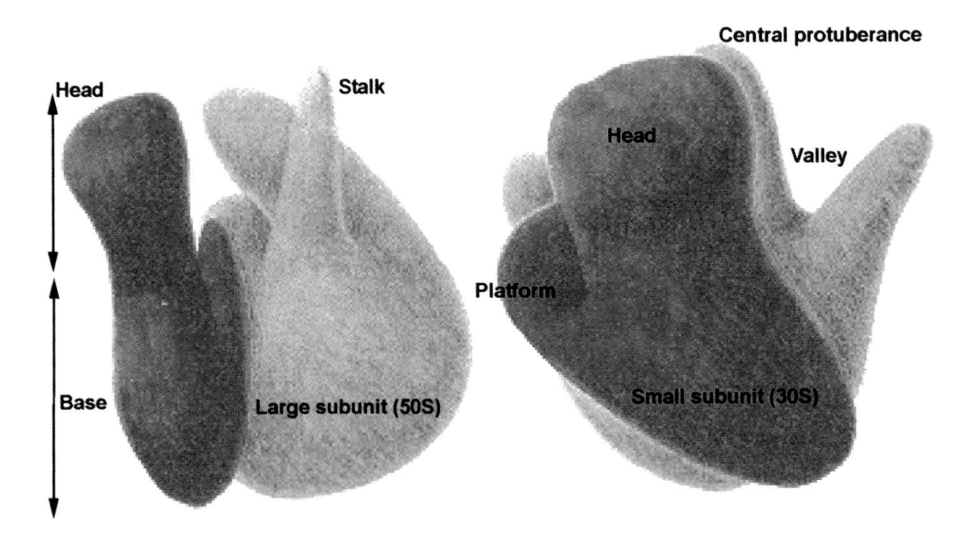

Figure 8.39 A three-dimensional model of the *E. coli* ribosome derived from electron microscopy. (Reproduced with permission from Voet, D., Voet, J.G & Pratt, C.W. *Fundamentals of Biochemistry*. John Wiley & Sons Inc, Chichester, 1999)

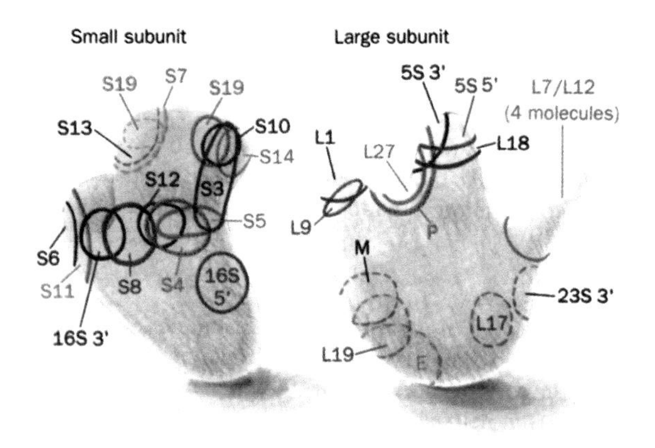

Figure 8.40 A positional map of surface epitopes for large and small *E. coli* ribosomal subunits. The small subunit is shown on the left and dashed lines indicate the positions for subunits lying of the reverse surface. The symbols 16S 3′ and 16S 5′ mark the two ends of the 16S rRNA molecule. In the large subunit P indicates the peptidyl transferase site; E marks the site of emergence for the nascent polypeptide from the 50S subunit and M specifies the ribosome's membrane anchor site. The 3′ and 5′ sites are indicated for 5S rRNA. (Reproduced with permission from Voet, D., Voet, J.G., and Pratt, C.W. *Fundamentals of Biochemistry*. John Wiley & Sons, Chicheter, 1999)

However, delineation of the catalytic sites of protein synthesis required higher levels of resolution – a major experimental problem – for an assembly with over 50 different proteins, several RNA molecules and a mass in excess of 2.5 MDa. Ribosome crystals were produced in the 1960s but yielded two-dimensional arrays of poor diffraction quality. Improved crystal quality stemmed from use of ribosomes derived from thermophiles (*Thermus thermophilus* and *Haloarcula marismortui*) where protein and nucleic acid stability was enhanced when compared with mesophiles and from advances in technology that introduced synchrotron radiation, high efficiency area detectors and crystal freezing techniques that limited radiation damage.

A structural basis for protein synthesis

The first structures for the ribosome were published in 2000 by Tom Steitz and Peter Moore and represented the culmination of decades of intensive effort. The structure of the large subunit was followed by structures for the small (30S) subunits from both *E. coli* and

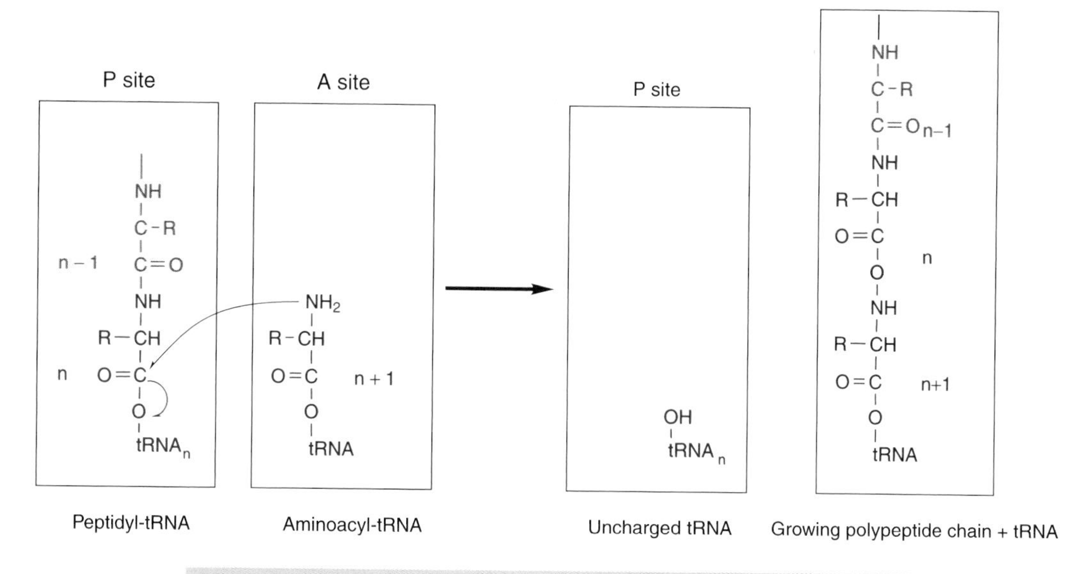

Figure 8.41 The peptidyl transferase reaction at the A and P sites

T. thermophilus. However, a broad understanding of the functional properties of prokaryotic ribosomes arose from many years of biochemical studies long before structural data was available. These studies established the self-assembly of ribosomes, inactivation of protein synthesis by antibiotics, the involvement of accessory proteins in initiation, elongation and termination, a role for GTP in protein synthesis, the nature of tRNA binding via codon/anticodon recognition and the presence of discrete binding sites within the ribosome.

An outline of protein synthesis

Translation of mRNA consists of three stages – initiation, elongation and termination – with the most important process being elongation where nascent polypeptides are extended by the addition of amino acids. Protein synthesis results in the addition of amino acid residues to a growing polypeptide chain in a direction that proceeds from the N to C terminals as mRNA is read in a 5′–3′ direction. Synthesis proceeds at surprisingly high rates for a complex reaction with up to 20 residues per second added to a growing chain.

During elongation the peptidyl transferase reaction catalyses movement and covalent linkage of an amino acid onto a growing polypeptide chain in a process involving two specific sites on the ribosome. The A site is the location for amino acyl tRNA binding to the ribosome and is proximal to the peptidyl (P) site. Displacement of extending polypeptide chains from the ribosome by high salt treatment leaves a peptidyl-tRNA species as the final 'residue'. This results from transfer of the peptidyl tRNA to the incoming amino acyl tRNA to form a peptidyl tRNA with one additional residue located in the A site. The next stage is translocation to create P site peptidyl tRNA, a vacant A site with uncharged tRNA transferred to the E or exit site (Figure 8.41).

Chain initiation

Methionine, a relatively uncommon residue, is frequently found as the first residue of a polypeptide chain modified by the addition of formic acid. In most cases the *N*-formylmethionine group is removed (Figure 8.42), but occasionally the reaction does proceed efficiently and methionine appears to be the first residue.

The initiation site is not solely defined by the AUG or start codon because a mRNA sequence will normally contain other AUG triplets for internal methionine residues. Additional structural elements such as the Shine–Dalgarno sequence located ~10 nucleotides upstream of the start codon are complementary to a

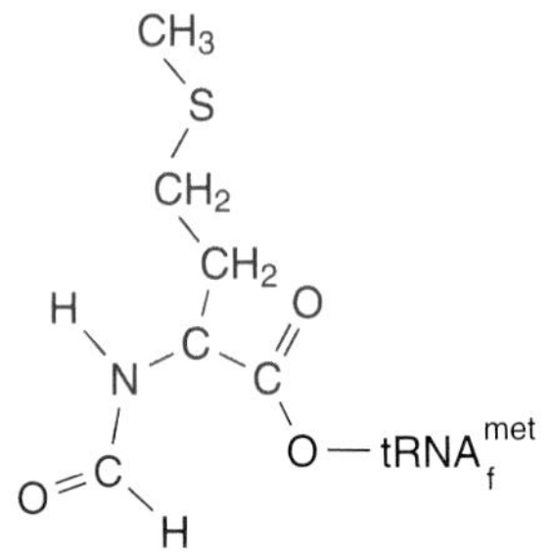

Figure 8.42 *N*-formylmethionine complexed to $tRNA_f^{met}$

pyrimidine-rich sequence found at the 3′ end of 16S rRNA of the 30S subunit, and facilitates ribosomal selection of the correct AUG codon (Figure 8.43).

Protein synthesis requires additional proteins known as initiation factors (IF) that exhibit dynamic and transient association with the ribosome (Table 8.7). Initiation starts after completion of polypeptide synthesis with the ribosome as an inactive 70S complex. IF-3 binding to the 30S subunit promotes complex dissociation and is assisted by IF-1 (Figure 8.44). In a dissociated state mRNA, IF-2, GTP and fMet-tRNA bind to the small subunit. IF-2 is a G protein and is obligatory for binding fMet-tRNA to the 30S subunit in a reaction that is unique in not requiring mRNA and interactions between codon and anticodon. In this state mRNA binds to the small subunit completing priming reactions and the complex is capable of binding the 50S subunit. The association of the 50S subunit results in conformational change, GTP hydrolysis by IF-2 and the release of all initiation factors. The result is a ribosome primed with fMet-tRNA (P site) and a vacant A site poised to accept the next amino acyl tRNA specified by the second codon in an event that marks the start of chain elongation.

Chain elongation

Elongation involves the addition of a new residue to the terminal carboxyl group of an existing polypeptide chain. Elongation is divided into three key events: aminoacyl tRNA binding, the peptidyl transferase reaction, and translocation. Elongation factors assist in many stages of these reactions (Figure 8.45).

Charged amino acyl tRNA is escorted to the vacant A site by an elongation factor -EF-Tu. EF-Tu is yet another G protein involved in protein synthesis and after depositing amino acyl tRNA at the A site hydrolysis of GTP releases the EF-Tu–GDP complex

	Sequence (Initiation codon highlighted)
araB	- U U U G G A U G G A G U G A A A C G A U G G C G A U U -
galE	- A G C C U A A U G G A G C G A A U U A U G A G A G U U -
lacI	- C A A U U C A G G G U G G U G A U U G U G A A A C C A -
lacZ	- U U C A C A C A G G A A A C A G C U A U G A C C A U G -
Q β phage replicase	- U A A C U A A G G A U G A A A U G C A U G U C U A A G -
ϕX174 phage A protein	- A A U C U U G G A G G C U U U U U U A U G G U U C G U -
R17 phage coat protein	- U C A A C C G G G G U U U G A A G C A U G G C U U C U -
Ribosomal S12	- A A A A C C A G G A G C U A U U U A A U G G C A A C A -
Ribosomal L10	- C U A C C A G G A G C A A A G C U A A U G G C U U U A -
trpE	- C A A A A U U A G A G A A U A A C A A U G C A A A C A -
trp leader	- G U A A A A A G G G U A U C G A C A A U G A A A G C A -
3′ end of 16S rRNA	3′ HO A U U C C U C C A C U A G - 5′

Figure 8.43 Translation initiation sequences aligned with the start codon recognized by *E. coli* ribosomes. The RNA sequences are aligned at the start (AUG) codon highlighted in blue. The Shine–Dalgarno sequence is highlighted in red and is complementary to a region at the 3′ end of the 16S rRNA. This sequence is shown in green and involves G-U pairing. (Reproduced with permission from Voet *et al.*, John Wiley & Sons, Ltd, Chichester, 1998)

Table 8.7 Soluble protein factors involved in *E. coli* protein synthesis

Factor	Mass (kDa)	Role
Initiation		
IF-1	9	Assist in IF-3 binding
IF-2	97	Binds initiator tRNA and GTP
IF-3	22	Dissociates 30S subunit from inactive ribosome and aids mRNA binding
Elongation		
EF-Tu	43	Binds amino acyl tRNA and GTP
EF-Ts	74	Displaces GTP from EF-Tu
EF-G	77	Promotes translocation by binding GTP to ribosome. Molecular mimic of EF-Tu + tRNA.
Termination		
RF-1	36	Recognizes UAA and UAG stop codons
RF-2	38	Recognizes UAA and UGA stop codons
RF-3	46	G protein (binds GTP) and enhances RF-1 and RF-2 binding

from the ribosome. At this stage the amino acyl tRNA is checked in a process known as 'proof-reading' and if incorrect a new tRNA is reloaded to the A site. The activity of EF-Tu is aided by EF-Ts – a protein that binds to EF-Tu as a binary complex to facilitate further rounds of GTP binding and further cycles of elongation (Figure 8.46).

The structure of Ef-Tu determined by crystallography in a complex with GDP and a slowly hydrolysing analogue of GTP known as guanosine-5′-(β, γ-imido)-triphosphate (GDPNP) reveals three domains connected by short flexible peptide linkers. The 210 residue N-terminal domain with a GTP/GDP binding site undergoes structural reorganization when GTP is hydrolysed, changing orientation with respect to the remaining two domains by ~90°. The Phe-tRNA–EF-Tu/GDPNP complex is stable and shows two macromolecules arranged into the shape of a 'corkscrew' (Figure 8.47). The 'handle' consists of EF-Tu and the acceptor region (-CCA) of Phe-tRNA. The remaining part of the Phe-tRNA molecule forms the 'screw' and by comparing individual tRNA and EF-Tu structures with the complex it is clear that structural perturbations are small. In the complex formed with tRNA most interactions involve the amino acyl region of the tRNA – the CCA arm – and side chains of EF-Tu. These interactions also explain the observation of poor binding between uncharged tRNA molecules and EF-Tu.

The peptidyl transferase reaction is the second phase of elongation and involves nucleophilic reactions between uncharged amino groups of A site amino acyl-tRNA and the carbonyl carbon of the final residue of P site tRNA arranged in close proximity in the 50S subunit. Evidence points towards a site and mechanism of protein synthesis performed entirely by RNA. Ribosomes consist predominantly of highly conserved rRNA (Figure 8.48),[1] the proteins of large and small subunits show little conservation of sequence, almost all of the proteins of the large subunit of the ribosome of *T. aquaticus* could be removed without loss of activity, and mutations conferring antibiotic resistance were localized to ribosomal RNA genes.

In the final phase of elongation (translocation), uncharged tRNA is moved to the E site whilst recently extended tRNA in the A site is moved to the P site. The A site becomes vacant and will accept the next charged tRNA. Ribosomes move along the mRNA chain in the 3′ direction so that a new codon occupies the A site. This reaction involves an additional G protein, EF-G, whose structure reveals a remarkable similarity to the amino acyl-tRNA/EF-Tu/GTP complex despite an absence of sequence homology and RNA. This is not

[1]The sequence of rRNA is so highly conserved that it is used in taxonomic studies to identify new species or to highlight evolutionary lineage.

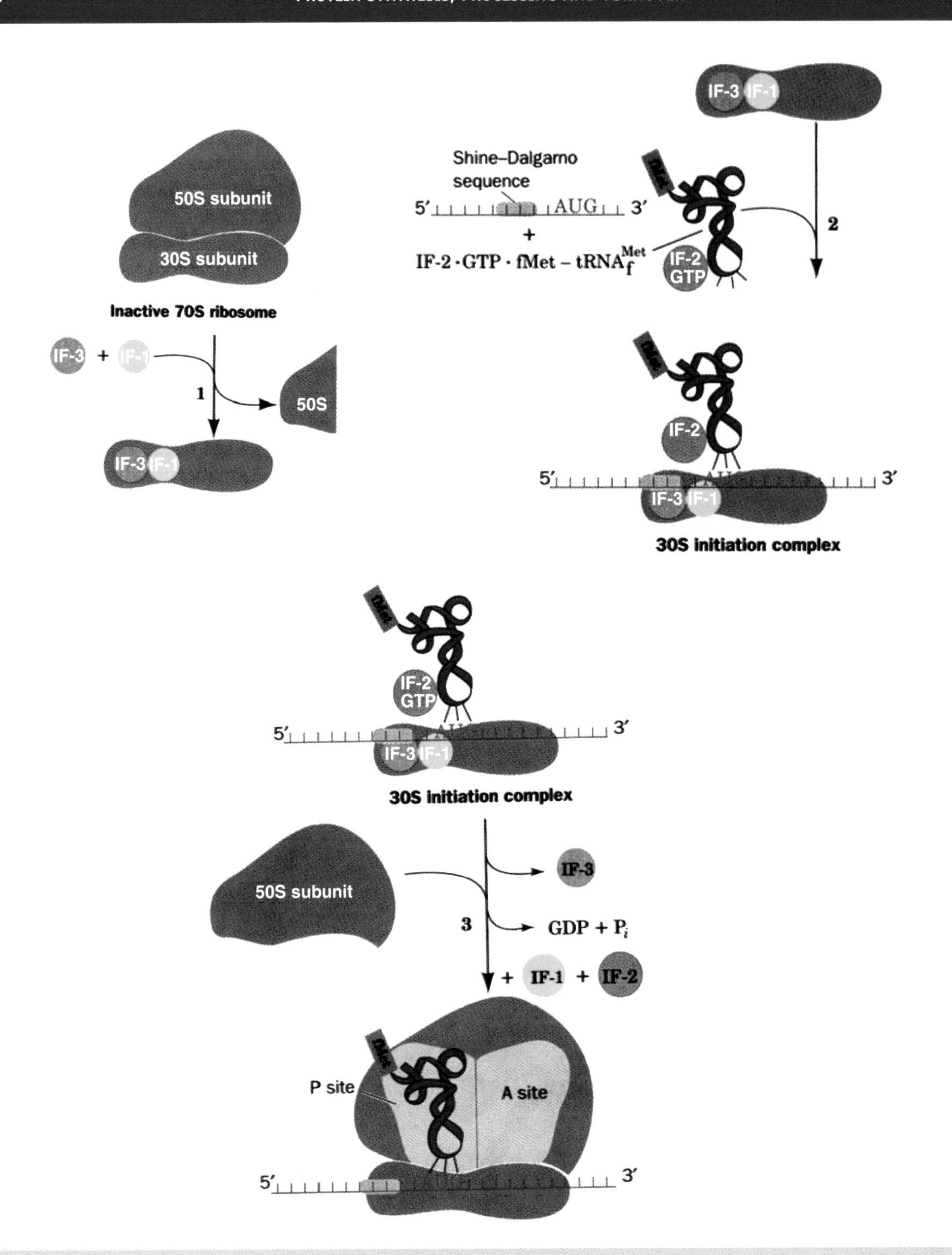

Figure 8.44 The initiation pathway of translation in *E. coli* ribosomes (reproduced with permission from Voet, D., Voet, J.G & Pratt, C.W. *Fundamentals of Biochemistry*. John Wiley & Sons, Ltd, Chichester, 1999)

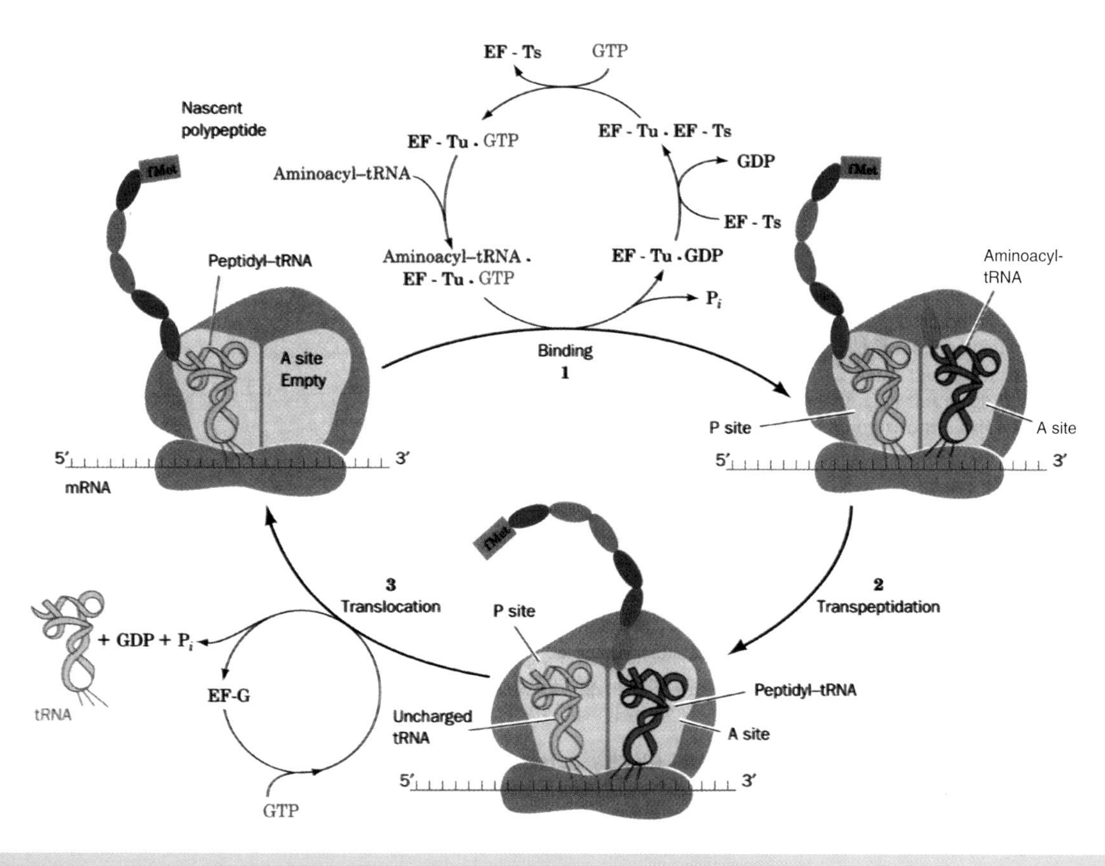

Figure 8.45 The elongation phase of translation. Only the A and P sites are shown in each stage of the elongation cycle (reproduced with permission from Voet, D., Voet, J.G & Pratt, C.W. *Fundamentals of Biochemistry*. John Wiley & Sons, Ltd, Chichester, 1999)

a coincidence but a strategy that involves competition between EF-G and the EF-Tu complex for the A site.

EF-G has five domains (Figure 8.49); domains 1 and 2 resemble the EF-Tu complex and the remaining three domains fold in a similar arrangement to the anticodon stem of tRNA. These features assist in promoting conformational changes in the ribosome by displacing peptidyl-tRNA from the A site and switching the A site to a low affinity for amino acyl tRNAs. It is an example of molecular mimicry and the 'switch' completes translocation.

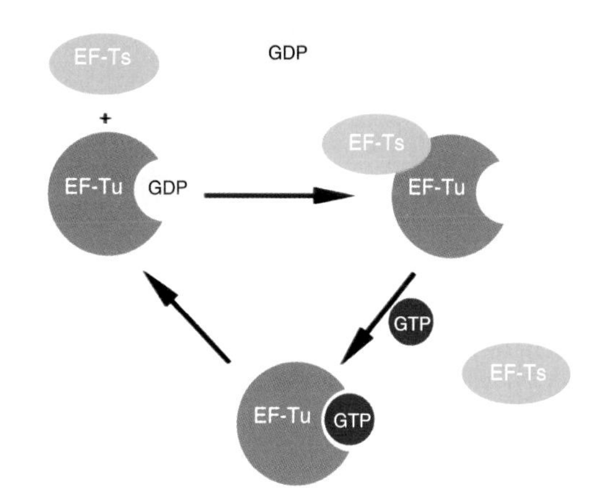

Figure 8.46 Regeneration of EF-Tu-GTP through the action of EF-Ts binding

Termination

Termination marks the end of protein synthesis and is mediated by three codons, UAA, UAG and UGA.

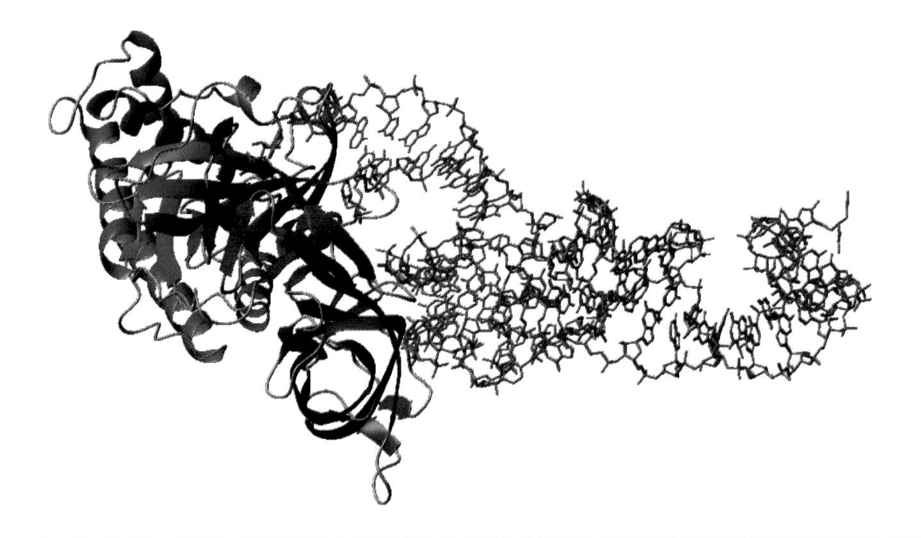

Figure 8.47 The structure of EF-Tu in a complex with Phe-tRNA (PDB: 1TTT). Domain 1 is shown in blue, domain 2 in red and domain 3 in green. The CCA arm of the Phe-tRNA is bound between the blue and red subunits

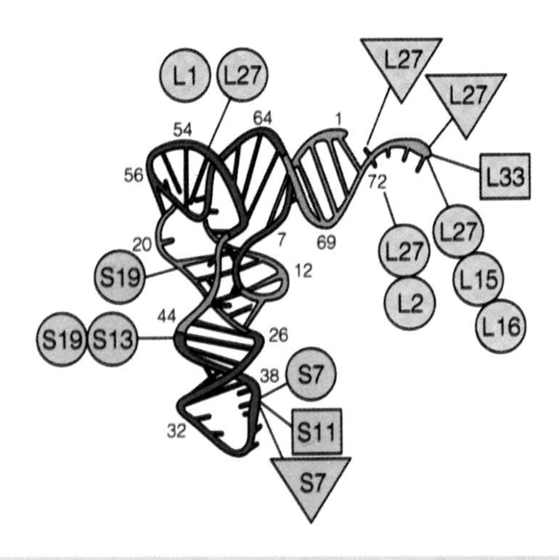

Figure 8.48 tRNA environment within the ribosome defined by cross-linking experiments. The cross links were from defined nucleotide positions within a tRNA to ribosomal proteins in the large and small subunits (denoted by prefix L and S). The A site was defined by triangular symbols, the P site by circles and the E site by squares. (Reproduced and adapted from Wower, J. *Biochimie* 1994, **76**, 1235–1246)

There are no tRNAs for termination codons but protein release factors RF-1, RF-2 and RF-3 identify the codons. RF-1 recognizes UAA and UAG whereas RF-2 identifies UAA and UGA. The third protein RF-3 is a G protein and in the presence of GTP stimulates RF-1 and RF-2 binding to the large subunit. With RF-3 and either RF-1 and RF-2 bound to the A site the peptidyl transferase reaction adds water to the end of the growing chain releasing the peptide into the cytosol along with an uncharged tRNA.

Antibiotics provide insight into protein synthesis

Some antibiotics block protein synthesis and are used to probe initiation, elongation and termination. Their effectiveness in halting prokaryotic protein synthesis coupled with their relative ineffectiveness in eukaryotic systems has allowed their use in medicines. Amongst the antibiotics used to interfere with protein synthesis in microbes are tetracycline, erythromycin, puromycin, streptomycin and chloramphenicol. These reagents block different parts of protein synthesis and from analysis of their chemical structures their effects on ribosome function can be rationalized. Chloramphenicol (Figure 8.50), for example, blocks the peptidyl transferase reaction by acting as a competitive inhibitor in which the secondary

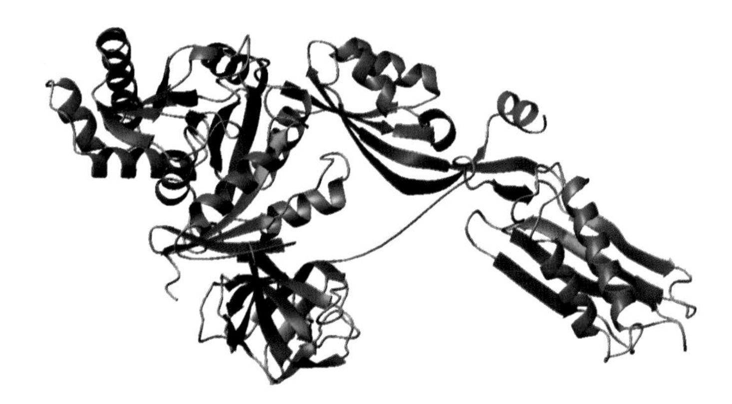

Figure 8.49 The five domains of EF-G. Domains 3–5 mimic the conformation of tRNA complexed to EF-Tu (PDB: 2EFG)

amide resembles the normal peptide bond. Similarly puromycin (Figure 8.50) contains within its structure a portion resembling the 3′ end of the amino acyl tRNA. As a result it enters the A site and is transferred to extending peptide chains causing premature release from the ribosome.

Affinity labelling and RNA 'footprinting'

Chemical probes such as dimethyl sulfate or carbodiimides attack accessible bases of rRNA but in the presence of ligands such as antibiotics, mRNA or tRNA some regions will remain unmodified and leave a 'footprint' when analyzed by electrophoresis. In 23S RNA A2451and A2439 were protected by the acyl region of tRNA at the A and P sites whilst the 3′ terminal protected G2252 and G2253. These bases were universally conserved (or very nearly so) across the phylogenetic spectrum and the results point to functional importance. Similarly regions of the 16S rRNA known as helix 44, the 530 loop and helix 34 are involved in the 30S decoding site. Genetic studies further identified two bases A1492 and A1493 from their universal conservation and requirement for viability. A few 'hot spots' or critical bases effectively modulate protein synthesis and these experiments allowed crystallographers to assess structures on the basis of the known importance of these bases.

Structural studies of the ribosome

The structure of the large ribosomal subunit of *Haloarcula marismortui* defined sites of protein synthesis and highlighted potential mechanisms for the peptidyl

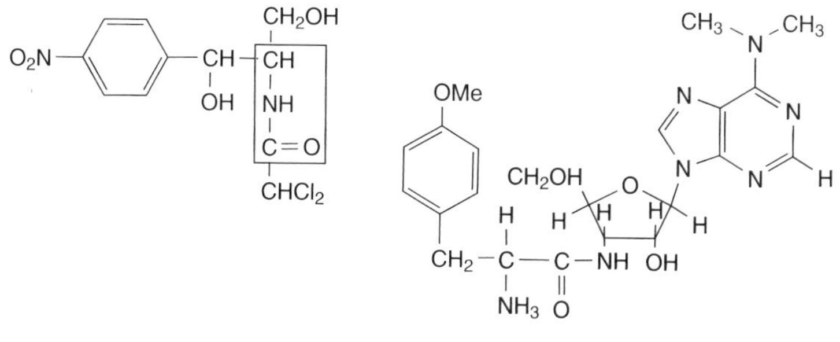

Figure 8.50 The structures of chloramphenicol and puromycin

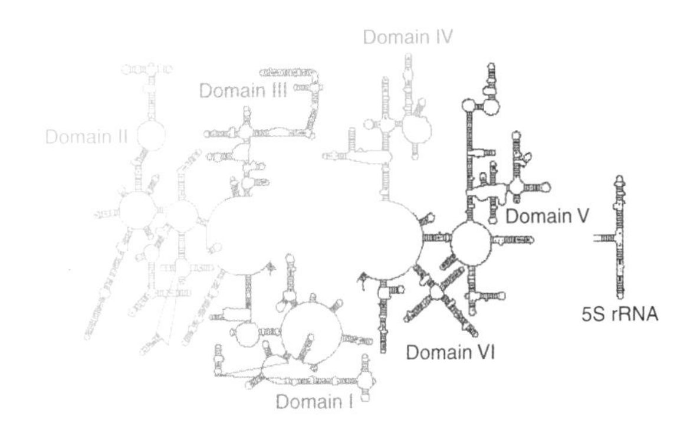

Figure 8.51 Computer-based secondary structure prediction for 23S and 5S rRNA sequences. (Reproduced with permission from Ramakrishnan, V. & Moore, P.B. *Curr. Opin. Struct. Biol.* 2001, **11**, 144–154. Elsevier)

transferase reaction with coordinates for RNA and protein representing over 90 percent of the subunit.

The secondary structure of 23S rRNA (Figure 8.51) shows regions of helix formed by base pairing as well as extended regions but provides little information on how the RNA folds to give the structure adopted by the large subunit. The 23S rRNA has six helical rich domains linked via extended loops.

The tertiary structure of 23S rRNA (Figure 8.52) showed domains that are *not* widely separated but are interwoven via helical interactions to form a compact monolithic structure. The 23S rRNA is a highly convoluted structure dictating the overall shape of the ribosome as well as defining the peptidyl transferase site.

Large subunit proteins (Figures 8.53 and 8.54) are located towards the ribosome's surface, often exposed to solvent, or on the periphery between the two subunits. Many proteins have multiple RNA binding sites and a principal function is directed towards stabilization of RNA folds. One surprising observation was that some proteins were not globular domains and for those located in crevices between rRNA helices were often extended structures. Many were basic proteins – an unsurprising observation in view of the large number of phosphate groups present within RNA.

Catalytic mechanisms within the large subunit

Using substrate analogues of the -CCA region of tRNA the location of the peptidyl-transferase site was shown to lie at the bottom of a deep cleft at the interface between large and small subunits. The peptidyl transferase reaction involves the bimolecular reaction between two substrates, namely the A site amino acyl-tRNA and the P site peptidyl-tRNA, and to enhance reactivity the molecules are constrained and in close proximity.

Transition state analogues based on puromycin pinpointed the location of the CCA arms of the A and P site tRNA molecules on the large subunit. In the presence of CCdA-phosphate-puromycin (the Yarus inhibitor, Figure 8.55) the peptidyl transferase reaction was inhibited after transfer of peptide chains from P site bound tRNA to the α amino group of puromycin. The peptidyl transferase site was based around nucleotides belonging to the central loop of 23*S* rRNA domain V a region called the 'peptidyl transferase loop' and known to bind CCdA-phosphate-puromycin. No proteins were located close to the puromycin complex and catalysis was mediated entirely by rRNA.

To facilitate formation of a tetrahedral intermediate an active functional group is required. The nearest suitable residue was the N3 atom of A2486 (A2451 in *E. coli*) located ~3 Å from the phosphoramide oxygen of the CCdA-phosphate-puromycin. No other functional group was within 5 Å of this reaction site and more significantly there were no atoms derived from polypeptide chains within 18 Å of this centre. This residue had previously been implicated in peptidyl transferase activity from genetic, footprinting and crosslinking studies.

Under normal conditions the N1 atom of adenine monophosphate has a pK of ~3.5 with the N3 centre observed to be an even weaker base, with a pK two units lower. The crystal structure (Figure 8.56) suggested a higher pK for the N3 atom of A2486 because the distance between the N atom and the oxygen is 3 Å and appears as a formal hydrogen bond – an interaction that would only occur if the N atom is protonated. The crystallization performed at pH 5.8 suggested a value for the p*K* of the N3 group above pH 6.0 to account for N3 protonation.

Many details of the catalytic process remain to be clarified but a charge relay network is believed to elevate the p*K* of the N3 atom of A2486 via a network of hydrogen bonds with G2482 and G2102

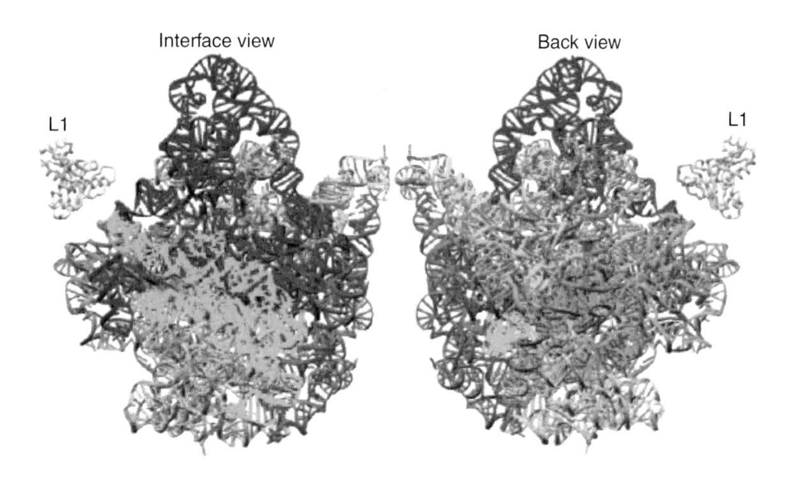

Figure 8.52 The tertiary structure of the 23S rRNA sequence. The colours utilize those shown in Figure 8.51 for the secondary structure. (Reproduced with permission from Ramakrishnan, V. & Moore, P.B. *Curr. Opin. Struct. Biol.* 2001, **11**, 144–154. Elsevier)

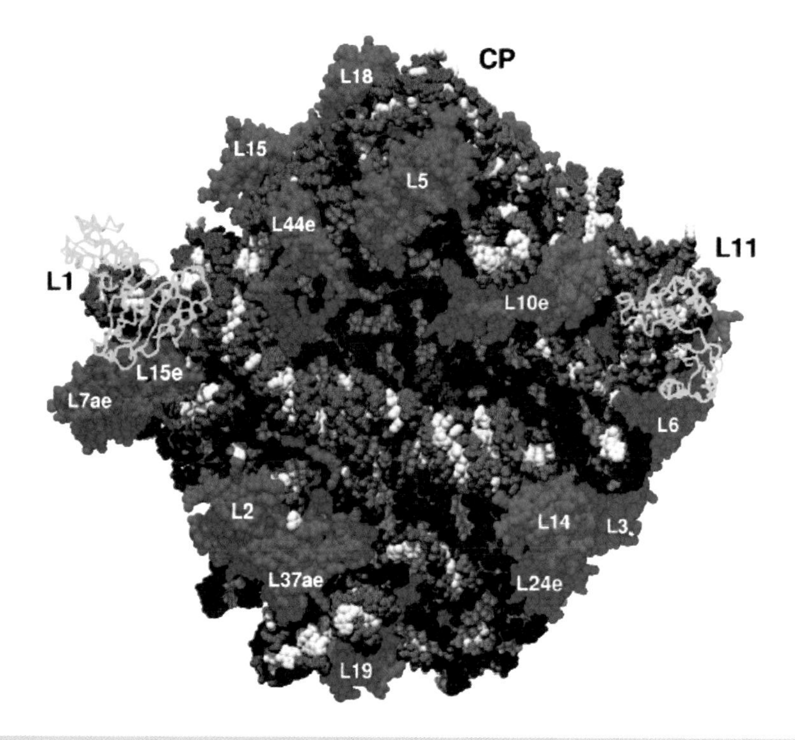

Figure 8.53 The structure of the large subunit of the ribosome of *H. marismortui*. A space-filling model of the 23*S* and 5*S* rRNA together with the proteins is shown looking down on the active site cleft. The bases are white and the sugar phosphate backbones are orange. The numbered proteins are blue with the backbone traces shown for the L1 and L11 proteins. The central protuberance is labelled CP. (Reproduced with permission from Nissen, P. *et al. Science* 2000, **289**, 920–930)

(Figure 8.57). A2486 and G2102 are conserved in the 23S rRNA sequences analysed from all three kingdoms.[1]

With this level of structural characterization it was proposed that peptidyl transferase involved the abstraction of a proton from the α amino group of the amino acyl tRNA by the N3 atom of A2486. In turn the NH_2 group of amino acyl tRNA acts as a nucleophile attacking the carbonyl group of peptidyl-tRNA. The protonated N3 stabilizes the tetrahedral carbon intermediate by hydrogen bonding to the oxyanion centre with subsequent proton transfer from the N3 to the peptidyl tRNA 3′ OH occurring as the newly formed peptide deacylates. The reaction mechanism described is not firmly established and future work will undoubtedly seek to experimentally test the proposed mechanism and role of A2486; a possible mechanism is shown in Figure 8.58.

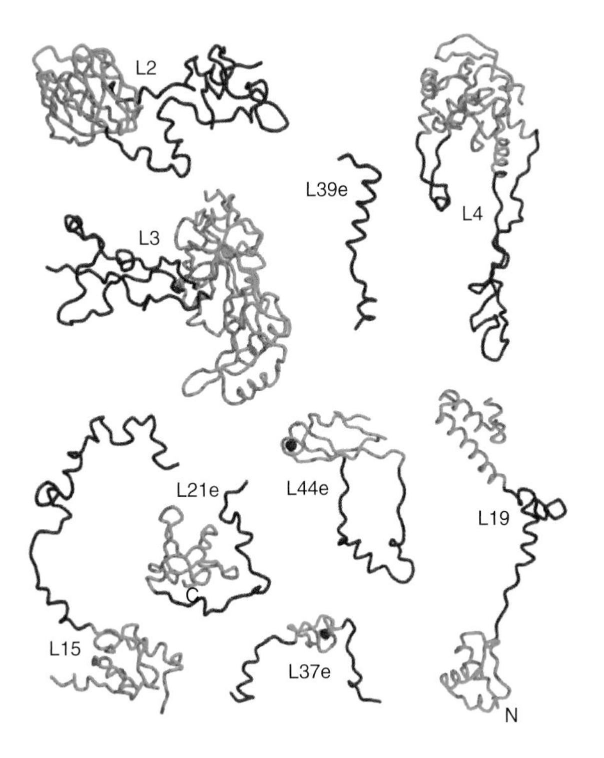

Figure 8.54 The globular domains and extended regions of proteins found in the large subunit. The long extensions would probably prove destabilizing in isolated proteins. Globular domains are shown in green with the extended region in red. (Reproduced with permission from Ramakrishnan, V. & Moore, P.B. *Curr. Opin. Struct. Biol.* 2001, **11**, 144–154. Elsevier)

[1]G2482 is conserved at a level of ~98 percent in all sequences but is replaced by an A in some archae sequences and deleted altogether in some eubacterial sequences.

The structure and function of the 30S subunit

Although the A and P sites are located on the large subunit the small subunit plays a crucial and obligatory role in protein synthesis. Protein synthesis is

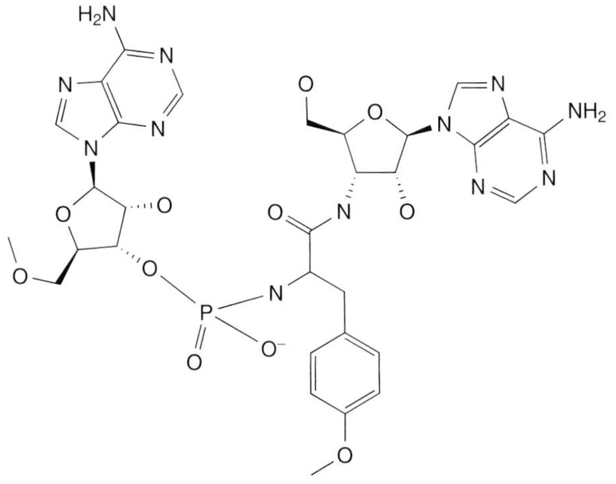

Figure 8.55 The transition state analogue CCdA-phosphate-puromycin

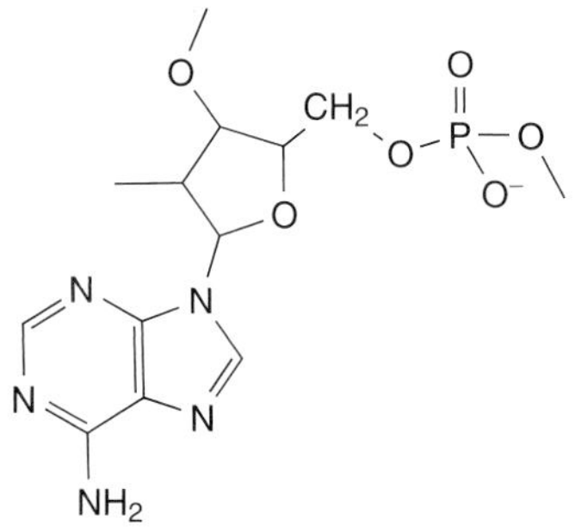

Figure 8.56 The N1 and N3 of adenine monophosphate; the N1 is shown in blue and the N3 in red

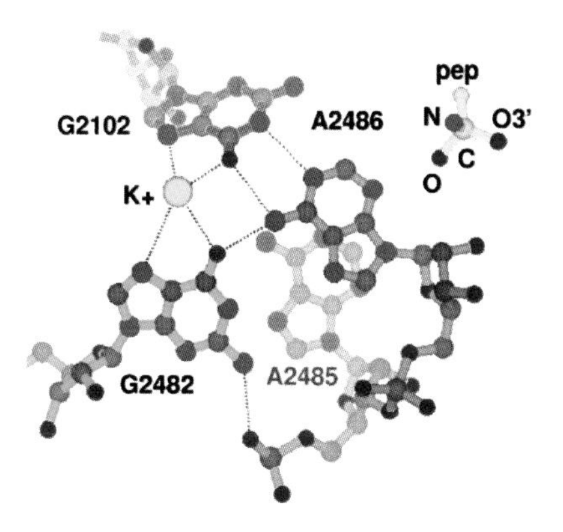

Figure 8.57 A skeletal representation with dashed hydrogen bonds showing G2482, G2102, and A2486, as well as the buried phosphate that may result in a charge relay through G2482 to the N3 of A2486. (Reproduced with permission from Nissen, P. *et al.* *Science* 2000, **289**, 920–930)

a multi-step process and starts with IF-3 binding to the small subunit. The 30S subunit plays a direct role in 'decoding' mRNA by facilitating base pairing between mRNA codon and the anticodon of relevant tRNAs.

The small subunit from *T. thermophilus* was crystallized in the 1980s but poor crystal diffraction properties limited use for many years until improvements in resolution occurred with removal of the S1 subunit from the ribosome prior to crystallization. Structures, initially at a resolution of 5.5 Å were followed by resolutions below ~3 Å and detailed studies of the 30S ribosomal subunit were largely the results of two groups headed by Ada Yonath and V. Ramakrishnan. The structure defined ordered regions of 16S rRNA and confirmed the general shape of the small subunit deduced previously using microscopy. All morphological features were derived from RNA and not protein.

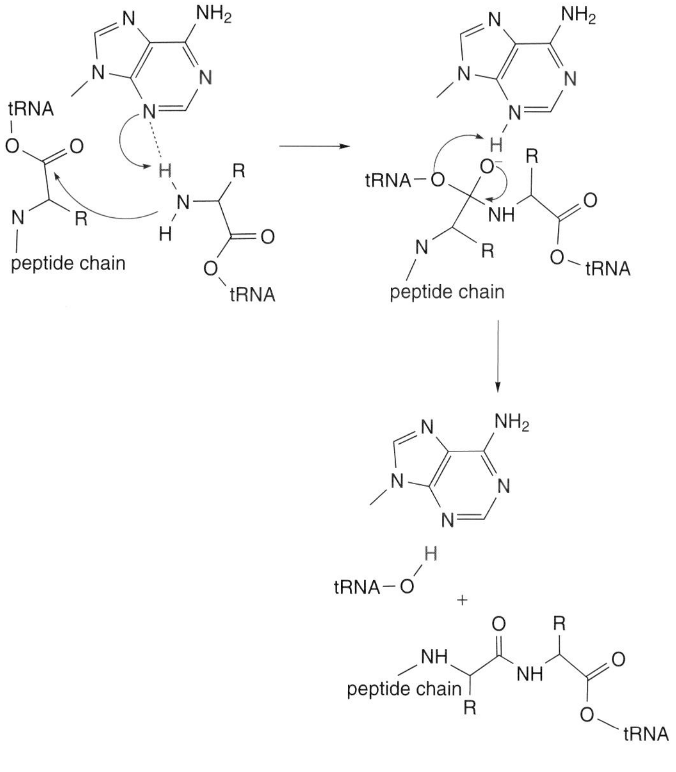

Figure 8.58 A possible mechanism for the peptidyl transferase involving the N3 atom of A2486 (adapted from Nissen, P. *et al.* *Science* 2000, **289**, 920–930)

The 16S RNA structure

The 16S rRNA contributes a significant proportion of the mass and volume of the 30S subunit and consists of approximately 50 elements of double stranded helix interspersed with irregular single stranded loops. The 16S RNA fold consists of four domains based on inter-helical packing and interactions with proteins (Figure 8.59). Anatomical features define the 30S structure with a head region containing a beak that points away from the large subunit. This structure is on top of a shoulder region whilst at the bottom a spur or projection is observed with a main interface defined by body and platform areas (Figure 8.60).

The 5′ domain of 16S rRNA forms the body region, the central domain and most of the platform of the 30S subunit. In contrast, the 3′ major domain constitutes the bulk of the head region whilst the 3′ minor domain forms part of the body at the subunit interface. The four domains of the 16S rRNA secondary structure radiate from a central point in the neck region of the subunit and are closely associated in this functionally important region of the 30S subunit.

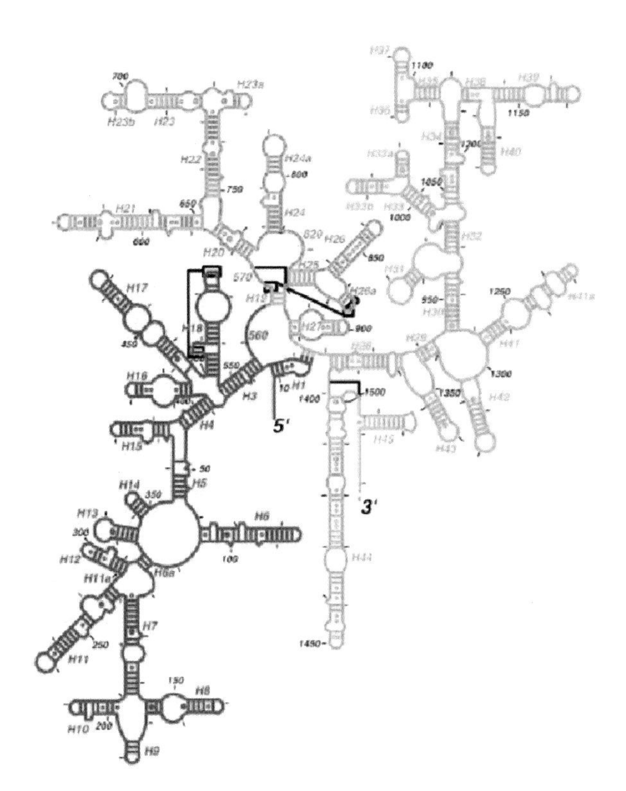

Figure 8.59 Secondary structure diagram of 16S RNA showing the various helical elements. The 5′ domain is shown in red; a central domain in green; a major domain of the 3′ region in orange-yellow and a minor domain in this region in cyan

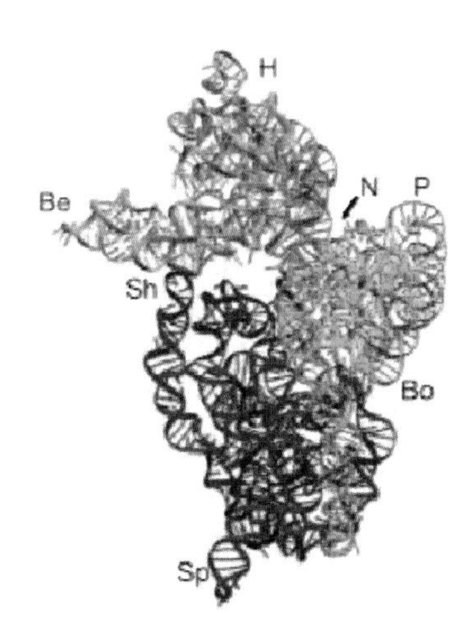

Figure 8.60 The tertiary structure of 16S RNA with the same colour scheme for the domains as in Figure 8.59. The model shows H, head; Be, beak; N, neck; P, platform; Sh, shoulder; Sp, spur; Bo, body regions of the 30S subunit from the perspective of the 50S subunit. (Reproduced with permission from Ramakrishnan, V. & Moore, P.B. *Curr. Opin. Struct. Biol.* 144–154. 2001, **11**, Elsevier)

Proteins of the small subunit

Small differences in composition of 30S subunits were noticed between *T. thermophilus* and the previously dissected *E. coli* ribosome. *Thermus* lacks subunit S21 but contains an additional short 26 residue peptide fragment. As a result of S1 removal the 30S structures contain only subunits S2–S20 along with the short 26 residue peptide fragment. Many proteins are located at junctions between helices of the RNA. For example, the S4 subunit binds to a junction formed by five helices in the 5′ domain whilst S7 binds tightly to a junction formed from four rRNA helices in the 3′ major domain. Both proteins are important in the

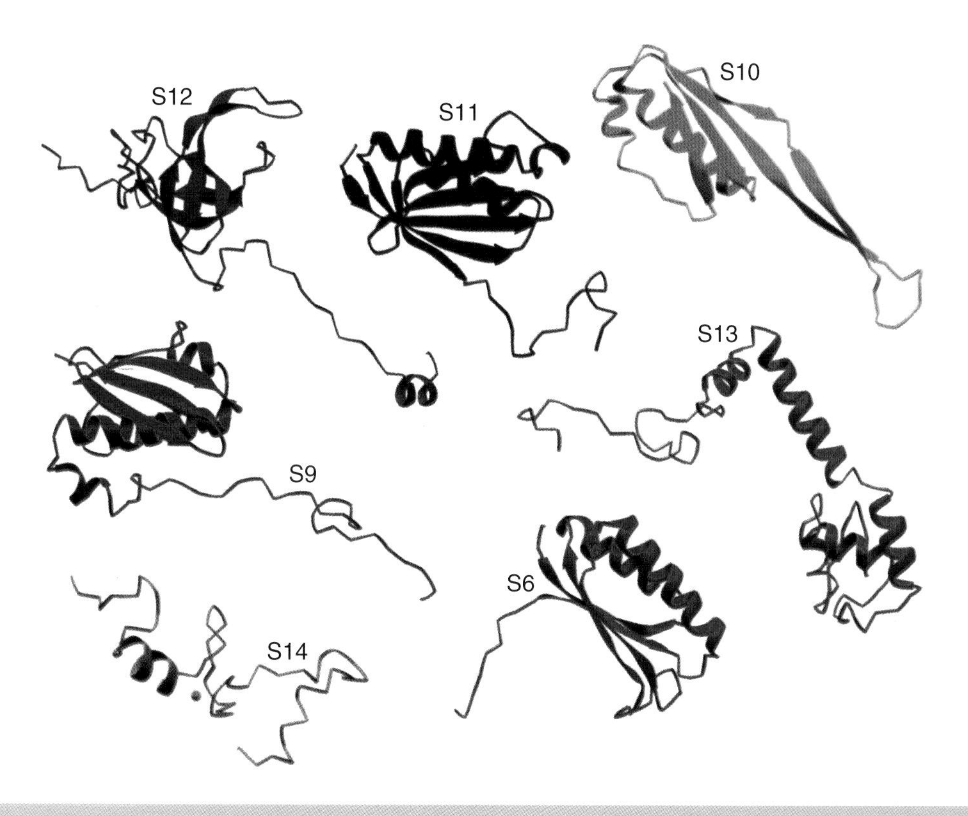

Figure 8.61 The globular domains and extended regions of some of the proteins found in the 30S subunit. Long extensions are probably destabilizing in isolated proteins. S14 is a Zn-binding protein, the cation is shown by a green sphere. (Reproduced with permission from Brodersen, D.E. *et al. Cold Spring Harbor Symp.* 2001, **66**, 17–32. CSHL Press)

assembly of the 30S subunit forming parts of the body and head, respectively. Table 8.8 summarizes the polypeptide structures of the small subunit.

Almost all of the proteins contain one or more globular domains and common topologies such as the β barrel are found in subunit S12 and S17. The packing of α-helices against an extended β-sheet is observed in several proteins such as S3, S10, S6 and S11. However, a defining characteristic is the relatively long extended regions found in subunits. These extensions are ordered, occur at the N or C terminal and may contain helical regions such as hairpin structures (S2) or C-terminal helices (S13). The extensions include loops or long β hairpins (S10 and S17) or 'tails' to proteins such as S4, S9, S11, S12, S13 and S19 (Figure 8.61). In almost all cases their role is to stabilize the RNA fold. In the 30S subunit the extensions reach far into cavities surrounded by RNA and make contact with several RNA elements. The extensions are well suited to this role since they are narrow, allowing close approach to different RNA elements, and they have basic patches to counter the highly charged sugar-phosphate backbone of RNA. Although protein–RNA interactions are obviously important, some of the protein subunits interact with each other via hydrophobic contacts. S3, S10 and S14 form a tight cluster held together by hydrophobic interactions. Other subunits interact via electrostatic and hydrogen bonding interactions and these include S4, S5 and S8.

Functional activity in the 30S subunit

The major function of the small subunit is decoding and matching the anticodon of tRNA with the mRNA

Table 8.8 Summary of the structural properties of polypeptide chains in the 30S subunit. (Adapted from Wimberley, B.T. *et al. Nature* 2000, **407**, 327–339. Macmillan).

Protein	Residues	No. of domains	Secondary structure in domains	Protein interaction	Unusual features
S2	256	2	α_2, $\alpha_1\beta_5\alpha_3$	None	Extended α hairpin
S3	239	2	α_2b_3, $\alpha_2\beta_4$	S10, S14	N-terminal tail
S4	209	3	Zn finger, α_4, $\alpha_3\beta_4$	S5	N-terminal Zn finger
S5	154	2	$\alpha_1\beta_3$, $\alpha_2\beta_4$	S4, S8	Extended β hairpin
S6	101	1	$\alpha_2\beta_4$	S18	C-terminal tail
S7	156	1	$\alpha_6\beta_2$	S9, S11	Extended β hairpin
S8	138	2	$\beta_2\alpha_3$, $\alpha_1\beta_3$	S5, S12, S17	
S9	128	1	$\beta\alpha_3\beta_4$	S7	Long C-terminal tail
S10	105	1	$\alpha_2\beta_4$	S3, S14	Long β hairpin
S11	129	1	$\alpha_2\beta_5$	S18, S7	Long N-terminal tail
S12	135	1	$\alpha_1\beta_5$	S8, S17	Long N-terminal tail and extended β-hairpin loops
S13	126	1	α_3	S19	Long C-terminal tail
S14	61	1	none	S3, S10	Zn module mostly extended
S15	89	1	α_4	none	
S16	88	1	$\beta_1\alpha_2\beta_4$	none	C terminal tail
S17	105	1	β_5	S12	C-terminal helix + β hairpin loops
S18	88	1	α_4	S6	β strand extends S11 sheet
S19	93	1	$\alpha_1\beta_3$	S13	
S20	105	1	α_3	none	

Partial structural information was available for S4–S8, and S15–S19 and aided their identification in crystals of 30S subunits.

codon. Within the 30S subunit the A, P and E sites were defined by RNA elements derived from different domains (45 different helical domains are found in 16S rRNA designated as H1–H45). The A and P sites are defined predominantly by RNA with helix 44, helix 34 and the 530 loop together with S12 forming part of the A site. In addition, the extended polypeptide chains of S9 and S13 intrude into the tRNA binding sites and may form interactions with tRNA. In contrast, the E site is largely defined by protein.

A combination of molecular genetics, sequence analysis and biochemical studies highlighted A1492 as an important base to the decoding process. The structure of the small subunit confirmed the importance of A1492 along with its neighbour and conserved base A1493. The crux of the decoding process is the ability to discriminate between cognate and near cognate tRNAs and this activity lies in the unique conformation of bases found at the A site. Discrimination against non-cognate tRNA results in two or three mismatches in base pairing at the codon–anticodon level with the high energetic cost making the process unfavourable.

The structure of the small subunit is split into domains, unlike the 50S subunit, with each possessing independent mobility and an involvement in conformational changes that influence ribosomal activity. The E site is predominantly protein and formed by the S7 and S11 subunits, a small interface between subunits forms the anticodon stem loop binding site whilst an extended β hairpin structure in S7 plays a role in dissociation of the vacant tRNA molecule from the ribosome.

Strong sequence conservation of rRNA suggests that RNA topology is likely to be maintained within the eukaryotic ribosomal subunits although there is clearly an insertion of elements with the formation of 5.8S, 18S and 28S rRNA. Yeast has a 18S rRNA 256 nucleotides longer than the 16S rRNA (*E. coli*) although the pattern of stem loops and major domains seen in 16S rRNA is mirrored by rRNA of yeast and other eukaryotes. However, the 40S and 60S subunits possess increased numbers of polypeptides relative to their *E. coli* counterparts, there are functional differences in the mechanism of initiation, elongation involves a greater variety and number of accessory proteins whilst the mechanisms of antibiotic inhibition are not shared with prokaryotes.

The remarkable confirmation that the ribosome is a ribozyme defined a new era in structural biology. The structures produced for the 50 and 30S subunits assimilates and unifies four decades of biochemical data on the ribosome and provides a wealth of new information about RNA and protein structure, their respective interactions and ribosome assembly. In 50 years our knowledge has progressed from the initial elucidation of the atomic structure of DNA to a near complete description of replication, transcription and translation.

Post-translational modification of proteins

The initial translation product may not represent the final, mature, form of the protein with some polypeptide chains undergoing additional reactions called post-translational modification. These reactions include processing to remove sequences normally at the ends of the molecule, the addition of new groups such as phosphate or sugars, and the modification of existing groups such as the oxidation of thiol groups. In almost all cases these modifications are vital to protein structure and function.

Proteolytic processing

The digestive enzymes chymotrypsin, trypsin and pepsin function in the alimentary tracts of animals to degrade proteins. These enzymes are initially translated as inactive forms called zymogens (chymotrypsinogen, trypsinogen and pepsinogen) to prevent unwanted degradation. Zymogens are converted into active enzyme by removal of a short 'pro' sequences. Such processes are important in the production of pancreatic proteases and mechanisms such as the blood clotting cascade (see Chapter 7).

Peptide hormones as examples of processing

Peptide hormones are frequently synthesized as longer derivatives. Angiotensin, a hormone involved in the control of vasoconstriction and insulin, with a role in the maintenance of blood glucose concentrations are 'processed' hormones. The 'pro' sequences may also have additional sequences located at the N-terminal to facilitate targeting to specific intracellular compartments. For example insulin, is synthesized as preproinsulin (Figure 8.62).

The 'pre' sequence transfers insulin across the endoplasmic reticulum (ER) membrane into the lumen where cleavage by proteases leaves the 'pro' insulin molecule (Figure 8.63). In this state folding starts with disulfide bridge formation stabilizing tertiary structure. A fragment known as the C peptide is released (Figure 8.64) and persists in secretory granules seen in pancreatic cells. The insulin remains inactive as a hormone and in high concentrations in the pancreas. The chain requires further modification and a second proteolytic reaction cuts the primary

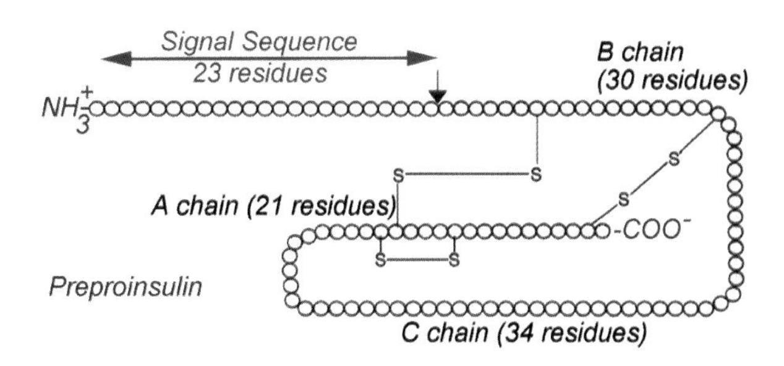

Figure 8.62 Preproinsulin contains 110 residues within a single polypeptide. Each circle represents a single amino acid residue

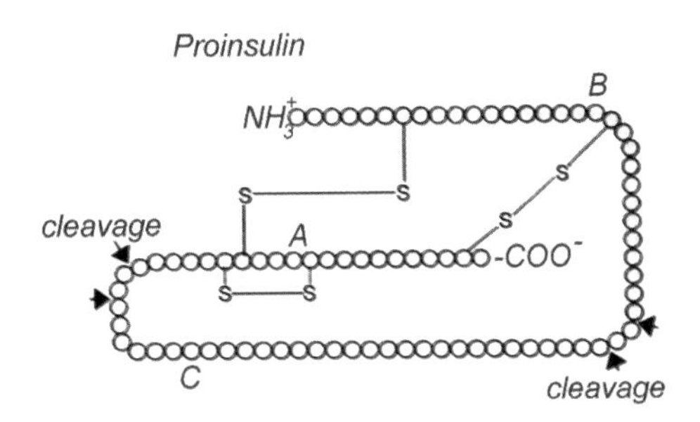

Figure 8.63 Signal sequence removal yields proinsulin. The A chain has an intramolecular disulfide bridge as well as two intermolecular disulfide bridges with the B chain

sequence at pairs of Lys-Arg and Arg-Arg residues between the A and B regions creating two separate polypeptides.

A very dramatic example of processing is the peptide hormone opiomelanocortin (POMC). The POMC gene is expressed in the anterior and intermediate lobes of the pituitary gland where a 285-residue precursor undergoes differential processing to yield at least eight hormones (Figure 8.65).

Disulfide bond formation

The formation of disulfide bonds between cysteine residues is a common post-translational modification resulting in strong covalent bonds. The covalent bond restricts conformational mobility in a protein and normally occurs between residues widely separated in the primary sequence. For example, in ribonuclease four disulfide bridges form between cysteine residues 26–84, 40–95, 58–110 and 55–72.

Disulfide bond formation occurs post-translationally in the ER of eukaryotes or across the plasma membrane of bacteria in proteins destined for secretion. The formation of disulfide bonds is often a rate-limiting step during folding, and sometimes the 'wrong' disulfide bond forms where there is a potential for multiple bridges. In all cells enzymes catalyse efficient formation of disulfide bonds and correct 'wrong' disulfide bonds by rapid isomerization. These enzymes are collectively called disulfide oxidoreductases.

Eukaryotes and prokaryotes catalyse disulfide formation with different groups of enzymes. In prokaryotes the Dsb family of enzymes are identified in Gram-negative bacteria and carry names such as DsbA, DsbB, DsbD etc. In *E. coli* DsbA contains a consensus motif Cys-Xaa-Xaa-Cys and shows a similar fold to thioredoxin despite a low sequence identity (<10 percent). In the cytoplasm of *E. coli* thioredoxin reduces disulfides whilst in the periplasm DsbA oxidizes thiol groups. Since both thioredoxin and DsbA contain the Cys-Xaa-Xaa-Cys motif the ability to oxidize disulfide bonds centres around the properties of this catalytic centre. In DsbA the first cysteine (Cys30) has a low pK of 3.5, compared with a normal value around 8.5, and the sidechain forms a reduced thiolate anion S^- representing a highly oxidizing centre that allows DsbA to catalyse disulfide formation.

In eukaryotes comparable reactions are performed by protein disulfide isomerase (PDI). Again the activity of PDI depends on the diagnostic motif Cys-Xaa-Xaa-Cys and when the active-site cysteines are present as a

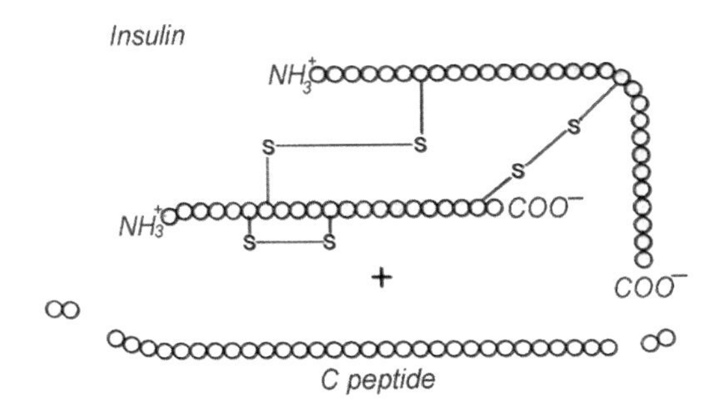

Figure 8.64 The processed (active) form of insulin plus proteolytic fragments

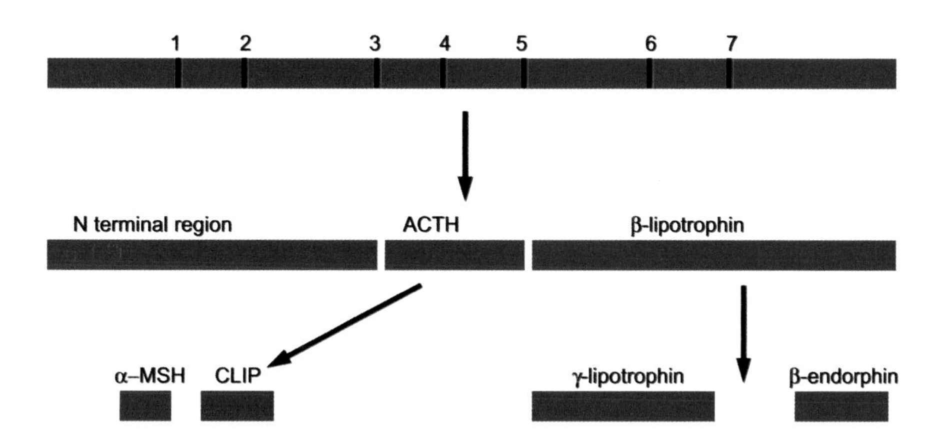

Figure 8.65 The processing of pro-opiomelanocortin yields additional bioactive peptides. Initial processing of pro-opiomelanocortin in the anterior and intermediate lobes of the pituitary yields an N-terminal fragment; ACTH, adrenocorticotrophic hormone (39 residues); and β-lipotrophin. Further processing of ACTH in the intermediate lobes yields a melanocyte stimulating hormone – a 13-residue peptide acetylated at the N-terminal and amidated at the C-terminal residue together with CLIP, a corticotropin-like intermediate lobe peptide of 21 residues. Further processing of β lipotrophin yields γ-lipotrophin and β endorphin. The peptides are 59 and 26 residues in length respectively with the β endorphin acetylated at the N terminal

disulfide the enzyme transfers disulfide bonds directly to substrate proteins acting as a dithiol oxidase. Under more reducing conditions with thiols in the active-site the enzyme reshuffles disulfides (isomerase) in target proteins.

Hydroxylation

Hydroxylation is another example of a post-translational modification. It is particularly important in the maturation of collagen where hydroxylation of proline and lysine residues found in the Gly-Xaa-Xaa motif occurs in procollagen in the ER as part of the normal secretory pathway. The reaction maintains structural rigidity in collagen enabling a role

HO H
C
H_2C CH_2 O
—N—CH—

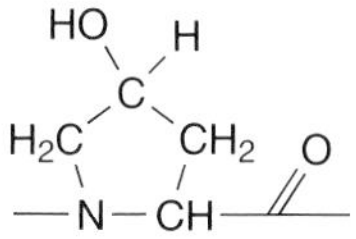

Figure 8.66 4-Hydroxyproline, a common post-translational modification observed in collagen

as a biological scaffold or framework. Proline is hydroxylated most commonly at the γ or fourth carbon by the enzyme prolyl-4-hydroxylase in reactions requiring oxygen, ascorbate (vitamin C) and α-ketoglutarate (Figure 8.66).

The requirement of vitamin C for effective folding and stability of collagen emphasizes the physiological consequences of scurvy. Scurvy is a disease arising from a lack of ascorbate in the diet, and was frequently experienced by sailors in the 16th and 17th centuries during long maritime voyages with the absence of fresh fruit or vegetables containing high levels of vitamin C. Scurvy can also occur under conditions of malnourishment and is sometimes seen in the more disadvantaged regions of the world where famine is prevalent. Hydroxylation of proline residues stabilizes collagen by favouring additional hydrogen bonding and in its absence the microfibrils are weaker. One observed effect is a weakening of connective tissue surrounding the teeth, a common symptom of scurvy. Proline residues are also hydroxylated at the β or C3 position with lower frequency.

In an entirely analogous manner lysine residues of collagen are hydroxylated in a reaction conferring increased stability on triple helices by allowing the

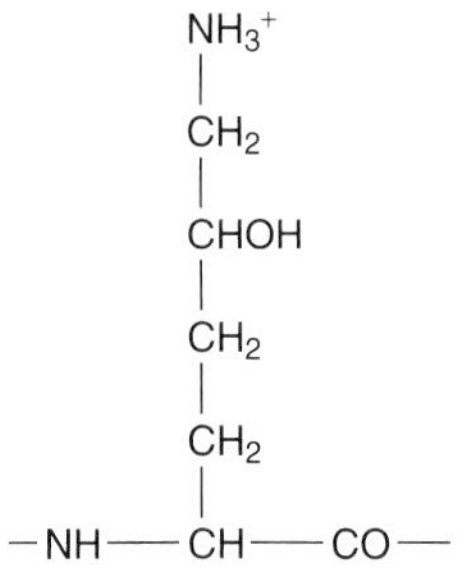

NH_3^+
|
CH_2
|
CHOH
|
CH_2
|
CH_2
|
—NH— CH — CO—

Figure 8.67 The hydroxylation of lysine at the C5 or δ position

subsequent attachment of glycosyl groups that form cross links between microfibrils. Lysine residues are hydroxylated at the δ carbon (C5) by the enzyme lysyl-5-hydroxylase (Figure 8.67).

Phosphorylation

The phosphorylation of the side chains of specific serine, threonine or tyrosine residues is a general phenomenon known to be important to functional activity of proteins involved in intracellular signalling. The addition of phosphoryl groups occurs through the action of specific enzymes (kinases) that utilize ATP as a donor whilst the removal of these groups is controlled by specific phosphatases. A generalized scheme for phosphorylation is

$$\text{Protein} + \text{ATP} \rightleftharpoons \text{protein-P} + \text{ADP}$$

and results in the products phosphoserine, phosphotyrosine, and phosphothreonine (Figure 8.68). More rarely other residues are phosphorylated such as His, Asp and Lys.

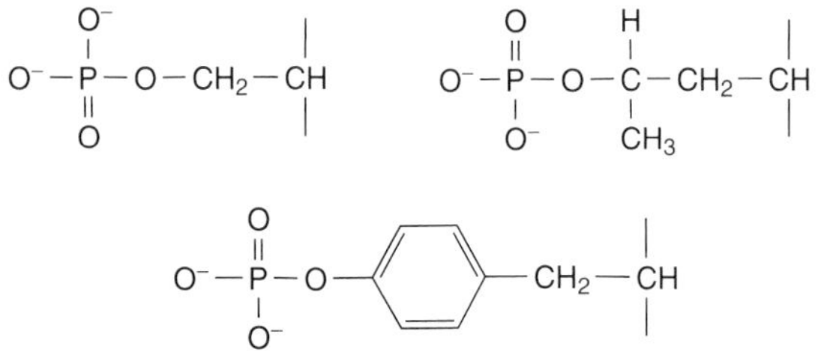

$O^- - P(=O)(O^-) - O - CH_2 - CH$

$O^- - P(=O)(O^-) - O - C(H)(CH_3) - CH_2 - CH$

$O^- - P(=O)(O^-) - O - C_6H_4 - CH_2 - CH$

Figure 8.68 The phosphorylation of serine, tyrosine and threonine residues

Glycosylation

Many proteins found at the cell surface are anchored to the lipid membrane by a complex series of glycosylphosphatidyl inositol (GPI) groups. These groups are termed GPI anchors and they consist of an array of mannose, galactose, galactosamine, ethanolamine and phosphatidyl inositol groups. All eukaryotic cells contain cell-surface proteins anchored by GPI groups to the membrane. These proteins have diverse functions ranging from cell-surface receptors to adhesion molecules but are always located on exterior surfaces.

The term GPI was first introduced for the membrane anchor of the variant surface glycoprotein (VSG) of *Trypanosoma brucei*, a protozoan parasite that causes sleeping sickness in humans (a disease that is fatal if left untreated and remains of great relevance to sub-Saharan Africa). GPI anchors are particularly abundant in protozoan parasites. The overall biosynthetic pathway of GPI precursor is now well understood with signals in the form of consensus sequences occurring within a polypeptide chain as sites for covalent attachment of GPI anchors. This site is often a region of ~20 residues located at the C-terminal, and specific proteases cleave the signal sequence whilst transamidases catalyse the addition of the GPI anchor to the newly generated C-terminal residue. The outline organization of GPI anchors (Figure 8.69) involves an ethanolamine group linked to a complex glycan containing mannose, galactose and galactosamine and an inositol phosphate before linking to fatty acid chains embedded in the lipid bilayer.

Alongside GPI anchors the principal post-translational modification involving glycosylation involves the addition of complex sugars to asparagine side chains (N-linked) or threonine/serine side chains (O-linked) in the form of glycosidic bonds. In N-linked glycosylation an oligosaccharide is linked to the side chain N of asparagine in the sequence Asn-X-Thr or Asn-X-Ser, where X is any residue except proline. The first sugar to be attached is invariably *N*-acetylglucosamine and strictly occurs co-translationally as

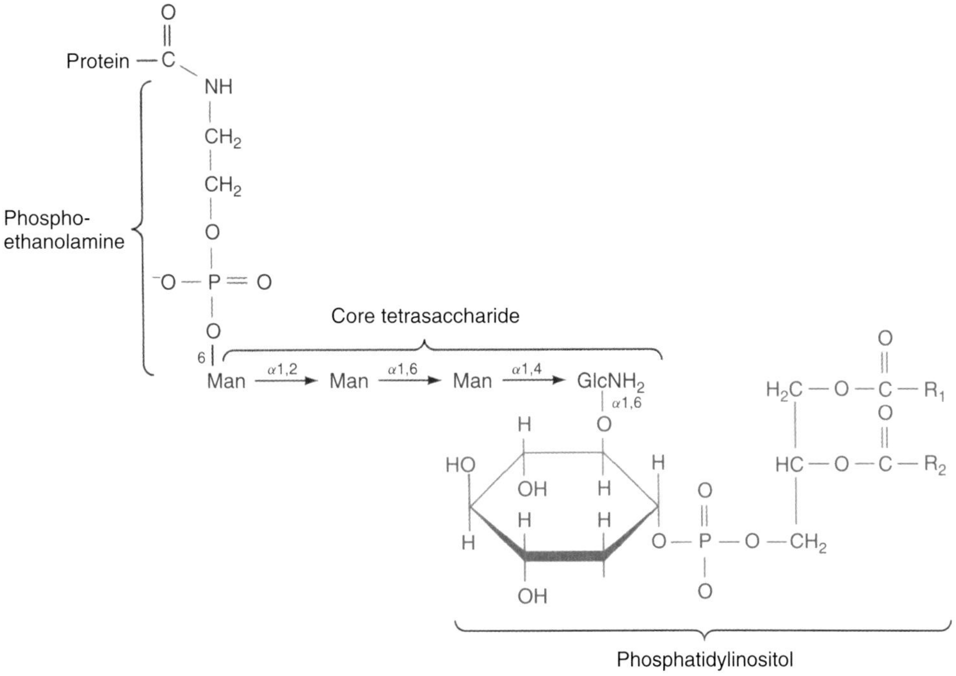

Figure 8.69 The attachment of GPI anchors to proteins together with the molecular organization of these complex oligosaccharides. GPI anchors are present in many proteins including enzymes, cell adhesion molecules, receptors and antigens. They are found in virtually all mammalian cell types and share a core structure of phosphatidylinositol glycosidically linked to non-acetylated glucosamine (GlcN). Glucosamine is usually found acetylated or sulfated form and thus the presence of non-acetylated glucosamine is an indication of a GPI anchor

the polypeptide chain is synthesized. This attachment is the prelude for further glycosylation which can involve nine mannose groups, three glucose groups and two *N*-acetyl-glucosamine groups being covalently linked to a single Asn side chain. Further processing removes some of these sugar groups in the ER and in the Golgi through the action of specific glucosidases and mannosidases, whilst in some instances fucose and sialic acid groups can be added by glucosyl transferases. These reactions lead to considerable heterogeneity and diversity in glycosylation reactions occurring within cells. However, all N-linked oligosaccharides have a common core structure (Figure 8.70).

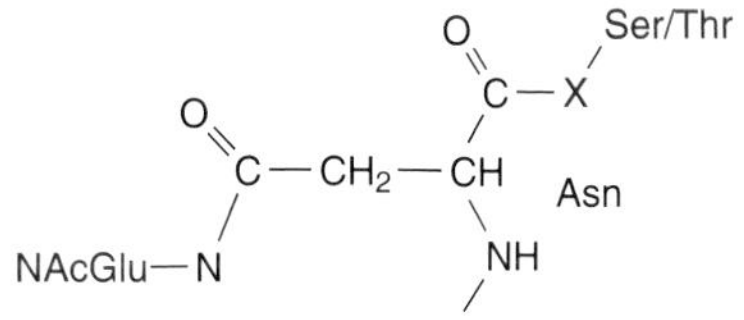

Figure 8.70 Common core structure for N-linked oligosaccharides

In O-linked glycosylation the most common modification involves a disaccharide core of β-galactosyl(1 → 3)-α-*N*-acetylgalactosamine forming a covalent bond with the side chain O of Thr or Ser. Less frequently, galactose, mannose and xylose form O-linked glycosidic bonds. Whilst N-linked glycosylation proceeds through recognition of Asn-containing motifs, within primary sequences O-linked glycosylation has proved more difficult to identify from specific Ser/Thr-containing sequences and appears to depend more on tertiary structure and a surface accessible site.

A single protein may contain N- and O-linked glycosylation at multiple sites and the effect of linking large numbers of oligosaccharide units is to increase molecular mass dramatically whilst also increasing the surface polarity or hydrophilicity. The function of added oligosaccharide groups has proved difficult to delineate and in some instances proteins can function perfectly well without glycosylated surfaces. Generally, many cell-surface proteins contain glycosylation sites and this has raised the possibility that oligosaccharides function in molecular recognition events although mechanistic roles have yet to be uncovered.

N- and C-terminal modifications

Post-translational modification of the N-terminal of proteins includes acetylation, myristoylation – the addition of a short fatty acid chain containing 14 carbon atoms, and the attachment of farnesyl groups. Many eukaryotic proteins are modified by acetylation of the first residue and the donor atoms are provided by acetyl-CoA and involve *N*-acetyltransferase enzymes. Acetylation is a common covalent modification at the N-terminal but other small groups are attached to the free amino group including formyl, acyl and methyl groups. The precise reasons for these modifications are unclear but may be correlated with protein degradation since chains with modified residues are frequently more resistant to turnover than unmodified proteins.

In contrast the attachment of myristoyl units has a clearly defined structural role. As a long aliphatic and hydrophobic chain with 14 carbon units it causes proteins to associate with membranes although these proteins become soluble when this 'anchor' is removed. Myristoylated proteins have properties typical of globular proteins and remain distinct from integral membrane proteins with which they are often associated. Important examples of myristoylated proteins include small GTPases that function in many intracellular signalling pathways.

Myristoylation is actually a co-translational modification, since the enzyme *N*-myristoyltransferase is often bound to ribosomes modifying nascent polypeptide as they emerge. The substrate is myristoyl-CoA with glycine being the preferential N-terminal amino group. When the residue after glycine contains Asn, Gln, Ser, Val or Leu myristoylation is enhanced, whilst Asp, Phe, or Tyr are inhibitory. The sequence specificity appears efficient with all proteins containing N-terminal glycine followed by a stimulatory residue observed to be myristoylated. Palmitoylation follows many of the patterns of myristoylation with the exception that a C16 unit is attached to proteins post-translationally in the cytoplasm. The reaction is catalysed by palmitoyl transferase and recognizes a sequence motif of Cys-Aliphatic-Aliphatic-Xaa where Xaa can be any amino acid. As progressively more proteins are purified it is clear that the range of modifications extents to iodination and bromination of tyrosine side chains, adenylation, methylation, sulfation, amidation and side chain decarboxylation.

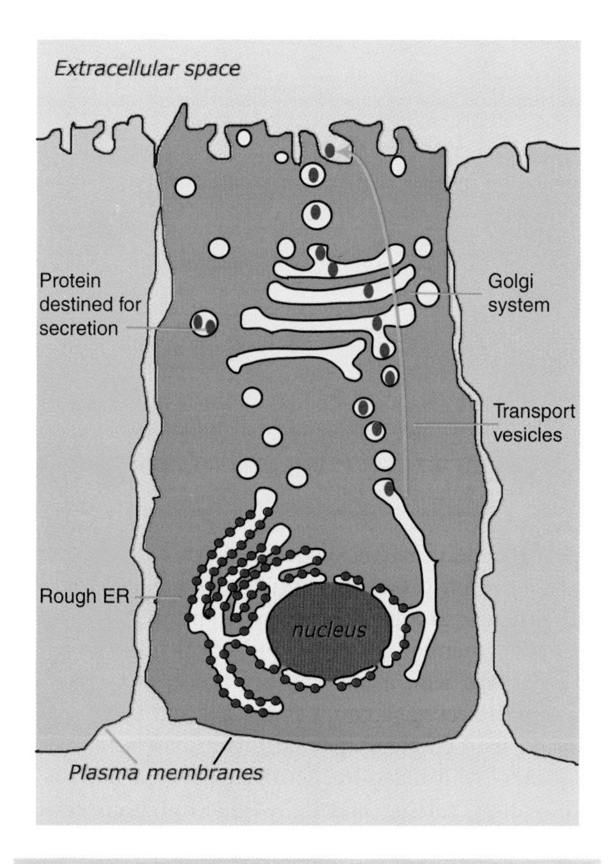

Figure 8.71 The movement of vesicles as part of the normal route for transfer of proteins from the ER to the Golgi apparatus and on to other destinations. Transfer through the Golgi occurs with protein processing before further vesicle budding from the trans Golgi region. Vesicles from the trans Golgi are either directed to the plasma membrane, for secretion or to lysosomes.

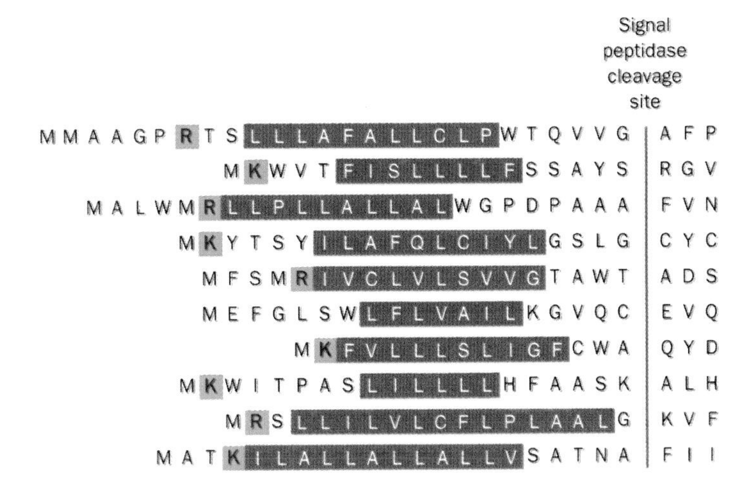

Figure 8.72 The N-terminal sequence of secretory preproteins from eukaryotes. The sequences show basic residues in blue and the hydrophobic cores in brown. The signal peptidase cleavage site is indicated

Protein sorting or targeting

The cell sorts thousands of proteins into specific locations through the use of signal sequences located normally at the N-terminal. The hypothesis that signal sequences direct proteins towards cell compartments was proposed by Gunter Blobel and David Sabatini in the early 1970s to explain the journey of secreted proteins from cytoplasmic sites of synthesis through the ER to the cell's exterior (Figure 8.71). Signal sequences direct proteins first to the ER membrane in a process that is best described as a co-translational event.

A fundamental part of sorting is the interaction of the ribosome with a macromolecular complex known as the signal recognition particle (SRP). The SRP identifies a signal peptide sequence on nascent polypeptide chains emerging from the ribosome and by association temporarily 'halts' protein synthesis. The whole ensemble remains 'halted' until the SRP–ribosome complex binds to further receptors (SRP receptors) in the ER membrane. Chain elongation resumes and the growing polypeptide is directed towards the lumen.

Proteins with signal sequences recognized by the SRP have a limited number of destinations that includes insertion of membrane proteins directly into the bilayer, transfer to the ER lumen as a soluble protein confined to this compartment, transfer via the ER lumen to the Golgi apparatus, lysosomal targeting, or secretion from the cell. Protein targeting to the nucleus, mitochondrion and chloroplast is based on a SRP-independent pathway.

The SRP-mediated pathway

The presence of additional residues at the N terminus of nascent polypeptide chains not found in the mature form of the protein provided a clue that signal sequences existed (Figure 8.72). Edman sequencing methods showed that the primary sequences of signal motifs lacked homology but shared physicochemical properties. The overall length of the sequence ranged from 10 to 40 residues and was divided into regions with different properties. A charged region between two and five residues in length occurs immediately after the N-terminal methionine residue. The basic region is followed by a series of residues that are predominantly hydrophobic such as Ala, Val, Leu, Ile, and Phe and are succeeded by a block of about five hydrophilic or polar residues.

Protein synthesis of ~80–100 residues exposes the signal peptide through the large subunit channel and results in SRP binding and the cessation of further extension until the ribosomal–mRNA–peptide–SRP complex associates with specific receptors in the

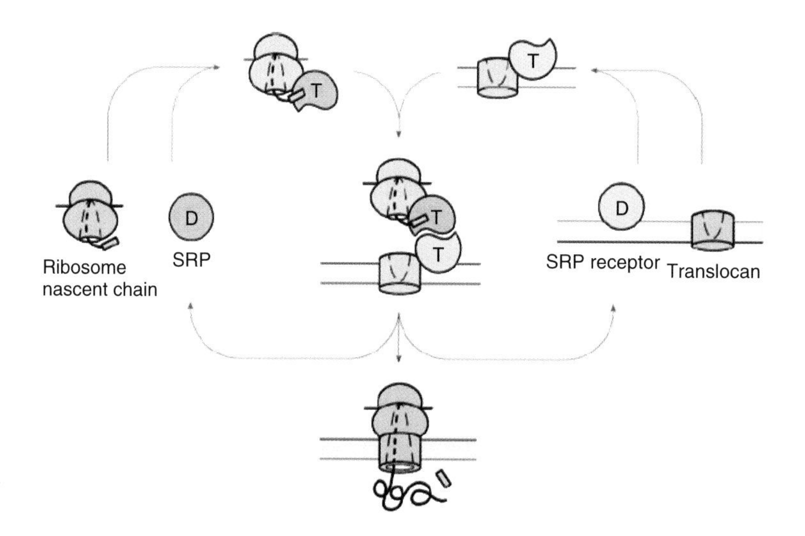

Figure 8.73 Co-translational targeting by SRP. Cycles of GTP binding and hydrolysis occur during the SRP cycle. GTPase activity is associated with SRP54 and the SRP receptor subunits. Activity is modulated by the ribosome and the translocon. The SRP complex is shown in blue and the receptor in green. SRP D reflects GDP binding, SRP T reflects a GTP-bound state. The signal sequence is shown in yellow for the nascent chain (reproduced with permission from Keenan, R.J. *et al. Ann. Rev. Biochem.* 2001, **70**, 755–75. Annual Reviews Inc.)

ER membrane. This is known as co-translational targeting (Figure 8.73) and is conveniently divided into two processes; recognition of a signal sequence and association with the target membrane.

The structure and organization of the SRP

SRPs are found in all three major kingdoms (archae, eubacteria and eukaryotes) with the mammalian SRP consisting of a 7S rRNA and six distinct polypeptides called SRP9, SRP14, SRP19, SRP54, SRP68 and SRP72 and named according to their respective molecular masses. With the exception of SRP72 all bind to the 7S rRNA and SRP54 has a critical role in binding RNA *and* signal peptide. The diversity of signal sequences plays a significant part in determining the mode of binding since introduction of a single charged residue destroys interaction with SRP54, whilst a block of hydrophobic residues remains critical to association. In SRP54 the M domain is important in molecular recognition of signal peptide. The name derives from a high percentage of methionine residues and is conserved in bacterial and mammalian homologues The homologous *E. coli* protein Ffh (Ffh = fifty four homologue) has ~16 percent of residues in the M domain as methionines – a frequency six times greater than its typical occurrence in proteins.

The *E coli* SRP has a simpler composition – the Ffh protein and a 4.5S RNA – and evolutionary conservation is emphasized by the ability of human SRP54 to bind 4.5S RNA. Bacterial SRPs are minimal homologues of the eukaryotic complex and are attractive systems for structural studies. The crystal structure derived for Ffh from *T. aquaticus* (Figure 8.74) revealed a C-terminal helical M domain forming a long hydrophobic groove that is exposed to solvent and is made from three helices together with a long flexible region known as the finger loop linking the first two helices. The dimensions of the groove are compatible with signal sequence binding in a helical conformation and, as predicted, the surface of the groove is lined entirely with phylogenetically conserved hydrophobic residues.[1]

Homologous domains from *E. coli* and mammalian SRP54 have similar structures (Figure 8.75),

[1]In *T. aquaticus*, a thermophile living above 70 °C, many of the conserved Met residues found in SRP54/Ffh are replaced by Leu, Val and Phe, possibly reflecting a need for increased flexibility to counter increased thermal mobility.

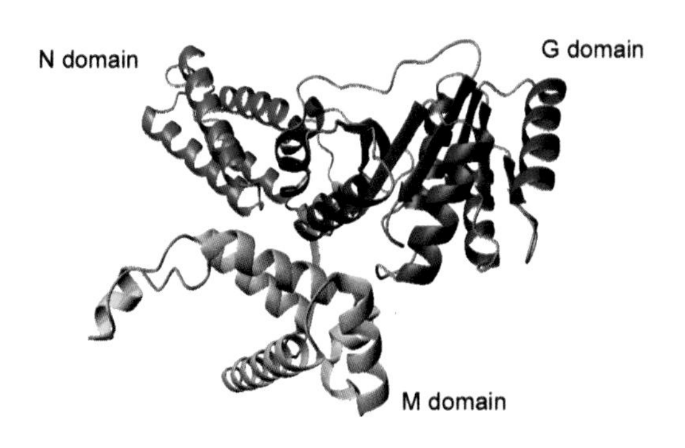

Figure 8.74 The structure of the full length FfH subunit from *Thermus aquaticus*. The M domain encompassing residues 319–418 is shown in yellow, the N domain, residues 1–86, in blue and the largest G domain, residues 87–307, in green (PDB: 2FFH)

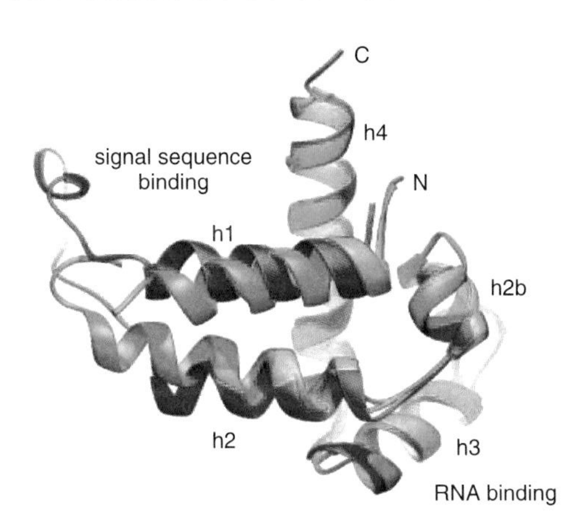

Figure 8.75 Superposition of *E. coli* (light blue), *T. aquaticus* (magenta), and human (green) M domains showing similar structures (reproduced with permission from Batey, R.T. *et al. J. Mol. Biol.* 2001, **307**, 229–246. Academic Press)

although the finger loop region adopts different conformations – a variation that reflects intrinsic flexibility. Additional N-terminal and GTPase domains interact with the SRP receptor.

SRP54 is a GTPase and the catalytic cycle involves GTP/GDP exchange and GTP hydrolysis although signal sequence binding does not require GTP hydrolysis. Ribosome association stimulates GTP binding by SRP54 and is followed by hydrolysis upon interaction with the membrane receptors. The SRP receptor contains two subunits designated α and β with the α subunit showing homology to the GTPase domain of SRP54 and interacting with the β subunit, a second GTPase, that shows less similarity to either SRP54 or the α subunit.

The SRP receptor structure and function

Mammalian SRP receptors are membrane protein complexes but the corresponding bacterial receptors are simpler containing a single polypeptide known as FtsY (Figure 8.76). FtsY is a GTPase loosely associated with the bacterial inner membrane that can substitute for the α subunit of the SRP receptor. FtsY is composed of N and G domains comprising the GTPase catalytic unit and adopting a classic GTPase fold based on four conserved motifs (I–IV) arranged around a nucleotide-binding site. Motif II is contained within a unique insertion known as the insertion box domain and this extends the central β sheet of the domain by two strands and is characteristic for the SRP GTPase subfamily.

The N and G domains of SRP54 (Ffh) interact with the α subunit of the SRP receptor (FtsY) through their respective, yet homologous, NG domains. The interaction occurs when each protein binds GTP and the interaction is weaker in the apo protein. Nucleotide dependent changes in conformation are the basis for the association between SRP and its cognate receptor and GTP hydrolysis causes SRP to dissociate from the receptor complex. SRP lacking an NG domain fails to target ribosome–nascent chain complexes to the membrane and conversely the NG domain of FtsY when fused to unrelated membrane proteins promotes association. The primary role of the SRP receptor is to 'shuttle' the ribosomal-mRNA–nascent chain from a complex with SRP to a new interaction with a membrane-bound complex known as the translocon. The translocon is a collection of three integral membrane proteins, Sec61α, Sec61β and Sec61γ, that form pores within the membrane allowing the passage of nascent polypeptide chains through to the ER lumen.

Signal peptidases have a critical role in protein targeting

After translocation across the ER membrane polypeptides reach the lumen where a peptidase removes the

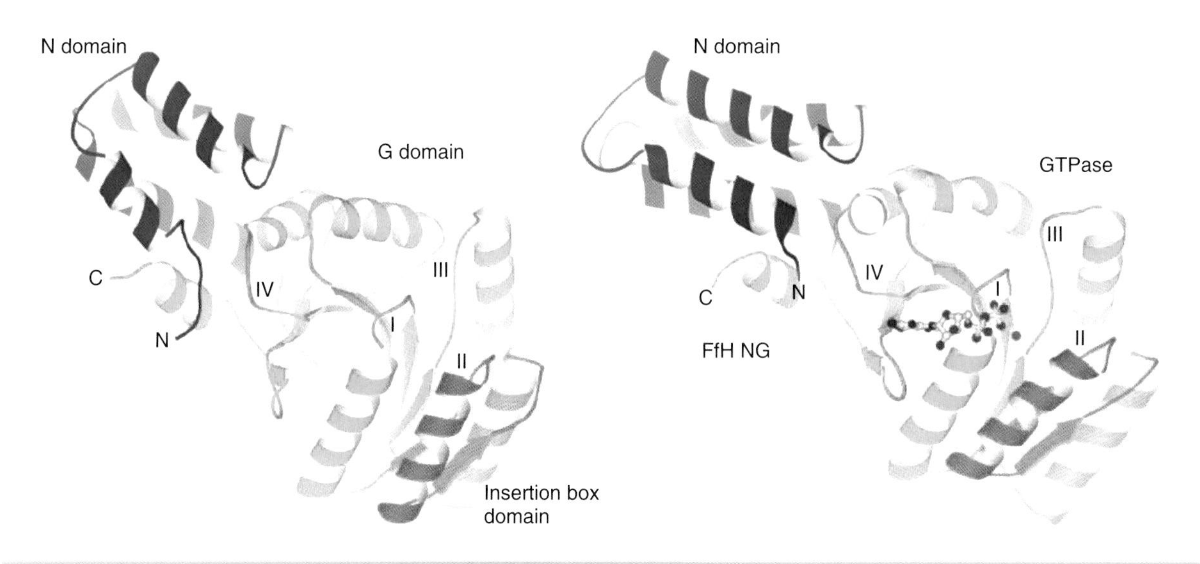

Figure 8.76 The structures of *E. coli* FtsY show homology to Ffh. The apo-form of the NG domain shows the N-terminal N domain (blue) packing tightly against the GTPase fold (green). The conserved insertion box domain (orange) is unique to the SRP family of GTPases. The four conserved GTPase sequence motifs are indicated (I–IV). (Reproduced with permission from Keenan, R.J. *et al. Ann. Rev. Biochem.* 2001, **70**, 755–75. Annual Reviews Inc.) The structure of Ffh is show for comparison in similar colours

'signal peptide' by cleavage at specific sites. Sites of cleavage are determined by local sequence composition after the 'polar' portion of the signal peptide. The ER signal peptidase shows a preference for residues with small side chains at positions −1 and −3 from the cleavage site. Consequently Gly, Ala, Ser, Thr and Cys are common residues whilst aromatic, basic or large side chains at the −3 position inhibits cleavage. Alanine is the most common residue found at the −1 and −3 positions and this has given rise to the Ala-X-Ala rule or −1, −3 rule for cleavage site identification.

Post-lumenal targeting is directed by additional sequence-dependent signals

When proteins reach the lumen other sequence dependent signals dictate targeting to additional sites or organelles. One signal pathway involves the KDEL sequence named after the order of residues found at the C terminal of many soluble ER proteins. In mammals proteins with the sequence Lys-Asp-Glu-Leu (i.e. KDEL) are marked for recovery from transport pathways. One example is PDI and altering this sequence causes the protein to be permanently secreted from the ER. The KDEL sequence binds to specific receptors located in the ER membrane and in small secretory vesicles that bud off from the ER en route for the Golgi. The regulation of this receptor is puzzling but it represents a very efficient mechanism for restraining and recovering ER proteins.

An interesting variation of this pattern of transport occurs in channel proteins such as the K/ATPase complex consisting of two different subunits each containing a KDEL motif. Individually these subunits are prevented from leaving the ER but their assembly into a functional channel protein leads to the occlusion of the KDEL motif from the receptor. Thus once assembled these subunits are exported from the ER and this represents an effective form of 'quality control' ensuring that only functional complexes enter the transport pathway.

After synthesis, partially processed proteins destined for the plasma membrane, lysosomes or secretion are observed in the Golgi apparatus – a series of flattened membrane sacs. Within the Golgi apparatus progressive processing of proteins is observed in the form of glycosylation and once completed proteins are sent to their final destination through the use of clathrin coated vesicles.

Protein targeting to the mitochondrion and chloroplast

Mitochondria and chloroplasts have small genomes and most proteins are nuclear coded. This requires that the majority of chloroplast and mitochondrial proteins are transferred to the organelle from cytoplasmic sites of synthesis. Both organelles have outer and inner membranes and targeting involves traversing several bilayers. The import of proteins by mitochondria and chloroplasts is similar to that operating in the ER with N-terminal sequences directing the protein to organelles where specific translocation pathways operate to assist movement across membranes. The mitochondria and chloroplast present unique targeting systems in view of the number of potential locations. For the mitochondrion this includes locations in the outer or inner membranes, a location in the inter-membrane space, or a location in the matrix. Unsurprisingly the number of potential locations even within a single organelle places stringent demands on the 'signal' directing nuclear coded proteins to specific sites.

Mitochondrial targeting

In the mitochondrion translocation of precursor proteins is an energy-dependent process requiring ATP utilization and complexes in the inner and outer membranes. For small proteins (<10 kDa) the outer membrane has pores that may allow entry but in most cases the targeting information for precursor proteins resides in an N-terminal sequence extension. Mitochondrial signal sequences vary in length and composition although they have a high content of basic residues and residues with hydroxyl side chains but lack acidic side chains. An important property of these signal sequences is the capacity to form amphipathic helices in solution and this is probably related to the requirement to interact with membranes, receptors *and* aqueous environments. The majority of mitochondrial proteins are synthesized in the cytosol and chaperones play a role in maintaining newly synthesized protein in an 'import competent' state that is not necessarily the native fold but a form that can be translocated across membranes.

Table 8.9 Proteins identified as components of the yeast outer membrane translocase system

Protein	Proposed function
Tom5	Component of GIP, transfer of preprotein from Tom20/22 to GIP
Tom6	Assembly of the GIP complex
Tom7	Dissociation of GIP complex
Tom20	Preprotein receptor, preference for presequence-containing preproteins
Tom22	Preprotein receptor, cooperation with Tom20, part of GIP complex
Tom37	Cooperation with Tom machinery (Tom70)
Tom40	Formation of outer membrane translocation channel (GIP)
Tom70	Preprotein receptor, preference for hydrophobic or membrane preproteins
Tom72	None (homologue of Tom70)

Adapted from Voos, W. *et al. Biochim. Biophys. Acta* 1999, **1422**, 235–254.

Translocation involves specific complexes in the membrane given the abbreviations Tom (translocation outer membrane) and Tim (translocation inner membrane). The first step of translocation involves precursor protein binding to a translocase system consisting of at least nine integral membrane proteins (see Table 8.9). Tom20, Tom22 and Tom70 are the principal proteins involved in import. Tom20 has a single transmembrane domain at the N-terminus with a large cytosolic domain recognizing precursor proteins from their targeting signals. Tom20 cooperates with Tom22 a protein with an exposed N terminal domain that also binds targeting sequences. This may indicate autonomous rather than cooperative function but import is driven by a combination of hydrophobic and electrostatic interactions between pre-sequence and receptor. Tom70, the other major receptor has a large cytosolic domain preferentially interacting with preproteins carrying 'internal' targeting information. Specific recognition by Tom 20, 22, and 70 leads to

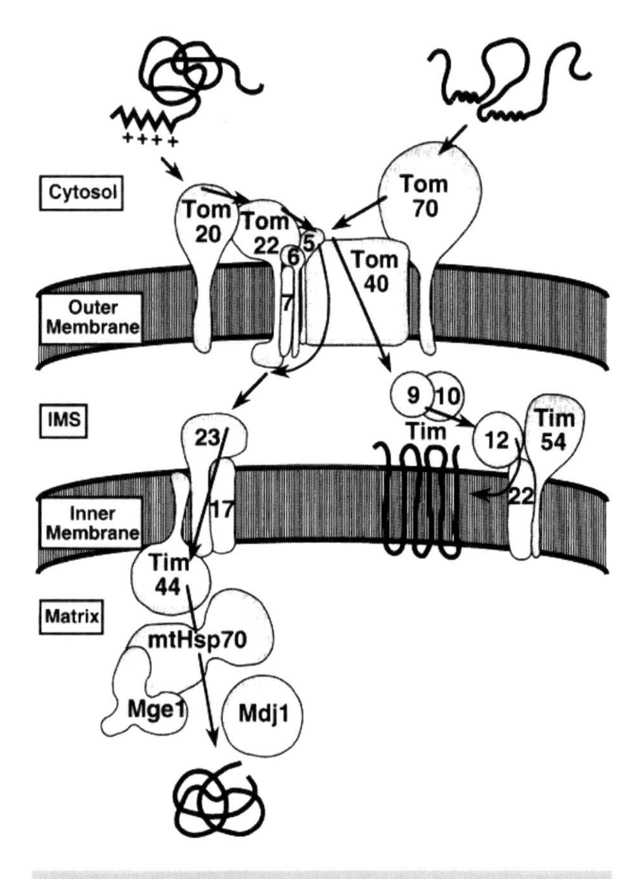

Figure 8.77 Schematic model of the mitochondrial protein import machinery of *S. cerevisiae*. (Reproduced with permission from Voos, W. *et al. Biochim. Biophys. Acta* 1999, **1422**, 235–254. Elsevier)

Table 8.10 Proteins identified as components of the yeast inner mitochondrial membrane translocase system

Protein	Possible function
Tim8	Cooperation with Tim9-Tim10
Tim9	Complex with Tim10, guides hydrophobic carrier proteins through inter membrane space.
Tim10	Complex with Tim9, guides hydrophobic carrier proteins through inter membrane space.
Tim12	Complex with Tim22 and Tim54, inner membrane insertion of carrier proteins.
Tim13	Cooperation with Tim9-Tim10
Tim17	Complex with Tim23, translocation of preproteins through inner membrane into matrix
Tim22	Inner membrane insertion of carrier proteins
Tim23	Complex with Tim17, translocation of preproteins through inner membrane into matrix
Tim44	Membrane anchor for mitochondrial Hsp70
Tim54	Complex with Tim22, inner membrane insertion of carrier proteins
Tim11	Unclear role.

Adapted from Voos, W. *et al. Biochim. Biophys. Acta* 1999, **1422**, 235–254.

a second step involving pre-sequence insertion into the Tom40 hydrophilic channel.

The inner mitochondrial membrane import pathway (Figure 8.77) completes the sequence of events started by the Tom assembly. The pathway is composed of Tim proteins (Table 8.10) with an inner membrane import channel formed by Tim23, Tim17 and Tim44. Tim23 is a membrane protein containing a small exposed acidic domain that may act as a potential binding site for sequences translocated through the outer membrane. Tim23 is hydrophobic and with Tim17 forms the central channel of the inner membrane import pathway. Tim44 is a peripheral protein that interacts with mitochondrial chaperones (Hsp70) to regulate protein folding and prevent aggregation. Since proteins are translocated in an extended conformation the role of chaperones is important to mediate controlled protein folding in the mitochondrial matrix.

Closely coupled to translocation across the inner membrane is removal of the N-terminal signal sequence by mitochondrial processing proteases (MPP) located in the matrix. An interesting, but unexplained, observation, and one that will be recalled by the careful reader, is that MPPs constituted the 'core' proteins of complex III found in mitochondrial respiratory chains.

The import pathway will target proteins into the matrix of mitochondria but to reach other destinations the basic import pathway is supplemented by additional sorting reactions.

The Rieske FeS protein is synthesized with an N-terminal matrix targeting sequence that is cleaved by the normal processing peptidase and *in vitro* this protein is detected in the matrix, a foreign location for this protein. *In vivo* additional sorting mechanisms target the protein back to the inner membrane. In contrast cytochrome c_1 has a dual targeting sequence. A typical *matrix* targeting sequence is supplemented by a second sorting signal that has a hydrophobic core and resembles bacterial export signals. For proteins in the inter-membrane space an incredibly diverse array of pathways exist. Cytochrome c is targeted to this location without obvious targeting sequences whilst yeast flavocytochrome b_2 (lactate dehydrogenase) is located in this compartment by a dual targeting sequence.

Targeting of proteins to the chloroplast

The chloroplast (Figure 8.78) uses similar principles of sorting and targeting to the mitochondrion. Targeting sequences contain significant numbers of basic residues, a high content of serine and threonine residues and are highly variable in length ranging from 20 to 120 residues. In direct analogy to their mitochondrial counterparts the translocating channel of the outer and inner chloroplast membranes are given the nomenclature Toc and Tic.

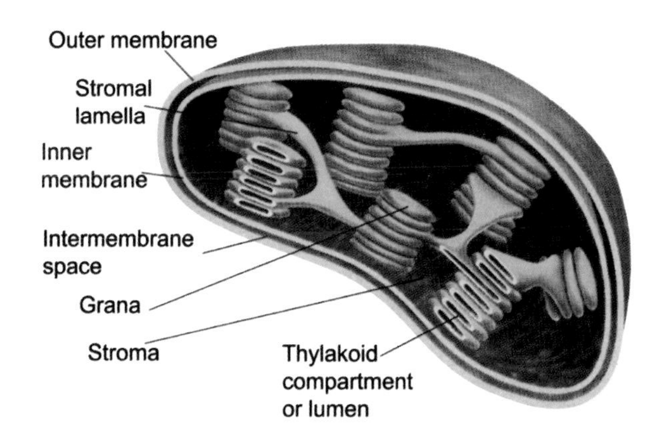

Figure 8.78 A schematic diagram of a chloroplast

Although *in vitro* the Toc and Tic translocation systems are separable the two complexes coordinate activities *in vivo* to direct proteins from the cytosol to the stroma. Two proteins, Toc159 and Toc75, act as receptors and form a conducting channel with a diameter of 0.8–0.9 nm that requires proteins to traverse outer membranes in an extended conformation. The Tic proteins involved in precursor import are Tic110, Tic55, Tic40, Tic22 and Tic20, but little is known about the mechanism of import or their structural features.

Stromal proteins such as ribulose bisphosphate carboxylase, ferredoxin or ferredoxin-NADP reductase do not require additional sorting pathways. They are synthesized with a single signal sequence that is cleaved by a stromal processing peptidase. Thylakoid proteins such as the light-harvesting chlorophyll complexes, reaction centres, and soluble proteins such as plastocyanin must utilize additional targeting pathways. Targeting is via a bipartite N-terminal signal sequence where the first portion ensures transfer into the stromal compartment whilst a second part directs the peptide to the thylakoid membrane where further proteolysis in the lumen removes the second signal sequence.

Nuclear targeting and nucleocytoplasmic transport

The final targeting system involves nuclear import and export. Nuclear proteins are required for unpacking, replication, synthesis and transcription of DNA as well as forming the structure of the nuclear envelope. There are no ribosomes in the nucleus so all proteins are imported from cytosolic sites of synthesis. This creates obvious problems with the nucleus being surrounded by an extensive double membrane. Proteins are imported through a series of large pores that perforate the nuclear membrane and result from a collection of proteins assembled into significant macromolecular complexes.

Specific nuclear import pathways were demonstrated with nucleoplasmin, a large pentameric protein complex ($M_r \sim 165$ kDa), which accumulates in the soluble phase of frog oocyte nuclei. Extraction of the protein followed by introduction into the cytosol of oocytes led to rapid accumulation in the nucleus. The rate of accumulation was far greater than expected on the basis of diffusion and involved active transport mechanisms and sequence dependent receptors. The sequence specific import was demonstrated by removing the tail regions from nucleoplasmin pentamers. This region contained a 'signal' directing nuclear import and in tail-less protein no accumulation within the nucleus was observed.

Nuclear proteins have import signals defined by short sequences of 4–10 residues that lack homology but show a preference for lysine, arginine and proline. Nuclear localization signals (NLS) are sensitive to mutation and linking sequences to cytoplasmic proteins directs import into the nucleus. Today it is possible to use computers to 'hunt' for NLS within primary sequences and their identification may indicate a potential role for a protein as well as a nuclear localization (Table 8.11). More complex bipartite sequences occur in proteins where short motifs of two to three basic residues are separated by a linker region of 10–12 residues from another basic segment.

Nucleoplasmin-coated gold particles defined the route of protein entry into the nucleus when electron-dense particles were detected by microscopy around pore complexes shortly after injection into the cytosol of oocytes. These observations suggested specific association between cytoplasmic-facing proteins of the nuclear pore complex and NLS containing protein. Nuclear transport is bi-directional (Figure 8.79) with RNA transcribed in the nucleus exported to the cytoplasm as ribonucleoprotein whilst the disassembly of the nuclear envelope during mitosis requires the re-import of all proteins. Many proteins shuttle continuously between the nucleus and cytoplasm and the average cell has an enormous level of nucleocytoplasmic traffic – some estimates suggest that over 10^3 macromolecules are transferred between the two compartments every second in a growing mammalian cell.

Table 8.11 Sequences identified to act as signals for localization to the cell nucleus

Protein	NLS sequence
SV40 T antigen	Pro Lys Lys Lys Arg Lys Val
Adenovirus E1a	Lys Arg Pro Arg Pro
Nucleoplasmin	Lys Arg Pro Ala Ala Thr Lys Lys Ala Gly Gln Ala Lys Lys Lys Lys

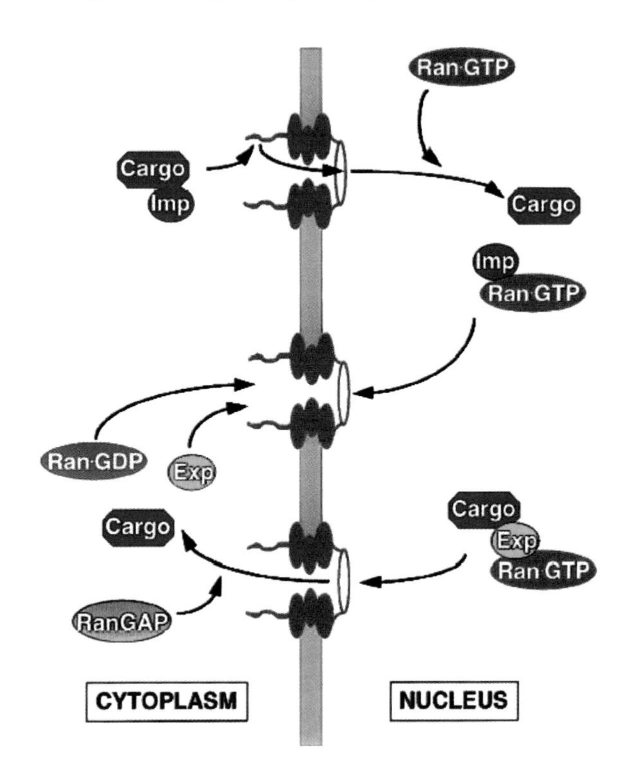

Figure 8.79 A schematic representation of nuclear import involving importins, cargo and Ran-GTP (reproduced with permission from Cullen, B.R. *Mol. Cell. Biol.* 2000, **20**, 4181–4187)

The first proteins discovered with a role in shuttling proteins from the cytosol to nucleus were identified from their ability to bind to proteins containing the NLS. The protein was importin-α (Figure 8.80) and homologues were cloned from other eukaryotes to define a family of importin-α-like proteins. Importin-α functions as a heterodimer with a second protein called importin-β. A typical scenario for nuclear transport involves cargo recognition via the NLS by importin-α and complex formation with importin-β followed by association and translocation through the nuclear pore complex in a reaction requiring GTP. In the

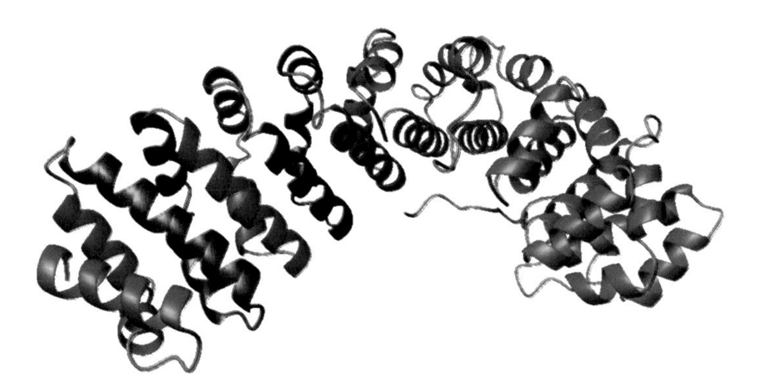

Figure 8.80 The super-helix formed by collections of armadillo motifs in mouse importin-α. A unit of three helices is shown in green (PDB: 1AIL)

nucleus the complex dissociates in reactions catalysed by monomeric G proteins such as Ran-GTP. Ran-GTP binds specifically to importin-β allowing it to be recycled back to the cytoplasm in events again linked to GTP hydrolysis. Importin-α and the cargo dissociate with the protein 'delivered' to its correct destination.

The structure of importin-α, determined in a complex with and without NLS sequence, reveals two domains within a polypeptide of ~60 kDa. A basic N-terminal domain binds importin-β and a much larger second domain composed of repeating structural units known as armadillo repeats (Figure 8.80). Each unit is composed of three α helices and the combined effect of each motif is a translation of ~0.8–1.0 nm and a rotation of ~30°, forming a right-handed super-helical structure.

Importin-β binds to the N-terminal of importin-α but also interacts with the nuclear pore complex and Ran GTPases. The interaction with Ran proteins is located at its N-terminal whilst interaction with importin-α is centred around the C-terminal region. Importin-β (Figure 8.81) also contains repeated structural elements – the HEAT motif (the proteins in which the motif was first identified are Huntingtin protein–Elongation factor (EF3)–A subunit of protein phosphatase and the TOR protein). The motif possesses a pattern of hydrophobic and hydrophilic residues and is predicted to form two helical elements of secondary structure. In a complex with the N-terminal fragment (residues 11 to 54) of the α subunit the HEAT motifs arrange in a convoluted snail-like appearance where the

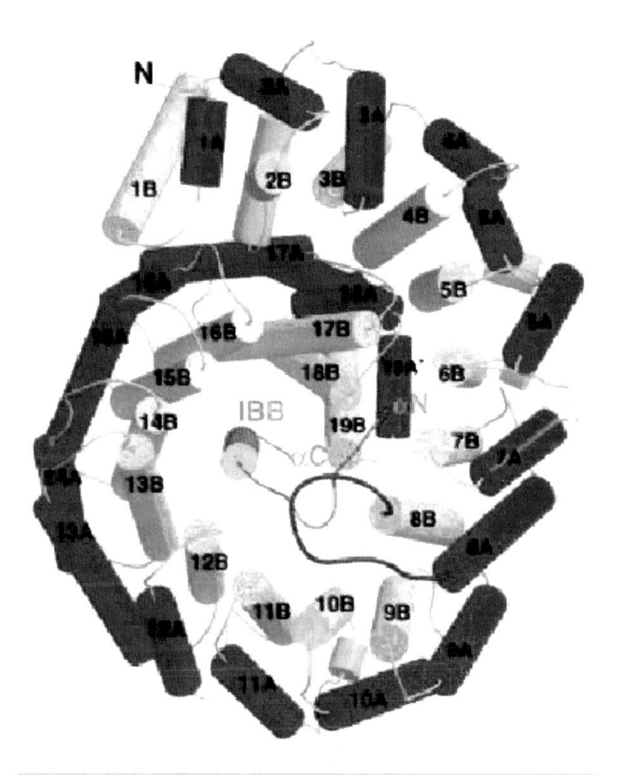

Figure 8.81 Structure of importin-β plus IBB domain from a view down the superhelical axis. A and B helices of each HEAT motif are shown in red and yellow respectively. The acidic loop with the DDDDDW motif in HR8 is shown in blue. (Reproduced with permission from Cingolani, G. *et al. Nature* 1999, **399**, 221–229. Macmillan)

extended IBB domain is located at the centre of the protein. Importin-β is composed of 19 HEAT repeats forming a right-handed super-helix where each HEAT motif, composed of A and B helices connected by a short loop, arranges into an outer layer of 'A' helices defining a convex surface and an inner concave layer of 'B' helices. The HEAT repeats vary in length from 32 to 61 residues with loop regions containing as many as 19 residues. HEAT repeats 7–19 bind importin-α with the N-terminal fragment of importin-α bound on the inner surface forming an extended chain from residues 11–23 together with an ordered helix from residues 24–51. The axis of this helix is approximately coincident to that of importin-β super-helix. The N-terminal region of the importin-α interacts specifically with a long acidic loop linking helices 8A and 8B that contains five aspartate residues followed by a conserved tryptophan residue. HEAT repeats 1–6 are implicated in the interaction with Ran.

The third protein with a critical role in nuclear–cyto plasmic transport is Ran – a GTPase essential for nuclear transport. Ran is a member of the Ras-like family of GTP binding proteins and switches between active GTP bound forms and an inactive state with GDP (Figure 8.82). Ran is similar to other monomeric GTPases but it also possesses a long C-terminal extension that is critical for nuclear transport function and is terminated by a sequence of acidic residues (DEDDDL).

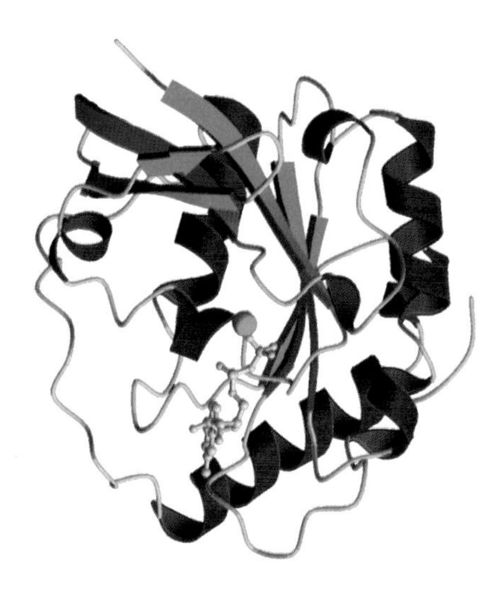

Figure 8.82 The structure of Ran-GDP

Ran consists of a six-stranded β sheet surrounded by five α helices. GDP and GTP are bound via the conserved NKXD motif with Lys123 and Asp125 interacting directly with the base, the ribose portion exposed to the solvent and the phosphates binding via a P loop motif and a large number of polar interactions with the sequence GDGGTGKT. The loop region acts as a switch through nucleotide-exchange induced conformational changes and dictates interactions with other import proteins such as importin-β.

Ran itself is 'regulated' by a cytosolic GTPase activating protein termed Ran GAP1 that causes GTP hydrolysis and inactivation. In the nucleus a chromatin-bound nucleotide exchange factor called RCC1 functions to exchange GDP/GTP. The interplay of factors controlling nucleotide exchange and hydrolysis generates gradients of Ran-GTP that direct nuclear–cytoplasmic transport.

Ran binds to the N terminal region of importin-β, but importantly the C-terminal of Ran is able to associate with other proteins largely through charged residues in the tail (DEDDDL). These charges mediate Ran-GTP/importin-β complexes interaction with Ran binding proteins in the nuclear pore complex but also offer a mechanism of dissociating complexes in the nucleus. Gradually many of the molecular details of protein import via nucleocytoplasmic transport machinery have become clear with structures for the above proteins and complexes formed between Ran and RCC1 as well as the ternary complex of Ran–RanBP1-RanGAP. Ran, importin-α and -β subunits are exported back to the cytosol for future rounds of nucleocytoplasmic transport.

Proteins identify themselves to transport machinery with two signals. In addition to the positively charged NLS some proteins have nuclear export signals (NES) based around leucine-rich domains.

The nuclear pore assembly

In higher eukaryotes the nuclear pore complex is a massive macromolecular assembly containing nearly 100 different proteins with a combined mass in excess of 120 MDa. In yeast this assembly is simpler with approximately 30 different polypeptides forming the

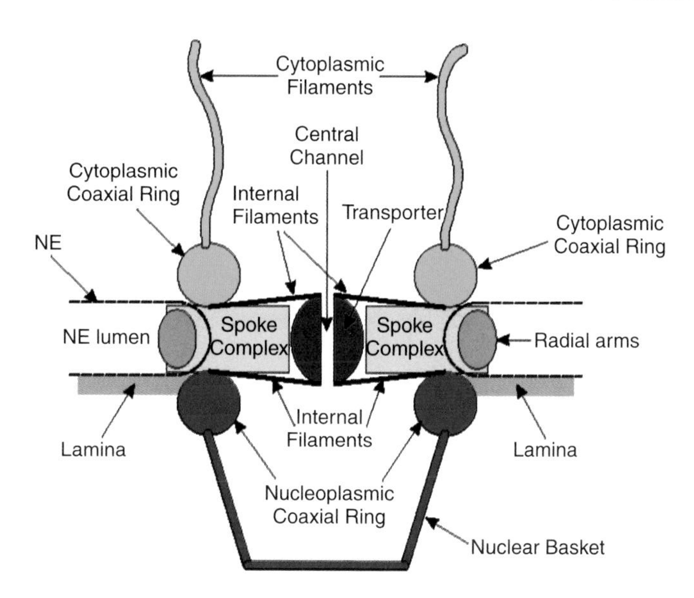

Figure 8.83 A cutaway representation of the nuclear pore complex (reproduced with permission from Allen, T.D. *et al. J. Cell Sci.* 2000, **113**, 1651–1659. Company of Biologists Ltd)

complex. The increased size of the vertebrate pore probably reflects complexity associated with metazoan evolution, but the basic structural similarities between the pore complexes of yeast and higher eukaryotes indicates a shared mechanism of translocation that can be discussed within a single framework.

The proteins of the nuclear pore complex are called nucleoporins or 'nups'. As the role of Ran, importins and other import proteins were uncovered over the last decade emphasis has switched to the structure and organization of the nuclear pore complex. Electron microscopy reveals the pore complex as a symmetrical assembly containing a pore at the centre that presents a barrier to most proteins. The nucleoporin complex (NPC) spans the dual membrane of the nuclear envelope and is the universal gateway for macromolecular traffic between the cytoplasm and the nucleus. The basic framework of the NPC consists of a central core with a spoke structure (Figure 8.83). From this central ring long fibrils 50–100 nm in length extend into the nucleoplasm and the cytoplasm whilst the whole NPC is anchored within the envelope by the nuclear lamina.

Clarification of nuclear pore organization came from studies of the yeast complex where genetic malleability coupled with genome sequencing revealed 30 nucleoporins that are conserved across phyla and thus allowed identification of homologues in metazoan eukaryotes. Once identified nups were cloned and localized within the envelope using a combination of immunocytochemistry, electron microscopy and cross-linking studies. The yeast nuclear pore complex has been redefined in terms of nucleoporin distribution. Nup358 is equated with the RanBP2 protein described previously whilst p62, p58, p54 and p45 are proteins found at the centre of the nuclear pore complex and gp210 and POM121 form all or part of the spoke assembly.

Protein turnover

Within the cell sophisticated targeting systems direct newly translated proteins to their correct destinations. However, proteins have finite lifetimes and normal cell function requires specialized pathways for degradation and recycling. The half-life of most proteins is measured in minutes although some, such as actin or haemoglobin, are more stable with half-lives in excess of 50 days. At some point all proteins 'age' as a result of limited proteolysis, covalent modification or non-enzymatic reactions leading to a decline in activity.

Figure 8.84 The primary sequence of ubiquitin from plants, animal and microorganisms shows only four changes in sequence at positions 19, 24, 28 and 57

These proteins are degraded into constituent amino acids and re-cycled for further synthetic reactions. The proteasome is a large multimeric complex designed specifically for controlled proteolysis found in all cells. In eukaryotes the ubiquitin system contributes to this pathway by identifying proteins destined for degradation whilst organelles such as the lysosome perform ATP-independent protein turnover.

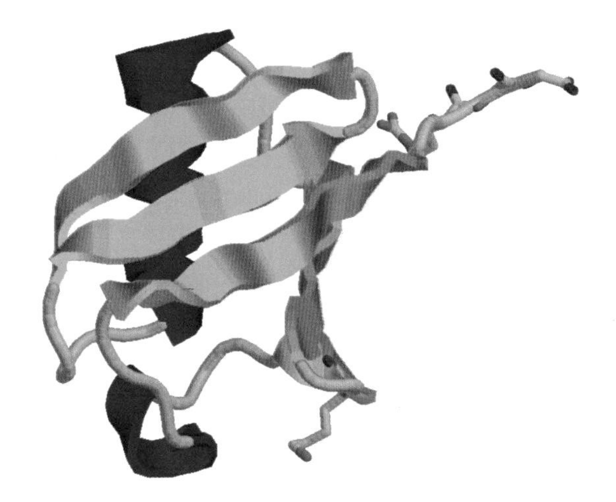

Figure 8.85 The structure of ubiquitin showing four significant β strands (1–7,10–17, 40–45, 64–72) and a single α helix (23–34) whilst shorter strand and helix regions exists from 48–50 and 56–59, respectively. The residues Arg 74, Gly75 and Gly76 are shown along with Lys48

The ubiquitin system and protein degradation

In eukaryotes ubiquitin (Figure 8.84) occupies a pivotal role in the pathway of protein degradation. It is a highly conserved protein found in all eukaryotic organisms but never in either eubacteria or archae. The level of homology exhibited by ubiquitin is high with only 3 differences in primary sequence between yeast and human protein. The lack of evolutionary divergence points to functional importance and suggests that most residues in ubiquitin have essential roles.

In view of its importance to protein degradation the structure of ubiquitin (Figure 8.85) is unspectacular containing five β strands together with a single α helix. The final three C-terminal residues are Arg-Gly-Gly and extend away from the globular domain into bulk solvent. This arrangement is critical to biological function as the C-terminal glycine forms a peptide bond with the ε-amino group of lysine in target proteins. Covalent modification by adding ubiquitin in a process analogous to phosphorylation has led to the terminology ubiquitination (also called ubiquitinylation). Multiple copies of ubiquitin attach to target proteins acting as 'signals' for degradation by the proteasome.

Many enzymes catalyse ubiquitin addition to proteins and include E1 enzymes or ubiquitin-activating enzymes, E2 enzymes, known as ubiquitin-conjugating enzymes, and E3 enzymes, which act as ligases. Ubiquitin is activated by E1 in a reaction hydrolysing ATP and forming ubiquitin adenylate. The activated ubiquitin binds to a cysteine residue in the active site of E1 forming a thiol ester. E2, the proximal donor of ubiquitin to target proteins, transfers ubiquitin to the acceptor lysine forming a peptide bond, although E3 enzymes also participate by forming complexes with target protein and ubiquitin-loaded E2. Figure 8.86 shows the ubiquitin-mediated degradation of cytosolic proteins.

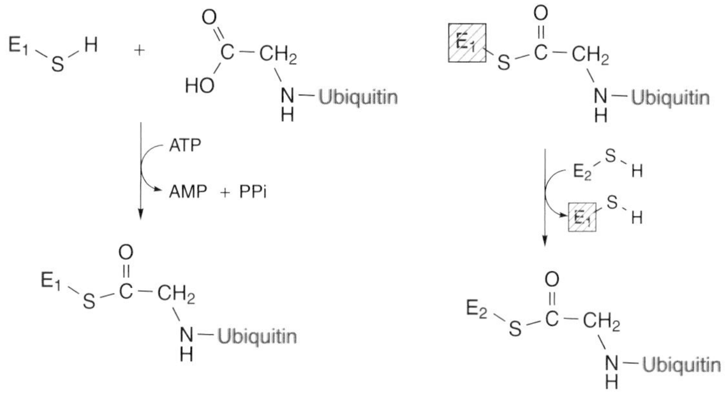

Figure 8.86 The first step in the ubiquitin-mediated degradation of cytosolic proteins is ATP dependent activation of E1 to form a thiol ester derivative with the C-terminal glycine of ubiquitin. It is followed by transfer of ubiquitin from E1 to the E2 enzyme, the proximal donor of ubiquitin to proteins

Multiple ubiquitination is common and involves peptide bond formation between the carboxyl group of Gly76 and the ε-amino group of Lys48 on a second ubiquitin molecule. Substitution of Lys48 with Cys results in a protein that does not support further ubiquitination and is incapable of targeting proteins for proteolysis. Four copies of ubiquitin are required to target proteins efficiently to the 26*S* proteasome with 'Ub_4' units exhibiting structural characteristics that enhance recognition. Ubiquitin does not degrade proteins but tags proteins for future degradation. A view of ubiquitin as a simple tag may be an over-simplification since a role for ubiquitin enhancing association between proteins and proteasome is implied by the observation that without ubiquitin proteins interact with the proteasome but quickly dissociate.

Alternative strategies exist for degrading proteins within cells. One strategy used by cells is to inactivate proteins by oxidizing susceptible side chains such as arginine, lysine and proline. This event is enough to 'mark' proteins for degradation by cytosolic proteases. A second method of protein turnover is identified with sequence information and 'short-lived' proteins contain sequences rich in proline, glutamate, serine and threonine. From the single letter code for these residues the sequences have become known as PEST sequences. Comparatively few long-lived proteins contain PEST-rich regions and the introduction of these residues into stable proteins increases turnover although the structural basis for increased ability is unclear along with the relationship to the ubiquitin pathway. A third mechanism of controlling degradation lies in the composition of N-terminal residues. A correlation between the half-life of a protein and the identity of the N-terminal residue gave rise to the 'N-end' rule. Semi-quantitative predictions of protein lifetime from the identity of the N-terminal residue suggests that proteins with Ser possess half-lives of 20 hours or greater whilst proteins with Asp have on average half-lives of ~3 min. The mechanism that couples recognition of the N-terminal residue and protein turnover is unknown but these correlations are also observed in bacterial systems lacking the ubiquitin pathway of degradation.

The proteasome

Intracellular proteolysis occurs via two pathways: a lysosomal pathway and a non-lysosomal, ATP-dependent, pathway. The latter pathway degrades most cell proteins and involves the proteasome first identified as an endopeptidase with multicatalytic activities from bovine pituitary cells. The multicatalytic protease was called the proteasome, reflecting its complex structure and proteolytic role.

The proteosome degrades proteins via peptide bond scission, and multiple catalytic functions are seen via

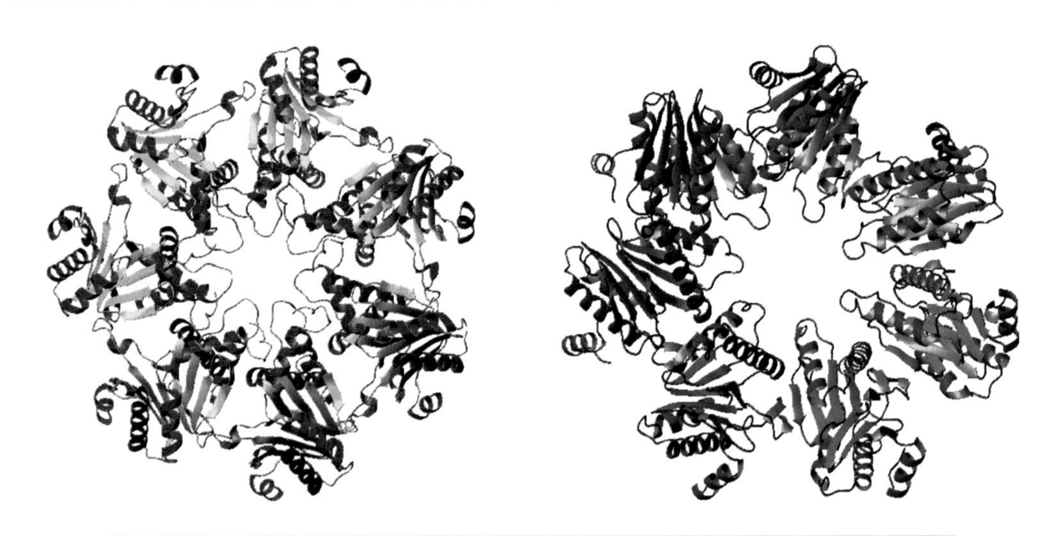

Figure 8.87 Superior views of α and β heptameric rings showing central cavity

the hydrolysis of many synthetic and natural substrates. The eukaryotic proteasome has activities described by terms such as 'chymotryptic-like' activity (preference for tyrosine or phenylalanine at the P1 position), 'trypsin-like' activity (preference for arginine or lysine at the P1 position) and 'post-glutamyl' hydrolysing activity (preference for glutamate or other acidic residues at the P1 position). The variety of catalytic activities created a confusing functional picture since proteasomes cleave bonds after hydrophobic, basic and acidic residues. In the proteasome the myriad of catalytic activities suggested a unique site. The proteasome was first identified in eukaryotes but comparable complexes occur in prokaryotes where the system can be deleted without inducing lethality. As usual, eukaryotic proteasomes have complex structures whilst those from prokaryotes are simpler.

The hyperthermophile *Thermoplasma acidophilum* proteasome contains two subunits of 25.8 kDa (α) and 22.3 kDa (β) arranged in a 20S proteasome containing 28 subunits: 14 α subunits and 14 β subunits arranged in four stacked rings. The two ends of the cylinder each consist of seven α subunits whilst the two inner rings had seven β subunits – each ring is a homoheptamer. The $\alpha_7\beta_7\beta_7\alpha_7$ assembly forms a three-chambered cylinder with two antechambers located on either sides of a central cavity (Figure 8.87). Sequence similarity exists between α and β subunits and there is a common fold based around an antiparallel array of β strands surrounded by five helices. Helices 1 and 2 are on one side of the β sandwich whilst helices 3, 4 and 5 are on the other. In the case of the α subunits an N-terminal extension of ~35 residues leads to further helical structure that fits into a cleft in the β strand sandwich. In the β subunits this extension is absent and the cleft remains 'open' forming part of the active site. The crystal structure of the 20S complex (Figure 8.88) showed a cylindrical assembly (length ~15 nm, diameter 11.3 nm) with the channel running the length of the assembly but widening to form three large internal cavities separated by narrow constrictions. The cavities between the α subunit and β subunit rings are the 'antechambers' (~4 × 5 nm diameter) with the third cavity at the centre of the complex containing the active sites. Between the antechamber and the central cavity access is restricted to approximately 1.3 nm by a loop region containing a highly conserved RPXG motif derived from the α subunit. In this loop a Tyr residue protrudes furthest into the channel to restrict access. Further into the channel side chains derived from the β subunits exist at the entrance to the central cavity and restrict the width to ~2.2 nm. Together, these systems form a gating system controlling entry of polypeptide chains to those that thread their way into the central cavity. A scheme of proteolysis involving protein unfolding and

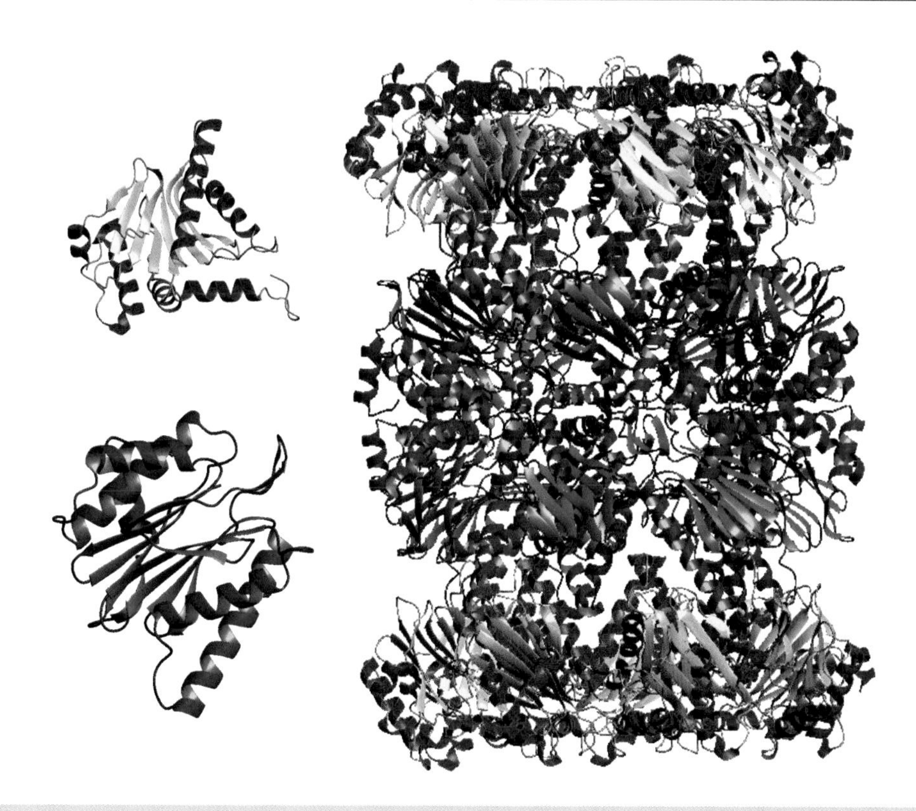

Figure 8.88 The three-dimensional structure of the 20S proteasome from the archaebacteria *T. acidophilum* (PDB: 1PMA). The structure – a 673 kDa protease complex has a barrel-shaped structure of four stacked rings. An individual α subunit is shown top left, an individual β subunit bottom left. The colour scheme is maintained in the whole complex and shows only the distribution of secondary structure for clarity

the formation of extended structure prior to degradation is likely from the organization of the proteasome.

The active site is located on the β subunit where the structure reveals a β sandwich with one side predominantly open. This side faces the inside pointing towards the central cavity. The α subunits possess a highly conserved N-terminal extension with residues 1–12 remaining invisible in the crystal structure possibly due to mobility whilst part of this extension is seen as a helix (residues 20–31) that occupies the cleft region in the β sandwich. Although the precise function of this extension is unknown its strategic location coupled with sequence conservation suggests an important role in translocation of substrate to the proteasome interior. The β subunits lack these N-terminal extensions but have pro-sequences of varying length that are cleaved during assembly of the proteasome. The most important function of this proteolytic processing is the generation of the active site residues.

The 14 identical catalytically active (β) subunits show highest activity for bond cleavage after hydrophobic residues but extensive mutagenesis involving all serines, two histidines, a single cysteine and two conserved aspartate residues in the β subunit failed to inhibit enzyme activity. The results suggested degradation was not associated with four classical forms of protease action, namely serine, cysteine, aspartyl and metalloproteinases. Further mutagenesis revealed that deletion of the N-terminal threonine or its replacement by alanine resulted in inactivation. Mechanistic studies showed that *N*-acetyl-Leu-Leu-norLeu-CHO (Figure 8.89) was a potent inhibitor of proteolysis, and crystallography of the proteasome-inhibitor adduct located the peptide bound via the

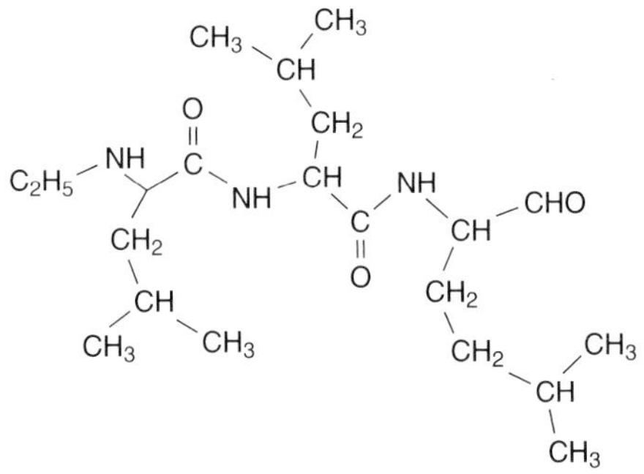

Figure 8.89 The potent inhibitor of proteasome activity *N*-acetyl-Leu-Leu-norLeu-CHO

aldehyde group to the –OH group of the N-terminal threonine.

Further insight into catalytic mechanism came with observations that the metabolite lactocystin derived from *Streptomyces* bound to the eukaryotic complex resulting in covalent modification of Thr1.[1] The results implied the involvement of N-terminal Thr residues in catalytic reactions. Initially the proteasome fold was thought to be a unique topology but gradually more members of this family of proteins have been uncovered with a common link being that they are all N-terminal nucleophile hydrolases or NTN hydrolases.

Fourteen genes contribute to the eukaryotic 20S proteasome (Figure 8.90). Each subunit has a similar structure but is coded by a separate gene. Crystallography and cryo-EM studies have shown that the patterns of subunit organization extend to eukaryotic proteasomes and when viewed in the electron microscope archaeal, bacterial, and eukaryotic 20S proteasomes form similar barrel-shaped particles ~15 nm in height and ~11 nm in diameter consisting of four heptameric rings (Figure 8.91).

The 20S assembly of the proteasome identified in eukaryotes, and first structurally characterized in *T. acidophilum*, is one component of a much larger complex found in eukaryotes. The 20S proteasome represents the core region of a larger assembly that is capped by a 19S regulatory complex. The attachment of a regulatory 'unit' to the 20S proteasome results in the formation of a 2.1 MDa complex called the 26S proteasome.

With two 19S regulatory complexes attached to the central 20S proteasome the 26S proteasomes from *Drosophila*, *Xenopus*, rat liver and spinach cells have similar shapes with an approximate length of 45 nm and a diameter of 20 nm. The significance of the 19S regulatory complex in association with the 20S assembly is that it confers ATP- and ubiquitin-dependent proteolysis on the proteasome; it is this pathway that performs most protein turnover within eukaryotic cells degrading over 90 percent of cellular proteins.

The organization of the 19S complex remains an active area of research and is relatively poorly understood in comparison to the 20S assembly. The complex from yeast contains at least 18 different proteins whose sequences are known and reversible interactions occur with proteins in the cytosol complicating compositional analysis. Within the group of 18 different subunits are six polypeptides hydrolysing ATP (ATPases) and showing homology to other members of the AAA superfamily. The AAA superfamily represents ATPases associated with various cellular activities and are found in all organisms where a defining motif is the presence of one or more modules of ~230 residues that includes an ATP-binding site and the consensus sequence A/G-x_4-G-K-S/T.

A general working model for the action of the proteasome envisages that the cap assembly binds ubiquitinated proteins, removes the ubiquitin at a later stage for recycling and unfolds the polypeptide prior to entry into the 20S core particle. The proteolytic active sites of the proteasome are found in the core particle and the cap assembly does not perform proteolysis. However, free 20S proteasome does not degrade ubiquitinated protein conjugates and shows lower activity suggesting that the 19S cap assembly is crucial to entry of substrate, activity and ubiquitin dependent hydrolysis.

Lysosomal degradation

Lysosomes contain hydrolytic enzymes including proteases, lipases, and nucleases. Any biomolecule trapped in a lysosome is degraded, and these organelles are viewed as non-specific degradative systems whilst the ubiquitin system is specific, selective and 'fine-tuned' to suit cellular demand. Lysosomes are often described as enzyme sacs and defects in lysosomal

[1] Surprisingly this inhibitor does not inactivate the *Thermoplasma* proteasome.

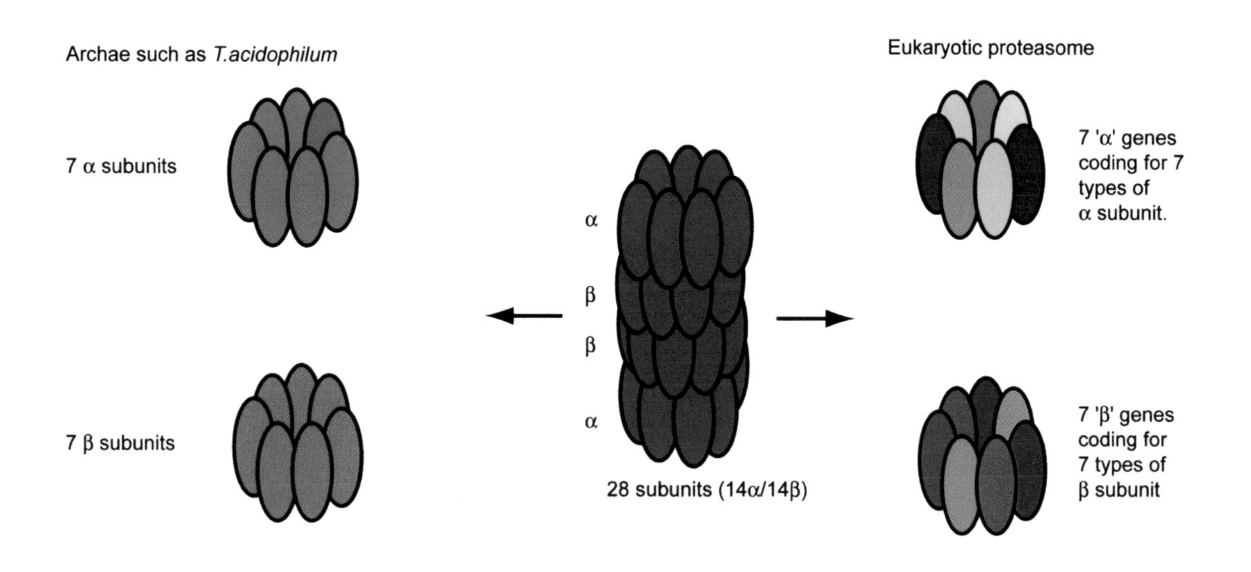

Figure 8.90 The structural organization of the proteasome.

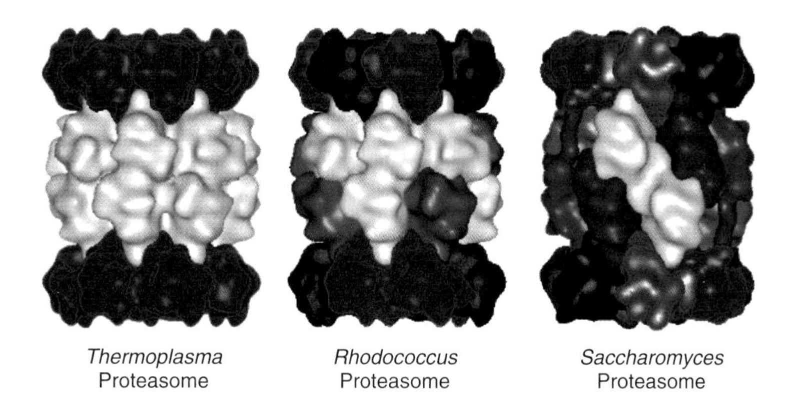

Figure 8.91 Cryo-EM derived maps of the assembly of 20S proteasomes from archaea, eubacteria and eukaryotic cells. The α type subunits are colored in red and the β type subunits in blue. The different shades of these colors indicate that two or seven distinct α and β-type subunits form the outer and inner rings of the *Rhodococcus* or *Saccharomyces* proteasome respectively (reproduced with permission from Vosges, D. *et al. Ann. Rev. Biochem.* 1998, **68**, 1015–1068. Annual Reviews Inc)

proteins contribute to over 40 recognizable disorders, collectively termed lysosomal storage diseases.

Individually, the disease states are rare but lead to progressive and severe impairment resulting from a deficiency in the activity of specific enzymes. The loss of activity impedes the normal degradative function of lysosomes and contributes to the accumulation of unwanted biomolecules. Included in the collection of identified lysosomal based diseases are: (i) Hurler syndrome, a deficiency in α-L-iduronidase required to degrade mucopolysaccharides; (ii) Gauchers' disease, a deficiency in β-glucocerebrosidase needed in the degradation of glycolipids; (iii) Fabry disease, caused by a deficiency in α-galactosidase, an enzyme degrading glycolipids; (iv) Pompe disease, a deficiency of α-glucosidase required to break down glycogen; and (v) Tay–Sachs disease, a lysosomal defect occurring as a result of a mutation in one subunit of the enzyme β-hexosaminidase A that leads to the accumulation of the GM2 ganglioside in neurones.

Apoptosis

Apoptosis is the process of programmed cell death. Whilst superficially this appears to be an unwanted process more careful consideration reveals that apoptosis is vital during development. A developing embryo goes through many changes in structure that can only be achieved by programmed cellular destruction. Similarly, the development of insects involves the reorganization of tissues between the larval and adult states. The concept of apoptosis is vital to normal development and although many details, such as control and initiation remain to be elucidated, the importance of a family of intracellular proteases called caspases is well documented.

Caspases are synthesized as zymogens with their activity inhibited by covalent modification, in which the active enzyme is synthesized as a precursor protein joined to additional death effector domains. This configuration is vital to achieve complete inhibition of activity since the cell is easily destroyed by the activity of caspases. Activation involves a large number of 'triggers' that excise the death effector domains, process the zymogen and form active caspases. Once activated caspases catalyse the formation of other caspases, leading to a proteolytic cascade of cellular destruction. Several methods of regulating caspase activity and hence apoptosis exist within the cell. Zymogen gene transcription is regulated whilst anti-apoptotic proteins such as those of the Bcl-2 family occur in cells and block activation of certain pro-caspases.

The first members of the caspase family identified were related to human interleukin-1 converting enzyme (ICE) and the product of the nematode cell-death gene *CED3*. To date 11 caspases have been identified in humans although mammals may possess 13 different caspases, *Drosophila melanogaster* contains seven and nematodes such as *Caenorhabditis elegans* contain three. Thus, there appears to have been an expansion in their number with increasing cellular complexity. The term caspase refers to the action of these proteins as cysteine-dependent aspartate-specific proteases. Their enzymatic properties are governed by specificity for substrates containing Asp and the use of a Cys285 sidechain for catalysing peptide-bond cleavage located within a conserved motif of five residues (QACRG). Whereas the activity of the proteasome governs the day to day turnover of proteins the activity of caspases signals the end of the cell with all proteins degraded.

Summary

Fifty years have passed since the discovery of the structure of DNA. This event marking the beginning of molecular biology expanded understanding of all events in the pathway from DNA to protein. This included structural and functional descriptions of proteins involved in the cell cycle, DNA replication, transcription, translation, post-translational events and protein turnover. In many instances determining structure has uncovered intricate details of their biological function.

The cell cycle is characterized by four distinct stages: a mitotic or M phase is followed by a G_1 (gap) phase that represents most of the cell cycle, a period of intense synthetic activity called the S phase, and finally a short G_2 phase as the cell prepares for mitosis.

Genetic studies of yeast identified mutants in which cell division events were inhibited. Control of the cell cycle is mediated by protein kinases known as Cdks where protein phosphorylation regulates cellular activity. Activity of Cdks is dependent on cyclin binding with optimal activity occurring for Cdk2 in a complex with cyclin A and phosphorylation of Thr160. The structure of this complex results in the critical movement of the T loop, a flexible region of Cdk2, governing accessibility to the catalytic cleft and the active site threonine.

Transcription is DNA-directed synthesis of RNA catalysed by RNA polymerase. Transcription proceeds from a specific sequence (promoter) in a 5′–3′ direction until a second site known as the transcriptional terminator is reached. In eukaryotes three nuclear RNA polymerases exist with clearly defined functions. RNA polymerase II is concerned with the synthesis of mRNA encoding structural genes.

Sequences associated with transcriptional elements have been identified in prokaryotes and eukaryotes. In eukaryotes the TATA box is located upstream of the transcriptional start site and governs formation of a pre-transcriptional initiation complex. The TATA box resembles the Pribnow box or −10 region found in prokaryotes.

Specific TATA binding proteins have been identified and structural studies reveal that basal transcription requires in addition to RNA polymerase the formation of pre-initiation complexes of TBP, TFIIB, TFIIE, TFIIF and TFIIH.

In eukaryotes transcription is followed by mRNA processing that involves addition of 5′ G-caps and 3′ polyA tails. Non-coding regions of mRNA known as introns are removed creating a coherent translation-effective mRNA. Introns are removed by the spliceosome.

The spliceosome contains snRNA complexed with specific proteins. Processing initial mRNA transcripts involves cutting at specific pyrimidine rich recognition sites followed by splicing them together to create mRNA that is exported from the nucleus for translation at the ribosome.

Translation converts mRNA into protein and occurs in ribonucleoprotein components known as the ribosomes. Ribosomes convert the genetic code, a series of three non-overlapping bases known as the codon, into a series of amino acids covalently linked together in a polypeptide chain.

All ribosomes are composed of large and small subunits based predominantly on highly conserved rRNA molecules together with over 50 different proteins. Biochemical studies identified two major sites known as the A and P sites. The P site (peptidyl) contains a growing polypeptide chain attached to tRNA. The A site (aminoacyl) contains charged tRNA species bearing a single amino acid that will be added to extending chains.

Protein synthesis is divided into initiation, elongation and termination. All stages involve accessory proteins such as IF1, IF2 and IF3 together with elongation and release factors such as EF1, EF2, EF3, RF1, RF2 and RF3.

Elongation is the most extensive process in protein synthesis and is divided into three steps. These processes are amino acyl tRNA binding at the A site, peptidyl transferase activity and translocation.

Structures for 50 and 30S subunits revolutionized understanding of ribosome function. The structure of the large subunit confirmed conclusively that the peptidyl transferase reaction is catalysed entirely by RNA; the ribosome is a ribozyme.

Initial translation products undergo post-translational modification that vary dramatically in type from oxidation of thiol groups to the addition of new covalent groups such as GPI anchors, oligosaccharides, myristic acid 'tails', inorganic groups such as phosphate or sulfate and larger organic skeletons such as heme.

The removal of peptide 'leader' sequences in the activation of zymogens is another important post-translational modification and converts inactive protein into an active form. Many enzymes such as proteolytic digestive enzymes, caspases and components of the blood clotting cascade are activated in this type of pathway.

N-terminal signal sequences share physicochemical properties and are recognized by a SRP. The SRP directs nascent chains to the ER membrane or cell membrane of prokaryotes. Signal sequences do not exhibit homology but have a basic N-terminal region followed by a hydrophobic core and a polar C-terminal region proximal to the cleavage site.

SRPs are found in the archae, eubacteria and eukaryotes. Mammalian SRP is the most extensively characterized system consisting of rRNA and six distinct polypeptides. The SRP directs polypeptide chains to the ER membrane and the translocon, a membrane bound protein-conducting channel.

Other forms of intracellular protein sorting exist within eukaryotes. Proteins destined for the mitochondria, chloroplast and nucleus all possess signals within their polypeptide chains. For the nucleus protein import requires the presence of a basic stretch of residues arranged either as a single block or as a bipartite structure anywhere within the primary sequence.

NLS are recognized by specific proteins (importins) that bind the target protein and shuttle the 'cargo' towards the nuclear pore complex. The formation of importin–cargo complexes is controlled by G proteins such as Ran.

Proteins are not immortal – they are degraded with turnover rates varying from minutes to weeks. Turnover is controlled by a complex pathway involving ubiquitin labelling.

Ubiquitin is a signal for destruction by the proteasome. The proteasome has multiple catalytic activities in a core unit based around four heptameric rings. The 20S proteasome from *T. acidophilum* has an $\alpha_7\beta_7\beta_7\alpha_7$ assembly forming a central channel guarded by two antechambers. The central chamber catalyses proteolysis based around the N terminal threonine residue (Thr1) where the side chain acts as a nucleophile attacking the carbonyl carbon of peptide bonds.

In prokaryotes the proteasome degrades proteins in ubiquitin independent pathways. In eukaryotes the 20S proteasome catalyses ubiquitin-dependent proteolysis only in the presence of a cap assembly.

Further pathways of protein degradation exist in the lysosome where defects are responsible for known metabolic disorders, such as Tay–Sachs disease, and by caspases that promote programmed cell death (apoptosis).

Problems

1. Using knowledge about the genetic code describe the products of translation of the following synthetic nucleotide ......AUAUAUAUAUAUA.... in a cell free system. Explain the results and describe the critical role of alternating oligonucleotides in determining the genetic code. Using the data presented in Table 8.6 how would this translation product differ using a system reconstituted from mitochondria.

2. Arrange the following macromolecules in decreasing order of molecular mass: tRNA, subunit L23, tetracycline, 23S rRNA, the large ribosomal subunit, 5S RNA, spliceosome protein U1A, *E. coli* DNA polymerase I, *E. coli* RNA polymerase. Obtain information from databases or information given within Chapter 8.

3. Recover the structure of the cyclinA: cdk2 binary complex from a protein database. Using any molecular graphics package highlight the following residues Lys266, Glu295, Leu255, Asp240, and Arg211 of cyclin A and Glu 8, Lys 9, Asp36, Asp38 Glu40 and Glu42 of cdk2. Describe the arrangement of these residues.

4. Puromycin binds to the A site and prevents further chain elongation. From the structure presented in Figure 8.54 account more fully for this observation.

5. The Yarus inhibitor is described as a transition state analogue that mimics the tetrahedral intermediate generated when the α amino group of the A site bound amino acyl tRNA attacks the carbonyl group of the ester linking a peptide to the peptidyl tRNA at the P site. Draw the normal assumed tetrahedral intermediate most closely linked to this inhibitor highlighting the tetrahedral carbon centre.

6. An N-terminal photo-affinity label has been used to map the tunnel of the large ribosomal subunit. Describe how you might perform this experiment and what results might you expect to obtain from such a procedure?

7. Unlike the mitochondria and chloroplast signal sequence nuclear localization signals are not removed. What might be the reasons underlying this observation?

8. Caspases are proteolytic enzymes with an active site cysteine. Discuss possible enzyme mechanisms in view of the known activity of serine proteases.

9

Protein expression, purification and characterization

To study the structure of *any* protein successfully it is necessary to purify the molecule of interest. This is often a formidable task especially when some proteins are present within cells at low concentration, perhaps as few as 10 molecules per cell. Frequently, this involves purifying a single protein from a cell paste containing over 10 000 different proteins. The ideal objective is to obtain a single protein retaining most, if not all, of its native (*in vivo*) properties. This chapter focuses on the methods currently employed to isolate and purify proteins.

Although functional studies often require small amounts of protein (often less than 1 ng or 1 pmol), structural techniques require the purification of proteins on a larger scale so that 10 mg of pure protein is sometimes needed. Taken together, the requirement for purity and yield often places conflicting demands on the experimentalist. However, the successful generation of high-resolution structures is testament to the increased efficacy of methods of isolation and purification used today in protein biochemistry. Many of these methods evolved from simple, comparatively crude, protocols into sophisticated computer-controlled and enhanced procedures where concurrent advances in bioinformatics and material science have supported rapid progress in this area.

The isolation and characterization of proteins

Two broad alternative approaches are available today for isolating proteins. We can either isolate the protein conventionally by obtaining the source cell or tissue directly from the host organism or we can use molecular biology to express the protein of interest, often in a host such as *Escherichia coli*. Today molecular biology represents the most common route where DNA encoding the protein of interest is inserted into vectors facilitating the high level expression of protein in *E. coli*.

Recombinant DNA technology and protein expression

Prior to the development of recombinant protein expression systems the isolation of a protein from mammalian systems required either the death of the organism or the removal of a small selected piece of tissue in which it was suspected that the protein was found in reasonable concentrations. For proteins such as haemoglobin removal of small amounts of blood does not present ethical or practical problems. Given that haemoglobin is found in a soluble state and at very high natural concentrations (~145 g l^{-1} in humans) it is not surprising that this protein was amongst the first to be studied at a molecular level. However, many proteins are found at much lower concentrations and in tissues or cells that are not easily obtained except post mortem.

Recombinant systems of protein expression have alleviated most problems of this type and allow the study of proteins that were previously inaccessible. Once DNA encoding the protein of interest has been obtained it is relatively routine to obtain proteins via recombinant hosts in quantities one could previously only dream about (i.e. mg–g amounts). This does not mean that it is always possible to express proteins in *E. coli*. For example, membrane proteins still present numerous technical problems in heterologous expression, as do proteins rich in disulfide bonds or those bearing post-translational modifications such as myristoylation and glycosylation. Eukaryotic cells carry out many of these modifications routinely and this has led to the development of alternative expression systems besides *E. coli*. Alternative expression hosts include simple eukaryotes such as *Saccharomyces cerevisiae* and *Pichia pastoris*, as well as more complex cell types. The latter group includes cultured insect cells such as *Spodoptera frugiperda* infected with baculovirus vectors, *Drosophila* cell-based expression systems and mammalian cell lines often derived from immortalized carcinomas.

However, the most common approach and the route generally taken in any initial investigation is to express 'foreign' proteins in *E. coli*. These methods have revolutionized studies of protein structure and function by allowing the expression and subsequent isolation of proteins that were formerly difficult to study perhaps as a result of low intrinsic concentration within cells or tissues. Of importance to the development of recombinant DNA technology was the discovery of restriction enzymes (see Chapter 7) and their use with DNA ligases to create new (recombinant) DNA molecules that could be inserted into a carrier molecule or vector and then introduced into host cells such as *E. coli* (Figure 9.1). The process of constructing and propagating new DNA molecules by inserting into host cells is referred to as 'cloning'. A clone is an exact replica of the parent molecule, and in the case of *E. coli* this results, after several cycles of replication, in all cells containing new and identical DNA molecules.

The generation of new recombinant DNA molecules involves at least five discernable steps (Figure 9.2):

1. *Preparation of DNA*. Today DNA is usually generated using the polymerase chain reaction (PCR). The technique, described in Chapter 6, involves the action of a thermostable DNA polymerase with forward and

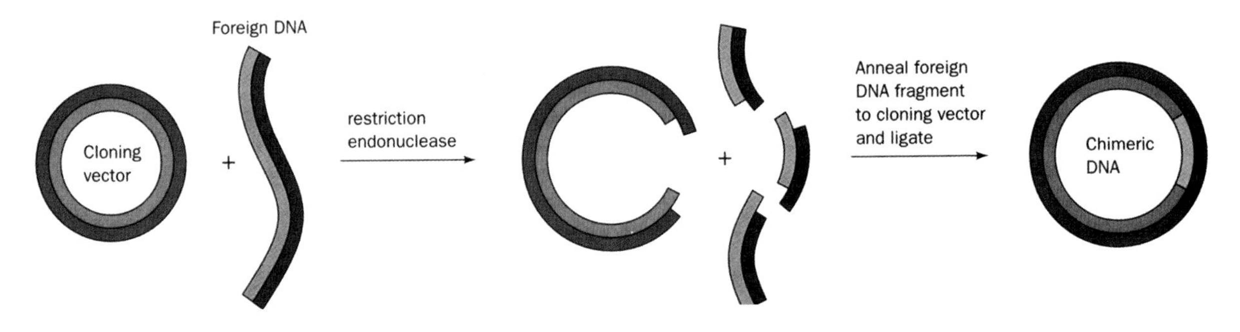

Figure 9.1 The use of restriction enzymes to generate DNA fragments and the use of DNA ligase to join together vector and insert together to form a new recombinant (chimeric) circular DNA molecule. The diagram shows the joining together of a cut vector and a cut fragment described as having sticky ends. This type of end occurs with most restriction enzymes and is contrasted with the smaller group of type II restriction endonucleases that produce blunt-ended fragments (reproduced with permission from Voet, D. Voet, J.G and Pratt, C.W. *Fundamentals of Biochemistry*. John Wiley & Sons Ltd., Chichester, 1999)

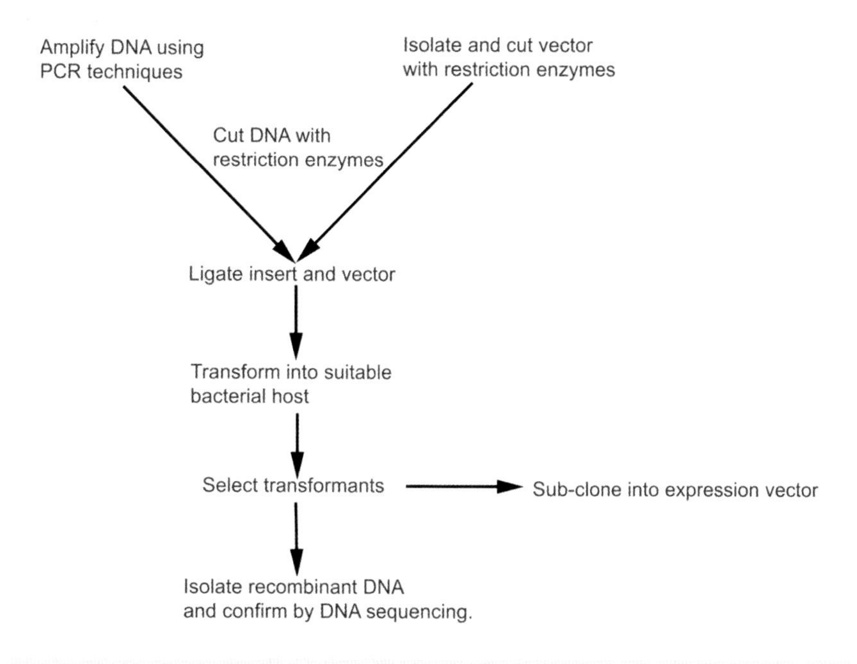

Figure 9.2 Generalized scheme for the creation of recombinant DNA molecules

reverse primers in a series of cycles involving denaturation, annealing and extension. The result is large quantities of amplified and copied DNA. Template DNA is preferably derived from cDNA via the action of reverse transcriptase on processed mRNA. This avoids the presence of introns in amplified sequences. In other instances DNA is derived from the products of restriction enzyme action.

2. *Restriction enzyme cleavage*. The generation of DNA fragments of specific length and with 'tailored' ends was vital to the success of recombinant DNA technology. This allowed the 'insert' (the name given to the DNA fragment) to be joined into a vector containing compatible sites (see Figure 9.1). Today the generation of PCR DNA fragments normally includes sites for restriction enzymes that facilitate subsequent cloning reactions and today a wide variety of restriction enzymes are commercially available.

3. *Ligation of DNA fragments*. As part of the overall cloning process insert and vector DNA are joined in a process called ligation. Without covalent linkage of pieces of DNA and the formation of closed circular DNA subsequent transformation reactions (see step 4) occur with very low efficiency.

4. *Transformation*. The propagation of recombinant DNA molecules requires the introduction of circular DNA (ligated vector + insert) into a suitable host cell. Replication of the cells generates many copies of the new recombinant DNA molecules. The whole process is called transformation and represents the uptake of circular DNA by host cells with the acquisition of new altered properties. The exact process by which cells take up DNA is unclear but two common methods used for *E. coli* involve the application of a high electric field (electroporation) or a mild heat shock. The result is that membranes are made 'leaky' and take up small circular DNA molecules. For *E. coli* the process of transformation requires special treatment and cells are often described as 'competent' meaning they are in a state to be transformed by plasmid DNA.

5. *Identification of recombinants*. After transformation cells such as *E. coli* can be grown on nutrient rich plates and it is here that marker genes found in many

vectors are most useful. Normal *E. coli* cells used in the laboratory lack resistance to antibiotics such as ampicillin, tetracycline or kanamycin. However, transformation with a vector containing one or more antibiotic resistance genes leads to resistant cells. As a result when cells are grown on agar plates containing an antibiotic only those cells transformed with a plasmid conferring resistance will grow. This selection procedure allows the screening of cells so that only transformed cells are studied further. By transforming with known amounts of plasmid DNA it is possible to calculate transformation efficiencies and it is not unusual for this value to reach 10^9-10^{10} transformants/μg DNA. Further identification of the recombinant DNA might involve isolation of the plasmid DNA and its digestion with restriction enzymes to verify an expected pattern of fragments. Definitive demonstration of the correct DNA fragment would involve nucleotide sequencing.

One of the first vectors to be developed was called pBR322 and it is the 'ancestor' of many vectors still used today for cloning. The vector pBR322 (Figure 9.3) contains distinctive sequences of DNA that includes an origin of replication (*ori*) gene as well as a gene (*rop*) that control DNA replication and the number of copies of plasmid found within a cell. In addition pBR322 contains two 'marker' genes that code for resistance to the antibiotics tetracycline and ampicillin. These useful markers are denoted by tet^R and amp^R.

The next step requires the transfer of DNA from the cloning vector to a vector optimized specifically for protein expression. A cloning vector such as pBR322 will allow several copies of the plasmid to be present in cells and may facilitate sequence analysis or analysis of restriction sites but they are rarely used for protein expression. Other common cloning vectors used widely in molecular biology and frequently described in scientific literature are pUC, pGEM®, pBluescript®. Each vector has many variants containing different restriction enzyme sites that allow the orientation of the insert in either direction as well as common features such as one or more antibiotic resistance genes, multiple cloning sites, an origin of replication, and a medium to high copy number.

In contrast, expression vectors are designed with additional features beneficial to protein production. The vectors retain many of the features of 'cloning vehicles'

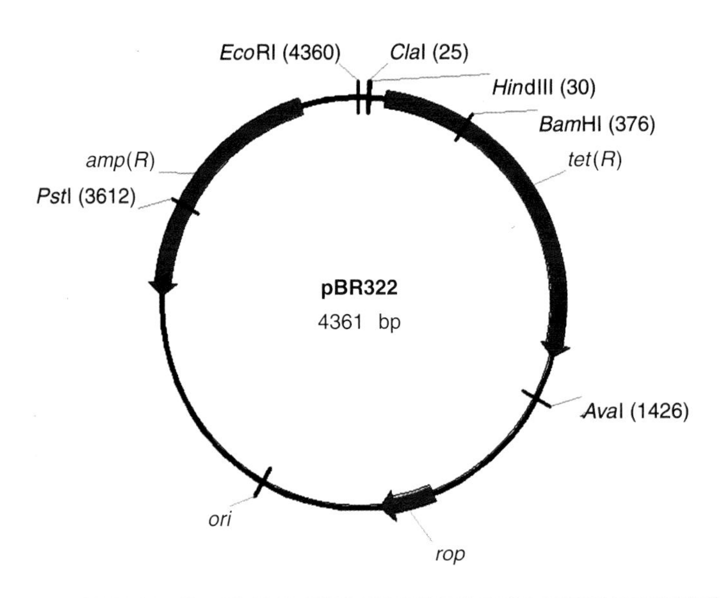

Figure 9.3 Plasmid vector pBR322. The small circular DNA molecule contains 4361 bp, the *rop*, *ori*, amp^R and tet^R genes as well as numerous sites for restriction enzymes such as *Eco*RI, *Bam*HI, *Ava*I, *Pst*I, *Cla*I and *Hin*dIII

such as multicloning sites, antibiotic resistance genes and the *rop* and *ori* genes but have in most cases:

1. *An inducible promoter*. Expressed proteins may prove toxic to the host organism. To limit this effect the activity of the promoter can be 'turned on' by an inducer. Although the mechanism of action of inducers varies from one system to another the most common system in use involves the promoter derived from the *lac* operon of *E. coli*. In many systems the sequence of this promoter has been modified to contain a 'consensus' sequence of bases for optimal activity (see Chapters 3 and 8). The *lac* operon (Figure 9.4) is a group of three genes – β-galactosidase, lactose permease and thiogalactoside transacetylase – whose activities are induced when lactose is present in the medium. The *lac* repressor is a homotetrameric protein with an N-terminal domain containing the HTH motif found in many transcription factors. It specifically binds DNA sequences in the operator region inhibiting transcription. A non-metabolized structural analogue of lactose, the natural inducer, is used to induce protein expression. Isopropylthiogalactose (IPTG) binds the lac repressor removing the inhibition of transcription (Figure 9.5). This system is the basis for most inducible expression vectors used in genetic engineering studies today.

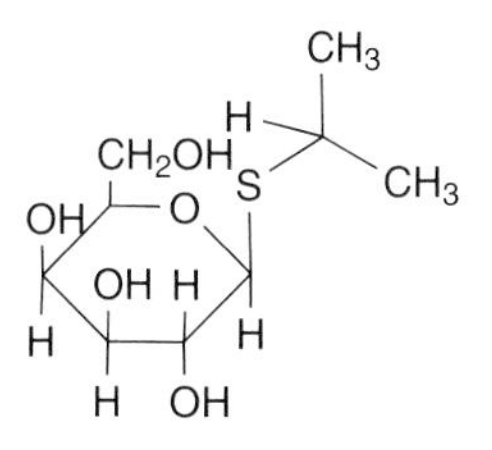

Figure 9.5 The structure of isopropylthiogalactose (IPTG)

2. *Restriction enzyme sites*. A large number of restriction sites are often clustered together in so called multi-cloning sites. These sites allow the insert to be placed 'in frame' with the start codon and in a correct relationship with the Shine–Dalgarno sequence. Two restriction enzymes, *Nco*I and *Nde*I, cut at the sequence C↓CATGG and CA↓TATG respectively. Careful inspection of this sequence reveals that ATG is part of the start codon and fragments cut with *Nco*I/*Nde*I will, if ligated correctly, be in frame with the start codon. This will yield the correct protein sequence upon translation.

3. *A prokaryotic ribosome binding site containing the Shine–Dalgarno sequence*. The Shine–Dalgarno sequence has a consensus sequence of AGGAGA located 10–15 bp upstream of the ATG codon.

Operator

lac operon

I *P* *O* *Y* *A*

Repressor binds to operator, preventing transcription of *lac* operon

Repressor

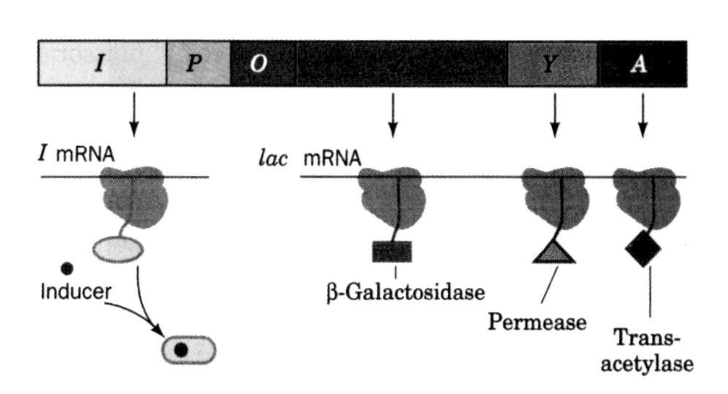

Figure 9.4 The expression of the lac operon. In the absence of inducer (IPTG) the repressor binds to the operator preventing transcription of the three genes of the lac operon. IPTG binds to the repressor allowing transcription to proceed. Most expression vectors contain the *lacI* gene together with the lac promoter as their inducible transcription system. (Reproduced with permission Voet, D. Voet, J.G and Pratt, C.W. *Fundamentals of Biochemistry* 1999. John Wiley & Sons Ltd., Chichester)

Successful expression requires the use of the host cells synthetic machinery and in particular formation of an initiation complex for translation. The 16S rRNA binds to purine rich sequences (AGGAGA) as a result of sequence complementarity enhancing formation of the initiation complex. The composition and length of the intervening sequences between the ribosomal binding site and the start codon are important with structures such as hairpin loops decreasing expression.

4. *A transcriptional terminator*. In an ideal situation mRNA is transcribed only for the desired protein and is then halted. To achieve this aim many expression vectors contain transcriptional terminators. Terminators are DNA sequences that cause the disassembly of RNA polymerase complexes, limit the overall mRNA length and are G-C rich and palindromic. The formation of RNA hairpin structures destabilizes DNA–RNA transcripts causing termination. In other cases termination results from the action of proteins like the *Rho*-dependent terminators in *E. coli*.
5. *Stop codons*. In frame stop codons prevent translation of mRNA by ribosomes and their presence as part of the 'insert' or as part of the expression vector is desirable to ensure the absence of extra residues in the expressed protein.
6. *Selectable markers*. One or more selectable markers such as the genes for ampicillin or tetracycline resistance allows the expression vector to be selected and maintained.

Extensive discussion of all of these features is beyond the scope of this book but one set of widely used expression systems are the pET-based vectors. These vectors are based on the bacteriophage T7 RNA polymerase promoter, a tightly regulated promoter, that allows high levels of protein expression. A large number of variants exist on a basic theme. The pET vector is approximately 4600 bp in size, contains either the Amp^{R} and Kan^{R} antibiotic resistance genes, as well as the normal *ori* and *rop* genes. The pET vectors (Figure 9.6) have regions containing sites for many Type II restriction enzymes and genes are cloned downstream under control of T7 RNA polymerase promoter. It might be thought that host RNA polymerases would initiate transcription from this promoter but it turns out that this promoter is highly selective for T7 RNA polymerase. One consequence is that background protein expression in the absence of T7 RNA polymerase is extremely low and this is desirable when expressed proteins are toxic to host cells. However, expression of foreign cDNA requires T7 RNA polymerase – an enzyme not normally found in *E. coli* – and to overcome this problem all *host* cells contain a chromosomal copy of the gene for T7 RNA polymerase. The addition of IPTG to cultures of *E. coli* (*DE3*) leads to the production of T7 RNA polymerase that in turn allows transcription of the target DNA.

One of the major advantages of bacterial cells is their ease of culture allowing growth in a sterile manner, in large volumes and with short doubling times (~40 minutes under optimal conditions). *E. coli* cells are easy to transform with vector DNA, especially when compared with eukaryotic cells, and growth media are relatively cheap. Large amounts of expressed protein can be obtained from *E. coli* and it is not unusual for the expressed protein to exceed 10 percent of the total cell mass. The success of genetic engineering has been demonstrated by production of large amounts of therapeutically important eukaryotic proteins using *E. coli* expression systems. Human insulin, growth hormones, blood coagulation factors IX and X as well as tissue plasminogen activator have all been successfully expressed and used to treat diseases such as diabetes, haemophilia, strokes and heart attacks.

Purification of proteins

Advances in recombinant DNA technology allow large-scale protein expression in host cells. Demonstration of protein expression is usually confirmed by SDS–polyacrylamide gel electrophoresis (see below) and is then followed by the question: how is the protein to be purified? An effective purification strategy requires protein recovery in high yield *and* with high purity. Achieving these two objectives can be difficult and is usually resolved in an empirical manner.

The first step in any purification strategy involves the fractionation of cells by mechanical disruption

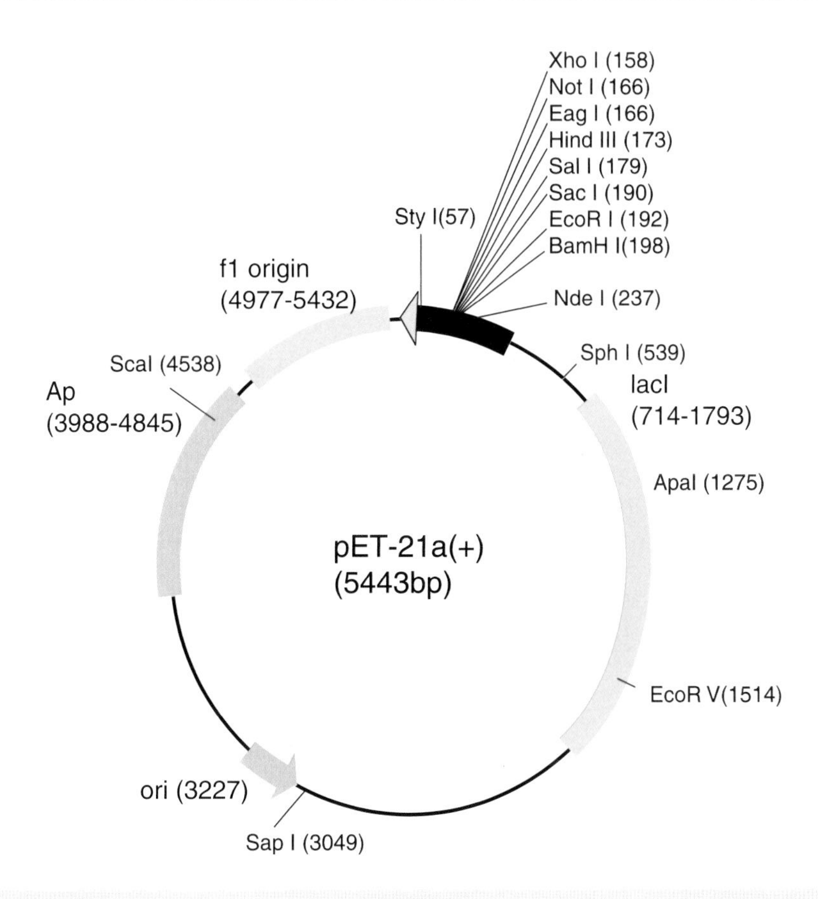

Figure 9.6 The pET series of vectors. Identifiable elements within the vector include the T7 promoter (311–327), T7 transcription start (310), T7 Tag coding sequence (207–239), Multiple cloning sites (*Bam*HI–*Xho*I) (158–203), His-Tag coding sequence (140–157), T7 terminator region (26–72), *lacI* coding sequence (714–1793), pBR322 origin of replication (3227), and the antibiotic resistance coding sequence (3988–4845)

or the use of chemical agents such as enzymes. Mechanical disruption methods can involve sonication where high frequency sound waves (ultrasound frequencies > 20 kHz) disrupt membrane structure. Other methods include mechanical shearing by rupturing membranes by passing cells through narrow orifices under extremely high pressure or grinding cells with abrasive material such as sand or glass beads to achieve similar results.

A different method of disruption involves the use of enzymes such as lysozyme to disrupt cell walls in Gram-negative bacteria like *E. coli* or chitinase to degrade the complex polysaccharide coats associated with yeast. Lysozyme acts by splitting the complex polysaccharide coat between *N*-acetylmuramic acid and *N*-acetylglucosamine units in the peptidoglycan layer (Figures 9.7 and 9.8). Multiple glycosidic bond cleavage weakens the cell envelope and the *E. coli* membrane 'sac' is easily broken by osmotic shock or mild sonication to release cytoplasmic contents that are separated from membranes by differential centrifugation.

Purification strategies normally exploit the physical properties of the protein as a means to isolation. If primary sequence information is available then predictions about overall charge, disulfide content, solubility, mass or hydrophobicity can be obtained. Bioinformatics may allow recognition of similar proteins

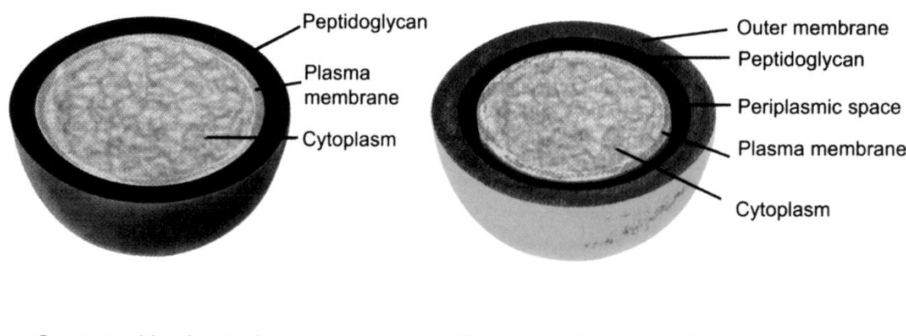

Figure 9.7 A diagram of the organization of the peptidoglycan layer in Gram-positive and Gram-negative bacteria. The peptidoglycan layer is a complex structure consisting of covalently linking carbohydrate and protein components and is more extensive in gram positive bacteria. Bacteria are defined as Gram-positive or -negative according to whether heat fixed cells retain a dye (crystal violet) and iodine after destaining with alcohol. Gram-positive bacteria allow the dye into the cells and heat fixing causes changes in the peptidoglycan layer that prevent the escape of the dye during destaining. In contrast the smaller peptidoglycan layer of Gram-negative bacteria allows the escape of dye during destaining (reproduced with permission Voet, D. Voet, J.G and Pratt, C.W. *Fundamentals of Biochemistry*. John Wiley & Sons Ltd., Chichester, 1999)

and therefore the implementation of comparable purification methods. However, in some circumstances the protein is 'new' and sequence information is absent. To obtain sequence information it is necessary to purify the protein yet to purify the protein it is often advantageous to have sequence information! Exploiting physical properties and judicious use of selective methods allows the purification of a protein from cell extracts containing literally thousands of unwanted proteins to a single protein (homogeneity) in a few steps.

Centrifugation

The principle of centrifugation is to separate particles of different mass. Strictly speaking this is not entirely accurate because centrifugation depends on other factors such as molecular shape, temperature and solution density. Heavy particles sediment faster than light particles – a fact demonstrated by mixing sand and water in a container and waiting for a few minutes after which time the sand has settled to the bottom of the container. This reflects our everyday experience that heavier particles are pulled downwards by the earth's gravitational force. However, a close inspection of the solution would reveal that small (microscopically small) grains remain suspended in solution. This also happens to macromolecules such as proteins where, as a result of Brownian diffusion, their buoyant density counteracts the effect of gravity.

Fractionation procedures (Figure 9.9) are important in cell biology and are used extensively to purify cells, subcellular organelles or smaller components for further biochemical analysis. Disruption of *E. coli* cells produces membrane and cytosolic fractions, each containing particles of varying mass. These particles are separated by centrifugation, a process enhancing rates of sedimentation by increasing the force acting on particles within solutions. Centrifugation involves the rotation of solutions in tubes within specially designed rotors at frequencies ranging from 100 revolutions per minute up to ~80 000 r.p.m. More commonly centrifuges are used at much lower g values and a force of 20 000 g is sufficient to pellet most membranes found in cells as well as larger organelles such as mitochondria, chloroplasts, and the ER systems of animal and plant cells.

In the preceding section centrifugation has been described as a preparative tool but it also finds an

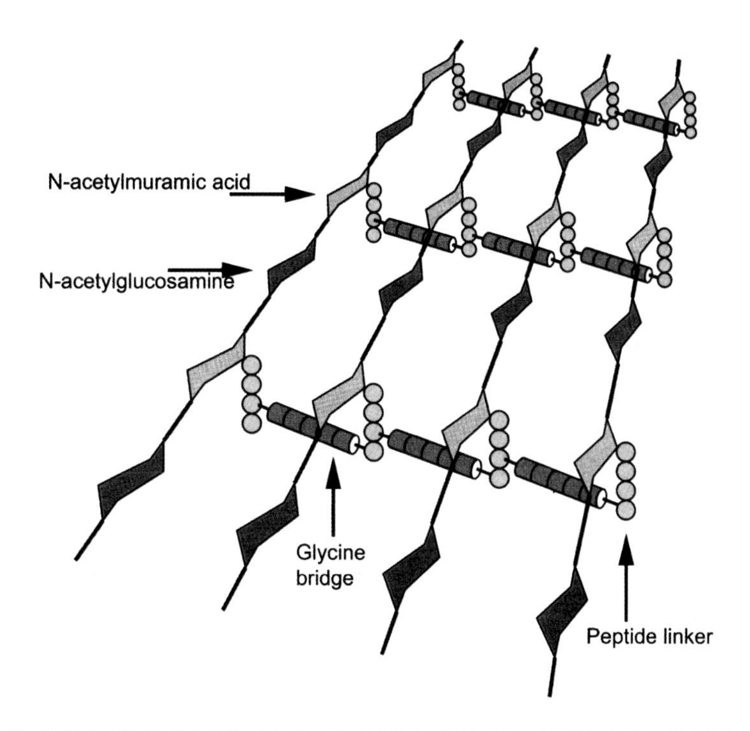

Figure 9.8 The arrangement of β(1–4) linked *N*-acetylglucosamine (NAG) and *N*-acetylmuramic acid (NAM) together with the linkage to the tetrapeptide bridge of L-Ala, D-isoglutamate, L-Lys and D-Ala peptidoglycan. The tetrapeptide bridges are joined together by five Gly residues that extend from the carboxyl group of one tetrapeptide to the ε-amino group of lysine in another. The isoglutamate residue is so called because the side chain or γ carboxyl group forms a peptide bond with the next amino group. The arrangement of units within the peptidoglycan layers has been studied particularly extensively for *Staphylococcus aureus* but considerable variation in structure can occur. The presence of D-amino acids renders the peptidoglycan layer resistant to most proteases (Reproduced with permission from Voet, D. Voet, J.G. Pratt, C.W. *Fundamentals of Biochemistry*. John Wiley & Sons Ltd., Chichester, 1999)

application as a precise analytical tool. Centrifuged solutions are subjected to strong forces that vary along the length of the tube or more properly the radius (r) about which rotation occurs. The centrifugal force acting on any particle of mass m is the product of a particle's mass multiplied by the centrifugal acceleration (from Newton's second law of motion). This centrifugal acceleration is itself the product of the radius of rotation and the angular velocity ($r\omega^2$). For any particle the *net* force acting on it will be a balance between centrifugal forces causing particles to pellet and buoyant forces acting in the opposite direction. Particles that are less buoyant than the solvent will sink whilst those that are lighter than the solvent will float. This results in the following expression describing the forces acting on a particle during centrifugation:

$$\text{Net force} = \text{centrifugal force} - \text{buoyant force} \quad (9.1)$$

$$= \omega^2 rm - \omega^2 rm_s \quad (9.2)$$

$$= \omega^2 rm - \omega^2 r\nu\rho \quad (9.3)$$

where m_s is the mass of solvent displaced by the particle, ν is the volume of the particle and ρ is the density of the solvent. If, for the moment, the second term on the right hand side of the equation is neglected the centrifugal force is simply the angular velocity multiplied by the radius of rotation. Usually, the value of the force applied to particles during centrifugation is compared to the earth's gravitational force ($\boldsymbol{g}$, $\sim 9.8\ \text{m}^2\ \text{s}^{-1}$) and the dimensionless quantity

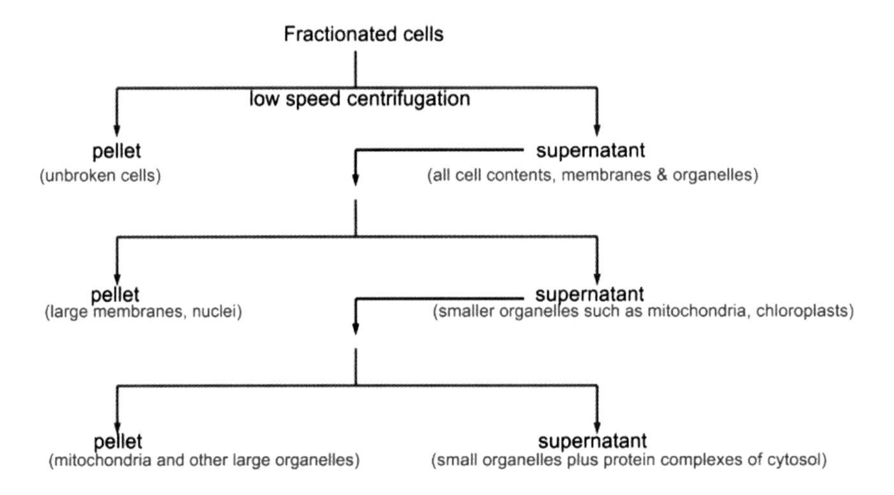

Figure 9.9 The fractionation of cells. After initial tissue maceration the solution is filtered to remove large debris and centrifuged at low speeds to remove unbroken cells (<1000 ***g***). Progressive increases in centrifugal forces pellet smaller cellular organelles and eventually at the highest speeds large protein complexes may be isolated

called the relative centrifugal force (RCF) is used.

$$\text{RCF} = \text{centrifugal force/force due to gravity} \quad (9.4)$$

$$= \omega^2 rm/m\,\mathbf{g} \quad (9.5)$$

$$= \omega^2 r/\mathbf{g} \quad (9.6)$$

To determine the maximum relative centrifugal force it is therefore only necessary to know the radius of the rotor and the speed of rotation. The radius is fixed for a given rotor and the speed of rotation is a user-defined parameter. Although RCF is a dimensionless quantity it is traditionally quoted as a numerical value; for example 5000 ***g*** meaning 5000 × the earth's gravitational force. Since ω the angular velocity, in radians per second, is the number of revolutions per second multiplied by 2π the RCF of a 10 cm radius rotor rotating at 8000 r.p.m. is

$$\text{RCF} = (2\pi \times 8000/60)^2 \times 0.1/9.8 \sim 7160\,\mathbf{g} \quad (9.7)$$

This is actually the ***g*** force at the end of the rotor tube. More careful consideration reveals that the ***g*** force will vary along the radius of the tube. In the above example the ***g*** force at a position 5 cm along the tube is approximately 3580 ***g*** and this has frequently led to the use of a term $\mathbf{g}_{av}$ to reflect average centrifugal force.

In analytical methods of centrifugation the second term of Equation 9.2, the buoyant force, must be considered. In most applications this term is 'invisible' since rotor speeds are such that membrane or large organelles are forced to sediment whilst the lighter particles remain in solution. An extension of this relationship allows the size or molecular mass of proteins to be determined.

During centrifugation at high speeds the initial acceleration of a particle due to the net force is relatively short in duration (~1 ns), after which time the particle moves at a constant velocity. This constant velocity arises because the solution exerts a frictional force (f) on the particle that is proportional to the sedimentation velocity ($\mathrm{d}r/\mathrm{d}t$). Once the steady state is reached we can rewrite Equation 9.3 as

$$f\,\mathrm{d}r/\mathrm{d}t = \omega^2 rm - \omega^2 r\nu\rho \quad (9.8)$$

In Equation 9.8 the volume of a particle is a difficult quantity to measure and is replaced by a term called the partial specific volume ($\bar{\nu}$). This is defined as the increase in volume when 1 g of solute is dissolved in a large volume of solvent. The quantity $m\nu$ is defined as the increase in volume caused by the addition of one molecule of mass m and it is equal to the volume of the particle.[1] Introduction of these terms allows Equation 9.8 to be simplified

$$f\,\mathrm{d}r/\mathrm{d}t = \omega^2 rm - \omega^2 rm\,\bar{\nu}\rho \qquad (9.9)$$

$$f\,\mathrm{d}r/\mathrm{d}t = \omega^2 rm\,(1 - \bar{\nu}\rho) \qquad (9.10)$$

and rearrangement leads to

$$S = \frac{\mathrm{d}r/\mathrm{d}t}{\omega^2 r} = \frac{m\,(1 - \bar{\nu}\rho)}{f} \qquad (9.11)$$

where the term $\mathrm{d}r/\mathrm{d}t/\omega^2 r$ is defined as the sedimentation coefficient (S called the Svedburg unit after the pioneer of ultra-centrifugation Theodor Svedburg). The Svedburg has units of 10^{-13} s, and as an example haemoglobin has a sedimentation coefficient of 4×10^{-13} s or 4 S. Since the mass m can be described as the molecular weight of the solute divided by Avogadro's constant (M/N_o) and by assuming the particle to be spherical we can utilize Stokes law to describe the frictional coefficient, f, as

$$f = 6\pi\,\eta\,r_s \qquad (9.12)$$

where r_s is the radius of the particle and η is the viscosity of the solvent (normally water).[2] This leads to the following equation

$$S = \frac{M(1 - \bar{\nu}\rho)}{N_o\,6\pi\,\eta\,r_s} \qquad (9.13)$$

that can be rearranged to give

$$M = SN_o\,(6\pi\,\eta\,r_s)/(1 - \bar{\nu}\rho) \qquad (9.14)$$

If we do not make any assumptions about shape we can use the relationship

$$D = k_B T/f \qquad (9.15)$$

[1]For most proteins $\bar{\nu}$ has a value of 7.4×10^{-4} m^3/kg or 0.74 ml/g.
[2]For non-spherical particles correction of this relationship must consider anisotropy where one axis may be much longer than the other two. An example would be a cylindrically shaped molecule. Hydrodynamic analysis can in some instances allow the shape of the molecule to be deduced or at least distinguished from the simple spherical situation.

where D is the diffusion coefficient, f is the frictional coefficient, k_B is Boltzmann's constant and T the absolute temperature. This leads to

$$M = SRT/D\,(1 - \bar{\nu}\rho) \qquad (9.16)$$

and allows the molecular weight to be calculated if S, the sedimentation velocity, is measured since all of the remaining terms are constants or are derived from other measurements. From Equation 9.11 S is defined as $\frac{\mathrm{d}r}{\mathrm{d}t}\frac{1}{\omega^2 r}$ or

$$S\,\mathrm{d}t = 1/\omega^2 \mathrm{d}r/r \qquad (9.17)$$

with integration of the above equation leading to the relationship

$$\ln r/r_0 = S\,t\,\omega^2 \qquad (9.18)$$

By measuring the time taken for the particle to travel between two points, r_o (at time $t = 0$) and r (at time t) and knowing ω the angular velocity (in radians/s) it is relatively straightforward to calculate S. Measurement of protein sedimentation rates is performed by recording changes in refractive index as the protein migrates through an optical cell. Table 9.1 lists the sedimentation coefficients for various proteins.

A more direct way of measuring molecular weight using ultracentrifugation is to measure not the rate of sedimentation but the equilibrium established after many hours of centrifugation at relatively low speeds. At equilibrium a steady concentration gradient will occur with the flow of proteins towards sedimentation balanced by reverse flow as a result of diffusion. It can be demonstrated that for a homogeneous protein the gradient is described by the equation

$$c(r)/c(r_0) = \exp M(1 - \bar{\nu}\rho)\omega^2\,(r^2 - r_0{}^2)/2RT \qquad (9.19)$$

where $c(r)$ is the concentration at a distance r from the axis of rotation and $c(r_0)$ is the concentration of protein at the meniscus or interface (r_0). From Equation 9.19 a plot of ln $c(r)$ against $(r^2 - r_0{}^2)$ will yield a straight line whose slope is $M(1 - \bar{\nu}\rho)\omega^2/2RT$, from which M is subsequently estimated (Figure 9.10).

Solubility and 'salting out' and 'salting in'

One of the most common and oldest methods of protein purification is to exploit the differential solubility of

Table 9.1 Sedimentation coefficients associated different proteins

Protein	Molecular mass (kDa)	Sedimentation coefficient ($S_{20,w}$)
Lipase	6.7	1.14
Ribonuclease A	12.6	2.00
Myoglobin	16.9	2.04
Concanavalin B	42.5	3.50
Lactate Dehydrogenase	150	7.31
Catalase	222	11.20
Fibrinogen	340	7.63
Glutamate Dehydrogenase	1015	26.60
Turnip yellow mosaic virus	3013	48.80
Large ribosomal subunit	1600	50

Notice the 'abnormal' value for fibrinogen and the absence of a simple correlation between mass and S value (after Smith in Sober H.A. (ed.) *Handbook of Biochemistry and Molecular Biology*, 2nd edn. CRC Press).

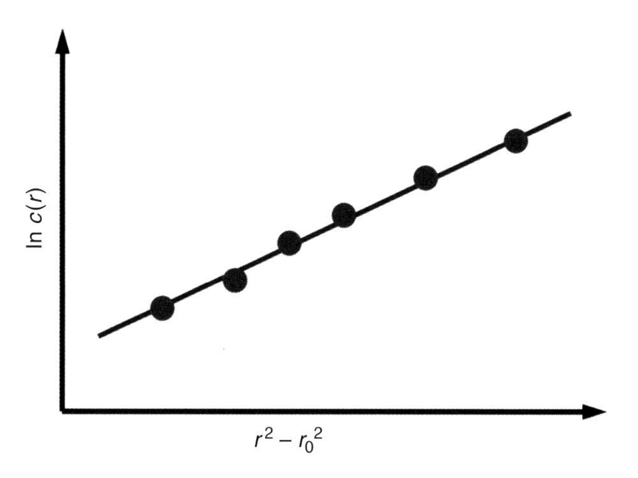

Figure 9.10 A plot of ln $c(r)$ against $(r^2 - r_0^2)$ yields a straight line that allows direct estimation of M, the protein molecular weight. The slope of the line is given by $M(1 - \bar{v}\rho)\omega^2/2RT$

proteins at various ionic strengths. In concentrated solutions the solubility of many proteins decreases differentially leading to precipitation whilst others remain soluble. This allows protein separation since as solubility decreases precipitation occurs with the precipitant removed from solution by centrifugation. The use of differential solubility to fractionate proteins is often used as one part in a strategy towards overall purification.

The solubility of proteins in aqueous solutions differs dramatically. Although membrane proteins are clearly insoluble in aqueous solution many structural proteins such as collagen are also essentially insoluble under normal physiological conditions. For small globular proteins with masses below 10 kDa intrinsic solubility is often high and concentrations of 10 mM are easily achieved. This corresponds to approximately 100 g l^{-1} or about 10 percent w/v. In solution globular proteins are surrounded by a tightly bound layer of water that differs in properties from the 'bulk' water component. This water, known as the hydration layer, is ordered over the surface of proteins and appears to interact preferentially with charged side chains and polar residues. In contrast, this water is rarely found close to hydrophobic patches on the surface of proteins. Altering the properties of this water layer effects the solubility of proteins. Similarly protein solubility is frequently related to the isoelectric point or pI. As the pH of the aqueous solution approaches the pI precipitation of a protein can occur. As the pH shifts away from the pI the solubility of a protein generally increases. This is most simply correlated with the overall charge of the protein and its interaction with water. Since the pI of proteins can vary from pH 2–10 it is clear that pH ranges for protein solubility will also vary widely.

The ionic strength (I) is defined as

$$I = 1/2\ \Sigma\ m_i z_i^2 = 1/2\ (m_+ z_+^2 + m_- z_-^2) \quad (9.20)$$

where m and z are the concentration and charge of all ionic species in solution. Thus a solution of 0.5 M NaCl has an ionic strength of 0.5 whilst a solution of 0.5 M $MgCl_2$ has an ionic strength of

1.5. The concept of ionic strength was introduced by G.N. Lewis to account for the non-ideal behaviour of electrolyte solutions that arose from the total number of ions present along with their charges rather than the chemical nature of each ionic species. A physical interpretation of properties of electrolytes from the studies of Peter Debye and Erich Hückel, although mathematical and beyond the scope of this book, assumed that electrolytes are completely dissociated into ions in solution, that concentrations of ions were dilute below 0.01 M and that on average each ion was surrounded by ions of opposite charge. This last concept was known as an ionic atmosphere and is relevant to changes in protein solubility as a function of ionic strength. Trends in protein solubility vary from one protein to another and show a strong dependence on the salts used to alter the ionic strength (Figure 9.11).

At low ionic strength a protein is surrounded by excess counter-ions of the opposite charge known as the ionic atmosphere. These counter-ions screen charges on the surface of proteins. Addition of extra ions increases the shielding of surface charges with the result that there is a decrease in intermolecular attraction followed by an increase in protein solubility as more dissolves into solution. This is the 'salting-in' phenomenon (Figure 9.12). However, the addition of

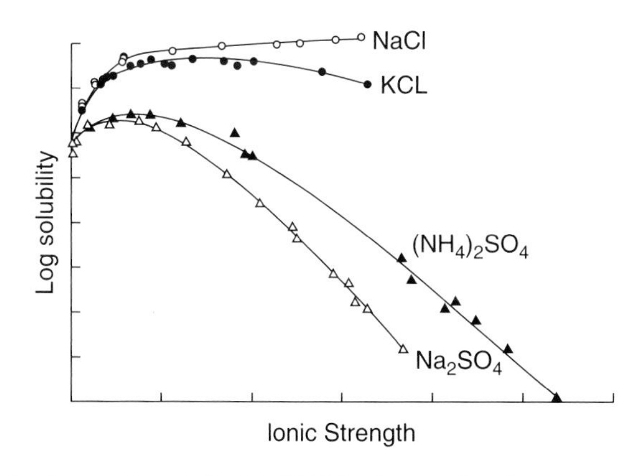

Figure 9.11 The solubility of haemoglobin in different electrolytes illustrates the 'salting in' and 'salting out' effects as a function of ionic strength. Derived from original data by Green, A.A. *J. Biol. Chem.* 1932, **95**, 47

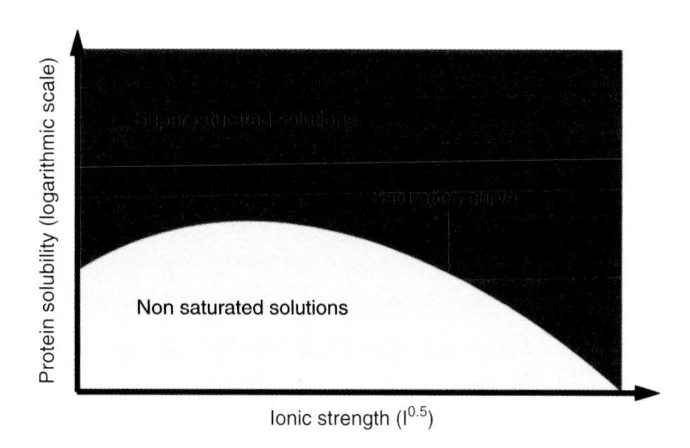

Figure 9.12 Model profile for solubility as a function of ionic strength. Saturation occurs when solid and solution phases are in equilibrium. 'Salting-out' is seen on the right-hand side of the diagram where there is a reduction in protein solubility as the concentration of salt increases whilst 'salting in' is apparent on the left-hand side of the diagram where there is an increase in protein solubility with ionic strength

more ions, reverses this trend. When the ionic concentration is very high (>1 M) each ion must be hydrated with the result that bulk solvent (water) is sequestered from proteins leading to decreases in solubility. If continued decreased solubility leads to protein aggregation and precipitation. Ionic effects on protein solubility were first recognized by Franz Hofmeister around 1888, and by analysing the effectiveness of anions and cations in precipitating serum proteins he established an order or priority that reflected each ions 'stabilizing' properties.

Cations: $N(CH_3)_3^+ > NH_4^+ > K^+ > Na^+$
$> Li^+ > Mg^{2+} > Ca^{2+} > Al^{3+}$
$>$ guanidinium

Anions: $SO_4^{2-} > HPO_4^{2-} > CH_3COO^-$
$>$ citrate $>$ tartrate $> F^- > Cl^- > Br^-$
$> I^- > NO_3^- > ClO_4^- > SCN^-$

The priority is known as the Hofmeister series (Table 9.2); the first ions in each series are the most stabilizing with substituted ammonium ions, ammonium itself and potassium being more stabilizing

Table 9.2 Properties of the Hofmeister series of cations and anions

Stabilizing ions	↔ Destabilizing ions
Strongly hydrated anions	↔ Strongly hydrated cations
Weakly hydrated cations	↔ Weakly hydrated anions
Kosmotropic	↔ Chaotropic
Increase surface tension	↔ No effect on surface tension
Decrease solubility of non-polar side-chains	↔ Increase solubility of non-polar side-chains
Salting out	↔ Salting in

than calcium or guanidinium, for example. Similarly for anions of the Hofmeister series sulfate, phosphate and acetate are more stabilizing than perchlorate or thiocyanate.

At the heart of this series is the effect of ions on the ordered structure of water. Ordered structure in water is perturbed by ions since they disrupt natural hydrogen-bond networks. The result is that the addition of ions has a similar effect to increasing temperature or pressure. Ions with the greatest disruptive effect are known as structure-breakers or chaotropes. In contrast, ions exhibiting strong interactions with water molecules are called kosmotropes. It is immediately apparent that molecules such as guanidinium thiocyanate are extremely potent chaotropes whilst ammonium sulfate is a stabilizing molecule or kosmotrope. Na^+ and Cl^- are frequently viewed as the 'border zone' having neutral effects on proteins.

Ammonium sulfate is commonly used to selectively precipitate proteins since it is very soluble in water thereby allowing high concentrations (>4 M). Under these conditions harmful effects on proteins such as irreversible denaturation are absent and NH_4^+ and SO_4^{2-} are both at the favourable, non-denaturing, end of the Hofmeister series. The mechanism of 'salting out' proteins resides in the disruption of water structure by added ions that lead to decreases in the solubility of non-polar molecules. By using ammonium sulfate it is possible to quantitatively precipitate one protein from a mixture. The remaining proteins are left in solution and such methods are widely used to purify soluble proteins from crude cell extracts.

Chromatography

Chromatographic methods form the core of most purification protocols and have a number of common features. First, the protein of interest is dissolved in a mobile phase normally an aqueous buffered solution. This solution is often derived from a cell extract and may be used directly in the chromatographic separation. The mobile phase is passed over a resin (called the immobile phase) that selectively absorbs components in a process that depends on the type of functional group and the physical properties associated with a protein. These properties include charge, molecular mass, hydrophobicity and ligand affinity. The choice of the immobile phase depends strongly on the properties of the protein exploited during purification.

The general principle of chromatography involves a column containing the resin into which buffer is pumped to equilibrate the resin prior to sample application. The sample is applied to the column and separation relies on interactions between protein and the functional groups of the resin. Interaction slows the progress of one or more proteins through the column whilst other proteins flow unhindered through the column. As a result of their faster progress through the column these proteins are separated from the remaining protein (Figure 9.13). In some instances protein mixtures vary in their interaction with functional groups; a strong interaction results in slow progress through the column, a weak interaction leads to faster rates of progress whilst no interaction results in unimpeded flow through the resin.

Pioneering research into the principles of chromatography performed by A.J.P. Martin and R.L.M. Synge in the 1940s developed the concept of theoretical plates in chromatography. In part this theory explains migration rates and shapes of eluted zones and views the chromatographic column as a series of continuous, discrete, narrow layers known as theoretical plates. Within each plate equilibration of the solute (protein) between the mobile and stationary (immobile) phases occurs with the solute and solvent moving through the

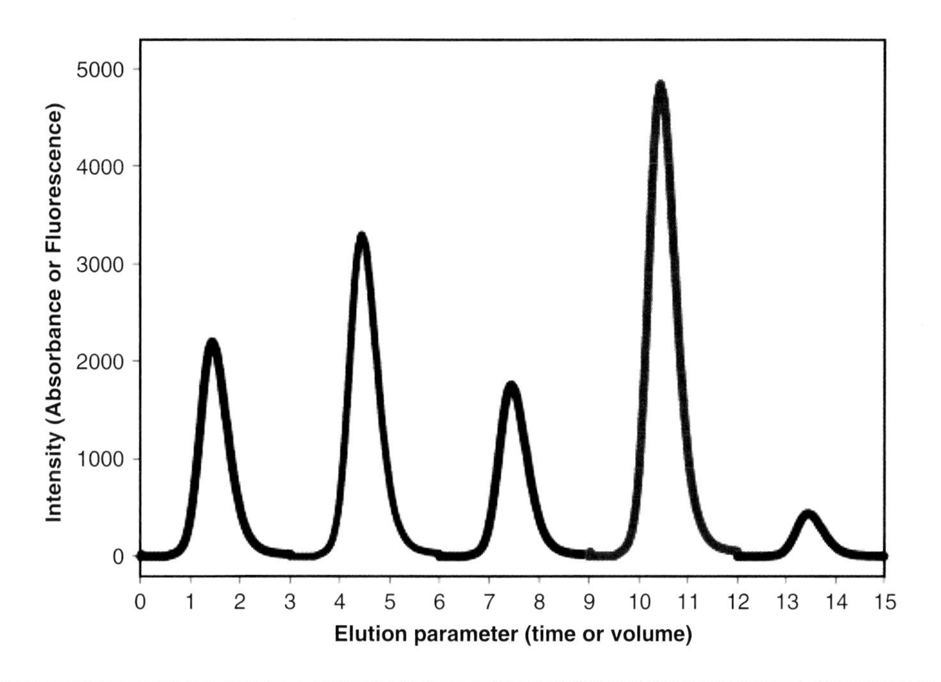

Figure 9.13 An 'ideal' separation of a mixture of proteins. The peaks representing different components are separated with baseline resolution. A less than ideal separation will result in peak overlap

column in a stepwise manner from 'plate' to 'plate'. The resolution of a column will increase as the theoretical plate number increases. Practically, there are a number of ways to increase the theoretical plate size, and these include making the column longer or decreasing the size of the beads packing the column. The value of increasing resolution is that two similarly migrating solutes may be separated with reasonable efficiency.

Today, sophisticated equipment is available to separate proteins. This equipment, controlled via a computer, includes pumps that allow flow rates to be automatically adjusted over a wide range from perhaps as low as 10 $\mu l\ min^{-1}$ to 100 $ml\ min^{-1}$. In addition, complex gradients of solutions can be constructed by controlling the output from more than one pump, and under some circumstances these gradients vastly improve chromatographic separations. The addition of the sample to the top of the column is controlled via a system of valves whilst material flowing out of the column (eluate) is detected usually via fluorescence- or absorbance-based detectors (Figure 9.14).

A major advance has been the development of resins that withstand high pressures (sometimes in excess of 50 MPa) generated using automated or high pressure liquid chromatography (HPLC) systems (Figure 9.15). Higher pressures are advantageous in allowing higher flow rates and decreased separation times. Previously, manual chromatographic methods relied on the flow of solution through the column under the influence of gravity and resulted in separations taking several days. Today it is possible to routinely perform chromatographic separations with high levels of resolution in less than 10 min.

Modern chromatographic systems contain many of the components shown schematically in Figure 9.14. Two or more pumps drive buffer onto columns at controlled rates usually in a range from 0.01–10 $ml\ min^{-1}$. The pumps are linked to the column via chemically resistant tubing that conveys buffers to the column. Between the pump and the column is a 'mixer' that adjusts the precise volumes dispensed from the pumps to allow the desired buffer composition. This is particularly important in establishing buffer gradients of ionic strength or pH. Buffer of defined composition is pumped through a series of motorized valves that

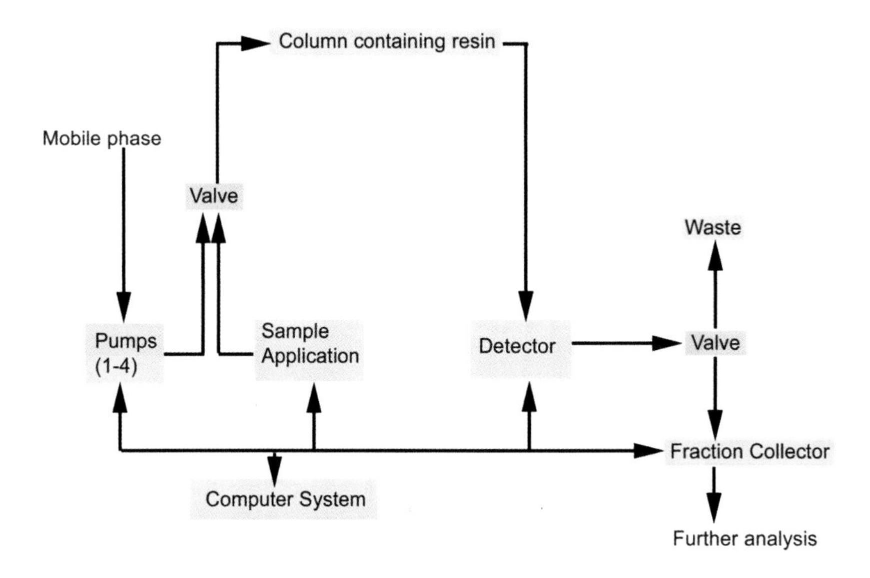

Figure 9.14 Schematic representation of chromatographic systems indicating flow of mobile phase and sample together with the flow of information between all devices and a computer workstation

Figure 9.15 A modern automated chromatography system embodying the flow scheme of Figure 9.14. (Reproduced courtesy of Amersham Biosciences)

allow addition of sample to the column or its diversion to a waste outlet. The column itself is made of precision-bored glass, stainless steel or titanium and contains chromatographic resin.

Elution of material from the column leads to a detection system that usually measures absorbance based around the near UV region at 280 nm for proteins. Frequently, detection systems will also measure pH and ionic strength of the material eluting from the column. The detection system is interfaced with a fraction collector that allows samples of precise volume to be collected and peak selection on the basis of rates of change of absorbance with time. In any system all of the components are controlled via a central computer and the user inputs desired parameters such as flow rate, fraction volume size, wavelength for detection, run duration, etc.

Ion exchange chromatography

Ion exchange chromatography separates proteins on the basis of overall charge and depends on the relative numbers of charged side chains. Simple calculations of the number of charge groups allow an estimation of the isoelectric point and an assessment of the charge at pH 7.0. Anionic proteins have a pI < 7.0 whilst cationic proteins have isoelectric points >7.0. In ion exchange chromatography anionic proteins bind to cationic groups of the resin in the process of *anion exchange* chromatography. Similarly, positively charged proteins bind to anionic groups within a supporting matrix in *cation exchange* chromatography.

The first ion exchangers developed for use with biological material were based on a supporting matrix of cellulose. Although hydrophilic and non-denaturing cellulose was hindered by its low binding capacity for proteins due to the small number of functional groups attached to the matrix, biodegradation and poor flow properties. More robust resins were required and the next generation were based around cross-linked dextrans, cross-linked agarose, or cross-linked acrylamide in the form of small, bead-like, particles. These resins offered significant improvements in terms of pH stability, flow rates and binding capacity. The latest resins are based around extremely small homogeneous beads of average diameter $<10\ \mu m$ containing hydrophilic polymers. One implementation of this technology involves polystyrene cross-linked with divinylbenzene producing resins resistant to biodegradation, usable over wide pH ranges (from 1–14) and tolerating high pressures (>10 MPa).

For cation exchange chromatography the common functional groups (Figure 9.16) involve weak acidic groups such as carboxymethyl or a strong acidic group such as sulfopropyl or methyl sulfonate. Similarly for anion exchange, the functional groups are diethylaminoethyl and the stronger exchange group of diethyl-(2-hydroxypropyl)amino ethyl. The binding of proteins to either anion or cation exchange resins will depend critically on pH and ionic strength. The pH of the solution remains important since it influences the overall charge on a protein. Fairly obviously, it is not sensible to perform ion exchange chromatography at a pH close to the isoelectric point of a protein. At this pH the protein is uncharged and will not stick to any ion exchange resin.

In ion exchange chromatography samples are applied to columns at low ionic strengths (normally $I < 0.05$ M) to maximize interactions between protein and matrix. The column is then washed with further solutions of constant pH and low ionic strength to remove proteins lacking any affinity for the resin. However, during any purification protocol many proteins are present in a sample and exhibit a range of interactions with ion exchange resins. This includes proteins that bind very tightly as well as those with varying degrees of affinity for the resin. These proteins are separated by slowly increasing the ionic strength of the solution. Today's sophisticated chromatography systems establish gradients where the ionic strength is increased from low to high levels with the ions competing with the protein for binding sites on the resin. As a result of competition proteins are displaced from the top of the column and forced to migrate downwards through the column. Repetition of this process occurs continuously along the column with the result that weakly bound proteins are eluted. Eventually at high ionic strengths even tightly bound proteins are displaced from the resin and the cumulative effect is the separation of proteins according to charge. Collecting proteins at regular intervals with a fraction collector and recording their absorbance or fluorescence establishes an elution profile and assists in locating the protein of interest.

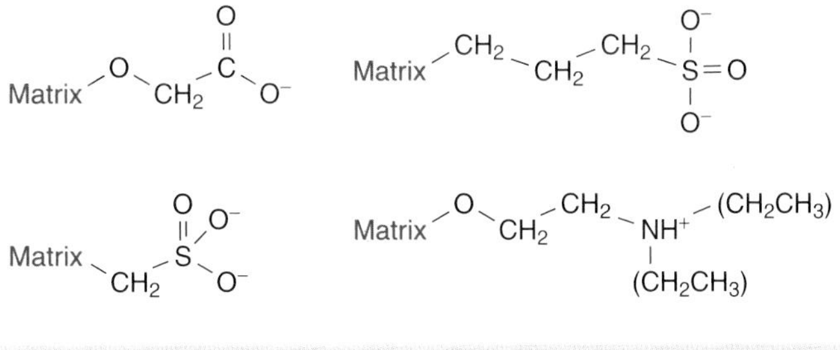

Figure 9.16 The common functional groups found in ion exchange resins. The terms strong and weak exchangers refer to the extent of ionization with pH. Strong ion exchangers such as QAE and SP are completely dissociated over a very wide pH range whereas with the weak exchangers (DEAE and CM) the degree of ionization varies and this is reflected in their binding properties

Size exclusion or gel filtration chromatography

Size exclusion separates proteins according to their molecular size. The principle of the method relies on separation by the flow of solute and solvent over beads containing a series of pores. These pores are of constant size and are formed by covalent cross-linking of polymers. High levels of cross-linking lead to a small pore size. In a mixture of proteins of different molecular mass some will be small enough to enter the pores and as a result their rate of diffusion through the column is slowed. Larger proteins will rarely enter these regions with the result that progress through the resin is effectively unimpeded. Very large proteins will elute in the 'void volume' and all proteins that fail to show an interaction with the pores will elute in this void volume irrespective of their molecular size. Migration rates through the column reflect the extent of interaction with the pores; small proteins spend a considerable amount of time in the solvent volume retained within pores and elute more slowly than those proteins showing marginal interactions with the beads. It is clear that this process represents an effective way of discriminating between proteins of different sizes (Figure 9.17).

In an ideal situation there is a linear relationship between the elution volume and the logarithm of the molecular mass (Figure 9.18). This means that if a size exclusion column is calibrated using proteins of known mass then the mass of an unknown protein can be deduced from its elution position. A reasonably accurate estimation of mass can be obtained if the unknown protein is of similar shape to the standard set. This occurs because the process of size exclusion or gel filtration is sensitive to the average volume occupied by the protein in solution. This volume is often represented by a term called the Stokes radius of the protein and this parameter is very sensitive to overall shape. So, for example, elongated molecules such as fibrous proteins show anomalous rates of migration. A rigorous description of size exclusion chromatography would view the separation as based on differences in hydrodynamic radius. Today, size exclusion chromatography is rarely used as an

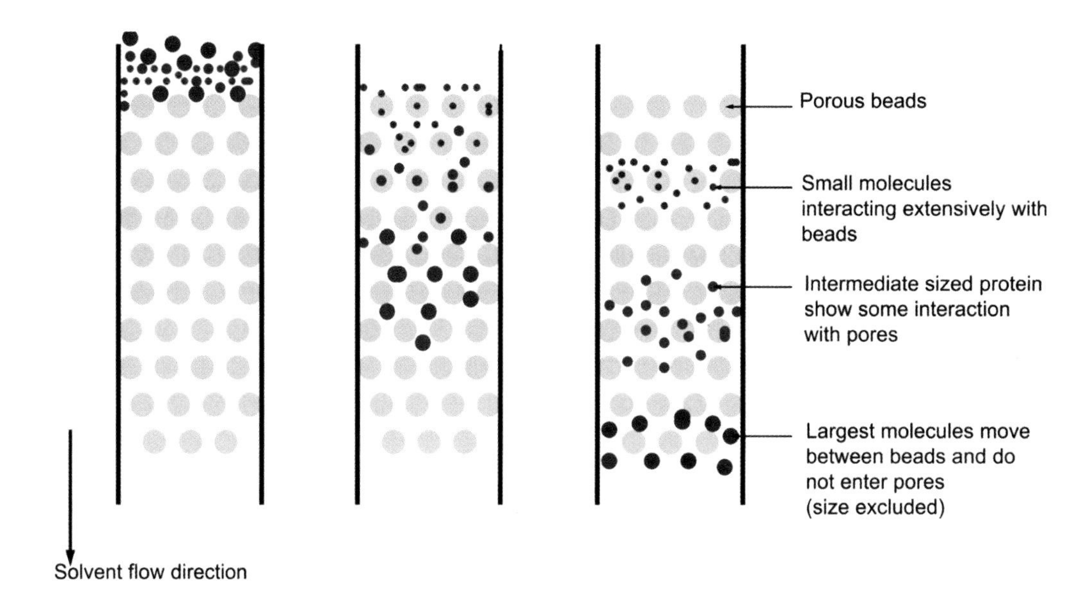

Figure 9.17 The principle of gel filtration showing separation of three proteins of different mass at the beginning, middle and end of the process. Resins are produced by cross linking agarose, acrylamide and dextran polymers to form different molecular size ranges from 100–10 000 (peptide–small proteins), 3000–70 000 (small proteins), 10^4–10^5 (larger proteins, probably multi-subunit proteins), and 10^5–10^7 (macromolecular assemblies such as large protein complexes, viruses, etc

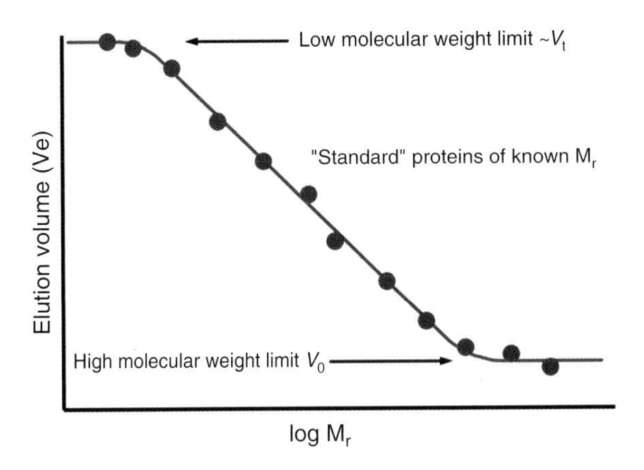

Figure 9.18 Estimation of molecular mass using size exclusion chromatography. The volume (or time) necessary to achieve elution from a column is plotted against the logarithm of molecular weight to yield a straight line that deviates at extreme (low and high) molecular weight ranges. These extremes correspond to the total volume of the column (V_t) and the void volume respectively (V_o)

analytical technique to estimate subunit molecular mass but it is frequently used as a preparative method to eliminate impurities from proteins. One common use is to remove salt from a sample of protein or to exchange the protein from one buffer to another.

Hydrophobic interaction chromatography

Hydrophobic interaction chromatography (HIC) is based on interactions between non-polar side chains and hydrophobic resins. At first glance it is not obvious how HIC is useful for separating and purifying proteins. In globular proteins the folded states are associated with buried hydrophobic residues well away from the surrounding aqueous solvent. However, close inspection of some protein structures reveals that a fraction of hydrophobic side chains remain accessible to solvent at the surface of folded molecules. In some cases these residues produce a distinct hydrophobic patch whose composition and size varies from one protein to another and leads to large differences in surface hydrophobicity. HIC separates proteins according to these differences. As a technique it exploits interactions between hydrophobic patches on proteins and hydrophobic groups covalently attached to an inert matrix. The most popular resins are cross-linked agarose polymers containing butyl, octyl or phenyl (hydrophobic) functional groups.

The hydrophobic interaction depends on solution ionic strength as well as chaotropic anions/cations. As a result HIC is usually performed under conditions of high ionic strength in solutions containing 1 M ammonium sulfate or 2 M NaCl. High ionic strength increases interactions between hydrophobic ligand and protein with the origin of this increased interaction arising from decreased protein solvation as a result of the large numbers of ions each requiring a hydration shell. The net result is that hydrophobic surfaces interact strongly as ions 'unmask' these regions on proteins.

Samples are applied to columns containing resin equilibrated in a solution of high ionic strength and proteins containing hydrophobic patches stick to the resin whilst those lacking hydrophobic interactions are not retained by the column. Elution is achieved by decreasing favourable interactions by lowering the salt concentration via a gradient or a sudden 'step'. Proteins with different hydrophobicity are separated during elution with the most hydrophobic protein retained longest on the column. However, the prediction of the magnitude of surface hydrophobicity on proteins remains difficult and the technique tends to be performed in an empirical manner where a variety of resins are tried for a particular purification protocol. One advantage of HIC is the relatively mild conditions that preserve folded structure and activity.

Reverse phase chromatography

A second chromatographic method that exploits hydrophobicity is reverse phase chromatography (RPC) although the harsher conditions employed in this method can lead to denaturation. The basis of RPC is the hydrophobic interaction between proteins/peptides and a stationary phase containing alkyl or aromatic hydrocarbon ligands immobilized on inert supports. The original support polymer for reverse phase methods was silica to which hydrophobic ligands were linked. Although still widely used as a base matrix it has a disadvantage of instability at alkaline pH and this restricts its suitability in biological studies. As a result, synthetic organic polymers based on polystyrene beads have proved popular and show excellent chemical stability over a wide range of conditions.

The separation mechanism in RPC depends on hydrophobic binding interactions between the solute molecule (protein or peptide) in the mobile phase and the immobilized hydrophobic ligand (stationary phase). The initial mobile phase conditions used to promote binding in RPC are aqueous solutions and lead to a high degree of organized water structure surrounding protein and immobilized ligand. The protein binds to the hydrophobic ligand leading to a reduction in exposure of hydrophobic surface area to solvent. The loss of organized water structure is accompanied by a favourable increase in system entropy that drives the overall process. Proteins partition between the mobile and stationary phases in a process similar to that observed by an organic chemist separating iodine between water and methanol. The distribution of the protein between each phase reflects the binding properties of the medium, the hydrophobicity of the protein and the overall composition of the mobile phase. Initial experimental conditions are designed to favour adsorption of the protein from the mobile phase to the stationary phase. Subsequently, the mobile phase composition is modified by decreasing its polarity to favour the reverse process of desorption leading to the protein partitioning preferentially back into the mobile phase. Elution of the protein or peptide occurs through the use of gradients that increase the hydrophobicity of the mobile phase normally through the addition of an organic modifier. This is achieved through the addition of increasing amounts of non-polar solvents such as acetonitrile, methanol, or isopropanol. At the beginning a protein sample is applied to a column in 5 percent acetonitrile, 95 percent water, and this is slowly changed throughout the elution stage to yield at the end of the process a solution containing, for example, 80 percent acetonitrile/20 percent water.

For both silica and polystyrene beads the hydrophobic ligands are similar and involve linear hydrocarbon chains (n-alkyl groups) with chain lengths of C2, C4, C8 and C18 (Figure 9.19). C18 is the most hydrophobic ligand and is used for small peptide purification whilst the C2/C4 ligands are more suitable for proteins. Although proteins have been successfully purified using RPC the running conditions often lead to denaturation and reverse phase methods are more valuable for purifying short peptides arising from chemical synthesis or enzymatic digestion of larger proteins (see Chapter 6).

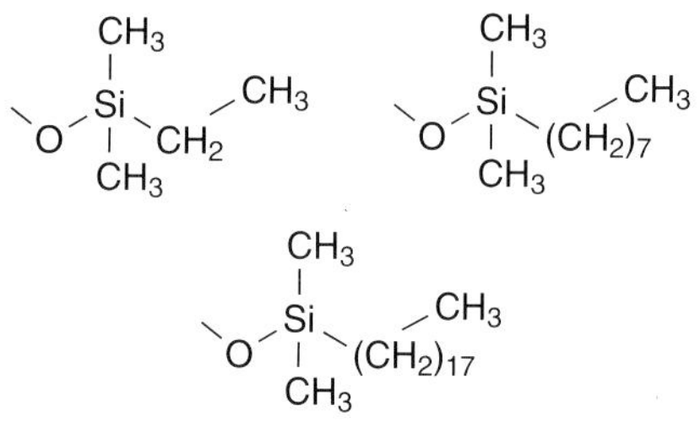

Figure 9.19 The hydrophobic C2, C8 and C18 ligands widely used in RPC

Affinity chromatography

An important characteristic of some proteins is their ability to bind ligands tightly but non-covalently. This property is exploited as a method of purification with the general principle involving covalently attachment of the ligand to a matrix. When mobile phase is passed through a column containing this group proteins showing high affinity for the ligand are bound and their rate of progress through the column is decreased. Most proteins will not show affinity for the ligand and flow straight through the column. This protocol therefore offers a particularly selective route of purification and affinity chromatography has the great advantage of exploiting biochemical properties of proteins, as opposed to a physical property.

Affinity chromatography relies on the interaction between protein and immobilized ligand. Binding must be sufficiently specific to discriminate between proteins but if binding is extremely tight (high affinity) it can leads to problems in recovering the protein from the column. Most methods of recovery involve competing for the binding site on the protein. Consequently, a common method of elution is to add exogenous or free ligand to compete with covalently linked ligand for the binding site of the protein. A second method is to change solution conditions sufficiently to discourage protein–ligand binding. Under these conditions the protein no longer shows high affinity for the ligand and is eluted. The last method must of course not use conditions that promote denaturation of the protein, and frequently it involves only moderate changes in solution pH or ionic strength. Enzymes are particularly suited to this form of chromatography since co-factors such as NAD, NADP, and ADP are clear candidate ligands for use in affinity chromatography.

One limitation in the development of affinity chromatography is that methods must exist to link the ligand covalently to the matrix. This modification must not destroy its affinity for the protein and must be reasonably stable i.e not broken under moderate temperature, pH and solution conditions used during chromatography. Recently, a great expansion in the number and type of affinity based chromatographic separations has occurred due to the development of expression systems producing proteins with histidine tags or fused to other protein such as glutathione-*S*-transferase or maltose binding protein (Table 9.3). Histidine tags can be located at either the N- or C-terminal regions of a protein and bind to metal chelate columns containing the ligand imidoacetic acid. Similarly glutathione-*S*-transferase is a protein that binds the ligand glutathione which itself can be covalently linked to agarose based resins.

Dialysis and ultrafiltration

The basis of dialysis and ultrafiltration is the presence of a semipermeable membrane that allows the flux of small (low molecular weight) compounds at the same time preventing the diffusion of larger molecules such as proteins. Both methods are commonly used in the purification of proteins and in the case of ultrafiltration particularly in the concentration of proteins.

Dialysis is commonly used to remove low molecular weight contaminants or to exchange buffer in a solution containing protein. A solution of protein is placed in dialysis tubing that is sealed at both ends and added to a much larger volume of buffer (Figure 9.20). Low molecular weight contaminants diffuse across the membrane whilst larger proteins remain trapped. Similarly, the buffers inside and outside the dialysis tubing will exchange by diffusion until equilibrium is reached between the compartments. This equilibrium is governed by the Donnan effect, which maintains electrical neutrality on either side of the membrane. For polyvalent ions such as proteins this means that ions of the opposite charge will remain in contact with protein and the solution ionic strength will not fall to exactly that in the bulk solution. For dilute solutions of protein coupled with higher exterior ion concentrations the Donnan effect becomes negligible but for high concentrations of protein or low ionic strength buffers it remains a significant effect on the colligative properties of ions. Despite this complication, buffer inside the dialysis tubing is gradually replaced by the solution found in the larger exterior volume. By replacing the exterior volume of buffer at regular intervals the process rapidly achieves the removal of low molecular weight contaminants or the exchange of solvent.

Ultrafiltration is used to concentrate solutions of proteins by the application of pressure to a sealed vessel whose only outlet is via a semipermeable membrane (Figure 9.20). The molecular weight limit for the semipermeable membrane can be closely controlled. Solvent and low molecular weight molecules pass through the membrane whilst larger molecules remain inside the vessel. As solvent is forced across the membrane proteins inside the ultrafiltration cell are progressively concentrated. Frequently, ultrafiltration is used to concentrate a protein during the final stages of purification when it may be required at high concentrations; in crystallization trials for example.

Polyacrylamide gel electrophoresis

The most common method of analysing the purity of an isolated protein is to use polyacrylamide gel electrophoresis (PAGE) in the presence of the detergent sodium dodecyl sulfate (SDS) and a reductant of disulfide bridges such as β-mercaptoethanol. The technique has the additional advantage of allowing the monomeric (subunit) molecular mass to be determined

Table 9.3 Ligands used for affinity chromatography together with the natural co-factor/substrate and the proteins/enzymes purified by these methods

Aim of affinity reaction	Ligand/natural substrate	Example
To bind nucleotide binding domains	Natural ligands are NADP or NAD. Most affinity-based methods covalently bind either ADP or AMP to resins and this is sufficient to bind the nucleotide binding domain	Dehydrogenases bind NAD or NADH and are frequently purified via this route
Separation of proteins based on the ability of some side chains to chelate divalent metal ions	Iminodiacetic acid covalently linked to gel binds metal ions such as Ni^{2+} or Zn^{2+} leaving a partially unoccupied coordination sphere. Histidine residues are the usual target but tryptophan and cysteine residues can also bind	Any recombinant protein containing a His-tag. Binding occurs via imidazole nitrogen and protonation destroys binding and hence is a route towards elution
Immobilized lectins such as concanavalin A	Binds α-D-mannose, α-D-glucose and related sugars via interaction with free OH groups	Separation of glycoproteins particularly viral glycoproteins and cell surface antigens
Glutathione binding proteins	Glutathione is the natural substrate and can be covalently linked to column	Fusion proteins containing glutathione-*S*-transferase
Affinity of monoclonal and polyclonal IgG for protein A, protein G or a fusion protein, protein A/G	Protein A is a protein of molecular weight 42 000 derived from *S. aureus*. It consists of six major regions five of which bind to IgG. Other affinity columns exploit the use of protein G – a protein derived from streptococci. Both proteins exhibit affinity for the Fc region of immunoglobulins	Separation of IgG subclasses from serum or cell culture supernatants
Reversible coupling of proteins containing a free thiol group	Thiol group immobilized on column matrix. Forms mixed disulfide with protein containing free–SH group. Reversed by addition of excess thiol or other reductant	Cysteine proteinases a family of enzymes with reactive thiols at their catalytic centres are purified by these methods
Affinity for oligonucleotides	Polyadenylic acid linked via N6 amino group or polyuridylic acid linked to matrix	Purification of viral reverse transcriptases and mRNA binding proteins

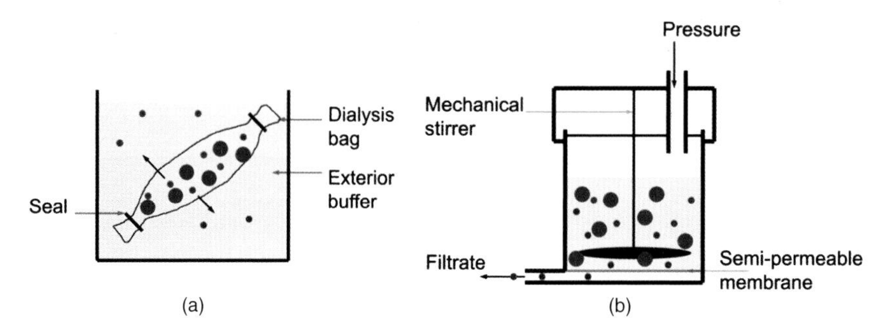

Figure 9.20 Dialysis (a) and ultrafiltration (b) assemblies used in protein purification

with reasonable accuracy. SDS–PAGE is widely used to assess firstly if an isolated protein is devoid of contaminating proteins and secondly whether the purified protein has the expected molecular mass. Both of these parameters are extremely useful during protein purification.

The principle of SDS–PAGE is the separation of proteins (or their subunits) according to molecular mass by their movement through a polyacrylamide gel of closely defined composition under the influence of an electric field. Looking at the individual components of this system allows the principles of electrophoresis to be illustrated and to emphasize the potential of this technique to provide accurate estimations of mass, composition and purity. The gel component is a porous matrix of cross-linked polyacrylamide. The most common method of formation involves the reaction between acrylamide and *N*, *N*′-methylenebisacrylamide (called 'bis'; Figure 9.21) catalysed by two additional compounds ammonium persulfate (APS) and *N*,*N*,*N*′,*N*′-tetramethylethylenediamine (TEMED). Polyacrylamide gels form when a dissolved mixture of acrylamide and the bifunctional 'bis' cross-linker polymerizes into long, covalently linked, chains. The polymerization of acrylamide is a free-radical catalysed reaction in which APS acts as an initiator. Initiation involves generating a persulfate free-radical that activates the quaternary amine TEMED, which in turns acts as a catalyst for the polymerization of acrylamide monomers. The concentration of acrylamide can be varied to alter 'resolving power' and this contributes to the definition of a 'pore' size through which proteins migrate under the influence of an electric field. A smaller pore size is generated by high concentrations of acrylamide with the result that only small proteins migrate effectively. This action of the gel is therefore similar to size exclusion chromatography where there is an effective filtration or molecular sieving process by pores.

The mobility of a protein through polyacrylamide gels is determined by a combination of overall charge, molecular shape and molecular weight. Native PAGE (performed in the absence of SDS) yields the mass of native proteins and is subject to deviations caused by non-spherical shape and residual charge as well as interactions between subunits. The technique is less widely used than SDS–PAGE. In the presence of SDS the parameters of shape and charge become unimportant and separation is achieved solely on the basis of protein molecular weight. The underlying reason for this observation is that SDS binds to almost all proteins destroying native conformation (Figure 9.22). SDS causes proteins to unfold, forming rod-like protein micelles that migrate through gels. Since almost all proteins form this rod-like unfolded structure with an excess of negative charge due to the bound SDS the effect of shape and charge is eliminated. SDS binds to proteins at a relatively constant ratio of 1.4 g detergent/g protein leading in a protein of molecular weight 10 000 (i.e ~100 residues) to approximately one SDS molecule for every two amino acid residues.

Acrylamide

Bisacrylamide

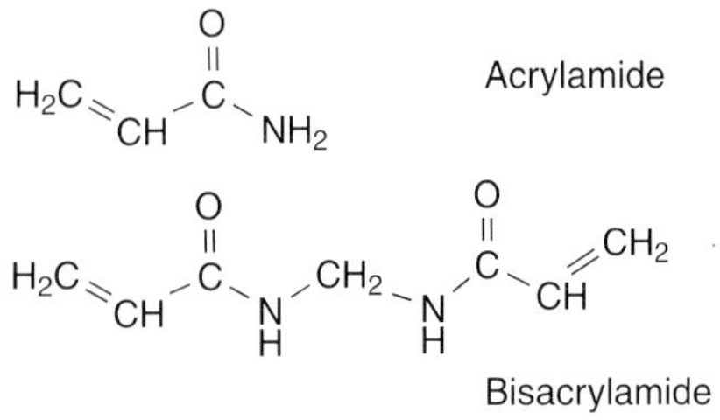

Figure 9.21 The structure of acrylamide and bisacrylamide

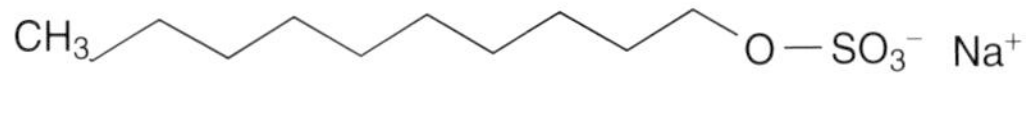

A consequence of detergent binding is to coat all proteins with negative charge and to eliminate the charge found on the native protein. As a result of SDS binding all proteins become highly negatively charged, adopt an extended rod-like conformation and migrate towards the anode under the influence of an electric field.

A protein mixture is applied to the top of a gel and migrates through the matrix as a result of the electric field with 'lighter' components migrating faster than 'heavier' components. Over time the component

CH_3 — $O-SO_3^-$ Na^+

Figure 9.22 Sodium dodecyl sulfate, also called sodium lauryl sulfate, has the structure of a long acyl chain containing a charged sulfate group; it is an extremely potent denaturant of proteins

proteins are separated and the resolving power of the techniques is sufficiently high that heterogeneous mixtures of proteins can be separated and distinguished from each other. In practice most gels contain two components – a short 'focusing' gel containing a low percentage of acrylamide (<5 percent) that assists in the ordering and entry of polypeptides into a longer 'resolving' gel (Figure 9.23). The above discussion has assumed a constant acrylamide concentration throughout the gel but it is now possible to 'tailor' gel performance by varying the concentration of acrylamide. This creates a gradient gel where acrylamide concentrations vary linearly, for example, from 10 to 15 percent. The effect of this gradient is to enhance the resolving power of the gel over a narrower molecular weight range. Although an initial investigation is usually performed with gels containing a constant acrylamide concentration (say 15 percent) increased resolution can be obtained by using gradient gels. The range of gradient usually depends on the mass of the polypeptides under investigation.

The migration of proteins through a gel will depend on the voltage/current conditions used as well as the temperature and it is most common to compare the mobility of an unknown protein or mixture of proteins with pure components of known molecular mass. For example, in Figure 9.24 the migration of proteins in an extract derived from recombinant *E. coli* is compared to the mobility of standard proteins whose molecular masses are known and range from 6 kDa to 200 kDa. The fractionation of the protein of interest can be followed clearly at each stage.

A large number of proteins ranging in mass from less than 10 kDa to greater than 100 kDa are seen in the starting material. Progressive purification removes many of these proteins and allows bands of similar monomeric mass to be identified.

After separation the gel is colourless but profiles such as those of Figure 9.24 result from gel staining that locates the position of each protein band. A number of different stains exist but the two most common dyes are Coomassie Brilliant Blue R250 (see Chapter 3) and silver nitrate. For routine use Coomassie blue staining is preferable and has a detection limit of approximately 100 ng protein. Coomassie Brilliant Blue reacts non-covalently with all proteins and the success of the staining procedure relies on

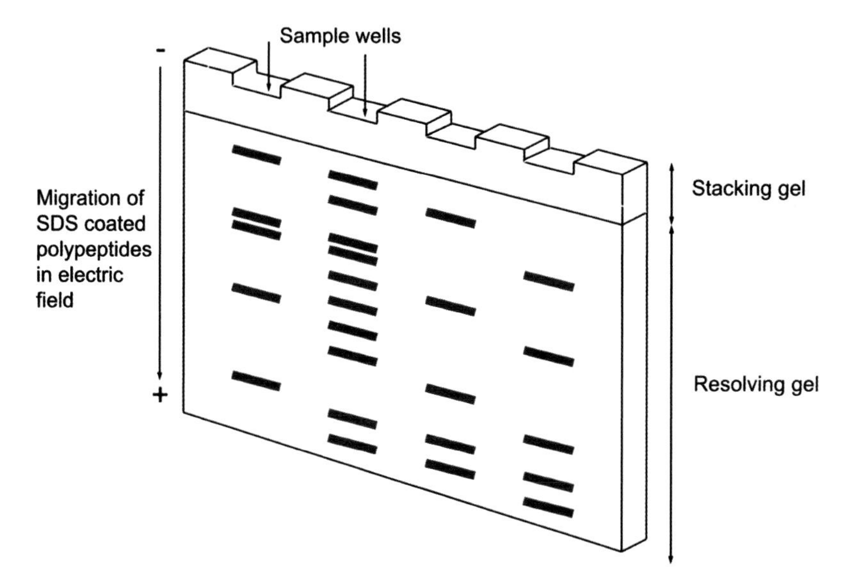

Figure 9.23 A conventional polyacrylamide gel. A top 'stacking' gel containing a lower concentration of acrylamide facilitates entry of SDS-coated polypeptides into the 'resolving' gel where the bands separate according to molecular mass. Separation is shown by the different migration of bands (shown in red) within the gel

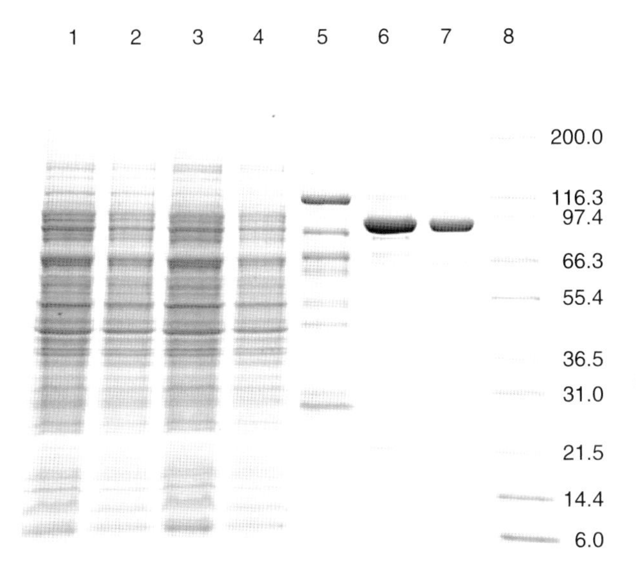

Figure 9.24 A 'real' gel showing the fractionation of a foreign protein expressed originally as a fusion protein from *E. coli* cell lysates. The gel was stained using Coomassie blue. From left to right the protein undergoes progressive purification to eventually yield a pure protein. The cell lysates are shown in lanes 1 and 2, lanes 3 and 4 show the non bound material from application to a glutathione affinity column whilst lane 5 represents the bound material eluted from this column. Lane 6 represents the protein released from the GST fusion by limited proteolysis whilst lane 7 shows the result of further purification by anion exchange chromatography. Lane 8 shows molecular weight markers over a range of 200–6 kDa (reproduced with permission from Chaddock, J.A. *et al. Protein Expression and Purification*. 2002, **25**, 219–228. Academic Press)

'fixing' by precipitation the migrated proteins with acetic acid followed by removal of excess dye using methanol–water mixtures. Silver nitrate has the advantage of increased sensitivity with a detection limit of ~1 ng of protein. The molecular basis for silver staining remains less than clear but involves adding silver nitrate solutions to the gel previously washed under mildly acidic conditions. Today, purpose-built machines have automated the running and staining of gels (Figure 9.25).

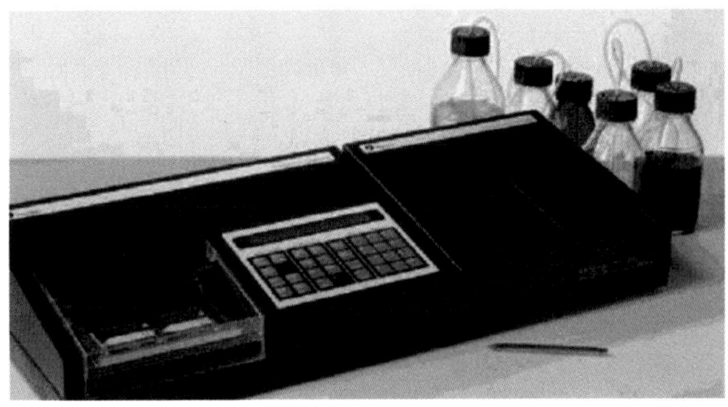

Figure 9.25 An automated electrophoresis system. The left-hand compartment performs electrophoretic separation of proteins on thin, uniformly made, gels whilst the second compartment performs the staining reaction to visualize protein components. The gels are run horizontally. (Reproduced courtesy of Amersham Biosciences)

Further use of polyacrylamide gels in studying proteins occurs in the techniques of Western blotting and two-dimensional (2D) electrophoresis. In the first area SDS–PAGE is combined with 'blotting' techniques to allow bands to be identified via further assays. These procedures involve enzyme-linked immunosorbent assays (ELISA), in which antibody binding is linked to a second reaction involving an enzyme-catalysed colour change. Since the reaction will only occur if a specific antigen is detected this process is highly specific. This combination of techniques is called Western blotting (Figure 9.26). Western blotting was given its name as a pun on the technique Southern blotting, pioneered by E.M. Southern, where DNA is transferred from an agarose gel and immobilized on nitrocellulose matrices. In Western blotting the protein is eluted/transferred from SDS–polyacrylamide gels either by capillary action or by electroelution. Electroelution results in the movement of all of the separated proteins out of the gel and onto a second supporting media such as polyvinylidene difluoride (PVDF) or nitrocellulose. Immobilized on this second supporting matrix the proteins undergo further reactions that assist with identification (Figure 9.27). A common technique reacts the eluted protein with previously raised antibodies. Since antibodies will only bind to a specific antigen this reaction immediately

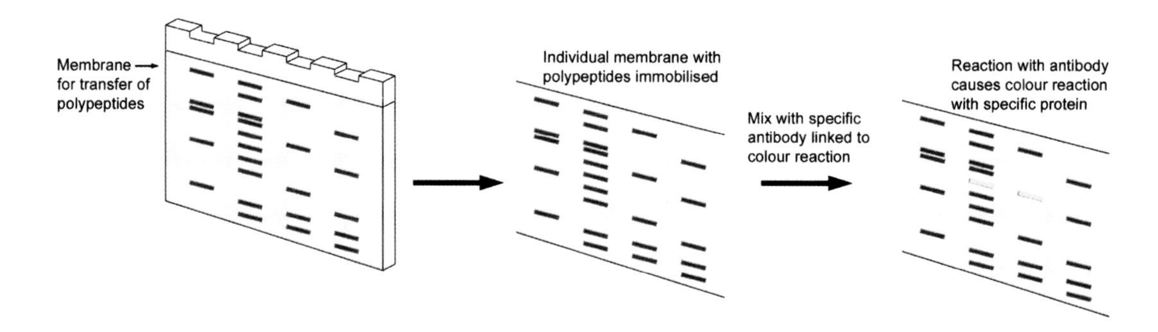

Figure 9.26 Western blotting and identification of a protein via cross reaction with a specific antibody. In the example above a single protein in sample wells 2 and 3 is identified

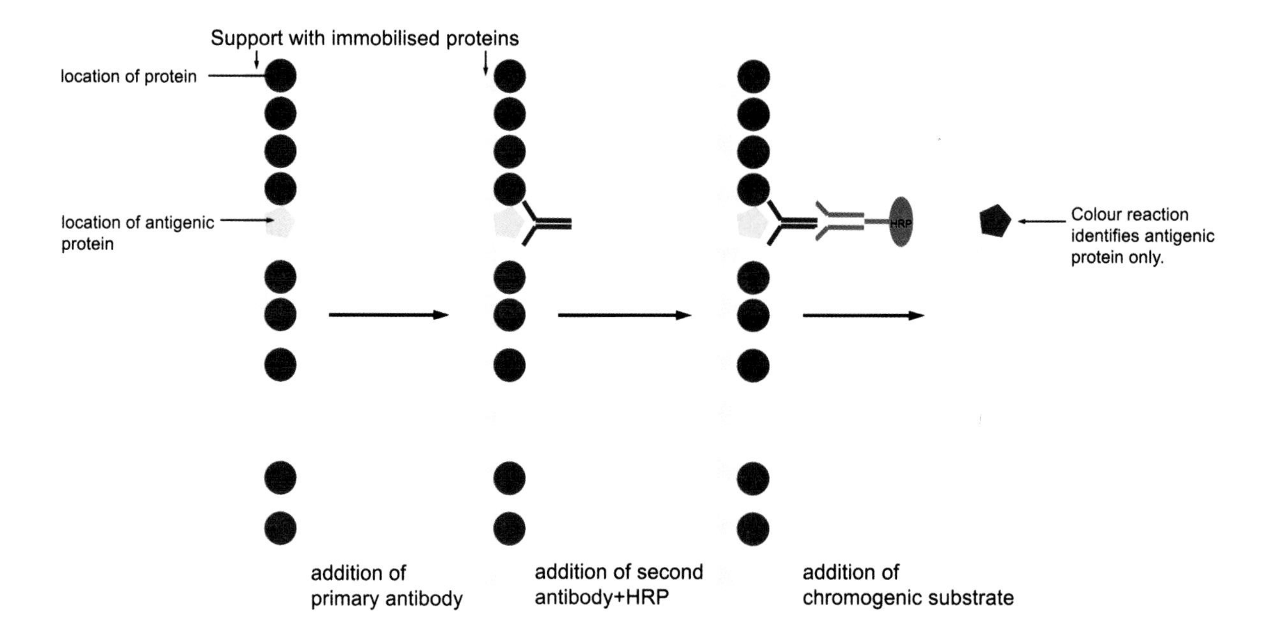

Figure 9.27 An antibody raised against the protein of interest is used to locate the polypeptide on the PVDF membrane. This membrane is essentially a replica of the polyacrylamide gel. A second antibody containing covalently bound horseradish peroxidase reacts with the first or primary antibody and in the presence of a chromogenic substrate (3,3′,5,5′-tetramethylbenzidine and hydrogen peroxide) causes a reaction that identifies the location of the polypeptide via the production of a coloured band. Care has to be taken to avoid non-specific binding, but to a first approximation the intensity of colour indicates the quantity of antigenic polypeptide as well as its location on the support. Since this matrix is a replica of the original SDS–polyacrylamide gel staining of the gel will enable identification of this band within a complex mixture of polypeptides as a result of the specific antigen–antibody reaction

identifies the protein of interest. However, by itself the reaction does not lead to any easily detectable signal on the supporting matrix. To overcome this deficiency the antibody reaction is usually carried out with a second antibody cross-reacting to the initial antibody and coupled to an enzyme. Two common enzymes linked to antibodies in this fashion are horseradish peroxidase (HRP) and alkaline phosphatase. These enzymes were chosen because the addition of substrates (benzidine derivatives and hydrogen peroxide for HRP together with 5-bromo-4-chloro-3-indoyl phosphate or 4-chloro-1-naphthol substrates with alkaline phosphatase) leads to an insoluble product and a colour reaction that is detected visually. The appearance of a colour on the nitrocellulose support allows the position of the antibody–antigen complex to be identified and to be correlated with the result of SDS–PAGE.

An alternative procedure to the reaction with antibodies is to identify proteins absorbed irreversibly to a nitrocellulose sheet using radioisotope-labelled antibodies. Iodine-125 is commonly used to label antibodies and the protein is identified from the exposure of photosensitive film. Together these two techniques allow a protein to be detected on the basis of molecular mass (SDS–PAGE) and also native conformation (Western blotting) at very low levels. These techniques are now the basis for many clinical diagnostic tests.

A second route of analysis involves spreading protein separation into a second dimension. This technique is called two-dimensional gel electrophoresis and separates proteins according to their isoelectric points in the first dimension followed by a second separation based on subunit molecular mass using SDS–PAGE.

The bioinformatics revolution currently underway in protein biochemistry is providing ever-increasing amounts of sequence data that requires analysis at the molecular level. The area of genomics has led naturally to proteomics – the wish to characterize proteins expressed by the genome. This involves analysis of the structure and function of all proteins within a single organism, including not only individual properties but also their collective properties through interactions with physiological partners. Proteomics requires a rapid way of identifying many proteins within a cell, and in this context 2D electrophoresis is advancing as a powerful method for analysis of complex protein mixtures (Figure 9.28). This technique sorts proteins according to two independent properties. The first-dimension relies on isoelectric focusing (IEF) and centres around the creation of a pH gradient under the influence of an electric field. A protein will move through this pH gradient until it reaches a point where its net charge is zero. This is the p*I* of that protein and the protein will not migrate further towards either electrode. The separation of protein mixtures by IEF is usually performed with gel 'strips' which are layered onto a second SDS gel to perform the 2D step. The result is a 2D array where each 'spot' should represent a single protein species in the sample. Using this approach thousands of different proteins within a cell or tissue can be separated and relevant information such as isoelectric point (p*I*), subunit or monomeric mass and the amount of protein present in cells can be derived.

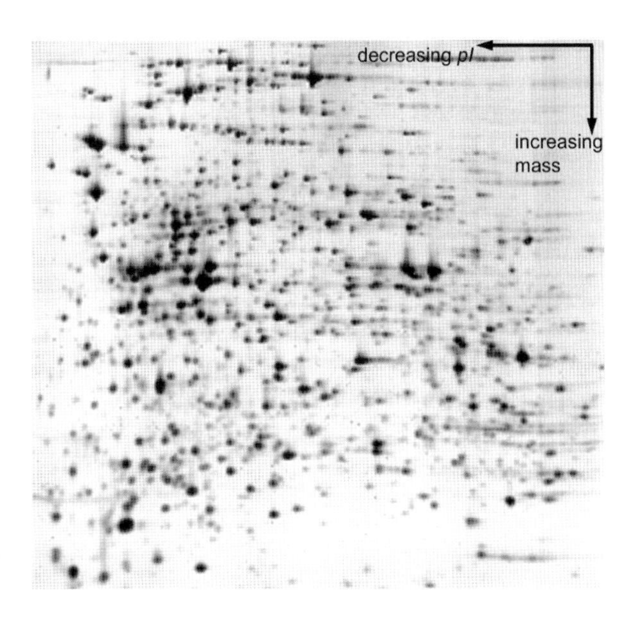

Figure 9.28 Separation of proteins using 2D electrophoresis. The proteins separate according to isoelectric point in dimension 1 and according to molecular mass in dimension 2. Often the proteins are labelled via the incorporation of radioisotopes and exposure to a sensitive photographic film allows identification of each 'spot'

Although the value of 2D electrophoresis as an analytical technique was recognized as long ago as

1975 it is only in the last 10 years that technical improvements allied to advances in computers have seen this approach reach the forefront of research. The observation of a single spot on 2D gels is sufficient to allow more detailed analysis by eluting or transferring the 'spot' and subjecting it to microsequencing methods based on the Edman degradation, amino acid analysis or even mass spectrometry. Computers and software are now available that allow digital evaluations of the complex 2D profile and electrophoresis results can be readily compared between different organisms. One obvious advantage of these techniques is their ability to detect post-translational modifications of proteins. Such modifications are not readily apparent from protein sequence analysis and are not easily predicted from genome analysis. 2D electrophoresis is also permitting differential cell expression of proteins to be investigated, and in some cases the identification of disease markers from abnormal profiles.

Mass spectrometry

Mass spectrometry has become a valuable tool in studying covalent structure of proteins. This has arisen because mass determination methods have become more accurate with the introduction of new techniques such as matrix assisted laser desorption time of flight (MALDI-TOF) and electrospray methods. This allows accurate mass determination for primary sequences as well as the detection of post-translational modifications, the analysis of protein purity and where appropriate the detection of single residue mutations in genetically engineered proteins.

In mass spectrometry the fundamental observed parameter is the mass to charge ratio (m/z) of gas phase ions. Although this technique has its origins in studies originally performed by J.J. Thomson at the beginning of the 20th century, over the last 20 years the technique has advanced rapidly and specifically in the area of biological analysis. The technique is important to protein characterization, as demonstrated by the award of the Nobel prize for Chemistry in 2002 to two pioneers of this technique, John Fenn and Koichi Tanaka, for the development of methods permitting analysis of structure and identification of biological macromolecules and for the development of 'soft' desorption–ionization methods in mass spectrometry.

All mass spectrometers consist of three basic components: an ion source, a mass analyser and a detector. Ions are produced from samples generating charged states that the mass analyser separates according to their mass/charge ratio whilst a detector produces quantifiable signals. Early mass spectrometry studies utilized electron impact or chemical ionization methods to generate ions and with the extreme conditions employed this frequently led to the fragmentation of large molecules into smaller ions (Table 9.4). Improvements in biochemical mass spectrometry arise from the development of 'soft' ionization methods that generate molecular ions without fragmentation (Table 9.4). In particular, two methods of soft ionization are widely used in studies of proteins and these are matrix-assisted laser desorption ionization (MALDI) and electrospray ionization (ESI).

An initial step in all mass spectrometry measurements is the introduction of the sample into a vacuum

Table 9.4 The various mass spectrometry methods have different applications

Ionization method	Analyte	Upper limit for mass range (kDa)
Electron impact	Volatile	~1
Chemical ionization	Volatile	~1
Electrospray ionization	Peptides and proteins	~200
Fast atom bombardment	Peptides, small proteins	~10
Matrix assisted laser desorption ionization (MALDI)	Peptides or proteins	~500

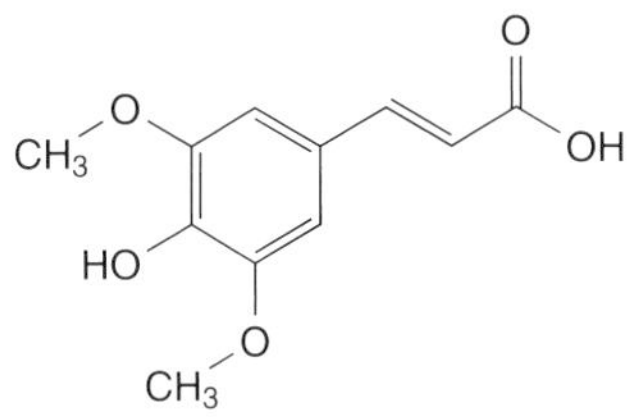

Figure 9.29 Structure of sinapinic acid used as matrix in MALDI methods of ionization. Sinapinic acid (3,5-dimethoxy-4-hydroxycinnamic acid) is the matrix of choice for protein of $M_r > 10\,000$, whilst for smaller proteins α-cyano 4-hydroxycinnamic acid is frequently preferred as a matrix

and the formation of a gas phase. Since proteins are not usually volatile it was thought that mass spectrometry would not prove useful in this context. Although early studies succeeded in making volatile derivatives by modifying polar groups the biggest advance has arisen through the use of alternative strategies. One of the most important of these is the MALDI-TOF method where the introduction of a matrix assists in the formation of gas phase protein ions (Figure 9.29). A common matrix is sinapinic acid (M_r 224) where the protein (analyte) is dissolved in the matrix at a typical molar ratio of ~1:1000. The protein is frequently present at a concentration of 1–10 μM. This mixture is dried on a probe and introduced into the mass spectrometer. The next stage involves the use of a laser beam to initiate desorption and ionization and in this area the matrix prevents degradation of the protein by absorbing energy. The matrix also assists the irradiated sample to vaporize by forming a rapidly expanding matrix plume that aids transfer of the protein ions into the mass analyser. Most commercially available mass spectrometers coupled to MALDI ion sources are 'time-of-flight' instruments (TOF-MS) in which the processes of ion formation, separation and detection occur under a very stringent high vacuum. The ionization of proteins does not cause fragmentation of the protein but generates a series of positively charged ions the most common of which occurs through the addition of a proton and is often designated as the $(M + H)^+$ ion. A second ionized species with half the previous m/z ratio denoted as $(M + 2H)^{2+}$ is also commonly formed. If Na^+ or K^+ ions are present in samples then additional molecular ions can form such as $(M + Na)^+$ and $(M + K)^+$ and will be observed with characteristic m/z ratios. The gaseous protein ions formed by laser irradiation are rapidly accelerated to the same (high) kinetic energy by the application of an electrostatic field and then expelled into a field-free region (called the flight tube) where they physically separate from each other according to their m/z ratios. A detector at the end of the flight tube records the arrival time and intensity of signals as groups of mass-resolved ions exit the flight tube. Small molecular ions will exit the flight tube first followed by proteins with progressively greater m/z ratios. Commonly, the matrix ions will be the first ions to reach the detector followed by low molecular weight impurities. To estimate the size of each molecular ion samples of known size are usually measured as part of an internal calibration. By estimating the TOF for these standard proteins the TOF and hence the mass for an unknown protein can be estimated with very high accuracy.

The mass of the intact protein is one of the most useful attributes in protein identification procedures. Although SDS–PAGE can yield a reasonably accurate mass for a protein such measurements are prone to error. Some proteins run with anomalous mobility on these gels leading to either over or under estimation of mass that can range in error from 5 to 25 percent. As a result, mass spectrometry has been used to refine estimations of molecular weight by excision of a protein from 1 or 2D gels. MALDI-TOF is the mass spectrometry technique of choice for rapid determination of molecular weights for gel 'spots' and can be performed on samples directly from the gel, from electroblotting membranes or from the protein extracted into solution. In fast-atom bombardment (FAB) a high-energy beam of neutral atoms, typically Xe or Ar, strikes a solid sample causing desorption and ionization. It is used for large biological molecules that are difficult to get into the gas phase. FAB causes little fragmentation of the protein and usually gives a large molecular ion peak, making it ideal for molecular weight determination. A third method of ion generation is electrospray ionization (ESI) mass spectrometry and involves a solution of protein molecules suspended in a volatile solvent that are sprayed at atmospheric pressure through a narrow channel (a capillary of diameter ~0.1 mm) whose outlet is maintained within an electric field operating at voltages of ~3–4 kV. This field causes the liquid to disperse into fine droplets that can pass down a potential and pressure gradient towards the mass analyser. Although the exact mechanism of droplet and ion formation remains unclear further desolvation occurs to release molecular ions into the gas-phase. Like the MALDI-TOF procedure the electrospray method is a 'soft' ionization technique capable of ionizing and transferring large proteins into the gas-phase for subsequent mass spectrometer detection. In the previous discussion the mass analyser has not been described in detail but three principal designs exist: the magnetic and/or electrostatic sector mass analyser, the time of flight (TOF) analyser and the quadrupole analyser. Each analyser has advantages and disadvantages (Table 9.5).

The principle of using mass spectrometry in protein sequencing is shown for a 185 kDa protein derived from SDS–PAGE and digested with a specific protease (Figure 9.30). One peptide species is 'selected' for collision (m/z 438) and the derived fragments of this peptide are measured to give the product mass spectrum. Some of these fragments differ in their mass by values equivalent to individual amino acids. The sequence is deduced with the exception that leucine and isoleucine having identical masses are not distinguished. This sequence information is used together with the known specificity of the protease to create a 'peptide sequence tag'. Such tag sequences can then be compared with databases of protein sequences to identify either the protein itself or related proteins.

Mass spectrometry has very rapidly become an essential tool for all well-funded protein biochemistry laboratories and the range of applications is increasing steadily as the proteomics revolution continues.

Table 9.5 Different methods of detection employed in mass spectrometry

Analyser	Features of system
Quadrupole	Unit mass resolution, fast scan, low cost
Sector (magnetic and/or electrostatic)	High resolution, exact mass
Time-of-flight (TOF)	Theoretically, no limitation for m/z maximum, high flux systems

How to purify a protein?

Armed with the knowledge described in preceding sections it should be possible to attempt the isolation of any protein with the expectation that a highly enriched pure sample will be obtained. In practice, numerous difficulties present themselves and often render it difficult to achieve this objective. Assuming a protein is soluble and found either in the cytoplasm or a subcellular compartment then perhaps the major problem likely to be encountered in isolation is the potentially low concentration of some proteins in tissues. The use of molecular biology avoids this problem and allows soluble proteins to be over-expressed in high yields in hosts such as *E. coli* (Table 9.6). If cDNA encoding the protein of interest has been cloned it is logical to start with an expression system as the source of biological material.

The first step will involve disrupting the host cells (*E. coli*) using either lysozyme or mild sonication followed by centrifugation to remove heavier components such as membranes from the soluble cytoplasmic fraction which is presumed to contain the protein of interest. At this stage with the removal of substantial amounts of protein it may be possible to assay for the presence of the protein especially if it has an enzymatic activity that can be readily measured.

Alternatively the protein may have a chromophore with distinctive absorbance or fluorescence spectra that can be quantified. However, it is worth noting that at this stage the protein is just one of many hundreds of soluble proteins found in *E. coli*. These measurements remain important in any purification because they define the initial concentration of protein. All subsequent steps are expected to increase the concentration of protein at the expense of unwanted proteins. Concurrently, most investigators would also run SDS–PAGE gels of the initial starting material to obtain a clearer picture of the number and size range of proteins present in the cell lysate (Figure 9.31). Again, successive steps in the purification would be expected to remove many, if not all, of these proteins leaving the protein of interest as a single homogeneous band on a gel.

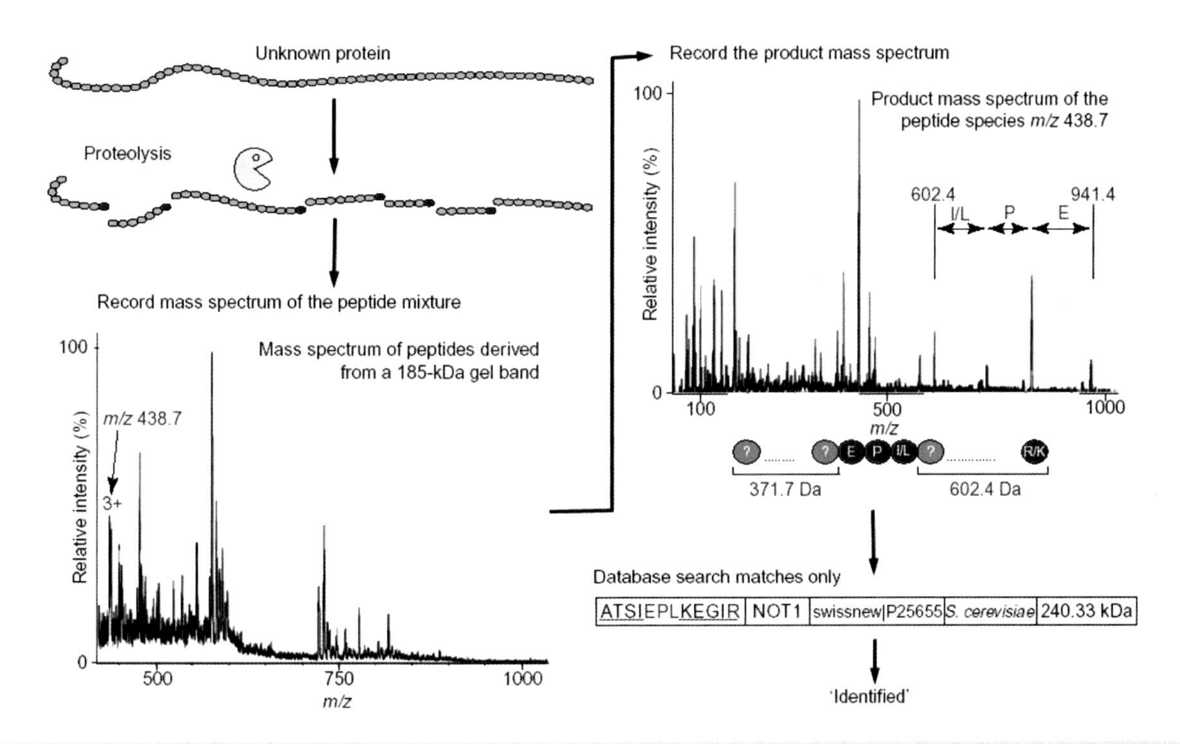

Figure 9.30 Mass spectrometric identification of an unknown protein can be obtained from determining the sequence of proteolytic fragments and comparing the sequence with available databases (reproduced and adapted with permission from Rappsilber, J. and Mann, M. *Trends Biochem. Sci.* 2002, **27**, 74–78. Elsevier)

Table 9.6 Example of the purification of an over-expressed enzyme from *E. coli*

Purification of the enzyme NADH-cytochrome b_5 reductase					
Purification step	Activity (units)	Protein (mg)	Specific activity (units/mg)	Recovery (% total)	Enrichment (over initial level)
Cell supernatant	26 247	320	82	100	1
Ammonium sulfate precipitation (50 %)	19 340	93	208	74	3
Ammonium sulfate precipitation (90 %)	18 627	72	259	71	3
Affinity	12 762	10	1276	49	16
Size exclusion	11 211	8	1400	43	17

The enzyme's activity is measured at all stages along with the total protein concentration. The total protein falls during purification but the specific activity increases significantly. Activity involved measurement of the rate of oxidation of NADH (Data adapted from Barber, M. and Quinn, G.B. *Prot. Expr. Purif.* 1996, 8, 41–47). The enzyme binds NAD and can therefore be purified by affinity chromatography using agarose containing covalently bound ADP

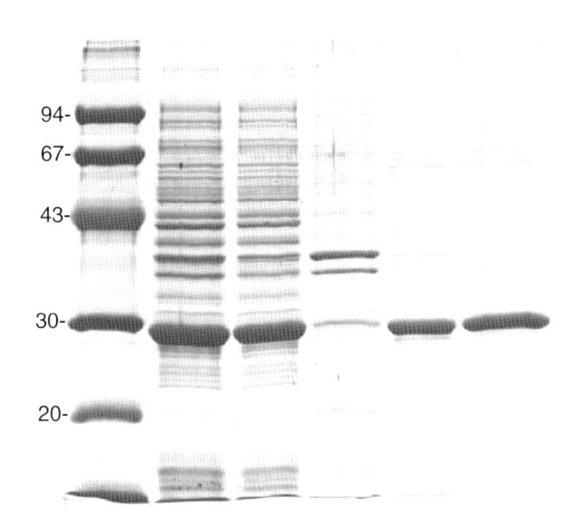

Figure 9.31 Progressive purification of a protein monitored by SDS-PAGE. Analysis was carried out on 12 % acrylamide gels with molecular mass markers shown in the first lane. The lanes contain from left to right total extract; polypeptide composition of supernatant after centrifugation at 14 000 g; polypeptide composition of pellet fraction after centrifugation at 14 000 g; material obtained after elution from an anion-exchange column and finally material obtained after gel filtration. The final protein is nearly pure, although a few impurities can still be seen (reproduced with permission from Karwaski, M.F. *et al. Protein Expression and Purification*. 2002, **25**, 237-240. Academic Press)

It is rare to achieve purification in one step using a single technique. More frequently the strategy involves 'capture' of the material from a crude lysate or mixture of proteins followed by one or more steps designed to enrich the protein of interest followed by a final 'polishing' stage that removes the last traces of contaminants.

It is not unreasonable to ask which method should be used first? In the absence of any preliminary information there are no set rules and the purification must be attempted in an empirical fashion by assaying for protein presence via enzyme activity, biological function or other biophysical properties at each step of the purification procedure. However, it is rare that purifications are attempted without some ancillary information obtained from other sources. Such information should generally be used to guide the procedure in the direction of a possible purification strategy. So, for example, if a sequentially homologous protein has been previously purified using anion exchange chromatography and size exclusion to homogeneity it is almost certain to prove successful again.

Membrane proteins present major difficulties in isolation. Removing the protein from the lipid bilayer will invariably lead to a loss of structure *and* function whilst attempting many of the isolation procedures described above with hydrophobic proteins is more difficult as a result of the conflicting solvent requirements of such proteins. However, it is by no means impossible to overcome these difficulties and many hydrophobic proteins have been purified to homogeneity by combining procedures described in the above sections with the judicious use of detergents. In many instances detergents allow hydrophobic proteins to remain in solution without aggregating and to be amenable to chromatographic procedures that are applicable to soluble proteins.

Summary

Purification requires the isolation of a protein from a complex mixture frequently derived from cell disruption. The aim of a purification strategy is the isolation of a single protein retaining most, if not all, biological activity and the absence of contaminating proteins.

Purification methods have been helped enormously by the advances in cloning and recombinant DNA technology that allow protein over-expression in foreign host cells. This allows proteins that were difficult to isolate to be studied where previously this had been impossible.

Methods of purification rely on the biophysical properties of proteins with the properties of mass, charge, hydrophobicity, and hydrodynamic radius being frequently used as the basis of separation techniques. Chromatographic methods form the most common group of preparative techniques used in protein purification.

In all cases chromatography involves the use of a mobile, usually aqueous, phase that interact with an inert support (resin) containing functional groups that enhance interactions with some proteins. In ion exchange chromatography the supporting matrix contains negatively or positively charged groups. Similar methods allow protein separation on the basis of hydrodynamic radius (size exclusion), ligand binding (affinity), and non-polar interactions (HIC and RPC).

Alongside preparative techniques are analytical methods that establish the purity and mass of the product. SDS–PAGE involves the separation of polypeptides under the influence of an electric field solely on the basis of mass. This technique has proved of widespread value in ascertaining subunit molecular mass as well as overall protein purity.

An extension of the basic SDS–PAGE technique is Western blotting. This method allows the identification of an antigenic polypeptide within a mixture of size-separated components by its reaction with a specific antibody.

2D electrophoresis allows the separation of proteins according to mass and overall charge. Consequently, through the use of these methods for individual cell types or organisms, it is proving possible to identify large numbers of different proteins within proteomes of single-celled organisms or individual cells.

Of all analytical methods mass spectrometry has expanded in importance as a result of technical advances permitting accurate identification of the mass/charge ratio of molecular ions. The most popular methods are MALDI-TOF and electrospray spectrometry. Using modern instrumentation the mass of proteins can be determined in favourable instances to within 1 a.m.u.

In combination with 1 and 2D gel methods mass spectrometry is proving immensely valuable in characterization of proteomes. The expansion of proteomics in the post-genomic revolution has placed greater importance on preparative and analytical techniques. When the methods described here are combined with the techniques such as NMR spectroscopy, X-ray crystallography or cryo-EM it is possible to go from gene identification to protein structure within a comparatively short space of time.

Problems

1. Calculate the relative centrifugal force for a rotor spun at 5000 r.p.m at distances of 9, 6 and 3 cm from the centre of rotation. Calculate the forces if the rotor is now rotating at 20 000 r.p.m.

2. Subunit A has a mass of 10 000, a p*I* of 4.7; subunit B has a mass of 12 000, a p*I* of 10.2, subunit C has a mass of 30 000, a p*I* of 7.5 and binds NAD; subunit D has a mass of 30 000, a p*I* of 4.5 and subunit E has a mass of 100 000, with a p*I* of 5.0. Subunits A, B and E are monomeric, subunits C and D are heptameric. A preliminary fractionation of a cell lysate reveals that all proteins are present in approximately equivalent quantities. Outline an efficient possible route towards purifying each protein. How would you assess effective homogeneity?

3. Why is it helpful to know amino acid composition prior to Edman sequencing?

4. A colleague has failed to sequence using Edman degradation an oligopeptide believed to be part of a larger polypeptide chain. Provide reasonable explanations why this might have happened. Suggest alternative strategies that could work. After further investigation a partial amino acid sequence Cys-Trp-Ala-Trp-Ala-Cys-$CONH_2$ is obtained. Does this highlight additional problems. Describe the methods you might employ to (i) identify this protein and (ii) isolate the complete protein?

5. Discuss the underlying reasons for the use of ammonium sulfate in protein purification.

6. It has been reported that some proteins retain residual structure in the presence of SDS. What are the implications of this observation for SDS–PAGE. What other properties of proteins might cause problems when using this technique?

7. In gel filtration the addition of polyethylene glycol to the running buffer has been observed to make proteins elute at a later stage as if they were of smaller size. Explain this observation. Suggest other co-solvents that would have a similar effect.

8. How would you confirm that a protein is normally found as a multimer?

9. Four mutated forms of protein X have been expressed and purified. Mutant 1 is believed to contain the substitution Gly>Trp, Mutant 2 Glu>Asp, Mutant 3 Ala>Lys, and Mutant 4 Leu>Ile. How would you confirm the presence of these substitutions in the product protein?

10. Describe the equipment a well-equipped laboratory would require to isolate and purify to homogeneity a protein from bacteria.

10

Physical methods of determining the three-dimensional structure of proteins

Introduction

Advances in biochemistry in the 20th century included the development of methods leading to protein structure determination. Atomic level resolution requires that the positions of, for example, carbon, nitrogen, and oxygen atoms are known with precision and certainly with respect to each other. By knowing the positions of most, if not all, atoms sophisticated 'pictures' of proteins were established as highlighted by many of the diagrams shown in previous chapters.

The impact of structural methods on descriptions of protein function includes understanding the mechanism of oxygen binding and allosteric activity in haemoglobin as well as comprehending the catalytic activity of simple enzymes such as lysozyme. It would extend further to understanding large enzymes such as cyclo-oxygenase, type II restriction endonucleases, and amino acyl tRNA synthetases. Structural techniques are also applied to membrane proteins such as photosynthetic reaction centres, together with complexes of respiratory chains. More recently, the role of macromolecular systems such as the proteasome or ribosome were elucidated by the generation of superbly refined atomic structures. All of these advances stem from understanding the arrangement of atoms within proteins and how these topologies are uniquely suited to their individual biological roles.

At the beginning of 2004 over 22 000 protein structures (22 348) were deposited in the Protein DataBank. Over 86 percent of all experimentally derived structures (~19 400) were the result of crystallographic studies, with most of the remaining structures solved using nuclear magnetic resonance (NMR) spectroscopy. Slowly a third technique, cryoelectron microscopy (cryo-EM) is gaining ground on the established techniques and is proving particularly suitable for asymmetric macromolecular systems. Although this chapter will focus principally on the experimental basis behind the application of X-ray crystallography and NMR spectroscopy to the determination of protein structure the increasing impact of electron crystallography warrants a discussion of this technique. This approach was originally used in determining the structure of bacteriorhodopsin and is likely to expand in use in the forthcoming years as structural methods are applied to increasingly complex structures.

Proteins: Structure and Function by David Whitford

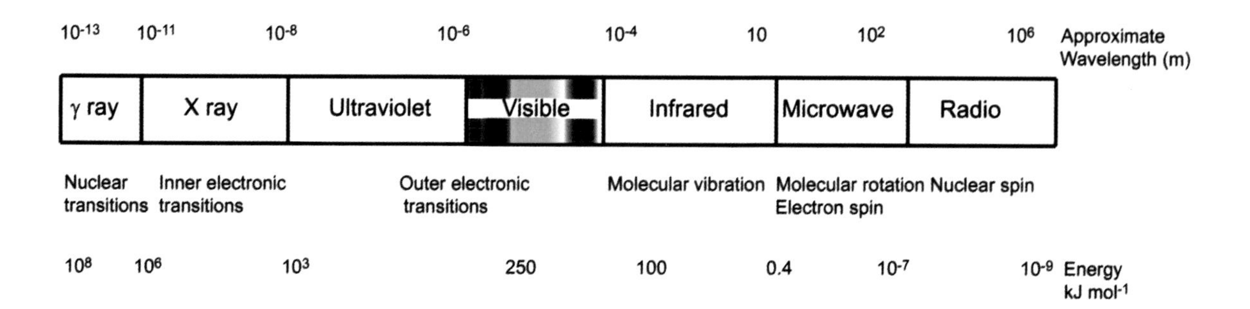

Figure 10.1 Bar showing the distribution of wavelengths associated different regions of the electromagnetic spectrum used in the study of protein structure

These methods are the only experimental techniques yielding 'structures' but other biophysical methods provide information on specific regions of a protein. Spectrophotometric methods such as circular dichroism (CD) provide details on the helical content of proteins or the asymmetric environment of aromatic residues. UV-visible absorbance spectrophotometry assists in identifying metal ions, aromatic groups or co-factors attached to proteins whilst fluorescence methods indicate local environment for tryptophan side chains.

The use of electromagnetic radiation

It is useful to appreciate the different regions of the electromagnetic spectrum involved in each of the experimental methods described in this chapter. The electromagnetic spectrum extends over a wide range of frequencies (or wavelengths) and includes radio waves, microwaves, the infrared region, the familiar ultraviolet and visible regions of the spectrum, eventually reaching very short wavelength or high frequency X-rays. The energy (E) associated with radiation is defined by Planck's law

$$E = h\nu \qquad \text{where } \nu = c/\lambda \tag{10.1}$$

where c is the velocity of light, and h, Planck's constant, has a magnitude of 6.6×10^{-34} Js, and ν is the frequency of the radiation. The product of frequency times wavelength (λ) yields the velocity – a constant of $\sim 3 \times 10^8$ m s^{-1}. From the relationship $c = \nu\lambda$ it is apparent that X-rays have very short wavelengths of approximately 0.15×10^{-9} m or less. At the opposite end of the frequency spectrum radio waves have longer wavelengths, often in excess of 10 m (Figure 10.1). The frequency scale is significant because each domain is exploited today in biochemistry to examine different atomic properties or motions present in proteins (Table 10.1).

Radio waves and microwaves cause changes to the magnetic properties of atoms that are detected by several techniques including NMR and electron spin resonance (ESR). In the microwave and infrared regions of the spectrum irradiation causes bond movements and in particular motions about bonds such as 'stretching', 'bending' or rotation. The ultraviolet (UV) and visible regions of the electromagnetic spectrum are of higher energy and probe changes in electronic structure through transitions occurring to electrons in the outer shells of atoms. Fluorescence and absorbance methods are widely used in protein biochemistry and are based on these transitions. Finally X-rays are used to probe changes to the inner electron shells of atoms. These techniques require high energies to 'knock' inner electrons from their shells and this is reflected in the frequency of such transitions ($\sim 10^{18}$ Hz).

All branches of spectroscopy involve either absorption or emission of radiation and are governed by a fundamental equation

$$\Delta E = E_2 - E_1 = h\nu \tag{10.2}$$

where E_2 and E_1 are the energies of the two quantized states involved in the transition. Most branches of spectroscopy involve the absorption of radiation with the elevation of the atom or molecule from a ground state to one or more excited states. All spectral

Table 10.1 The frequency range and atomic parameters central to physical techniques used to study protein structure

Technique	Frequency range (Hz)	Measurement
NMR	~0.6 – 60 × 10^7	Nucleus' magnetic field
ESR	~1–30 × 10^9	Electron's magnetic field
Microwave	~0.1–60 × 10^{10}	Molecular rotation
Infrared	~0.6–400 × 10^{12}	Bond vibrations and bending
Ultraviolet/visible	~7.5–300 × 10^{14}	Outer core electron transitions
Mossbauer	~3–300 × 10^{16}	Inner core electron transitions
X-ray	~1.5–15 × 10^{18}	Inner core electron transitions

lines have a non-zero width usually defined by a bandwidth measured at half-maximum amplitude. If transitions occur between two discrete and well-defined energy levels then one would expect a line of infinite intensity and of zero width. This is never observed, and Heisenberg's uncertainty principle states that

$$\Delta E \Delta t \approx h/2\pi \qquad (10.3)$$

where ΔE and Δt are the uncertainties associated with the energy and lifetimes of the transition. Transitions do not occur between two absolutely defined energy levels but involve a series of sub-states. Thus if the lifetime is short (Δt is small) it leads to a correspondingly large value of ΔE, where ΔE defines the width of the absorption line (Figure 10.2).

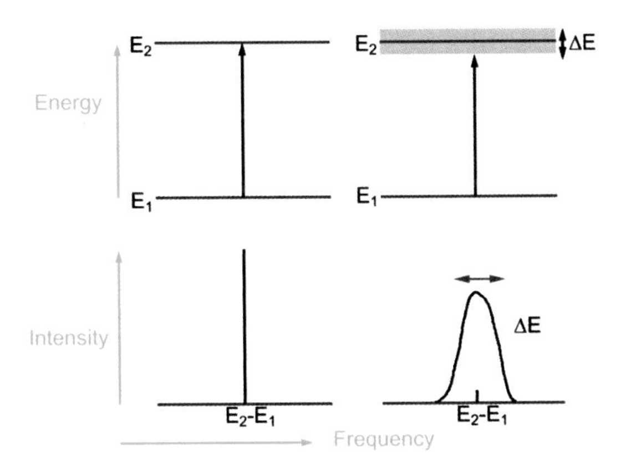

Figure 10.2 Theoretical absorption line of zero width and a line of finite width (ΔE)

X-ray crystallography

X-ray crystallography is the pre-eminent technique in the determination of protein structure progressing from the first low-resolution structures of myoglobin to highly refined structures of macromolecular complexes. X-rays, discovered by Willem Röentgen, were shown to be diffracted by crystals in 1912 by Max von Laue. Of perhaps greater significance was the research of Lawrence Bragg, working with his father William Bragg, who interpreted the patterns of spots obtained on photographic plates located close to crystals exposed to X-rays. Bragg realized 'focusing effects' arise if X-rays are reflected by series of atomic planes and he formulated a direct relationship between the crystal structure and its diffraction pattern that is now called Bragg's law. All crystallography since Bragg has centred around a basic arrangement of an X-ray source incident on a crystal located close to a detector (Figure 10.3). Historically, detection involved sensitive photographic films but today's detection methods include charged coupled devices (CCD) and are enhanced by synchrotron radiation, an intense source of X-rays.

Bragg recognized that sets of parallel lattice planes would 'select' from incident radiation those wavelengths corresponding to integral multiples of this wavelength. Peaks of intensity for the scattered X-rays are observed when the angle of incidence is equal to the angle of scattering and the path length difference is equal to an integer number of wavelengths. From diagrams such as Figure 10.4 it is relatively

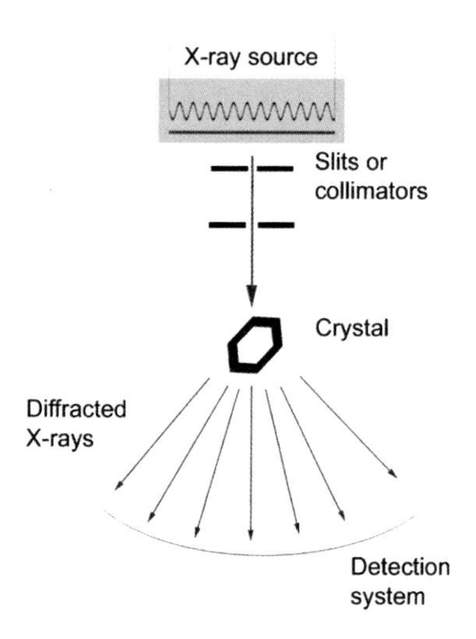

Figure 10.3 The basic crystallography 'set-up' used in X-ray diffraction. Monochromatic X-rays of wavelength 1.5418 Å, sometimes called Cu Kα X-rays, denote the dislodging of an electron from the K shell and the movement of an electron from the next electronic shell (L). After passing through filters to remove Kβ radiation the X-rays strike the crystal

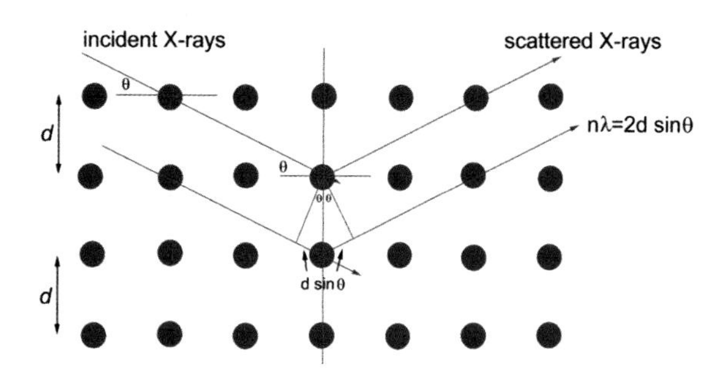

Figure 10.4 X-rays scattered by a crystal lattice

straightforward to establish the path difference ($n\lambda$) using geometric principles as

$$n\lambda = 2d \sin\theta \qquad (10.4)$$

This equality, a quantitative statement of Bragg's law, allows information about the crystal structure to be determined since the wavelength of X-rays is closely controlled. From the Bragg equation diffraction maxima are observed when the path length difference for the scattered X-rays is a whole number of wavelengths. The arrangement of atoms can be redrawn to make this equality more clear. The path difference is equivalent to

$$d \cos\theta_i - d \cos\theta_r = n\lambda \qquad \text{(where } n = 1, 2, 3, \ldots) \qquad (10.5)$$

This formulation is readily extended into three dimensions and is called the Laue set of equations (Figure 10.5). The Laue equations must be satisfied for diffraction to occur but are more cumbersome to deal with and lack the elegant simplicity of the Bragg equation. For each dimension Laue equations are written as

$$a(\cos\alpha_i - \cos\alpha_r) = h\lambda \qquad \text{(where } h = 1, 2, 3, \ldots) \qquad (10.6)$$

$$b(\cos\beta_i - \cos\beta_r) = k\lambda \qquad \text{(where } k = 1, 2, 3, \ldots) \qquad (10.7)$$

$$c(\cos\gamma_i - \cos\gamma_r) = l\lambda \qquad \text{(where } l = 1, 2, 3, \ldots) \qquad (10.8)$$

where a, b and c refers to the spacing in each of the three dimensions (shown by d in Figure 10.5).

Within any crystal the basic repeating pattern is the unit cell (Figure 10.6) and in some crystals more than one unit cell is recognized. In these instances the simplest unit cell is chosen governed by a series of selection rules. The unit cell can be translated (moved sideways but not rotated) in any direction within the crystal to yield an identical arrangement (Figure 10.7).

The unit cell, the basic building block of a crystal, is repeated infinitely in three dimensions but is characterized by three vectors (a, b, c) that form the edges of a paralleliped. The unit cell is also defined by three angles between these vectors (α, the angle between b and c; β, the angle between a

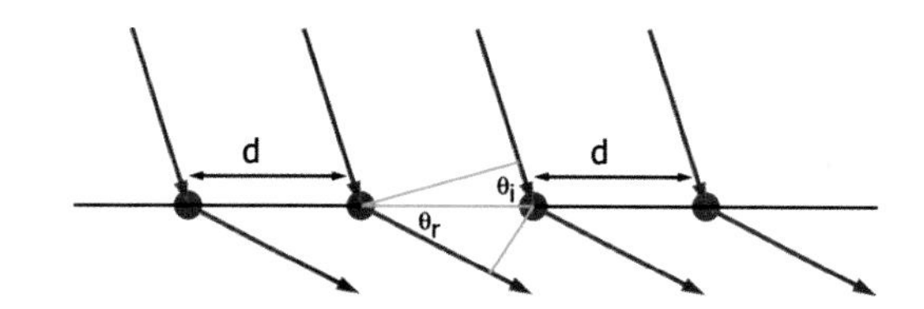

Figure 10.5 The Laue equations

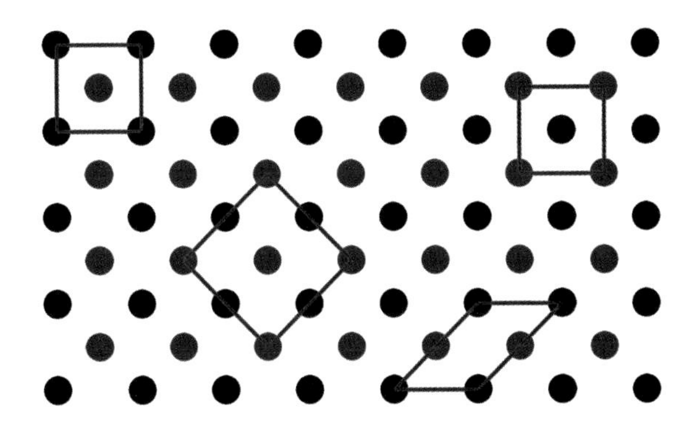

Figure 10.6 Possible unit cells in a two-dimensional lattice

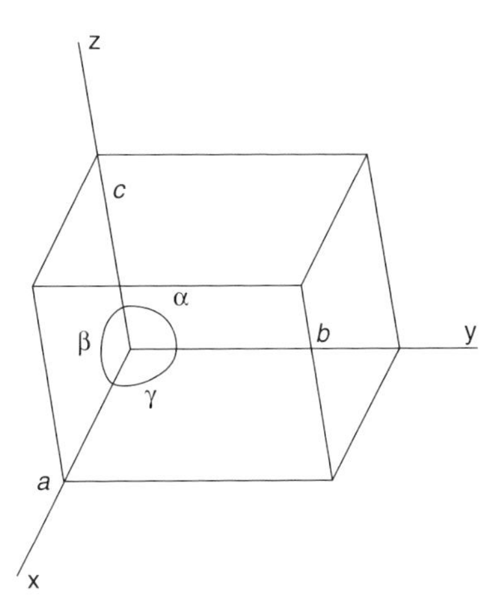

Figure 10.8 The angles and vectors defining any unit cell

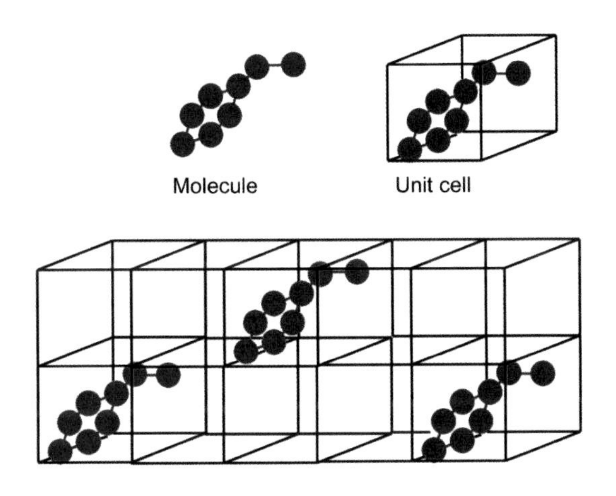

Figure 10.7 A unit cell for a simple molecule

and c; γ, the angle between a and b, Figure 10.8). The recognition of different arrangements within a unit cell was recognized by Auguste Bravais during the 19th century.[1] In two dimensions there are five distinct Bravais lattices, whilst in three dimensions the number extends to 14, usually classified into crystal types. Any crystal will belong to one of seven possible designs and the symmetry of these systems introduces constraints on the possible values of the unit cell parameters. The seven crystal systems are triclinic, monoclinic, orthorhombic, tetragonal, rhombohedral, hexagonal and cubic (Table 10.2). The description of unit cells and lattice types owes much to the origins of crystallography within the field of mineralogy.

In biological systems the unit cell may possess internal symmetry containing more than one protein molecule related to others via axes or planes of symmetry. A series of symmetry operations allows the generation of coordinates for atoms in the next unit cell and includes operations such as translations in a plane, rotations around an axis, reflection (as in a mirror), and simultaneous rotation and inversion. Collection of symmetry operations that define a particular crystallographic arrangement are known as space groups, and with 230 recognized space groups published by the International Union of Crystallography crystals have been found in most, but not all, arrangements.

Scattering depends on the properties of the crystal lattice and is the result of interactions between the incident X-rays and the electrons of atoms within the crystal. As a result metal atoms such as iron or copper and atoms such as sulfur are very effective at scattering X-rays whilst smaller atoms such as the proton are ineffective. The end result of X-ray diffraction experiment is not a picture of atoms, but rather a map of the distribution of electrons in the

[1]First described by Frankenheim in 1835 who incorrectly assigned 15 different structures as opposed to the correct number of 14, recognized by Bravais which to this day carries only his name.

Table 10.2 The parameters (a, b, c and α, β, γ) governing the different crystal lattices together with some of their simpler arrangements

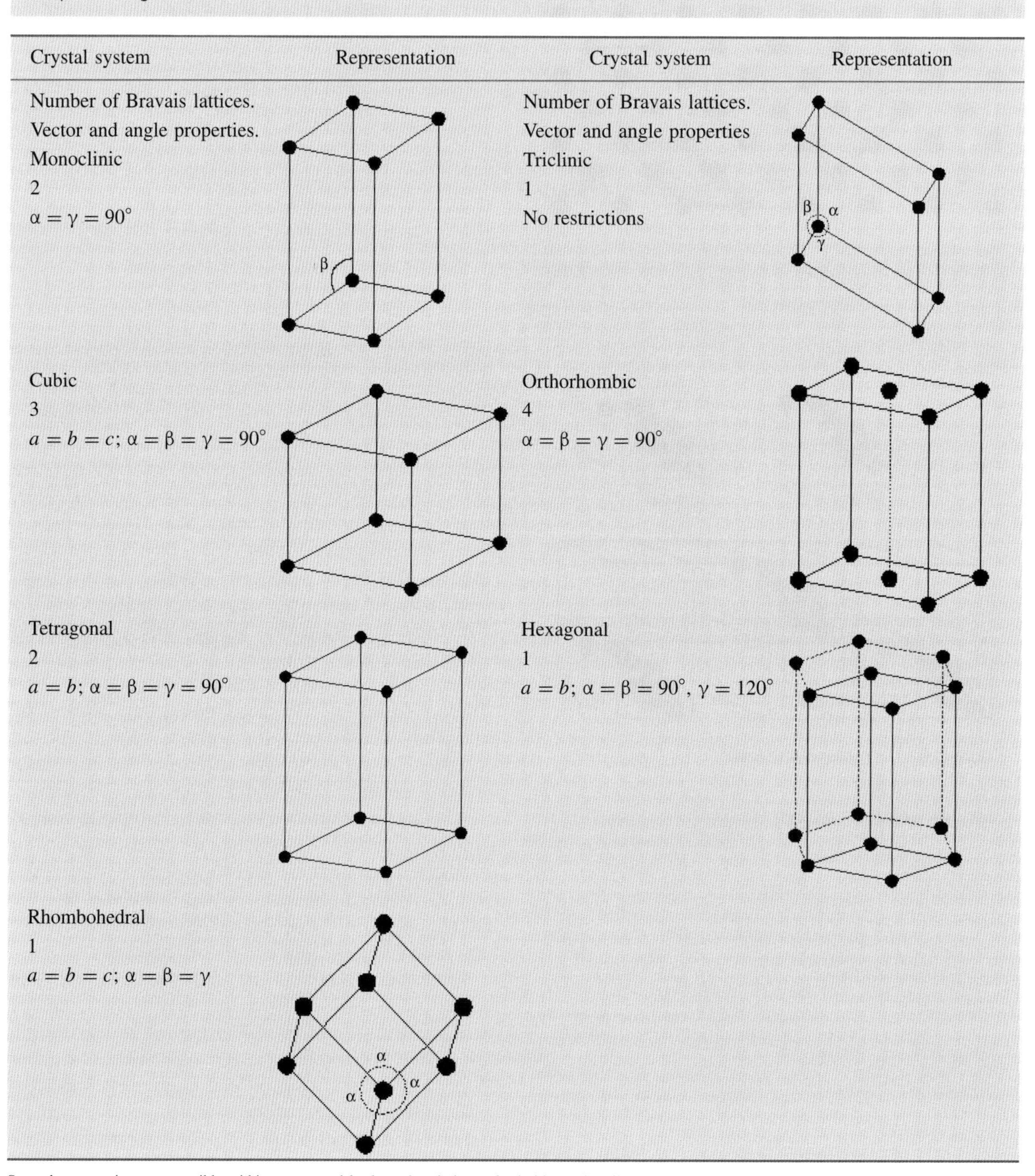

Crystal system	Representation	Crystal system	Representation
Number of Bravais lattices. Vector and angle properties. Monoclinic 2 $\alpha = \gamma = 90°$		Number of Bravais lattices. Vector and angle properties Triclinic 1 No restrictions	
Cubic 3 $a = b = c$; $\alpha = \beta = \gamma = 90°$		Orthorhombic 4 $\alpha = \beta = \gamma = 90°$	
Tetragonal 2 $a = b$; $\alpha = \beta = \gamma = 90°$		Hexagonal 1 $a = b$; $\alpha = \beta = 90°$, $\gamma = 120°$	
Rhombohedral 1 $a = b = c$; $\alpha = \beta = \gamma$			

Several permutations are possible within groups and leads to descriptions of primitive unit cells, face-centred containing an additional point in the center of each face, body-centred contains an additional point in the centre of the cell and centred, with an additional point in the centre of each end.

molecule -it is an electron density map. However, since electrons are tightly localized around the nucleus of atoms the electron density map is a good approximation of atomic positions within a molecule.

Diffraction of X-rays from a single molecule perhaps containing only one or a few electron-dense centres would be very difficult to measure and to distinguish from ambient noise. One advantage of a crystal is that there are huge numbers of molecules orientated in an identical direction. The effect of ordering is to enhance the intensity of the scattered signals (reflections). The conditions for in phase scattering can be viewed as the reflection of X-rays from (or off) planes passing through the collections of atoms in a crystal. A consideration of Bragg's law ($n\lambda = 2d \sin\theta$), i.e. the relationship between scattering angle (θ) and the interplanar spacing (d) shows that if the wavelength (λ) is increased the total diffracted intensity becomes less sensitive to the spacing or to changes in angle. The resulting diffraction pattern will be less sensitive and fine detail will be obscured. At a fixed wavelength (the normal or traditional condition) a decrease in planar spacing will require higher angles of diffraction to observe the first peak in the diffracted intensity. This inverse relationship between spacing within the object and the angle of diffraction leads to the diffraction space being called 'reciprocal space'.

Initial applications of X-ray crystallography were limited to small biological molecules or repeating units found in fibrous proteins like collagen. In 1934 J. D. Bernal and Dorothy Hodgkin showed that pepsin diffracted X-rays; observations consistent with the presence of organized and repeating structure. Despite this success structure determination of large proteins seemed a long way away. In 1947 Dorothy Hodgkin solved the structure of vitamin B_{12}, at that time a major experimental achievement, but further application of these methods to proteins proved very difficult. The partial success for smaller molecules relied on a 'trial and error' approach.

Thus at the beginning of the 1950s it seemed that X-ray crystallography was going to falter as a structural technique until Max Perutz demolished this barrier by introducing the method of isomorphous heavy atom replacement. In a typical diffraction pattern the irradiation of a protein crystal with monochromatic X-rays results in the detection of thousands of spots or reflections. These 'spots' are the raw data of crystallography and arise from all atoms within the unit cell. A complete analysis of the diffraction pattern (Figure 10.9) will allow the electron density map, and by implication the position of atoms to be deciphered.

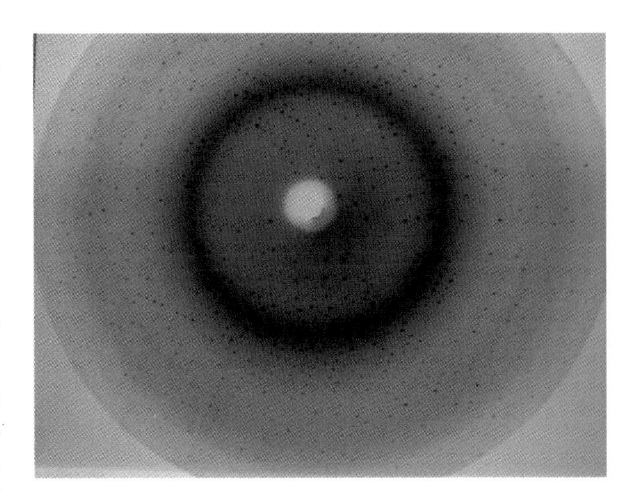

Figure 10.9 Protein diffraction patterns

Since all of the atoms within a unit cell contribute to the observed diffraction pattern it is instructive to consider how the properties of a wave lead to the location of atoms within proteins in the unit cell of crystals. A wave consists of two components – an amplitude f and a phase angle ψ (Figure 10.10). The wave can be further described as a vector ($\mathbf{f}$) of magnitude f and phase angle ψ, and using the relationship between vector algebra and complex numbers the vector becomes the product of real and imaginary components in a complex number plane.

From Figure 10.11

$$\mathbf{f} = f \cos\psi + \mathrm{i} f \sin\psi \tag{10.9}$$

leads to

$$\mathbf{f} = f(\cos\psi + \mathrm{i}\sin\psi) = f e^{\mathrm{i}\psi} \tag{10.10}$$

This minor bit of algebra combined with trigonometry becomes important when one considers that all atoms contribute a scattered wave to the diffraction pattern.

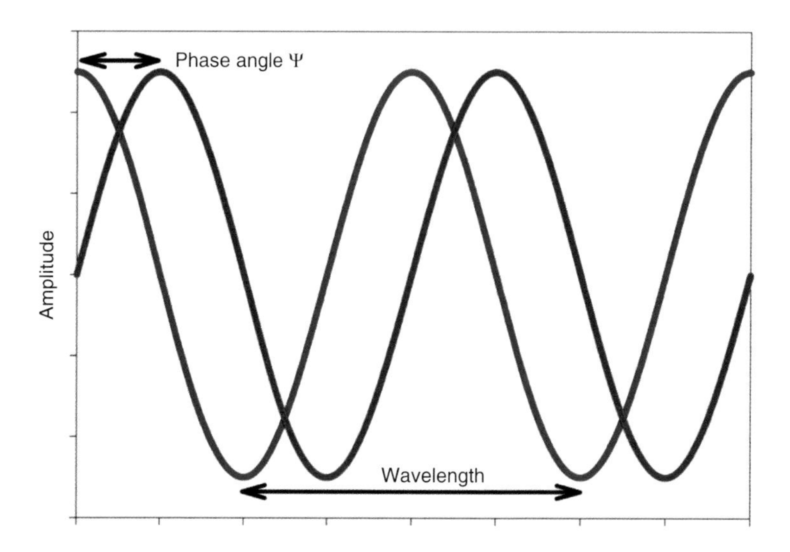

Figure 10.10 A wave described by its amplitude and phase angle. A cosine and sine wave are simply related by a phase shift of 90° or π/2 radians

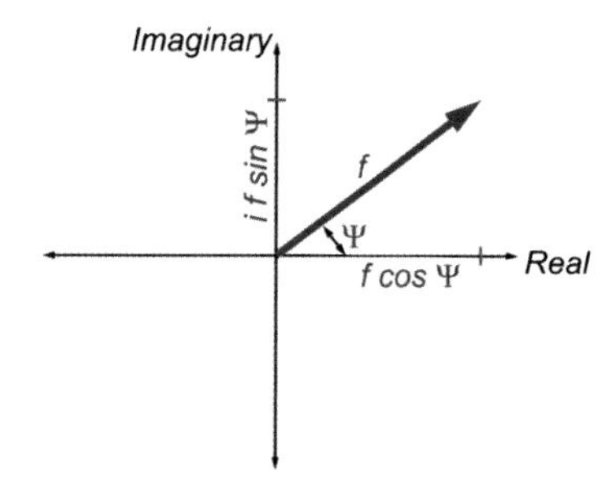

Figure 10.11 Vector representation of a wave and its expression as a complex number

Diffracted waves of intensity (amplitude, f) and phase (ψ) are summed together and described by the vector $\mathbf{F}_{\mathrm{hkl}}$ known as the structure factor. Equation 10.10 becomes

$$\mathbf{F}_{\mathrm{hkl}} = \sum f \cos\psi + \sum \mathrm{i} f \sin\psi \qquad (10.11)$$

and leads to

$$\mathbf{F}_{\mathrm{hkl}} = F_{\mathrm{hkl}}(\cos\varphi_{\mathrm{hkl}} + \mathrm{i}\sin\varphi_{\mathrm{hkl}}) = F_{\mathrm{hkl}}\mathrm{e}^{\mathrm{i}\varphi_{\mathrm{hkl}}} \qquad (10.12)$$

F_{hkl} is the square root of the intensity of the observed (measured) diffraction spot often called I_{hkl}, whilst the φ_{hkl} term represents the summation of all phase terms contributing to this spot.

The next problem is to relate the structure factor $\mathbf{F}_{\mathrm{hkl}}$ to the three-dimensional distribution of electrons in the crystal – the distribution of atoms. The structure factor is the Fourier transform of the electron density and is a vector defined by intensity F_{hkl} *and* phase φ_{hkl}. The value of the electron density at a real-space lattice point (x, y, z) denoted by $\rho\ (x, y, z)$ is equivalent to

$$\rho(x, y, z) = 1/V \sum_{\mathrm{hkl}=-\infty}^{+\infty} \mathbf{F}_{\mathrm{hkl}}\mathrm{e}^{-2\pi\mathrm{i}(hx+ky+lz)} \qquad (10.13)$$

where ρ is the value of the electron density at the real-space lattice point (x, y, z) and V is the total volume of the unit cell. This is rearranged using Equation 10.12 to give

$$\rho(x, y, z) = 1/V \sum_{\mathrm{hkl}=-\infty}^{+\infty} F_{\mathrm{hkl}}\mathrm{e}^{\mathrm{i}\varphi_{\mathrm{hkl}}}\mathrm{e}^{-2\pi\mathrm{i}(hx+ky+lz)} \qquad (10.14)$$

where φ_{hkl} is the phase information.

To obtain '3D pictures' of molecules the crystal is rotated while a computer-controlled detector produces two-dimensional electron density maps for each angle of rotation and establishes a third dimension. A rotating Cu target is the source of X-rays and this generator is normally cooled to avoid excessive heating. X-rays pass via a series of slits (monochromators) to the crystal mounted in a goniometer. The crystal is held within a loop by surface tension or attached by glue to a narrow fibre and can be rotated in any direction. Nowadays crystals are flash cooled to ~77 K (liquid nitrogen temperatures) with benefits arising from reduced thermal vibrations leading to lower conformational disorder and better signal/noise ratios. Cryocooling also limits radiation damage of the crystal and it is possible to collect complete data sets from a single specimen. Finally, a detector records the diffraction pattern where each spot represents a reflection and has parameters of position, intensity and phase. For a typical crystal there may be 40 000 reflections to analyse, and although this remains a time-consuming task computers facilitate the process. In general all of the diffraction apparatus is controlled via computer interfaces. The data collected is a series of frames containing crystal diffraction patterns as it rotates in the X-ray beam. From these frames data analysis yields a list of reflections (positions) and their intensities (amplitudes) but what remains unknown are the relative phases of the scattered X-rays.

A useful scheme with which to understand the occurrence of reflections is the Ewald construction. If a sphere of radius $1/\lambda$ centred around the crystal is drawn then the origin of the reciprocal lattice lies where the transmitted X-ray beam passes straight through the crystal and is described as at the edge of the Ewald sphere (Figure 10.12). Diffraction spots occur only when the Laue or Bragg equations are satisfied and this condition occurs when the reciprocal lattice point lies exactly on the Ewald sphere. At any one instant the likelihood of observing diffraction is small unless the crystal is rotated to bring more points in the reciprocal lattice to lie on the Ewald sphere by rotation of the crystal (Figure 10.13).

A detection device perpendicular to the incoming beam records diffraction on an arbitrary scale with the most obvious attributes being position and intensity of the 'spots'. The intensity is proportional to the square of the structure factor magnitude according to the relationship

$$I_{hkl} = k|F_{obs}|^2 \tag{10.15}$$

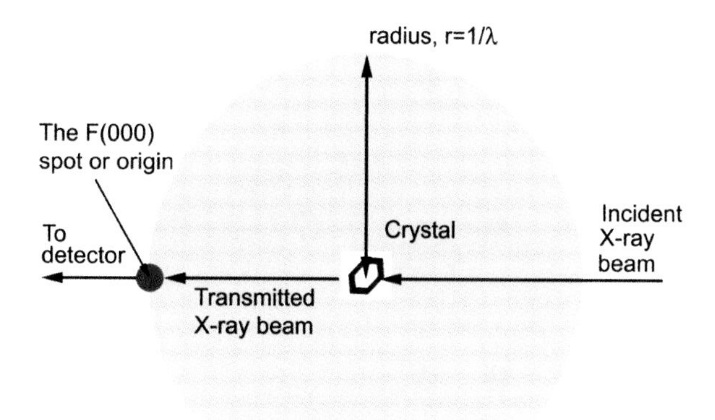

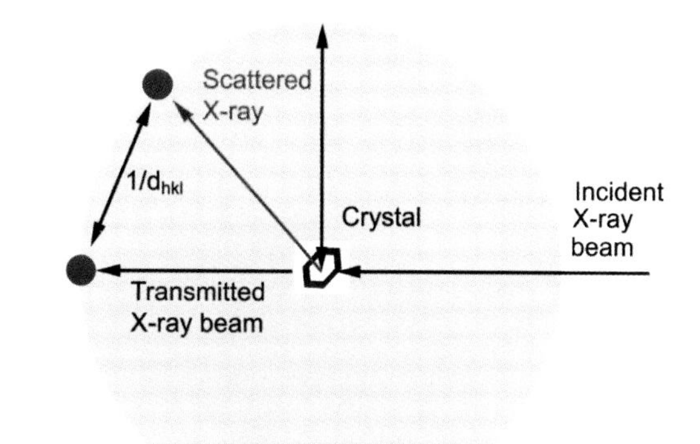

Figure 10.12 Scattering of X-rays by atoms within a crystal

where k is a constant that depends on several factors such as the X-ray beam energy, the crystal volume, the volume of the unit cell, the angular velocity of the crystal rotation, etc.

From Equation 10.14 the relationship between intensity, phase and the fractional coordinates of each atom (x, y and z) is clearly emphasized and from this equation it is clear that if we know the structure we can generate $\mathbf{F}_{hkl}$. However, in crystallography the aim is to determine the position of atoms, i.e. x, y and z from $\mathbf{F}_{hkl}$ – the inverse problem. A complete description of each reflection – its position, intensity *and* phase – is represented by the structure factor, $\mathbf{F}_{hkl}$ and if this

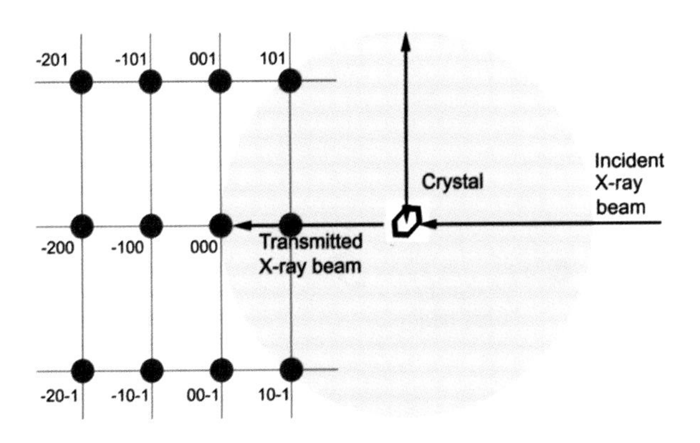

Figure 10.13 Rotation of the crystal brings more planes (collections of atoms) into the Ewald plane

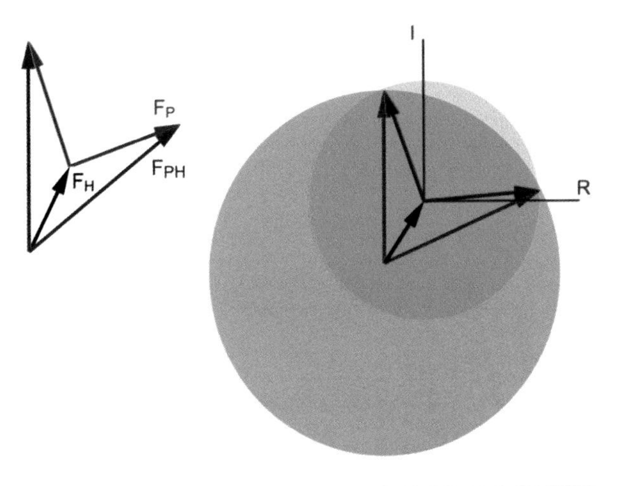

Figure 10.14 A vector diagram showing intensity and phases attributed to protein, heavy atom and derivative. The Harker construction emphasizes two possible phases for F_P. Diffraction data from a second derivative identify a unique solution. The diagram is constructed by drawing a circle with a radius equal to the amplitude of F_P and centred at the origin (shown by brown/green shading). The circle indicates a vector obtained for all of the possible phase angles for F_P. A second circle with radius F_{PH} centred at a point defined by F_H (pink in the figure above). Where the two circles intersect represents possible values for F_P (magnitude and phase) that satisfy the equation $F_{PH} = F_H + F_P$ while agreeing with the measured amplitude F_{PH}

parameter is determined accurately enough then sufficient information exists to generate an atomic structure for proteins.

The phase problem

When crystallographers worked on the structures of simple molecules it was possible to make 'guesses' about the conformation of a molecule. By calculating a diffraction pattern for the 'guess' the theoretical and experimentally determined profiles were compared. If the guess placed atoms in approximately the right position then the calculated phases were almost correct and a useful electron density map could be computed by combining the observed amplitudes with the calculated phases. In this empirical fashion it was possible to successively refine the model until a satisfactory structure was obtained. These direct methods work well for small molecule crystal structures but not proteins.

The solution to the phase problem determines the value of φ_{hkl} and overcomes a major bottleneck in the determination of protein structure by diffraction methods. The achievement of Max Perutz in solving the 'phase problem' in initial studies of myoglobin was significant because it pointed the way towards a generalized method for macromolecular structure determination. Perutz irradiated crystals of myoglobin soaked in the presence of different heavy metal ions. Isomorphous replacement required that metal ions were incorporated into a crystal without perturbing structure and is sometimes difficult to achieve. In 1954 Perutz and co-workers calculated a difference Patterson $(F_{PH} - F_P)^2$ using the amplitudes of a mercury 'labelled' haemoglobin crystal and the amplitudes of a native, but isomorphous, haemoglobin crystal. The scattering due to the 'light' atoms (from the protein) was mathematically removed leaving a low level of background noise with the residual peaks on the difference Patterson map showing the vectors existing between heavy atoms. These maps define the positions of the heavy atoms and allow structure factors to be calculated. This assumes that scattering from protein atoms is unchanged by complex formation with heavy atoms. With the proviso that the heavy atom does not alter the protein then the structure factor for the derivative crystal (F_{PH}) is equal to the sum of the protein structure factor (F_P) and the heavy atom structure factor (F_H), or

$$\mathbf{F}_{PH} = \mathbf{F}_P + \mathbf{F}_H \qquad (10.16)$$

The structure factor is a vector and leads to a representation called the Harker construction (Figure 10.14). Since the length and orientation of one side ($\mathbf{F}_H$) is known along with the magnitude of $\mathbf{F}_{PH}$ and $\mathbf{F}_P$ there are two possible solutions for the phase of $\mathbf{F}_P$.

Heavy metal atom derivatives must give minimal perturbations of protein structure and retain the lattice structure of the unmodified crystal. The derivatives must also perturb reflection intensities sufficiently to allow calculation of phases. Protein crystals are usually prepared with heavy metal atoms such as uranium, platinum or mercury introduced at specific points within the crystal with thiol groups representing common high affinity sites. Amongst the compounds that have been used are potassium tetrachloroplatinate(II), *p*-chloromercuribenzoate, potassium tetranitroplatinate(II), uranyl acetate and *cis*-platinum (II) diamine dichloride.

Modern diffraction systems are highly specialized exploiting advances in computing, material science, semiconductor technology and quantum physics to allow the generation of accurate and precise protein structures. All of the originally tedious calculations are performed computationally whilst all equipment is controlled via microprocessors. A further enhancement of X-ray diffraction has been the use of highly intense or focused beams such as those obtained from synchrotron sources. This has allowed new solutions to the phase problem involving the use of synchrotron radiation at multiple wavelengths. Today one of the methods of choice in X-ray crystallography is called multiple anomalous dispersion (MAD). Just as the use of heavy metal derivatives allowed the extrapolation of phase information and permitted the determination of electron density from observed diffraction data the use of multiple wavelengths near the absorption edge of a heavy atom achieves a comparable effect. This is commonly performed by replacing methionine with selenomethionine during protein expression. Nowadays crystallography centres are based around the location of synchrotron sources and the speed and quality of structure generation has improved dramatically (Figure 10.15).

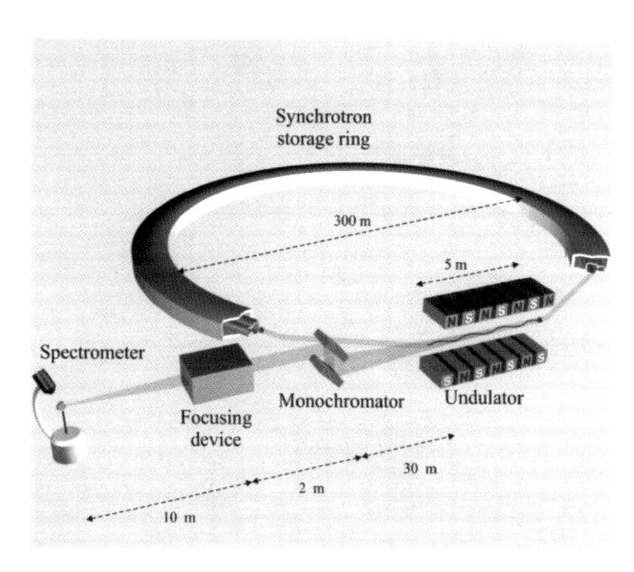

Figure 10.15 The arrangement of data collection centres about a synchrotron storage ring (reproduced with permission from Als-Nielsen, J. & McMorrow, D. *Elements of Modern X-ray Physics*. John Wiley & Sons)

Fitting, refinement and validation of crystal structures

The initial electron density map (Figure 10.16) does not resolve individual atoms and in the early stages several 'structures' are compatible with the data. Higher resolution allows the structure to be assigned. Interpretation of electron density maps requires knowledge of the primary sequence since the arrangement of heavy atoms for the side chains dictates the fitting process. Resolution is not a constant even between similarly sized proteins because crystal lattices have motion from thermal fluctuations and mobility contributes to the final overall resolution. One estimate of mobility within a crystal is the B factor (Debye–Waller factor) that reflects spreading or blurring of electron density. The B factor represents the mean square displacement of atoms and has units of $Å^2$.

Structures before refinement are often at a resolution >4.5 Å where only α helices are observed and the identification of side chains is unlikely. At higher resolutions of 2.5–3.5 Å the polypeptide backbone is

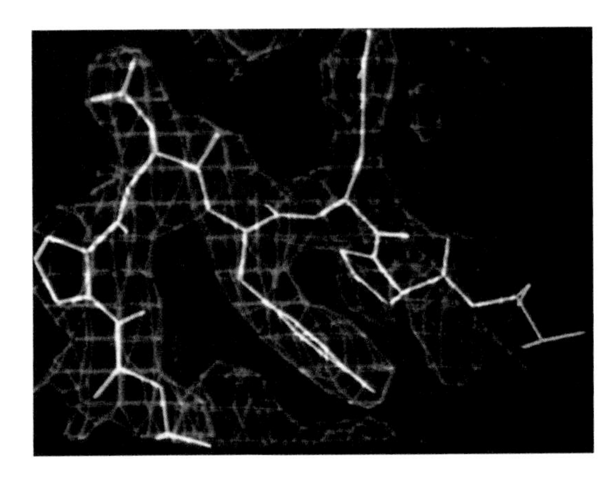

Figure 10.16 An electron density map

traced via electron density located primarily on carbonyl groups and helices, strands and aromatic side chains such as tryptophan are defined. Around 2.0 Å almost all of the structure will be identified including the conformations associated with side chains. Specialized computer programs fit electron density maps and the process is assisted by assuming standard bond lengths and angles. Refining models in an iterative fashion progressively improves the agreement with experimental data. A structure is judged by the crystallographic R-factor, defined as the average fractional error in the sum of the differences between calculated structural factors (F_{cal}) and observed structural factors (F_{obs}) divided by the sum of the observed structural factors.

$$R = \sum |F_{obs} - F_{cal}| / \sum F_{obs} \qquad (10.17)$$

A value of 0.20 is often represented as an R factor of 20 percent and 'good' structures have R-factors ranging from 15 to 25 percent or approximately $1/10^{th}$ the resolution of the data. A structure of resolution 1.9 Å is expected to yield an R factor of <0.19. One result of protein structure determination is the generation of a file that lists x, y, and z coordinates for all heavy atoms whose locations are known. These files are deposited in protein databanks and the PDB files listed throughout this book represent the culmination of this analysis.

Protein crystallization

One of the slowest steps in protein crystallography is the production of protein crystals. The methods employed in crystal production rely on the ordered precipitation of proteins. The first protein (urease) was crystallized by James Sumner as long ago as 1926, and was followed by the crystallization of pepsin in 1930 by John Northrop. In the next 20 years over 40 additional proteins were crystallized, including lyzosyme, trypsin, chymotrypsin, catalase, papain, ficin, enolase, carbonic anhydrase, carboxypeptidase, hexokinase and ribonuclease, although structure determination had to wait until the advances of Perutz and Kendrew (Figure 10.17).

Crystallization requires the ordered formation of large (dimensions greater than 0.1 mm along each axis), stable crystals with sufficient long-range order to diffract X-rays. Structures produced by X-ray diffraction are only as good as the crystals from which they are derived.

To form a crystal protein molecules assemble into a periodic lattice from super-saturated solutions. This involves starting with solution of pure protein at a concentration between 0.5 and 200 mg ml^{-1} and adding reagents that reduce protein solubility close to the point of precipitation. These reagents perturb protein–solvent interactions so that the equilibrium shifts in favour of protein–protein association. Further concentration of the solution results in the formation of

Figure 10.17 Examples of protein crystals. Top row: azurin from *Pseudomonas aeruginosa*, flavodoxin from *Desulfovibrio vulgaris*, rubredoxin from *Clostridium pasteurianum*. Bottom row: azidomethemerythrin from the marine worm *Siphonosoma funafatti*, lamprey haemoglobin and bacteriochlorophyll a protein from *Prosthecochloris aesturii*. The beauty of the above crystals is their colour arising from the presence of light absorbing co-factors such as metal, heme, flavin or chlorophyll (reproduced with permission from Voet, D., Voet, J.G. & Pratt, C.W. *Fundamentals of Biochemistry* John Wiley & Sons Ltd. Chichester, 1999)

nucleation sites, a process critical to crystal formation and is the first of three basic stages of crystallization common to all systems. It is followed by expansion and then cessation as the crystal reaches a limiting size.

Two experimental methods used to form crystals from protein solutions are vapour diffusion and equilibrium dialysis. Vapour diffusion is the standard method used for protein crystallization. It is suitable for use with small volumes, easy to set-up and to monitor reaction progress. A common format involves 'hanging drops' containing protein solution plus 'precipitant' at a concentration insufficient to precipitate the protein (Figure 10.18). The drop is equilibrated against a larger reservoir of solution containing precipitant and after sealing the chamber equilibration leads to supersaturating concentrations that induce protein crystallization in the drop.

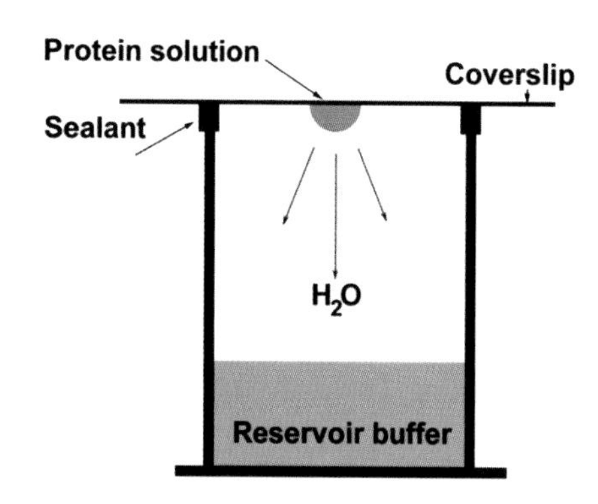

Figure 10.18 The 'hanging drop' or vapour diffusion method of protein crystallization. As little as 5 μl of concentrated solution (protein + solvent) may be suspended on the coverslip

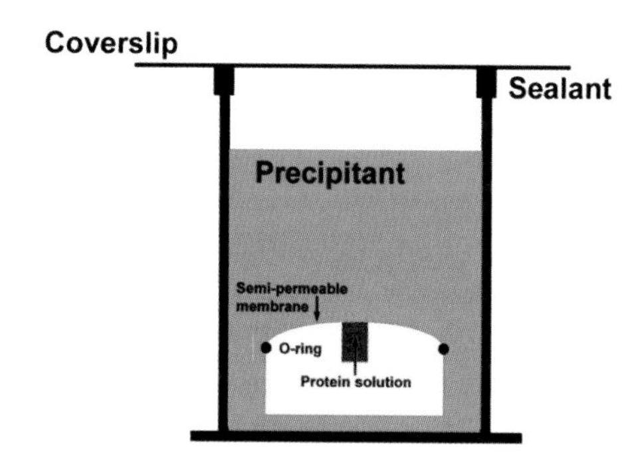

Figure 10.19 Equilibrium dialysis can be achieved with many different 'designs' although the basic principle involves the separation of protein solution from the precipitant by a semipermeable membrane. Diffusion across the membrane promotes ordered crystallization

A second method is equilibrium dialysis and is used for crystallization of proteins at low and high ionic strengths (Figure 10.19). Small volumes of protein solution are placed in a container separated from precipitant by a semi-permeable membrane. Slowly the precipitant causes crystal formation within the well containing the protein solution.

To perform large numbers of crystallization trials it is common to use robotic systems to automate the process and to eliminate labour-intensive manipulations whilst crystallization is performed in temperature-controlled rooms, free of vibration, leading to crystals appearing over a period of 4–10 days. When the above techniques work successfully X-ray diffraction yields unparalleled resolution for protein structures. There is little doubt that the technique will continue to represent the main method of protein structure determination in the exploration of new proteomes.

Nuclear magnetic resonance spectroscopy

NMR spectroscopy as a tool for determining protein structure arose more recently than X-ray crystallography. With the origins of X-ray diffraction lying in the discoveries of Röentgen, Laue and the Braggs at the beginning of the 20th century it was not until 1945 that a description of the NMR phenomenon was given by Felix Bloch and Edward Purcell. NMR spectra are observed upon absorption of a photon of energy and the transition of nuclear spins from ground to excited states. The observation of signal associated with these transitions and the use of radio frequency irradiation to elicit the response marked the start of NMR spectroscopy.

This discovery would have remained insignificant except for subsequent observations showing that nuclear transitions differed in frequency from one nucleus to another but also showed subtle differences according to the nature of the chemical group. So for the proton this property meant that proteins exhibited many signals with, for example, the methyl protons resonating at different frequencies to amide protons which in turn are different to the protons attached to the α or β carbons. In 1957 the first NMR spectrum of a protein (ribonuclease) was recorded but progress as a structural technique remained slow until Richard Ernst described the use of transient techniques. Transient signals produced after a pulse of radio frequency radiation are converted into a normal spectrum by the mathematical process of Fourier transformation – this technique would pave the way towards important advances, particularly multi-dimensional NMR spectroscopy, that are now the basis of all biomolecular structure determination.

Table 10.3 Spin properties and abundance of important nuclei in protein NMR studies

Nuclei	Spin	Abundance (% total)	Magnetogyric ratio (γ) ($\times 10^7$ $T^{-1}s^{-1}$)	Ratio (γ_n/γ_H)
^{1}H	1/2	99.985	26.7520	1.00
^{2}H	1	0.015	4.1066	0.15
^{13}C	1/2	1.108	6.7283	0.25
^{14}N	1	99.630	1.9338	0.07
^{15}N	1/2	0.370	2.712	0.10
^{18}O	5/2	0.037	3.6279	0.14
^{31}P	1/2	100.000	10.841	0.41

NMR phenomena

Underlying the NMR phenomenon is a property of all atomic nuclei called 'spin'. Spin describes the nature of a magnetic field surrounding a nucleus and is characterized by a spin number, I, which is either zero or a multiple of 1/2 (e.g. 1/2, 2/2, 3/2, etc.). Nuclei whose spin number equals zero have no magnetic field and from a NMR standpoint are uninteresting. This occurs when the number of neutrons and the number of protons are even. Nuclei with non-zero spin numbers have magnetic fields that vary considerably in complexity (Table 10.3). Spin 1/2 nuclei represent the simplest situation and arise when the number of neutrons plus the number of protons is an odd number.[2] The most important spin 1/2 nucleus is the proton with a high natural abundance (~100 %) and its occurrence in all biomolecules.

For nuclei such as ^{12}C the most common isotope is NMR 'silent' and the 'active' spin 1/2 nucleus (^{13}C) has a low natural abundance of ~1.1 percent. Similarly ^{15}N has a natural abundance of ~0.37 percent. However, advances in molecular biological techniques enable proteins to be expressed in host cells grown on media containing labelled substrates. Growing bacteria, yeast or other cell cultures on substrates containing ^{15}N or ^{13}C (ammonium sulfate and glucose are common sources) allows uniform enrichment of proteins in

[2] A third possibility exists when the number of neutrons and the number of protons are both odd and this leads to the nucleus having an integer spin (i.e. I = 1, 2, 3, etc.).

these nuclei. Biomolecular NMR spectroscopy requires proteins enriched with ^{13}C or ^{15}N or ideally both nuclei.

Although the shape of the magnetic field of a nucleus is described by the parameter I (spin number) the magnitude is determined by the magnetogyric ratio (γ). A nucleus with a large γ has a stronger magnetic field than a nucleus with a small γ. The proton (^1H) has the largest magnetogyric ratio and coupled with its high natural abundance this contributes to its popularity as a common form of NMR spectroscopy.

When samples are placed in magnetic fields nuclear spins become polarized in the direction of the field resulting in a net longitudinal magnetization. In the normal or macroscopic world two bar magnets can be aligned in an infinite number of arrangements but at an atomic level these alignments are governed by quantum mechanics and the number of possible orientations is $2I + 1$. For spin 1/2 nuclei this gives two orientations that in the absence of an external magnetic field are of equal energy. For spin 1/2 nuclei such as ^1H, ^{13}C, or ^{15}N application of a magnetic field removes degeneracy and the energy levels split into parallel and antiparallel orientations (Figure 10.20).

Spins aligned parallel with external magnetic fields are of slightly lower energy than those aligned in an antiparallel orientation. The concept of two energy levels allows one to envisage transitions between lower and higher energy levels analogous to that seen in other forms of spectroscopy. The difference in population ($n_{\mathrm{upper}}/n_{\mathrm{lower}}$) between each level is governed by Boltzmann's distribution

$$n_{\mathrm{upper}}/n_{\mathrm{lower}} = \mathrm{e}^{-(\Delta E/k_{\mathrm{B}}T)} \tag{10.18}$$

When $\Delta E \sim k_{\mathrm{B}}T$, as would occur with two closely spaced energy levels, the ratio $n_{\mathrm{upper}}/n_{\mathrm{lower}}$ approaches 1. At thermal equilibrium the number of nuclei in the lower energy level slightly exceeds those in the higher energy level. As a result of this small inequality it is possible to elicit transitions between states by the application of short, intense, radio frequency pulses.

Instead of considering a single spin it is more useful to consider the magnetic ensemble of spins. In the presence of the applied magnetic field (B_0) spin polarization occurs and a vector model of NMR views the net magnetization lying in the direction of the z-axis. Irradiation of the sample by a radiofrequency (rf) field denoted as B_1 along the x-axis rotates the macroscopic magnetization into the xy plane (Figure 10.21).

Most frequently the pulse length is calculated to tip 'magnetization' through 90° ($\pi/2$ radians). The xy plane lies perpendicular to the magnetic field and causes transverse magnetization to precess under the influence of the applied magnetic field. Precession in the xy plane at the Larmor frequency of the nuclei under investigation induces a current in the detector coil that is the *observable* signal in all NMR experiments.

Transverse magnetization decays exponentially with time in the form of a signal called a free induction decay (FID) eventually reaching zero. All NMR spectra (such as the one shown in Figure 10.22) are derived by converting the FID signal of intensity versus time into a profile of intensity versus frequency via Fourier transformation (FT).

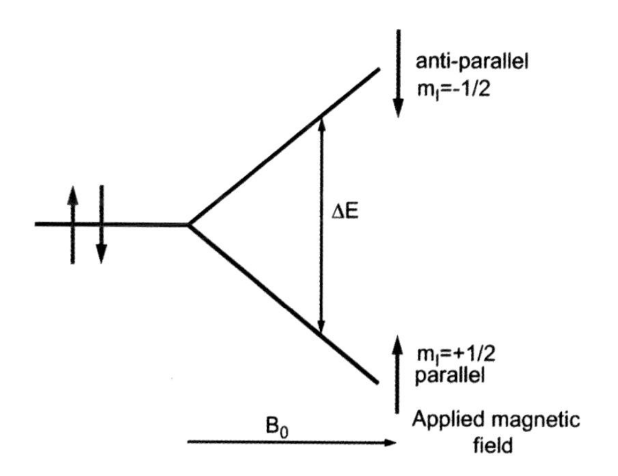

Figure 10.20 An energy level diagram reflecting the alignment of spin 1/2 nuclei in applied magnetic fields

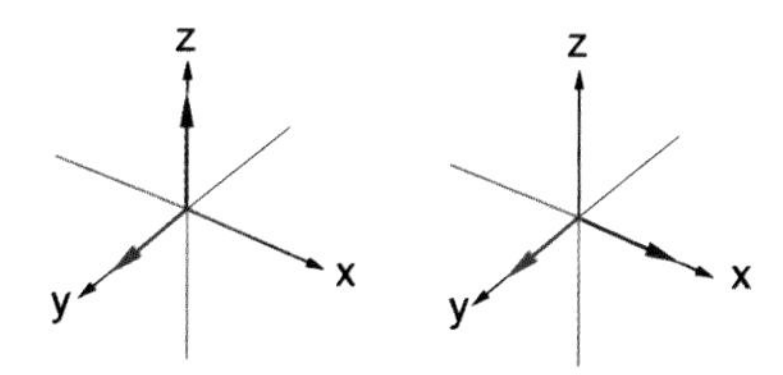

Figure 10.21 Application of a radiofrequency pulse (blue arrow) along the *y*-axis rotates macroscopic magnetization (red arrow) into the *xy* plane

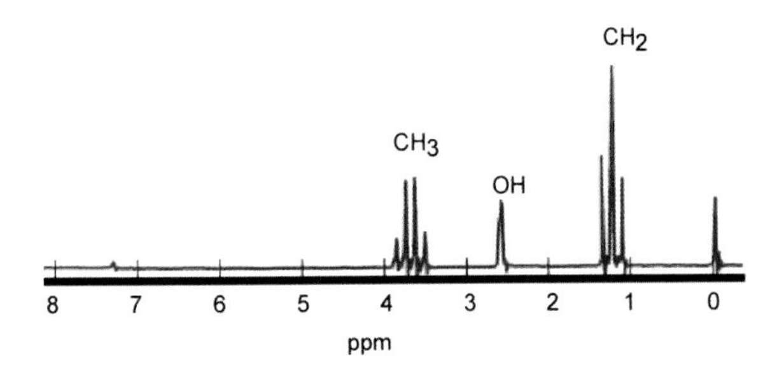

Figure 10.22 A simple NMR spectrum for the molecule ethanol

Simple NMR experiments involve repetitive application of rf pulses with a suitable recovery period between pulses allowing the return of magnetization to equilibrium. Each FID is acquired, stored on computer and added to the previous FID permitting signal averaging. In this manner high signal to noise ratio spectra are acquired on protein samples in a few minutes. Unlike most forms of spectroscopy where the incident radiation is slowly scanned through the spectral range FT-NMR involves the application of rf pulses that excite all nuclear spins simultaneously.

In the construction of magnets for use in NMR spectroscopy the magnetic field is achieved through the use of superconducting materials operating at liquid helium temperatures. Magnets are described in terms of their field strength with designations such as 14.1 T (where 1 T or Tesla is equivalent to 10^4 Gauss and 1 G is equivalent to the earth's magnetic field). The Larmor frequency is obtained from the relationship $\omega_o = -\gamma B_0$ where $\omega_o = 2\pi\nu_0$ and ν_0 is the Larmor frequency. A quick calculation of the Larmor frequency of the proton (where B_0, the field strength = 14.1 T; γ, the magnetogyric ratio of $^1H = 26.7520 \times 10^7$ T s^{-1}) shows that ν_0 (^{1}H) occurs at a frequency of ~600 MHz. As a result spectrometers are often referred to as 400, 600, 750, 800 or 900s reflecting the proton Larmor frequency in MHz at a given magnetic field strength.

Parameters governing NMR signals

The use of NMR spectroscopy as a tool to determine protein structure is based around several related parameters that influence the observation of signals. These parameters include the chemical shift (δ), spin-spin coupling constants (J), the spin lattice or T_1 relaxation time (sometimes denoted as R_1 the spin lattice relaxation rate = $1/T_1$), the spin–spin or T_2 relaxation time ($R_2 = 1/T_2$), the peak intensity and the nuclear Overhauser effect (NOE). Since these effects are vital to all aspects of NMR spectroscopy a brief description is given to highlight their respective importance.

The peak intensity represented formally by the integrated area reflects the number of nuclei involved in the signal. The ^{1}H NMR spectrum of ethanol illustrates this by containing three resonances due to the hydroxyl group, the methylene group and the methyl group. The integrated areas under each line are in the ratio of 1:2:3. Within proteins methyl resonances exhibit intensities three times greater than the Hα proton. Although the peak height is frequently used as an indicator of intensity many parameters can modify peak height so the integrated area is always the best indicator of the number of protons forming each peak.

The chemical shift denotes the position of a resonance along a frequency axis and is uniquely sensitive to the environment in which a nucleus is located. The chemical shift is defined relative to a standard such as 2,2-dimethyl-2-silapentane-5-sulfonate (DSS) and is quoted in p.p.m. units. Since resonant frequencies are directly proportional to the static field minor variations between instruments make it very difficult to compare spectra obtained on different spectrometers. To avoid this problem all resonances are measured relative to a standard defined as having a chemical shift of 0 p.p.m. For a resonance the chemical shift (δ) is measured as

$$\delta = (\Omega - \Omega_{ref})/\omega_0 \times 10^6 \qquad (10.19)$$

where Ω and Ω_{ref} are the offset frequencies of the signals of interest and reference respectively. The resonance attributable to the protons of water occurs at ~4.7 p.p.m. at ~20 °C.

The fundamental equation of NMR, $\omega = -\gamma B_o$, suggests that all nuclei subjected to the same magnetic field will resonate at the same frequency. This does not occur – nuclei experience different fields due to their local magnetic environment. This allows the equation to be recast as

$$\omega = \gamma(B_o - \sigma B_o) \qquad (10.20)$$

where σ represents a screening or shielding constant that reflects the different magnetic environments found in molecules. Another way of envisaging this effect is to use an effective field at the nucleus B_{eff} that is reduced by an amount proportional to the degree of shielding.

$$B_{eff} = B_0(1 - \sigma) \tag{10.21}$$

Spin–spin coupling constants or J values are defined by interactions occurring through bonds. These scalar interactions are field independent and occur as a result of covalent bonds between nuclei linked by three or less bonds. The scalar coupling constants contribute to the fine structure observed for resonances in ^{1}H NMR spectra of small molecules although they are rarely resolved in studies of proteins. Scalar coupling leads to the splitting of the methylene signal of ethanol into a quartet as a result of interactions with each of the three protons of the methyl group. Similarly the methyl proton resonance is seen as a triplet due to its interactions with each methylene proton. As the molecular weight increases this fine structure is frequently obscured by line broadening effects.

A correlation exists between the magnitude of the spin–spin coupling constant (^3J) and the torsion angles found for vicinal protons in groups of the polypeptide chain. The Karplus equation derived from theoretical studies of ^{1}H–C–C–^{1}H couplings suggests a relationship of the form

$$^3J = A\cos^2\theta + B\cos\theta + C \tag{10.22}$$

and in proteins the $^3J_{NH\alpha}$ coupling constant and torsion angle ϕ are related by

$$^3J_{NH\alpha} = 6.51\cos^2\theta - 1.76\cos\theta + 1.60 \quad (\text{where } \theta = \phi - 60) \tag{10.23}$$

If the coupling constant $^3J_{NH\alpha}$ is measured with sufficient accuracy the torsion angle can be estimated and used as a structural restraint. Many coupling constants between nuclei are measured using heteronuclear and homonuclear methods (Table 10.4).

The longitudinal relaxation time (T_1) reflects the rate at which magnetization returns to the longitudinal axis after a pulse and has the units of seconds. If M_o is the magnetization at thermal equilibrium then at time t after a pulse the recovery of M_o is expressed as

$$M_z = M_o(1 - e^{-t/T_1}) \tag{10.24}$$

where T_1 is the longitudinal relaxation time. In solution T_1 is correlated with the overall rate of tumbling of a macromolecule but is also modulated by internal molecular motion arising from conformational flexibility. Both T_1 and T_2 depend on the correlation time τ_c a factor closely linked with molecular mass, and estimated assuming a spherical protein of hydrodynamic radius r in a solution of viscosity, η, via the Stokes equation as

$$\tau_c = 4\pi\eta r^3/3k_BT \tag{10.25}$$

Macromolecules such as proteins have longer correlation times than small peptides. A typical value for a 100 residue spherical protein is 3–5 ns.

The transverse or spin–spin relaxation time (T_2) describes the decay rate of transverse magnetization in the xy plane. T_2 is always shorter than T_1 and is correlated with dynamic processes occurring within a protein. T_2 governs resonance linewidth and decreases with increasing molecular mass. For

Table 10.4 One, two and three bond coupling constants

Coupling	Magnitude (Hz)	Coupling	Magnitude (Hz)
^{1}H–^{13}C (^1J)	110–130	^{13}Cα –^{13}CO (^1J)	55
^{1}H–^{15}N (^1J)	89–95	^{13}Cα –Hα and ^{13}Cβ –Hβ (^1J)	130–140
H–C–C–H vicinal (^3J)	2–14	HN–HA (^3J)	2–12
H–C–H geminal (^3J)	−12–−15	^{13}Cα –^{15}N (^1J)	45
^{15}N–^{13}CO	15	^{13}Cα –^{15}N (^2J)	70

Lorentzian lineshapes the linewidth at half maximum amplitude is

$$\Delta\nu_{1/2} = 1/\pi T_2 \qquad (10.26)$$

with decreases in T_2 leading to broader lines. NMR spectroscopy of large proteins is often described as 'limited by the T_2 problem'. Many factors influence T_2 and the most important are molecular mass, temperature, solvent viscosity, and exchange processes.

Probably the most important measurable parameter in NMR experiments for the determination of protein structure is the nuclear Overhauser effect (NOE). This is the fractional change in intensity of one resonance as a result of irradiation of another resonance. As a result of dipolar or 'through space' interactions the irradiation of one resonance perturbs intensities of neighbouring resonances. The NOE is expressed as

$$\eta = (I - I_o)/I_o \qquad (10.27)$$

where I_o is the intensity without irradiation and I is the intensity with irradiation. The NOE effect is rapidly attenuated by distance and declines as the inverse sixth power of the distance between two nuclei.

$$\eta \propto r^{-6} \qquad (10.28)$$

The NOE phenomenon, like the relaxation times T_1 and T_2, varies as a function of the product of the Larmor frequency and the rotational correlation time. Armed with an appreciation of these parameters it is possible to extract much information on the structure and dynamics of regions of the polypeptide chain in proteins.

Practical biomolecular NMR spectroscopy

NMR signals are obtained by placing samples into strong, yet highly homogeneous, magnetic fields. These magnetic fields arise from the use of superconducting materials (niobium–tin and niobium–titanium alloy wires) wound around a 'drum' maintained at temperatures of ~4 K. High field strengths (11.7–21.5 T) arise from the influence of current flowing through the wires and the constant temperature ensures the field strength does not fluctuate. Samples contained within a quartz tube are lowered via a cushion of compressed air into the centre (bore) of the magnet where a 'probe' is maintained at a consistent temperature (±0.1 °C), usually between 5 and 40 °C. Electronics located in the probe allow excitation of nuclei and detection of signals and are linked to computer-controlled devices that permit the generation of rf pulses of defined timing, phase, amplitude and duration whilst allowing the detected signal (FID) to be amplified, filtered and subjected to further processing (Figure 10.23). This processing includes storage of the raw data as well as Fourier transformation.

Experiments may continue for 3–4 days and require protein stability for this period. To observe magnetization involving exchangeable protons such as amides (NH) in proteins it is necessary to perform experiments in water. This brings additional problems of intense signals due to the high concentration of water protons (110 M) that far exceeds the concentration of 'signals' from the protein (~10^{-3} M). Sophisticated methods of solvent (water) suppression allied to post-acquisition processing eliminate these signals effectively from protein spectra.

Chemical shifts reflect the magnetic microenvironments of groups. The amide (NH) proton of a polypeptide backbone has a chemical shift between 8.0 and 9.0 p.p.m., methyl groups have chemical shifts between 0 and 2 p.p.m. whilst the Hα proton has a value between 4.0 and 4.6 p.p.m. ^{1}H chemical shifts were derived from an analysis of short unstructured model peptides of the form Gly-Gly-X-Ala, where X was each of the 20 residues (Table 10.5).

Within *folded* proteins some chemical shifts deviate from their expected values reflecting magnetic environments that depend on conformation. This is seen in the ^{1}H NMR spectrum of ubiquitin where peaks above 9.0 and below 0 p.p.m. reflect tertiary structure although many resonances are found close to their 'random coil' conformations. The dispersion over a much wider frequency range offers the possibility of identifying resonances in a protein – a process called assignment.

The assignment problem in NMR spectroscopy

The assignment problem for a protein of ~100 residues and perhaps 700 protons requires identifying which resonance belongs to a particular proton. The assignment problem remains the crux of determining protein

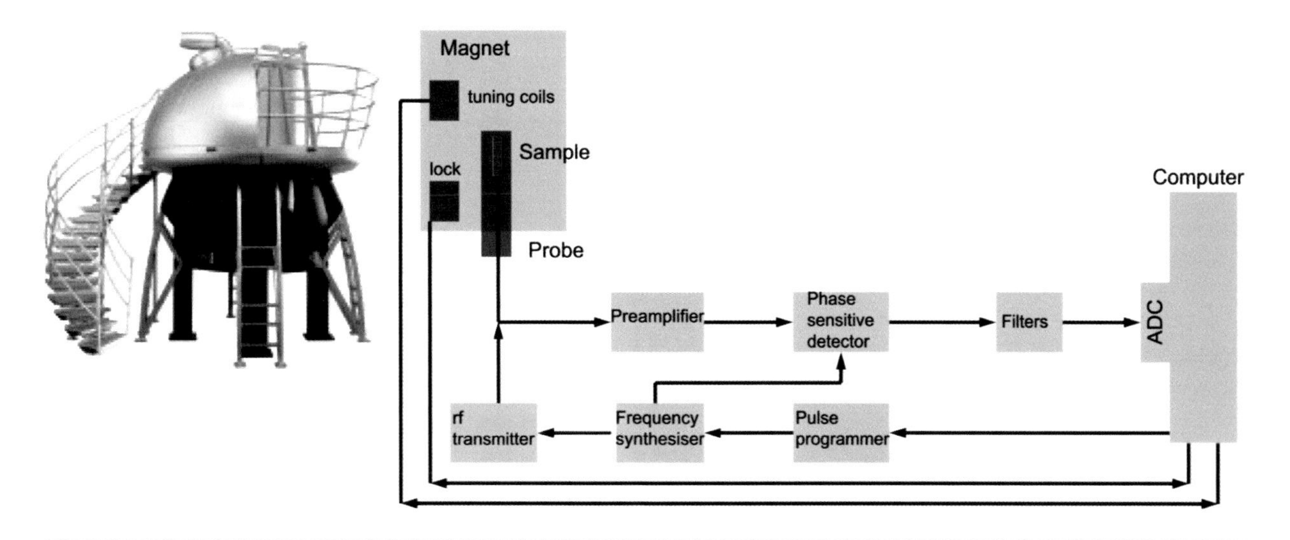

Figure 10.23 A state-of-the-art NMR spectrometer operating at 900 MHz and a block diagram representing a NMR spectrometer. The major components are the magnet containing the probe together with 'shim' or tuning coils to maintain homogeneity and electronics to 'lock' the field at a given frequency. Outside of the magnet are radio frequency transmitters with pulses designed via a pulse programmer and a frequency synthesizer together with a detection system of numerous amplifiers, filters and analogue–digital converters. All systems are controlled via a computer (Reproduced courtesy Varian Inc)

structure by NMR spectroscopy and is perhaps analogous to the phase problem of X-ray crystallography. Without first 'assigning' the protein we cannot move on to derive restraints important in the derivation of molecular structure. The starting point involves assigning a signal to a specific atom or group of atoms (e.g. HA proton, CH_3 protons, etc.). Spectral overlap in 1D spectra usually prevents complete assignment even in small proteins and as shown in the ^{1}H NMR spectra of ubiquitin (Figure 10.24) – a protein with 76 residues – many overlapping peaks preclude absolute identification.

Historically the major advance in this area was to spread conventional 1D spectra into a second dimension that reduces spectral overlap. In the last two decades these methods have expanded dramatically and almost all structural investigations of proteins start with the acquisition of 2D spectra. A normal 1D NMR spectrum involves a pulse followed by measurement of the resulting FID. 2D NMR spectroscopy involves the application of successive pulses that lead to magnetization transfer between nuclei. The mechanism of transfer proceeds either by a through bond (scalar mechanism) or by a through space (dipolar) interaction. A 2D experiment consists of four time periods; the preparation, evolution, mixing and acquisition periods often preceeded by a *relaxation* delay. The preparation period consists of a single pulse or a series of pulses and delays and its purpose is to create the magnetization terms that will make up the indirect dimension. These terms evolve during the *evolution* or t_1 period. The evolution of magnetization during the t_1 period is followed by a second pulse that initiates the mixing period and results in magnetization transfer to other spins, normally via through bond or through space interactions; transfer can also occur via chemical exchange. The decay of the FID is detected during the *acquisition* or t_2 period. The whole pulse scheme is repeated to allow for signal averaging and other instrumental factors including relaxation. The experiment is then repeated by incrementing the t_1 period to build up a series of FIDs recorded at different t_1 intervals. A 2D experiment contains two time variables, t_1 and t_2, and Fourier transformation of this dataset, $S(t_1, t_2)$ yields a 2D contour plot, $S(\omega_1, \omega_2)$, where the precession

Table 10.5 The ^{1}H chemical shifts of the amino acids residues in a random coil conformation

	Chemical shift (p.p.m.)			
Residue	HN	HA	HB	others
Ala	8.25	4.35	1.39	–
Asp	8.41	4.76	2.84, 2.75	–
Asn	8.75	4.75	2.83, 2.75	7.59, 6.91 (sc amide)
Arg	8.27	4.38	1.89, 1.79	1.70 (HG), 3.32 (HD), 7.17, 6.62 (sc NH)
Cys	8.31	4.69	3.28, 2.96	–
Gln	8.41	4.37	2.13, 2.01	2.38 (HG), 6.87,7.59 (sc NH_2)
Glu	8.37	4.29	2.09, 1.97	2.31,2.28 (HG)
Gly	8.39	3.97	–	–
His	8.41	4.63	3.26, 3.20	8.12 (2H), 7.14 (4H)
Ile	8.19	4.23	1.90	1.48, 1.19 (HG=CH_2), 0.95 (HG=CH_3), 0.89 (HD)
Leu	8.42	4.38	1.65	1.64 (HG), 0.94,0.90 (HD)
Lys	8.41	4.36	1.85, 1.76	1.45 (HG), 1.70 (HD), 3.02 (HE), 7.52 (sc NH_3)
Met	8.42	4.52	2.15, 2.01	2.64 (HG), 2.13(HE)
Phe	8.23	4.66	3.22, 2.99	7.30 (2,6H), 7.39 (3,5H), 7.34 (4H)
Pro	–	4.44	2.28, 2.02	2.03 (HG), 3.68,3.65 (HD)
Ser	8.38	4.50	3.88, 3.88	–
Thr	8.24	4.35	4.22	1.23
Trp	8.09	4.70	3.32, 3.19	7.24(2H), 7.65(4H), 7.17(5H), 7.24(6H), 7.50(7H), 10.22 (indole NH)
Tyr	8.18	4.60	3.13, 2.92	7.15 (2,6H), 6.86 (3,5H)
Val	8.44	4.18	2.13	0.97,0.94 (HG)

Adapted from *NMR of Proteins and Nucleic Acids*. Wuthrich, K. (ed.) Wiley Interscience, 1986.

frequencies occurring during the evolution and detection periods determine peak positions (ω_1 and ω_2) in the 2D plot.

In general, three homonuclear ^{1}H-NMR experiments are used for assignment of proteins. The COSY (correlated spectroscopy) and TOCSY[3] (total correlated spectroscopy) describe magnetic interactions between scalar coupled nuclei – resonances linked via 'through bond' interactions (Figure 10.25). When the COSY experiments are performed on protein dissolved in H_2O, cross peaks are observed in 2D spectra reflecting, for example, connectivity between NH and HA protons.

In ^{1}H NMR through bond interactions (Figure 10.26) as a result of coherence transfer are limited to two or three bonds and are restricted to *intra*-residue correlations. The four-bond interaction between the $HN_{(i+1)}$,$HA_{(i)}$ is too weak to be observed. Cross peaks can also occur between HA and HB protons via ^{3}J couplings and it is soon apparent that characteristic patterns of connectivity are observed for different residues (Figure 10.27).

Performing experiments in D_2O ($^{2}H_2O$) removes most HN signals since protons exchange for deuterons leading to the loss of the left-hand side of the 2D spectra. These protons are described as labile.

[3]The TOCSY experiment is sometimes called HOHAHA (Homonuclear Hartmann-Hahn) after the discoverers of an original solid state experiment demonstrating cross-polarization between nuclei.

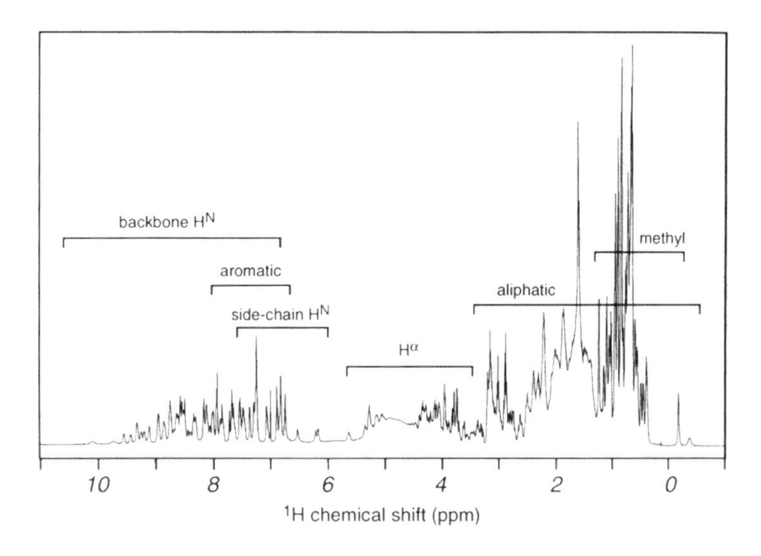

Figure 10.24 The ^{1}H NMR spectra of ubiquitin, a protein containing 76 residues. (Reproduced with permission from Cavanagh, J. *et al. Protein NMR Spectroscopy: principles and practice*. Academic Press, 1996)

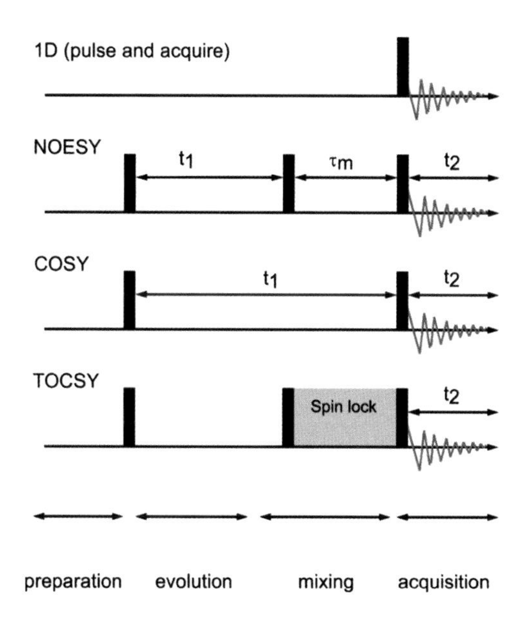

Figure 10.25 Schematic representing 1 and 2D homonuclear pulse sequences used in protein NMR spectroscopy. The pulse schemes have been given names such as COSY (correlation spectroscopy), TOCSY (total correlation spectroscopy) and NOESY (nuclear Overhauser effect spectroscopy) that define the basic mode of magnetization transfer

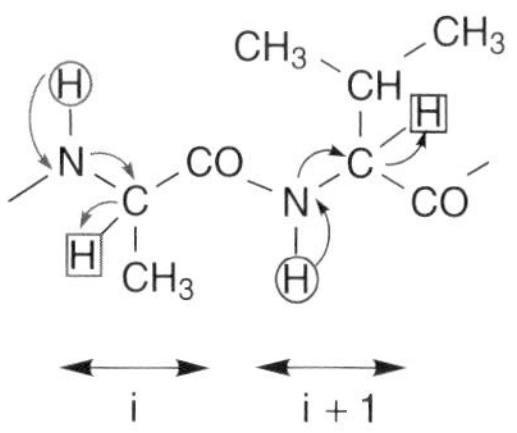

Figure 10.26 Through bond interactions in residue i and i + 1 of a polypeptide chain (... Ala-Val...). The amide proton and HA proton of the same residue are linked via a three-bond coupling

The TOCSY experiment relies on cross-polarization, and during a complicated pulse sequence the application of a 'spin locking' field leads to all spins becoming temporarily equivalent. In a condition of "equivalence" magnetization transfer occurs when the mixing time is equivalent to 1/2 J. For proteins containing many residues with different side chains there is no single value for 1/2 J that allows all connectivities to be observed. Instead TOCSY experiments are repeated with different mixing times and the ideal situation involves observing complete spin systems from the HN to the HA and HB and on to the remaining resonances.

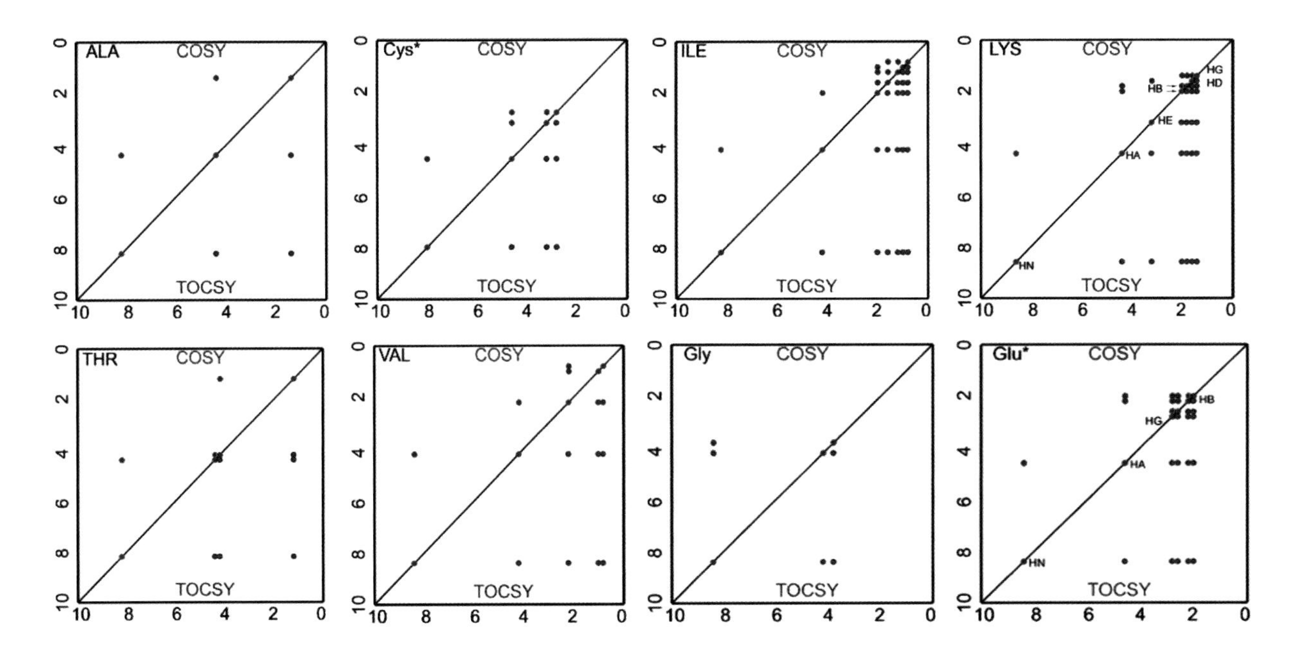

Figure 10.27 Patterns of connectivity seen in 2D COSY and TOCSY spectra of eight of the 20 amino acid residues. *The pattern shown by Cys is similar for all other residues in which the spin system involves a HN–HA and two non-degenerate HB protons. This is likely to include the aromatic residues Phe, Tyr and Trp as well as Asp, Asn and Ser. Similarly, the profile shown by Glu is shared by Gln and Met. The pattern does not take any account of peak fine structure that occurs in COSY type experiments

Figure 10.28 The pattern of sequential NOEs expected within a polypeptide chain for three successive residues

Connectivity patterns are sometimes diagnostic for individual residues and having identified spin system type it is necessary to establish its location within the primary sequence (Figure 10.28). This is achieved through NOESY experiments. The 2D homonuclear NOESY experiment relies on *through space* interactions between nuclei (^{1}H) separated by less than 6 Å. By establishing sequential connectivity between resonances of a known spin system type it is possible to identify clusters of residues that are unique within the primary sequence. In Figure 10.28 valine and alanine exhibit sequential connectivity and if these residues occur sequentially only once in the primary sequence then these residues are identified unambiguously. If the dipeptide Val-Ala occurs in more than one region of the polypeptide chain then the identity of residues $i-1$ and $i+2$ becomes important in establishing conclusive sequence specific assignments.

Some sequential NOEs are indicative of particular units of secondary structure (Table 10.6). For example, the regular periodicity of helices leads to the close approach of HA and HN between residues $(i,i+2)$, $(i,i+3)$ and $(i,i+4)$ that do not occur in antiparallel or parallel β strands. More significantly the observation of $d_{\alpha N}$ $(i,i+4)$ is strongly indicative of a regular α

Table 10.6 Regular secondary structure gives characteristic NOEs

Interaction	α helix	3_{10} helix	Antiparallel β strand	Parallel β strand	Type I turn	Type II turn
$d_{\alpha N}$	2.7	2.7	2.8	2.8	2.8	2.7
$d_{\alpha\beta}$	2.2–2.9	2.2–2.9	2.2–2.9	2.2–2.9	2.2–2.9	2.2–2.9
$d_{\beta N}$	2.2–3.4	2.2–3.4	2.4–3.7	2.6–3.8	2.2–3.5	2.2–3.4
$d_{\alpha N}(i,i+1)$	3.5	3.4	2.2	2.2	3.4	2.2
$d_{NN}(i,i+1)$	2.8	2.6	4.3	4.2	2.6	4.5
$d_{\beta N}(i,i+1)$	2.5–2.8	2.9–3.0	3.2–4.2	3.7–4.4	2.9–4.1	3.6–4.4
$d_{\alpha N}(i,i+2)$	4.4	3.8	–	–	3.6	3.3
$d_{NN}(i,i+2)$	4.2	4.1	–	–	3.8	4.2
$d_{\alpha N}(i,i+3)$	3.4	3.3	–	–	3.1–4.2	3.8–4.7
$d_{\alpha\beta}$ (i,i + 3)	2.5–4.4	3.1–5.1	–	–	–	–
$d_{\alpha N}(i,i+4)$	4.2	–	–	–	–	–

Distances involving β protons vary due to the range of distances possible. The first three sets of distances refer to intra-residue connectivity. All distances are in Å

helix whilst the presence of only $d_{\alpha N}$ (i,i + 2) and $d_{\alpha N}$ (i,i + 3) may indicate more tightly packed 3_{10} helices. Similarly, intense NOEs between $d_{\beta N}$ (i,i + 1) indicates a β strand. Since both strands and helices tend to persist for several residues the observation of blocks of sequential NOEs with these connectivities allows secondary structure to be defined.

The NOESY spectrum not only resolves sequentially connected residues but also represents the basis of protein structure determination using NMR data. Integrating the volumes associated with the cross peaks in NOESY spectra quantifies this interaction and from intra-residue and sequential NOEs allows the volumes to be converted into distances, since regular secondary structure is associated with relatively fixed separation distances (see Table 10.6). Some NOE cross peaks arise as a result of long-range interactions (i.e between residues widely separated in the primary sequence) and these cross peaks play a major role in determining the overall fold of a protein since by definition they represent separation distances of less than 6 Å. Using this approach Kurt Wuthrich pioneered protein structure determination for BUSI IIA – a proteinase inhibitor from bull seminal plasma in 1984 – and it marked a landmark in the progression of NMR spectroscopy from analytical tool to structural technique (Figure 10.29).

Table 10.7 Some of the first proteins whose structures were determined by homonuclear NMR spectroscopy

Protein	Date	Mass
BUSI IIA	1985	6050
Lac repressor headpiece	1985	5500
BPTI	1987	6500
Tendamistat	1986	8000
BDS-I	1989	5000
Human complement protein C3a	1988	8900
Plastocyanin	1991	10 000
Thioredoxin	1990	11 700
Epidermal growth factor	1987	5800

Between 1984 and 1990 structures for many small (<10 kDa) proteins were determined via 2D NMR techniques (Table 10.7). To compare the power of crystallography and NMR spectroscopy Wuthrich and

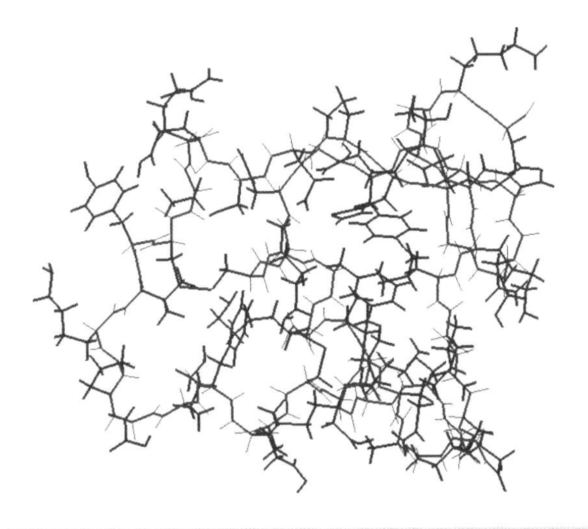

Figure 10.29 The structure of BUSI IIA – the first protein structure determined using NMR spectroscopy. Backbone atoms are shown in blue, side chains in green

Robert Huber independently determined the structure of tendamistat, a protein inhibitor of α-amylase with 74 residues (Figure 10.30). The results showed emphatically that crystal and solution state structures were comparable in almost every respect and testified to the ability of NMR spectroscopy in deriving 3D structures.

However, larger proteins containing more than 100 residues presented new and increasingly difficult problems that arose from the greater number of resonances. This resulted in spectral overlap with increased correlation times producing broader linewidths. The answer was heteronuclear NMR spectroscopy.

Heteronuclear NMR spectroscopy

Resonance assignment is vital to protein structure determination. Additional spin 1/2 nuclei allow alternative assignment strategies and new pulse techniques exploiting heteronuclear scalar couplings present in $^{13}C/^{15}N$ enriched proteins (Figure 10.31). Experiments involved the transfer of magnetization from ^{1}H to ^{13}C and/or ^{15}N through large one-bond scalar couplings. The magnitude of these couplings (30–140 Hz) is much greater than the ^{1}H–^{1}H ^{3}J couplings (2–10 Hz).

The simplest heteronuclear 2D experiments correlate the chemical shift of the ^{15}N nucleus with its attached proton. In 2D ^{15}N–^{1}H heteronuclear spectra cross peaks are spread out according to the nitrogen chemical shift

Figure 10.30 The structures of tendamistat derived by NMR and crystallography are superimposed for the polypeptide backbone. The NMR structure is shown in red and the crystallographic structure is shown in blue. (PDB files: 2AIT and 1HOE). With the minor exception of the N terminal region the two structures agree very closely. Tendamistat is a protein based largely on β sheet (right)

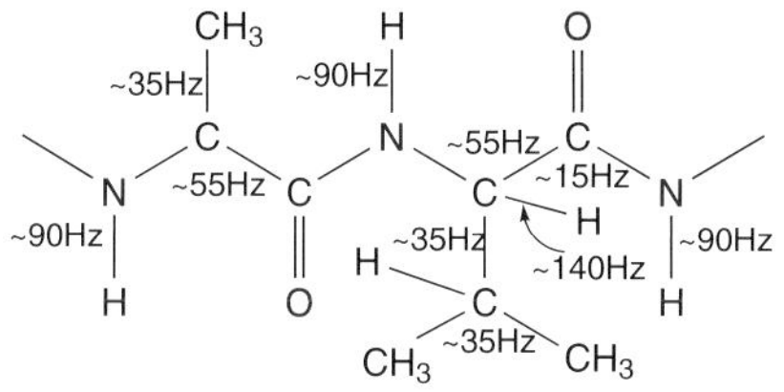

Figure 10.31 Heteronuclear coupling between ^{13}C, ^{1}H and ^{15}N within proteins together with the magnitude of these one bond coupling constants (^{1}J)

as well as the proton chemical shift (Figure 10.32). Many residues have characteristic ^{15}N and ^{13}C chemical shifts (Tables 10.8 and 10.9). For example, Gly, Ser and Thr residues show ^{15}N chemical shifts below 115 p.p.m. along with the side chain amides of Gln and Asn. Observation of a cross peak in heteronuclear spectra in these regions is most probably attributed to one of these residues.

Table 10.8 Nitrogen chemical shifts for the backbone and side chain amides of residues found in proteins

Residue	^{15}N	Residue	^{15}N
Ala	125.04	Leu	122.37
Arg	121.22	Lys	121.56
Asn sc	119.02	Met	120.29
Asp	119.07	Phe	120.69
Cys	118.84	Pro	–
Glu	120.23	Ser	115.54
Gly	107.47	Thr	111.9
Gln sc	120.46	Trp indole	122.08
His	118.09	Tyr	120.87
Ile	120.35	Val	119.31

Determined from position in a sequence AcGGXGG-NH_2 in 8 M urea. Adapted from *J. Biomol. NMR* 2000, **18**, 43–48.

Heteronuclear 2D versions of the NOESY, COSY and TOCSY are often recorded but spectral overlap can remain a problem especially in NOE spectra where both intra- and inter-residue cross peaks are observed. This problem was resolved by further pulse sequences that involved three time variables (t_3, t_2 and t_1) and where Fourier transformation leads to a cube or 3D plot. The major advantage of these techniques is that 2D spectra are extended into a third dimension usually the ^{15}N or ^{13}C chemical shift. A panoply of new techniques based on heteronuclear J couplings massively enhanced NMR spectroscopy as a tool for 'large' protein structure determination. Ambiguities seen previously were eliminated by new pulse sequences that identified spin systems with high sensitivity. These

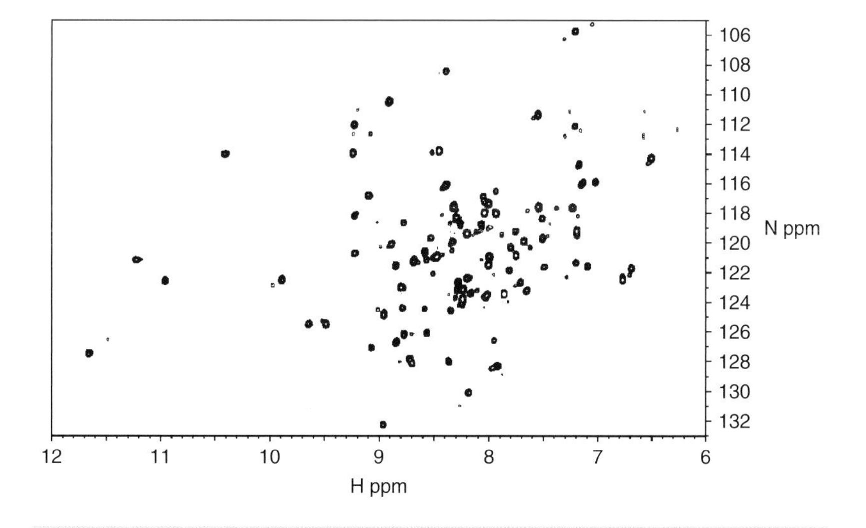

Figure 10.32 A 2D heteronuclear ^{15}N–^{1}H correlation spectra for a protein

Table 10.9 ^{13}C chemical shifts for the backbone and side chain amides of residues found in proteins

Residue	CO or C′	C_α	C_β	C_γ	C_δ	Other
Ala	175.8	50.8	17.7			
Arg	175.0	54.6	28.8	25.7	41.7	
Asn	173.1	51.5	37.7			175.6 (amide)
Asp	174.2	52.7	39.8			178.4 (carboxylate)
Cys	175.7	57.9	26.0			
Gln	174.0	54.1	28.1	32.2		179.0 (amide)
Glu	174.8	54.9	28.9	34.6		182.8 (carboxylate)
Gly	172.7	43.5				
His	172.6	53.7	28.0			135.2 (C2), 118.7 (C4), 130.3 (C5)
Ile	174.8	59.6	36.9	25.4, 15.7	11.3	
Leu	175.9	53.6	40.5	25.2	23.1, 21.6	
Lys	174.7	54.4	27.5	23.1	31.8	
Met	175.0	53.9	31.0	30.7		15.0 (C_ε)
Phe	176.0	57.4	37.0			136.2 (C1), 130.3 (C2/C6), 130.3 (C3/C5), 128.6 (C4)
Pro	175.2	61.6	30.6	25.5	48.2	*trans* configuration,
		61.3	33.1	23.2	48.8	*cis* shows small differences with exception of C_β
Ser	172.6	56.6	62.3			
Thr	172.7	60.2	68.3	20.0		
Trp	176.7	56.7	27.4			C3 (108.4), 112.8 (C7), 137.3 (C8), 127.5 (C9)
Tyr	176.0	57.4	37.0			128.0 (C1), 130.0 (C2/C6), 117.0 (C3/C5), 156.0 (C4)
Val	174.9	60.7	30.8	19.3, 18.5		

The chemical shifts were determined in linear pentapeptides of the form GGXGG. Data adapted from Wuthrich, K. *NMR of Proteins and Nucleic Acids*. John Wiley & Sons, 1986.

techniques carry names that highlight the different correlation made during experiments. For example a 3D HNCO correlates the NH of residue i with the CO of the proceeding residue (i − 1). Triple resonance experiments (Table 10.10) greatly increase the number of assignments, and armed with a larger number of assignments comes the possibility of deriving greater numbers of structural restraints from NOEs and coupling constant data.

The use of multi-dimensional NMR methods has seen larger proteins assigned and it is now feasible to attempt to completely assign proteins of molecular mass in excess of 30 kDa. As of 2003 the NMR derived structures (coordinates) of over 2500 proteins have been deposited in the Protein Databank. Of these proteins the vast majority have masses below 15 000 and a search of databases suggests that structures for proteins with more than 150 residues are increasingly steadily. In 2002 a backbone assignment for a 723-residue protein (malate synthase) was reported and bears testament to the effectiveness of these new multi-dimensional NMR methods. Assignment is, however, only the first step in structure determination and for any protein (large or small) it is necessary to obtain

Table 10.10 A small selection of common triple resonance experiments used for sequential assignment in proteins

Experiment	Observed correlation	Magnetization transfer	J couplings
HNCA	$^{1}H^{N}_{i}$–$^{15}N_{i}$–$^{13}C^{\alpha}_{i}$ $^{1}H^{N}_{i}$–$^{15}N_{i}$–$^{13}C^{\alpha}_{i-1}$	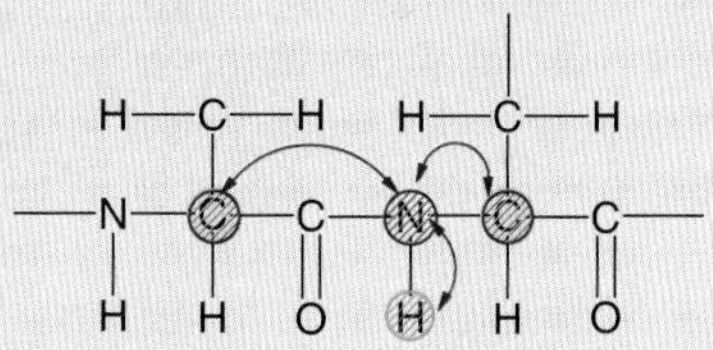	$^{1}J_{NH}$, $^{1}J_{NC}$, $^{2}J_{NC\alpha}$
HNCO	$^{1}H^{N}_{i}$–$^{15}N_{i}$–$^{13}CO_{i-1}$	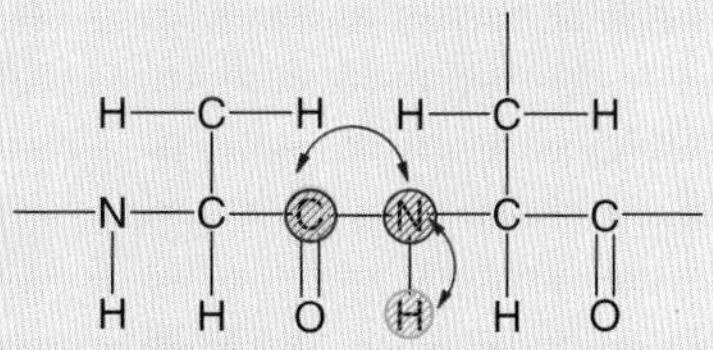	$^{1}J_{NH}$, $^{1}J_{NCO}$
CBCANH	$^{13}C^{\beta}_{i}/^{13}C^{\alpha}_{i}$–$^{15}N$–$^{1}H^{N}_{I}$ $^{13}C^{\beta}_{i}/^{13}C^{\alpha}_{i}$–$^{15}N_{i+1}$–$^{1}H^{N}_{I+1}$	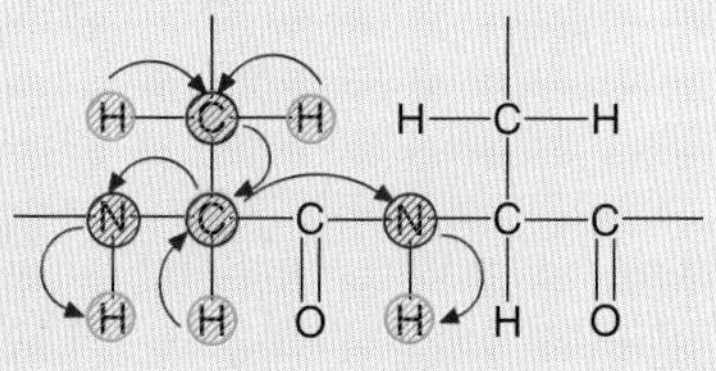	$^{1}J_{CH}$, $^{1}J_{C\alpha C\beta}$, $^{1}J_{NC\alpha}$, $^{2}J_{NC\alpha}$, $^{1}J_{NH}$
CBCA(CO)NH	$^{13}C^{\beta}_{i}/^{13}C^{\alpha}_{i}$–$^{15}N_{i+1}$–$^{1}H^{N}_{I+1}$	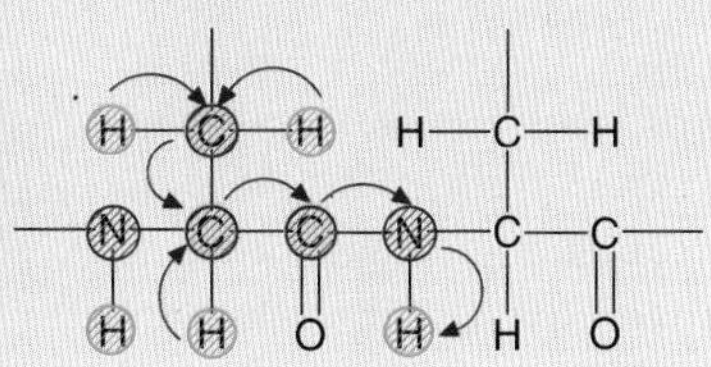	$^{1}J_{CH}$, $^{1}J_{C\alpha C\beta}$, $^{1}J_{C\alpha CO}$, $^{1}J_{NC}$, $^{1}J_{NH}$

a sufficient number of structurally significant NOEs or other conformational restraints to define structure with reasonable precision.

Generating protein structures from NMR-derived constraints

The information derived from NMR spectroscopy about protein conformation can be divided into two major classes. One class of conformational restraints reflect angular information whilst a second class are distance dependent. It is self evident that if sufficient angles and distances between atoms are defined then, in theory at least, it is possible to define overall conformation. The majority of conformational restraints are obtained via homo- and heteronuclear NOESY experiments and convert the NOE cross peak volumes derived from the relationship

$$NOE_{ij} \propto 1/r_{ij}^{6} \qquad (10.29)$$

into distances that represent the separation between two nuclei i and j. Most schemes categorize NOE cross peak volumes into three classes of restraints (strong, medium and weak with sometimes a fourth group termed very weak). Strong NOEs represent distances between nuclei that are close together ranging from 1.9 to 2.9 Å; medium NOEs represent distances from 2.9–3.5 Å and weak NOEs represent distances from 3.5–5.0 Å. Very weak NOEs are generally assumed to include separation distances in a range 5.0–6.5 Å. In effect these distances represent upper and lower limits for the separation distances. Until recently NOEs were the only experimental restraints used to determine protein structure since the number of torsion angles measurable with accuracy was often limited. However, heteronuclear NMR methods have permitted increased numbers of restraints based on torsion angles derived from the measurement of coupling constants between backbone atoms as well as some side chains.

Most approaches (see Figure 10.33) derive protein structures from NMR data using distance geometry and simulated annealing. In all cases the methods satisfy the conformational data yielding a structure consistent with experimental data and parameters governing bond lengths and angles. Unfortunately almost all NMR restraints include a range of possible values. So for example, an NOE cross peak volume may be consistent with distances ranging between 3.5 and 5.0 Å whilst a $^3J_{HNHA}$ coupling constant of <5 Hz is consistent with a torsion angle (ϕ) between 0 and $-120°$. When this is repeated for every NOE and torsion angle many protein conformations are consistent with the data and there is no uniquely defined structure.

The approach used to calculate structures from NMR data involves the generation of an ensemble of 'low energy' conformations all consistent with the data that are progressively refined during calculation. Structure calculations based on experimental data derived from NMR spectroscopy yield a family of closely related structures all of which are consistent with the input data. The three-dimensional structures of interleukin-1β and thioredoxin derived using NMR spectroscopy are shown in Figure 10.34, and like their crystallographic counterparts these structures are models of clarity providing a detailed insight into the mechanism of protein function.

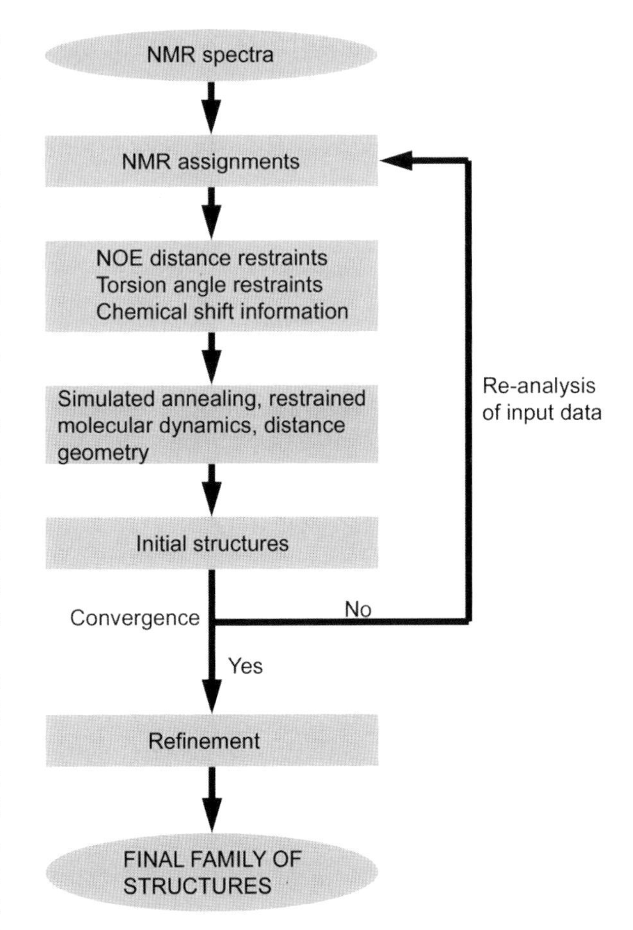

Figure 10.33 The approach used to derive protein structures from torsion angle and distance restraints obtained via NMR spectroscopy

Structure calculations yield families of related conformations all of low overall energy. Although a mean structure calculated from a family of 20–50 closely related structures is often shown this structure is of no more relevance than any of the other structures of low energy generated via structure calculation. The generation of a mean structure allows a root mean square deviation to be calculated from differences in the centre of mass of atoms in a standard (mean) structure with that in family of molecules. A small difference in position for each atom between the mean and test structures leads to a low rmsd value and

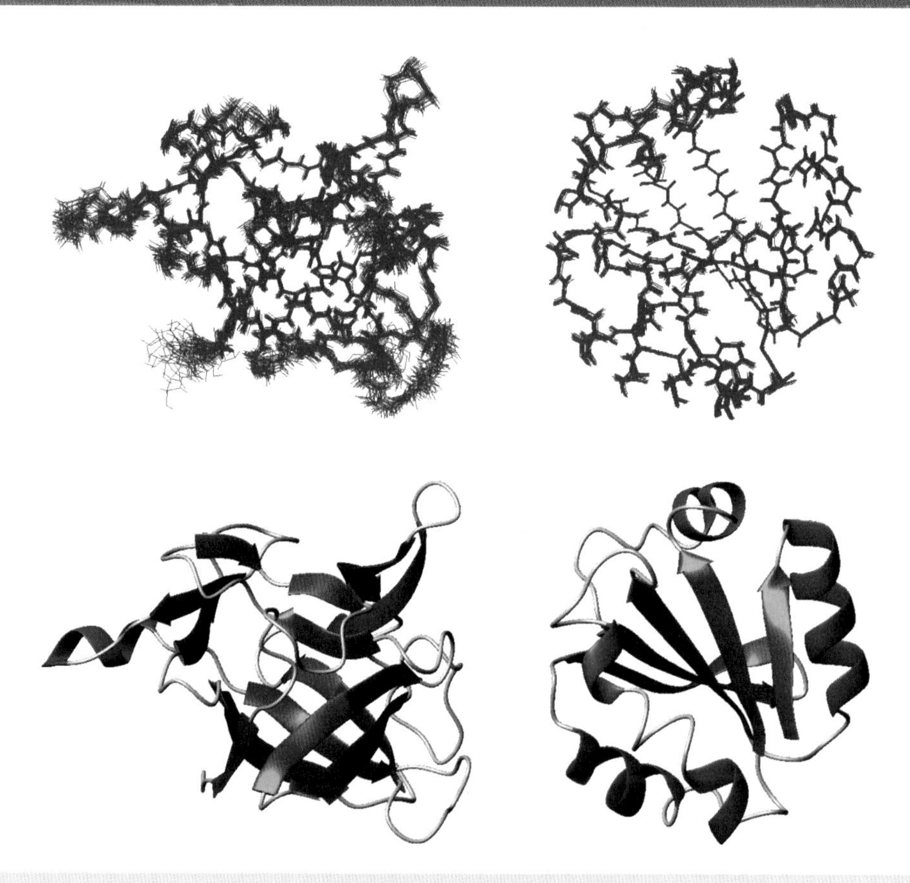

Figure 10.34 The solution structures of interleukin-1β and thioredoxin derived using NMR spectroscopy. The structures show the 50 lowest energy structures for the polypeptide backbone together with the secondary structure distribution of the lowest energy structure (bottom row) (PDB: 7I1B and 1TRU). Interleukin-1β contains ~150 residues and has a mass of ~17 500, whilst thioredoxin contains 105 residues and has a mass of ~11 500. Reproduced courtesy of FEI Company www.feicompany.com

implies structures are similar.

$$\mathrm{rmsd} = \sqrt{\frac{1}{N}\sum_{i=1}^{N}(r_{\mathrm{mean}} - r_{\mathrm{test}})^2} \qquad (10.31)$$

The precision of NMR structures is ultimately related to the number of experimental restraints defining conformation. If there are no restraints between side chains then structure is poorly defined. Most frequently backbones are defined with highest precision followed by the side chains. In high quality NMR derived structures as many as 50 restraints per residue may be obtained leading to a backbone rmsd of 0.3–0.5 Å. Advances in molecular biology, instrumentation, heteronuclear pulse sequences and computational methods are now permitting structures to be determined using NMR spectroscopy of equal precision to those obtained by X-ray diffraction. However, one group of proteins for which NMR and to a lesser extent X-ray diffraction struggle to provide detailed structures are large asymmetric complexes or macromolecular assemblies.

Cryoelectron microscopy

Visualizing structure has provided enormous impetus to understanding biological processes, and electron microscopy has provided many views of cells and subcellular organelles through the use of a transmission electron microscope (Figure 10.35; TEM).

Pioneering work using electron microscopy to study viral organization in the tail assembly of bacteriophage T4 by Aaron Klug laid the foundation for determining 3D structure from series of 2D images or projections of an object's electron density formed at an image plane (Figure 10.35). Three-dimensional information could be recovered if a number of views of the object were recorded at different angles of observation. This is usually done with a tilting stage that elevates specimens to angles of ~70°. For structures of high rotational symmetry such as helices it is possible to use just a single or small number of orientations to reconstruct a 3D model. The discovery had wider implications since it was exploited in the field of medicine as a medical imaging technique called computerized axial tomography (CAT). This work also allowed Henderson and Unwin to determine the orientation of helices in bacteriorhodopsin and pointed the way forward towards the use of electron microscopy to study proteins at higher levels of atomic resolution. From continuous development of instrumentation, improved specimen preparation, massive advances in data processing and increased computational power electron microscopy has become a technique capable of providing structural information. The technique is useful for proteins not amenable to study by X-ray crystallography or NMR.

Early electron microscopy studies involved 'fixing' biological preparations using cross-linking agents followed by staining with heavy metal (electron dense) compounds; treatments that could distort structure by introducing artifacts. This problem persisted until Kenneth Taylor and Robert Glaeser reported electron diffraction data recorded at a temperature of ~100 K from frozen thin films containing hydrated catalase crystals. In one swoop the technique advanced allowing biological 'units' to be imaged under representative conditions. The technique became known as cryo-electron microscopy (cryo-EM). Sample preparation

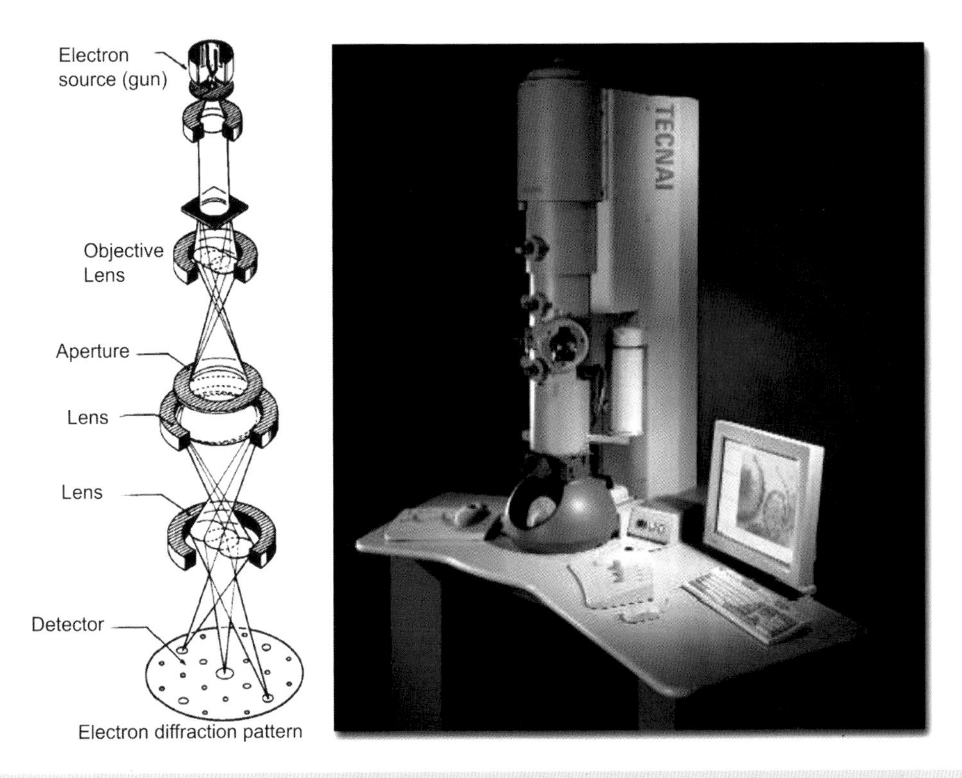

Figure 10.35 An electron diffraction pattern together with a picture of a modern TEM used for structural analysis. (Reproduced by permission of FEI Company.)

is important to cryo-EM and plunging a thin layer supported on a carbon grid into a bath of ethane cooled by liquid nitrogen leads to rapid and efficient cooling. Liquid ethane, a very efficient cryogen, produces a metastable form of water called the vitrified state. The vitrified state is amorphous and lacks the crystalline ice-like lattice. Although metastable the vitrified form is maintained indefinitely at low temperatures (~77 K) and all cryo-EM methods use this approach. Images are recorded at ~100 K using 'low dose' techniques (0.1 electrons $Å^{-2}$) to identify areas of interest and to limit radiation damage. Areas of interest are then captured at greater magnification with a higher dose of electrons (5–10 $Å^{-2}$). The data has a low signal/noise ratio and requires extensive averaging and processing in deriving structural information.

The early successes of cryo-EM centred mainly on viral structure determination and many viral capsids have been defined. One example involved determination of the hepatitis B capsid protein structure by the groups of R.A. Crowther and A. Steven (Figure 10.36). At a resolution of 7.6 Å this study identified elements of secondary structure within subunits.

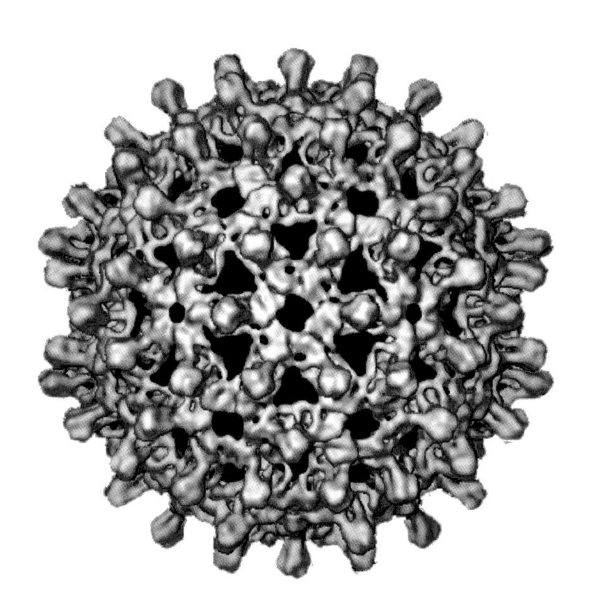

Figure 10.36 The reconstructed 3D structure of the hepatitis B capsid (adapted with permission from Baumeister, W. & Steven, A.C. *Trends Biochem. Sci.* 2000, **25**, 625–631. Elsevier)

Single-particle analysis deals with large proteins or protein complexes and is growing in its application to biological problems largely through the efforts of Joachim Frank in the study of ribosome structure. In this method macromolecular complexes are trapped in random orientations within vitreous ice and images from all possible orientations are combined and averaged to provide enhanced views.

From a sample preparation point of view single particle analysis is straightforward – it requires a well-dispersed, homogeneous complex frozen in amorphous ice and aligned over the holes of a carbon grid. Structure determination from differently oriented particles requires the combination of data from many images using methods that distinguish 'good' from 'bad' particles, determine the orientation relative to a reference point and evaluate the quality of the resulting data. These methods are intensely computational and beyond the scope of this text. Although the method has not produced data comparable in resolution to X-ray diffraction or NMR spectroscopy methods the technique has allowed ribosomal subunits to be determined at resolutions approaching 7.0 Å. Significantly, the results of crystallography and/or NMR can often be assimilated into the cryo-EM data to build models based on more than one structural technique.

The power of electron crystallography has been demonstrated for bacteriorhodopsin but success has also been achieved with a variety of proteins including tubulin, light harvesting complexes associated with the reaction centres of photosynthetic organisms, aquaporin a transport protein of the red blood cell membrane and the acetylcholine receptor of neurones.

Tubulin is a key protein of the microtubule assembly regulating intracellular activity particularly cytoskeletal function and eukaryotic cell division where it facilitates the separation of chromatids as part of mitosis. In this role the ability of microtubules to polymerize and depolymerize is important. The microtubule contains 13 protofilaments arranged in parallel to form a hollow structure approximately 24 nm in diameter. The protofilaments are composed of alternating α- and β-tubulin monomers each composed of a single polypeptide chain (the α and β subunits) of ~55 kDa. Within the microfibril each protofilament is staggered

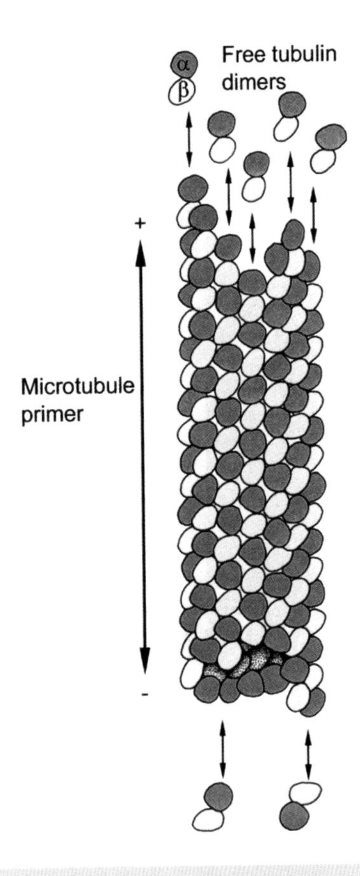

Figure 10.37 The arrangement, assembly and disassembly of tubulin dimers within a microtubule. Assembly is primarily from a preformed microtubule primer. The growing (+) end represents rapid addition of dimers when compared with rates of removal whilst at the (−) end the converse is true. Assembly is enhanced at 37 °C and by the presence of GTP whilst the absence of nucleotide and lower temperatures drives disassembly (adapted from Darnell *et al. Molecular Cell Biology*, 2nd edn. Scientific American Books, 1990.)

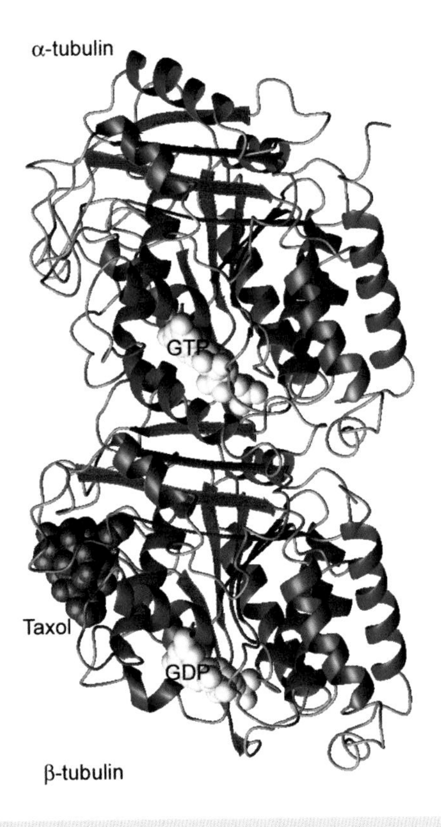

Figure 10.38 The structure deduced for the αβ tubulin dimer determined at a resolution of 0.37 nm by electron crystallography (PDB: 1TUB reproduced with permission from Nogales, E. *et al. Nature* 1998, **391**, 199–203. Macmillan Publishers Ltd)

with respect to the next to give a characteristic helical pattern (Figure 10.37).

Assembly involves GTP binding to α and β subunits. Subunit addition brings β -tubulin into contact with the α subunit promoting GTP hydrolysis on the interior β-tubulin with GTP bound to α-tubulin unhydrolyzed during polymerization. In the presence of Zn ions purified tubulin assembles into a two-dimensional ordered sheet ideal for electron crystallography (Figure 10.38). In these sheets the protofilaments are similar in appearance to those in microtubules with the exception that they are associated in an antiparallel array. The polymerization reactions make studies using NMR spectroscopy unfeasible whilst the inherent mobility of the filaments has prevented crystallization suitable for X-ray diffraction.

The electron density profile determined directly from cryo-EM studies for the αβ dimer allowed the sequence of each subunit to be fitted directly with little ambiguity. Each monomer was a compact structure containing a core of β strands assembled into a sheet surrounded by α helices. A conventional Rossmann fold was found at the N-terminal region and contained a nucleotide-binding region. An intermediate domain contained a four-stranded β sheet and three helices along with a Taxol-binding site. A third domain of two antiparallel helices crossing the N- and intermediate regions represented a binding surface for additional motor proteins. A Mg ion was bound near the GTP binding site whilst a binding site for Zn ions was identified at the interface between protofilaments.

Cryo-EM methods are now firmly established as a third method for macromolecular structure determination. The list of successes is continuing to expand with increasingly detailed pictures of the ribosome, the thermosome, viral capsids, transport proteins, amyloid fibres and receptors obtained using cryo-EM methods. There is little doubt that cryo-EM will increase in importance as a method of three-dimensional structure determination for macromolecular complexes.

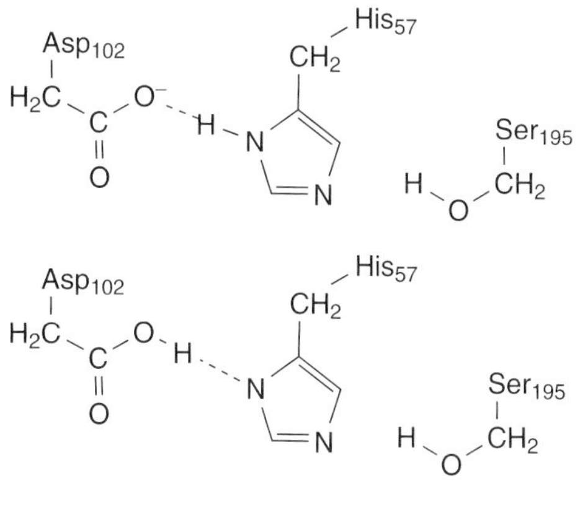

Figure 10.39 The exchangeable proton in the catalytic site of serine proteases was shown to remain attached to His57 and was not transferred to Asp102

Neutron diffraction

Neutron diffraction requires the crystallization of proteins followed by measurement of the diffraction of neutrons. Neutron beams were traditionally generated *via* atomic reactors but increasingly sources are derived *via* synchrotron beam lines. Until the introduction of synchrotron radiation a major problem was achieving sufficiently high neutron fluxes to enable adequate data collection times. Although less widely used the technique has proved very useful for locating hydrogen atoms within protein crystals something that is difficult to achieve with X-rays.

The usefulness of neutron diffraction to biological structure determination in the post-genomic era involves its ability to probe dynamic properties involving protons. Hydrogen exchange underscores many biological reactions and neutron diffraction is adept at locating protons that exchange for deuterium within enzyme active sites. Such observations help to establish catalytic mechanisms and the exchange of hydrogen for deuterium leads to a large change in neutron scattering factors.

Two noteworthy examples where neutron diffraction has proved of value are the catalytic mechanisms of serine proteases and lysozyme. In serine proteases such as trypsin the catalytic mechanism revolves around the catalytic triad of invariant Ser, His and Asp residues (Figure 10.39). The nucleophilic Ser residue forms two tetrahedral intermediates during bond cleavage and is assisted by increased imidazole basicity caused by hydrogen bonding to the aspartate. Hydrogen bonding could arise from protonation of the imidazole group or protonation of aspartyl side chains.

Neutron protein crystallography showed that the proton remained attached to the imidazole ring of His 57 and not the carboxylate of Asp 102 and clarified the reaction mechanism. Similarly in lysozyme, the deuterium atom was found to reside on the carboxyl side chain of Glu 35, rather than Asp 52, again emphasizing a key aspect of the catalytic mechanism.

Optical spectroscopic techniques

Absorbance

Transitions between different electronic states occur in the ultraviolet, visible and near infrared regions of the electromagnetic spectrum and are widely used for studying protein structure. Absorbance involves transitions of outer shell electrons between various electronic states and is governed by the rules of quantum mechanics. The absorbance of light excites an electron from the ground state to a higher excited state (Figure 10.40).

Absorbance is governed by several rules; the frequency of the incident radiation must match the

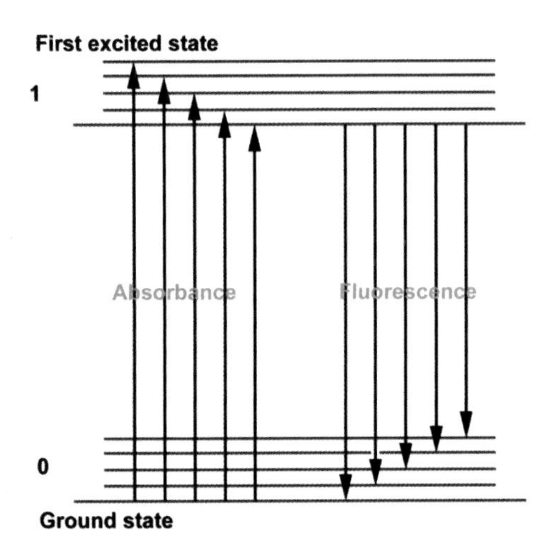

Figure 10.40 Absorbance and fluorescence resulting from excitation and emission between the ground and first excited state

quantum of energy necessary for transition from ground to excited states (i.e. $h\nu = \Delta E$). When the resonance condition is satisfied transitions can occur if further complex selection rules are obeyed. Selection rules are divided into high probability or allowed transitions and forbidden transitions of much lower probability. Within the latter set of transitions are spin-forbidden and symmetry-forbidden transitions. Spin forbidden transitions involve a change in spin multiplicity defined as $(2S + 1)$ where S is the electron spin number (analogous to I, the nuclear spin quantum number). Spin multiplicity reflects electron pairing (see Table 10.11). For a favourable transition there is no change in multiplicity ($\Delta S = 0$).

Table 10.11 Spin multiplicity for atoms and molecules

No. of unpaired electrons.	Electron spin S	$(2S + 1)$	Multiplicity
0	0	1	Singlet
1	1/2	2	Doublet
2	1	3	Triplet
3	3/2	4	Quartet

Symmetry-forbidden transitions reflect redistribution of charge during transitions in a quantity called the transition dipole moment. Differences in dipole moment arise from the different electron distributions of ground and excited states and when these differences approximate to zero the transition is forbidden. To express this slightly differently absorbance requires a change in dipole moment.

Absorbance is normally observed by peaks of finite intensity and definitive linewidth in plots against frequency. The intensity is reflected by a Boltzmann distribution that describes the population of each energy level. In molecules atomic interactions lead to the association of electrons with neighbouring nuclei whilst the nuclei move relative to each other in the form of quantized vibrational or rotational motion. As a result molecules possess ground and excited states that split into a variety of sub-states, each corresponding to different vibrational modes, and then to further rotational sub-states. Absorbance of light leads to transitions from the lowest ground state to a variety of energy levels in the first excited state. In the absence of chemical reactions or energy transfer the electrons return to the ground state. Absorbance spectra of proteins, when recorded at room temperature, consist of several closely related transitions each of similar, but non-identical, frequency leading to distinct bandwidths, reflecting the statistical sum of all individual transitions. Despite broad bands absorbance spectra are frequently characterized by maxima diagnostic of a particular transition and chromophore (e.g. A_{280} for proteins). In proteins transitions often involve aromatic side chains (Figure 10.41) but in the near UV/visible regions (350–750 nm) co-factors such as heme, flavin, or chlorophyll will also give intense absorbance bands. Using molecular orbital theory electrons are defined according to the orbitals in which they reside as either σ, π or n (non-bonding) with the corresponding anti-bonding orbitals denoted as σ^*, π^* or n*. Transitions between $\sigma \rightarrow \sigma^*$ lie in the far UV region (large energy difference) and are not normally observed by optical methods, but transitions between $\pi \rightarrow \pi^*$ and $n \rightarrow \pi^*$ are frequently observed in the UV and visible region of the electromagnetic spectrum. Physicists and chemists plot the spectra of molecules in the form of intensity versus frequency, where the frequency is expressed as

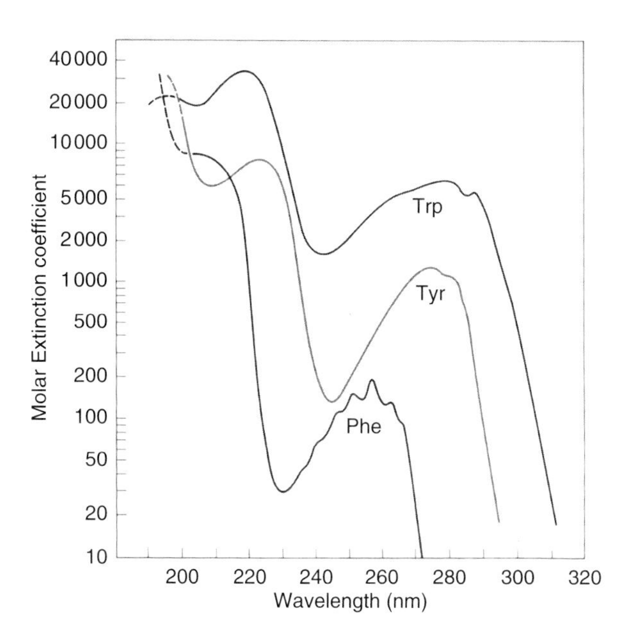

Figure 10.41 Detailed absorbance spectra in the UV visible region for the aromatic amino acids Tyr, Trp and Phe. In proteins a summation of these profiles is observed reflecting the number of residues (reproduced and adapted with permission from Wetlaufer, D.B. *Adv. Prot. Chem.* 1962, **17**, 304–390. Academic Press)

a wavenumber. The wavenumber ($\bar{\nu}$) is defined as

$$\bar{\nu} = 1/\lambda = \nu/c \qquad (10.31)$$

with molecular transition frequency (λ) expressed in cm^{-1}, and where c is the velocity of light. In contrast, biochemists usually plot intensity versus wavelength, although in view of the relationship $c = \nu\lambda$ it is straightforward to convert these equations.

Fluorescence

Fluorescence is the process by which electronically excited molecules decay to the ground state via the emission of a photon without any change in spin multiplicity. Emission is detected by spectrofluorimeters and occurs at longer wavelengths than the corresponding absorbance band (Figures 10.42 and 10.43).

The quantum yield of fluorescence emission is defined as the ratio of photons emitted through

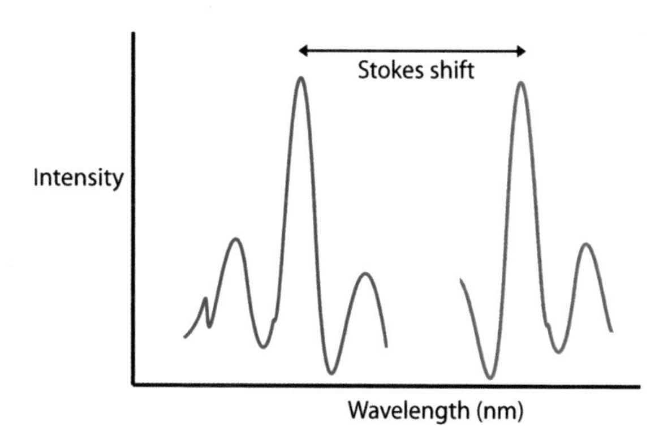

Figure 10.42 Absorbance and fluorescence spectra for a chromophore. The difference between the peaks of maximum intensity is called the Stokes shift

fluorescence to the number of photons absorbed. Its maximum value is 1, although other processes contribute to deactivation of excited molecules and lower values are observed.

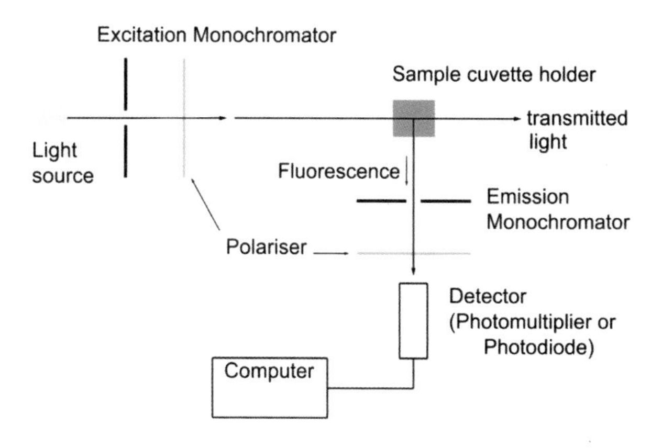

Figure 10.43 A basic instrumental arrangement for detection of fluorescence. The light source is a deuterium- or xenon-based lamp with the wavelength created by the use of diffraction gratings. Light is passed through one or more adjustable excitation monochromators and strikes a sample cuvette. Fluorescence, measured at right angles to the incident beam, is recorded by a photomultiplier sensitive between ~200–1000 nm. Polarizers can be placed into the light path of both excitation and emission beams for measurement of fluorescence polarization

In proteins the principal fluorophore is the indole side chain of tryptophan with a contribution that exceeds the other aromatic residues, and dominates the fluorescence of proteins between 300 and 400 nm. Folded states of proteins show different fluorescence spectra to unfolded states due to the influence of local environment and solvent. In the folded state tryptophan emission shows a blue shift towards 330 nm from a 'free' or unfolded value of ~350 nm. Changes in intensity and wavelength reflect the influence of local environment on emission. Other aromatics side chains or molecules such as heme may quench fluorescence whilst collisional quenching by small molecules such as oxygen, iodide and acrylamide is an important effect.

Collisional quenching leads to excited state deactivation and is a nanosecond process occurring via diffusion-controlled encounters. Aromatic molecules on protein surfaces are quenched more effectively than buried groups with the fluorescence intensity decaying exponentially with time. The intensity at time t is given by

$$I = I_o e^{-t/\tau} \tag{10.33}$$

where I_o is the intensity at time $t = 0$ and τ is the mean lifetime of the excited state. The mean lifetime is therefore the time decay for the fluorescence intensity to fall to 1/e. Measurement of fluorescent lifetimes is possible with sensitive photon-counting fluorimeters and is reflected by time constants ranging from a few picoseconds to hundreds of nanoseconds. These measurements shed light on molecular motion and fluorophore environments.

Collisional quenching can determine the accessibility of fluorophores in proteins via the Stern–Volmer analysis. It is modelled as an additional process deactivating the excited state. Collisional quenching depopulates the excited state leading to decreased fluorescence lifetimes. The dependence of emission intensity, F, on quencher concentration [Q] is given by the Stern–Volmer equation

$$F_o/F = \tau_o/\tau = 1 + k_q\tau_o[Q] \tag{10.34}$$

$$= 1 + K_{SV}[Q] \tag{10.35}$$

where τ and τ_o are the lifetime in the presence and absence of quencher, k_q is the bimolecular rate constant for the dynamic reaction of the quencher with the fluorophore and the product $k_q\tau_o$ is called the Stern–Volmer constant or K_{SV}. A low value for K_{SV} indicates that residues (fluorophores) have low solvent exposure and are buried within a protein.

In many proteins tyrosine fluorescence is effectively quenched by tryptophan as a result of energy transfer occurring primarily from dipole–dipole (acceptor–donor) interactions where the magnitude is governed by the distance, intervening media and orientation. Semiquantitative models of fluorescence resonance energy transfer allow estimation of distances between donor and acceptor. Energy transfer involving a dipolar (through space) coupling mechanism of the donor excited state and the acceptor is known as Forster energy transfer and varies as the inverse sixth power of the intervening distance.

Fluorescence polarization measurement associated with emission intensity and fluorescence anisotropy provide insight into the mobility of fluorophores in proteins. Fluorescence polarization occurs when linearly polarized light is used to excite molecules. Molecules oriented in parallel to the direction of propagation are preferentially excited; for fixed or rigid molecules arranged in parallel this will yield maximum polarization whilst motion associated with the fluorophore leads to randomized orientations and depolarization. Depolarization is also sensitive to the rotational motion or 'tumbling' of molecules. This Brownian motion is described by a rotational correlation time (τ_c) that to a first approximation is estimated assuming spherical proteins of known radius. The correlation time reflects a relaxation process and if it is shorter than the fluorescence decay time leads to molecular reorientation between the events of absorbance and emission.

The degree of fluorescence polarization, P, is defined as

$$P = (I_\parallel - I_\perp)/(I_\parallel + I_\perp) \tag{10.37}$$

where $I_\parallel$ is the fluorescence intensity measured with polarization parallel to the absorbed plane-polarized radiation, and $I_\perp$ is the fluorescence intensity measured perpendicular to the absorbed radiation. A rigid system leads to a maximum value of 1.0 whilst a system exhibiting complete depolarization due to molecular tumbling has a value of −1. These values are theoretical and are rarely, if ever, observed in solution. Typical rotational correlation times for proteins in solution are of the order of 1–100 ns.

Frequently, proteins show partial polarization or are completely unpolarized as a result of tumbling. In addition, internal motion, protein concentration, solvent viscosity, temperature, as well as resonance energy transfer can lead to substantial alterations in polarization. The most common method of analysing fluorescence polarization involves the Perrin equation

$$1/P - 1/3 = (1/P_o - 1/3)(1 + 3\tau/\tau_c) \qquad (10.38)$$

where τ is the fluorescence lifetime, τ_c is a correlation time reflecting the rotational relaxation rate due to molecular tumbling and is equivalent to

$$\tau_c = 4\pi\eta r^3/k_B T \qquad (10.39)$$

where η is the solvent viscosity and r is the hydrodynamic radius. P_0 represents the intrinsic polarization observed under conditions promoting least molecular motion such as high solvent viscosity and low protein concentration. For non-spherical proteins the equation should be re-written as $\tau_c = 3\eta V/RT$ where V is the volume.

Fluorescence polarization depends on the rotational correlation time *and* the fluorescence lifetime and is difficult to use if there is no knowledge of the lifetime (τ) or the limiting value of P_o. Fluorescence anisotropy decays as the sum of exponentials and is used in time resolved measurements. Anisotropy is defined as the ratio of the difference between the emission intensity parallel to the polarization of the electric vector of the exciting light ($I_\parallel$) and that perpendicular ($I_\perp$) divided by the total intensity (I_T), where $I_T = I_\parallel + 2I_\perp$

$$A = (I_\parallel - I_\perp)/(I_\parallel + 2I_\perp) \qquad (10.40)$$

The emission anisotropy (A) is related to the correlation time of the fluorophore (τ_c) through the Perrin equation

$$A_o/A = 1 + \tau/\tau_c \qquad (10.41)$$

where A_o, is the limiting anisotropy of the fluorophore and depends on the angle between the absorption and emission transition dipoles, and τ is the fluorescence lifetime. Fluorescence anisotropy is used to probe tryptophan residues in proteins as well as motion of covalently attached fluorophores. Measurement of tryptophan anisotropy reveals that proteins often exhibit additional motion associated with Trp residues and has been attributed to rotation of the indole side chain about the CA–CB bond with models involving precession of the side chain within a cone.

Green fluorescent protein

The green fluorescent protein (GFP) is responsible for the natural green 'colour' exhibited by the jellyfish, *Aequorea victoria* and other coelenterates, where it gives cells a characteristic glow visible on the surface of sea at sunset. Details of the structure of GFP (Figure 10.44) have revealed an 11-stranded antiparallel β sheet forming a barrel-like structure of diameter 3.0 nm and length 4.0 nm and that has been likened to a 'lantern'. GFP is a remarkably stable protein, resistant to proteases and denatured only under extreme conditions such as 6 M guanidine hydrochloride at 90 °C. The lantern functions to protect the 'flame' which is positioned in a single α helix located towards the centre of the β barrel. The 'flame' responsible for the intrinsic fluorescence of GFP is *p*-hydroxy-benzylideneimidazolinone and results from cyclization of the tripeptide Ser65-Tyr66-Gly67 and 1,2-dehydrogenation of the Tyr.

GFP absorbs blue light around 395 nm preferentially but also exhibits a smaller absorbance peak at 475 nm (ε of ~30 000 and 7000 $M^{-1}cm^{-1}$ respectively). In the coelenterate *A. victoria* GFP absorbs blue light by fluorescence energy transfer from a second protein aequorin converting the incident 'blue' light via energy transfer into green light. *Aequorea* are luminescent jellyfish observed to glow around the margins of their umbrellas as a result of light produced from photogenic cells in this region. Exposure to physical stress causes the intracellular calcium concentrations to increase leading to an excited state of aequorin generated through Ca^{2+} binding to the pigment coelenterazine. The light produced by aequorin generates GFP fluorescence by energy transfer, although the physiological basis of these reactions is unclear (Figure 10.45).

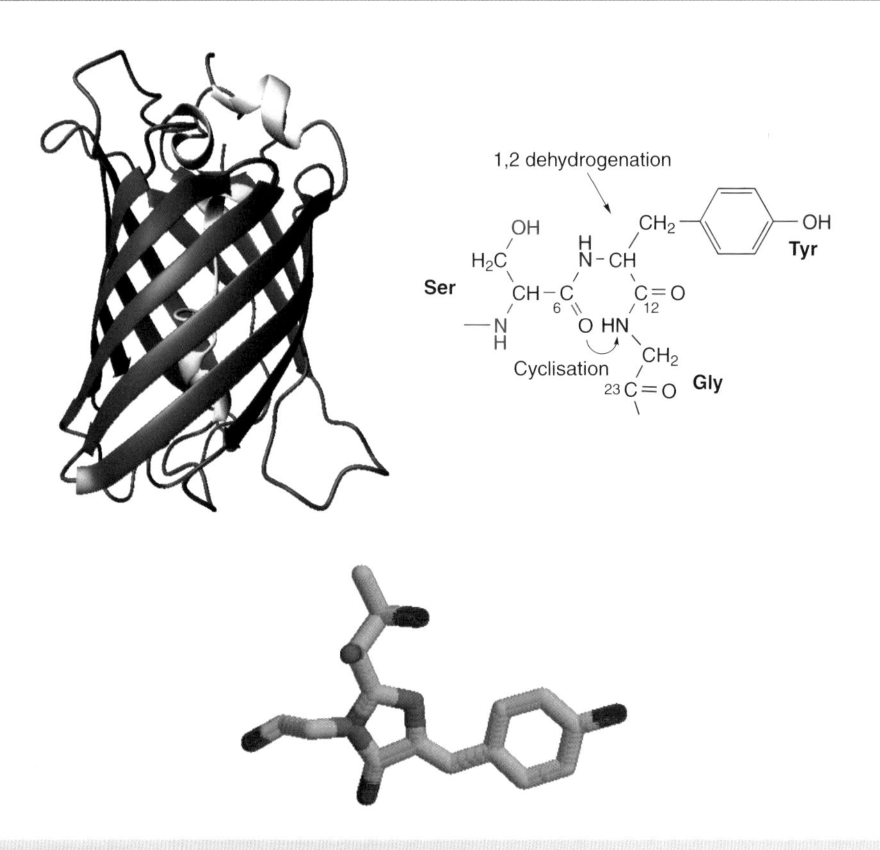

Figure 10.44 The structure of GFP showing 11-stranded β barrel (PDB:1EMA). The cyclic *p*-hydroxybenzylideneimidazolinone fluorophore

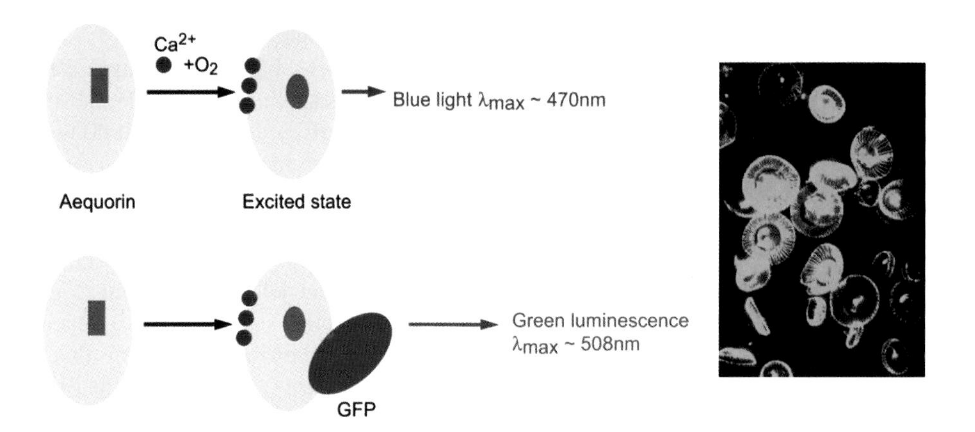

Figure 10.45 GFP fluorescence. *In vitro* mixing of aequorin and Ca^{2+} produces blue light ($\lambda_{max} \sim 470$ nm). The blue fluorescent form of aequorin represents an excited state that decays to the ground state. In the presence of GFP energy transfer occurs with a green luminescence observed comparable to that seen in the living animal (right) (reproduced with permission from Yang F., Moss L.G. & Phillips G.N. *Nat. Biotechnol.* 1996, **14**, 1246–1251)

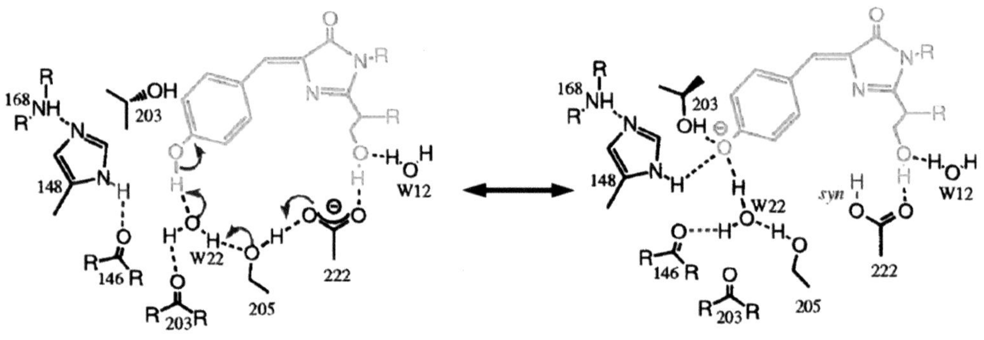

395 nm absorbance (protonated)

475 nm absorbance (deprotonated)

Figure 10.46 The structure at the active site of GFP associated with the 398 and 475 nm absorbance bands (reproduced with permission from Brejc, K. *et al. Proc. Natl Acad. Sci.* 1997, **94**, 2306–2311)

Absorbance of GFP around 398 nm leads to a fluorescence emission peak at 509 nm with a quantum yield between 0.72–0.85, and hence the green colour observed *in vivo* and *in vitro*. The intensity ratio between the absorbance peaks is sensitive to factors such as pH, temperature, and ionic strength, suggesting the presence of two different forms of the chromophore. It has been shown that the two forms differ in their protonation states with the 398 nm band reflecting a protonated fluorophore whilst the 475 nm band reflects deprotonation of this group (Figure 10.46).

The unique fluorescence of GFP allows the expressed protein to be fused to other domains creating a chimeric protein with a fluorescent tag that can be measured and has been used in N- and C-terminal fusions to follow gene expression, protein–protein interactions, cell sorting pathways, and intracellular signalling.

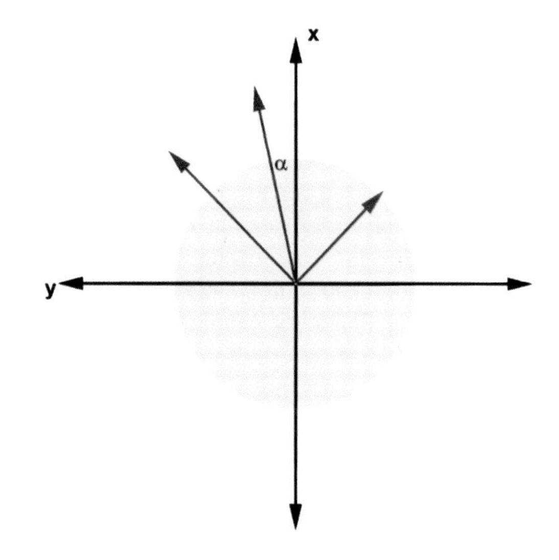

Figure 10.47 The resultant rotation of ε when ε_R and ε_L make unequal angles with the *x*-axis. A levorotatory chiral centre exists causing a rotation of angle α

Circular dichroism

Circularly polarized light travels through optically active media with different velocities due to different indices of refraction for right (dextro) and left (laevo) components (Figure 10.47). This is called optical rotation and the variation of optical rotation with wavelength is known as optical rotary dispersion (ORD). ORD is normally measured at a specific wavelength and temperature and is usually denoted by a parameter α, the angle of rotation of polarized light.

The right and left circularly polarized components are also absorbed differentially at some wavelengths due to differences in extinction coefficients for the two polarized components (Figure 10.48). The addition of the left and right components yields ε at every point from the relationship $\varepsilon = \varepsilon_L + \varepsilon_R$. When this light is

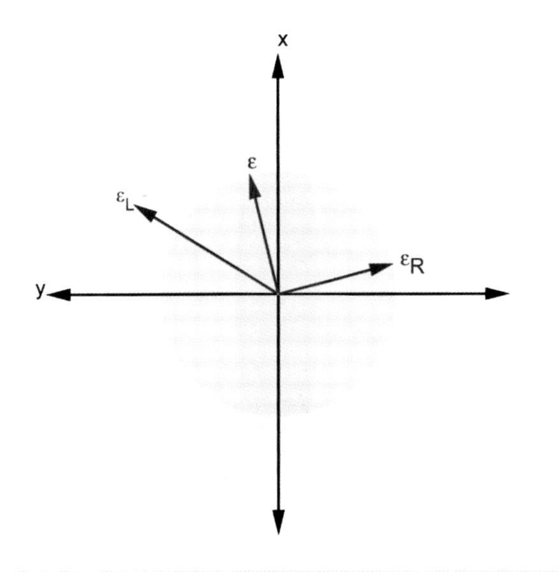

Figure 10.48 The variation of ε for a circular dichroic sample; the vector traces out an ellipse

passed through an optically active medium the left and right circularly polarized waves are rotated at different speeds due to small differences in molar absorptivity for the left and right components. The differential absorption of the left and right circularly polarized light gives rise to the technique of circular dichroism (CD).

CD spectroscopy measures differences in absorbance of right and left circularly polarized light. In proteins the optically active absorption bands tend to be located in the far-UV or near-UV regions of the spectrum and involve $\pi \rightarrow \pi^*$, $n \rightarrow \pi^*$, and $\sigma \rightarrow \sigma^*$ transitions. The chromophores responsible for these transitions are either intrinsically asymmetric or more frequently asymmetric as a result of their interactions with the protein. CD spectroscopy is of considerable importance to the study of proteins as a quantitative tool in the estimation of secondary structure. This arises because regularly repeating helix or sheet produce different CD signals. In a protein CD signals reflect the sum of the secondary structure components and it has proved possible to estimate the proportions of helix, strand or turn in a protein of unknown structure with remarkable precision.

Historically, ellipticity is the unit of CD and is defined as the tangent of the ratio of the minor to major elliptical axis. An axial ratio of 1:100 will therefore result in an ellipticity of 0.57°. In CD spectroscopy the measured parameter is the rotation of polarized light expressed with the units of millidegrees. Modern CD instrumentation is capable of precision down to thousandths of a degree (1 millidegree). There are a number of ways of expressing the CD signal of a sample although frequently many incorrect forms are used in the literature. The observed CD signal (S) is expressed in millidegrees and is normally converted to $\Delta\varepsilon_m$ or $\Delta\varepsilon_{mrw}$, where $\Delta\varepsilon_m$ is the molar CD extinction coefficient and $\Delta\varepsilon_{mrw}$ is the mean residue CD extinction coefficient, respectively. In any CD experiment it is vital to know the protein concentration extremely accurately to avoid significant error. It can be shown that

$$\Delta\varepsilon_m = S/(32980Cl) \qquad (10.42)$$

where C is the concentration in mol dm^{-3} (or M) and l is the path length in cm. The light path is usually a value between 0.1 and 1.0 cm. The units of $\Delta\varepsilon_m$ are therefore $M^{-1}cm^{-1}$ and the analogy with the molar extinction coefficient determined via absorbance measurement is clear. Alternatively, expressing the ellipticity as a mean residue CD extinction coefficient leads to

$$\Delta\varepsilon_{mrw} = Smrw/(32980Cl) \qquad (10.43)$$

where the mean residue weight (mrw) is the molecular weight divided by the number of residues. In this instance C is expressed in mg ml^{-1}. CD intensities are sometimes reported as molar ellipticity ($[\theta]_M$) or mean residue ellipticity ($[\theta]_{mrw}$). These terms are calculated from the following equalities

$$[\theta]_M = S/(10Cl) \qquad (10.44)$$

where C is again the concentration in mol dm^{-3} or

$$[\theta]_{mrw} = Smrw/(10Cl) \qquad (10.45)$$

Both $[\theta]_M$ and $[\theta]_{mrw}$ have the units degrees cm^2 dmol^{-1}. $[\theta]$ and $\Delta\varepsilon$ may be inter-converted using the relationship

$$[\theta] = 3298\Delta\varepsilon \qquad (10.46)$$

CD extinction coefficients are the most logical unit since they are direct analogs of extinction coefficient in absorbance measurements and lead to values of $\Delta\varepsilon_{mrw}$ in an approximate range ± 20 whilst the corresponding values for $\Delta\varepsilon$ range from ± 3000. Using the above relationships it will be apparent that values of $[\theta]_{mrw}$ values are in the range $\pm 70\,000$.

In the determination of secondary structure content one approach is to assume that spectra are linear combinations of each contributing secondary structure type ('pure' α helix, 'pure' β strand, Figure 10.49) weighted by its relative abundance in the polypeptide conformation. Unfortunately the problem with this approach is that there are no standard reference CD spectra for 'pure' secondary structure. More significantly synthetic homopolypeptides are poor models of secondary structure and most homopolymers do not form helices nor is there a good example of a 'model' β strand. An empirical approach involved determining the experimental CD spectra of proteins for which the structures are already known. Using knowledge of the structure the content of helix, turn and strand is defined accurately and by using a database of reference proteins these methods prove accurate and reliable when applied to unknown samples. A number of different mathematical procedures have been adopted using reference proteins of varying size and secondary structure content and all of these methods give similar results.

A variation of CD occurs when samples are placed in magnetic fields maintained at low temperatures (<77 K). Under these conditions all molecules exhibit CD spectra and the technique is called magnetic circular dichroism (MCD). In most cases the spectra of proteins are too complex to interpret fully but for metalloproteins the MCD technique provides a powerful tool with which identify ligands and metal ions.

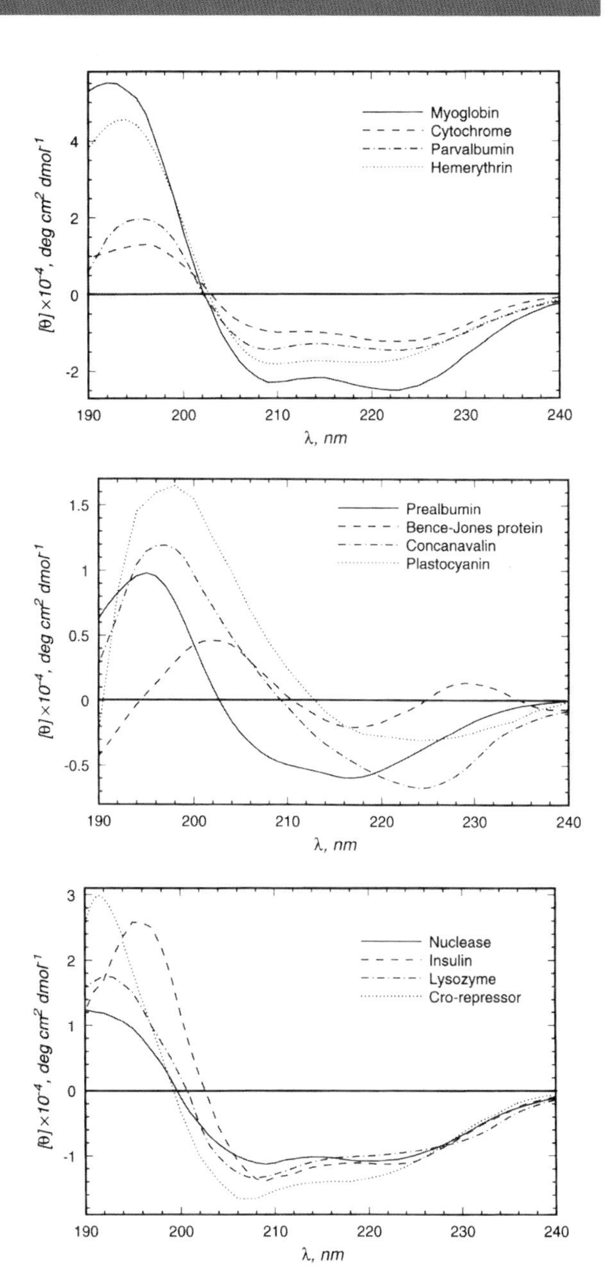

Figure 10.49 CD spectra of proteins with diverse secondary structure content. Top: extensively α helical proteins; centre: proteins rich in β strands; bottom: protein with (α + β) structure (reproduced with permission from *Circular Dichroism* – the conformational analysis of biomolecules, Fasman, G.D (ed) Kluwer Academic 1996.)

Vibrational spectroscopy

The vibrational spectra of proteins are extremely complex and lie at lower frequencies than electronic spectra within the infrared (IR) region. It is useful to divide the IR region into three sections, the near, middle and far IR regions with the most informative zone located at wavenumbers between 4000 and 600 cm^{-1} (Table 10.12). Historically, IR

Table 10.12 The relationship between wavenumber and wavelength for the near, middle and far IR regions

Region	Wavelength range (μm)	Wavenumber range (cm^{-1})
Near	0.78–2.5	12 800–4000
Middle	2.5–50	4000–200
Far	50–1000	200–1000

It is worth noting that 750 nm is located at the far red end of the visible spectrum. This is 0.75 μm or a wavenumber of $1/0.75 \times 10^{-4}$ cm^{-1} or 13 333 cm^{-1}.

spectroscopy uses wavenumbers to denote frequency. In any molecule the atoms are not held in rigid or fixed bonds but move in a manner that is reminiscent of two bodies attached by a spring. Consequently all bonds can either bend or stretch about an equilibrium position that is defined by their standard bond lengths (Figure 10.50).

Exposure to IR radiation with frequencies between 300 and 4000 cm^{-1} leads to the absorption of energy and transition from the lowest vibrational state to a higher excited state. In a simple diatomic molecule there is only one direction of stretching or vibration and this leads to a single band of IR absorption. Atoms that are held by weak bonds require less energy to reach excited vibrational modes and this is reflected in the frequencies associated with IR absorbance. As molecules increase in complexity more modes of vibrations exist with the appearance of more complex spectra. For a linear molecule with n atoms, there are $3n - 5$ vibrational modes whilst if it is non-linear it will have $3n - 6$ modes. Water is a non-linear molecule containing three atoms and has three modes of vibration in IR spectra (Figure 10.51).

There is one more condition that must be met for a vibration to be IR active and this requires changes to the electric dipole moment during vibration ($d\mu/dr \neq 0$). This arises when two oppositely charged atoms move positions. By treating a diatomic molecule as a simple harmonic oscillator Hooke's law can be used to calculate the frequency of radiation necessary to cause a transition. For a simple harmonic oscillator the

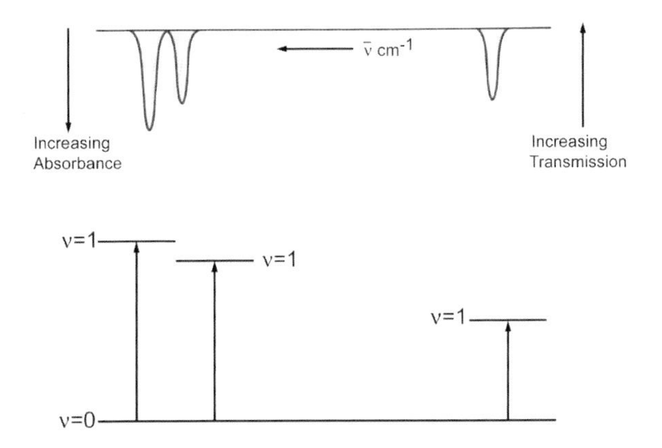

Figure 10.51 The IR spectrum of water showing three fundamental modes of vibration. At higher resolution each peak will show fine structure as a result of simultaneous transitions between different rotational levels

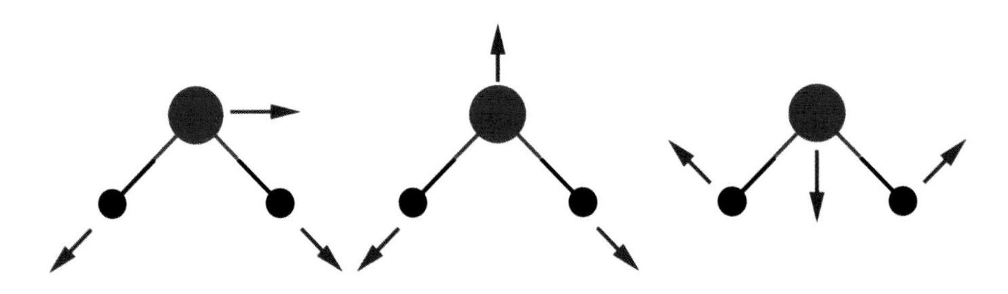

Figure 10.50 Bond stretching and bending in a simple molecule (H_2O). Ignoring fine structure arising from simultaneous transitions the IR spectrum of water has three peaks at 3943, 3832 and 1648 cm^{-1}

frequency of vibration (ν) is given by

$$\nu = \frac{1}{2\pi}\sqrt{\frac{k}{\mu}} \tag{10.47}$$

where k is the force constant determined by bond strength, and μ is the reduced mass. The term μ, the reduced mass is equivalent to $(m_1 + m_2)/m_1 m_2$ and leads to

$$\nu = \frac{1}{2\pi}\sqrt{\frac{k\, m_1 m_2}{m_1 + m_2}} \tag{10.48}$$

From the above equation it can be seen that a strong bond between atoms leads to bands at higher frequency; IR spectra of C≡C bonds absorb between 2300 and 2000 cm^{-1}, C=C bonds occur at 1900–1500 cm^{-1} whilst C–C occur below 1500 cm^{-1} and the spectral region becomes crowded and complex. Despite complexity it is called the fingerprint region and is diagnostic for small biomolecules.

Fourier transform IR spectroscopy measures the vibrations associated with functional groups and highly polar bonds within proteins. In proteins these groups provide a biochemical 'fingerprint' made up of the vibrational features of all contributing components. Vibrational spectra associated with proteins are complex but the common components contributing to the IR spectrum include at least seven amide bands denoted as amide I–amide VII representing different vibrational modes associated with the peptide bond. For most observation on proteins only the first three amide bands denoted as amide I, amide II and amide III are of significant interest. Of these the amide I band is most widely used for secondary structure analyses. Amide I is the result of C=O stretching of the amide group coupled to the bending of the N–H bond and the stretching of the C–N bond. These vibrational modes give IR bands between 1600–1700 cm^{-1} and are sensitive to hydrogen bonding and to the secondary structure environment. Amide I bands centred around 1650 to 1658 cm^{-1} are usually believed to be characteristic of groups located within α helices. Unfortunately, disordered regions of polypeptides as well as turns can also give amide I bands in this region and this can complicate assignment. In contrast β strands give highly diagnostic bands in the region located from 1620–1640 cm^{-1}. Parallel and antiparallel β strands are sometimes distinguishable with antiparallel regions showing a large splitting of the amide I band due to the interactions between strands.

Water (H_2O) is a very strong infrared absorber with prominent bands centred at wavenumbers of ~3800–3900 cm^{-1} (H–O stretching band), 2125 cm^{-1} (water association band) and one at 1640 cm^{-1} (the H–O–H bending vibration) lying in a major spectral region of interest (the conformationally sensitive amide I vibrations). Despite the high s/n ratios, accurate frequency determination, speed and reproducibility of modern instrumentation it is frequently necessary to perform measurements in aqueous solution around 1640 cm^{-1} in D_2O where no strong bands are close to the amide I frequency. Measurements performed on the amide I band in D_2O and hence on deuterium substituted proteins are often distinguished as amide I′. One advantage IR spectroscopy shares with CD is that the technique is readily performed on relatively small amounts of material (often ~100 μl) and represents a good starting point for new structural investigation of a protein. For both methods increasingly successful attempts to extract quantitative information on protein secondary structure have been made. In IR spectroscopy analysis of the amide I bands has proved successful and has permitted deconvolution of spectra into components attributable to helical, strand and turn elements of secondary structure. As a result IR like CD provides a means of estimating the secondary structure present in proteins.

Raman spectroscopy

Raman spectroscopy is concerned with a change of frequency of light when it is scattered by molecules. The Raman technique owes its name to Chandrasekhara Venkata Raman who discovered in 1928 that light interacts with molecules via absorbance, transmission *or* scattering. Scattering can occur at the same wavelength when it is known as Rayleigh scattering or it can occur at altered frequency when it is the Raman effect (see Figure 10.52).

Raman's discovery was that when monochromatic light is used to irradiate a sample the spectrum of scattered light shows a pattern of lines of shifted frequency known as the Raman spectrum. The shifts are independent of the excitation wavelength and are

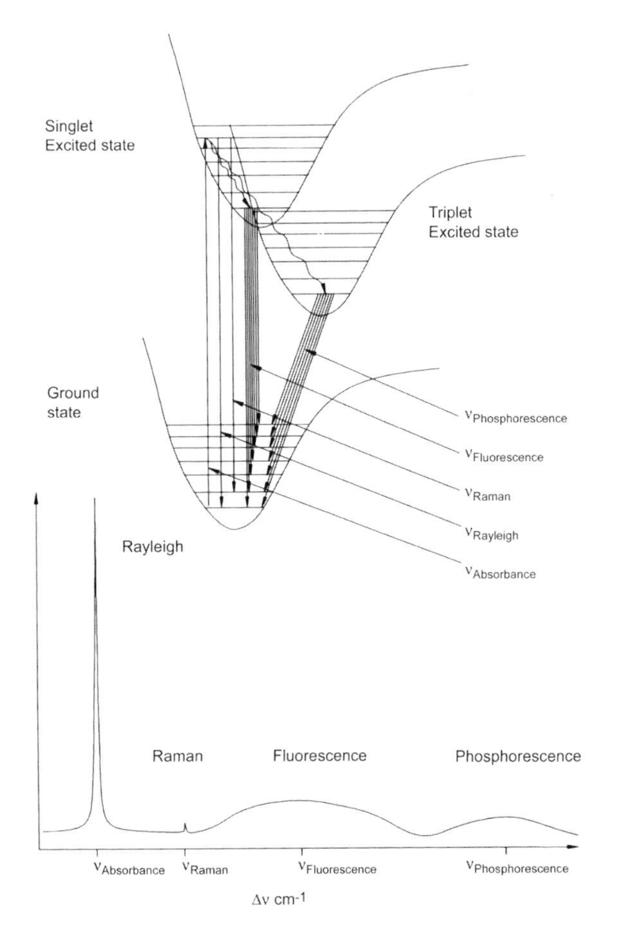

Figure 10.52 Relationship between Raman, Rayleigh, absorbance and fluorescence (adapted from Wang, Y. & van Wart H.E. *Methods Enzymol.* 1993, **226**, 319–373. Academic Press)

characteristic for the molecule under investigation. Collision of a photon with a molecule leads to elastic and inelastic scattering; the former contributes to the Rayleigh line whilst inelastic collisions cause quantum transitions of lower and higher frequency and are responsible for the negative and positive shifts observed in Raman spectroscopy corresponding to vibrational and rotational transitions within scattering molecules. The two types of Raman transitions are sometimes called Stokes lines. These changes are normally observed directly in the IR region of the spectrum but they can also be observed as frequency shifts within the more accessible UV or visible regions. Raman signals from aqueous protein samples are easily measured with signals from water being relatively weak. A much greater problem derives from fluorescence associated with proteins that swamps smaller Raman signals.

A variation on the basic technique of Raman spectroscopy is called resonance Raman spectroscopy. The technique is often applied to metalloproteins where electronic transitions associated with chromophore ligands to metal centres produce enhancements to the basic Raman effect. Vibrational modes associated with the ligand are dramatically enhanced often by factors of 10^3 or more and have proved extremely useful in examining the structural organization of metal centres within proteins. Since these centres are often found at the catalytic core of an enzyme resonance Raman has facilitated understanding the chemistry occurring within these sites. The techniques of fluorescence, Raman and absorbance are related via excitation of an electron to an excited state followed by its decay back to the ground state (Figure 10.52).

ESR and ENDOR

Electron spin resonance (ESR) spectroscopy measures a comparable process to NMR spectroscopy with the exception that changes in *electron* magnetic moment are measured as opposed to changes in nuclear spin properties. Despite this difference much of the basic theory of NMR applies to ESR spectroscopy.[4] Electrons possess a property called spin that leads to the generation of a magnetic moment when an external magnetic field is applied. Spin is quantized with an electron spin quantum number $M_s = \pm 1/2$. In the presence of an applied magnetic field degeneracy is removed and leads to transitions between the two states (Figure 10.53). The resonance condition is expressed as

$$\Delta E = h\nu = g\beta H \tag{10.49}$$

where g is a dimensionless constant (often called the Lande g factor and is equal to 2.00023) and β is the

[4]The terms ESR and EPR are used interchangeably and both terms are acceptable and used by authors in the subject literature. EPR = electron paramagnetic resonance.

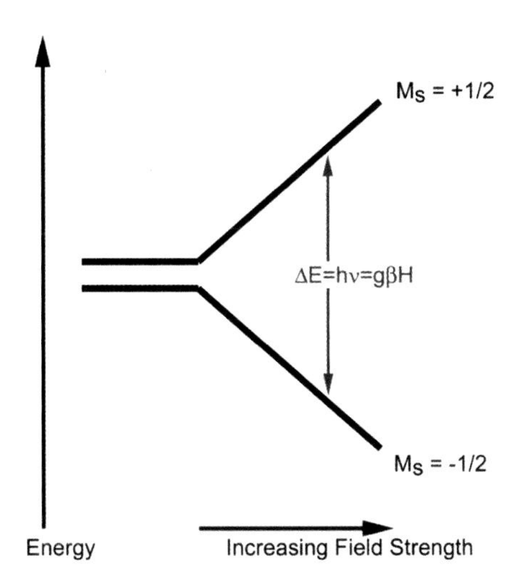

Figure 10.53 The resonance condition for ESR spectroscopy. Such a transition would give rise to a single line. For instrumental reasons ESR spectra are presented as first derivative spectra

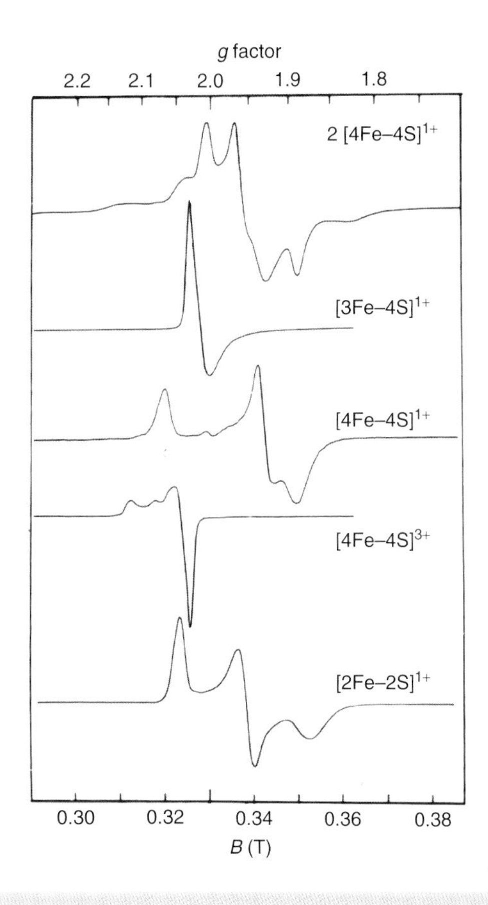

Figure 10.54 Representative ESR spectra for iron–sulfur clusters found in ferredoxins. From top to bottom the spectra reveal the properties of the iron centres in proteins from *Clostridium pasteurianum*, *Desulfovibrio gigas*, *Bacillus stearothermophilus*, *Chromatium vinosum* high potential protein and *Mastogocladus laminosus*. All spectra were recorded at X-band frequencies at a temperature between 10 and 20 K

Bohr magneton. The electron magnetic moment is 600 times greater than that of the proton. ESR experiments are normally carried out at a field strength of 3400 G (0.34 T) and a frequency of ~9 GHz (known as the X-band) that lies in the microwave region of the electromagnetic spectrum. In ESR the frequency is usually fixed and spectra are acquired by sweeping the applied magnetic field whilst measuring concomitant changes in resonance absorption.

A fundamental requirement for ESR is the presence of one or more unpaired electrons. In proteins this occurs frequently for transition metals such as Fe, Co, Mn, Ni and Cu with unpaired d electrons. Routinely ESR spectra are recorded at low temperatures, usually below 50 K, through the use of liquid helium cooled probes. Rapid spin lattice relaxation rates at elevated temperatures are slowed at lower temperatures leading to increased signal intensity and less saturation. Aqueous samples absorb microwaves strongly and freezing results in the formation of ice a less efficient microwave absorber.

Iron exists in three different oxidation states in biological systems. Usually the ferrous (d^6) and ferric (d^5) configurations are encountered but occasionally enzymes such as peroxidases or cytochromes P450 have a ferryl (d^4) state. The ferric iron form occurs as an unpaired high spin ($S = 5/2$) state or a partially paired low spin ($S = 1/2$) state that are distinguished using ESR spectrometry. In ferredoxins, metalloproteins containing iron–sulfur clusters, ESR has revealed a variety of different clusters {[Fe-Cys4], [2Fe-2S], [3Fe-3S] and [4Fe-4S] centres}, interacting redox centres, redox state transitions, as well as ligand identity, coordination and geometry (Figure 10.54).

Table 10.13 International and national facilities for X-ray crystallography and NMR spectroscopy

Facility and role	Web address
Synchrotron sources	
Daresbury UK. Crystallography and CD	http://www.srs.ac.uk/srs/
Cornell USA. Crystallography	http://www.chess.cornell.edu/
European facility Grenoble France. X-ray and Neutron diffraction	http://www.esrf.fr/
NMR facilities	
Madison USA. Ultra high field NMR spectroscopy	http://www.nmrfam.wisc.edu/
Frankfurt Germany. EU funded large scale facility for NMR and ESR	http://www.biozentrum.uni-frankfurt.de/LSF/

Although ESR spectroscopy has been most widely used to characterize iron-containing proteins it has also been used to study: (i) the metal centre of Ni^{2+} (d^8) in enzymes like urease, hydrogenases and carbon monoxide dehydrogenase; (ii) the wide variety of Cu^{2+} (d^9) centres such as those found in superoxide dismutase, tyrosinase and haemocyanin; (iii) Mo centres such as that found in enzymes like xanthine oxidase; and (iv) Mn centres found within photosynthetic systems involved in oxygen evolution.

Recent developments in ESR spectroscopy utilize advantages of pulse methods for acquiring spectra as opposed to sweeping the magnetic field. ESR spectroscopy has traditionally been a continuous wave (CW) technique but techniques such as ESEEM (electron spin echo envelope modulation) or ENDOR (electron-nuclear double resonance spectroscopy) probe electron–nuclear interactions in metalloproteins unsuitable for study using crystallography or NMR. The techniques can derive relaxation times and information on metal–ligand mobility, distance and orientation.

Armed with several of the above spectroscopic tools the structure of almost all proteins can be gradually determined. The time scale from isolation of a protein to determination of its three-dimensional structure has decreased dramatically from several years to perhaps a few months. In many cases acquisition of experimental data takes only a few days whilst analysis and interpretation of the raw data is often a longer process. Unfortunately the cost of instrumentation required to perform these studies at the highest level has become extremely prohibitive. Ultra high field NMR spectrometers, powerful electron microscopes and synchrotron beam lines are beyond the scope of most individual institutions, with a single instrument costing millions of pounds. Increasingly this instrumentation is being organized around core national, and sometimes internationally coordinated centres of excellence where these facilities serve many research groups (Table 10.13).

Summary

By exploiting different regions of the electromagnetic spectrum and the interaction of atoms with radiation considerable information can be acquired on the structure and dynamics of proteins.

The plethora of techniques available for studying proteins means that if one technique is unsuitable there is almost certainly another method that can be applied.

X-ray crystallography and multi-dimensional NMR spectroscopy yield detailed pictures of protein structure at an atomic level. X-ray crystallography relies on the diffraction of X-rays by electron dense atoms constrained within a crystal. Some proteins fail to crystallize and this represents the major limitation of the technique.

The 'phase problem' is overcome through the use of isomorphous replacement but a basic requirement is that addition of heavy metal atoms does not alter protein structure. The structures derived for the proteasome, the ribosome and viral capsids serve to emphasize the success of the technique.

Membrane proteins or proteins with extensive hydrophobic domains are notoriously difficult to crystallize, although several structures now exist within the Protein Databank.

NMR spectroscopy measures nuclear spin reorientation in an applied magnetic field most frequently for spin 1/2 nuclei. In proteins this centres around the ^{1}H but with isotopic labelling can also include ^{15}N and ^{13}C nuclei.

A major hurdle in NMR spectroscopy is the assignment problem – identifying which resonances belong to a given residue. Sophisticated multidimensional heteronuclear NMR methods have been developed to facilitate this process based around the interaction of nuclei 'through bonds' and 'through space'.

Solution structure is defined from the use of a combination of torsion angle and distance restraints. The former are obtained from J coupling experiments whilst the r^{-6} dependence of the NOE is the basis of distance constraints.

NMR structures are usually presented as a family or ensemble of closely related protein topologies. Although most structures for proteins determined by NMR spectroscopy have masses below ~20 kDa the size limit is increasing steadily with the introduction of new methods.

The vast majority of structures deposited in the Protein Databank (>95 percent) have been determined using crystallography or NMR spectroscopy, but slowly a third technique of cryo-EM is gaining prominence.

Cryo-EM methods have the advantage of requiring little sample preparation and are particularly suitable to very large protein complexes. The technique relies on electron diffraction by particles immersed in a frozen lattice. Single particle analysis is the current 'hot' area and involves trapping macromolecular complexes in random orientations within vitreous ice and combining thousands of images to provide an enhanced picture of the system.

Optical techniques based around electronic transitions provide information on the absorbance and fluorescence of chromophores found in proteins usually aromatic residues. Tryptophan exhibits a significantly greater molar extinction coefficient when compared to either tyrosine or phenylalanine with the result that most of a protein's absorbance or fluorescence reflects the relative abundance of Trp within a polypeptide chain.

Time-resolved measurements allow changes in fluorescence or absorbance to be followed on very short time scales sometimes in the sub-nanosecond range. This allows measurements of protein mobility such as the motion of aromatic side chains.

CD involves the measurement of differential absorption of right and left circularly polarized light as a function of wavelength. In proteins the far UV region from 260–180 nm is dominated by the CD signals due to elements of secondary structure such as helix, turns and strands.

Proteins composed extensively of helical structure have very different CD spectra to those containing proportionally more β strands. CD spectra allow the secondary structure content of unknown proteins to be predicted with reasonable accuracy.

IR spectroscopy of proteins has been used to measure changes in the vibrational states associated with the amide bond. Amide bonds in helices, turns and strands show characteristic transitions in a region known as the amide I band between 1600 and 1700 cm^{-1}. Experimental spectra are fitted as combinations of helices and strands and can be used to estimate secondary structure content in a comparable manner to CD spectroscopy.

Problems

1. Why are fluorescence maxima observed at longer wavelengths than absorbance maxima?
2. Why do peaks in absorbance spectra become narrower when the temperature is lowered to 77 K?

3. What is the wavelength range of IR studies conducted at 300–4000 cm^{-1}.

4. Show that an axial ratio of 1:100 leads to an observed rotation of circularly polarized light of 0.57°.

5. The absorbance of a 10 μM solution of tryptophan in buffer at 280 nm is 0.06. The buffer alone gives an absorbance of 0.004 at 280 nm. Assuming a path length of 1 cm calculate the extinction coefficient of tryptophan. What are the expected absorbance values for pathlengths of 1, 2 and 20 mm?

6. Using the data provided in Table 10.4 calculate the Larmor frequency of the following nuclei ^{1}H, ^{13}C and ^{15}N at field strengths of 20, 18.1, 16, 11.7 and 8 T.

7. Sketch the high resolution 1D ^{1}H-NMR spectra of alanine, glycine and threonine.

8. Draw the expected pattern of connectivities in 2D homonuclear TOCSY and COSY spectra for the residues not included in Figure 10.27. (ignore fine structure for cross peaks).

9. Explain how heavy metal derivatives of proteins might be detected using non-denaturing electrophoretic techniques. What might be the limitations to this technique? Using Harker diagrams show how the use of a second heavy metal derivative leads to a unique solution for the intensity and phase.

10. The structure of cytochrome c′ from *Chromatium vinosum* has been determined and the data has been deposited in protein databanks. Locate the relevant PDB file and from the primary citation identify the methods used to determine the structure together with the unit cell dimensions, the number of molecules per unit cell, the resolution achieved. Describe the structure determined and any new features discovered.

11

Protein folding *In vivo* and *In vitro*

Introduction

To address the mechanisms of protein folding it is necessary to understand thermodynamic and kinetic parameters governing the formation of the native state. Folding is a reversible reaction defined most simply as

$$U \underset{k_2}{\overset{k_1}{\rightleftharpoons}} N \qquad (11.1)$$

where the equilibrium lies to the right ($K_{eq} > 1 = [N]/[U]$) for the native state (N) to be defined as more stable than the unfolded state (U). Quite clearly the respective rates of the forward (k_1) and reverse (k_{-1}) reactions are related to the value K_{eq} and to understand folding requires estimating the magnitude of these rate constants.

Although many proteins fold *in vitro* at comparable rates to those observed *in vivo* the environment and the presence of 'helper' proteins play critical roles in folding within the cell. The realization that some proteins are assisted to find the correct folded structure by molecular chaperones has expanded the area of protein folding.

Protein misfolding is no longer viewed as an experimental artefact. It results from gene mutation and underpins diseases such as Alzheimer's, collagen based diseases such as osteogenesis imperfecta, cystic fibrosis, familial amyloidotic polyneuropathy – a comparatively rare disease arising from mutations in the protein transthyretin – as well as more familiar diseases such as emphysema.

Studies of protein folding have revolutionized our ideas of disease transmission. Until recently it was thought that nucleic acid uniquely carried information 'directing' the transmission of a disease. This occurred either through a mutation in the host genome, integration of viral DNA into a genome or by infection directed by a pathogen's genome. Today there is compelling evidence that some diseases arise solely from a protein and a misfolded one at that!

Factors determining the protein fold

Globular proteins fold into conformations of ordered secondary and tertiary structure where hydrophobic side chains are buried on the inside of the protein and the polar/charged side chains are solvent accessible. Interactions governing the formation of secondary and tertiary structure involve the formation of hydrogen bonds, disulfide bridges, charge–pair interactions and non-polar or hydrophobic effects. The cumulative effect of these forces is that in *any* folded protein the magnitude of favourable interactions outweighs the sum of the unfavourable ones.

These interactions are disrupted by extremes of temperature, immersing proteins in acidic or alkaline solutions or adding solvents such as alcohol in a process known as denaturation that results in a loss

Proteins: Structure and Function by David Whitford

of activity and ordered structure. The disordered state is often described as a 'random coil' although this term should be used sparingly since unfolded states lack truly randomized structure. Denaturation results in a loss of compactness with the unfolded 'state' fluctuating between ensembles of iso-energetic, disordered and extended conformations.

In vitro studies of folding subject proteins to extreme conditions in the expectation that a progressive loss of native structure will be observed. Measuring parameters associated with the kinetics of folding or the stability of the native state allows the process of denaturation to be quantified. Loss of folded structure is usually measured by changes in absorbance or fluorescence although other techniques such as circular dichroism and NMR spectroscopy are increasingly being applied. In most cases folding is a cooperative process arising from simultaneous formation of multiple interactions within a polypeptide chain (Figure 11.1). Individually, each interaction is weak but their cooperative formation drives polypeptide chains towards folded states.

Denaturants like sodium dodecyl sulfate (SDS) are very effective at disrupting protein structure at low concentrations. This is useful for techniques such as SDS–polyacrylamide gel electrophoresis (PAGE) but less desirable in folding studies where a progressive and reversible effect is required. Denaturants such

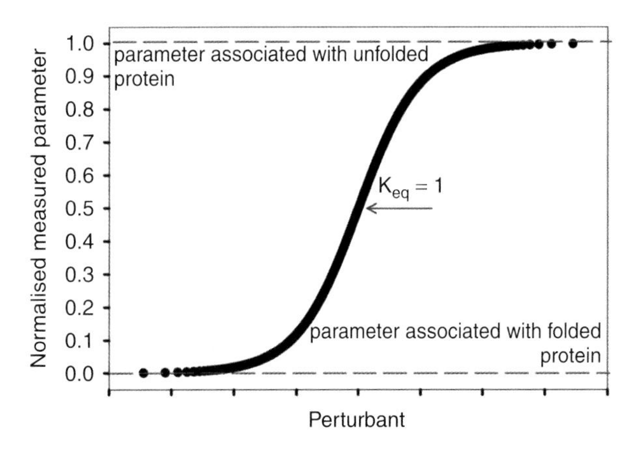

Figure 11.1 The denaturation of a protein showing a cooperative profile. The perturbant can be increasing concentrations of denaturant or increasing temperature

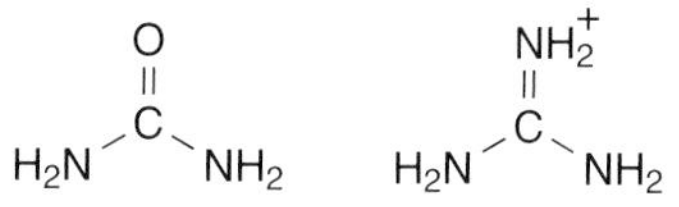

$H_2N-C(=O)-NH_2$ urea

$H_2N-C(=NH_2^+)-NH_2$ guanidinium cation

Figure 11.2 Urea is neutral whilst the charged guanidine is normally associated with chloride or thiocyanate anions

as the neutral diamide urea and guanidine (most commonly used as a chloride or thiocyanate salt) bind weakly to protein surfaces with typical affinities of $\sim$100 mM^{-1} and cause progressive loss of structure.

Urea and guanidine hydrochloride (Figure 11.2) are widely used as denaturants during studies of protein folding with their action arising from disruption of a large number of weak interactions. In view of their ability to unfold proteins there are more interactions with the unfolded state than with native forms. Both reagents are classified as chaotropes and in general effects are based on increases in the solubility of polar and non-polar regions of proteins. This effect can be measured by an increase in the partitioning of individual amino acids between water and denaturant. Studies with model compounds initiated by Charles Tanford in the 1960s showed that solubility of amino acids increased with elevated concentrations of urea and appeared to correlate with the size (accessible surface area) of a non-polar side chain. A linear correlation between accessible side chain surface area (Figure 11.3, shown in Å^2) and the free energy of transfer was apparent from model studies. However, the action of urea and guanidinium salts as denaturants is complex and preferential solubility of non-polar side chains does not account for all of the effects of these denaturants.

Denaturation caused by extremes of pH is better understood and involves protonation of side chains where these interactions underpin tertiary structure. Below pH 5.0 and above pH 10.0 many proteins denature as a result of a loss of stabilizing interactions. A destabilizing effect will include ionization of a buried side chain that is uncharged in the folded state. The presence of a charged group on the inside of a protein surrounded by hydrophobic residues is destabilizing and shifts equilibria towards the unfolded protein.

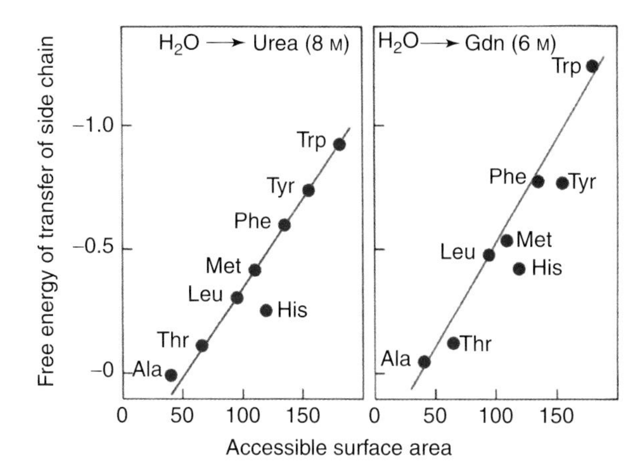

Figure 11.3 Denaturants such as urea and guanidinium chloride increase the solubility of amino acids measured by an increased transfer of amino acid from water to denaturant

Since Lys, Arg, Asp and Glu residues are frequently charged and lie on the surface of proteins the above effect is often exhibited by His and Tyr side chains as they are frequently accommodated within the interior.

Protein folding is a thermodynamically favourable process with a decrease in free energy from unfolded to folded states. This is shown by free energy profiles where the unfolded state (reactant) reaches the folded state (product) via an activated or transition state in a representation that is analogous to that seen for unimolecular reactions. The reaction is described as a two state system when only the folded and unfolded states are identified at any point during the reaction (Figure 11.4).

For a simple folding reaction (see Equation 11.1) the equilibrium constant is defined as the ratio of the concentration of products to reactants (unfolded). In a two state process and with a means of estimating the concentration of either products or reactant during the course of the reaction the free energy associated with unfolding is estimated from the equilibrium constant (K_{eq}) according to

$$\Delta G = RT \ln K_{eq} \quad (11.2)$$

where

$$K_{eq} = [F]/[U] \quad (11.3)$$

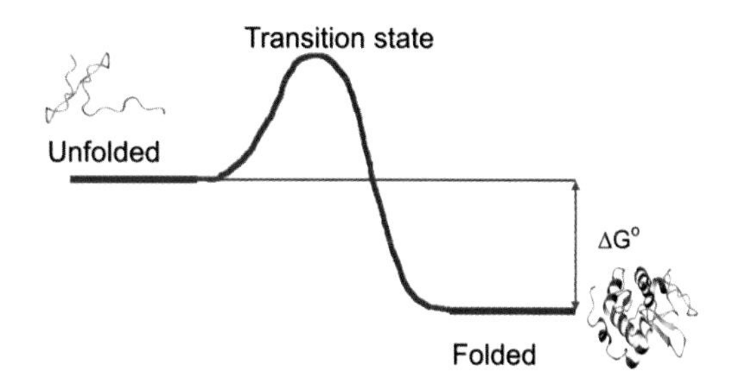

Figure 11.4 A simple reaction profile for two state reaction. The reaction can be described in terms entirely analogous to reaction kinetics where ΔG is the conformational stability of the folded protein ($G_U - G_F$). For folding to occur the change in free energy must be negative

At any point along a cooperative folding curve the concentration of [U] is determined and since the system is two state the fraction of folded protein (f_F) plus the fraction of unfolded protein (f_U) must be equal to unity ($f_F + f_U = 1$). The observed quantity (y_{obs}), usually absorbance or fluorescence, is equivalent to

$$y_{obs} = y_F.f_F + y_U.f_U \quad (11.4)$$

where y_U and y_F are the values characteristic of unfolded and folded states, respectively. Since $f_F = 1 - f_U$ this leads to the following equality

$$f_U = (y_F - y_{obs})/(y_F - y_U) \quad (11.5)$$

and a re-formulation of the equilibrium constant for protein folding as

$$K_{eq} = (y_F - y_{obs})/(y_{obs} - y_U) \quad (11.6)$$

$$\Delta G = -RT \ln (y_F - y_{obs})/(y_{obs} - y_U) \quad (11.7)$$

Transition midpoint temperatures, T_m, define thermal unfolding curves but are not useful estimates of conformational stability since they vary widely from one protein to another (Figure 11.5). However, the T_m remains useful for comparing closely related sequences such as homologous proteins or single site mutants. The unusually named 'cold shock' proteins from mesophilic *Bacillus subtilis* (*Bs*-CspB) and its thermophilic counterpart *B. caldolyticus* (*Bc*-Csp) contain 67 and 66 residues respectively with a high sequence homology

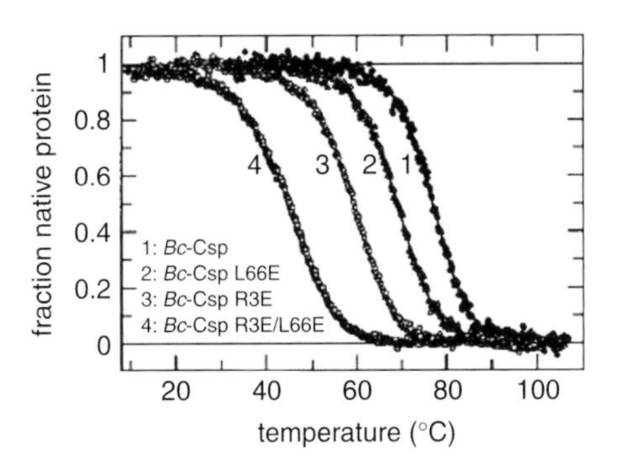

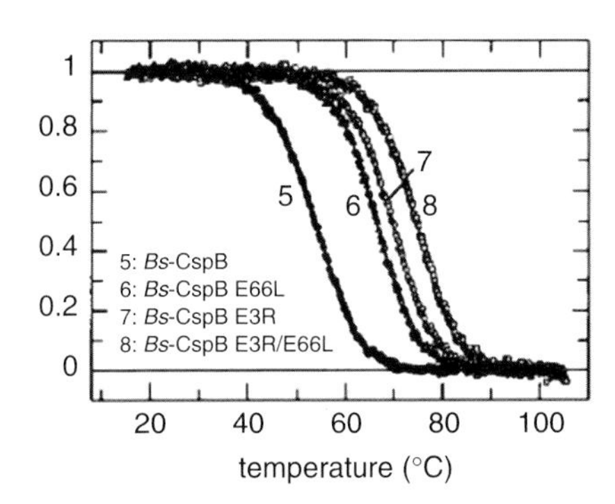

Figure 11.5 Thermal stability curves. Left: Unfolding transitions of wild type *Bc*-Csp and three destabilized variants. Right; Unfolding transitions of wild type *Bs*-CspB and three stabilized variants. The fractions of native protein obtained after a two-state analysis of the data are shown as a function of temperature with the continuous lines showing the results of the analysis (reproduced with permission from Perl, D. *et al. Nat. Struct. Biol.* 2000, **7**, 380–383. Macmillan)

seen by a difference in only 12 residues. The proteins have similar 3D structures and all of the different residues are located on the molecular surface. These residues contribute to a transition midpoint temperature of nearly 77 °C for *Bc*-Csp, approximately 23° higher than that observed in *Bs*-CspB. Using site directed mutagenesis the 12 residues were replaced in *Bc*-Csp by the corresponding residues found in *Bs*-CspB. Only two out of the 12 residues made significant contributions to thermal stability. These residues were Arg3 and Leu66, with each contributing to enhanced electrostatic (Arg) and hydrophobic (Leu) interactions. A recurring theme for elevated thermostability is increased numbers of charged residues and hydrophobic side chains whilst the number of polar uncharged residues usually decreases.

Analysis of thermal induced unfolding curves centres around substitution of the equation

$$\Delta G_U = \Delta H_U - T\Delta S_U \quad (11.8)$$

Since $K_{eq} = e^{-\Delta G/RT}$ this leads to

$$K_{eq} = e^{-\Delta H_U/RT + \Delta S_U/R} \quad (11.9)$$

where ΔH and ΔS are the enthalpy and entropy associated with unfolding. The van't Hoff analysis of the variation in equilibrium constant with temperature

$$\ln K_{eq} = -\Delta H/RT + \Delta S/R \quad (11.10)$$

is widely used in chemistry to estimate reaction enthalpy and entropy. Over a narrow temperature range a straight line defines the enthalpy and entropy in plots of ln K_{eq} versus $1/T$. These plots in protein unfolding transitions are non-linear, indicating that ΔH is not temperature independent. In proteins ΔH and ΔS are temperature dependent with non-negligible variations over a typical range of 70–100 K.

The heat capacity is defined as the change in enthalpy with temperature

$$C_p = \partial H/\partial T = T\partial S/\partial T \quad (11.11)$$

and during protein unfolding the product and reactants have different heat capacities leading to significant changes in enthalpy. Recasting Equation 11.11 leads to

$$\partial H/\partial T = C_p(U) - C_p(F) = \Delta C_p \quad (11.12)$$

$$\Delta H(T_2) = \Delta H(T_1) + \Delta C_p\,(T_2 - T_1) \quad (11.13)$$

where $C_p(U)$ and $C_p(F)$ are the heat capacities of the unfolded and native states respectively and ΔC_p is the change in heat capacity accompanying unfolding. From these equations calculating ΔG at any temperature T requires knowledge of ΔC_p and ΔH. The enthalpy is best estimated at a single temperature, and T_m provides the most convenient estimation point in the thermal unfolding curve since at this point

$$\Delta G(T_m) = 0 = \Delta H_m - T_m\Delta S_m \quad (11.14)$$

To calculate ΔG at any temperature T is given by

$$\Delta G(T) = \Delta H_{\mathrm{m}}\,(1 - T/T_{\mathrm{m}}) - \Delta C_{\mathrm{p}}\,[(T_{\mathrm{m}} - T) + T\ln\,(T/T_{\mathrm{m}})] \quad (11.15)$$

where T_{m} is estimated from the unfolding curve; ΔH_{m} is estimated from the gradient of the straight line of $\ln K_{\mathrm{eq}}$ versus $1/T$ around T_{m} – at this point $\Delta H_{\mathrm{m}} = T_{\mathrm{m}}\,\Delta S$, since $\Delta G = 0$, and this leaves ΔC_{p} as the only parameter remaining to be estimated.

Although generally true that proteins with a high T_{m} are more stable than those with lower T_{m} values this parameter is not a definitive measure of stability. Changes in protein stability reflected by small changes in T_{m} are approximated through the relationship

$$\Delta\Delta G_{\mathrm{U}} \approx T_{\mathrm{m}}\,\Delta S_{\mathrm{m}} \approx \Delta H_{\mathrm{m}}\,\Delta T_{\mathrm{m}}/T_{\mathrm{m}} \quad (11.16)$$

If either the entropy or enthalpy associated with unfolding is unknown an estimate of $\Delta\Delta G_{\mathrm{U}}$ may be obtained with reasonable precision from Equation 11.16. Intuitively it is clear that as the temperature increases a protein unfolds. The basic equations of state relating to protein folding reveal that as T increases the term $T\Delta S$ dominates to such an extent that it becomes greater than ΔH. At high temperatures the entropy of the unfolded state dominates and favours unfolding.

The heat capacity change is most accurately measured by differential scanning calorimetry (DSC). DSC enables direct measurement of T_{m}, ΔH and ΔC_{p} from increases in heat transfer that occur with unfolding as a function of temperature. The instrument is based around an adiabatic chamber where one cell contains the sample (protein plus solvent) at a concentration of $\sim$1 mg ml^{-1} whilst another cell acts as a reference and is normally filled with an identical volume of solvent (Figure 11.6). Both cells are heated with the temperature difference between the two cells constantly measured as a complex feedback loop increases or decreases the sample cell's power input via a heater in an effort to keep the temperature difference close to zero. Since the masses and volumes of the two cells are matched the power added or subtracted by the system is a direct measure of the difference between the heat capacity of sample and reference solutions – in other words the heat capacity of the protein. Despite first impressions it is surprisingly difficult to achieve perfect matching of sample and reference cells with the result that experimental scans are observed with a baseline offset. This 'constant' is usually subtracted from the data to give a corrected profile and accurate estimates of C_{p} (Figure 11.7).

Although ΔC_{p} can be estimated from a single thermogram it is generally obtained from a series of profiles obtained at different pH values. The T_{m} decreases as the pH is lowered from small enthalpy changes and a plot of ΔH_{cal} against T_{m} has a slope of ΔC_{p}. From Equation 11.11 integration of the heat

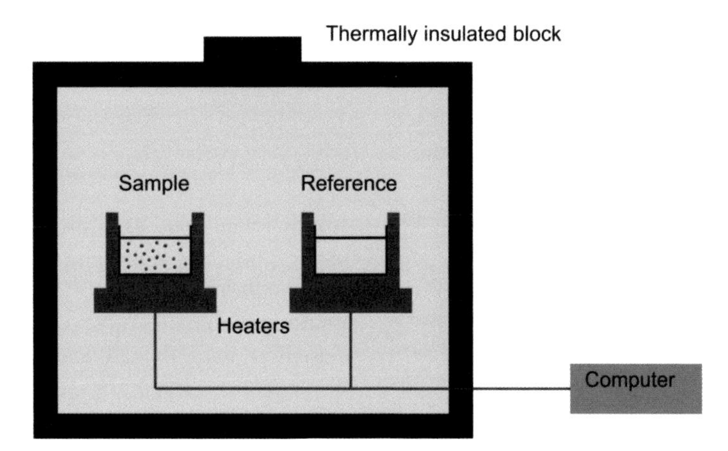

Figure 11.6 In a differential scanning calorimeter the sample cell contains protein plus solvent whilst the reference cell lacks protein. Volumes of $\sim$0.5 ml are used in each cell

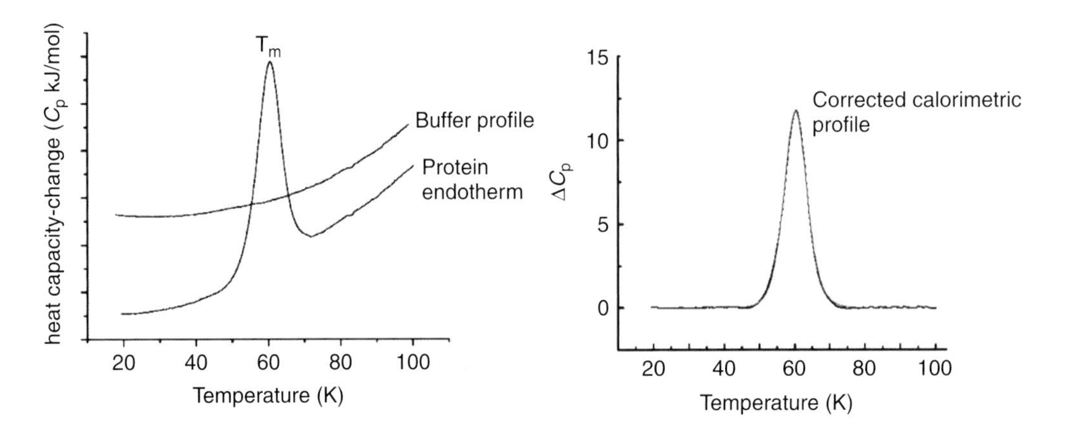

Figure 11.7 DSC profile for protein unfolding showing pre- and post-denaturational regions and baseline correction to achieve accurate estimation of $\Delta C_{p(U-F)}$

capacity curve yields the enthalpy (ΔH) and measures the total heat absorbed by the protein during the process of unfolding. Calorimetric studies of protein unfolding show that the heat capacities associated with unfolded states are larger than those of native states, although both are positive values.

In small soluble domains the value of ΔC_p ranges between 7000 and 10 000 J mol^{-1} K^{-1}. Systematic analysis of these values in many proteins suggests that a rule of thumb can be constructed for 'estimating' ΔC_p – it involves multiplying the number of residues in a protein by 50 J mol^{-1} K^{-1}. It is common to compare the enthalpy obtained from the van't Hoff plot (ΔH_{vh}) with that determined calorimetrically (ΔH_{cal}). A ratio ($\Delta H_{vh}/\Delta H_{cal}$) of 1.0 confirms that the process is two state whilst a value of less than 1 may be used as evidence of an intermediate and departure from a simple two-state analysis. A value of $\Delta H_{vh}/\Delta H_{cal} > 1$ often indicates polymerization at some point in the heating process.

Over a wide temperature range the variation in equilibrium constant is defined by a curve as opposed to a straight line. This curve has two points where the equilibrium constant is 1 and the concentrations of folded and unfolded forms are equal (Figure 11.8). The first point is the thermal denaturation temperature (T_m) whilst the second point reflects a low temperature denaturation point. The ΔG for unfolding is dominated by the entropy component at high and low temperatures

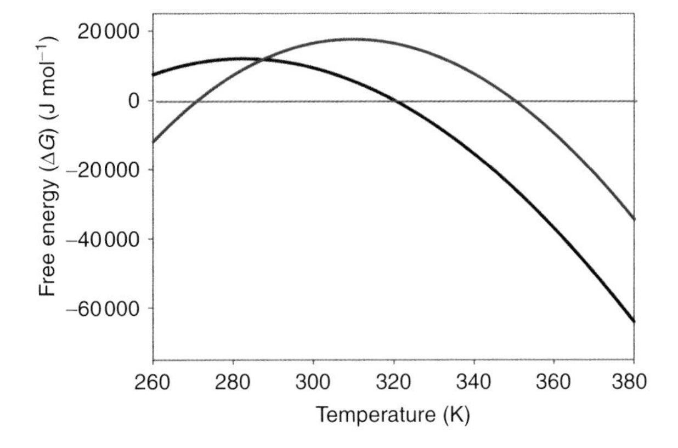

Figure 11.8 The temperature dependence of protein unfolding. The two curves show the variation in ΔG with temperature in two systems with different T_m, ΔH and ΔC_p

although the cold denaturation point lies below the freezing point of water and is not routinely accessible.

An alternative is to promote unfolding using denaturants such as urea or guanidine hydrochloride (GdnHCl) with the advantage that unfolding curves are generally simpler to interpret. In direct analogy to T_m a midpoint in the unfolding transition is given the symbol C_m and can be determined with reasonable accuracy ($\pm$0.05 M). For similar proteins such as those involving single site-specific mutations this approach offers a valid means of comparison but different proteins exhibit a wide range of C_m and unfold under very different conditions. A logical step is to extrapolate estimations of protein stability to zero denaturant concentration. This allows conformational stability of proteins to be compared in the absence of denaturant ($\Delta G_{u(H_2O)}$).

In view of the correlation between denaturant concentration and accessible surface area changes in free energy are related to denaturation via the 'linear extrapolation method' (Figures 11.9–11.11). This model assumes without any strong theoretical basis that ΔG changes linearly with denaturant concentration, according to the equation

$$\Delta G_U = \Delta G_{U(H_2O)} - m_{eq}[D] \qquad (11.17)$$

where ΔG_U is the energy of unfolding in the presence of denaturant, D is the molar concentration of denaturant and m_{eq} is a constant reflecting the association between denaturant and protein. Values of m_{eq} (see Table 11.1) reflect the shape of denaturation curves, with high values leading to sharp transitions between folded and unfolded states. Values for m_{eq} are

Table 11.1 Experimentally determined values of m_{eq} and ΔC_p for selected proteins

Protein	m_{eq} (J mol^{-1} M^{-1})	ΔC_p (J mol^{-1} K^{-1})
Ubiquitin	7 300	5 690
Cytochrome b$_5$	8 950	6 000
Barnase	18 400	6 900
Thioredoxin	13 850	6 950
Lysozyme	14 480	6 610
Ribonuclease T1	10 700	5 300
Metmyoglobin	15 500	7 820
α-chymotrypsin	17 150	12 650
Phosphoglycerate kinase	40 580	31 380
Staphylococcal nuclease	28 580	9 700

All m_{eq} values were determined using GdnHCl as denaturant. Values of m calculated with urea are most frequently observed to be smaller than those obtained using GdnHCl.

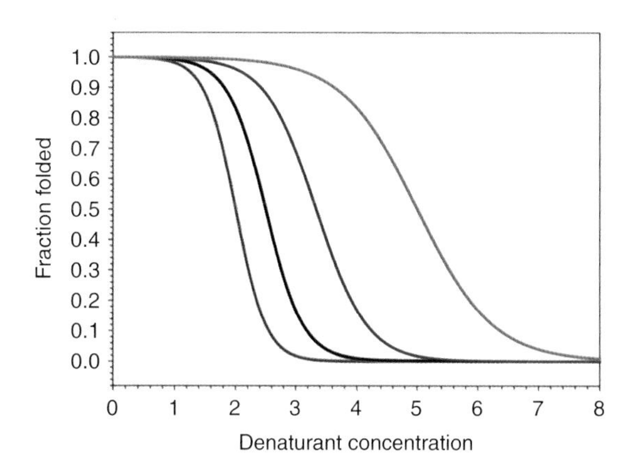

Figure 11.9 Four ideal denaturation curves representing a protein with a conformational stability $\Delta G_{(H_2O)} = 20$ kJ mol^{-1}. The curve reflects different values of m_{eq} ranging from 10 000 (red) 8000 (blue), 6000 (black) and 3000 (green) (J mol^{-1} M^{-1})

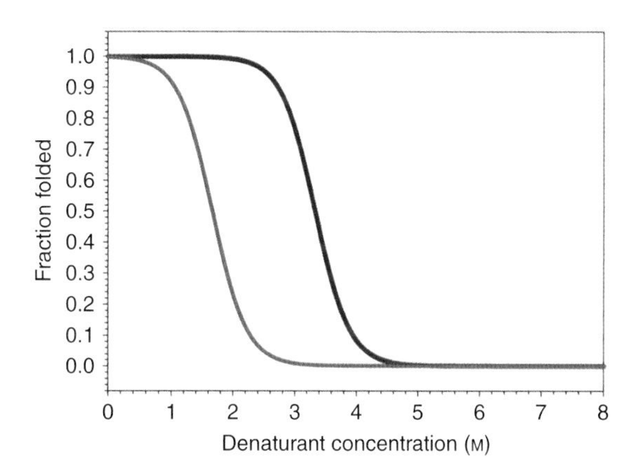

Figure 11.10 The denaturation of two proteins with different stabilities but comparable m_{eq} values. The first protein (red) has a stability of 30 kJ mol^{-1} whilst the second (green) has a stability of 12 kJ mol^{-1}. Denaturation leads to C_m values of ~1.8 M and 3.5 M. The two systems represent a relatively unstable protein and one of moderate stability

positive, with typical values ranging between 2000 and 15 000 J mol^{-1} M^{-1}.

A plot of ΔG versus denaturant concentration gives the conformational stability in the absence

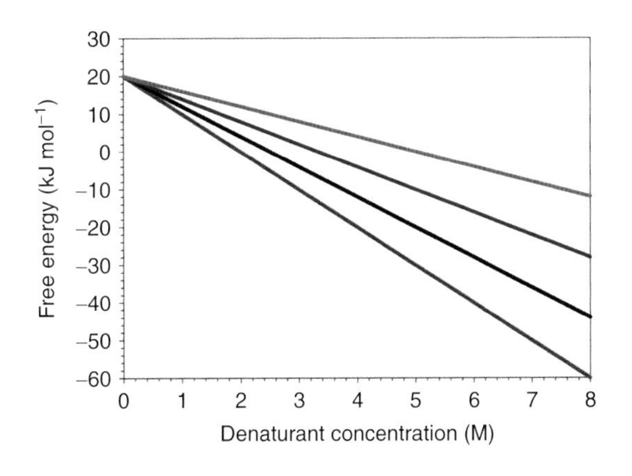

Figure 11.11 The linear extrapolation method applied to the data of Figure 11.9

of denaturant and experimental studies suggest that using either urea and guanidine hydrochloride and the same 'test' protein yields a common intercept at zero denaturant and a constant value of $\Delta G_{U(H_2O)}$ (Figure 11.11).

A conclusion from many denaturation studies is that overall stability of a protein (ΔG, $G_U - G_F$) is small with the folded state being marginally more stable than the unfolded state. This leads to values for protein conformational stability ranging between 10 and 75 kJ mol^{-1}, with most proteins exhibiting values towards the lower end of this range (Table 11.2).

Table 11.2 The conformational stability of proteins

Protein	Chain length	Conformationally stability (kJ mol^{-1})
λ repressor	80	12.5
Cytochrome c	104	74.0
Cold shock protein CspB	67	12.6
SH3 domain of α spectrin	62	12.1
CI2	64	29.3
U1A spliceosomal protein	102	38.9
Ubiquitin	76	15.9
CD2	98	33.5

Ribonuclease defines the 'outline' of protein folding in vitro

Ribonuclease has contributed much to our understanding of protein folding *in vitro* through the landmark studies of Christian Anfinsen who posed the question: 'what is the origin of the information necessary for folding?' Ribonuclease, with 124 amino acid residues and four disulfide bridges located between cysteines 26–84, 40–95, 58–110, and 65–72, catalyses the hydrolysis of RNA. Reduction of the disulfide bridges to thiols by mercaptoethanol in the presence of urea results in protein unfolding and a concomitant loss of activity.

Anfinsen noticed that when ribonuclease was oxidized (by standing in air) and the urea removed by dialysis that enzyme activity slowly recovered as a result of protein folding, the reformation of tertiary structure and most importantly the active site. Repeating these reactions in the presence of denaturant (oxidation of the thiols in 8 M urea) led to the regeneration of less than 1 percent of the total enzyme activity. Urea preventing correct disulfide pairings resulting in a 'scrambled' ribonuclease whilst in its absence correct disulfide bridge formation allowed the folded and thermodynamically most stable state to be reached. These classic studies showed that *all* of the information necessary for protein folding resides *within* the primary sequence. The intervening years have demonstrated the generality of Anfinsen's results and although a few caveats must be included the maxim 'sequence defines conformation' remains relevant today.

One of the first caveats in this picture of protein folding is the protease subtilisin E from *B. amyloliquefaciens*. The enzyme is synthesized as a preproenzyme of ~380 residues to permit secretion and to avoid premature proteolysis. Folding studies involving the removal of the pro sequence showed that its loss was coupled to a lack of subsequent protein refolding. In contrast unfolding the full length preproprotein resulted in a folded enzyme. The results were interpreted along the lines that the pro-sequence participated in folding reactions guiding subtilisin towards the native state. In its absence the native state could not be reached and emphasized that it is not necessarily the 'final' protein sequence that encodes folding information. The second caveat arose with the observation that 'helper' proteins assist folding *in vivo* by preventing molecular aggregation. These helper proteins, molecular chaperones, derive their name from their ability to prevent unwanted interactions between newly synthesized chains and other proteins.

Factors governing protein stability

The native state is the most stable form of a protein but what factors contribute to this stability? From the previous section it might be thought that disulfide bridges or covalent bonds in general are major determinants of conformational stability. However, in the unfolded protein these bonds remain intact and do not contribute to conformational stability. To put this in a slightly different way the covalent bonds contribute equally to the stability of the folded and unfolded proteins and it is necessary when addressing this concept to focus on non-covalent interactions.

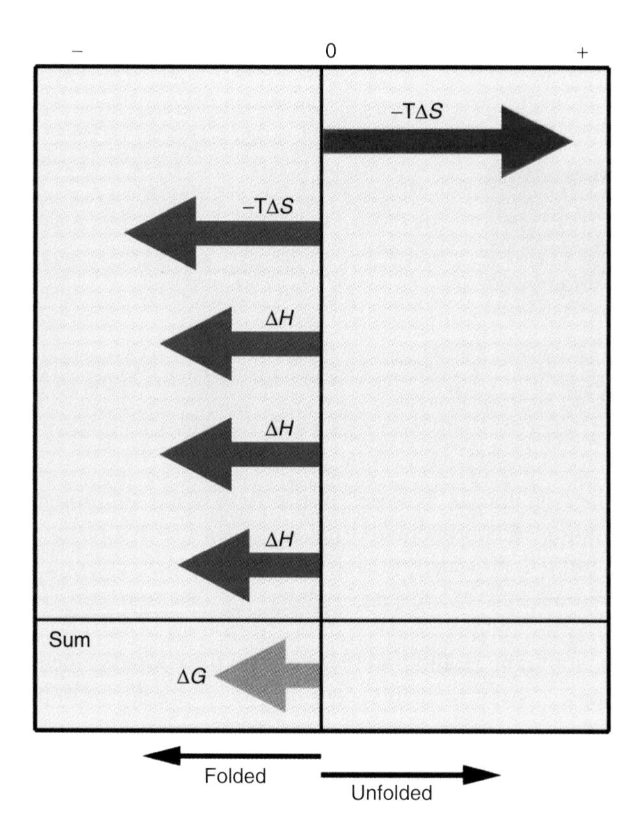

Figure 11.12 Contributions to the free energy of folding of soluble proteins

Non-covalent interactions include hydrogen bonds, hydrophobic forces and interactions between charged groups that cumulatively contribute to increased stability of the native state (Figure 11.12). The entropic contribution to folding is generally unfavourable since it involves transitions from large numbers of unstructured conformations to a single ordered structure. The decrease in entropy works against protein folding and this conformational entropy is largely responsible for favouring the unfolded state. To fold proteins must overcome conformational entropy from entropic and enthalpic contributions that derive from interactions between charged groups, hydrophobic effects, hydrogen bonding and van der Waals interactions.

A major contribution to thermodynamic stability arises from the hydrophobic interaction. Protein folding results in the burial of hydrophobic side chains away from water and their interaction with similar side chains on the inside of the molecule. In thermodynamic terms ordered water molecules specifically arranged in unfolded state are released from this state and are free to move with the result that the entropy increases and makes favourable contributions to the overall free energy change. Favourable enthalpic terms include the formation of stabilizing interactions in the native state via charge interactions and hydrogen bonding. The magnitude of ΔS and ΔH varies from one protein to another but the result appears to be largely compensatory with ΔG values confined to a narrow range generally between 15 and 40 kJ mol^{-1}.

Folding problem and Levinthal's paradox

The Ramachandran plot reveals that many dihedral angle combinations are compatible with stable secondary structure with the result that in an average protein of 100 residues the number of possible conformations is enormous. Some idea of the magnitude of this value is obtained by assuming that only three conformations exist for each residue in a protein of 100 residues and that each conformation can be sampled in 10^{-13} s or 0.1 ps. To sample all conformations would

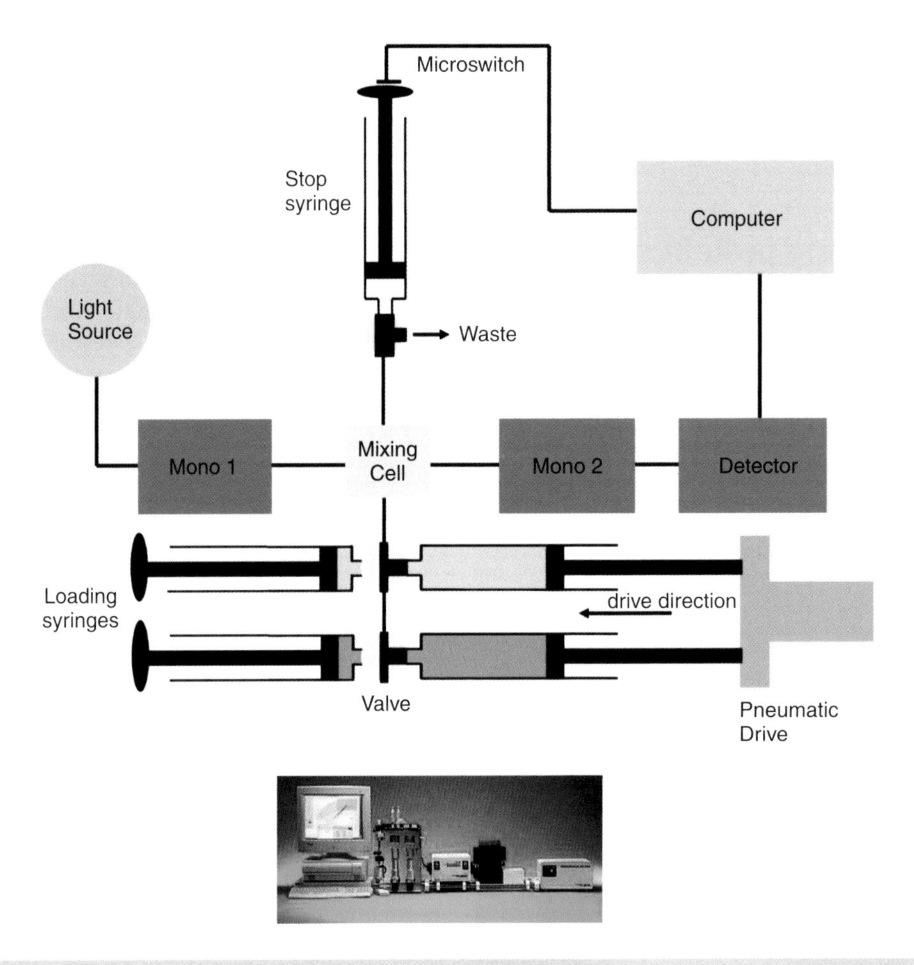

Figure 11.13 Schematic arrangement of a stopped flow system (lower figure reproduced with permission Applied Photophysics Limited)

take $3^{100} \times 10^{-13}$ s – an impossibly long period – that emphasizes that proteins do not sample *all* possible conformations.[1] This problem commonly known as Levinthal's paradox has focussed attention on the kinetics of protein folding.

Kinetics of protein folding

A consideration of Levinthal's paradox quickly eliminates random models of protein folding but a relevant question to ask is what is the time scale of protein folding? Experimentally, these studies involve placing unfolded protein into a buffer system that promotes folding and by mixing rapidly the reaction can be followed on time scales ranging from 1 ms to ~10 min. A common method is the stopped flow kinetic technique where reactants combine in a mixing cell in a process followed using changes in fluorescence or absorbance. In modern instrumentation it is possible to measure changes in CD signals with time but these experiments are technically difficult and less sensitive. In stopped flow experiments (Figure 11.13) pressure-driven syringes force reactants into a mixing cell where the reactants mix displacing resident solution within

[1] There are far more than three conformations per residue, some proteins contain more than 100 residues and 10^{-13} is an unrealistically short period in which to sample each conformation.

a dead time of ~1 ms. This initiates measurement of folding reactions and kinetic profiles usually follow single exponential decays that are fitted to obtain the observed rate constant (k_{obs}) and the maximum amplitude (A_i)

$$k_{obs} = A_i e^{-kt} \quad (11.15)$$

For a two-state unfolding system kinetic analysis is straightforward since the equilibrium constant K_{eq} is simply the ratio of forward (k_f) and reverse (k_u) rates defining the folding and unfolding reactions. Combining Equations 11.2 ($\Delta G_{U(H_2O)} = -RT \ln K_{eq}$) and 11.14 ($\Delta G_U = \Delta G_{U(H_2O)} - m_{eq}[D]$) with the relationship

$$K_{eq} = k_{fw}/k_{uw} \quad (11.16)$$

allows the dependence of the folding and unfolding rates on denaturant concentration to be expressed as

$$\ln k_u = \ln k_{uw} + m_u[D] \quad (11.17)$$

$$\ln k_f = \ln k_{fw} + m_f[D] \quad (11.18)$$

where D is the molar concentration of denaturant.[2] Plotting the logarithm of the rate of folding against denaturation concentration yields a linear relationship that when combined with the dependence of unfolding rates on denaturant concentration leads to a characteristic profile known as a chevron plot (Figure 11.14). C_m is defined by the point where $k_u = k_f$ and the unfolding and refolding rates in the absence of denaturant are extrapolated by extending each limb of the plot to zero denaturant. At any point the observed rate is calculated from

$$k_{obs} = k_u + k_f \quad (11.19)$$

$$k_{obs} = k_{uw} \exp(m_u[D]) + k_{fw} \exp(m_f[D]) \quad (11.20)$$

where

$$m_{eq} = (m_u - m_f)/RT \quad (11.21)$$

The parameters m_u and m_f are kinetic m values to distinguish from the equilibrium m value (m_{eq}) and are used to estimate the position occupied by the transition state along the reaction coordinate from U to F often

[2]The parameters $m_{ku} = RTm_u$ and $m_{kf} = RTm_f$ and k_{uw}/k_{fw} is the rate of unfolding/folding in the absence of denaturant.

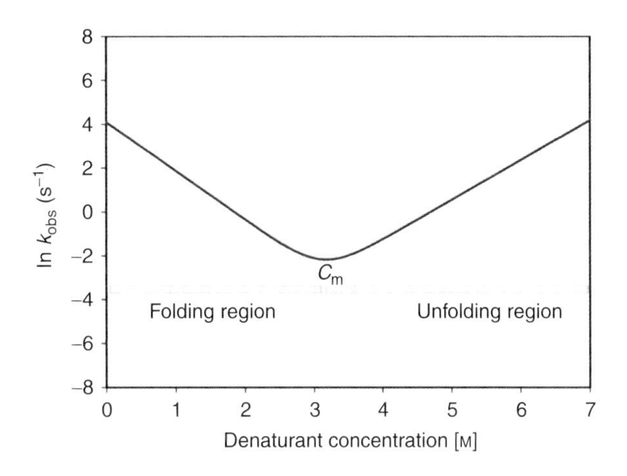

Figure 11.14 A chevron plot describes kinetics of protein folding for a simple soluble domain as a function of denaturant concentration

given the symbol α and equal to $m_f/(m_u + m_f)$. A value of 0.5 suggests a transition state midway between folded and unfolded states.

Barnase and CI2 are two proteins that have been actively studied by Alan Fersht to provide quantitative insight into the mechanism of folding. CI2 is a 64-residue inhibitor of chymotrypsin or subtilisin with a structure based around a single α helix from residues 12 to 24, and five β strands arranged in parallel and antiparallel sheets (Figure 11.15). The α helix lies between strands 2 and 3 and CI2 forms a single folding unit that exhibits two state kinetics and a $t_{1/2}$ of 13 ms at room temperature for the major phase of protein folding. Smaller contributions attributed to *cis–trans* peptidyl proline isomerization are observed but remain a minor element in the overall folding pathway that lacks intermediates in kinetic or equilibrium-based measurements. CI2 represents the simplest kinetic pathway for folding and one widely observed to occur in other small soluble domains.

Barnase is a small, monomeric, soluble, extracellular ribonuclease secreted by *B. amyloliquefaciens* containing 110 residues and lacking disulfide bonds. The protein exhibits reversible unfolding under a wide variety of denaturing conditions. The structure (Figure 11.16) shows a five-stranded β sheet with three helices (α_1–α_3) located between loops and outside of

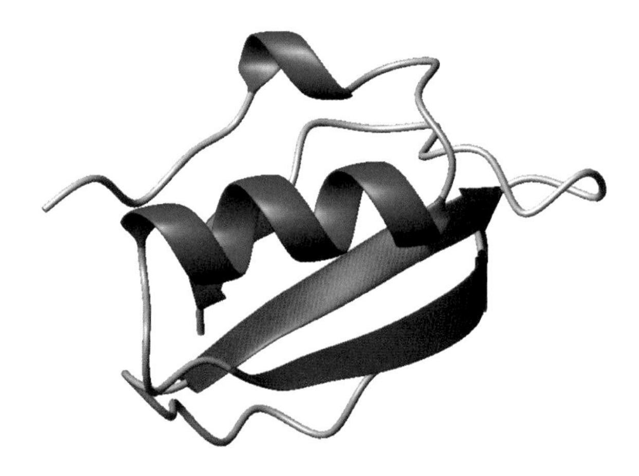

Figure 11.15 The structure of CI2 (PDB: 2CI2). Some strands are poorly defined

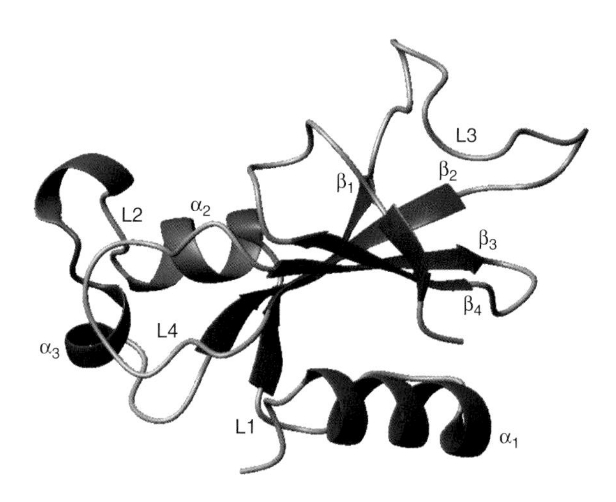

Figure 11.16 The structure of barnase showing distribution of secondary structural elements. L1-L4 are loop regions whilst β strands extend from residues 50-55, 70-76, 85-91, 94-99 and 106-108

the sheet region. The major helix (α_1) is located at the N-terminal and extends from residues 6–18 where its interaction with the β sheet forms the major hydrophobic core. A smaller core is formed from the interactions between loop2, α_2 and α_3 helices and the first β strand.

As a system for studying folding and stability barnase represents an excellent paradigm. With the α and β structures in different regions of the molecule, as opposed to a more mixed arrangement, it is possible to study either fragments of the protein or individual mutations confined to one secondary structural element. To understand protein folding, and in particular the pathway involved, it is necessary to determine the structure and energetics of the initial unfolded state, all intermediate states, the final folded states and importantly the transition states occurring along this pathway.

One approach that has shed considerable light on the mechanism of protein folding as well as the non-covalent interactions involved in these pathways introduced small mutational changes to barnase that removed specific interactions. By measuring changes in stability (i.e. ΔG) alongside kinetic measurements estimating the activation energy ($\Delta G^{\ddagger}$) and other properties associated with the transition state it is possible to begin to describe the energetic consequences of mutation.

The difference in conformational stability between the folded states of the wild type and mutant forms is difficult to measure directly but can be exploited via relationships that invoke Hess' law. The basis of the method is a mutational cycle (Figure 11.17) in which the introduction of specific mutations into barnase is used to estimate kinetic and thermodynamic parameters. The mutation acts as a sensitive reporter of local events during the pathway of protein folding. The mutational cycle measures the free energy of

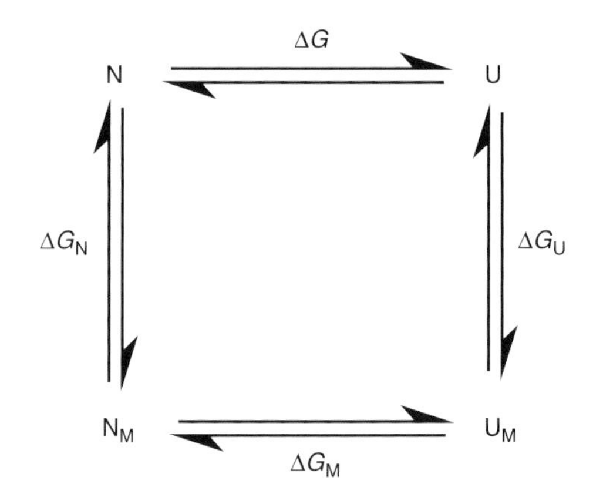

Figure 11.17 Application of Hess's law to the unfolding free energy of wild type and mutant proteins

unfolding for wild type (N) and mutant (M) proteins in the horizontal reactions whilst the vertical reactions represent experimentally indeterminable or 'virtual' free energy changes. They are the free energy changes that arise in the folded or unfolded protein as a result of mutation.

Hess's Law states that the overall enthalpy change for a chemical reaction is independent of the route by which the reaction takes place as long as the initial and final states are identical (Figure 11.17). It is simply a statement of the law of energy conservation. Using the mutational cycle approach it was shown that the energetic effects of mutation in barnase are evaluated from the relationship

$$\Delta G + \Delta G_{\mathrm{U}} = \Delta G_{\mathrm{N}} + \Delta G_{\mathrm{M}} \quad (11.22)$$

Re-arranging leads to

$$\Delta G - \Delta G_{\mathrm{M}} = \Delta G_{\mathrm{N}} - \Delta G_{\mathrm{U}} \quad (11.23)$$

where $\Delta\Delta G_{\mathrm{N-U}}$ is the change in free energy due to mutation (M) and the experimental quantity one would wish to access directly.

Mutations do not have a single effect on one 'folding' parameter but can alter solvent exposure, hydrophobic contacts, hydrogen bonding – the list is almost endless. As a result of the large number of parameters altered by a mutation it is very difficult to ascribe the changes to a single energetic parameter. One way to eliminate this problem is to borrow a trick from physical organic chemistry and use the Brønsted equation. The Brønsted equation relates changes in rate constant k and equilibrium constant K to a parameter β as a consequence of altering a non-reacting functional group

$$\log k = \text{constant} + \beta \log K \quad (11.24)$$

In physical organic chemistry this involves making series of derivatives whilst in the area of protein folding this involves the creation of mutants. In each case the effect on rate and equilibrium constants for folding are measured. For protein folding this equation was modified to

$$\Delta G^{\ddagger} = \text{constant} + \phi_{\mathrm{U}} \Delta G_{\mathrm{U}} \quad (11.25)$$

in view of the relationship of free energy terms $\Delta G^{\ddagger}$ and ΔG_{U} to log k and log K respectively whilst the parameter ϕ replaced β from the Brønsted relation. This scheme allows two measurable reactions to be compared since the constant can be cancelled by subtraction to give

$$\Delta G^{\ddagger}{}_{\mathrm{U}} - \Delta G^{\ddagger}{}_{\mathrm{N}} = \text{constant} + \phi_{\mathrm{U}}\ \Delta G_{\mathrm{U}} - \text{constant} + \phi_{\mathrm{U}}\ \Delta G_{\mathrm{N}} \quad (11.26)$$

$$\Delta\Delta G^{\ddagger} = \phi_{\mathrm{U}}\ \Delta\Delta G_{\mathrm{U}} \quad (11.27)$$

$$\phi_{\mathrm{U}} = \Delta\Delta G^{\ddagger}/\Delta\Delta G_{\mathrm{U}} \quad (11.28)$$

where $\Delta\Delta G_{\mathrm{U}}$ is the difference in conformational stability between folded and unfolded protein. A value of ϕ of 0 (measured from the unfolded state and in the direction of folding) implies that the structure at the site of mutation in the transition state is comparable to the structure in the unfolded state. Alternatively a ϕ value of 1 suggests that the structure in the vicinity of the mutation in the transition state is as folded as that found in the native structure. Fractional values of ϕ imply a mixture of states.

Kinetic analysis of the refolding of barnase was consistent with one major intermediate and a linear scheme $\mathrm{U} \rightleftharpoons \mathrm{I} \rightleftharpoons \mathrm{N}$, where the rate-limiting step was conversion of I to N. The $t_{1/2}$ for the observable step of refolding was ~30 ms, although the kinetics of the step preceding I occur within the dead time of stopped flow studies. Since ΔG_{U} and the forward and reverse rates were measured a partial energy diagram for barnase folding was constructed where kinetic and thermodynamic analysis exposed the properties of the transition state and major, late folding, intermediate in wild type and mutant forms of the protein (Figure 11.18).

Residues mutated at the C-terminal end of the α_1 helix show values of ϕ of ~1.0 for the transition and folded states. This implies that this region of the helix-ordered structure is formed rapidly. Interestingly, residues such as Thr6 at the beginning of the helix show intermediate values suggesting disorder is present before formation of native structure. In a similar fashion residues in loop 3 rapidly form structure comparable to the folded state whilst loops 1,2 and 4 remain disordered throughout the folding process until the later stages. Residues at the centre of the β sheet form rapidly but those located at the edges do not fold until later. More significantly, residues

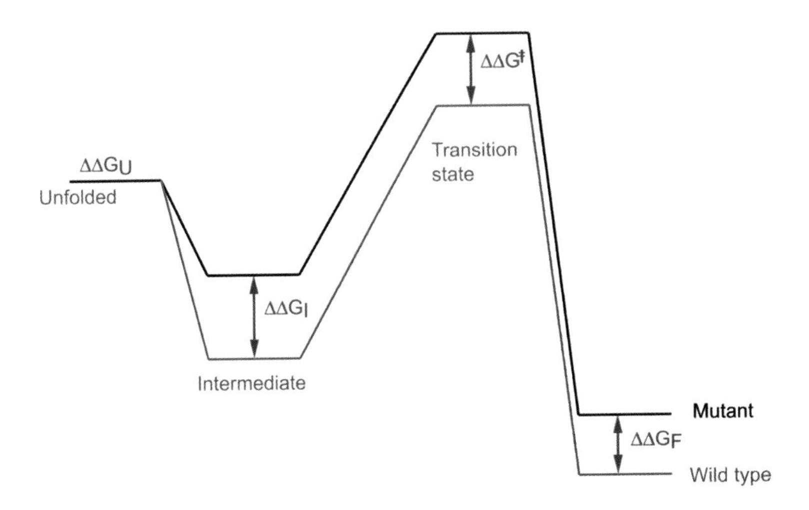

Figure 11.18 Partial free energy diagram describing the folding of wild type and mutant forms of barnase. Although not strictly necessary for simplicity the unfolded forms of the wild type and mutant barnase are shown as iso-energetic. The remaining energy terms describe the differences between the intermediate, transition state and folded forms of wild type and mutant barnase

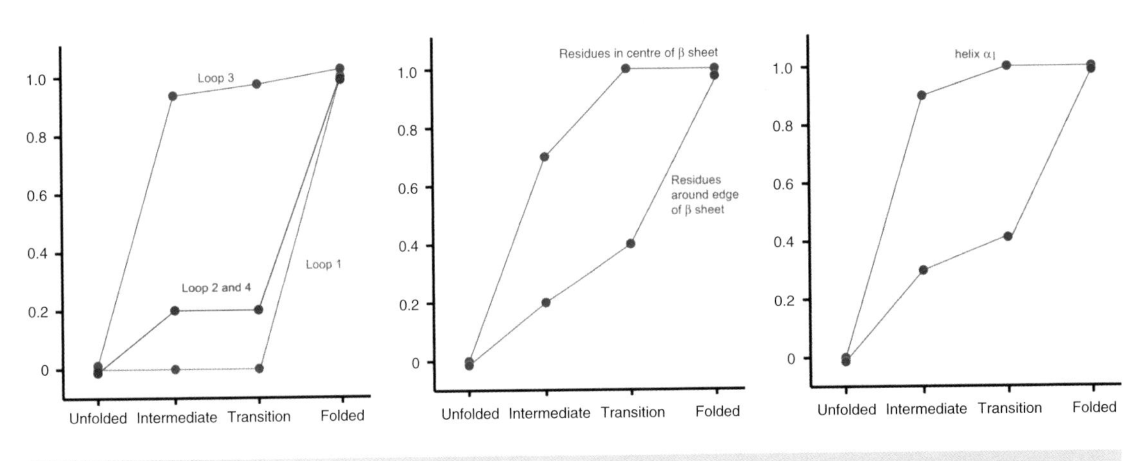

Figure 11.19 ϕ-values found for mutation of residues in the loop, β sheet and helix regions of barnase

located in the core region between the helix and sheet showed ϕ values dependent on location. Residues at the centre of the core exhibit ϕ values close to 1 in the intermediate and subsequent states whilst those located at the periphery of this core showed a gradation of values (Figure 11.19). In this fashion an extraordinary detailed picture of the pathway and energetics of protein folding and its sensitivity to mutation was established for barnase.

Models of protein folding

Exceptionally rates of protein folding are observed with time constants less than 5 ms but most domains with less than 100 residues fold over slower time scales ranging from 10 ms to 1 s. In most small domains folding is a cooperative process but for a smaller group of proteins folding is characterized by the transient population of detectable intermediates. These proteins offer the possibility of characterizing partially folded states or sub-domains. Initial studies of protein folding focused extensively on helical domains or domains containing mixtures of helix and strand (α/β proteins). Gradually, it has been shown that small β sheet proteins also fold in a highly cooperative manner whilst larger β sheet domains especially those with complex topology frequently exhibit intermediates during folding. Thus, models of protein folding must account for processes spread over a wide range of time scales as well as the presence in some cases of intermediates or partially folded structures.

Helix formation

The formation of segments of helical structure occurs with time constants around 1 μs. The reactions proceed with a slow initial stage as the polypeptide chain struggles to form hydrogen bonds between donor and acceptor separated by three or four residues in a disordered chain. This is followed by faster rates that reflect progressively rapid extension of helices from an ordered structure sometimes called a nucleation site. Most experiments on helix formation have been derived using synthetic homopolypeptides and a major problem is that whilst considerable data has been accumulated for helical regions the formation of β structure relies on more than one element of secondary structure for stabilization and there is no satisfactory model system.

Although helices have regular repeating hydrogen bonds coupled with a uniformity of bond lengths and angles this periodicity masks their marginal stability. In an isolated state most helices will unfold and only synthetic polyalanine helices are reasonably described as stable. Initiation of helix formation is a slow and unfavourable process. This arises because five residues must be precisely positioned to define the first hydrogen bond between residues 1 and 5. Hydrogen bonds between residue i, i + 4 characterize regular α helices whilst the intervening three residues must be arranged to form a turn as well as arranged in a geometry favouring the formation of future hydrogen bonds. There is a large entropic penalty in forming the first turn whilst addition of a subsequent residue to the helix is far less severe in terms of energetic cost. At the same time dipoles associated with the peptide bond in a helix influence the energetics since for the first turn their alignment lies in the same direction (parallel) and is an unfavourable arrangement. With the completion of the first turn subsequent formation of peptide dipoles separated by 3 to 4 residues leads to a head to tail arrangement that may assist helix propagation.

The molten globule

During studies of protein folding it was recognized that intermediate states showing many of the properties of the native fold could be identified. This state became known as the molten globule and is characterized by a compact structure, containing most elements of secondary structure as helices, turns and strands, although many long range or tertiary contacts are lacking. The molten globule has a hydrophobic core consistent with burial of non-polar side chains away from the solvent.

Studies of α-lactalbumin using circular dichroism (CD) under denaturing low pH (<4) conditions revealed formation of a stable intermediate state that was distinguishable from the native *and* unfolded states by possessing far-UV CD spectra resembling the native fold but an unfolded-like near-UV CD spectrum. In other words it appeared to possess considerable secondary structure but no tertiary structure. This state was given the name molten globule and similar forms have been identified during the folding of other proteins. Proteins showing molten globule-like states include many familiar soluble proteins such as myoglobin, α-lactalbumin, β-lactoglobulin, BPTI, cytochrome c and azurin. These proteins differ considerably in their secondary structure content as well as tertiary folds. In contrast, some proteins fold without evidence of molten globule intermediates, and lysozyme is one interesting example, especially since lysozyme and α-lactalbumin are homologous proteins.

Although the molten globule state was identified by its stability and presence at equilibrium under partially denaturing conditions similar states have been detected in kinetic experiments. In particular stopped-flow CD studies appear to show kinetic intermediates populated during folding possessing high levels of secondary structure yet lacking complete side chain packing and extensive tertiary structure. These studies have led to the suggestion that molten-globule like intermediates form as part of the overall process of reaching the native state and have provided a picture in which the formation of elements of secondary structure occurs prior to consolidation of the tertiary fold.

Hydrophobic interactions

Hydrophobic interactions are intrinsically weak forces between non-polar side chains that lead to dramatic ordering of water. In globular proteins these interactions lead to charged/polar side chains residing on the surface of a protein whilst non-polar side chains are buried on the inside of the molecule. The fluorescence probe anthraquinone naphthalene sulfonate (ANS) binds effectively to exposed hydrophobic surfaces and exhibits a characteristic λ_{max} that changes when the probe is placed in a polar environment. By measuring changes in fluorescence of ANS during protein folding in a rapid mixing experiment it is possible to monitor the formation of hydrophobic cores. From many studies of folding involving proteins with different topologies it has been recognized that different events can occur during the formation of the native state. These events include some or all of the following reactions:

1. Formation of all native contacts in a highly cooperative transition;
2. Folding via a molten globule intermediate;
3. Folding via kinetic intermediates resembling molten globules;
4. Condensation of hydrophobic cores prior to formation of secondary and tertiary structure.

A major problem for experimentalists and theoreticians has been to reconcile all of these different views within a single scheme of protein folding allowing formation of the folded state on time scales between 10 ms and 1 s.

A hierarchic model has been proposed where structure is determined by local interactions with conformational preferences among short sections of peptide chain guiding the polypeptide towards forming larger units of secondary structure. An obvious weakness with this model is that few short peptides in isolation form helices and none will form β strands. Although at first these elements of secondary structure are only metastable it is envisaged that interactions with neighbouring units reinforce stability. Early stages of protein folding are characterized by the interaction of secondary structure via a purely diffusional model in which elements collide to allowing favourable interactions and stable association. These interactions accumulate cooperatively to form larger collections of secondary structure that can acquire stabilizing tertiary structure and long-range order at longer time intervals. The detection of intermediates or transition states containing secondary but not tertiary structure is perhaps supportive of a hierarchic model whilst the fact that secondary structure can be predicated from sequence is also used as a supportive line of evidence.

Recently, considerable attention has focused on a concept called contact order. Although the length of a protein does not show a simple correlation with rates of protein folding the topology of a protein does seem to influence this process. To reflect the concept of topology the term contact order (CO) can be defined as the average sequence distance between all pairs of contacting residues normalized by the total sequence length of the protein and is described as

$$CO = \frac{1}{LN}\sum^{N} \Delta S_{i,j} \qquad (11.29)$$

where N is the total number of contacts, $\Delta S_{i,j}$ is the sequence separation in residues between contacting residues i and j, and L is the total number of residues within the protein. Providing the structure of a protein has been determined at a reasonable level of resolution the CO can be calculated. A correlation exists between CO and observed rates of protein folding. This leads, for example, to proteins that are extensively α helical having low CO and folding more quickly than proteins containing higher proportions of β strand. These proteins are characterized by higher CO. The observation that proteins with low CO tend to fold rapidly could suggest that local contacts are the most effective route towards rapidly reaching the native state.

Irrespective of the precise details during the early events of protein folding a common objective achieved by all schemes is to restrict the conformational space sampled by the protein and therefore to avoid the worst excesses of Levinthal's paradox. Restricting

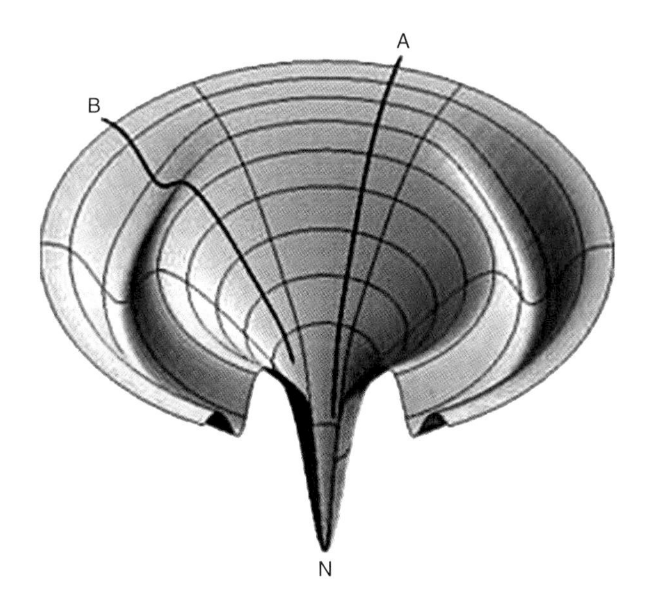

Figure 11.20 The route towards the folded conformation for protein A and B differ. Real energy landscapes are undoubtedly much more complex in profiles

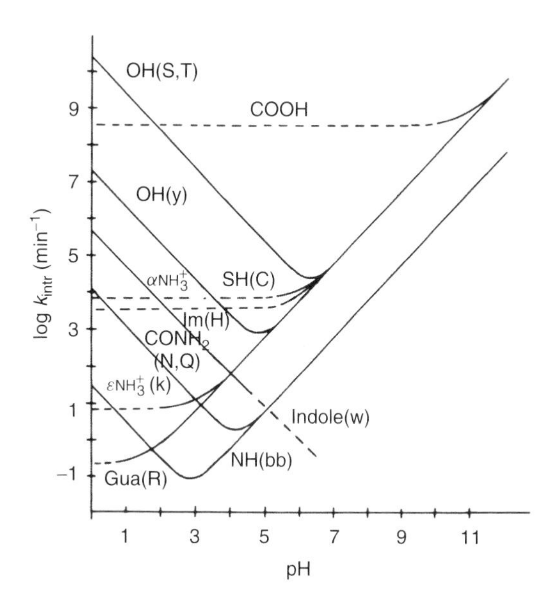

Figure 11.21 Rates of amide exchange for different functional groups in proteins. Backbone amide groups show minimum exchange at pH 3.5 with time constants of ~1 h whilst at pH 10 the rate is increased leading to exchange on the ms time scale

the number of conformations leads to remarkably rapid folding. A new picture of protein folding represented by contoured energy landscapes directs the unfolded protein towards the final (low energy) conformation by a 'funnel' (Figure 11.20). A three-dimensional plot suggests several routes may be used in reaching the native fold with local energy minima existing along pathways. The shape of the funnel directs all conformations towards the native state yet avoids the need to sample all of the possible conformations.

A new view of protein folding is one of energy landscapes, multiple pathways and multiple intermediates where folding leads towards the most stable state in a process influenced by kinetic events.

Amide exchange and measurement of protein folding

The measurement of amide exchange rates is a powerful way of assessing the protection (resistance to exchange) of these groups as a protein folds from denatured to native states. Protection occurs because the amide is involved a hydrogen bond or is buried within the hydrophobic interior of a protein. In NMR experiments based around the observation of a proton the substitution of H by ^2H (D) leads to a loss of a signal. Exchange rates are dominated by pH and proteins show profiles where the rate has a minimum around pH 3.5 but increases either side of this minimum due to acid and base catalysis (Figure 11.21). The intrinsic rate of exchange (k_{ex} or k_{intr}) reflects a summation of acid- and base-catalysed rates and differs according to the type of functional group.

$$k_{intr} = k_{OH}[OH^-] + k_H[H^+] + k_w \qquad (11.30)$$

The rates for each of the functional groups found in proteins have different pH optima and are strongly influenced by pH, local chemical environment, solvent, side chain identity, neighbouring residues and temperature. In proteins NH exchange can be extremely slow and may take months or years to reach completion. A protection factor is described via a parameter, θ_p, where

$$\theta_p = k_{intrinsic}/k_{obs} \qquad (11.31)$$

and a slow rate of exchange is seen by values of θ_p of 10^6-10^7.

A theory of amide exchange in proteins developed by Kai Linderström-Lang involves open conformations where the protein is exposed to the solvent in equilibrium with a closed structure.

$$\text{closed} \underset{k_{clo}}{\overset{k_{op}}{\rightleftharpoons}} \text{open} \xrightarrow{k_{intrinsic}} \text{hydrogen exchange}$$

The closed state is often equated with folded molecules and the open state with denatured proteins. Protected amide hydrogens are 'closed' to exchange but become accessible to exchange through formation of the 'open' state at rates comparable to those observed for unstructured peptides. Two kinetic regimes are recognized for hydrogen exchange in the above scheme. The first regime is called the EX1 mechanism and occurs when $k_{intrinsic} \gg k_{clo}$. The observed rate is therefore approximated by the rate of protein 'opening' such that $k_{obs} = k_{op}$. In contrast the more common mechanism found in proteins occurs when $k_{intrinsic} \ll k_{clo}$ and the observed rate (k_{obs}) is therefore equivalent to the product $k_{op}k_{intrinsic}$. This is called the EX2 mechanism and occurs when a protein is relatively stable and exchange rates are comparatively slow.

In proteins the most important exchangeable proton is attached to the secondary amide of the polypeptide backbone. Amide exchange rates in proteins are readily measured in protein uniformly enriched in ^{15}N in a 'pulse hydrogen' exchange experiment. The principle of the experiment involves transferring a protein from water-supported buffers to those based on ^2H_2O (D_2O) and leads to a loss of cross-peak intensity in 2D NMR spectra as a result of proton/deuteron (H/D) exchange (Figure 11.22). By observing the rate of change in the intensity of cross peaks amide exchange rates are estimated.

Denatured protein in 6 M guanidinium HCl in solutions of ^2H_2O is allowed to fold in solutions containing ^2H_2O and leads to protein 'labelled' with deuterium. Placing the samples at high pH favours rapid exchange for accessible amides and is followed by 'quenching' and a lowering of the solution pH. Folding is allowed to continue to completion and the method samples amides that are rapidly 'removed' from exchange reactions by forming hydrogen bonds, or more probably from burial on the inside of the protein. After folding has been completed the pattern of NH and ND labels is analysed by 2D NMR spectroscopy (Figure 11.23). Increasing the refolding time (t_f) allows a greater number of exchangeable sites to be protected. Using this approach it was possible to sample the protection of over 50 different amide groups in the enzyme lysozyme. The results showed two groups of amides existed; approximately 50 percent were completely protected from exchange within about 200 ms whilst the remaining group remained accessible even after 1 s. Further analysis of rapidly protected amides revealed a location within a domain based around four α helices whilst the more accessible amides were contained in the β sheet rich domain.

This picture of amide protection provides a view of protein folding in which each domain of lysozyme acts as a separate folding unit (Figure 11.24). Studies with other proteins reinforced this view and domains are often defined as 'folding units'. It might be argued that the rates of protection simply reflect the units of secondary structure but working against this idea is the observation that 3_{10} helices found in each domain showed different rates of protection. The different technique used in the study of the kinetics of protein folding are listed in Table 11.3.

Kinetic barriers to refolding

Kinetic studies highlight the presence of transient intermediates during folding that represent partially folded forms along the pathway between denatured and folded states (Figure 11.25). In 'off' pathway intermediates it may be necessary to unfold before completing a folding pathway. When this type of reaction is performed *in vitro* the rate of reaction is slow. *In vivo* a different picture emerges with specific proteins catalysing the elimination of 'unwanted' folds.

Additional barriers to efficient protein folding exist. One major kinetic barrier to protein folding involves *cis–trans* isomerization of the amide bond preceding

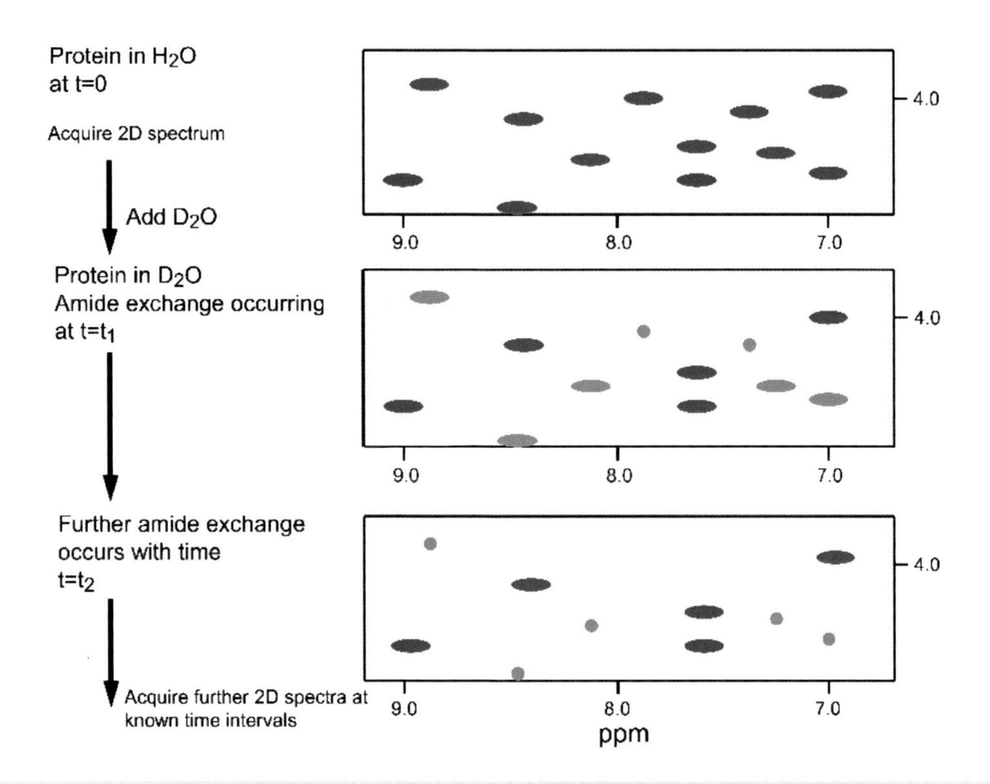

Figure 11.22 The principal of measuring H/D exchange by NMR spectroscopy. Initial cross-peaks are shown in red. Those exhibiting altered intensities are shown in orange and eventually exchange leads to a complete loss of intensity

Table 11.3 Techniques used to study kinetics of protein folding

Method	Structural information
Intrinsic fluorescence	Primarily environment of Trp and Tyr residues. Can also include environment of co-factors such as heme or flavin
Absorbance	Environment of aromatic groups or other conjugated systems
Near UV CD	Asymmetry of aromatic residues within tertiary structure
Far UV CD	Formation of secondary structure
H/D exchange NMR	Formation of persistent hydrogen bonds or amide groups protected from solvent
H/D exchange MS	Detection of folding intermediates and different populations
FTIR	Formation of hydrogen bonds
ANS fluorescence	Accessibility of hydrophobic residues

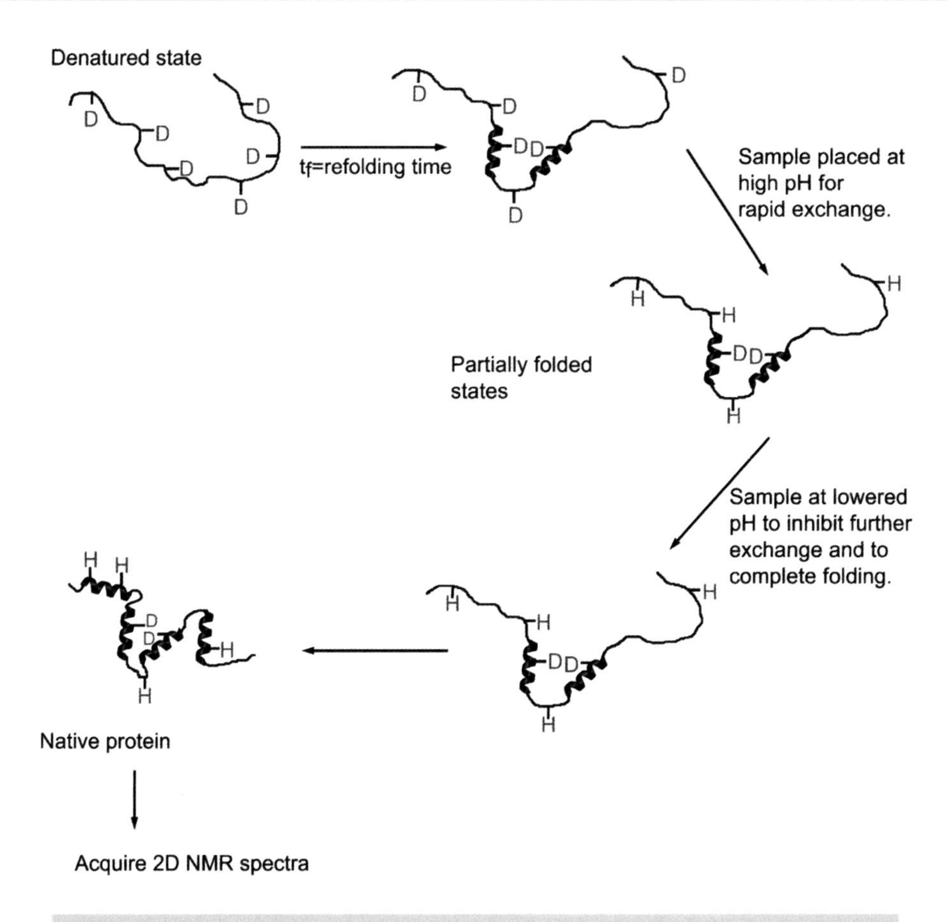

Figure 11.23 The principle of pulsed H/D exchange in a protein during folding

proline residues. Proline has unique properties among the 20 naturally occurring building blocks of proteins stemming from the cyclic side chain and the absence of an amide proton (NH). The peptide bond immediately preceding proline is unlike other peptide bonds in that the ratio of *cis–trans* isomers is ~4. Unfolding removes many of the constraints placed on peptide bonds and leads to randomization of prolyl peptide bonds so that in a protein containing five Pro residues it is expected that on average one prolyl peptide bond will exist in the less favourable *cis* conformation. The potential for proline *cis–trans* isomerization to inhibit the rate of protein folding arose from studies by Robert Baldwin showing that kinetically heterogeneous mixtures of molecules existed in refolding denatured ribonuclease. An explanation for this observation provided by John Brandts centred around the *cis–trans* isomeric state of prolyl peptide bonds. The refolding of a proportion of protein with prolyl peptide bonds in an unfavourable conformation is characterized by slow rates of formation of the native state superimposed on normal rates of folding. Kinetic heterogeneity is observed and arises because molecules with bonds arranged favourably fold rapidly whilst a fraction with unfavourable *cis* peptide bonds will fold more slowly. As a result protein refolding involving *cis–trans* isomerization is characterized by one or more slow refolding phases.

This view has been supported through the use of model peptides containing proline residues and the observation of slow refolding rates in proteins including insulin, tryptophan synthase, T4 lysozyme, staphylococcal nuclease, chymotrypsin inhibitor CI2,

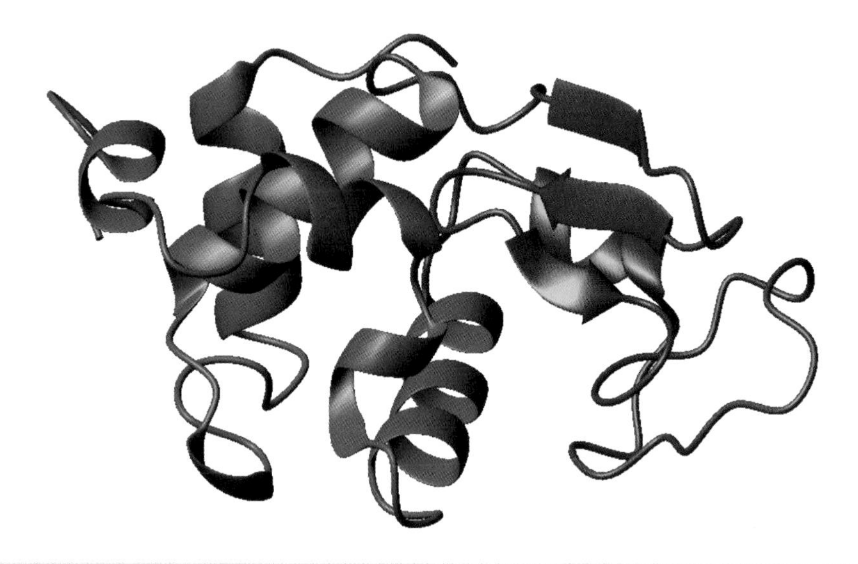

Figure 11.24 Structure of lysozyme showing the rapidly formed helix rich domain (blue) along with the slowly 'protected' sheet-rich region (green)

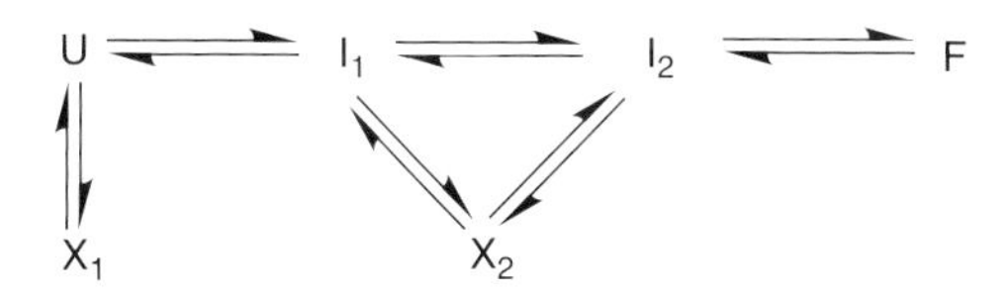

Figure 11.25 On and off pathway intermediates during protein folding. I_1 and I_2 represent on pathway intermediates such as the molten globule like state whilst X_1 and X_2 represent off pathway forms of the protein

barnase, yeast iso-1 and iso-2 cytochromes c, pepsinogen, thioredoxin as well as calcium binding proteins such as parvalbumin and calbindin. One direct observation is the refolding of three homologous parvalbumins (Ca binding proteins found in muscle). Two of the parvalbumins contain a single proline residue and exhibit refolding reactions with a slow component whilst this phase is absent in a third parvalbumin lacking proline.

Cis–trans proline isomerization is one of the major causes of kinetic heterogeneity in protein refolding and deviations away from a purely two-state system. The presence of such effects during refolding are normally confirmed by site directed mutagenesis, by kinetic analysis using 'double mixing' stopped flow techniques and from analysis of the temperature dependency associated with each folding reaction. Those associated with *cis–trans* isomerization of Xaa–Pro peptide bonds are characterized by activation energies ranging 80 to 100 kJ/mol.

In vivo protein folding

Proteins fold to form the native state *in vitro* but do similar pathways or mechanisms apply *in vivo*? Translation occurs at rates of ~20 residues per second, yet many small soluble proteins fold completely within 100 ms. The disparity between optimal translation rates and folding suggests that *in vivo* limitations exist in the latter process.

Peptidyl prolyl isomerases (PPI) catalyse one of two important but slow reactions associated with folding. The reaction is *cis–trans* isomerization of prolyl peptide bonds. The first enzyme was isolated from pig kidney in 1984 but during this period other studies looking at the binding of the immunosuppressant drug cyclosporin A (a vital immunosuppressant during transplant surgery preventing the rejection of donor organs) discovered its intracellular target and named this protein cyclophilin. Subsequent protein sequencing showed that PPI isolated from pig kidney and cyclophilin were identical. The precise basis for cyclosporin A binding to PPIs remains to be established but it may be unrelated to the activity of the immunosuppressant in decreasing T-cell proliferation and activity. Peptidyl prolyl isomerase overcomes kinetic barriers presented by incorrectly oriented peptide bonds during protein folding.

Additional prolyl isomerases exist within cells and are grouped within one of at least three major protein families. Besides cyclophilins FK506-binding proteins and parvulins represent different families that are not homologous but share a common biochemical function. A plethora of PPIs exist in the human genome with at the last count 11 cyclophilins, 18 FK506 binding proteins and two parvulins encoded.

The second 'slow' reaction of protein folding is the formation of disulfide bonds. This reaction is catalysed by the enzyme protein disulfide isomerase (PDI). Since both reactions (*cis–trans* proline isomerization and disulfide bond formation) were shown to inhibit rapid protein folding *in vitro* the presence of enzymes within cells to specifically catalyse these 'slow' reactions is compelling evidence that similar mechanisms exist for folding *in vitro* and *in vivo*. Protein disulfide isomerase contains the conserved active site motif of Cys-Xaa-Xaa-Cys found in thioredoxin and shares a similar role to the Dsb family of proteins found in *E. coli*. Together with molecular chaperones these systems allow proteins to fold effectively in the cell.

Chaperones

The fundamental experiments of Anfinsen showed that denatured ribonuclease folds *in vitro* with the primary sequence directing folding to the native state. It was widely assumed that folding of all newly synthesized proteins *in vivo* would proceed similarly. Nascent polypeptide chains would not require additional assistance to fold efficiently. With hindsight this assumption seems foolish, and gradually observations established that specialized proteins within the cytosol assist formation of native states.

One of the first observations was that *E. coli* containing a defective operon called *groE* could not assemble wild type bacteriophage λ. This was despite the fact that all protein components of mature λ were encoded by the viral genome. A second observation was that in the mitochondrial matrix translocated polypeptide associated with a protein (Hsp60). The name reflected its mass and that production increased approximately two-fold after heat shock. The name is misleading because experiments showed that deletion of the Hsp60 gene in yeast was lethal and the protein was essential to the cell under all growth conditions. This is expected for a protein playing an important role in the folding of mitochondrial proteins.

The heat-shock or cell-stress response (i.e. changes in the expression of intracellular proteins) is a common event, with increased protein production an essential survival strategy that allows responses to diverse stimuli, including heat or cold, osmotic imbalance, toxins and heavy metals as well as pathophysiological signals. The proteins synthesized in response to such environmental stresses are collectively called heat-shock proteins (or HSPs), stress proteins or, more commonly today, molecular chaperones. The response of cells to stress is a primitive mechanism that appears to be evolutionarily ancient as well as essential to survival. The ancient nature is demonstrated by significant sequence homology in molecular chaperones from a wide range of organisms. Systematic work over the last 20 years has identified many proteins with chaperone-like activity. Chaperonins refer to a sub-group of these proteins that function as components of multimeric systems.

Many newly synthesized proteins reach their folded states *in vivo* spontaneously and without assistance, in processes analogous to the folding of ribonuclease *in vitro*. However, folding efficiency may be limited by side-reactions such as aggregation (Figure 11.26) promoted by transiently exposed hydrophobic surfaces. In *E. coli* aggregation of proteins is readily detected by SDS–PAGE after heat shock. Cells respond to heat shock and the production of significant amounts of unfolded protein by the synthesis of new systems designed to promote refolding. These systems are the molecular chaperones.

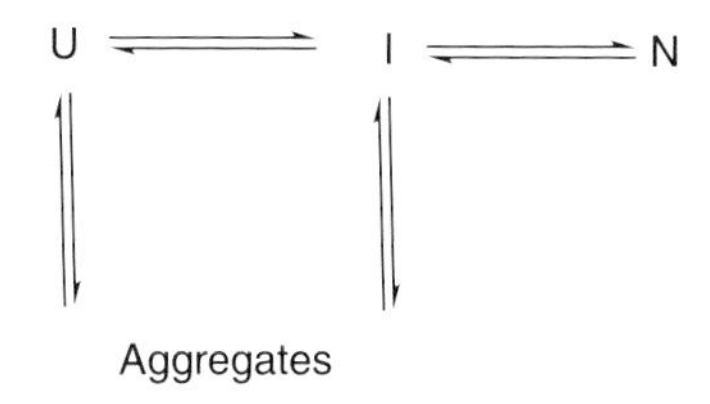

Figure 11.26 Possible routes of aggregation for unfolded and partially-folded protein. Chaperones limit the extent of aggregation reactions by binding to U and I forms

Several molecular chaperone systems have been characterized in *E. coli* including the GroES-GroEL, DnaK-DnaJ-GrpE and ClpB molecular chaperone systems. The acronym Hsp (heat-shock protein) is still used to describe these chaperones and the literature is full of terms such as Hsp60, Hsp70 and Hsp100 for families of homologous proteins. Today these terms are used interchangeably and the Hsp60 name could refer to GroEL – one component of the best-characterized chaperone.

Molecular chaperones are distributed in all cells ranging from archaebacteria through to complex eukaryotic cells. Although many chaperone systems have not been structurally characterized the GroEL-ES, the thermosome of *Thermoplasma acidophilum* and small HSPs from *Methanococcus jannaschii* represent exceptions.

The GroEL-ES system

The GroEL-ES chaperone system is essential for *E. coli* growth under all conditions with mutations in the *GroE* operon proving lethal. The GroEL-ES complex of *E. coli* has formed the basis of understanding chaperone function as a result of detailed structural studies using crystallography and electron microscopy. *In vivo* GroES is composed of seven identical subunits (M_r~10 000) whilst GroEL is composed of 14 larger subunits each with a mass of ~60 000. This leads to Gro-EL being called chaperonin60 (cpn60) and Gro-ES chaperonin10 (cpn10).

A model for the organization of GroEL obtained from negative staining EM studies showed a double ring cylindrical structure with a diameter of ~14 nm and height ~16 nm. The seven-fold symmetrical complexes were of comparable size to the ribosome with a central cavity of ~6 nm that bound polypeptide prior to ATP binding or GroES attachment. The first detailed structure for GroEL was determined by Paul Sigler and provided beautiful detail of the toroidal architecture indicated from electron microscopy. GroEL had a substantial central cavity in a cylindrical structure composed of two stacked heptameric rings arranged with seven-fold rotational symmetry. The rings are arranged back to back, contacting each other through a region known as the equatorial interface that is one of three distinct domains found in each GroEL subunit. The remaining domains are the apical and intermediate domains; these zones are simply designated as E, I and A domains (Figure 11.27).

GroEL contains a single polypeptide chain of 550 residues starting at the E domain extending to the intermediate region before the central part of the sequence makes up the A domain. The sequence then returns to form part of the I domain before terminating in the E domain. The E domain is a helical rich region that provides the ATP/ADP binding site whilst the apical domain shows greater mobility and has a lower percentage of regular secondary structure. Schematically the GroEL subunit is a tripartite structure (Figure 11.28).

The structure of GroES was determined independently of GroEL to confirm seven-fold symmetry arranged in a dome-shaped architecture. The dome contained seven subunits each of 110 residues and formed a core β barrel structure with two prominent loop structures (Figure 11.29).

One remarkable feature of the GroEL structure is a massive conformational change caused by binding co-chaperonin (GroES) and nucleotides. Revealed by cryo-EM studies of the ternary complex formed between GroEL/ADP/GroES image reconstructions showed a small elongation of the cylinder upon ATP binding but a very substantial elongation at one end of the GroEL complex upon GroES binding. This was accompanied by increases in diameter at the end cavities with the GroES ring observed as a

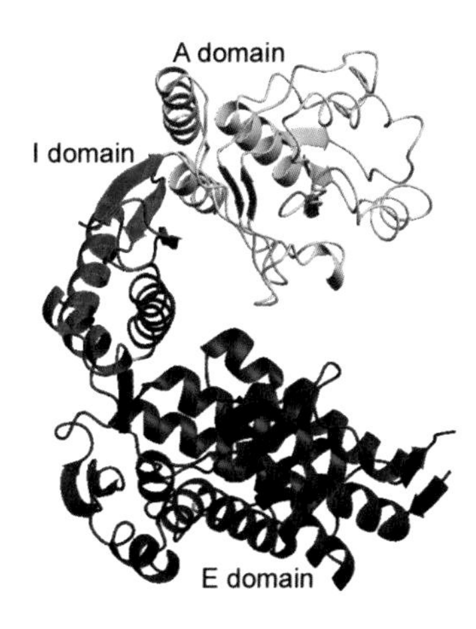

Figure 11.27 The monomeric structure of GroEL showing relative positions of E, I and A domains (PDB: 1GRL)

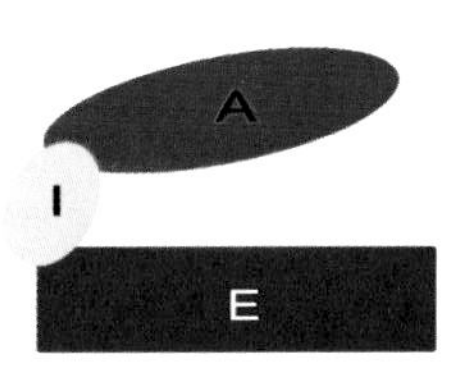

Figure 11.28 The tripartite structure of a GroEL subunit showing a schematic arrangement of apical (A), intermediate (I) and equatorial domains

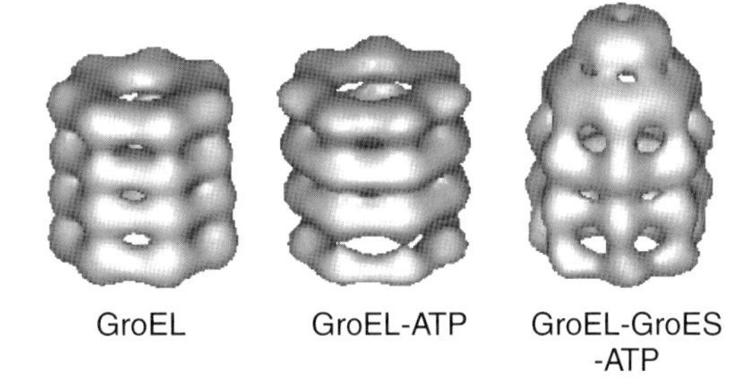

Figure 11.30 3D reconstructions of the surfaces of GroEL, GroEL-ATP and GroEL-GroES-ATP from cryo-EM. The GroES ring is seen as a disc above the GroEL and contributes to the bullet-like shape. (Reproduced with permission from Chen, S. *et al. Nature* 1994, **371**, 261–264. Macmillan)

flat, disc-like object above the neighbouring GroEL heptamer (Figure 11.30). In the presence of adenine nucleotides GroES binds to the stacked rings of GroEL forming a 'bullet-like' asymmetric structure with GroES attached to only one of the heptameric rings of GroEL (the *cis* ring).

The chaperonin-assisted catalysis of protein folding proceeded through cycles of ATP binding and hydrolysis. Nucleotide binding modulates the interaction between GroEL and GroES binding. ATP binding to GroEL occurs initially with low affinity ($K_d \sim 4$ mM)

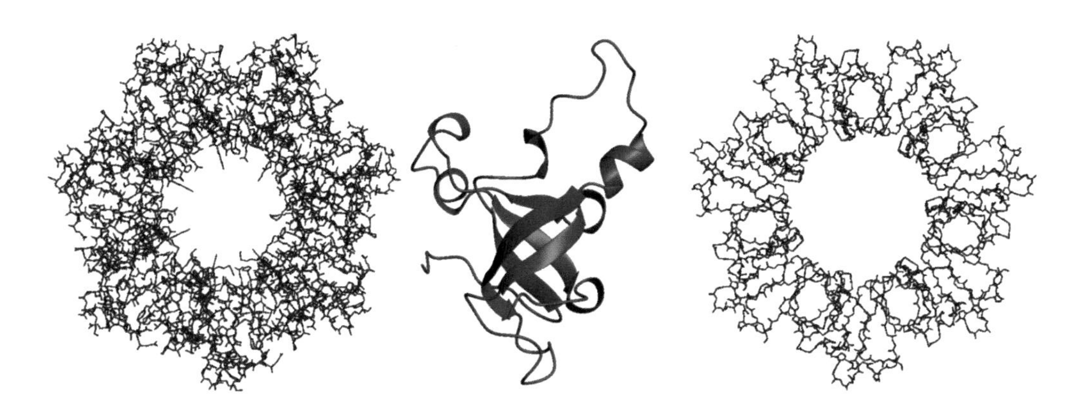

Figure 11.29 The structure of GroES within the heptameric ring seen in side and superior views. One subunit within the ring is shown in red. The arrangement of secondary structure showing the β barrel core is shown from the side in the centre for an individual subunit (PDB: 1G31)

to one subunit but as a result of positive cooperativity binding occurs with substantially higher affinity to the remaining subunits ($K_d \sim 10\ \mu M$). The cooperative nature of ATP binding establishes GroEL as an allosteric enzyme. In GroEL with a total of 14 subunits in two heptameric rings it is observed that tight nucleotide binding occurs only to one of the two available rings. ATP binding to the remaining ring is inhibited but promotes GroES association and a change in conformation for the *cis* ring subunits. The change in conformation is central to the switch in binding modes from an allosteric T state with low affinity for ATP and high affinity for non-native chains to an R state that shows the opposite trends.

The molecular basis for some of the conformational changes have been understood by comparing the structures of Gro-EL complexes in the presence of Gro-ES as well as ADP, ATP and AMP-PNP, using cryo-EM methods. Although the resolution of these studies was relatively low by crystallographic standards the large size of the complex meant that levels of resolution were sufficient to identify key catalytic features. Nucleotide binding in the equatorial domain pocket leads to substantial changes in the GroEL heptameric ring structures and movement in particular for the apical domain. The result of domain movement is the existence of several distinct structures.

Crystallographic analysis of the GroEL-GroES-$(ADP)_7$ complex showed a similar bullet shape and by comparing this structure with the un-liganded GroEL structure the significant conformational perturbations were shown to arise from 'en bloc' movements of I and A domains with respect to the E domain in the *cis* ring capped by GroES (Figure 11.31). The overall architecture of GroEL and the GroEL-GroES-$(ADP)_7$ complex emphasized the bullet shape

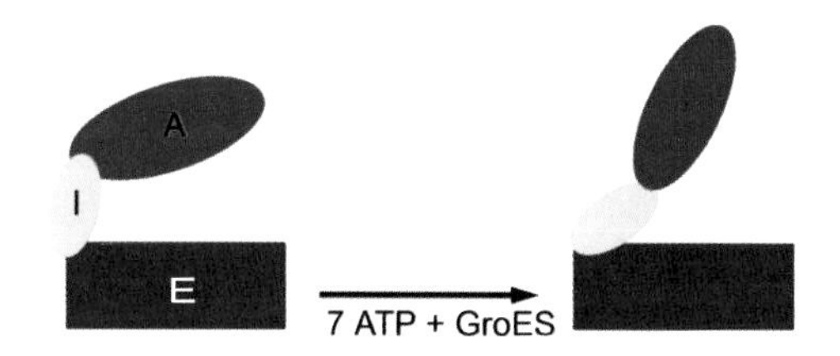

Figure 11.31 Domain movement in GroEL as a result of GroES and ATP binding

(Figure 11.32) and allowed insight into the mechanism of binding partially folded proteins and preventing protein aggregation.

The reorganization of the *cis* ring of GroEL results from domain re-arrangement involving intermediate and apical domains with the I domain swinging downwards towards the E domain by approximately 25°. One effect of this movement is to close the occupied nucleotide (ATP) binding site located on the top inner surface of the equatorial domain. The A domain shows greater movements, swinging ~60° upwards relative to the equator, but also twisting about its long axis by ~90° to form new interfaces with neighbouring A domains. These movements lead to interactions between the A domain of GroEL and mobile loops on GroES. In comparison with the A domain the *cis* equatorial (E) domains do not show large conformational shifts, with smaller inward movement of the *cis* assembly by ~4°. Since these regions interact with the neighbouring heptameric ring there is a complementary outwards tilt in the E domains of the *trans* ring. These movements have important functional consequences in the overall cycle of catalysis. In contrast to the large conformational changes occurring in GroEL, the structure of the GroES ring within the complex is similar to that observed in the standalone structure. A minor exception to this rule would include mobile and disordered loops that become structured within the complex as a result of interaction with the A domain.

The chaperonin catalytic cycle

With the structure of the 'inactive' GroEL and 'active' GroEL-GroES-ATP complexes known it was possible to attempt to understand how this macromolecular complex assists in protein folding. At the simplest level the complex catalyses cyclic binding and release of target polypeptides. This process is divided into four distinct stages or phases (Figure 11.33). Phase I involves polypeptide binding; phase II the release of the polypeptide into the central channel and the initiation of folding; phase III involves hydrolysis of ATP, reorganization of the *cis* heptameric ring and the start of product (GroES, folded peptide and ADP) release; finally, phase IV involves ATP binding to GroEL subunits of the *trans* ring providing the trigger

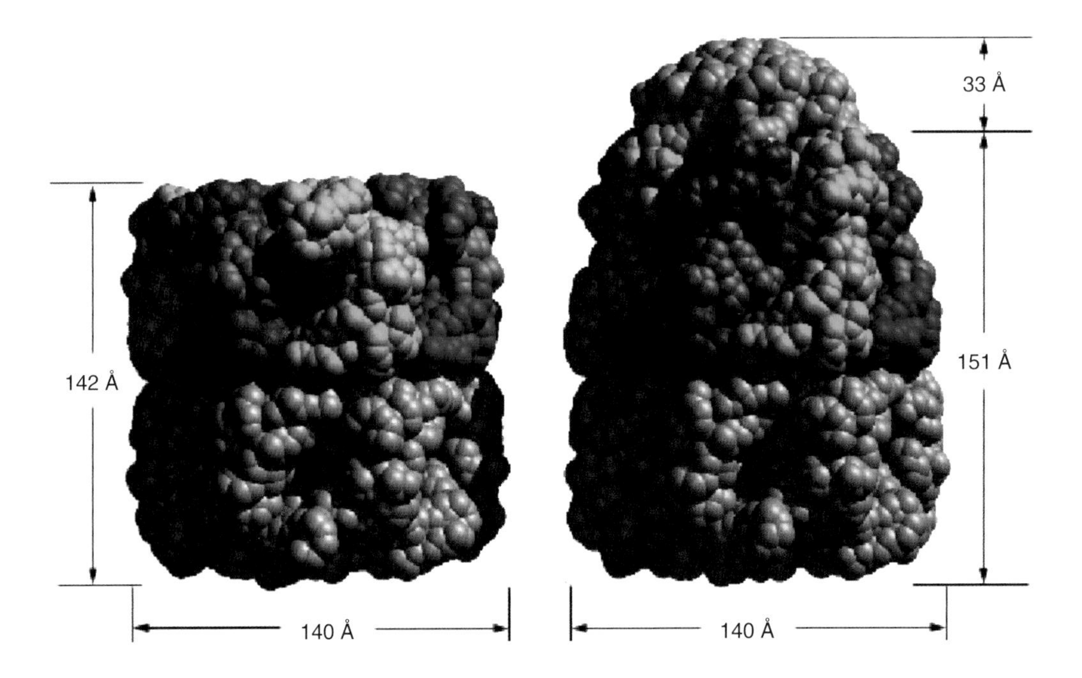

Figure 11.32 Structures of GroEL and GroEL-GroES-$(ADP)_7$ complex derived by X-ray crystallography. Different colours denote the individual subunits in the upper heptameric ring with the apical, intermediate and equatorial regions show in progressively darker hues. In the lower ring all subunits are shown in yellow. GroES is shown in grey (reproduced with permission Sigler, P.B. *et al. Annu. Rev. Biochem.* 1998, **67**, 581–608. Annual Reviews Inc)

to discharge GroES and entrapped folded polypeptide from the opposite side of the structure.

The cavity within GroEL provides a 'reaction vessel' in which the polypeptide chain can be 'incubated' until folding is complete. This vessel is called the 'Anfinsen cage' and estimations of internal volume place an upper limit of $\sim 85\,000$ Å^3 which is consistent with a spherical protein of 70 kDa.

In the first phase unfolded polypeptide binds to the apical domain largely through hydrophobic interactions and the direct involvement of at least nine residues (determined by site-directed mutagenesis) on each subunit. Eight of these residues have hydrophobic side chains and point inwards into the cavity creating a hydrophobic ring that binds target polypeptides. The central channel of GroEL functions as two separate cavities, one in each ring, as a result of a disordered 24 residue C-terminal region in each of the seven ring subunits that effectively 'blocks' communication between channels.

The second stage of the catalytic cycle involves nucleotide binding. From observations of GroEL-assisted refolding experiments it was noted that GroEL alone inhibits refolding whilst in the presence of K^+ ions, Mg-ATP and GroES efficient folding to the native state was seen. This led to a proposal that at least two distinct conformations of the complex existed; one that binds unfolded polypeptides tightly and ATP weakly and another in which the binding properties are reversed. Nucleotide-modulated conformational changes of GroEL are inherent to the protein folding cycle. ATP binding preceded GroES binding with the latter event occurring at near diffusion controlled rates (4×10^7 $\mathrm{M}^{-1}\,\mathrm{s}^{-1}$) in the presence of ATP, but approximately 100 times more slowly in the presence of ADP.

The third stage involves completion of polypeptide folding coupled with its release into the cavity of the *cis* ring of the GroEL-ES complex. Binding of unfolded polypeptide to the A domain has suggested

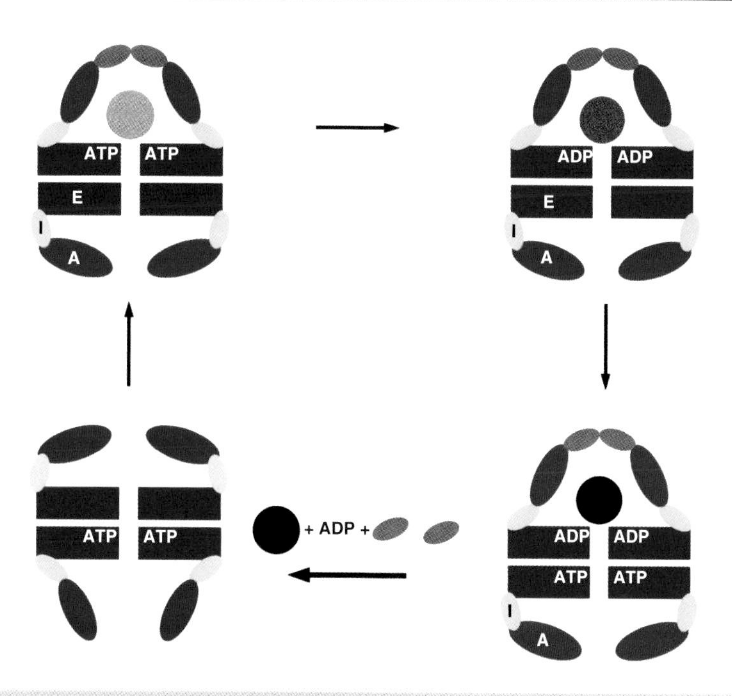

Figure 11.33 The chaperonin catalytic cycle (phases II–IV). ATP and GroES (purple subunits) bind to the GroEL ring causing change in conformation. ATP is trapped inside complex. In this state non-native polypeptide chains will no longer bind to apical domain but are released into central channel. ATP on *cis* ring is hydrolysed. Protein folds in central cavity. ATP binds to *trans* ring causing conformational change in opposite ring that releases GroES, ADP and folded peptide

the involvement of residues Leu234, Leu237, Val259, Leu263, and Val264. Mutation of any of these residues inhibits correct protein folding and in some instances proves lethal to *E. coli*. ATP and GroES binding cause 'en bloc' movements of the E, I and A domains and lead to competition for these sites within the complex and the rotation and displacement of the heptameric rings in opposite directions. As a result, unfolded chains are no longer bound and are released into the cavity of the *cis* ring where folding occurs. The non-polar side chains of residues responsible for binding the non-native polypeptide through hydrophobic interactions in the cavity of the un-liganded ring are now buried in the *cis* ring assembly. This leads to their replacement by polar residues on the walls of the cavity. This change is very significant because it now drives folding where the released polypeptide chain is able to re-initiate folding in an enlarged cavity whose lining is now hydrophilic.

The fourth phase represents the dissociation of the ligands (GroES and ADP) from the complex. The disassembly of the *cis* ring complex is triggered by hydrolysis of ATP to ADP with the latter forming weaker interactions at the nucleotide binding site as a result of the loss of a phosphate group. Weaker interactions breakdown the GroEL–GroES complex with the susceptibility to disassemble increased further by binding ATP to the opposite ring. This event also initiates the start of another cycle of binding of the non-native polypeptide chain and GroES association.

In summary, the GroEL-ES complex represents a remarkable molecular machine designed to ensure protein folding occurs efficiently and without the potential for complicating side reactions. As with many important protein complexes it appears that nature has seized the 'design' and used it in other chaperonin-based systems. Whilst GroEL and GroES are respectively the prototypical chaperonin and co-chaperonin there exist in other cells chaperonins that form similar ring-shaped assemblies. The most convincing evidence derives from the structure of the corresponding complex from hyperthermophilic bacteria.

It is likely that only a fraction of all cytosolic proteins found in *E. coli*, or indeed in other cells, are processed via molecular chaperonins. Some proteins will be too large to be accommodated within the central cavity of chaperonins whilst other nascent polypeptides, particular small soluble domains, can fold 'safely' without aggregating and without the intervention of other proteins. Estimates of the number of proteins acting as substrates for GroEL vary but at least 300 translated polypeptides including essential enzymes of the cell have been identified as using this system. About one-third of these proteins unfold within the cell and repeatedly return to GroEL for conformational maintenance. It appears that GroEL substrates consist preferentially of two or more domains of the α/β class containing mixtures of helices and buried β sheets with extensive hydrophobic surfaces. Such proteins are expected to fold slowly and to be aggregation-prone whilst the hydrophobic binding regions of GroEL are well suited to interaction with non-native states of α/β proteins.

The thermosome of T. acidophilum

The characterization of the GroEL complex as a toroidal structure based on two heptameric rings capped by a dome like GroES ring defined a major group of chaperonins found widely throughout cells. Systems based on this design are called Type I chaperonins. Sequence homology studies show that eukaryotic cells have similar protein complexes. These complexes are located within the mitochondria and chloroplasts with Hsp60, and Cpn60 representing two prominent systems. This similarity is not surprising in view of the endosymbiont origin of these organelles from eubacteria.

In *T. acidophilum* comparable reactions are catalysed by thermosomes that show variations in organization and are example of Type II chaperonins. The structure of the thermosome from *T. acidophilum* was the first Type II chaperonin to be described in detail. It is a hexadecamer composed of two eight membered rings. Each ring contained an alternating series of two polypeptide chains denoted as $(\alpha\beta)_4(\alpha\beta)_4$. Group II chaperonins in archaea and in the eukaryotic cytosol have been shown to use the same mechanism as type I chaperonins, with the binding of substrate to a central cavity and ATP dependent substrate release. The thermosome shares many of the properties of the GroEL rings although the lid structure is derived from the secondary structure elements of the apical domains and does not involve GroES like subunits to cap a central cavity.

Strands S12 and S13 and most of the N-terminal part of helix H10 protrude toward the pseudo eight-fold molecular axis at the ends of the particle to block the entrance to the central cavity and form a lid domain. The β strand (S13) from each A domain forms a circularly closed β sheet of eight strands and leaves a central pore of diameter $\sim$2.0 nm. The triangle of structural elements, strands S12 and S13 and helix H10, pack against corresponding elements in the neighbouring two subunits forming a hydrophobic core with nine non-polar side chains and a threonine residue from each subunit (Figure 11.34). The core is hydrophobic with a third of all lid segment residues buried completely and these interactions assist in maintaining the lid structure in a conformation comparable to that observed in the GroEL-ES complex.

Membrane protein folding

The preceding sections have dealt largely with the folding of soluble proteins *in vitro* and the assistance of macromolecular assemblies in governing folding *in vivo*. As with much of biochemistry the details of folding in soluble proteins has largely outpaced understanding of the same reactions in membrane bound systems. A key observation made in 1982 by H. Gobind Khorana was that bacteriorhodopsin could be denatured and refolded in a similar fashion to ribonuclease. This confirmed that membrane proteins possess all of the information necessary for folding and formation of the native state within their primary sequences. In the case of bacteriorhodopsin this involved the formation

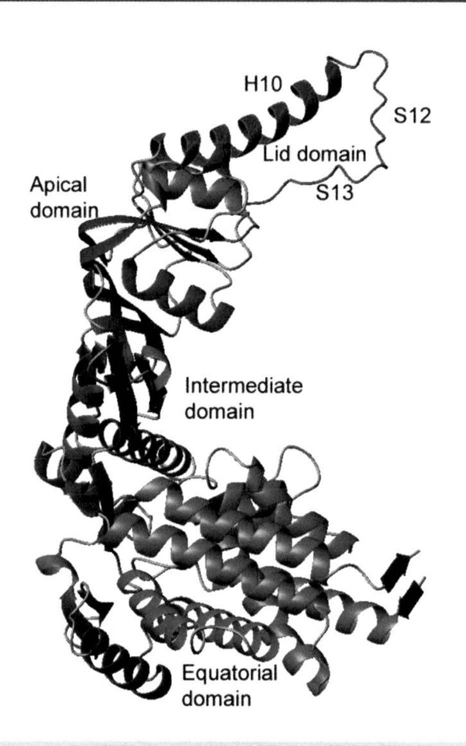

Figure 11.34 The structural arrangement of the thermosome E, I and A domains with the extended helix-strand-strand regions of H10, S12 and S13 shown in red (PDB: *1AG6*)

of seven transmembrane helices, but similar reactions have not been demonstrated for all membrane proteins. Most data currently available refers to the β-barrel rich domains of porins and the helix-rich domains of rhodopsin-like proteins.

Soluble proteins represent the minimum free energy conformation, as defined in Anfinsen's original experiments, and membrane proteins are no exception. However, the mechanism of reaching this folded state *in vitro*, and especially *in vivo*, is less obvious. In soluble proteins thermodynamic studies of the folding process in response to denaturing agents provided information on forces stabilizing tertiary structure. This approach is not successful for membrane proteins as they are very resistant to denaturation. Bacteriorhodopsin does not unfold completely in SDS, retaining helical structure, and denaturation arises primarily from subunit dissociation, loss of interactions between secondary structure elements and from unfolding of domains located either in the aqueous phase or at interfaces.

As a membrane protein bacteriorhodopsin is representative of proteins with seven transmembrane helices – a very large family – with the added advantage of a chromophore that is exquisitely sensitive to changes in physical properties. In the purple membrane bacteriorhodopsin is found as a two-dimensional crystalline lattice where the protein aggregates as collections of trimers. Two thermal transitions with enthalpies of denaturation (ΔH_d) of ~30 and ~400 kJ/mol have been linked to a dissociation of the lattice and dispersion into individual trimers at 80 °C followed by irreversible disruption of trimers via the loss of helix–helix interactions between monomers at 100 °C.

One framework for describing membrane protein folding is the two-stage model proposed by Donald Engelman that involves in the first step the independent formation of hydrophobic helices of ~25 residues upon insertion into the membrane. Stage II involves the ordering of structure where interhelix interactions drive proteins towards the native state. Insertion and ordering are believed to be separate and independent events. Supporting evidence for this model was obtained using fragments of the polypeptide chain of bacteriorhodopsin where five helices (A–E) assembled in the absence of the rest of the protein. This confirmed the ability of fragments to form stable secondary structure in membranes.

These studies emphasize the thermodynamics of protein folding, but an equally relevant aspect of the overall process is the kinetic parameters associated with formation of the native state. Folding kinetics of bacteriorhodopsin measured using CD and absorbance spectrophotometry indicate a rate limiting step is the formation of transmembrane helices with the formation of one or more helices allowing helix–helix packing and rapid completion of protein folding. Completion occurs upon retinal binding and a pathway defined by a series of intermediates with near native secondary structure but lacking the tertiary organization is observed (Figure 11.35).

Intermediate I_1 is formed within a few hundred ms whilst the transition from I_1 to I_2 is rate limiting and equated with the formation of transmembrane helices.

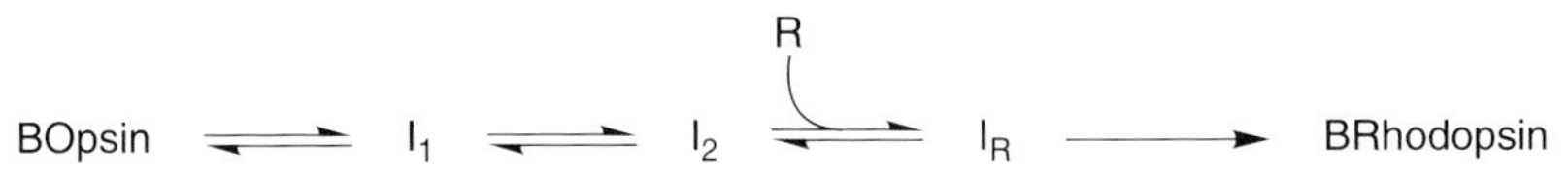

$$\text{BOpsin} \rightleftharpoons \text{I}_1 \rightleftharpoons \text{I}_2 \xrightleftharpoons{\text{R}} \text{I}_\text{R} \longrightarrow \text{BRhodopsin}$$

Figure 11.35 Sequential folding pathway for assembly of bacteriorhodopsin from co-factor (R) and apo-protein (bacterio-opsin)

It occurs slowly on a time scale of 10–100 s. These kinetics distinguish the process from those seen in soluble proteins that are frequently complete within 1 s. It seems unlikely that these reactions could occur so slowly *in vivo*, although data is lacking.

Fairly obviously, folding schemes derived from studies of helical membrane proteins may not be of great relevance to systems based around the β barrel, such as porins. Despite different structures experimental evidence suggests similarities extend to the processes of insertion and assembly. Porin OmpA is monomeric and this simplifies protein folding since there is no assembly into trimers. OmpA is synthesized with a signal sequence that is cleaved in the periplasm to leave a mature protein that inserts into membranes. *In vitro* OmpA can be completely refolded in artificial lipid bilayers and because of the moderate hydrophobicity of the individual strands of the β barrel these proteins are easily extracted from membranes in an unfolded state with urea or guanidinium chloride. In this state OmpA has been shown by CD spectroscopy to be unfolded at high denaturant concentrations. Rapid denaturant dilution in the presence of lipid leads to folded, membrane-inserted, conformations with several 'events' identified during a folding pathway distinguished by time constants of several minutes *in vitro*. The formation of individual β strands is unlikely and has not been described for any isolated sequence, suggesting that insertion of OmpA into the membrane may involve partial folding and the interaction of several strands. Despite a slow series of reactions the broad similarity to the folding of bacteriorhodopsin – a protein with very different topology – is reassuring and suggests that an outline of the basic pathway of membrane protein folding has been uncovered. Gradually, the folding of membrane proteins is becoming understood and it is clear that no new principles are involved in the formation of the native state (see Table 11.4).

Translocons and in vivo membrane protein folding

The details of membrane protein folding *in vivo* involve co-translational insertion of nascent proteins into the ER membrane at sites termed translocons. Translocons consist of membrane proteins that form a pore into which newly synthesized polypeptide chains enter as part of the normal cell trafficking pathways. Helical membrane proteins are targeted to the ER membrane in eukaryotic cells and to the plasma membrane in bacterial cells by a signal sequence composed of approximately 20–25 residues. Interaction between the signal sequence and the signal recognition particle (SRP) arrests translation and promotes binding to receptors with the direction of polypeptide chains into the translocon followed by integration into the membrane.

Uncovering the organizational structure of the translocon relied on yeast genetic studies of protein secretion (*SEC*) genes. Using electron microscopy it was possible to identify the cytological event perturbed by lesions and one set of experiments identified mutants (*SEC61*) defective in transport into the ER with unprocessed proteins accumulating in the cytosol. Subsequent cloning of the mammalian version of *SEC61* established a close homology with yeast and between prokaryote and eukaryote. The *SEC61* gene encoded the main channel-forming subunits within the translocon.

The translocon or Sec61 complex consists of three integral membrane proteins Sec61α, Sec61β and Sec61γ. Sec61α has many transmembrane helices whilst Sec61β and γ contain only single helical segments crossing the bilayer. In 1993 elegant reconstitution experiments by Tom Rapoport showed that it was possible to mimic protein translocation in artificial membranes containing the incorporated Sec61 complex although the catalytic efficiency of the reconstituted

Table 11.4 Some of the membrane proteins whose folding has been studied

Protein	Structure and function	Protein denaturation	Formation of folded protein
Bacteriorhodopsin	7 TM helices and retinal co-factor; light-driven protein pump	Apoprotein fully denatured in trifluoroacetic acid; apoprotein partly denatured in SDS, leaving ~55 % of native helix	Transfer from organic acid to SDS and then folded by mixing with lipid vesicles
LHC-II	3 TM α-helices and 1 short amphipathic helix at membrane surface. Photosynthetic light-harvesting protein	Apoprotein partly denatured in SDS, leaving about 30 % of native helix content	Reconstituted with thylakoid membrane extracts by freeze–thaw. Reconstituted in micelles containing pigments, lipid and detergent
E. coli DGK	Unknown structure	Apoprotein slightly denatured by SDS, leaving ~85 % of native helix content	Refolded in DM micelles
E. coli OmpA	Membrane domain of 170 amino acids in 8-stranded β barrel. Exact function unknown, but likely channel	Protein completely denatured in urea	Folding in lipid vesicles by mixing with urea-denatured state
E. coli OmpF	16-stranded β barrel. Trimer forms pore in outer membrane	Completely denatured in urea or GdnHCl	Poor folding in lipid vesicles on mixing the urea-denatured state. Folding increased if detergent used in mixed micelles

Adapted from Booth, P. *et al. Biochem. Soc. Trans.* 2001, **29**, 408–413. Abbreviations: TM, transmembrane; DM, dodecylmaltoside; GdnHCl, guanidine hydrochloride; LDS, lithium dodecylsulphate; OG, octylglucoside; PG, phosphatidyl-D,L-glycerol; DGDG, digalactosyl diacylglycerol; LHC-II, light harvesting complex of higher plants; DGK, diacylglycerol kinase

complex was low. With structural characterization of the SRP and its receptor greater emphasis has been placed on the organization of the translocon (Figure 11.36). Electrophysiological studies emphasize the presence of an ion channel in the translocon whilst studies with fluorescent labels attached to residues within a nascent polypeptide chain indicate changes in environment and conformation. Much work remains to

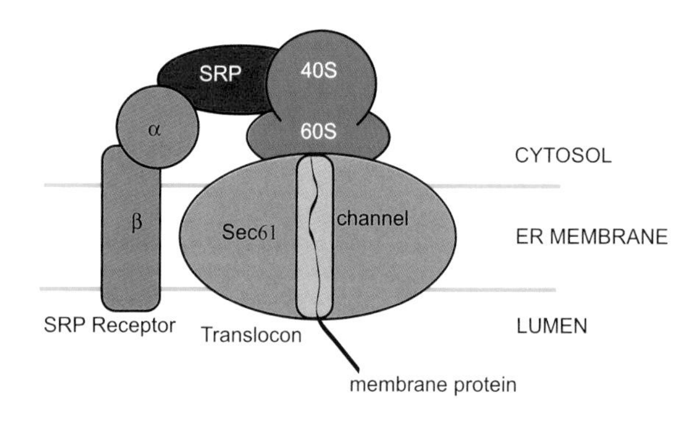

Figure 11.36 A model for the organization of the translocon. The signal recognition particle (SRP)-ribosome complex binds to receptor (α/β) in the ER membrane. The translocon consists of the Sec61 proteins (α, β, γ) and accessory proteins. Translocation requires GTP hydrolysis and the growing polypeptide chain is pushed across the membrane towards the lumen

be done in the characterization of the translocon and its role in membrane protein folding.

A key event in membrane protein folding is identification of a signal peptide by the SRP, but membrane proteins may contain many blocks of 20–25 non-polar residues each reminiscent of such a peptide. This could lead to unwanted transfer into the ER lumen and therefore requires specific recognition mechanisms. The translocon complex must also deal with integral proteins that differ considerably in topology. One answer to this problem lies in the presence of sequences within membrane proteins that control their assembly within the ribosome–translocon complex. These sequences are known as topogenic signals and include conventional N-terminal signal peptides, signal-anchor sequences, reverse signal-anchor sequences, and start-transfer/stop-transfer sequences.

Proteins with more than one transmembrane segment are possibly threaded into membranes through the effect of start- and stop-transfer sequences. In eukaryotes the main features of these transfer sequences are groupings of hydrophobic residues flanked by residues with positively charged side chains. The effect of these sequences is to promote and arrest the transfer of residues through the translocon (Figure 11.37). This process aids the formation of transmembrane helices but is also vital to the formation of important loop regions that frequently link such domains.

The mechanism by which transmembrane domains exit the translocon pore to reach the bilayer remains unclear. Evidence suggests that transmembrane helices fold within this pore and then exit to the bilayer either individually or in pairs. This scheme would lead to the final helix packing reactions occurring in the membrane bilayer after translocation.

Protein misfolding and the disease state

With a greater understanding of folding *in vivo* and *in vitro* has come the realization that diseases arise as a consequence of protein misfolding. Mutation within coding regions of genes can result in the insertion of a stop codon and the failure to synthesize full-length, folded, protein. Alternatively, mutations change the identity of one or more residues leading to a protein with altered folding properties. Ultimately, this research is traced back to studies of sickle cell anaemia identifying mutant haemoglobin as the basis for disease. Similar origins for disease (Table 11.5) occur in protein folding where mutations cause defective folding, aberrant assembly and incomplete processing.

Intracellular sorting and the emergence of defective protein trafficking

A key requirement for cellular function is that proteins are targeted to the correct compartment. In eukaryotic cells compartmentalization requires specialized sorting pathways and the ER and Golgi apparatus form part of the endocytic pathway where a key event is the budding (endocytosis) of membranes as part of the trafficking process. Defective protein folding often results in an inability to transfer mutant domains through the endocytic pathway.

One of the best examples described to date occurs in cystic fibrosis. Cystic fibrosis is very common in Caucasian populations with an incidence of approximately 1 in 20. In a homozygous state this results in a defective CFTR protein and a disease characterized by an inability to transport chloride across membranes effectively and problems achieving correct ion balance.

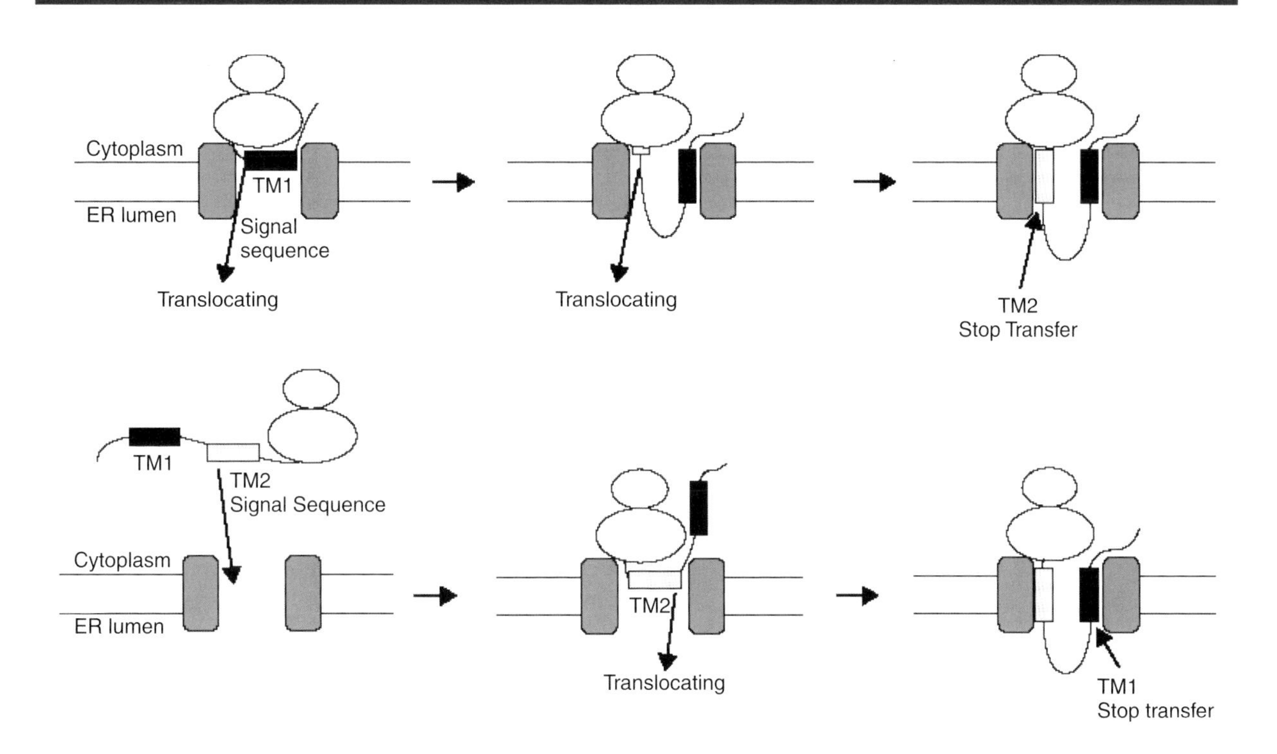

Figure 11.37 The function of topogenic sequences in membrane protein insertion. In the first mechanism TM1 functions as the signal sequence to initiate translocation of the carboxy-terminal region whilst TM2 functions as a stop transfer sequence halting translocation until the next start signal. Alternative mechanisms appear to exist such as that shown in the second pathway where TM2 initiates translocation of the N-terminal region and TM1 acts as the stop transfer signal (Reproduced with permission from Dalbey, R.E, Chen, M.Y., Jiang, F. & Samuelson, J.C. *Curr. Opin. Cell Biol.* 2000, **12**, 435–442. Elsevier)

Over 600 different mutations have been identified in the CFTR protein but the most common mutation involves deletion of one residue (Phe508). This does not lead to a truncated form of CFTR but instead results in a full-length protein missing a single residue. The mutant CFTR protein (ΔF508) is expressed and the translated product is easily detected. However, the protein is not processed further and is unable to complete post-translational glycosylation and does not transfer through the Golgi apparatus. Most of the protein becomes ubiquitinated and marked for destruction by the proteasome but a small amount ($\sim$1 percent) reaches the surface where it retains anion selectivity and conductance properties comparable to the wild type protein but has a decreased probability of ion channel opening. The result highlights how small 'defects' lead to major physiological impairment.

Protein aggregation and amyloidosis

Enormous interest has focussed on the appearance of large protein aggregates within tissues in diverse disease states. The appearance of protein deposits in close association with neurofibrillary networks in the cells of brain tissue has long been known to occur in Alzheimer's disease. Protein aggregation also underlies other diseases and may occur in tissues such as the liver and spleen, besides the brain. The protein aggregates are linked by formation of elongated fibrils (amyloid) and diseases showing this property are collectively grouped together by the term amyloidosis.[3]

[3]The term 'amyloid' was originally used to describe the resemblance of protein aggregates associated with disease states to the appearance of starch chains (amylose).

Table 11.5 Diseases arising from 'folding' defects

Disease	Protein or precursor involved in disease
Cystic fibrosis	CFTR
Primary systemic amyloidosis	Immunoglobulin light chain
Medullary carcinoma of the thyroid	Calcitonin
Osteogenesis imperfecta	Collagen (type I procollagen)
Lung, colon and other forms of cancer	p53 transcription factor
Maple syrup disease	α-ketoacid dehydrogenase complex
Amyotrophic lateral scoliosis	Superoxide dismutase
Creutzfeldt–Jakob disease, vCJD, scrapie, fatal familial insomnia	Prion protein
Alzheimer's disease	β-amyloid protein
Cataracts	Crystallins
Atrial amyloidosis	Atrial natriuretic factor (ANF)
Senile systemic amyloidosis, Familial amyloidosis	Transthyretin
Tay–Sachs disease	β-hexosaminidase
Hereditary emphysema	α-antitrypsin
Retinitis pigmentosa	Rhodopsin
Hereditary non-neuropathic systemic amyloidosis	Lysozyme
Transmissible spongiform encephalopathy (TSEs)	Prion protein
Huntington's chorea	Huntingtin
Familial hypercholesterolaemia	LDL receptor

Amyloidosis is apparent in many clinical conditions besides Alzheimer's and includes familiar diseases such as Parkinson's disease, type II diabetes together with less common conditions such as the spongiform encephalopathies. Some of these diseases are genetically determined (i.e. inherited) whilst others are acquired (sporadic) and others may be infectiously transmitted. In each amyloid disease a different protein or fragment aggregates, forming a fibril, and in systemic forms of amyloidosis aggregation leads to the deposition of kg quantities of protein. Ultimately, the progressive deposition of fibrils will cause death, especially when it occurs in a vital organ.

The association between formation of protein fibrils and irreversible protein aggregation was deduced from a number of studies but particularly those giving rise to a condition known as familial amyloidotic polyneuropathy (FAP). The disease was recognized accurately in the 1950s in a Portuguese family as an inherited condition. This disease centres around mutations in transthyretin, a protein involved in binding the hormone thyroxine.[4] Transthyretin is found in a binary complex with retinol binding protein (RBP) and acts as a soluble homotetrameric protein (Figures 11.38 and 11.39).

Each subunit contains 127 residues and a series of eight β strands that adopt a β barrel conformation. Seven of the strands (A–H) are seven to eight residues in length whilst the D strand is shorter and only three residues long (Figure 11.40). The eight strands form two sheets; the first is formed between strands *DAGH* and the second between *CBEF*. Unsurprisingly, the high content of β structure arranged in a barrel like topology contributes to significant stability.

Analysis of the molecular defects occurring in the transthyretin gene in individuals with FAP identified over 80 different mutations. The clinical symptoms of FAP usually begin from the third to fourth decades and are characterized by local neurologic impairment leading to wider autonomic dysfunction and death within 10 years. A common mutation noted in Portuguese, Japanese and Swedish kindreds is the transition Val30>Met. The identification and prevalence of this mutation led to an analysis of the structural and functional properties of Val30Met transthyretin especially since the mutant protein was known to be amyloidogenic.

Residue 30 lies on the inside of the monomer and the bulkier Met side chain displaces one set of four β strands from the remaining set of four. However, the mutant protein was folded with a conformation

[4]Transthyretin was formerly called prealbumin as a consequence of its migration pattern on gels just ahead of albumin.

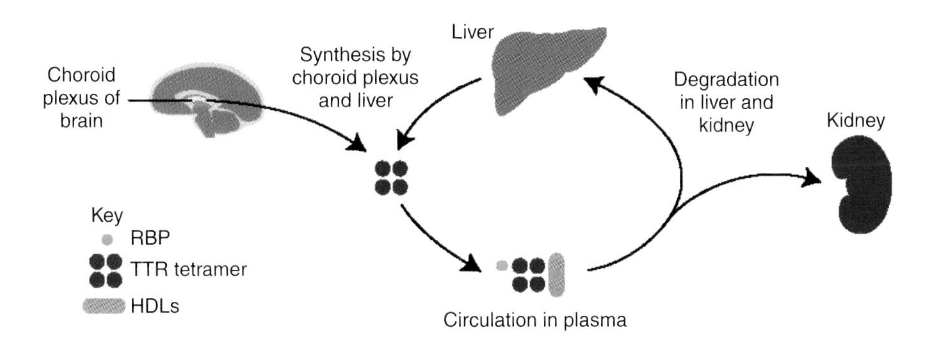

Figure 11.38 Synthesis, distribution and uptake of transthyretin. Transthyretin tetramer circulates in plasma bound to retinol-binding protein (RBP) providing a transport function for vitamin A and thyroxine with a small proportion binding high-density lipoproteins (reproduced with permission from Saraiva, M.J.M. *Expert Rev. Mol. Med.* Cambridge University Press, Cambridge, 2002)

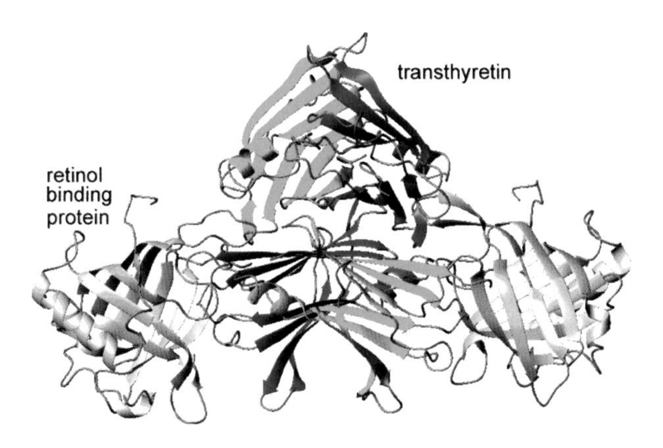

Figure 11.39 The complex of transthyretin and retinol binding protein. RBP is shown in yellow with the tetramer of transthyretin shown in blue and green. The larger RBP binds to a site formed by two of the subunits (PDB: 1RLB)

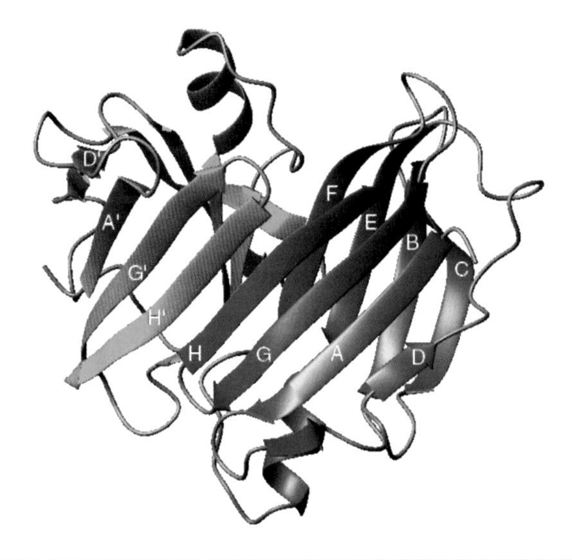

Figure 11.40 Transthyretin structure showing strands A–H and the sandwich formed between DAGH and CBEF in each monomer. Strands H and H′ interact at the interface (PDB: 1BMZ)

very similar to the wild type. This initially puzzling result was interpreted with the recognition that wild type protein aggregates to form fibrils under some conditions–an event associated with senile systemic amyloidosis, where fibril deposition in the heart leads to cardiomyopathy at the age of ~80. The results emphasized that amyloidosis is associated with *all* forms of the protein.

Mutations are viewed as exacerbating disease by promoting more active forms of amyloidosis. Structural studies of wild type protein at low pH (~4.5) show transthyretin dissociation to a monomeric but amyloidogenic intermediate of different tertiary structure involving rearrangement in the vicinity of the C strand–loop–D strand region. Dissociation of the tetramer is an unfavourable event under normal physiological conditions but these rates are

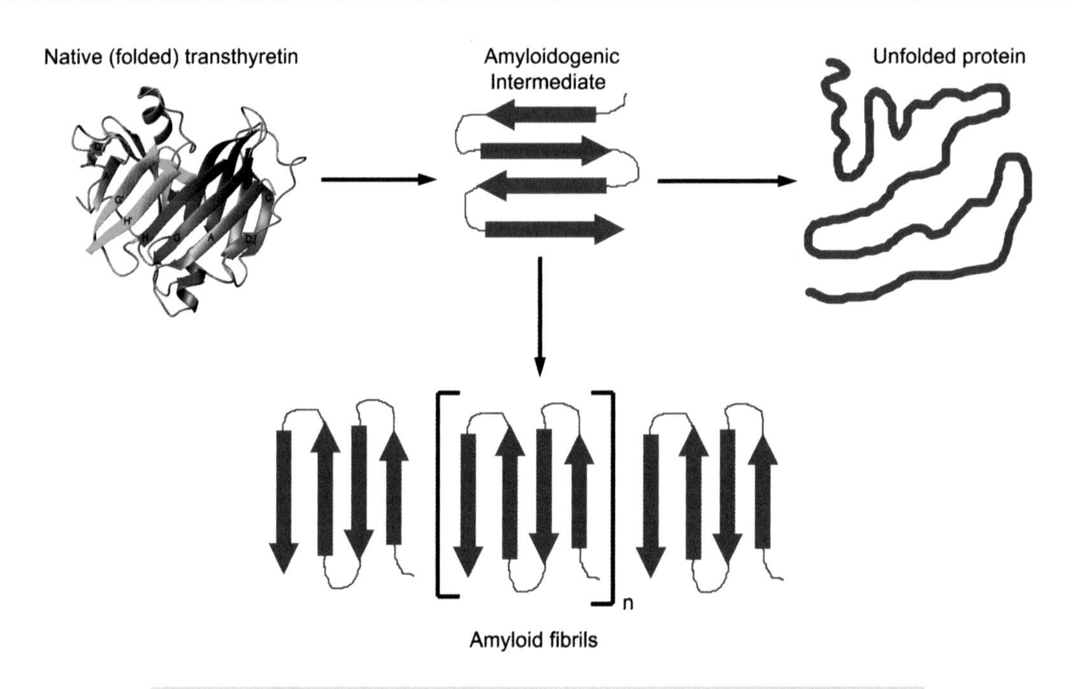

Figure 11.41 Scheme for initial formation of transthyretin-based fibrils

enhanced at low pH and for the Val30Met mutant a lower activation barrier increased the likelihood of amyloidogenic intermediates. Other mutants such as Leu55Pro confirmed this trend and led to the conclusion that mutants associated with FAP are less stable than wild type protein and form the amyloidogenic intermediate more quickly.

Individuals with the FAP mutations are kinetically and thermodynamically predisposed to amyloid fibril formation. Mutation increases the steady-state concentration of the amyloidogenic intermediate with the result that individuals with these defects are prone to amyloid formation and disease. The biophysical data mirrored the pathological outlook of the disease. Onset of the disease in the eighth, third and second decades correlated with the relative stabilities of wild-type, V30M and L55P forms of the transthyretin with the mutations shifting equilibria in favour of amyloidogenic intermediates and decreasing the age of appearance of FAP.

Fibril formation by the association of transthyretin results in comparable aggregates to those observed in other diseases and suggested a similar structure for all amyloid fibrils (Figure 11.41). Alzheimer's disease represents aggregates of a short fragment known as the A_β peptide. Since the A_β peptide and transthyretin do not share any sequence homology and are not related in any way the formation of similar fibril structures was remarkable and significant. The structure of amyloid fibrils examined using X-ray diffraction and cryo-EM methods reveal a β helix structure. In fibrils resulting from six precursor proteins present in different diseases a common structure existed and yet none of these proteins have sequence homology.[5] The morphology and properties of amyloid fibrils were based on a β scaffold with strands running transversely across a helix axis. In the direction of the helix axis the strands align to form β sheets. Four discrete chains are wound into the β helix (Figure 11.42).

[5]The systems studied were the A_β peptide; the λ immunoglobulin light chain (monoclonal protein systemic amyloidosis); lysozyme (hereditary lysozyme amyloidosis); calcitonin (medulary carcinoma of the thyroid); insulin (diabetes and insulin related amyloid disease); β_2 microglobulin (haemodialysis related amyloidosis), together with synthetic peptides of transthyretin (FAP) and the prion protein (transmissible spongiform encephalopathies).

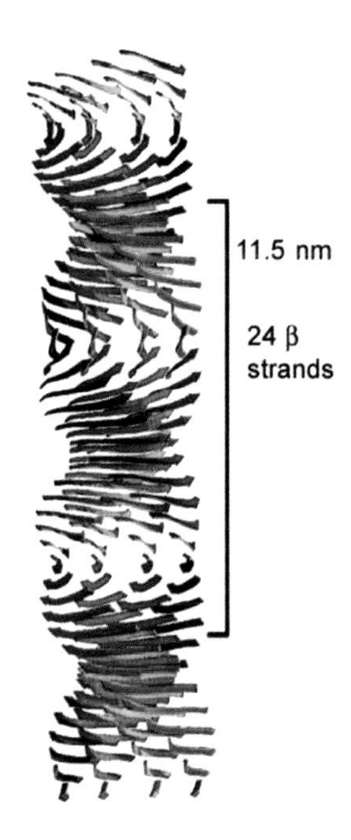

Figure 11.42 Model of a generic amyloid fibril structure. The structure of the amyloid fibrils is made up of four β sheets. Sheets run parallel to the axis of the protofilament with component β strands perpendicular. The twisting of β strands yields a series of parallel strands around a common helical axis that coincides with the axis of the protofilament and leads to a pitch of 11.55 nm containing 24 β strands (reproduced with permission from Sunde, M. *et al.*, *J. Mol. Biol.* 1997, **273**, 729–739. Academic Press)

These structures can be induced in protein not known to form amyloid fibrils in any documented disease phenotype. The SH3 domain (Figure 11.43) of the p85α subunit of phosphatidylinositol-3′-kinase was induced to form amyloid aggregates. The normal structure of the 84 residue domain is based around a β sandwich of five strands but under low pH conditions this domain unfolds in a reaction promoting fibril formation.

Mutating or 'mistreating' soluble proteins induces aggregation and the formation of fibrils. Fibrils derived from unfolded SH3 domains cannot form by simply linking together several units of the native fold because the diameter of the protein is too large to fit properly into the fibril structure. The SH3 domain must unfold to adopt a longer and thinner shape.

The assembly of β strands into a helix is a feature of amyloid fibrils and, although unusual, is known to occur in structural databases, with examples in folded proteins including alkaline protease, pectate lyases and the p22 tailspike protein. Collectively, these studies point to protein misfolding leading to aggregation and amyloid deposits.

As these studies progressed it became apparent that an obscure collection of neurodegenerative conditions resulted in a recognizable event, protein aggregation. These diseases were to become known as the prion-based diseases.

Prions and protein folding

Disease transmission requires the intervention of genetic material (DNA or RNA) in order to establish an infection. Most commonly, this event centres about bacterial or viral infections and the conversion of genetic information into protein with unpleasant consequences to the host. Gradually, convincing evidence has been acquired that a non-nucleic acid based agent is involved in transmission of a restricted number of diseases. These 'agents' were called 'proteinaceous infectious particles' or prions and were identified, after exhaustive analysis, as the active components in communicable and inherited forms of disease.

These ideas met with considerable scepticism and even hostility from some sections of the scientific and medical communities but there is good evidence that prions lie at the heart of neurodegenerative disorders associated with a pathological collection of diseases known as transmissible spongiform encephalopathies (TSEs). The common link between these diseases is development of gross morphological changes to tissue within the brain characterized by the formation of vacuoles and a sponge-like appearance (Figure 11.44). Disease progression is accompanied by the appearance of amyloid plaques within the tissue whilst in some states structural changes occur to cells leading to a loss of synaptic contacts. Three disease states showing these properties are scrapie, Creutzfeldt–Jakob disease (CJD) and Kuru.

Elucidation of the role of prions stems from attempts to understand the biochemical basis of these three obscure, and at first glance, unrelated diseases. Scrapie

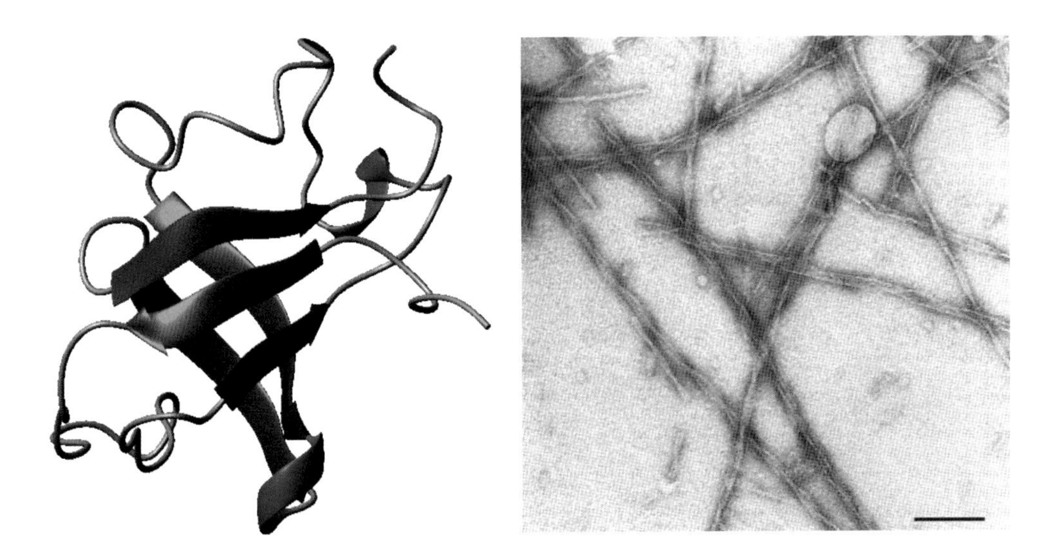

Figure 11.43 The normal fold of SH3 domains showing a sandwich based around five β strands (PDB:1PNJ) and fibrils induced to form via denaturation of SH3 domains at low pH. The solid scale bar is of length 100 nm (reproduced with permission from Guijarro, J.I. *et al. Proc. Natl Acad. Sci. USA* 1998, **95**, 4224–4228)

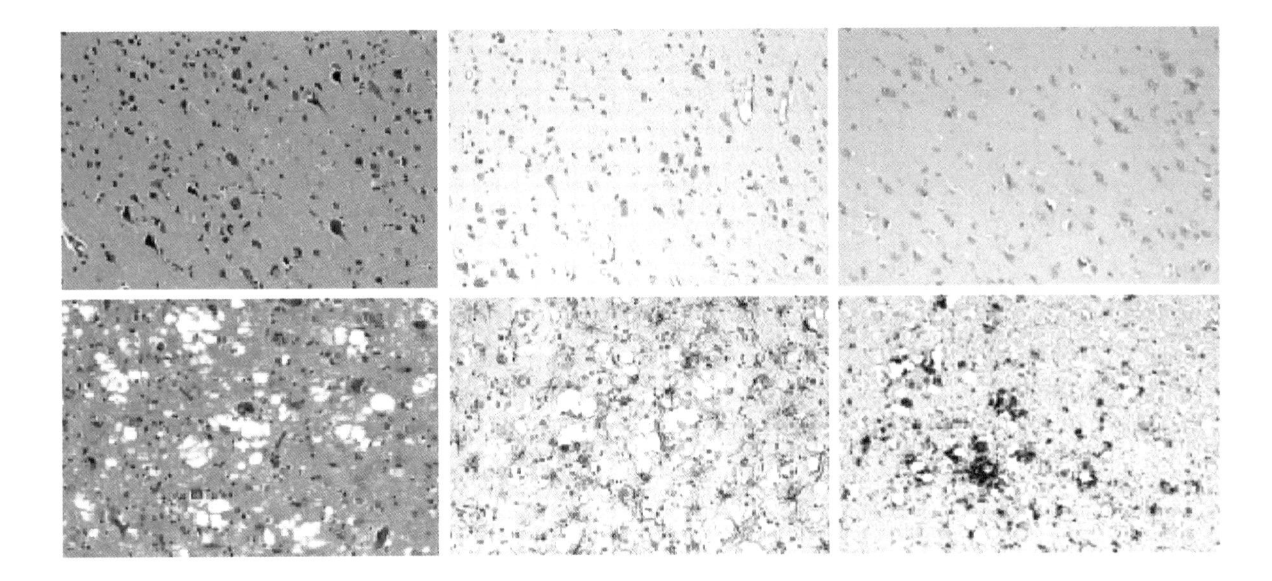

Figure 11.44 Neuropathological features associated with transmissible spongiform encephalopathies. Histological staining and immunohistochemical analysis of frontal cortex samples from a 'control' brain (top row) and of a patient suffering from CJD (lower row). Brain sections were stained with haematoxylin–eosin (left panel), with antibodies against glial fibrillary acidic protein (GFAP, middle) and with antibodies against the prion protein (PrP, right). Neuronal loss and vacuoles are visible in the H-E stain, proliferation of cross reactive astrocytes and prion protein deposits are detectable in the immunostains of the CJD brain samples (reproduced with permission from Aguzzi, A., Montrasio, F., & Kaeser, P.S. *Nature Rev* 2002, **2**, 118–126. Macmillan)

occurs in sheep and is defined by a progressive loss of motor coordination that leads to an inability to stand unsupported. Animals develop an intense itch that leads to wool scraping from skin and gives the disease its name. Scrapie has been recognized since the 17th century in the United Kingdom, with similar conditions noted more recently in animals such as mink, deer and elk.

CJD was first identified in the 1920s by H.G Creutzfeldt and then independently by A. Jakob. Patients exhibit dementia with poor motor coordination, poor perception and reasoning. The incidence of CJD is low, probably less than 1 case in two million, and typically occurs in individuals above 50 years of age. It is found throughout the world with little obvious association with economic, racial, or social patterns. Most frequently the disease arises sporadically, but in 10–15 percent of cases the disease is inherited in an autosomal dominant manner, and very rarely CJD is spread by iatrogenic transmission where contamination occurs most probably via donated tissue.

Kuru is an unusual disease confined to tribal people of the highlands of Papua, New Guinea where it is called the 'laughing death'. It was comprehensively described in 1957 as a fatal disease marked by ataxia and progressive dementia. A link was made between disease transmission and ritual cannibalism where tribes people honoured their dead by eating body parts including brain. A decline in cannibalism led to a decline in Kuru although the precise causative agent remained at that time unknown.

The three diseases are linked by the observation that transmission occurs when diseased brain extracts are injected into healthy animals. Of these diseases scrapie is experimentally accessible and initial studies attempted to identify the causative agent. Exhaustive analysis of scrapie infected brain tissue purified a single protein devoid of nuclei acid that was infectious when injected into hamster brains. It was called the prion protein (PrP) and the pioneering work in this area was attributable to the work of Stanley Prusiner and his extraordinary quest to define and characterize these agents.

Prion purification and the demonstration of infectivity raised a number of questions. How was PrP encoded? Was undetected DNA lurking in the 'pure' protein preparation? What is the origin of PrP? Major insight into the molecular events underpinning the generation of PrP arose with N terminal sequencing of the isolated protein. This sequence allowed the synthesis of degenerate oligonucleotide probes that hybridized to chromosomes found in a wide range of mammals including mice, hamsters and most importantly humans. The probes did not hybridize to the isolated PrP eliminating the possibility of trace amounts of DNA associated with this fraction. A gene encoding PrP was found on the short arm of chromosome 20 in humans where it coded for a glycoprotein of mass 33–35 kDa.

Most individuals never develop any form of the disease yet the prion gene is found in the human genome. Prusiner suggested one explanation. PrP was produced in two forms – a normal form and an abnormal one that generated disease. Immediate support for this view came with the demonstration that PrP found in infected brains was resistant to proteolysis whilst the normal form remained sensitive to proteases. The two forms are often called cellular PrP (normal) and scrapie PrP (infectious) – abbreviated as PrP^c and

Table 11.6 Structural differences between PrP^C and PrP^{Sc}

Property	PrP^c (cellular)	PrP^{Sc} (scrapie)
Polypeptide chain	1–231	1–231
Protease resistant	No	Stable core residues 90–231
Disulfide bridges	Yes (179–214)	Yes
Solubility	Soluble when expressed -GPI	Very insoluble except to all but the strongest 'solvents'
Aggregation state	Monomeric	Multimeric
Stability	25–40 kJ mol^{-1}	More stable than PrP^c
Structure	α helical	Increase in proportion of β strands

Adapted from Cohen, F. & Prusiner, S.B. *Annu. Rev. Biochem.* 1998, **67**, 793–819.

PrP^{Sc} (Table 11.6). The different proteolytic sensitivity of PrP^c and PrP^{Sc} suggested that each form of the protein possessed a different conformation.

Protein expression studies from the Syrian hamster, mice and humans showed monomeric species with little tendency to form aggregates whilst the full-length *infectious* conformer readily formed aggregates. PrP is composed of ~250 residues with a signal sequence of 22 residues, and close to the C-terminal residue is a GPI membrane anchor. Additional post-translational modifications include two N-linked glycosylation sites and a single disulfide bridge. In the absence of the GPI anchor the protein is soluble and does not partition in the membranes; this occurs when the protein is expressed in prokaryotic systems. As a result of its solubility it has been possible to subject the protein to detailed structural analysis.

The solution structure of a fragment containing residues 121–231 for the mouse PrP^c protein showed a folded domain with three α helices and two short β-strands (Figure 11.45). Structures of the analogous human protein (residues 23–231) and longer fragments from the mouse and Syrian hamster indicated that the prion protein consisted of a structured domain from residues 121–231 and a less ordered N-terminal domain. Very little secondary structure was detected in the N-domain. One reason for the absence of structure in the N-terminal region is seen in the primary sequence of the prion protein (Figure 11.46) where a series of five octapeptide repeats occur with the sequence PHGGGWGQ. The presence of this repeat is an unusual feature of the sequence.

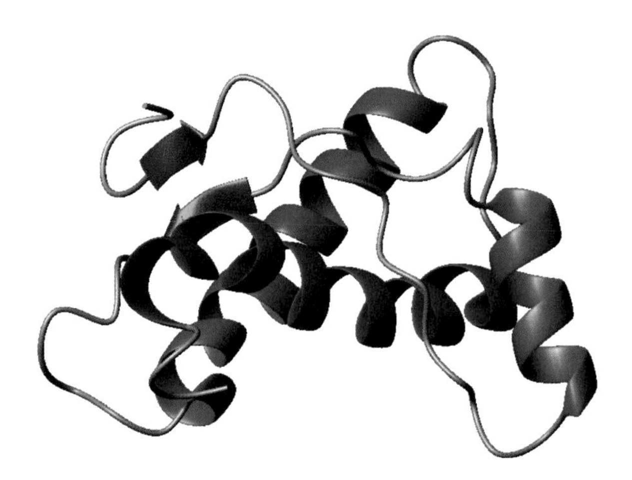

Figure 11.45 The structure of 121–231 fragment of the mouse PrPc determined using NMR spectroscopy (PDB: 1AG2)

Prions become infectious particles capable of transmitting disease (scrapie) by corrupting the conformation of other protein molecules and initiating a series of events that lead to the formation of protein fibrils or amyloid deposits within tissues.

Two additional TSEs, Gerstmann–Straussler–Scheinker (GSS) disease and fatal familial insomonia (FFI), also cause spongiform changes in the brain, neuronal loss, astrocytosis and amyloid plaque formation. GSS disease is an inherited neurodegenerative condition that offered the possibility of identifying mutations. Five different TSEs originate from different mutation sites within the prion gene (Table 11.7).

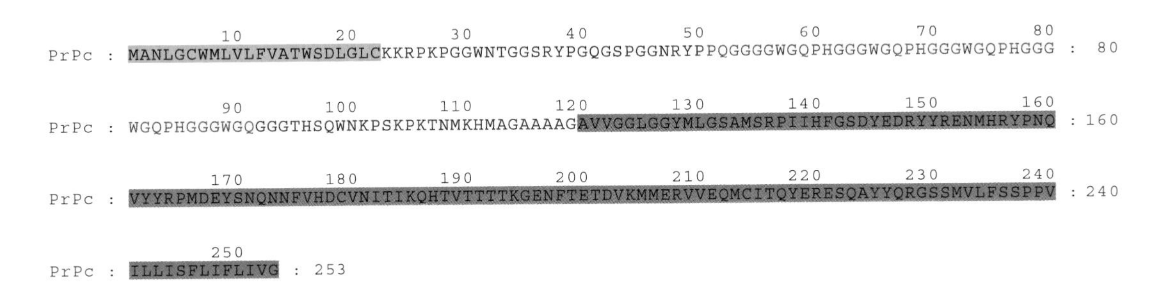

Figure 11.46 Primary sequence of the prion protein (PrP^c). The N-terminal signal sequence is highlighted by the green block whilst the blue block represents the structured domain of 5 helices and three strands. Shown in yellow is the unstructured region and within this sequence are five octapeptide repeats highlighted by red text

Table 11.7 Mutations of the *PNRP* gene associated with inherited forms of transmissible spongiform encephalopathy

Mutation	Disease plus phenotype
Insertion of 24, 48, 96, 120, 144, 168, 192, or 216 base pairs between codons 51 and 91 of PNRP gene leads to repeat of octapeptide motif in PrP	CJD, GSS, or atypical dementias
Pro102Leu, Pro105Leu, Ala117Val, Gly131Val, His187Arg, Asp202Asn, Phe198Ser, Gln212Pro, Glu217Arg	GSS: classical ataxic form plus other phenotypes
Tyr145stop, Thr183Ala	Alzheimer-like dementia
Asp178Asn	CJD (where residue 129 is V on mutant allele)
Asn178Asn	FFI (where residue 129 is M on mutant allele)
Val180Ile, Val203Ile, Arg208His, Val210Ile, Glu211Gln, Glu200Lys, Met232Arg	CJD

FFI = fatal familial insomnia; CJD = Creutzfeldt–Jakob disease; GSS = Gerstmann–Straussler–Scheinker syndrome.

Most mutations (Figure 11.47) occur in the structured C-domain of PrP and GSS is caused primarily by the substitution of Pro102>Leu, whilst in CJD the prevalent mutation is at residue 178 and involves the replacement of Asp by Asn. A slight complication to this genetic pattern is the existence of polymorphisms within the *PNRP* gene where Met or Val is found at residue 129. By itself this change does not result in disease but when combined with mutation at residue 178 the polymorphism contributes to either FFI (Met) or CJD (Val).

The identification of all diseases as the pathological consequence of mutations within the *PNRP* gene has unified this area of study but still leaves the awkward question of how does the prion cause a change in conformation and spread the disease? A major problem in this area has been the unambiguous demonstration of conversion of PrP^{c} into PrP^{sc}.

Strong evidence that conversion of prion protein from normal to abnormal states was responsible for neurodegenerative disease was obtained using 'knockout' mice where the *PNRP* gene was deleted. With the normal *PNRP* gene present the injection of mice with PrP^{Sc} resulted in the transmission of the disease; mice showed familiar symptoms of ataxia and cerebellar lesions within approximately 150 days. When knockout mice were subjected to the same experiment they did not develop the disease. In other words the disease requires native PrP protein encoded by the relevant gene to propagate into the development of amyloid fibrils. Two hypotheses known as the 'refolding' and 'seeding' models attempt to explain prion propagation (Figure 11.48).

Summary

The folding of individual polypeptide chains from less structured states to highly organized topologies is vital to biological function. *All* of the information directing protein folding resides within the primary sequence.

Thermodynamically the folded state is the global energy minimum with the free energy decreasing in the transition from unfolded to native protein. Native proteins are marginally more stable than unfolded states with estimates of conformational stability ranging from 10–70 kJ mol^{-1}.

A number of quantifiable factors have been shown to influence protein stability in globular proteins. These factors include conformational entropy, enthalpic contributions from non-covalent interactions and the hydrophobic interaction. When the sum of all favourable interactions outweighs the sum of unfavourable interactions a protein is stable.

Protein denaturation is the loss of ordered structure and occurs in response to elevated temperature,

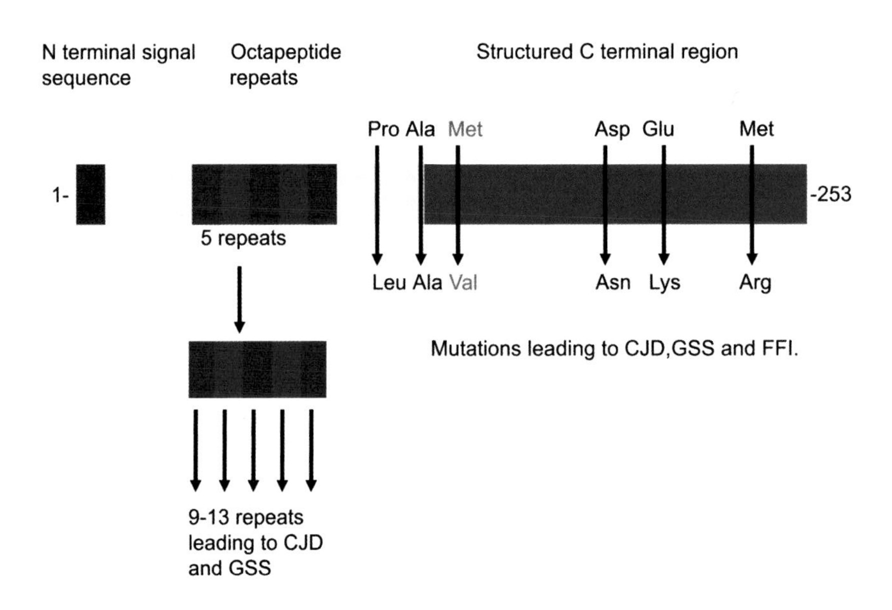

Figure 11.47 Schematic representing mutations and insertions in prion protein. Amplification of the number of octapeptide repeats leads to CJD or GSS

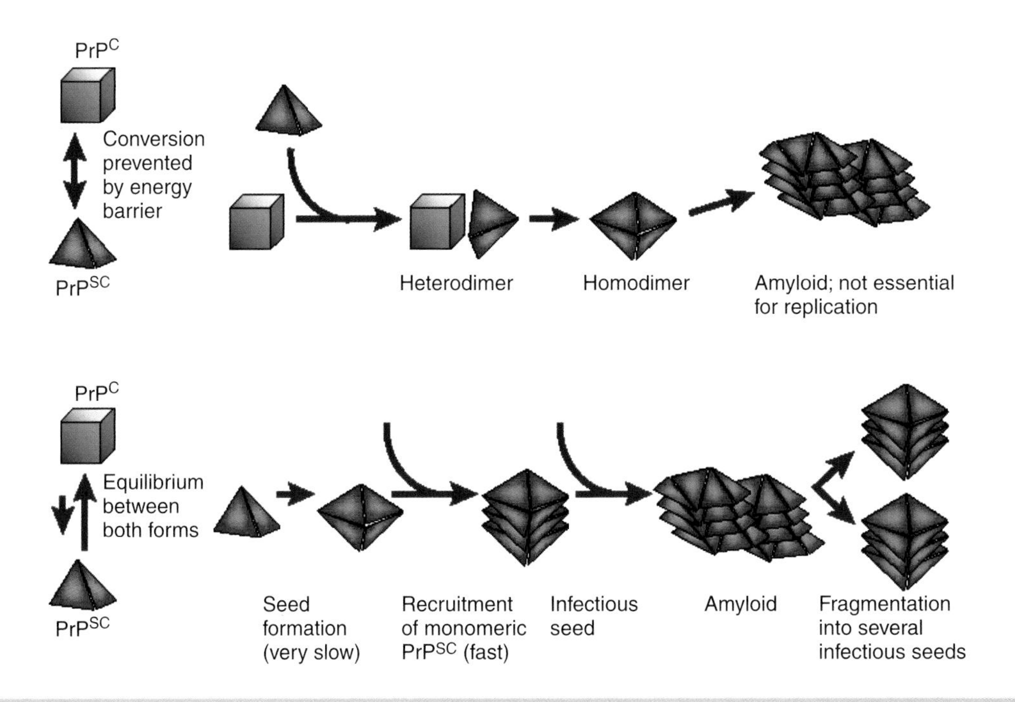

Figure 11.48 The 'protein-only' hypothesis and two popular models for the conformational conversion of PrP^{C} into PrP^{Sc} (reproduced with permission from Aguzzi, A., Montrasio, F., & Kaeser, P.S. *Nat. Rev.* 2002, **2**, 118–126. Macmillan)

extremes of pH or the addition of reagents such as urea or guanidine hydrochloride. Denaturation involves the disruption of interactions such as hydrophobic interactions, salt bridges and hydrogen bonds that normally stabilize tertiary structure.

Experimental measurements of protein stability include differential scanning calorimetry, absorbance and fluorescence optical methods, NMR spectroscopy and circular dichroism.

Kinetic methods allow protein folding to be followed as a function of time. Small soluble proteins (<100 residues) fold within 1 s and the observation of slower folding rates is normally associated with disulfide bond formation or cis-trans proline isomerization. To overcome slow *in vivo* folding specific enzymes catalyse reactions such as peptide prolyl isomerization and protein disulfide isomerization.

Critical events in the formation of the native fold are the acquisition of ordered stable secondary structure, the formation of hydrophobic cores, and the exclusion of water from the protein interior.

For proteins such as lysozyme, barnase and chymotrypsin inhibitor 2 the reaction pathway has been defined to identify properties of intermediates and transition states. In barnase it has proved possible to quantify the folding process in a pathway containing at least one intermediate between denatured and ordered states. A hydrophobic core and the formation of the β sheet are two early events in the folding pathway. In lysozyme each domain folds separately within the observation of multiple kinetic pathways for each folding unit.

Chaperones are multimeric systems found in all cells that prevent incorrect protein folding based around a toroidal structure of seven, eight or nine subunits. The toroidal structure is capped at either end and forms a cavity binding unfolded polypeptide via hydrophobic surfaces with ATP-driven conformational changes propelling the peptide towards the native state.

The kinetics of membrane protein folding *in vitro* are much slower than rates of folding for globular domains but the principles governing formation of the native state remain similar.

Incorrect folding leads to a loss of activity that is the basis of many disease states. Misfolding leads to incorrect protein trafficking within the cell and mutant proteins of *CFTR* are a good example.

Incorrect protein folding is also a key event in amyloidosis – the accumulation of long irreversibly aggregated protein within fibrils. Amyloidogenic diseases include many neurodegenerative disorders such as Alzheimer's, transmissible spongiform encephalopathies and hereditary forms of systemic amyloidosis. Fibrils from a range of amyloid proteins have a common structure based on collections of β strands that align into parallel sheets twisted into helical conformations known as β helices.

Neurodegenerative disorders such as CJD, BSE and scrapie are linked via the spongiform appearance of neuronal tissue and the accumulation of amyloid deposits. The disease arises from changes in the prion protein generating conformations with different secondary and tertiary structure. The abnormal form promotes amyloidosis by inducing other prion proteins to change conformation. These events occur spontaneously at very low frequency leading to sporadic occurrences of disease but are facilitated by mutations within the *PNRP* gene.

Problems

1. Draw the *cis* and *trans* conformations associated with a Xaa–Pro peptide bond.
2. The 3_{10} helix is named because there are three residues per turn and 10 atoms including hydrogen between the donor and acceptor atoms forming an intra-chain hydrogen bond. Describe the α helix using such notation. How does the helix compare with the 3_{10} helix? Repeat this analysis for the π helix.
3. Experimental studies of protein folding reveal the following trends of denaturation.
   Calculate the conformational stability and estimate any other relevant parameters.

Absorbance	Denaturant concentration (M)	Absorbance	Denaturant concentration (M)
0.985	0	0.375	1.8
0.985	0.1	0.325	1.9
0.985	0.2	0.285	2.0
0.980	0.3	0.220	2.1
0.975	0.4	0.175	2.2
0.970	0.5	0.140	2.4
0.960	0.6	0.120	2.5
0.945	0.8	0.115	2.6
0.890	1.0	0.110	2.7
0.760	1.2	0.105	2.8
0.625	1.4	0.105	2.9
0.510	1.6	0.100	3.0

4. Assume that in order to reach the native state a protein needs to sample only 10 conformations per residue with each conformation taking 0.1 ns. Estimate how long it might take for a 100 residue protein to fold. Now assume that the each block of 10 residues can fold independently of the remaining residues and by only sampling three conformations per residue. How long does folding take to occur? Comment on the two values?

5. Explain how disulfide bonds affect protein stability. What would you expect to be the effect on protein stability of introducing a disulfide bond using mutagenesis? What would be the effect of removing a disulfide bond on protein stability?

6. What factors contribute to the observation of heterogeneous folding kinetics of proteins? How would you attempt to evaluate the importance of these factors.

7. 'You cannot unscramble an egg'. Discuss this statement in the light of protein denaturation, refolding, modification and the known presence of chaperones.

8. Explain what might be the consequence of deleting the octapeptide repeats of PrP on prion infection in for example a host such as the mouse.

12

Protein structure and a molecular approach to medicine

Introduction

The introduction to this book made clear that one major reason for studying protein structure and function is to obtain insight into diseases afflicting modern civilizations. These diseases range in diversity from cancer to cholera, from human immunodeficiency virus (HIV) to hepatitis, from malaria to myocardial infarction to name a few. One potential impact of improved structural knowledge is the ability to treat many diseases with highly specific drugs. For bacterial and viral infections immunization has been refined to allow the successful introduction of live, but attenuated, biological samples or components of the active protein. Inoculation elicits an immune reaction that confers lifelong or substantial immunity and has played a major role in improved health care. In this manner diseases such as smallpox have been eradicated from all parts of the world whilst polio and measles have been drastically limited in distribution.

For diseases such as cholera treatments have been devised that limit routes of infection. Cholera spreads via a faecal–oral infection route and the presence of the bacterium *Vibrio cholerae*. By establishing clean water supplies infection routes via contaminated water or food is limited. Infection leads to colonization of the intestines and the production of toxin that results in chronic diarrhea which will, if left untreated, result in death as a result of fluid loss leading to shock and acidosis. Victims die of dehydration unless water and salts are replaced.

Cholera represents one example of how disease origin is traced back to the action of one protein in this case a toxin. By studying the structure and function of the cholera toxin we gain insight into the molecular events that underpin disease. The toxin is a hexameric protein of 87 kDa (subunit composition of AB_5) that binds to a ganglioside known as GM1 via the B subunits and transfers the A subunit across membranes via receptor mediated endocytosis. Inside the cell reductive cleavage of a disulfide bond releases a 195 residue fragment from the A subunit that is responsible for pathogenesis.

The fragment catalyses the transfer of ADP-ribose from NAD^+ to the side chain of Arg187 in the G_α subunit of the heterotrimeric G_s protein; a process known as ADP-ribosylation (Figure 12.1). GTP binding leads to dissociation into G_α and $G_{\beta\gamma}$ each of which activates further cellular components. G protein activation is usually transitory because G_α is a GTPase that hydrolyses GTP to GDP and promotes G protein re-association. G_α-GTP activates adenylate cyclase with the cAMP acting as a second messenger and regulating the activity of cAMP dependent

Proteins: Structure and Function by David Whitford

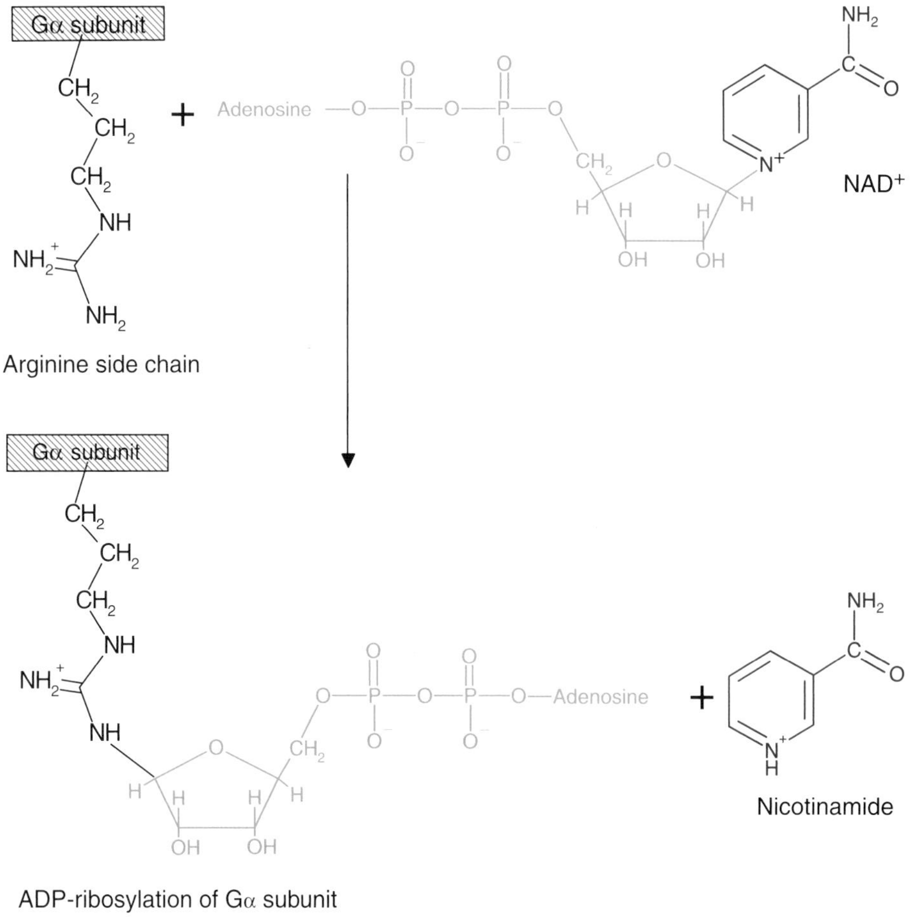

Figure 12.1 Cholera toxin promotes ADP-ribosylation of the G_α subunit of heterotrimeric G proteins

protein kinase. ADP ribosylation of the G_α subunit has a critical effect; it continues to activate adenylate cyclase whilst inhibiting the GTPase activity of G_α. Adenylate cyclase remains in a permanently activated state along with G_α and cellular cAMP levels increase dramatically. Intestinal cells respond to increased cAMP levels by activating sodium pumps and the secretion of Na^+. To compensate chloride, bicarbonate and water are also secreted but the net effect is the loss of enormous quantities of fluid and dehydration.

A molecular approach to the study of cholera elucidated the basis of toxicity and showed that dehydration is counteracted by replenishment of fluids containing salts. Significantly, other bacterial toxins such as *E. coli* heat stable enterotoxins and the pertussis toxin from the bacterium *Bordetella pertussis* act via similar mechanisms. Pertussis is responsible for whooping cough in young babies and is effectively countered in economically advantaged countries by vaccination of young babies around 6 months of age using an inactivated fragment of the whole toxin.

The fundamental events contributing to two diseases have been elucidated through their common mode of action and effective therapeutic strategies have been devised – rehydration and vaccination. However, this approach is not always possible. Increasingly the techniques of molecular and cell biology are providing insights into the underlying causes of disease giving rise to the new field of molecular medicine. Molecular medicine became a possibility with the completion of genome sequencing projects and the identification of many inherited diseases arising from mutated genes. Gene screening is now established as an important diagnostic tool with DNA extracted from biological samples and 'searched' for defects. Defective genes can be as small as a simple base change or may be larger involving missing, duplicated or translocated segments of genes.

The availability of 'predictive' gene tests has increased dramatically in the last decade. Currently genetic tests are available for the most common diseases such as cystic fibrosis, neurodegenerative disorders such amylotrophic lateral sclerosis (Lou Gehrig's disease), Tay–Sachs disease, Huntington's disease, familial hypercholesterolaemia (a disease causing catastrophically high cholesterol levels and myocardial infarction), childhood eye cancer (retinoblastoma), Wilms' tumour (a kidney cancer), Li–Fraumeni syndrome, familial adenomatous polyposis (an inherited predisposition to form pre-cancerous polyps and associated with colon cancer), and *BRCA*1 (a gene implicated in familial breast cancer susceptibility). The list could be extended for many more pages. Over 3000 mutated genes are known to be involved in inherited disorders or human disease.

Alongside molecular medicine there have been major advances in biotechnology to produce protein-based drugs for the treatment of disease. Today human insulin is given as fast and slow acting forms in the treatment of Type I diabetes arising from damage to the β cells of the islets of Langerhans found in the pancreas. Human insulin is produced by recombinant DNA methods and avoids many problems faced previously where insulin was extracted from the pancreas of pigs or cows.

Molecular medicine will occupy a pivotal role in clinical practice throughout the 21st century. This chapter deals with a selective group of proteins implicated in disease states where there is a growing understanding of the biological problem. As such it represents a series of case studies, by no means complete, described to provide a flavour of modern approaches to molecular medicine.

Sickle cell anaemia

The discovery of Ingram in 1956 that sickle cell anaemia arose from the substitution of glutamate by valine at position 6 of the β subunit of haemoglobin was the first example of a molecular defect contributing to human disease. It represents a key event, perhaps even the origin, of molecular medicine. Mutations in haemoglobin were identified because the protein is obtained from individuals in sufficient quantities for peptide sequencing; this situation rarely extends to other diseases where samples are normally only obtained via invasive biopsies or post-mortem.

Substitution of valine by glutamate alters the surface properties of the β subunit leading to aggregates of protein in the deoxy form of haemoglobin *S* (HbS). The hydrophobic side chain of Val6 projects out into solution fitting precisely into a small hydrophobic pocket on the surface of a neighbouring β subunit formed by Phe85 and Leu88. The pocket favours aggregation whilst oxygenation removes this pocket by conformational change. In normal haemoglobin the charged side chain of glutamate is not accommodated within a non-polar surface region. Suddenly the molecular basis of sickle cell anaemia is clear.

Long polymers or fibres of deoxy HbS ~20 nm in diameter extend the length of the erythrocyte deforming a normal biconcave cell into a sickle shape where distortion leads to cell rupture (Figure 12.2). The greatest danger of aggregation occurs in the capillaries where oxygen is 'off loaded' to tissues and the concentration of deoxy-HbS is higher. Precipitation prevents blood flow through the capillaries causing tissue necrosis and life threatening complications.

Sickle cell anaemia occurs when an individual carries two mutated forms of the β subunit gene. A single copy leads to the sickle cell trait and is generally asymptomatic. Generally, 'harmful' mutations are present at very low frequency and it is unusual for a mutant gene to be present at high frequency unless it confers a specific evolutionary advantage. In certain regions of West and Central Africa approximately 25 percent of the population exhibit heterozygosity, or the sickle cell trait. This is unexpected since harmful mutations are normally selectively weeded out by

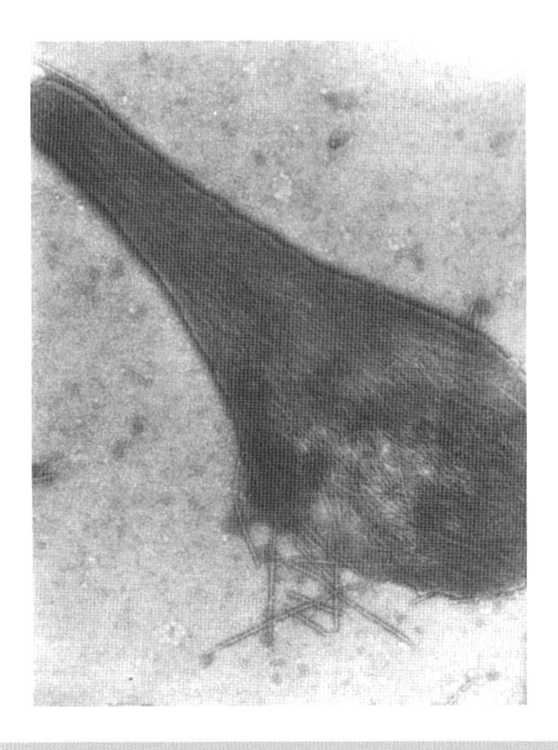

Figure 12.2 Electron micrograph of deoxy-HbS fibres spilling out from a ruptured erythrocyte

evolution. It is clear that the sickle cell trait correlates closely with the prevalence of malaria. Malaria kills over 1 million individuals each year mostly young children – a staggering and regrettable statistic – and is caused by the mosquito-borne protozoan parasite *Plasmodium falciparum*. The parasite resides in the erythrocyte during its life cycle and causes adhesion of red blood cells to capillary walls. Individuals with the sickle cell trait possess red blood cells containing abnormal haemoglobin and these tend to sickle when infected by the parasite. The result is that infected cells flow through the spleen and are rapidly removed from circulation. In other words individuals with sickle cell trait deal with the parasite more effectively than normal individuals – the consequence is that the mutant gene persists in malaria-endemic regions.

The gene encoding the β subunit is located on chromosome 11 – a region containing many other globin genes and known as the β globin cluster. Polymorphisms in this region are equated with disease severity and several distinct haplotypes are identified in sickle cell anaemia and given names reflecting their origins within Africa, such as Senegal haplotype, the most benign form of sickle cell disease, the Benin haplotype and the Central African Republic haplotype–one of the severest forms of sickle cell anaemia.

Screening of individuals is routinely performed using electrophoresis, where the sickle cell haemoglobin shows altered migration properties when compared with the normal protein. Tests are cheap, quick and reliable and a positive result points towards further diagnostic investigation. Tests provide prospective parents with realistic information on the inheritance of the trait. When both parents carry the sickle cell trait there is a 25 percent chance of the offspring having sickle cell anemia, a 50 percent chance of the sickle cell trait and a 25 percent chance of a normal individual. Despite good screening procedures the treatment of sickle cell anaemia is problematic and highlights a major dilemma of molecular medicine. Diseases can be identified but frequently there is no simple 'cure'. Before the introduction of antibiotics individuals with sickle cell anemia frequently died in childhood. Modern care involves treatment with penicillins and significantly reduces infant mortality.

Viruses and their impact on health as seen through structure and function

Viral infections are well known to all of us since they are responsible for the common cold as well as many more serious diseases. The term virus is derived from the Latin word for poison and seems entirely appropriate in view of the devastating illnesses caused in humans from viral infections. The diseases attributed to viruses are vast and include influenza (flu), measles, mumps, smallpox, yellow fever, rabies, poliomyelitis, as well as acquired immune deficiency syndrome (AIDS), haemorrhagic fever and certain forms of cancers.

The existence of sub-microscopic infectious agents was suspected by the end of the 19th century when the Russian botanist Dimitri Ivanovsky showed that sap from tobacco plants infected with mosaic disease contained an agent capable of infecting other tobacco plants even after passing through filters known to retain bacteria. Plant and animal cells are not the only living cells subject to viral attack. In 1915 Frederick Twort isolated filterable entities that destroyed bacterial cultures and produced cleared areas on bacterial lawns. The entities were bacteriophages, often abbreviated to phage, that attach to bacterial cell walls introducing genetic material. The result is rapid replication of new viral progeny with cell lysis releasing new infectious particles.

Crystallization of tobacco mosaic virus (TMV) by W.M. Stanley in 1934 represented one of the first systems to be crystallized but also emphasized the high inherent order present in viruses. Subsequently TMV has been a popular experimental subject. One important conclusion from studying TMV was that viruses were constructed using identical subunits packed efficiently around helical, cubic or icosahedral patterns of symmetrical organization. To clarify the terminology for virus components a number of terms were introduced. The capsid denotes the protein shell that encloses the nucleic acid and is based on a subunit structure. The subunit is normally the smallest functional unit making up the capsid. The capsid together with its enclosed nucleic acid (either DNA or RNA) constitutes the nucleocapsid. In some viruses the nucleocapsid is coated with a lipid envelope that can be derived from host cells. All of these structures form part of the virion – the complete infective virus particle.

Viruses lack normal cellular structure and do not perform typical cell-based functions such as respiration, growth and division. Instead they are viewed as 'parasites' that at one stage of a life cycle are free and infectious but once entered into a living cell are able to sequester the host cell's machinery for replication. The conversion of viral genetic material into new viruses or the incorporation of the viral DNA into the host genome can occur via a variety of different routes and depends on the organization of the viral genome.

The Baltimore classification (Table 12.1) is based on nuclei acid type and organization and uses the relationship between viral mRNA and the nucleic acid found in the virus particle. In this system viral mRNA is designated as the plus (+) strand and a RNA or DNA sequence complementary to this strand is designated as the minus (−) strand. Production of mRNA requires the use of either DNA or RNA (−) strands as templates and viruses can exist as double-stranded DNA, single-stranded + or − DNA, double-stranded RNA, and either (−) or (+) single strands of RNA. Almost all viruses can be classified in this scheme. RNA-based viruses form DNA using the enzyme reverse transcriptase and represent retroviruses. Retroviruses are involved in many diseases, transforming cells into cancerous states by the conversion of genetic material (RNA) into DNA followed by integration into the host cell genome (proviral state) and the production of transcripts and translation products via host cell machinery.

The two most important retroviruses in terms of their effect on man are the human influenza virus (flu) and HIV. These viruses are responsible for the two greatest pandemics affecting humans over the last 100 years.

Viruses, like bacterial infective agents, act as antigens in the body and elicit antibody formation in infected individuals. As a result vaccines can be developed against viruses most frequently based on a component of the virus coat or an inactivated form of the virus. Although diseases such as smallpox have been effectively eliminated through the use of vaccines this is not always an effective treatment route. For influenza and HIV this is particularly difficult and lies at the heart of the pandemics caused by these agents. These retroviruses are of considerable relevance to health care in the modern era and are described as a result of their importance in two major diseases affecting the world today.

HIV and AIDS

HIV has been the subject of intense research since 1983 when Luc Montagnier, Robert Gallo and colleagues identified it as the causative agent contributing to a complex series of illnesses grouped under the term 'acquired immune deficiency syndrome' (AIDS). The disease was originally recognized in the United States at the beginning of the 1980s as one affecting primarily homosexual men and led to a systematic decrease in the ability of the body to fight infection arising from a progressive loss of lymphocytes. The decline in the immune response led to the name AIDS. Lymphocyte loss rendered the body defenseless against infection, with individuals succumbing to rare

Table 12.1 A scheme of classifying viruses according to their mode of replication and genomic material sometimes called Baltimore classification

Genome organization	Examples	Details of replication
ds DNA	Adenoviruses Herpesviruses Poxviruses	Adenoviruses replicate in the nucleus using cellular proteins. Poxviruses replicate in the cytoplasm making their own enzymes for replication
ss (+) sense DNA	Parvoviruses	Replication occurs in the nucleus, involving the formation of a (−) sense strand, which serves as a template for (+) strand RNA and DNA synthesis
ds RNA	Reoviruses Birnaviruses	Viruses have segmented genomes. Each genome segment is transcribed separately to produce monocistronic mRNAs
ss (+) sense RNA	Picornaviruses Togaviruses	(a) Polycistronic mRNA, e.g. picornaviruses means naked RNA is infectious. Translation results in the formation of a polyprotein product, which is cleaved to make mature proteins (b) Complex transcription, e.g. togaviruses. Two or more rounds of translation are necessary to produce the genomic RNA
ss (−) sense RNA	Orthomyxoviruses Rhabdoviruses	Must have a virion particle RNA directed RNA polymerase (a) Segmented, e.g. orthomyxoviruses. First step in replication is transcription of the (−) sense RNA genome by the virion RNA-dependent RNA polymerase to produce monocistronic mRNAs, which also serve as the template for genome replication (b) Non-segmented, e.g. rhabdoviruses. Replication occurs as above and monocistronic mRNAs are produced
ss (+) sense RNA with DNA intermediate in life-cycle	Retroviruses	Genome is (+) sense but does not act as mRNA. Instead acts as template for reverse transcription
ds DNA with RNA intermediate	Hepadnaviruses	Viruses also rely on reverse transcription. Unlike retroviruses this occurs inside the mature virus particle. On infection the first event to occur is repair of gapped genome, followed by transcription

ds = double stranded, ss = single stranded.

opportunistic infections or diseases such as pneumonia and tuberculosis (TB). In some cases rare forms of skin cancer (Kaposi's sarcoma) were associated with advanced HIV infection along with other cancers.

In the immune system a defining event in HIV-mediated disease is a fall in the number of circulating helper T cells and this feature is used as a marker for immunodeficiency. Once the absolute T-cell count falls below a threshold of 200 cells per mm^3 in the peripheral blood individuals are vulnerable to AIDS-defining opportunistic infections. The virus preferentially infects helper T cells, as well as other cells such as macrophages and dendritic cells, because of the presence of CD4, a cell surface protein that acts as a high affinity receptor for the virus. The interaction between CD4 and virus mediates entry of the virus into the host cell and is a critical step in the overall process of HIV pathogenesis.

Subsequently, the disease was shown to exist in other populations; it was not restricted to homosexuals nor was it confined to the United States (Figure 12.3). HIV infection is responsible for significant mortality in *all* sexually active populations and is present throughout the world. In many countries the absence of good health care systems has meant that AIDS has become the fourth leading cause of mortality throughout the world and is responsible for major decreases in life expectancy. Infection with HIV also occurs from intravenous drug use and is also observed as a result of blood transfusions using infected (unscreened) donated samples.

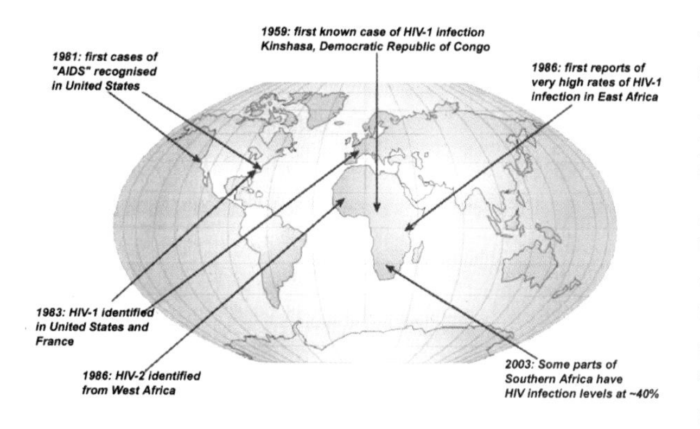

Figure 12.3 The sites of key events in the development of the HIV epidemic. This world map highlights the first known cases of infection with HIV-1 and HIV-2 in West and sub-Saharan Africa

The spectre of progressive HIV infection spread predominantly through sexual intercourse has dominated health care issues in all areas of the world. In 2003 it was estimated by the World Health Organization that over 40 million individuals were affected by HIV infection – a sombre and immensely problematic issue. There is little doubt that HIV infection and its sequelae will be responsible for massive increases in mortality rates, especially in sexually active populations and will also affect the young as a result of fetal or neonatal transmission. The pandemic nature of the disease means that dealing with HIV-infected individuals is a major health care issue in all areas of the world.

The HIV genome

HIV is a member of the lentivirus genus that includes retroviruses with complex genomes. An early objective in studying HIV was to sequence the genome to understand the size, function and properties of the virally encoded proteins. The HIV genome is small (~9.4 Kb) and encodes three structural proteins (MA, CA, and NC), two envelope proteins (gp41, gp120), three enzymes (RT, protease and integrase), and six accessory proteins (Tat, Rev, Vif, Vpr, Nef & Vpu) (Figures 12.4 and 12.5).

These viruses exhibit cone-shaped capsids covered with a lipid envelope derived from the membrane of the host cell. Puncturing the exposed surface are glycoproteins called gp120 – a trimeric complex of proteins anchored to the virus via interactions with a transmembrane protein called gp41. Both gp120 and gp41 are derived from the *env* gene product. Inside this layer is a shell, often called the matrix, composed of the matrix protein (MA). The matrix protein lines the inner surface of the viral membrane and a conical capsid core forms from the major capsid protein (CA). This complex is at the centre of the virus and surrounds two copies of the unspliced viral genome that are stabilized by interaction with the nucleocapsid protein (NC). One major region of the HIV genome is the *pol* gene (Figure 12.5); its product is a polyprotein of three essential enzymes. These enzymes are protease, reverse transcriptase and integrase.

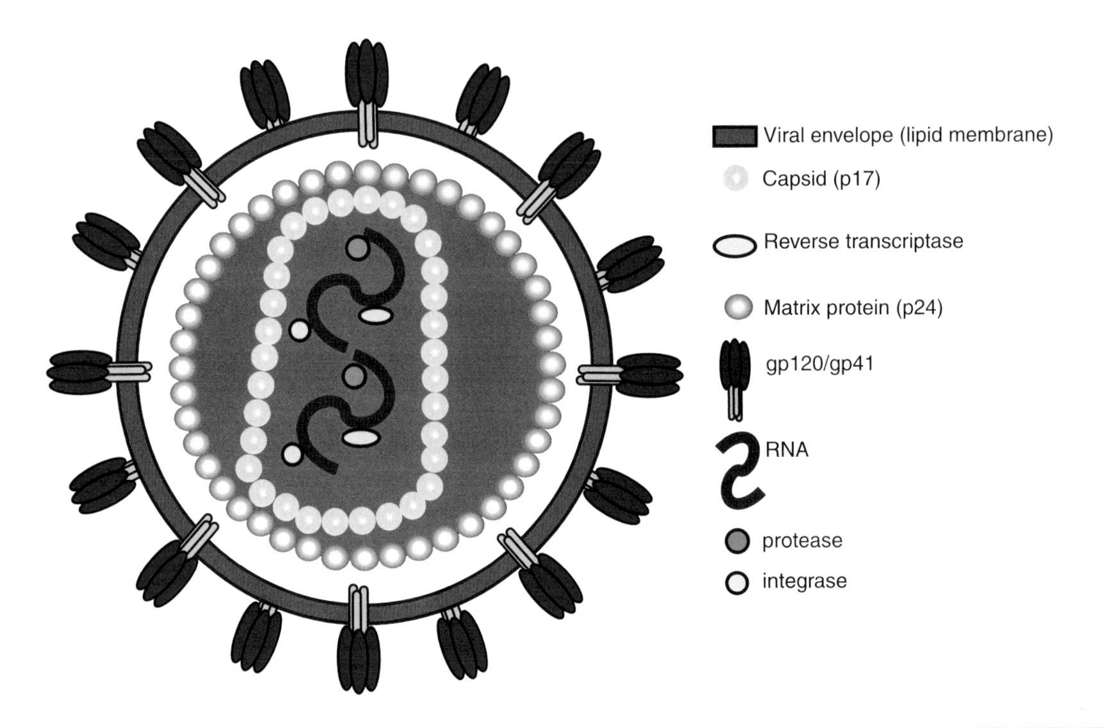

Figure 12.4 A picture of the organization of protein and RNA within the intact human immunodeficiency virus

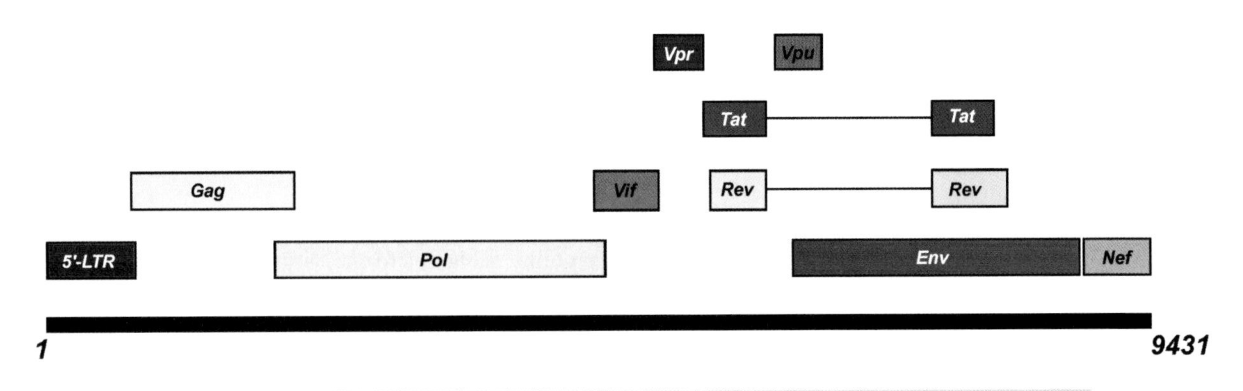

Figure 12.5 Organization of the HIV genome and the proteins derived from the genome

Of the accessory proteins three function within the host cell by facilitating transcription (Tat), assisting in transport of viral DNA to the cell nucleus (Rev), and acting as a positive factor for viral expression (Vif). These proteins are not found within the virion. The other three proteins are packaged with the virus and play important roles in regulating viral promoters (Vpr), acting as a negative regulatory factor for viral expression (Nef) and promoting viral release and CD4 degradation (Vpu) (Table 12.2).

The first phase of infection results in virus binding to $CD4^+$ lymphocytes. The helper T cells are the primary

Table 12.2 Functional role of HIV encoded gene products

Protein	Gene	Role
MA (p17)	*Gag*	Matrix protein
CA (p24)	*Gag*	Matrix protein
NC (p7)	*Gag*	Nucleocapsid protein
gp120	*Env*	Extracellular glycoprotein binds to CD4 receptor
gp41	*Env*	Transmembrane protein
Reverse transcriptase (RT)	*Pol*	Conversion of RNA > DNA
Integrase	*Pol*	Insertion of DNA into host genome.
Protease	*Pol*	Processing of translated polyprotein
Rev	*Rev*	Regulate expression of HIV proteins
Nef	*Nef*	Negative regulatory factor
Vpr	*Vpr*	Viral (HIV) promoter regulator
Vpu	*Vpu*	Viral release and CD4 degradation
Vif	*Vif*	Viral infectivity factor
Tat	*Tat*	Transcriptional activator

The three viral encoded enzymes play a major role in the HIV replication cycle and the formation of new virions and have formed the basis of most drug-based treatments of HIV infection. In part this has arisen from the highly resolved structural models available for reverse transcriptase, the integrase and protease of HIV.

target and binding leads to a series of intracellular events that causes their destruction with a half-life of less than two days. During initial stages of infection the viral genome becomes integrated into the chromosome of the host cell and is followed by a later phase of regulated expression of the proviral genome leading to new virion production. The stages of viral infection require the concerted involvement of viral proteins; this includes, for example, gp120-mediated binding to $CD4^+$ cells, conversion of viral RNA into DNA by reverse transcriptase, integration of viral DNA into the host chromosome by integrase, nuclear localization of viral DNA by Vpr, and processing of viral mRNAs by Tat, Rev and Nef.

The structure and function of Rev, Nef and Tat proteins

Nef is a single polypeptide chain of 206 residues originally recognized as a 'negative factor' where its absence caused low viral loads in cells. After viral infection the protein is expressed in high concentrations and inoculation of Rhesus monkeys with a Nef-deletion strain in the closely related SIV (simian immunodeficiency virus) did not lead to disease. Nef has at least two distinct roles during infection; it enhances viral replication, and by an unknown mechanism it reduces the number of CD4 receptors found on the surface of infected lymphocytes. The protein is unusual in having a myristoylated N-terminal residue indicating membrane association; removal of this modification inhibits activity. The structure of the core domain of Nef (residues 71–205) has been determined by NMR and crystallographic methods, with both techniques suggesting a fold containing a helix-turn-helix (HTH) motif with three α helices, a five-stranded antiparallel β sheet, an unusual left-handed poly-L-proline type-II helix and a 3_{10} helix.

The role of Rev is to regulate expression of HIV proteins by controlling the export rate of mRNAs from the nucleus to the cytoplasm where translation occurs. The Rev protein contains 116 residues and within this sequence are nuclear localization and export sequences

(NLS and NES) consistent with its role in mRNA export, but a structure has not been determined.

The final non-structural protein synthesized by the virus and packaged within the core is Tat–a transcriptional activator that enhances transcription of integrated proviral DNA. Transcription starts at the HIV promoter–a region of DNA located within the 5′ long terminal repeat region with the promoter binding RNA polymerase II and other transcription factors. Transcription frequently terminates prematurely due to abortive elongation and the Tat protein overcomes this problem by binding to stem–loop sites on the nascent RNA transcript. Description of the Tat protein is complicated by translation via two exons leading to a protein containing between 86 and 101 residues; the first exon encodes residues 1–72 and the second exon translates as residues 73–101. Structural studies have utilized an 86-residue fragment despite most HIV strains containing full-length protein, but have yet to yield a detailed picture for Tat. The protein is divided into four distinct regions; an N-terminal cysteine-rich domain, a core region, a glutamine-rich segment and a basic zone. The basic region contains the sequence $R_{49}KKRRQRRR_{57}$ essential for recognition and binding to RNA.

The role of Vif, Vpr and Vpu

A similar lack of structural information pertains to Vif, Vpr and Vpu. These three proteins range in size from 15 to 23 kDa and structures are unavailable, although considerable functional data has been acquired. It is known that Vpr, a 96-residue accessory protein present early in the viral life cycle, plays an important regulatory role during replication. Vpr localizes to the nucleus and is implicated in cell cycle arrest at the G_2/M interface. Vpr also enhances transcription from the HIV-1 promoter structure and cellular promoters.

Vif stands for viral infectivity factor and is a basic protein (23 kDa) whose precise function is unknown. Naturally occurring strains of HIV strains bearing mutations in Vif are known to replicate at significantly lower levels when compared with the wild-type virus. In some instances this is associated with low viral loads, high CD4 T-cell counts and slower progression to advanced forms of the disease. These results emphasize the importance of Vif and homologous proteins have been recognized in other lentiviruses from a highly conserved – SLQSLA motif located between residues 144–149. Vif is required during later stages of virus formation since virions produced in the absence of Vif cannot complete proviral DNA integration into host chromosomes. Unlike Vpr, Vif is primarily found in the cytoplasm as a membrane-associated protein.

Vpu is a small integral membrane protein that regulates virus release from the cell surface and degradation of CD4 in the endoplasmic reticulum. These two distinct biological activities represent independent mechanisms attributed to the presence of distinct structural domains within Vpu. A transmembrane domain is associated with virus release from the cell whereas CD4 degradation involves interaction with the cytoplasmic domain.

In contrast to the above proteins considerable understanding of HIV has come from structural studies of the remaining components of the virion particularly the enzymes and envelope proteins. Molecular biology has allowed individual proteins, or more frequently domains derived from whole proteins, to be cloned, expressed and subjected to detailed structural analysis. These studies assist in providing a picture of their roles during the viral life cycle and understanding the interdependence of structure and function in these proteins represents a key effort in the attempts to combat the disease. Alongside a large body of data for the reverse transcriptase and protease there also exists structural information for gp120, the integrase, domains of the transmembrane protein gp41 (the TM domain) and the matrix, capsid and nucleocapsid proteins.

Structural proteins of HIV

Structural proteins are derived from the *Gag* gene product by proteolytic cleavage to yield the MA, CA and NC proteins (Figure 12.6). With a combined mass of ~55 kDa the Gag polyprotein yields a matrix (MA) protein of 17 kDa, the capsid protein (CA) of 24 kDa and the nucleocapsid protein (NC) of ~7 kDa. The structure of the MA protein consists of five α helices, two short 3_{10} elements and a three-stranded β sheet. Although monomeric in solution several lines of evidence indicate trimeric structures represent 'building blocks' for the matrix shell within a mature virion.

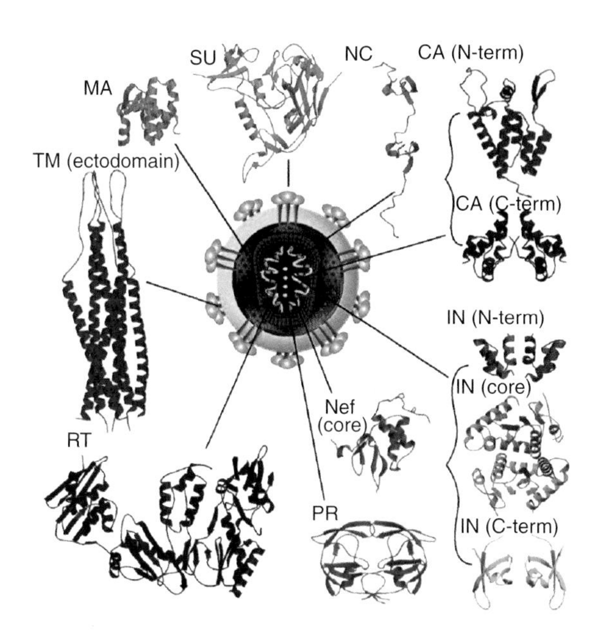

Figure 12.6 The structure of HIV. Drawing of the mature HIV virion surrounded by ribbon representations of the structurally characterized viral proteins and fragments. The protein structures have been drawn to approximately the same scale whilst the TM ectodomain shown is that determined for the closely related SIV (reproduced with permission from Turner, B.G. & Summers, M.F. *J. Mol. Biol.* 1999, **285**, 1–32. Academic Press)

The capsid protein occurs at ~2000 copies per mature virion forming a cone-shaped, dense, structure at the virus centre that encapsulates RNA as well as enzymes, NC protein and accessory proteins. Two copies of unspliced RNA are packaged in a ribonucleoprotein complex formed with the NC protein. The CA protein has proved difficult to isolate and no structure is available for the intact protein. In an N-terminal domain seven helices, two β hairpins and an exposed partially ordered loop are observed with the domain forming an arrowhead shape. This domain associates with similar regions as part of the condensation of the capsid core.

The nucleocapsid or NC protein is involved in recognition and packaging of the viral genome. Retroviral NC proteins contain the Cys-X_2-Cys-X_4-His-X_4-Cys array characteristic of Zn finger proteins, and in HIV the NC protein contains two such elements. The structure of this small protein is defined by these two Zn finger elements where reverse turns form a metal binding centre that leads via a loop to a C terminal 3_{10} helix. In the NC protein the two metal-binding centres may interact although this is thought to be a flexible, dynamic interaction and both play a role in binding RNA via conserved hydrophobic pockets.

Reverse transcriptase and viral protease

Two of the three enzymes, reverse transcriptase and protease, are amenable targets for drug therapy largely as a result of detailed structural analysis. Reverse transcriptase is responsible for the conversion of RNA to DNA whilst the protease catalyses processing of the initial polyprotein (Gag-Pol) product into separate functional proteins.

HIV protease The protease was the first HIV-1 protein to be structurally characterized and is translated initially as a Gag-Pol fusion protein; its role involves proteolytic cleavage of the polyprotein to release active components. In an autocatalytic reaction analogous to that seen in zymogen processing the protease liberates subunits found in the mature virus. The Gag precursor forms the structural proteins whilst Pol produces enzymes – reverse transcriptase, integrase, and protease.

As a protease the HIV enzyme is unique is catalysing bond scission between a tyrosine/phenylalanine and a proline residue with no human enzyme showing a similar specificity. Despite unique specificity the HIV protease is structurally similar to proteases of the pepsin or aspartic protease family and this has assisted characterization. Historically, pepsin was the second enzyme to be crystallized (after urease) and structures are available for pepsins from different sources. All reveal bilobal structures with two aspartate residues located close together in the active site. These residues activate a bound water molecule for use in proteolytic cleavage of substrate.

The bilobal structure of pepsin is based around a single polypeptide chain of ~330 residues where one lobe contains the first 170 residues and the remaining residues are located in a second lobe. Each lobe is related by an approximate two-fold axis of symmetry

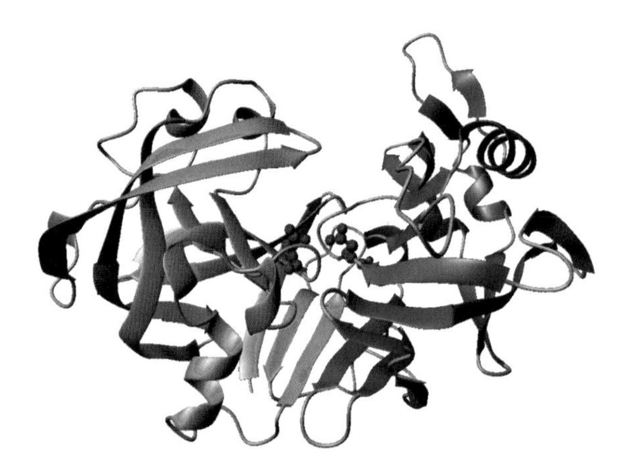

Figure 12.7 The overall topology of pepsin showing strand-rich secondary structure arranged within two lobes. Secondary structure elements for residues 1–170 of the first lobe are shown in orange with the remaining secondary structure shown for the second lobe in blue. Asp32 and 215 of the active site are shown in red

and at the centre, between each lobe, is the active site (Figure 12.7). It seems probable that this topology arose by gene duplication because each lobe contributes a single Asp side chain to the active site found within an Asp-Thr/Ser-Gly- motif preceded by two hydrophobic side chains. In pepsin the catalytic Asp residues are at positions 32 and 215. Each side chain interacts via a complex network of hydrogen bonds (Figure 12.8) with at least one molecule of water. The

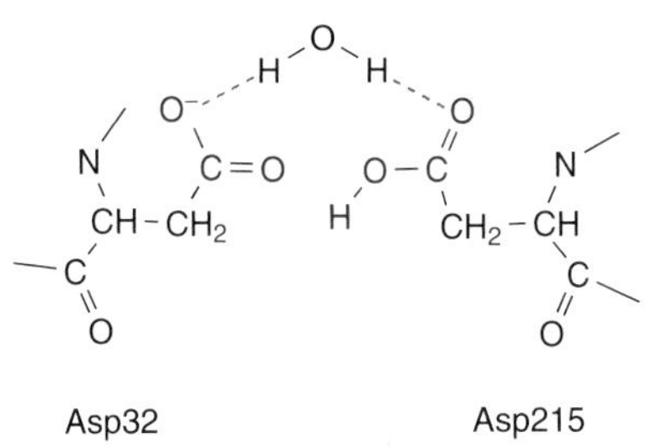

Asp32 Asp215

Figure 12.8 The interaction between Asp32 and 215 in pepsin involves a bound water molecule and two protonation states for the Asp side chain. A water molecule bridges between two oxygen atoms, one on each of the carboxyl groups of the side chain

side chains exhibit very different protonation profiles; one aspartate has a low pK value around 1.5 whereas the other exhibits a higher than expected value of 4.8. These values vary from one enzyme to another but the principle of two different pK values for aspartyl side chains is conserved in all proteins.

Pepsin shows optimal activity around pH 2.0 and substrates bind to the active site of pepsin with the C=O group of the scissile bond situated between the Asp side chains with the oxygen atom in a position comparable to that of the displaced water molecule. The unionized side chain of Asp215 protonates the carbonyl group allowing the displaced water molecule, hydrogen bonded to the side chain of Asp32, to initiate a nucleophilic attack on the carbonyl carbon atom by transferring a proton to Asp 32 (Figure 12.9). The resulting proton transfer forms a tetrahedral intermediate that promptly decomposes with protonation of the NH group.

HIV protease uses a slightly different trick to achieve the same catalytic function. The protein is not a single polypeptide chain but is a symmetrical homodimer with each chain containing 99 residues (Figure 12.10). In itself this organization supports a picture of aspartyl proteases arising from duplication of an ancestral gene. Crystallography shows the structure of this dimer as two identical chains each subunit containing a four-stranded β sheet formed by the N- and C-terminal strands together with a small α helix. The enzyme's active site is unusual in that it is formed at the interface of the two subunits. However, again the active site is based around a group of three residues in each subunit with the sequence Asp25-Thr26-Gly27 containing the catalytically relevant acidic side chain that is responsible for bond cleavage. The residues in the second subunit are often denoted as Asp25′, Thr 26′ and Gly27′.

The cavity region is characterized by two flexible flaps at the top of each monomer formed by an antiparallel pair of β strands linked by a four residue β turn (49–52). These flaps are also seen in pepsin and guard the top of the catalytic site with the active site Asp residues located at the bottom of this cleft. These 'flap' regions are flexible, and from studies of HIV protease in the presence of inhibitors it was observed that significant conformational changes occur in this region. Some backbone C_α atoms show

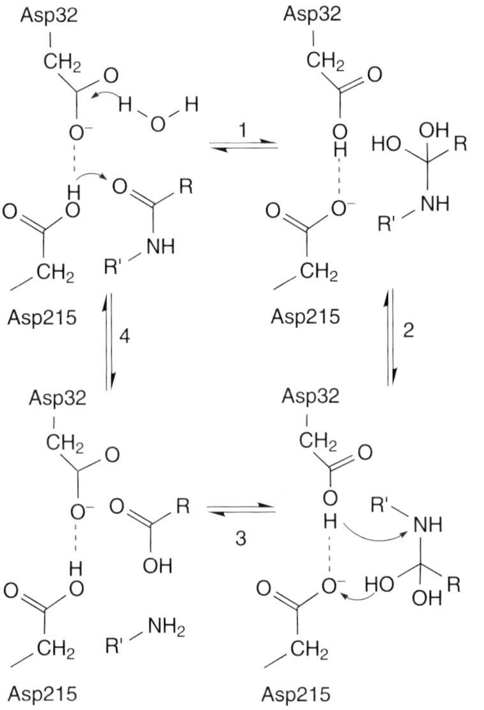

Figure 12.9 The catalytic activity of aspartyl proteases such as pepsin. The reaction mechanism involves the two aspartyl side chains interacting directly with substrate and a bound water molecule. Asp32 accepts a proton from water whilst Asp215 donates a proton to the substrate (1), a tetrahedral intermediate formed breakdown via the donation and acceptance of protons (2) with the resulting product yielding a free amino group and a free carboxyl group within the substrate (3). The dissociation of products allows the pepsin to start the reaction cycle again (4)

displacements of ~0.7 nm when the bound and free states of the enzyme are compared around the active site region.

Catalysis is mediated by twin Asp25 residues characterized by pK values that deviate considerably from a normal value around 4.3. One carboxyl group exhibits an unusually low pK of 3.3 whilst the other displays an unusually high pK of 5.3. The catalytic diad interacts with the amide bond of the target substrate in a manner analogous to that describe for pepsin. Inhibiting the protease represents one potential method of fighting HIV infection since this will prevent processing of the Gag-Pol polyprotein and the generation of proteins necessary for new virus formation. Many crystal structures of HIV-1 protease together with inhibitor complexes are known as part of drug development aimed at catalytic inhibition. In view of the preference of HIV protease for Tyr-Pro and Phe-Pro sequences within target substrates inhibitors have been designed to mimic the transition state complex and to prevent, by competitive inhibition, the turnover of the protease with native substrate (Figure 12.11).

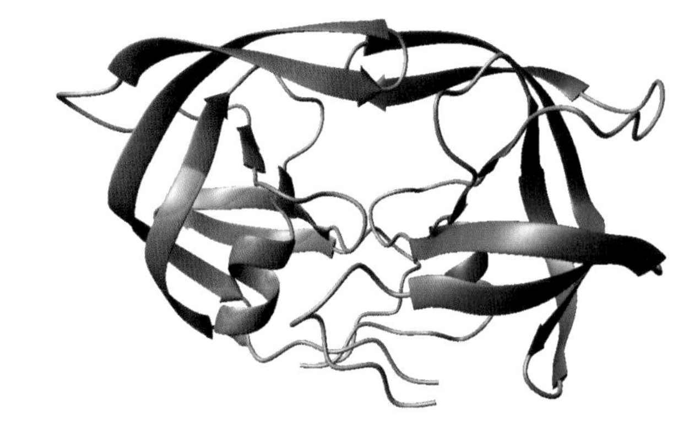

Figure 12.10 The HIV protease homodimer. The two monomers are shown in orange and blue

The protease inhibitors are known as peptidomimetic drugs because of their imitation of natural peptide substrates. These drugs mimic the enzyme's transition state tetrahedral complex and attempt to reproduce the hydrophobic side chain of phenylalanine, the restricted mobility of prolyl side chains and a carbonyl centre that is not easily protonated.

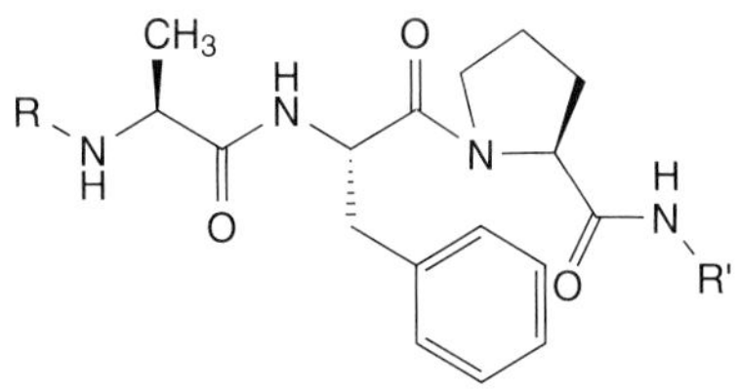

Figure 12.11 Ala-Phe-Pro sequence containing the target peptide bond (substrate) for HIV protease. The bond to be cleaved is shown in red

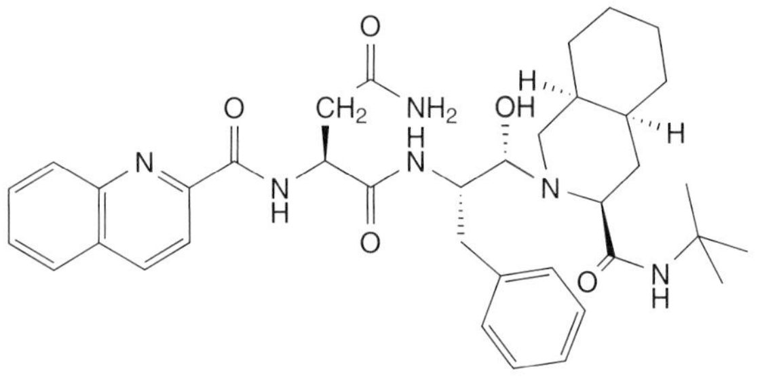

Figure 12.12 The structure of saquinavir

Several drugs were rapidly brought from the laboratory through clinical trials and into effective use in 1995 and included ritonavir, indinavir, nelfinavir and saquinavir (Figure 12.12). The presence of additional functional groups such as OH groups close to the target peptide bond coupled with aromatic groups results in inhibitors that bind with high affinity ($K_I < 1$ nM) and decrease viral protease activity by acting as competitive inhibitors. Although these drugs are responsible for remarkable increases in life expectancy they do not eliminate virus from the body. In general, the protease inhibitors are given when treatment with drugs aimed against reverse transcriptase fails or may be given in combination with other drug therapies. Unsurprisingly, these drugs have side effects but have proved effective at limiting viral proliferation. A major problem with this approach is resistance due to mutations in HIV protease that prevent the inhibitor from binding with optimal efficiency. Mutations within HIV protease lead to a loss of inhibitor binding but protein that retains catalytic activity. The mutations arise from the high error rate associated with activity of reverse transcriptase.

Reverse transcriptase The other major target for antiviral drug therapy is reverse transcriptase. The viral genome of RNA is transcribed into DNA by the action of reverse transcriptase followed by the integration of viral DNA into host cell chromosomes by the integrase. Reverse transcriptase is derived from the Gag-Pol precursor by proteolytic cleavage producing two homodimers of mass 66 000 (p66). The p66 subunit is bifunctional with a polymerase activity located in the larger of two domains and RNase activity located in a second smaller domain (called RNaseH). The mature, i.e. active reverse transcriptase is unusual in that it requires proteolytic removal of one of the RNase domains from a p66 subunit to form a heterodimer of p66 and p51 subunits. This heterodimer represents the functional enzyme of HIV.

Inhibiting the action of reverse transcriptase should slow down new virus production and by implication disease progression. To understand how this might be achieved it was necessary to understand the structure and function of the enzyme, as well as its interaction with substrates before effective inhibitors could be designed. From the structure of p66 (Figure 12.13) and p51 (Figure 12.14) subunits in the presence of DNA and inhibitors as well as unliganded states a number of important facts have emerged. The polymerase domain is composed of 'finger', 'palm', 'thumb' and 'connecting' regions and this is reminiscent of the organization of other RNA polymerases. Significantly the p66 subunit shows structural similarity to the Klenow fragment of *E. coli* DNA polymerase I suggesting that conservative patterns of structure and organization exist within this class of enzymes.

Studies of both p66 and p51 subunits suggested that binding DNA does not perturb complex structure significantly, but that p51 and p66 differ in their

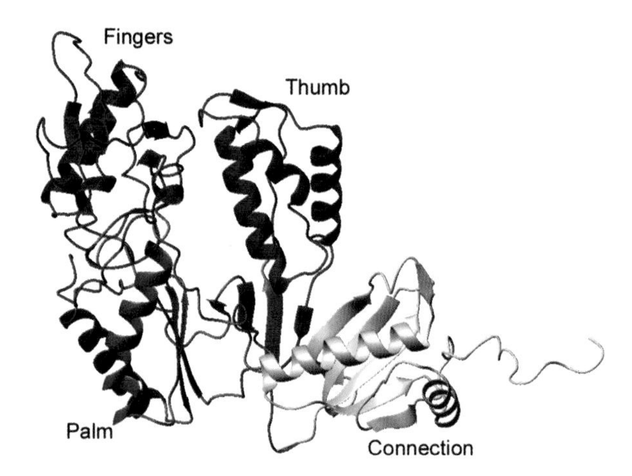

Figure 12.13 The organization of reverse transcriptase p66 subunit is shown. The finger domain is shown in red, the thumb in dark red, the palm in orange and the connecting region in gold. The active site aspartate residues lie at the base of the cleft between thumb and fingers

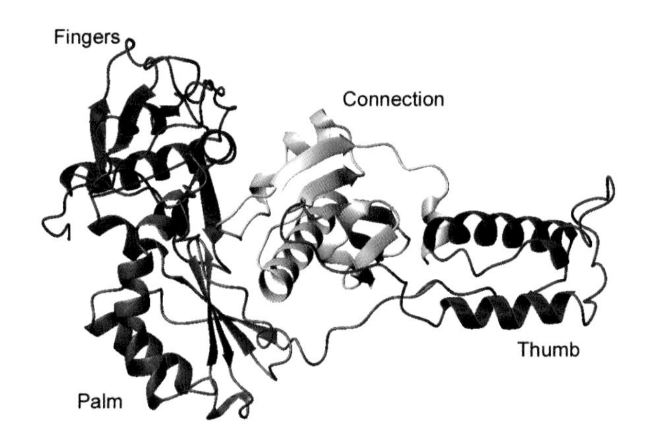

Figure 12.14 The p51 subunit of reverse transcriptase. The different conformations are clearly seen with the relative orientations of sub-domains forming a closed conformation in p51 and an open state in p66

structures largely as a result of the reorientation of domains or sub-domains within each subunit. The 'fingers' sub-domain is composed of β strands and three α helices whilst the 'palm' sub-domain possesses five β strands hydrogen bonded to four β strands at the base of the "thumb" region. The remainder of the 'thumb' region is predominantly α helical and links to a 'connection' sub-domain between the polymerase and RNaseH domains. The 'connection' domain is based around a large extended β sheet with two solitary helices. In p66 subunits the sub-domains are arranged in an open configuration exposing the active site containing three critical Asp side chains at residues 110, 185 and 186. In contrast, structural analysis of the p51 subunit reveals that, despite identical sequences, the polymerase sub-domains are packed differently with the fingers covering the palm and by implication burying the catalytically important active site residues. Unsurprisingly, functional studies reveal the p51 domain to be catalytically inactive although it interacts with the RNaseH domain of the p66 subunit and is important to the enzyme's overall conformation.

Inhibitors of reverse transcriptase fall into two classes. The first group based around the natural nucleotide substrate bind to the active site pocket whilst the second group, generally described as non-nucleoside inhibitors, function by binding at sites remote from the catalytic centre. Taken together these inhibitors represent two very different mechanisms of action. The first, and archetypal, reverse transcriptase inhibitor is 3′-azido-3′ deoxythymidine (AZT) and was the first HIV antiretroviral to be licensed for use. It contains an azido group in place of a 3′ OH group in the deoxyribose sugar. In a comparable fashion to dideoxynucleotides AZT acts as a chain terminator because it lacks a 3′ OH group. When the drug enters cells it is phosphorylated at the 5′ end and incorporated into growing DNA chains by the action of reverse transcriptase on the RNA template (HIV genome). The absence of a 3′ OH ensures that polymerization cannot continue. Although it might be expected that host DNA polymerases would be inhibited with severe consequences it appears fortuitous that host cell polymerases do not bind these inhibitors with high affinity. As a result AZT is effective against viral replication by inhibiting only reverse transcriptase activity. Competitive inhibition has proved extremely effective in the primary treatment of HIV with AZT, introduced in 1987 and complemented by additional nucleoside inhibitors, the accepted treatment option (Figure 12.15).

The non-nucleoside inhibitors (Figure 12.16) bind to a hydrophobic pocket remote from the active site. The effect of binding is to shift three β strands of the palm sub-domain containing the active site aspartyl residues towards a conformation seen in the p51 subunit. In other words, these inhibitors favour

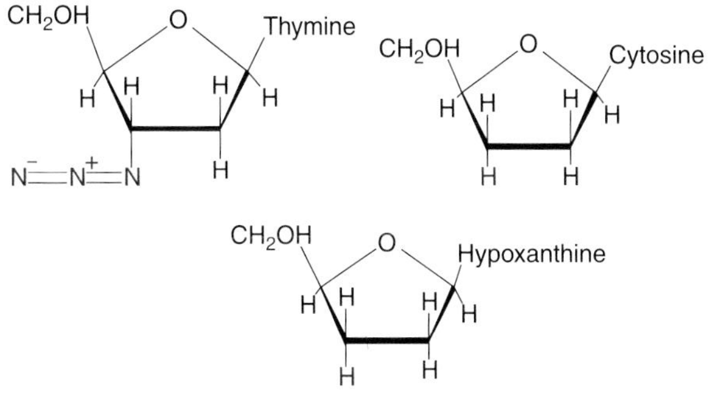

Figure 12.15 Active site (nucleoside inhibitors) of reverse transcriptase. Besides AZT, 2′,3′-dideoxycytidine (ddC) and 2′3′-dideoxyinosine (ddI) are other commonly used nucleoside analogues for fighting HIV infection

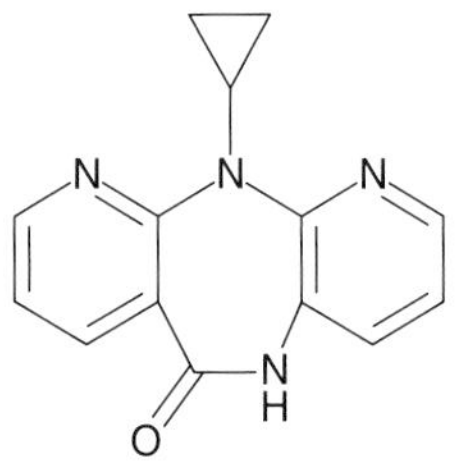

Nevirapine

Figure 12.16 Nevirapine, a non-nucleoside inhibitor of HIV reverse transcriptase

formation of inactive states by distorting the catalytic site as the basis for pharmacological action.

To date these drugs represent the most effective route of controlling HIV infection and are particularly important in the continual fight against viral proliferation. The combined use of nucleoside and non-nucleoside inhibitors in a cocktail known as highly active antiretroviral therapy (HAART) drastically reduces viral load (the number of copies of virus found within blood) and is associated with improved health. Unfortunately the spectre of drug resistance means that additional drugs or alternative regimes are needed to effectively combat the virus over a sustained period of time.

The introduction of antiviral drugs has drastically improved the quality of life and life expectancy of individuals infected with HIV but the prohibitive cost of providing these medicines has restricted treatment to countries with advanced forms of health care. This is particularly regrettable since HIV infection rates and deaths attributable to advanced HIV disease are most prevalent in the less economically advantaged areas of the world. Currently, many millions of individuals infected with HIV are unable to obtain effective life extending treatment.

The surface glycoproteins of HIV and alternative strategies to fighting HIV

An alternative approach to fighting HIV infection is to counter the ability of the virus to enter cells via the interaction between surface viral glycoproteins and the host cell receptors. The surface of HIV is coated with spikes that represent part of the envelope glycoprotein gp120. This exposed surface glycoprotein is anchored to the virus via interactions with the transmembrane protein gp41. Both proteins are synthesized initially as a precursor polyprotein – gp160. The envelope proteins gp41 and gp120 have been characterized as part of systematic attempts to understand the mechanism of entry into cells.

Viral entry is initiated by gp120 binding to receptors such as CD4, a protein containing immunoglobulin-like folds, expressed on the surface of a subset of T cells and primary macrophages. The gp120 protein binds to CD4 with high affinity exhibiting a dissociation constant (K_d) of ~4 nM and is a major route of HIV entry into cells although additional membrane proteins (the chemokine receptors CXCR4 and CCR5) can also operate in viral entry.

Structural studies of gp120 are aimed at understanding the molecular basis of protein–protein interactions. However, gp120 has proved a difficult subject because extensive glycosylation has prevented crystallization. Studies have therefore involved partially glycosylated core domains containing N- and C-terminal deletions and truncated loop regions. Immunoglobulin-like variable domains exist within gp120 in the form of five regions (V1–V5) and are found in all HIV strains. These variable zones are interspersed with five conserved regions. Four of these variable regions are located as surface-exposed loops containing disulfide links at their base. Despite substantial deletions the core gp120 binds to CD4 and presents a credible version of the native interaction.

The gp120 core structure has 'inner' and 'outer' domains linked by a four-stranded β sheet that acts as a bridge. The inner domain contains two α helices, a five-strand β sandwich and several loop regions, whilst the larger outer domain comprises two β barrels of six and seven strands, respectively. The core is often described as heart shaped with the N- and C-terminals located at the inner domain (top left-hand corner, Figure 12.17), the V1 and V2 loop regions located at the base of the inner domain and the V3 loop region located at the base of the outer domain. The V4 and V5 loop regions project from the opposite side of the outer domain.

The inner domain is directed towards the virus and from mutagenesis studies the interaction site between gp120 and CD4 is defined broadly at the interface

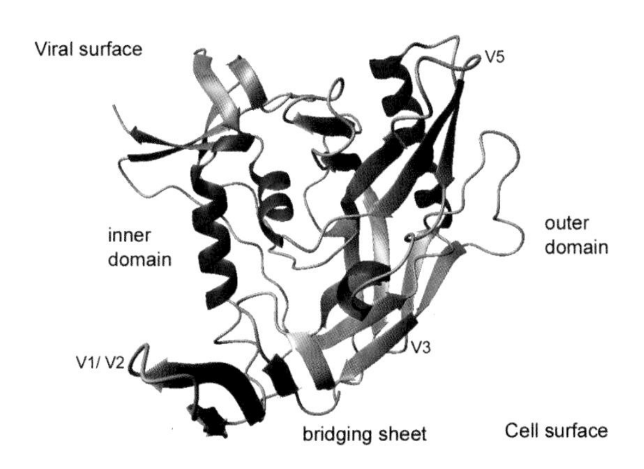

Figure 12.17 The structure of the gp120 core determined in a complex with a neutralizing antibody and a two domain fragment of CD4. Only the gp120 is shown for clarity. The bridging sheet is probably in contact with the cell surface with the N-terminus lying closer to the virus envelope surface (PDB: 1G9N)

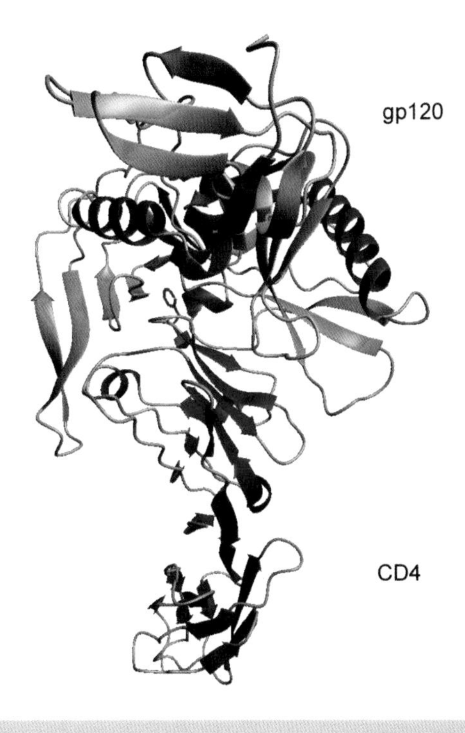

Figure 12.18 Interactions between the core of gp120 and the N-terminal immunoglobulin domain of CD4. The side chain of Phe43 of CD4 is shown capping a large hydrophobic cavity on gp120

of all three domains. This region contains conserved residues required for CD4 binding as well as variable residues. CD4 interacts with the variable residues through hydrogen bonds to main-chain atoms, but in addition cavities exist within the binding surface that are not in contact with CD4 (Figure 12.18). Within these cavities are many conserved residues and their mutation (as a result of error-prone RT activity) perturbs the interaction between CD4 and gp120 despite an absence of direct interactions in the crystal structures. The binding of CD4 to gp120 also results in interactions with chemokine co-receptors, and this site of interaction is distinct from that involved in CD4 binding. Current evidence suggests the co-receptor binding site is based around the bridging sheet between inner and outer domains, but on the opposite side of the molecule.

Initial attempts to construct a vaccine against HIV focused on raising specific antibodies against the gp120 component of the virus since this represents the first step of molecular recognition. Unfortunately immune recognition of the CD4 binding site is poor because of the mix of conserved and variable regions forming the binding site, the V1 and V2 regions mask the CD4 binding site prior to binding, and the CD4-bound state may represent a transient conformation of gp120. Vaccine development based on gp120 has not completed clinical trials.

The structure and function of gp41

The other major envelope protein is gp41, a transmembrane protein interacting with gp120, which consists of an N-terminal ectodomain, a transmembrane region and a C terminal intraviral segment within a single polypeptide chain of 345 residues. gp41 interacts with the matrix protein and its principal role is to mediate fusion between the viral and cellular membrane following gp120-based receptor binding.

At the N-terminal region a hydrophobic, glycine-rich, 'fusogenic' peptide is found and all structures presented to date have been of the ectodomain lacking this region. The ectodomain is a symmetrical trimer with each monomer consisting of two antiparallel α helices connected by an extended loop (Figure 12.19).

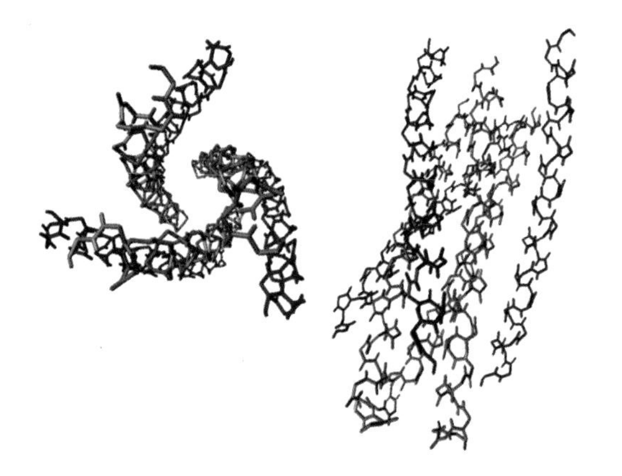

Figure 12.19 Part of the ectodomain region of gp41 is a coiled coil structure consisting of three pairs of helical segments

The combined action of gp120 and gp41 is the first contact point between virus and host cell. Unsurprisingly, if this stage can be inhibited then infection by HIV will be limited. Extensive proteolysis of a recombinant ectodomain revealed the existence of stable subdomains and a trimeric helical assembly composed of two discontinuous peptides termed N36 and C34 that are resistant to denaturation. The N- and C-peptides originate from the N-terminal and C-terminal regions of the gp41 ectodomain. The N-terminal derived peptides form a central trimeric 'coiled-coil' with the helical C-peptides surrounding the coiled coil in an antiparallel arrangement. This structural model was confirmed by X-ray crystallography for both HIV-1 and SIV peptides and is described as a trimer of hairpins (Figure 12.20).

In the trimer the C-peptides bind to the outside of the coiled-coil core interacting with a conserved hydrophobic groove formed by two helical N-terminal peptides. Although the structure of the intervening sequence of gp41 is unclear it would be required to loop around from the base of the coiled coil to allow the C-peptide to fold back to the same end of the molecule.

The mechanism of viral fusion to cells is slowly being unraveled and it may be very similar to fusion

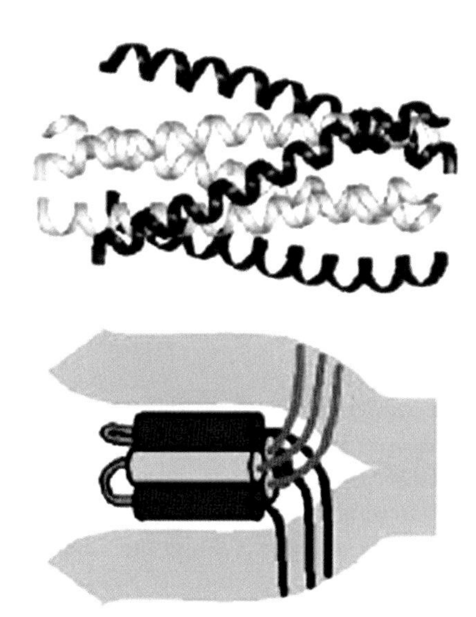

Figure 12.20 The trimer of hairpins exhibited by gp41 core domains (Reproduced by with permission from Eckert, D.E. & Kim, P.S. *Annu. Rev. Biochem.* 2001, **70**, 771–810. Annual Reviews Inc)

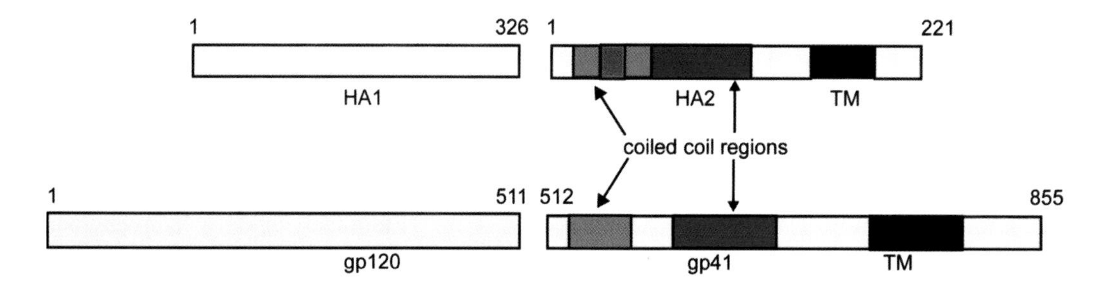

Figure 12.21 The organization of haemaglutinnin and HIV gp41/gp120 based around interacting helical regions or coiled coils

processes exhibited by the influenza virus during host cell infection. Some similarities exist between the sequences of gp41 and haemagglutinin (Figure 12.21) and may indicate potentially similar mechanisms exist *in vivo*.

However, this analogy cannot be taken too far, because HIV, unlike influenza virus fuses directly with cell membranes in a pH-independent manner. The trimer of hairpins is most probably the fusogenic state and is supported by the observation that gp41-derived peptides or synthetic peptides inhibit HIV-1 infection. Peptides derived from the C-terminal region are much more effective than those derived from N-peptides with inhibition noted at nM concentrations. These peptides act by binding close to the N-peptide region and preventing the hairpin structure. The observed inhibition of HIV by these peptides offers another route towards the development of antiviral drugs. Drugs that target additional steps in the viral life cycle, such as viral entry, are valuable in providing alternative therapies that may prove less toxic and less susceptible to viral resistance than current regimes.

The influenza virus

Influenza (or 'flu) is often considered a harmless disease causing symptoms slightly worse than a cold. Common symptoms are pyrexia, myalgia, headache and pharyngitis and the infection ranges in severity from mild, asymptomatic forms to debilitating states requiring several days of bed rest. The virus spreads by contact specifically through small particle aerosols that infiltrate the respiratory tract and with a short incubation period of 18 to 72 hours virus particles appear in body secretions soon after symptoms start causing further infection. In the respiratory tract the virus infects epithelial cells causing their destruction and susceptibility to infections such as pneumonia.

Modern medicines alleviate the worse symptoms of flu but this disease has killed millions in recent history and continues to present a major health problem. Periodically virulent outbreaks of influenza occur and one of the worst events occurring around 1918 and known as Spanish flu is conservatively estimated to have killed 20 million people. This flu pandemic resulted in more fatalities than all of the battles of

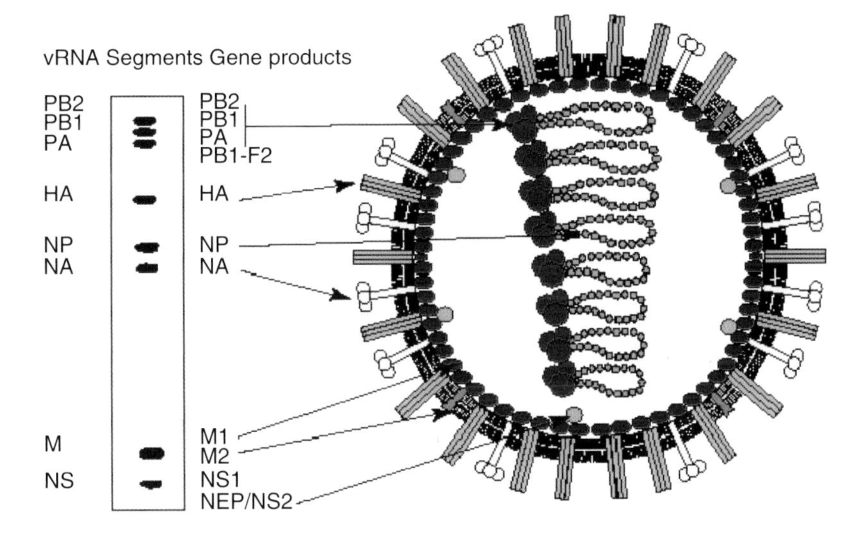

Figure 12.22 The influenza A virus. Nucleoprotein (green) associates with each RNA segment to form RNP complexes that contain PB1, PB2 and PA (red). The viral glycoproteins, haemagglutinin (blue) and neuraminidase (yellow), are exposed as trimers and tetramers. Matrix protein (purple) forms the inner layer of the virion along with the transmembrane M2 protein (orange). Non-structural proteins are also found (grey). (Reproduced with permission from Ludwig, S. *et al Trends Mol. Medicine* **9**, 46–52, 2003. Elsevier)

Table 12.3 The coding of different viral proteins and their functions by the segmented RNA genome

Segment	Size	Name	Function
1	2341	PB2	Transcriptase; Cap binding
2	2341	PB1	Transcriptase; Elongation
3	2233	PA	Transcriptase; Protease activity?
4	1788	HA	Haemagglutinin; membrane fusion
5	1565	NP	Nucleoprotein RNA binding; part of transcriptase complex; nuclear/cytoplasmic transport of vRNA
6	1413	NA	Neuraminidase: release of virus
7	1027	M1	Matrix protein: major component of virion
		M2	Integral membrane protein – ion channel
8	890	NS1	Non-structural: nucleus; role in on cellular RNA transport, splicing, translation. Anti-interferon protein.
		NS2	NS2 Non-structural: function unknown

World War I combined and emphasizes the serious nature of influenza outbreaks. Despite drugs and the use of vaccination influenza propagates around the world at regular intervals. There remains genuine fear of further major and lethal outbreaks striking again with a severity comparable to the events of 1918. When another pandemic occurs there is little doubt that we are in a much better position to deal with influenza than previous generations. This position of strength arises because the influenza virus has been extensively studied especially in relation to structural organization and interaction with cells.

The virus is a member of the orthomyxovirus family based around a (-) strand of RNA composed of eight segments in a total genome size of ~14 kb flanked by short 5′ and 3′ terminal repeat sequences (Figure 12.22). The segments bind to a nucleoprotein (NP) and associate with a polymerase complex of three proteins (PA, PB1 and PB2) that constitutes a ribonucleoprotein particle (RNP) and is located inside a shell. The shell is formed by the matrix (M1) protein and is itself surrounded by a lipid membrane that is derived from the host cell but contains two important glycoproteins – haemagglutinin (HA) and neuraminidase (NA) – together with a transmembrane-channel protein (M2). Two non-structural components (NS1 and NS2) are also encoded by the viral genome to make a total of 10 proteins[1] (Table 12.3).

Viral classification is based around the M and NP antigens and is used to determine strain type – A, B or C. Most major outbreaks of influenza are associated with virus types A or B with type B strain causing milder forms of the disease. The external antigens haemagglutinin and neuraminidase show greater sequence variation.

Viral infection and entry into the cell

The virus binds to sialic-acid-containing cell surface receptors via interactions with haemagglutinin entering the cell by endocytosis where fusion of the viral and endosomal membranes allows the release of viral RNA. After transfer to the nucleus the viral genome is transcribed and amplified before translation of mRNA in the cytoplasm. Late in the infection cycle the viral genome is transported out of the nucleus in RNP complexes and packaged with the structural proteins. Subsequent release of new virions completes a cycle

[1]For many years this was thought to be the complete complement of viral proteins but in 2002 further examination of the genome revealed an alternate reading frame in PB1 that allowed production of an eleventh protein (F2).

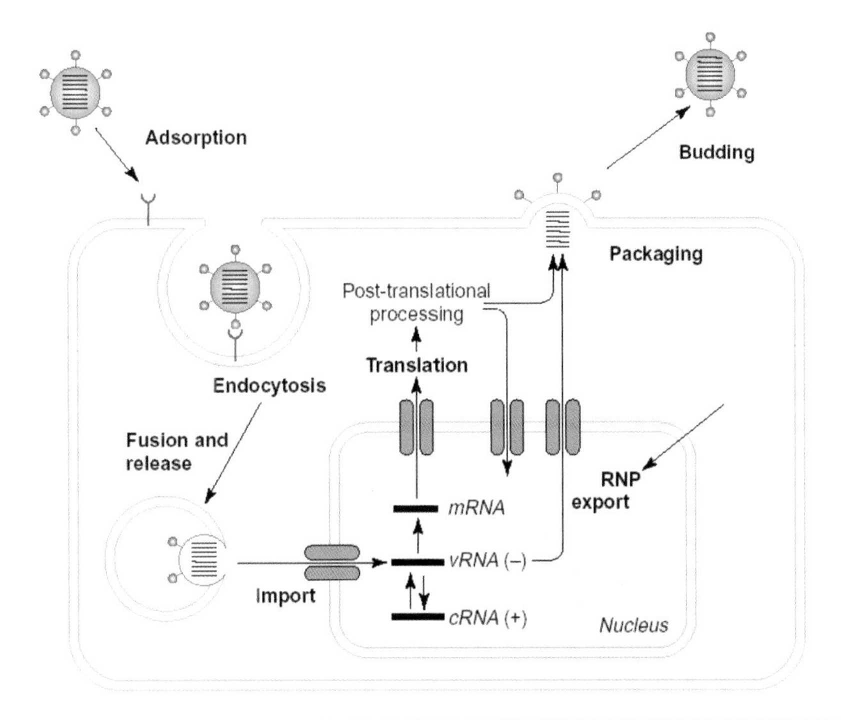

Figure 12.23 The lifecycle of the influenza A virus and the pathway of viral entry and exit from host cells. (Reproduced with permission from Ludwig, S. *et al Trends Mol. Medicine* **9**, 46–52, 2003 Elsevier)

that occurs over a time span between 6 to 16 h (Figure 12.23). Of critical importance to viral entry are the characteristic "spikes" of haemagglutinin and together with the second major envelope protein, neuraminidase, these proteins have been the targets for drug treatments to fight "flu".

Haemagglutinin structure and function

Haemagglutinin was identified from the ability of flu viruses to agglutinate red blood cells via the interaction between HA and sialic acid groups present on cell surface receptors. Haemagglutinin is anchored to the membrane by a short tail but limited proteolysis with enzymes such as bromelain or trypsin yield a soluble domain (Figure 12.24). The soluble domain was amenable to crystallographic analysis and in 1981 a structure of haemagglutinin determined by John Skehel, Donald Wiley and Ian Wilson provided the first detailed picture of any viral membrane protein.

Haemagglutinin is a cylindrically shaped homotrimer approximately 135 Å in length. It is synthesized as a precursor polypeptide (HA0) that associates to form a trimer but is first cleaved by proteases to give two disulfide linked subunits -HA1 and HA2. HA1 and HA2 are necessary for membrane fusion and virus infectivity and generally the polypeptide chains are separated by just one residue, a single arginine residue, removed from the C terminus of HA1 by carboxypeptidase activity. It is the HA1/HA2 assembly that is commonly, but slightly inaccurately, referred to as a monomer and haemagglutinin contains three HA1 and three HA2 chains. The membrane-anchoring region is located at the C terminal region of HA2 but structural studies were performed on tail less protein.

The crystallographic structure shows the free ends, at the cleavage site between HA1 and HA2, are separated by 20 Å and within the structure two distinct regions were readily identified; a globular head region responsible for binding sialic acid residues on host cell

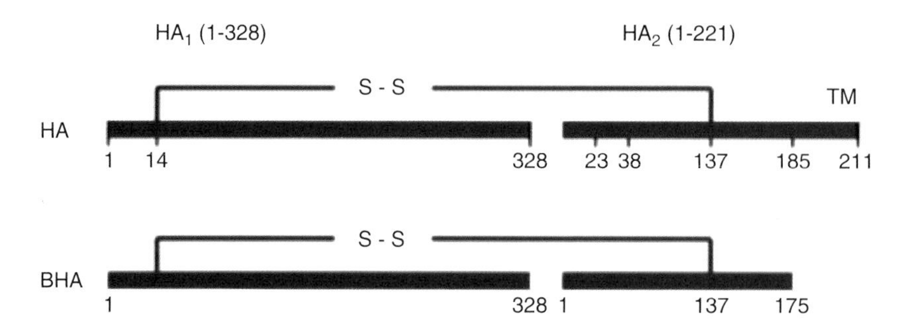

Figure 12.24 The membrane bound and protease treated form of haemagglutinin showing HA1 and HA2 regions. BHA refers to bromelain (protease) treated haemagglutinin

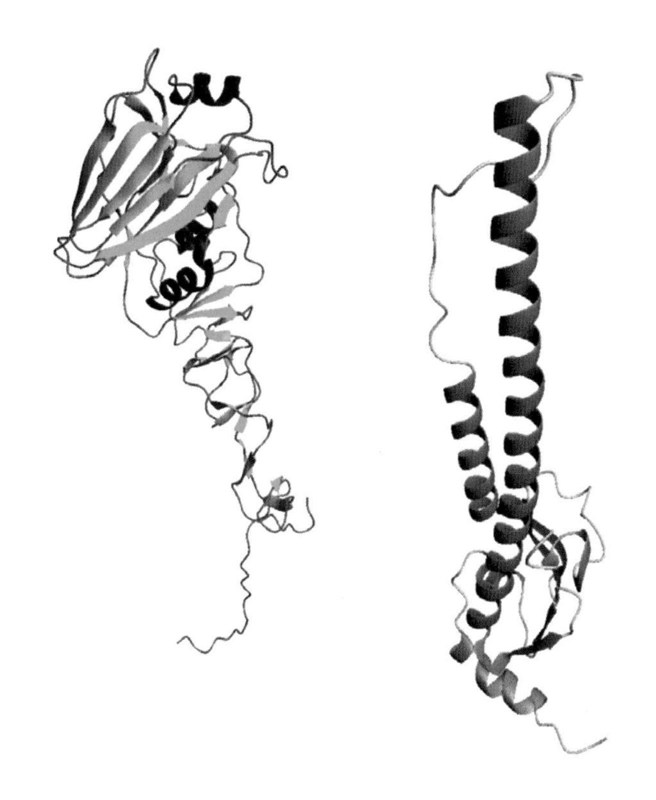

Figure 12.25 The individual HA1 and HA2 units of haemagglutinin

receptors and a longer extended stem-like region that contains a fusogenic sequence (Figure 12.25).

The three monomers assemble into a central α-helical coiled-coil region that forms the stem-like domain whilst at the top are three globular heads containing sialic acid-binding sites and responsible for cell adhesion or binding. The stem region consisting of a triple helix extends for ~76 Å from the membrane surface and is formed predominantly by the HA2 polypeptide. The globular region is based mainly on the HA1 portion of the molecule and each globular domain consists of a swiss roll (jelly roll) motif of eight anti-parallel β-strands by residues 116–261. At the distal tips receptor (sialic acid) binding pockets were identified. The binding sites reside in small depressions on the surface and were detected from crystallographic studies in the presence of different sialic acid derivatives. Amongst the residues forming the active sites are Tyr98, Ser136 Trp153, His183, Glu190, Leu194, Tyr195, Gly225 and Leu226 with site directed mutagenesis suggesting that Tyr98, His183, and Leu194 are particularly important in forming critical hydrogen bonds with the substrate (Figure 12.26).

Sialic acid groups bind with one side of the pyranose ring facing the base of the site whilst the axial carboxylate, the acetamido nitrogen, and the 8- and 9-OH groups of the sugar form hydrogen bonds with polar groups of conserved residues. Histidine 183 and glutamate 190 form hydrogen bonds with the 9-OH group, tyrosine 98 forms hydrogen bonds with the 8-OH group whilst the 5-acetamido nitrogen forms a hydrogen bond with the backbone carbonyl of residue 135 and the methyl group makes van der Waals contact with the phenyl ring of tryptophan 153. The 4-OH group projects out of the site and does not appear to participate in binding interactions.

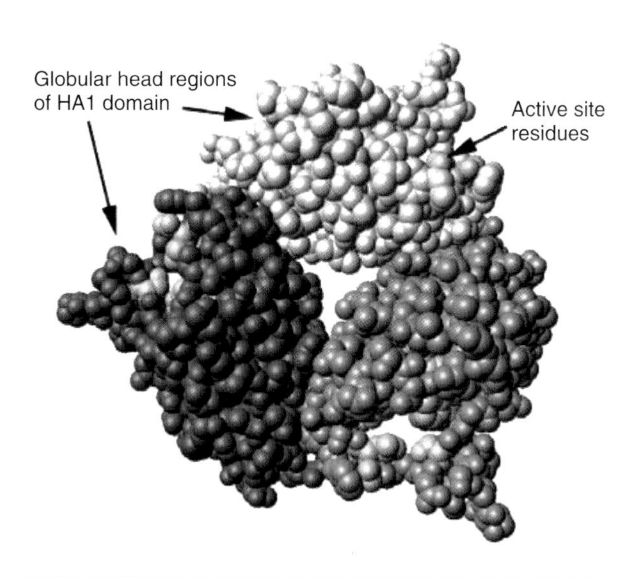

Figure 12.26 The active site residues of haemagglutinin are shown in the spacefilling representations of the globular HA1 domain. The view is from above the globular trimer

A key observation in understanding the structure and function of haemagglutinin was the demonstration that the molecule underwent a large conformational change when subjected to low pH. The HA proteolytic fragment does not bind to membranes and is functionally inactive but lowering the pH to a value comparable to that occurring in endosomes (~pH 5–6) during viral entry caused haemagglutinin to become a membrane protein. Electron microscopy showed protein aggregation into "rosettes" with an elongated and thinner structure. These observations suggested major conformational changes and performing crystallography on the 'low pH' form resolved the extent and location of structural perturbations (Figures 12.27 & 12.28).

At low pH the fusion region or peptide was located at the end of an elongated coiled-coil chain with domains near the C-terminal interacting with those at the N terminus. The low pH form is the thermodynamically stable form of HA2 and the initial state is best described as a metastable configuration. Exposure to low pH converts this metastable form into the thermodynamically stable form of HA2.

In total the structures of the ectodomain of haemagglutinin have been determined for the single-chain precursor (HA0), the metastable neutral-pH conformation and the fusion active, low pH-induced conformation. These structures provide a framework for interpreting results of experiments on receptor binding, the cyclical generation of new and re-emerging epidemics as a result of HA sequence variation as the process of membrane fusion during viral entry.

Neuraminidase structure and function

Neuraminidase is vital to viral proliferation. The enzyme is a glycosyl hydrolase cleaving glycosidic linkages between N-acetyl neuraminic acid and adjacent sugar residues by destroying oligosaccharides units present on host cell receptors. Viral neuraminidase

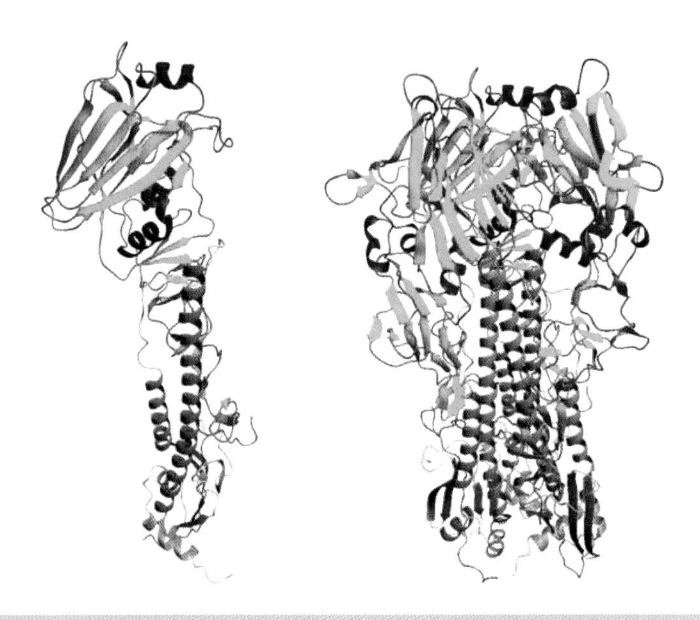

Figure 12.27 The structure of the HA1/HA2 'monomer' and a complete assembly of trimeric haemagglutinin (PDB:2HMG)

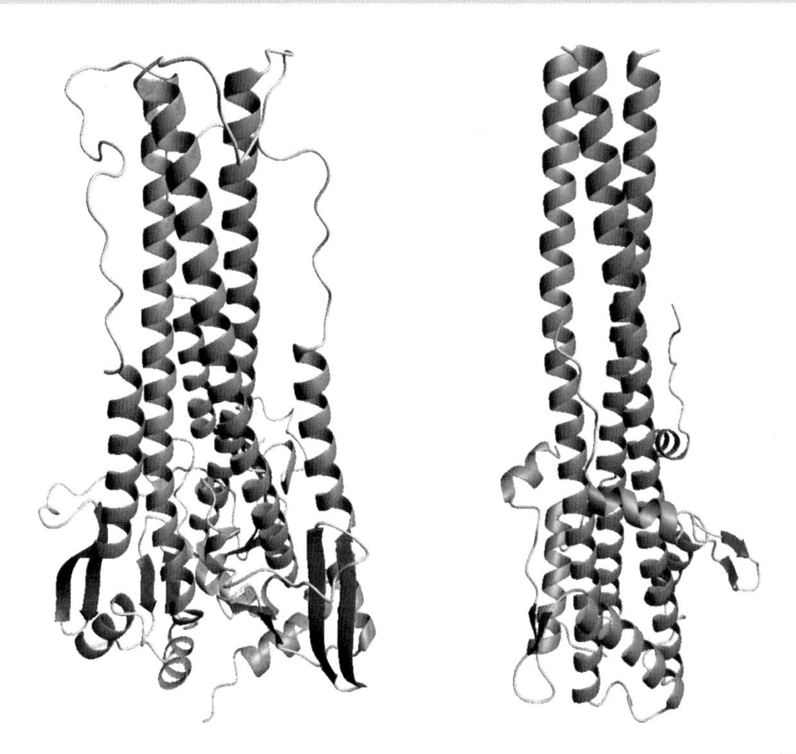

Figure 12.28 The changes in conformation associated with the coiled-coil regions of HA2 in the transition from neutral pH form to active haemagglutinin. To ascertain the change in conformation calculate the number of turns associated with the main helices

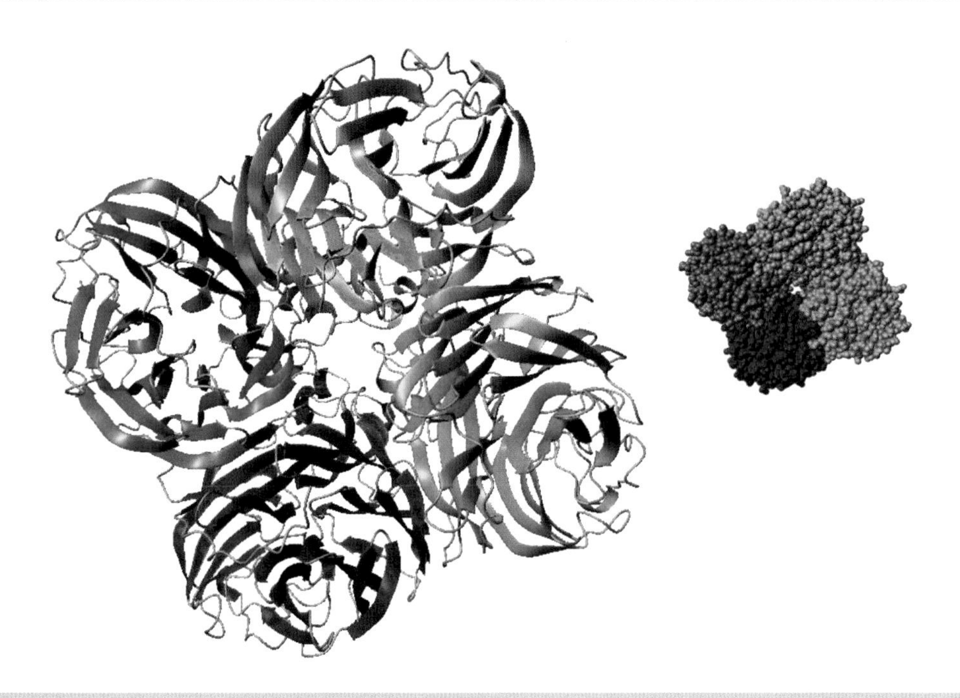

Figure 12.29 The box-like arrangement of the tetrameric state of neuraminidase shown alongside a spacefilling model of the same view

causes a loss of agglutination and allows subsequent infection of other host cells.

The amphipathic protein is a single polypeptide chain of ~470 residues with a hydrophobic membrane binding domain of ~35 residues connected to a larger polar domain of ~380 residues via a hypervariable stalk of ~50 residues. Limited proteolysis removes the N terminal membrane anchor and facilitated crystallization. The structure of the large fragment of neuraminidase revealed a homo-tetramer with monomers consisting of six topologically related units (Figure 12.29). Each unit contained a four-stranded sheet arrayed like the petals of a flower or more frequently described as one of the blades of a propeller (Figure 12.30).

Neuraminidase is an example of the β propeller motif. The four strands connect by reverse turns with the first strand of one sheet interacting across the top of the motif to the fourth strand of the adjacent motif. Disulfide bonds between adjacent strands contribute to a distortion of regular sheet conformation and introduce a distinct twist to each blade.

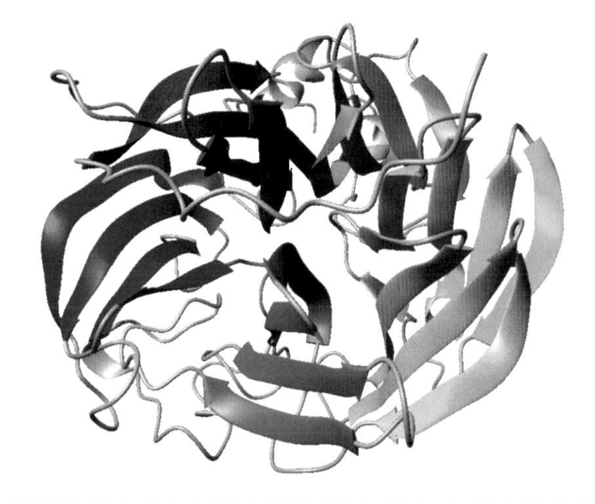

Figure 12.30 The neuraminidase monomer. Whilst the structure of the larger soluble fragment has been determined the structure of the membrane spanning region is unknown. The structure shows one monomer with six topologically related units (blades) composed of four β strands; each blade is shown in a different colour. (PDB:2BAT)

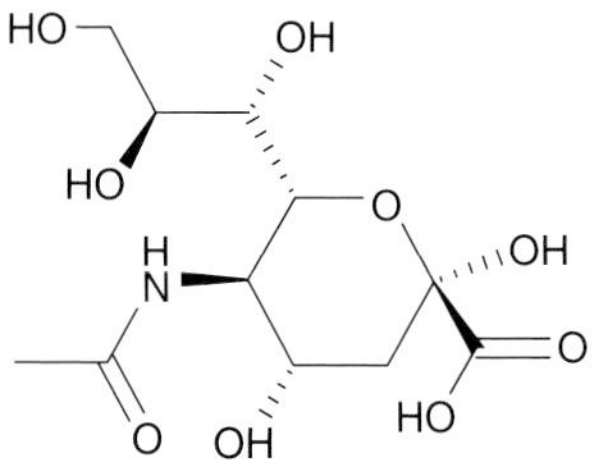

Figure 12.31 The structure of sialic acid (*N*-acetylneuraminic acid, Neu5Ac)

The monomeric units assemble into a box-shaped head with dimensions of ~100 Å × 100 Å × 60 Å attached to a slender stalk region. The tetrameric enzyme has circular 4-fold symmetry stabilized in part by metal ions bound on the symmetry axis with each subunit interacting via an interface dominated by hydrophobic interactions and hydrogen bonds.

Definition of the active site of neuraminidase proved important in designing inhibitors that interfere with neuraminidase activity and numerous crystal structures exist for enzyme-inhibitor complexes. Initial studies focused on a transition state analogue, 2-deoxy-2, 3-dehydro-*N*-acetylneuraminic acid (Neu5Ac2en), that inhibited the enzyme at micromolar concentrations but further refinement yielded inhibitors effective at nanomolar levels (Figures 12.31 & 12.32).

One of these compounds 4-guanidino-Neu5Ac2en (Zanamivir) inhibits viral proliferation as a result of significant numbers of hydrogen bonds formed between drug and *conserved* residues in the active site of neuraminidase (Figure 12.33).

Strategies to combat influenza pandemics

Influenza infection leads to a pronounced antibody response against haemagglutinin and neuraminidase based on the surface epitopes presented by the infecting virus. Ordinarily individuals have protective immunity but in the case of flu frequent changes in the sequences of these two antigens compromises the normal immune response. Vaccines against flu viruses are

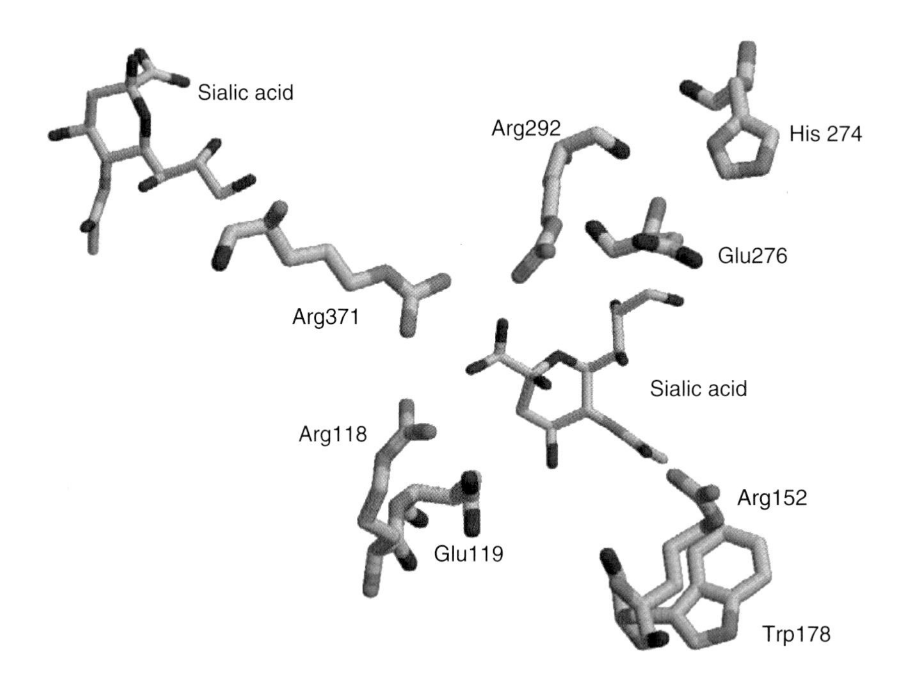

Figure 12.32 The active site of neuraminidase showing substrate bound to active site (PDB:1MWE). A second sialic acid binding site was determined on the surface of neuraminidase

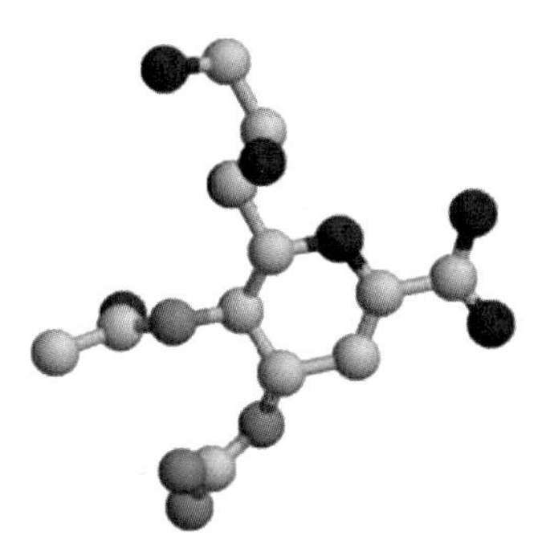

Figure 12.33 The molecular structure of Zanamivir. A potent inhibitor of neuraminidase that is now accepted as clinical treatment for flu in most countries

recommended by the WHO[2] to contain haemagglutinin components from two A strains and a single B strain and to be representative of viruses currently circulating the world. Unfortunately although vaccines are developed from purified HA the virus alters the sequence of residues around the epitopes without impairing protein function. This occurs *via* point mutations and leads to altered surface properties that eliminate cross-reaction with antibodies developed against previous forms of the virus. Two distinct mechanisms contribute to variation in surface properties of the HA. Point mutations cause antigenic drift whilst a second, potentially more serious, mechanism for variation is antigenic shift where the exchange of genes between influenza viruses creates new strains.

Haemagglutinin has at least fifteen different antigenic sub-types (H1–H15) whilst neuraminidase has nine antigenic subtypes (N1–N9). Most subtypes of type A virus infect avian species with sub-types H1, H2 and H3 infecting humans. The HA1 domains show greater variability than HA2 consistent with their role as binding sites for receptors and antibodies. With the recognition of these sub-types viral outbreaks are classified according to the virus class (usually A or B), the species from which the virus was isolated, its geographical region, the year of isolation and the H and N sub-group. The 20th century witnessed three influenza pandemics; Spanish influenza (defined as an H1N1 virus) in 1918, Asian influenza (defined as an H2N2 virus) in 1957, and Hong Kong influenza (H3N2 virus) in 1968. These global outbreaks caused severe illness with high mortality rates. DNA sequencing showed that the HA and NA genes of the H1N1 viruses emerged from an avian reservoir with subtle changes in primary sequence leading to altered antigenic properties. In contrast Asian and Hong Kong outbreaks were attributed to hybrid viruses arising from new combinations of avian and human viral genes. Aquatic birds are believed to be the source of influenza viruses affecting all other animal species. Recently in 1997, a highly pathogenic avian H5N1 influenza virus was transmitted directly from poultry to humans in Hong Kong. Of the 18 people infected, six died and this heightened worries about new pandemics in the future.

Influenza and HIV represent two viruses with an enormous impact on human health. They are not unique in their mechanism of infection but in both cases they have an unfortunate ability to mutate rapidly causing changes in antigenicity or loss of drug activity. This property creates pandemics and limits our ability to fight infection through vaccines and drugs. It is very unlikely that these viruses will remain the only threat to human health. The next decades are likely to see the emergence of new strains and new viruses, perhaps involving those crossing the species boundary, whose action will need to be countered through exploitation of knowledge about the structure and function of their constituent proteins.

Neurodegenerative disease

In comparison with influenza neurodegenerative diseases are less common, less easily transmitted and most of us will escape neurodegenerative disorders entirely. Despite a low incidence in all population groups there is evidence to suggest that these diseases are of greater importance to human health than previously thought. Many of these diseases have clear genetic links that enable insight into their molecular origin.

Prion based diseases are linked by the proliferative formation of insoluble aggregates of protein arising from mutations within the *PrP* gene. Creutzfeldt-Jakob disease, fatal familial insomnia, kuru and scrapie were linked to the PrP gene and lead to

[2]WHO = World Health Organization.

neurodegenerative conditions. All of these conditions lead to abnormal brain tissue and the presence of amyloid deposits. Collectively the term amyloidogenic disease stressed the link between protein misfolding and the development of neurodegenerative disorders.

BSE and new variant CJD

Scrapie was first reported in 1732 as a disease infecting sheep in England and has been present continuously since this time. It did not appear to effect bovine livestock. However, in the 1980s a new disease spread through cattle in the United Kingdom at an alarming rate producing symptoms normally associated with TSEs. The disease in cows led to abnormal gait and posture, poor body condition, nervous behaviour, decreased milk production and a progressive loss of mass. Diseased animals lost the ability to support themselves with death occurring 2–25 weeks after the appearance of symptoms. A feature of the disease was brain tissue of spongy appearance bearing large vacuoles. The disease was christened bovine spongiform encephalopathy (BSE) but its alarming rise in frequency yielded another sobriquet of 'mad cow disease' (Figure 12.34). The appearance of BSE raised a number of important questions. What was the origin of the disease, how was it transmitted and given the role of cattle in the human food chain was there a possibility of human infection?

BSE is a chronic degenerative disease and with the observation of scrapie like symptoms it was logical to look for a prion-like particle. At the time (mid 1980s) the concept of prions was controversial with the pioneering work of Prusiner slow to gain acceptance. The appearance of the disease in cattle was puzzling given the absence of simple routes of transmission but epidemiological data suggested BSE arose from animal feed containing contaminated protein sources. BSE had a substantial impact on the livestock industry in the United Kingdom with approximately 0.3 % of the total UK herd infected and continued to rise until 1992.

In 1988 the introduction of a total ban on the use of rendered ruminant based feed to cattle minimized transmission and was combined with the large scale slaughtering of infected cattle. In 1989 these preventative actions were supplemented by a ban on human

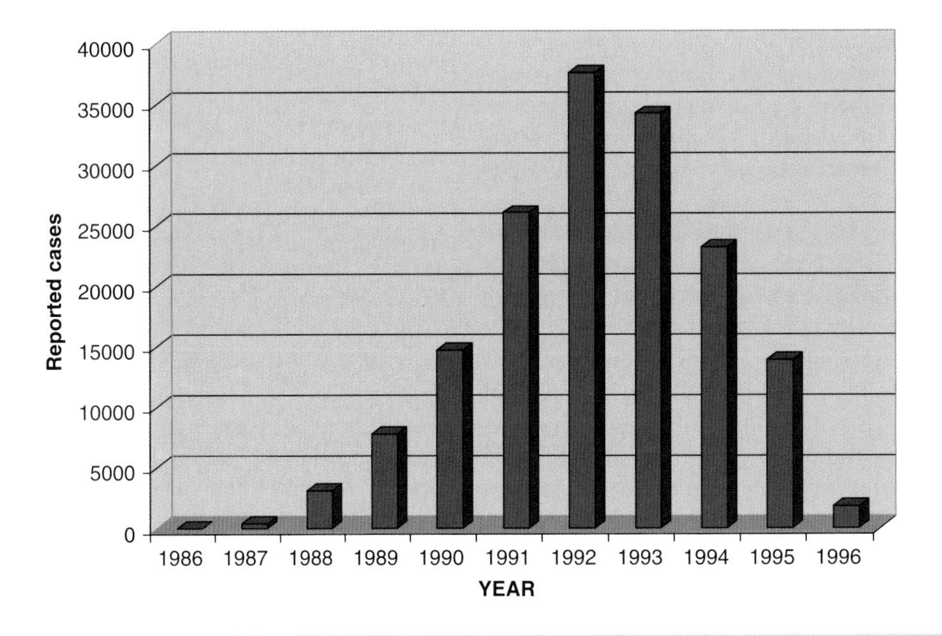

Figure 12.34 The yearly totals for clinical diagnosis of BSE in the UK herd (data derived from BSE Inquiry Report, Crown copyright 2000)

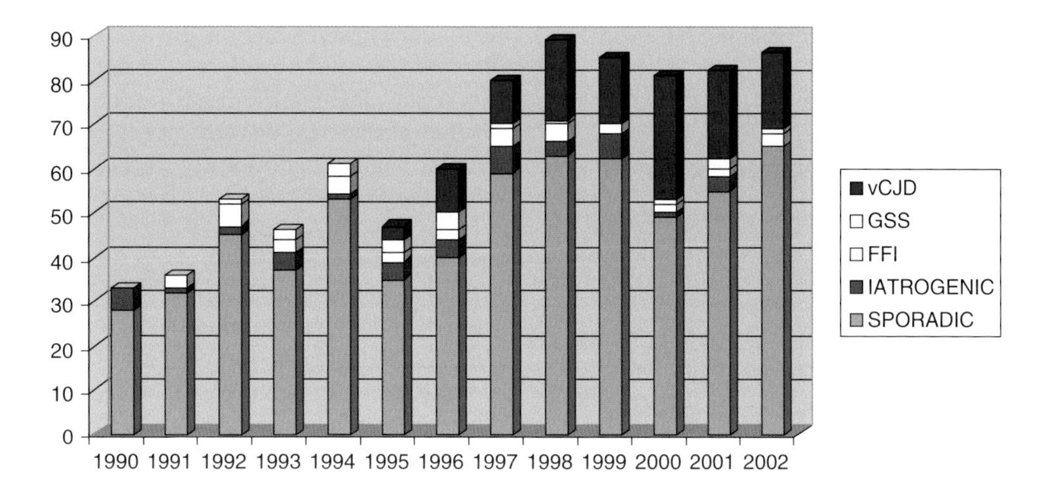

Figure 12.35 The incidence of all CJD cases in the United Kingdom (Data derived from CJD Surveillance unit, Edinburgh http://www.cjd.ed.ac.uk). From 1995 the number of cases of vCJD rise reaching a peak in 2000 before showing small decreases in the subsequent years

consumption of tissues shown to contain high levels of the infectious agent (when injected into mice brains) and therefore posing a potential danger to humans. Unfortunately this was not the end of the story because the long incubation period of BSE[3] ranging from 2 to 8 years meant that human consumption of infected cattle occurred throughout the 1980's.

In March 1996 the UK Spongiform Encephalopathy Advisory Committee announced the identification of 10 new cases of Creutzfeldt-Jakob disease (Figure 12.35). CJD is a rare disease and the observation of 10 cases was startling. Although direct evidence was lacking it seemed possible that exposure by ingestion to BSE agents was responsible for increase numbers of CJD cases. In 1994 initial speculation that BSE might be the causative agent of new cases of CJD was aired and despite a lack of definitive evidence the idea gained credibility and created widespread unease about the use of beef as a culinary product.

The 10 cases were unusual in terms of their high frequency and the clinical and pathological phenotypes associated with the condition. The new forms of CJD exhibited an onset of symptoms at a young age with teenagers or young adults most susceptible. The disease was associated with neurodegeneration and brain tissue showed lesions characteristic of spongiform encephalopathies. The new forms of CJD were called variant CJD (vCJD) and stemmed from exposure to BSE before the introduction of a specified bovine offal ban in 1989. Definitive proof of the link between BSE and vCJD required comparing both diseases at a molecular level to demonstrate that the etiological agents were prion proteins.

Comparing four different types of CJD (genetic, sporadic, iatrogenic, and "new variant") as well as BSE using SDS-PAGE showed that prion proteins exhibit distinct banding patterns when studied via Western blotting methods after treatment with proteinase K, a proteolytic enzyme that normally effectively degrades proteins into very small fragments but in the case of infectious prions produced protease resistant fragments. The pattern and intensity of classic forms of CJD (genetic, sporadic, or iatrogenic) were similar but variant CJD yielded a unique banding pattern suggesting a different genotype. The pattern was comparable to the profile seen in BSE infected organisms. Further analysis of the DNA sequence of individuals with vCJD showed that all were heterozygous at residue 129 (Valine/Methionine) suggesting that this transition conferred a genetic predisposition for contracting vCJD

[3] The time span between an animal becoming infected and then showing the first signs of disease.

whereas the homozygous (Valine/Valine) state was possibly resistant to infection.

Alzheimer's disease – a neurodegenerative disorder

Amyloidosis or protein aggregation is associated with Alzheimer's disease although like prion-based diseases little impact has been made towards effective treatment. An understanding of the molecular events involved in disease pathogenesis offers the possibility of future therapy.

Alzheimer's disease first recognized by Alois Alzheimer in the first decade of the 20th century is a slow progressive form of dementia normally seen in the later stages of life and resulting in intellectual impairment. Today a common diagnostic marker for physicians is memory impairment accompanied by a decline in language and decision-making ability, judgement, attention span, cognitive functions and personality. Diagnosis is usually confirmed by post mortem studies and the microscopic examination of brain tissue where a diagnostic feature is the presence of neurofibrillary tangles, aggregates of a single protein within brain tissue. The brain is seen to contain plaques and although plaques are detected in *all* brain tissue with age their number is drastically elevated in individuals with Alzheimer's disease.

Two types of Alzheimer's disease are recognized. Early onset disease results in symptoms before the age of 60 and has a genetic basis seen by its presence within families and an autosomal dominant form of inheritance. It accounts for less than 10 % of all forms of Alzheimer's. In contrast late onset forms are more common but the genetic basis is less well defined. Late onset disease develops in people above 65 with an incidence of 5–10 %.

The amyloid plaques found in Alzheimer's disease are composed of a 4.2 kD peptide called the Aβ protein. The plaques result from the transformation of soluble Aβ protein monomers into insoluble aggregates. The peptide called Aβ because of its amyloid properties and a partial β-stranded structure occurring within a 28-residue portion of the sequence is found in at least two forms; a 40 residue peptide and a 42 residue

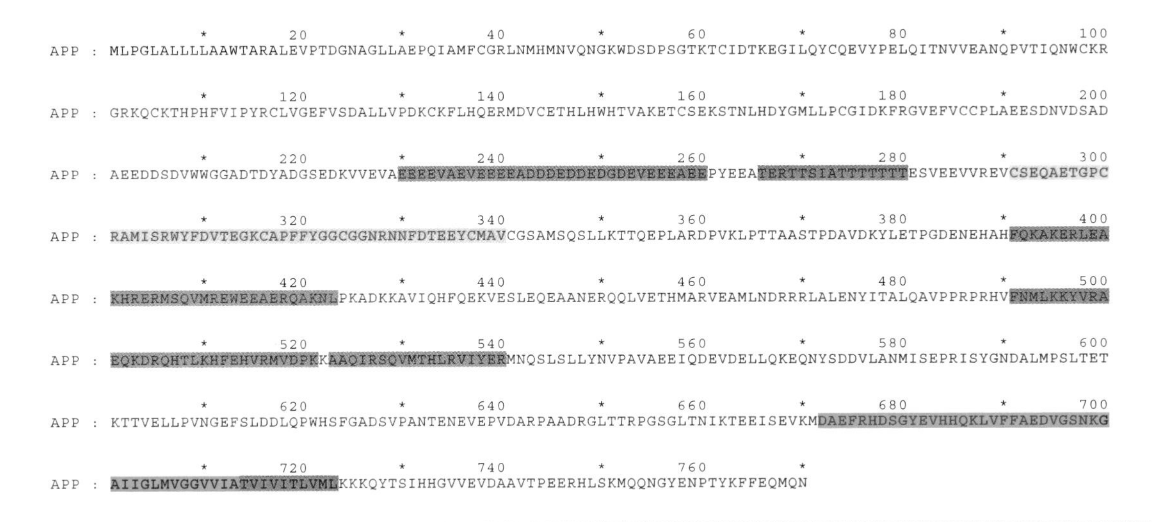

Figure 12.36 The primary sequence of the APP protein and its relationship to the Aβ peptide and other domains. The Aβ peptide extends from residues 672-711/713 (green) and overlaps with the single membrane spanning region from 700–723 (blue lettering and red block). The yellow block indicates a signal sequence (1–17) whilst the cyan block indicates the location of a domain showing homology to BPTI or Kunitz type inhibitors (291–341). The magenta blocks highlight poly-glutamate rich domains 230–260, threonine rich regions 265–282, heparin binding regions 391–424 and 491–522 and collagen binding domain 523–541

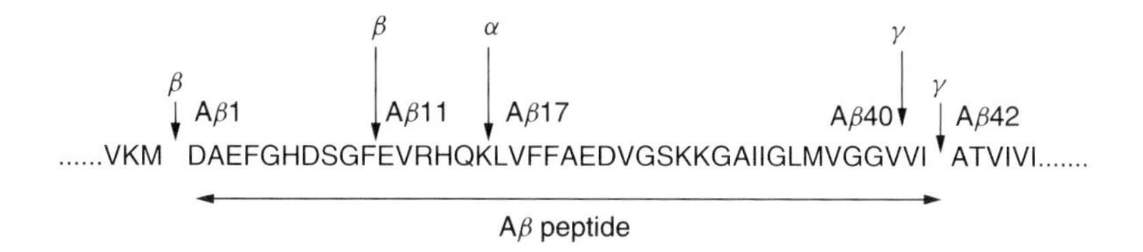

Figure 12.37 Schematic diagram of the Aβ peptide derived from the amyloid precursor protein. The amino acid sequence is shown with single letter codes and the α, β and γ secretase cleavage sites are shown along the polypeptide chain. The α secretase cleaves at the 17th residue whilst the β secretase cuts at the 1st and more rarely 11th residues and the γ secretase cuts at the 40th or 42nd residue

peptide designated as $A\beta_{1-40}$ and $A\beta_{1-42}$ respectively. Sequencing and database analysis showed that the peptides derived from a longer protein expressed in many tissues – the amyloid precursor protein (APP)- and DNA sequencing indicated a protein of ~770 residues with multiple domains (Figure 12.36).

APP undergoes extensive post-translational modification as a glycoprotein but despite domain recognition a clearly defined function for APP is absent although a role in cell adhesion is currently favoured. The APP gene was mapped to chromosome 21, contained 19 exons, with alternative splicing mechanisms giving rise to multiple transcripts (APP395, APP563, APP695, APP751, APP770). APP695 is the major isoform in neuronal tissue with APP751 dominating elsewhere.

The Aβ amyloid peptide (residues 672-711/713) is derived from a region of the protein that includes a small extracellular region together with its membrane spanning sequence (700–723). Understanding the origin of the Aβ peptide lies at the heart of discovering the molecular pathology of Alzheimer's disease. The amyloid peptide located *within* the full length protein arises from endoproteolysis of APP (Figure 12.37). The amyloid peptide was formed by cutting at residues 670 and 711/713 at sites associated with enzymes called β and γ secretase whilst the enzyme α secretase cuts APP at residue 687 within the amyloidogenic peptide. The properties and identity of these enzymes were initially obscure (hence their names) but slowly each system has been shown to occupy an important role in APP processing.

The demonstration that endoproteolysis of APP was associated with two different enzymes suggested that defining these 'activities' in more detail as well as the associated genes/proteins would assist understanding the molecular events contributing to disease. β secretase was identified as a new member of the aspartic protease family with the unusual property of being membrane tethered. The α secretase is a member of the ADAM family of proteins (standing for a disintegrin and metalloprotease) and finally γ secretase began to give up its secrets by suggesting similarity to a membrane protein called presenillin 1.

Although initially called β secretase the enzyme purified by different groups was given alternative names of BACE (β-site APP cleaving enzyme) and memapsin-2 (membrane-associated aspartic protease 2) and both are widely quoted in the literature. The enzyme, a single polypeptide chain of 501

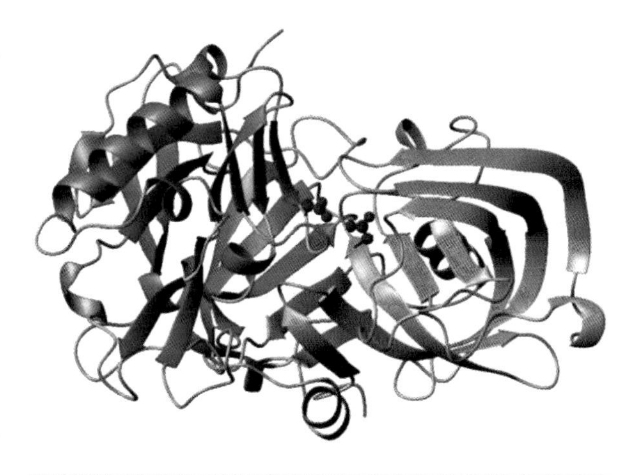

Figure 12.38 The organization of secondary structure into two lobes and the active site aspartate residues in β secretase (PDB:1M4H) shows similarities to previous aspartyl proteases

residues synthesized as a preproprotein, contains a single transmembrane region from residues 458 to 478 followed by a short cytoplasmic facing region of ~20 residues. The majority of residues are located in a large extracellular domain from residues 45–457 and this region of the enzyme has been studied to provide information on kinetics, substrate specificity and structure (Figure 12.38).

As an aspartyl protease the enzyme contains two conserved aspartate within a bilobal structure and exhibits an optimal pH for catalytic activity at pH 4.0 with substrates bound with K_m values between 1–10 mM and turnover rates ranging from 0.5–3 s^{-1}. The active site of β secretase is more open and less hydrophobic than corresponding sites in other human aspartic proteases but by defining the active site of this enzyme through substrate specificity and transition state analogues offers the hope of designing effective inhibitors. Inhibitors of β secretase are important pharmaceutical drug targets since they would prevent amyloid peptide formation.

The genetic basis of Alzheimer's disease has assisted understanding of disease pathology. Missense mutations are present in the APP gene in familial Alzheimer's disease and these mutations are frequently found in close proximity to the α-, β-, or γ secretase processing sites. In a Swedish pedigree a double mutation Lys → Asn and Met → Leu near the β-secretase cleavage site (671–672) leads to elevated production of the Aβ peptide (both Aβ40 and Aβ42) and therefore presumably accounts for the early onset of the disease. Defective genes are also responsible for early onset disease with mutations to presenilin 1 and presenilin 2 observed in families and leading to increased deposition of the Aβ peptide. Over 70 mutations are known most located in presenilin 1.

Despite a lack of a clearly defined function for APP the Aβ peptide aggregates into larger amyloid fibrils, destroying nerve cells and decrease levels of neurotransmitters. The correct balance of neurotransmitters is critical to the brain's activity and the levels of acetylcholine, serotonin, and noradrenaline are all altered in Alzheimers disease. With both structural defects and chemical imbalance in the brain Alzheimer's disease effectively disconnects areas of neuronal tissues that normally work cohesively together.

An interesting question is why have neurodegenerative diseases become more prominent in recent times? Undoubtedly part of the answer lies with longer life expectancy. In 1900 it is estimated that only 1 % of the world's population was above 65 years of age. By 1992 that figure had risen to nearly 7 % and in 2050 it is estimated that ~25 % of the population will be >65 years of age. The introduction of antibiotics, improved living conditions particularly in relation to water quality and food supply coupled with effective health care has meant that at least in the economically advantaged regions of the world average lifetimes for men and women are well above 80 years of age. One consequence of longer life expectancy is that diseases such as neurodegenerative disorders can develop over many years where previously an individual's short lifetime would preclude their appearance. Susceptibility to neurodegenerative diseases may be an inevitable consequence associated with longevity but as molecular details concerning these disease states are uncovered there is optimism about favourable treatment options in the future.

p53 and its role in cancer

Cancer involves the uncontrolled growth of cells that have lost their normal regulated cell cycle function. In many instances the origin of diseases lie in the mutation of genes controlling cell growth and division so that cancer is accurately described as a cell cycle based disease. Mutations alter the normal function of the cell cycle, with three classes of genes identified closely with cancer. Oncogenes 'push' the cell cycle forward, promoting or exacerbating the effect of gene mutation. In contrast, tumour suppressor genes function by normally applying the 'brakes' to cell growth and division. Mutation of tumour suppressor genes causes the cell to lose their ability to control coordinated growth. Finally, repair genes keep DNA intact by preserving their unique sequence. During the cell cycle mutagenic events result in DNA modification that in most circumstances are repaired unless the mutation resides in genes coding for repair enzymes.

One of the most important systems in the area of cancer and cell cycle control is the protein p53, and details of the action of this protein were largely the result of the research by Bert Vogelstein, David Lane and Arnold Levine. The p53 gene is located on the short arm of chromosome 17 where the open reading frame codes for a protein of 393 residues via 11 exons. The protein identified before the gene was originally characterized through co-purification with the large T antigen in SV40 virus-transformed cells and had a mass of ~53 kDa. Since the presence of the protein appeared to correlate with viral transformation of cells p53 was originally labelled as an oncogene. By the late 1980s it was clear that the gene product of cloned p53 was a mutant form of the protein and that normal (wild-type) p53 was the product of a tumour suppressor gene. Further modification of ideas concerning p53 function occurred with the demonstration that the protein caused G_1 cell cycle arrest but also activated genes by binding to specific DNA sequences – it was a transcription factor. The discovery that one transcriptional target of p53 was an inhibitor (p21) of cyclin dependent kinases (Cdk) provided a direct link to the cell cycle since p21 complex formation with cdk2 prevents cell division and represents one point of control. Mutant p53 does not bind effectively to DNA with the consequence that p21 is not produced and is unavailable to act as the 'stop signal' for cell division.

p53 has been called the 'Guardian of the Genome' and it is normally present in cells at low concentration in an inactive state or one of intrinsically low activity. When the DNA of cells is damaged p53 levels rise from their normal low levels and the protein is switched 'on' to play a role in cell cycle arrest, transcription and apoptosis. p53 regulates the cell cycle as a transcription factor so that when damage to DNA is detected cell cycle arrest occurs until the DNA can be repaired. Once repaired the normal cell cycle can occur but this process serves to ensure that damaged DNA is not replicated during mitosis and is not 'passed' on to daughter cells. If repair cannot take place apoptosis occurs (Figure 12.39).

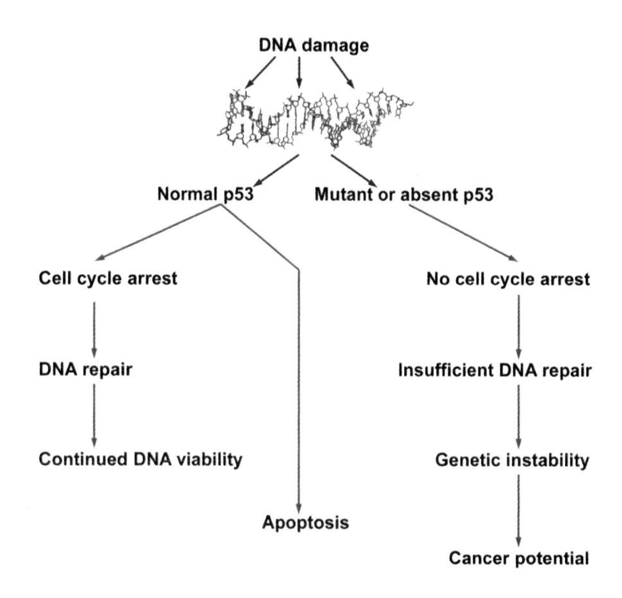

Figure 12.39 An outline of possible p53 actions with normal and damaged cells

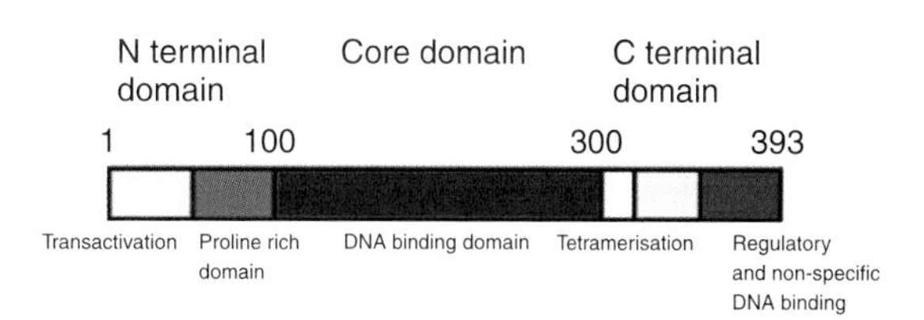

Figure 12.40 The organization of domains within the tumour suppressor p53

With an important role within cells the structural organization of p53 was of considerable interest to many different fields ranging from cell biology to molecular medicine. The protein contains a single polypeptide chain that is divided into three discrete domains (Figure 12.40). These domains present a modular structure that has facilitated their individual study in the absence of the rest of the protein and are concerned with transcriptional activation, DNA binding and tetramerization. An N-terminal transactivation domain extends from residues 1–99 and is followed by the largest domain–a central core region from residues 100–300 that binds specific DNA sequences. The third and final region of p53 is a C-terminal domain (residues 301–393) that includes both a tetramerization domain (from residues 325–356) and a regulatory region (363–393) (Figure 12.41). As a result of the two functional activities in this region p53 is sometimes described as containing four as opposed to three composite domains.

The presence of discrete domains linked by flexible linkers creates a conformationally dynamic molecule and p53 has proved difficult to crystallize as the

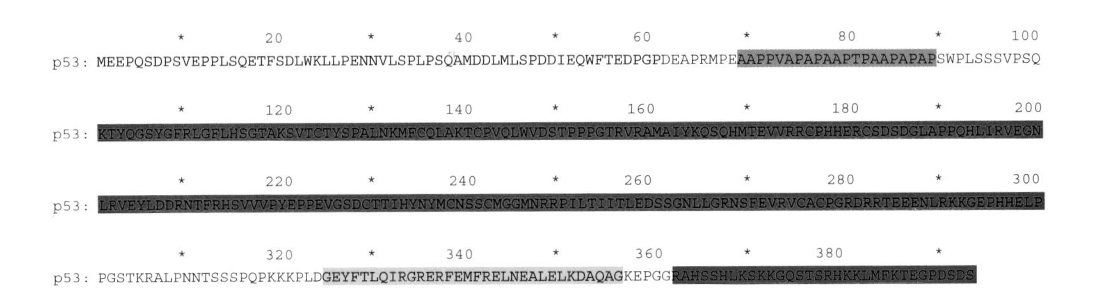

Figure 12.41 The primary sequence of p53 showing the partition of the primary sequence into discrete domains, using the colour codes of Figure 12.40

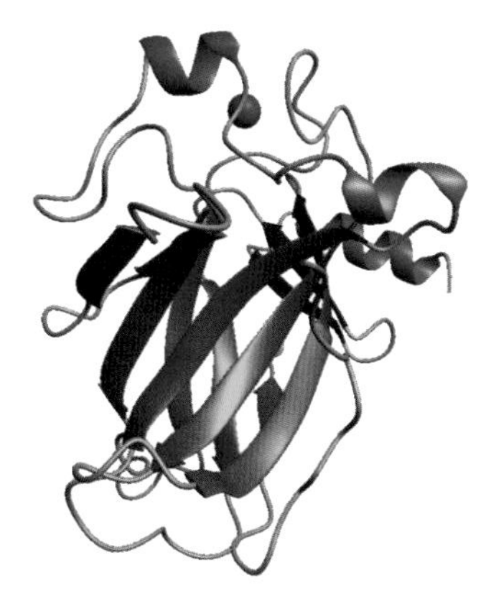

Figure 12.42 The structure of the isolated core domain of p53 showing β sandwich structure together with a bound Zn ion (purple)

full length molecule. This almost certainly results from molecular heterogeneity and structural insight into p53 followed a modular approach determining the structure of isolated domains. One incisive set of results were obtained by Nikolai Pavletich using proteolytic digestion of the full-length protein to yield a protease-resistant fragment of ~190 residues (residues 102–292) that corresponded to the central portion of p53 involved in sequence-specific DNA-binding. DNA binding was inhibited by metal chelating agents and unsurprisingly within the core domain a Zn ion was identified (Figure 12.42). The central core extending from residues 100–300 contains the DNA binding residues and also those residues representing mutational hot spots (see below) within the p53 gene.

As a DNA binding protein the core domain binds to both the major and minor groove of the double helix through the presence of charged side chains arranged precisely in conserved loops that link elements of β strands (Figures 12.43 and 12.44). Numerous arginine and lysine side chains form specific interactions with DNA. Most DNA binding proteins are based on the helix-turn-helix (HTH), HLH or leuzine zipper motifs, but p53 is essentially a β sandwich formed by the interaction of antiparallel four- and five-stranded elements of β sheet.

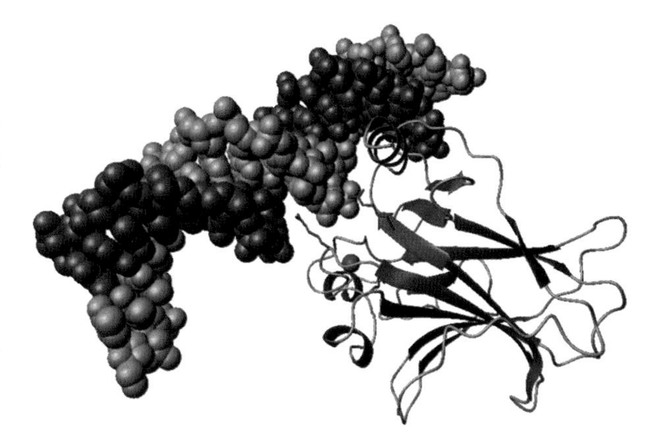

Figure 12.43 The structure of p53 core domain bound to DNA showing role of Arg175, Arg273 and Gly245 (red wireframe) in these interactions as well as the position of a bound Zn ion. The Zn ion is bound in a tetrahedral environment formed by the side chains of three cysteine residues together with a single histidine arranged as part of two loop regions and a single helix

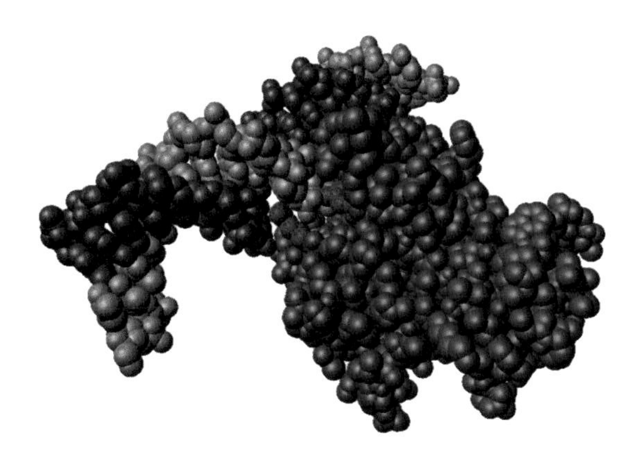

Figure 12.44 A spacefilling representation of p53-DNA complex showing precise interaction with major and minor grooves of helix (PDB:1TUP)

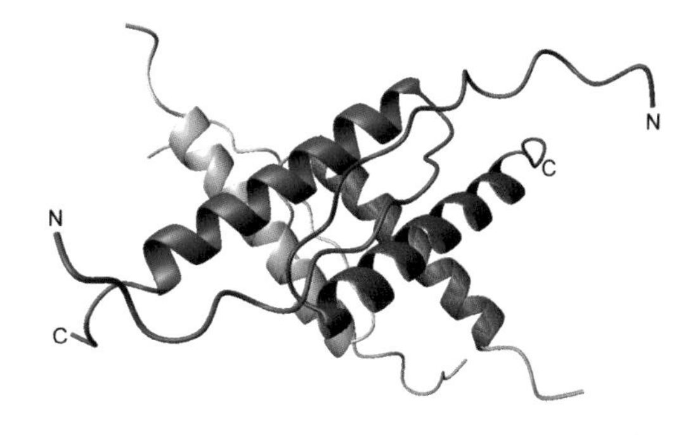

Figure 12.45 The structure of the isolated tetrameric region of p53. The blue and green helices pair together as one dimer with the yellow and orange helices forming the second dimer. The N- and C-terminals are indicated, and point in the case of the N-terminal domain towards the DNA binding core with the C-terminal region pointing towards the remaining parts of p53. The extended region before each helix resembles a β strand and two neighbouring strands on adjacent monomers also interact

The N-terminal domain consists of two contiguous transcriptional–activation regions (residues 1–42 and 43–63) together with an adjacent proline/alanine rich region (residues 62–91).

p53 binds to DNA with high affinity when assembled into a tetrameric complex. This reaction is directed by a tetramerization domain formed by the region between residues 320 and 360 in each polypeptide chain. It is joined to the previously described DNA binding domain via a long flexible linker. The structure of this domain determined in isolation as four 42 residue peptides consists of a dimer of dimers with a combined mass of 20 kDa. Each monomer consists of a regular helix (335–353) and an extended region resembling a poorly defined β strand conformation (residues 326–333) with the two elements linked via a tight turn based around Gly334. An antiparallel interaction of helices in each monomer promotes the formation of a pair of dimers. The interface between the pair of dimers is mediated solely by helix–helix contacts with the overall result being a symmetric, four-helix bundle (Figure 12.45). It has proved possible to mutate the tetramerization domain to favour dimer formation over that of the tetramer. Extensive mutagenesis combined with folding and stability studies suggest that two residues Leu340 and Leu348 promote tetramer formation via interactions of non-polar side chains.

A total of nine residues were identified as key residues in governing stability: Phe328, Ile332, Arg337, Phe338, Met340, Phe341, Leu344, Ala347 and Leu348. All of these residues are invariant in mammalian p53s with the exception of the minor transition Phe338Tyr, and point to important roles for these residues.

The realization that anomalous p53 activity is a common denominator in many human cancers was a major advance. In approximately 50 percent of tumours p53 is inactivated as a result of mutations within the p53 gene (Table 12.4). In other tumours the inactivation of p53 occurs indirectly through binding of viral proteins or through mutations in genes encoding proteins that interact with p53. These mutations alter the flow of information between p53 and the rest of the cellular apparatus.

Most p53 mutations occur in somatic cells and therefore only induce disease in the individuals carrying the defective gene. Mutations in germ line cells (egg and sperm) affect succeeding generations and for p53 an inherited defect known as Li–Fraumeni syndrome has been identified where the gene encoding p53 possesses a mutation. The mutation of p53 is associated with soft-tissue sarcomas and osteosarcomas (bone) where individuals with this inherited condition develop many different cancers from a young age. Since mutations in

Table 12.4 p53 malfunctions that may lead to human cancers

Mechanism of p53 inactivation	Typical tumours	Effect of inactivation
Mutation of residues in DNA binding domain	Colon, breast, lung, bladder, brain, pancreas, stomach, oesophagus and many others.	Prevents p53 binding to specific target DNA sequences and activating adjacent genes.
Deletion of C terminal domain	Occasional tumours at many different sites	Prevents the formation of p53 tetramers.
mdm2 gene multiplication	Sarcomas, brain	Enhances p53 degradation.
Viral infection	Cervix, liver, lymphomas	Products of viral oncogenes bind to p53 causing inactivation and enhanced degradation.
Deletion of $p14^{ARF}$ gene.	Breast, brain, lung and others especially when p53 is not mutated.	Failure to inhibit mdm2 with the result that p53 degradation is uncontrolled
Mislocalisation of p53 to cytoplasm.	Breast, neuroblastomas	p53 functions only in the nucleus.

Adapted from Vogelstein, B. *et al. Nature* 2000, **408**, 307–310.

p53 create genetic instability by the inability to repair damaged DNA or a failure to perform apoptosis defective DNA accumulates making individuals with the Li–Fraumeni syndrome very susceptible to cancer.

Of the mutations identified within p53 almost all are located in DNA binding and tetramerization domains and a vast array of different mutations have been identified. Databases exist where mutations of p53 found within tumours have been categorized; over 14 000 tumour-associated mutations of p53 have been described as of 2003. Within this vast panoply of mutations exist 'hot spots' along the gene. Three mutational hot spots are residues 175 (Arg), 248 (Arg) and 273 (Arg) of the protein. A distribution of all mutations identified to date suggest approximately 20 percent of mutations involve these three residues with residue 248 being the site of highest mutation (~7.5 percent). Along with a further three residues (245/Gly, 249/Arg and 282/Arg) these residues will account for 30 percent of all p53 mutations found in tumours.

If humans have an in-built tumour suppressor gene such as p53 why is cancer so common? Ignoring inherited conditions the answer lies in the influence of environmental conditions on p53. Mutagens alter the DNA sequence of p53 leads to base changes that are eventually reflected in an altered and functionally less active protein. One potent example of this environmental influence is the effect of cigarette smoking on p53. There is no doubt today about epidemiological data showing the overwhelmingly strong correlation between lung cancer and cigarette smoking, yet despite emphatic data a molecular basis for cancer development has taken longer to establish. One poisonous ingredient of cigarette smoke is a class of organic molecules known as benzopyrenes. Exposure of the p53 gene sequence to these mutagens caused alterations in base sequence most frequently the substitution of G by T. These mutagens covalently cross-link with DNA at the G residues found at codons 175, 248 and 273. Since p53 is a tetramer the mutation of just one subunit perturbs the activity throughout the complex resulting in impaired DNA binding.

Another mechanism for cancer lies in the inactivation of p53 by the human adenovirus and papilloma virus. These viruses bind to p53 inactivating the protein function and preventing a successful repair of damaged DNA. At residue 72 of p53 two alleles exist with either arginine or proline found in the wild type protein. Initial studies suggested these variants were functionally equivalent but current evidence suggests the Arg72 variant is more susceptible to viral mediated proteolysis. For the human papillomavirus (HPV18) – a major factor in cervical carcinoma – this involves the preferential interaction of Arg72 p53 with the E6 protein of the virus *in vitro* and *in vivo*. From these studies it was suggested that cancer of the cervix is approximately seven times higher in individuals homozygous for the Arg72 allele.

Other cancers involve amplification of mdm2 – a protein that inactivates p53 by binding to the transcriptional domain. High levels of mdm2 cause a permanent inactivation of p53 and an increased likelihood that mutations will be passed on to daughter cells. In view of its role 'controlling' p53 mdm2 is widely believed to represent the next avenue for intensive cancer research.

The amount of information existing on all aspects of p53 function and mutant protein expression in human cancers is vast and reflects a key role in pathogenesis. p53 is just one component of a network of events that culminate in tumour formation. Although inactivation of p53 either through direct or indirect mechanisms is most probably responsible for the majority of human cancers it must be remembered that genes other than p53 are targets for carcinogens. The identification and subsequent characterization of these genes/proteins will increase the molecular approach to the treatment of cancer.

Emphysema and α_1-antitrypsin

Emphysema is a genetic, as well as an acquired disease, characterized by over-inflation of structures in the lungs known as alveoli or air sacs. This over-inflation results from a breakdown of the walls of the alveoli and this results in decreased respiratory function. As a result breathlessness is a common sign associated with emphysema, and together with chronic bronchitis, emphysema comprises a condition known as chronic obstructive pulmonary disease (COPD). Frequently COPD is associated with lung disease arising from smoking but in some instances emphysema has been identified as a genetically inherited condition.

Emphysema is defined by destruction of airways distal to the terminal bronchiole with destruction of alveolar septae and the pulmonary capillary region leading to decreased ability to oxygenate blood. The destruction of the alveolar sacs is progressive and becomes worse with time. The body compensates with lowered cardiac output and hyperventilation. The low cardiac output leads to the body suffering from tissue hypoxia and pulmonary cachexia. A common occurrence is for sufferers to develop muscle wasting and weight loss and they are often identified as 'pink puffers'. In the United States it is estimated that two-thirds of all men and one-quarter of all women have evidence of emphysema at death with total sufferers approaching $\sim$2 million. As part of COPD it is responsible for the fourth leading cause of death in the United States and represents a major demand on health service care.

As a genetic condition emphysema has been attributed to a deficiency in a protein called α_1-antitrypsin. Unlike the common form of emphysema seen in individuals who have smoked for many years this form of the disease is often labelled as α_1-antitrypsin deficiency form of emphysema. The disease can occur at an unusually young age (before 40) after minimal or no exposure to tobacco smoke and has been shown to arise in families. The latter observation pointed to a genetic link.

The biological role of α_1-antitrypsin is as an enzyme inhibitor of serine proteases and despite its name its main target for inhibition is the enzyme elastase. It is a member of a group of proteins known as serpins or serine protease inhibitors. Collectively these proteins inhibit enzymes by mimicking the natural substrate and binding to active sites. Serpins have been identified in a wide variety of viruses, plants, and animals with more than 70 different serpins identified. Members of the serpin family have very important roles in the inflammatory, coagulation, fibrinolytic and complement cascades. Consequently malfunctioning serpins have been shown to lead to several forms of disease including cirrhosis, thrombosis, angio-oedema, emphysema as described here and dementia.

The site of action of α_1-antitrypsin is the serine protease elastase. Elastase is commonly found in neutrophils and is structurally homologous to trypsin or chymotrypsin (see Chapter 7). Neutrophils represent the primary defence mechanism against bacteria by phagocytosis of pathogens followed by degradation and in this latter reaction elastase is involved. At sites of inflammation hydrolytic enzymes of neutrophil origin are detected and this includes the enzyme elastase.

α_1-antitrypsin is the major serine protease inhibitor present in mammalian serum. It is synthesized in the liver and represents the major protein found in the α_1-globulin component of plasma. In the serum the concentration of α_1-antitrypsin is very high ($\sim$2 g l^{-1}) and this reflects its important role. A deficiency in certain individuals of α_1-antitrypsin was identified in the 1960s by detailed analysis of gel electrophoretic profiles but as with most diseases attention has recently been directed towards identifying the gene and its location within the human genome.

The gene is called the PI (protease inhibitor) gene and is located on the long arm of chromosome 14. Genetic screening of individuals presenting with early onset emphysema has allowed many mutations to be identified. The normal allele is the M allele and does not result in the disease. Genetic analysis of the M allele has shown that several different forms designated as M1, M2, M3, etc. can be distinguished with M1 representing approximately $\sim$70 percent of the population and distinguished by two variants; Val213 and Ala213. These two forms represent nearly all of the M1 phenotype and occur in a ratio of $\sim$2:1. Alternative alleles are attributed to an increased incidence of disease in individuals possessing one or more copies and the two most common are the S and Z forms. The Z allele is most frequently associated with codon 342 and results in the non-conservative substitution of glutamate by lysine as a result of a single base change from GAG to AAG. This mutation is estimated to be present in the north European population at $\sim$1 in 60. Individuals homozygous for the Z allele (often written as PI ZZ) are at greatly increased risk of emphysema *and* liver disease. The second form is the S allele and is associated with the substitution of glutamate by valine at codon 264. Individuals homozygous for the S allele (PI SS) do not display symptoms of the disease but individuals with one copy of the Z allele and one copy of the S allele (PI SZ) may develop emphysema. Over 70 different mutations have been identified and it appears that these mutations can be grouped into three major categories. Some mutations disrupt the serpin function of the protein, whilst rarer mutations are described as null or silent mutations and a third group appear to cause drastic reduction in the amount of α_1-antitrypsin present in the serum as opposed to its activity.

The single polypeptide chain of α_1-antitrypsin contains 394 amino acid residues after removal of a 24-residue signal sequence. The presence of three glycosylation sites leads to higher than expected molecular weight (M_r $\sim$52 000) when serum fractions are analysed by gel electrophoresis but the carbohydrate is not necessary for catalytic function. Within the polypeptide chain the sequence Pro357-Met358-Ser359-Ile360 represents a 'substrate' for target protease. This region of the polypeptide effectively forms a 'bait' for the cognate protease, in this case leucocyte elastase, and the reaction between serpin and protease leads to complete inactivation.

The structure of α_1-antitrypsin reveals three sets of β sheets designated A–C and nine helical elements. The basis of inhibition revolves around interaction between cognate serine protease and serpin by the presentation of a region of the serpin polypeptide chain that resembles natural substrates of the protease. Binding occurs at an exposed region of the polypeptide chain known as the reactive centre loop a region held at the apex of the protein and based around the critical Met358 residue (Figure 12.46). The mobile reactive centre loop presents a peptide sequence as the substrate for the target protease. Docking of the loop of the serpin with the protease is accompanied by cleavage at the P_1–P'_1 bond and the enzyme is locked into an irreversible and extremely stable complex. In the case of α_1-antitrypsin the P_1–P'_1 bond (i.e. the residues either side of the bond to be cleaved) lies between Met358 and Ser359 in the loop located at the apex of the protein.

Upon protease binding large conformational rearrangement occurs in which the two major structural transitions occur. The reactive centre loop is split between Met358 and Ser359 and the A_β sheet is able to

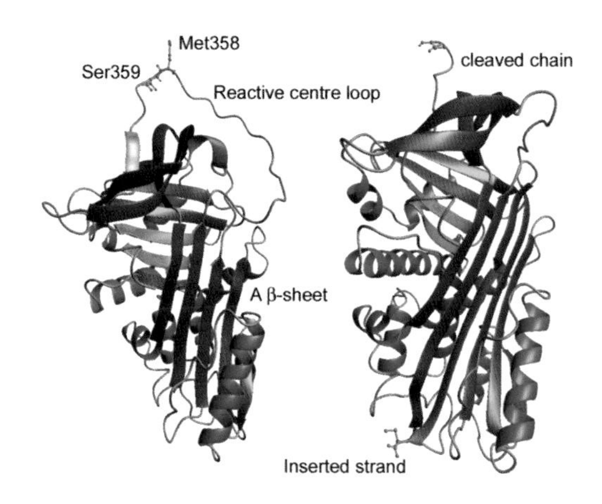

Figure 12.46 Two conformational states of α_1-antitrypsin showing conformational changes upon cleavage. Residues 350–394 are coloured orange to distinguish from the remainder of the molecule where the strands are shown in green and the helices in blue (PDB:2 PSI and 7API)

Figure 12.47 Space filling model of interaction between serpin and protease. Met358 and Ser359 can be seen at either end of the serpin (green)

open to accommodate much of the reactive centre loop region as a fifth β strand. This conformational change is vital for efficient inhibition and leads to Met358 being located at the foot of the molecule (Figure 12.47). In this conformation the α_1-antitrypsin and elastase are irreversibly bound in a stable complex that prevents further reactivity. Conformational mobility of these regions results in sensitivity to mutation and significantly the Z allele of α_1-antitrypsin is located within the reactive loop region (effectively as P'_4).

In many individuals with α_1-antitrypsin deficiency liver disease is detected as an associated condition. An explanation for this association has now been found. In the homozygous ZZ condition only about 15 percent of the normal level of α_1-antitrypsin is secreted into the plasma. The normal site of synthesis is the liver and it was observed that the remaining 85 percent accumulates in the ER of the hepatocyte. Although much is degraded the remainder aggregates forming insoluble intracellular inclusions. These aggregates are responsible for liver injury in approximately 12–15 percent of all individuals with this genotype. It is also observed that about 10 percent of newborn ZZ homozygotes develop liver disease leading to fatal childhood cirrhosis. Studies based on the structure of α_1-antitrypsin have demonstrated the molecular pathology underlying protein accumulation.

The principal Z mutation in antitrypsin results in the substitution of glutamate by lysine at residue 362 and allows a unique molecular interaction between the reactive centre loop of one molecule and the 'gap' in the A-sheet of another serpin. Effectively the reactive centre loop of one α_1-antitrypsin inserts into the A_β sheet region of a second molecule in a process known as loop to sheet polymerization. In mutant α_1-antitrypsin this occurs favourably at 37 °C and is very dependent on concentration and temperature. The polymerization is blocked by the insertion of a specific peptide into the A-sheet of α_1-antitrypsin.

α_1-antitrypsin is an acute phase protein and for mutant protein it will undergo a significant increase in association with temperature as might arise during bouts of inflammation. The result is harmful protein inclusions in the hepatocyte that may overwhelm the cell's degradative mechanisms, especially in the newborn. The control of inflammation and pyrexia in ZZ homozygote infants is very important to overall management of the disease. The serpins show considerable homology and the mechanism of loop–sheet polymerization is the basis of deficiencies associated also with mutations of C1-inhibitor (the activated first component of the complement system), antithrombin III, and α_1-antichymotrypsin implicated in other disease states. Collectively, these diseases could be called serpinopathies.

It has been shown that replacement of deficient protein by intravenous infusion of purified α_1-antitrypsin can limit disease progression. Unlike many of the previously described conditions this treatment represents an effective protocol that limits much of the necrosis of lung tissue associated with hereditary forms of emphysema. This form of emphysema is currently treated with protein isolated from plasma fractions, but with characterization of the cDNA encoding the protein α_1-antitrypsin has been expressed to allow high level production in the milk of transgenic sheep. Although this form of α_1-antitrypsin has yet to complete clinical trials it will offer a better alternative to conventionally purified protein. The underlying genetic condition will always remain but considerable improvement results from treatment with purified α_1-antitrypsin. The high concentration of α_1-antitrypsin in serum means that on average a patient may require over 200 g of highly purified proteins per year as the basis of a therapeutic regime.

For individuals with this genetic condition a lifetime avoidance of smoking is highly recommended since the effects of smoking as a primary cause of emphysema are well documented. In this context it has been shown that one effect of smoking can be to cause oxidation of the Met358 residue within α_1-antitrypsin. The oxidation of the methionine side chain to form a sulfoxide results in permanent irreversible damage to α_1-antitrypsin and eliminates its biological role as a serpin. A loss of inhibition allows unchecked elastase activity and promotes alveolar sac destruction that lead to emphysema. When combined with a genetic tendency towards emphysema it is clear that this situation must be avoided to prevent extreme lung damage. Genetically linked forms of emphysema represented one of the first molecular defects to be extensively characterized and the highly detailed structural and functional studies of serpins such as α_1-antitrypsin highlight the diagnostic, preventative and therapeutic powers of molecular medicine.

Summary

Molecular medicine represents a multidisciplinary approach to understanding human disease through the use of structural analysis, pharmacology, gene technology and even gene therapy. In 1955 sickle cell anaemia was identified as arising from a single amino acid change within haemoglobin. Many other diseases arise from similar genetic mutations.

Unprecedented progress has been made in understanding the molecular basis of many diseases including new, emerging, ones such as HIV and vCJD alongside familiar conditions such as influenza. Understanding at a molecular level offers the prospect of better disease diagnosis and in some cases improved therapeutic intervention.

Viruses represent a persistent human health problem, being responsible for many diseases. Vaccination using attenuated viruses or purified proteins effectively eradicated infections such as smallpox and polio but threats from rapidly mutating viruses such as influenza and HIV continue.

HIV continues to contribute to significant, worldwide, mortality in the 21st century whilst influenza returns periodically in the form of severe outbreaks or pandemics. In each virus the ability to mutate surface antigens rapidly compromises therapeutic drug action or antibody directed immune response.

The protein p53 has been called the 'Guardian of the genome'. Its major role is to prevent the perpetuation of damaged DNA by causing cell cycle arrest or apoptosis. The protein is a conformationally flexible molecule consisting of an N-terminal transcriptional activation domain, a central or core DNA binding domain and a third C terminal domain that contains multiple functions including the ability to promote tetramerization.

The core DNA binding domain is based on a β sheet scaffold that binds to major and minor grooves of the double helix through the presence of charged side chains arranged precisely in conserved loops that link elements of β strands.

Cancer and p53 are inextricably linked through the demonstration that many tumours have mutations in p53. Most, but not all, are located in the central DNA binding domain with mutational hot spots involving residues participating in DNA binding. The cell requires exquisite control pathways to modulate the activity of p53 and corruption of these pathways can often lead to the development of cancer.

α_1-antitrypsin is a serpin that acts to limit the unregulated activity of neutrophil elastase. The protein inhibits elastase by forming a tightly bound complex. It is found at high concentrations in the globulin fraction of serum where its abundance helped to identify individuals deficient in serum α_1-antitrypsin.

Deficiency arises from a gene mutation and is an inherited condition leading to emphysema as a result of damage caused to the alveolar lining of the lungs. Mutant α_1-antitrypsin fails to bind elastase effectively.

Mutated forms of α_1-antitrypsin show a tendency to exhibit loop–sheet polymerization – a process that causes the irreversible association of serpins.

Problems

1. Why should individuals with sickle cell anemia avoid intense exercise or exposure to high altitudes?
2. Vaccines have been developed against smallpox, measles and many other viral based diseases. So what is the problem with developing a vaccine for flu and HIV?
3. The ionization of maleic acid (COOH–CH_2–CHOH–COOH) occurs with pK values of 1.9 and 6.2. Discuss the reasons for these values and what is the relevance to the activity of aspartic proteases.
4. Using the known properties of the active site of neuraminidase describe why Zanamivir is an efficient inhibitor of enzyme activity especially when compared with Neu5Acen. Can you suggest potentially other beneficial drug design.
5. Is vCJD a new disease or merely an old one firmly recognized and identified?
6. Identify the conformationally sensitive strands to inhibitor binding and active site aspartates in HIV protease.

Epilogue

It is 50 years since the discovery of the double helical structure of DNA, an event that marks the beginning of molecular biology. The structure of DNA represents the seed that germinated, grew rapidly, flowered and bore fruit. It has led to descriptions of the flow of information from gene to protein. Molecular biology, expanded from a small isolated research area into a major, all embracing, discipline. As this book is being published close to the 50th anniversary it is timely to reflect that during this period biochemistry has been marked by several defining discoveries based on the structure and function of proteins. The list of significant discoveries could provide the bulk of the content of another book but a brief and selective description of these events might include:

- *The elucidation of the structure of myoglobin and haemoglobin by Perutz and Kendrew using X-ray crystallography.* This work paved the way for all future crystallographic studies of proteins and established the basis for allostery.

- *The structure of the first enzyme, lysozyme, by Phillips.* Structural characterization defined an active site and the geometry of residues that facilitated biological catalysis and pointed towards molecular enzymology.

- *Protein folding is encoded entirely by the primary sequence.* Anfinsen proved that proteins could fold to reach the native state whilst in cells large macromolecular complexes known as chaperones were shown much later to assist folding by forming environments known as the Anfinsen cage that limit unfavourable protein interactions.

- *Crystallization of the photosynthetic reaction centre by Michel, Diesenhofer and Huber.* The helical structure of transmembrane segments was confirmed and showed that membrane proteins could be crystallized and subjected to the same high resolution methods applied previously to soluble proteins.

- *The architecture of the ribosome.* A daunting experimental problem revealed catalysis of the peptidyl transferase was performed by RNA and the ribosome is a ribozyme. The structure of the ribosome confirmed that biological catalysis could proceed in the absence of protein-based enzymes redefining our traditional view of nucleic acid and proteins.

- *p53 and the molecular basis of cancer.* The demonstration that over 60 percent of all tumours are associated with mutations in p53 elucidated a direct link between molecular defects in a protein and subsequent development of cancer.

- *The prion hypothesis.* Consequently the prion hypothesis showed that some proteins ‘corrupt’ native conformations promoting protein aggregation; a hallmark of many neurodegenerative disorders.

By reading the preceding 12 chapters, where these discoveries are described in more detail, I hope the reader will gain an impression of the rapidly expanding and advancing area of protein biochemistry. This area offers the potential to revolutionize treatment of human health and to eradicate diseases in all avenues of life. The next 50 years will see the complete description of

Proteins: Structure and Function by David Whitford

URL	Description
www.rcsb.org/pdb	The repository for the deposition of biomolecular structures mainly nucleic acids and proteins
www.hgmp.mrc.ac.uk	The human genome mapping project centre in the UK
www.nhgri.nih.gov	The human genome mapping research institute of the NIH
www.expasy.ch	The ExPASy (Expert Protein Analysis System) proteomics server of the Swiss Institute of Bioinformatics (SIB)
www.ebi.ac.uk	European Bioinformatics Institute containing access to databases and software for studying proteins
www.ncbi.nlm.nih.gov	A national resource for molecular biology information in USA
www.srs.ebi.ac.uk	Sequence retrieval system for a wide variety of databases
rebase.neb.com/rebase/rebase.html	Restriction enzyme database
bimas.dcrt.nih.gov/sql/BMRBgate.html	Biomolecular Nuclear Magnetic Resonance databank
tolweb.org/tree/phylogeny.html	Tree of Life web page
www.ncbi.nlm.nih.gov/Omim	On line Mendelian Inheritance in Man
www.biochem.ucl.ac.uk/bsm/biocomp/index.html	A collection biocomputing resources at University College London
www.ensembl.org	A database annotating sequenced genomes
www.embl-heidelberg.de/predictprotein/predictprotein.html	Protein prediction server for secondary structure
www-nbrf.georgetown.edu/pirwww/pirhome3	Protein Information resource centre
archive.uwcm.ac.uk/uwcm/mg/hgmd0.html/	Human gene mutation database
gdbwww.gdb.org/	Another genome database
www.ornl.gov/hgmis	US Human Genome Project Information

See home page of this book for latest web-based or on-line resources.

other proteomes and promises the possibility of equally devastating discoveries to match those of the previous 50 years.

The following universal resource locators (URL) represent web addresses of sites that offer very useful information of relevance to the subject of this book. The author is not responsible for their content and due to the transitory nature of the web these sites may not always be accessible, maintained or presenting the latest information. However, all (as of April 2004) had potentially useful information that provides an ideal learning resource.

Glossary

α helix regular unit of secondary structure shown by polypeptide chains characterised by 3.6 residues per turn in a right-handed helix with a pitch of 0.54 nm.

β sheet collection of β strands assembled into planar sheet like structure held together by hydrogen bonds.

$3d_{10}$ helix a form of secondary structure containing three residues per turn and hydrogen bonds separated by 10 backbone atoms.

Ab initio from the beginning (*Latin*).

Acid–base catalysis reactions in which the transfer of a proton catalyses the overall process.

Acidic solution a solution whose pH is less than pH 7.0.

Active site part of an enzyme in which the amino acid residues form a specific three-dimensional structure containing the substrate binding site.

ADP adenosine diphosphate.

Affinity chromatography. separation of proteins on the basis of the specific affinity of one protein for an immobilized ligand covalently attached to an inert matrix.

AIDS disease ultimately resulting from infection with HIV, sometimes called advanced HIV disease.

Allosteric effectors molecules which promote allosteric transitions in a protein.

Allostery with respect to proteins, a phenomenon where the activity of an enzyme's active site is influenced by binding of an effector to a different part of the enzyme.

Alzheimer's a disease characterized by formation of protein deposits with the brain and classified as a neurodegenerative disorder.

Amino acid an organic acid containing an amino group, carboxyl group, side chain and hydrogen on a central α carbon. The building blocks of proteins.

Amphipathic for a molecule, the property of having both hydrophobic and hydrophilic portions. Usually one end or side of the molecule is hydrophilic and the other end or side is hydrophobic.

Ampholyte a substance containing both acidic and basic groups.

Amyloid accumulation of protein into an insoluble aggregate often a fibre in tissue such as the brain.

Anabolism process of synthesis from small simple molecules to large often polymeric states.

Antigen a substance that can elicit a specific immune response.

Anaerobe organism capable of living in the absence of oxygen.

Antibody protein component of immune system produced in response to foreign substance and consisting of two heavy and two light chains.

Anticodon sequence of three nucleotide bases found in tRNA that recognizes codon via complementary base pairing interactions.

Antigenic determinant a specific part of an antigen that elicits antibody production.

Antiparallel running in opposite directions as in DNA strands or β sheets.

Apoprotein protein lacking co-factor or coenzyme. Apoprotein often show decreased or negligible activity.

Apoptosis programmed cell death.

Archae one of the two major groupings within the prokaryotes, also called archaebacteria.

Asymmetric centre a centre of chirality. A carbon atom that has four different substituents attached to it.

Proteins: Structure and Function by David Whitford

ATP adenosine triphosphate.

ATPase an enzyme hydrolysing ATP to ADP and Pi.

Association constant/Affinity constant given by 'K', in the oxygen binding reaction of myoglobin, K is calculated as the concentration of myoglobin bound to oxygen divided by the product of the free oxygen concentration and the free myoglobin concentration.

B cells lymphatic cells produced by B lymphocytes.

Backbone part of the polypeptide chain consisting of N–C_α–C portion (distinct from side chain)

Bacteriophage virus that specifically infects a bacterium.

Basic solution a solution whose pH is greater than pH 7.0.

BSE bovine spongiform encephalopathy, a prion-based disease first seen in cattle in the UK.

bp base pair, often used to describe length of DNA molecule.

Buffer a mixture of an acid and its conjugate base at a pH near to their pK. Buffer solution composed of an acid and conjugate base resist changes in pH.

Cahn–Ingold–Prelog system of unambiguous nomenclature of molecules with one or more asymmetric centres via a priority ranking of substituents. Also known as RS system.

Cap a 7-methyl guanosine residue attached to the 5′ end of eukaryotic mRNA.

Capsid protein coat or covering of nucleic acid in viral particles.

Catabolism metabolic reactions in which larger molecules are broken down to smaller units. Proteins are broken down into amino acids.

cDNA complementary DNA derived from reverse transcription of mRNA.

Chaotrope a substance that increases the disorder (chaos). Frequently used to describe agents that cause proteins to denature. An example is urea.

Chaperonins proteins which assist in the assembly of protein structure. They include heat-shock proteins and GroEL/ES of *E. coli*.

Charge–charge interactions interactions between positively and negatively charged side chain groups.

Chiral possessing an asymmetric center due to different substituent groups and exist in two different configurations.

Chloroplasts plant organelles performing photosynthesis.

Clathrate structures hydrophobic molecules that dissolve in aqueous solutions form regular icelike structures called clathrate structures rather than the hydration shells formed by hydrophilic molecules.

Cloning process of generating an exact copy.

Coding strand analogous to sense strand.

Co-factor a small organic molecule or sometimes a metal ion necessary for the catalytic activity of enzymes.

Coiled coil arrangement of polypeptide chains where two helices are wound around each other.

Configuration arrangement of atoms that cannot be altered without breaking and reforming bonds.

Conformational entropy a protein folding process, which involves going from a multitude of random-coil conformations to a single, folded structure. It involves a decrease in randomness and thus a decrease in entropy.

Codon sequence of three nucleotide bases found in mRNA that determines a single amino acid.

Conformation proteins and other molecules occur in different spatial arrangements because of rotation about single bonds that leads to a variety of different, close related states.

Conservative amino acid changes Mutations in coding sequences for proteins which convert a codon for one amino acid to a codon for another amino acid with very similar chemical properties.

Cooperative transition a transition in a multipart structure such that the occurrence of the transition in one part of the structure makes the transition likelier to happen in other parts.

Covalent bonds the chemical bonds between atoms in an organic molecule that hold it together are referred to as covalent bonds.

Cristae invaginations of inner mitochondrial membrane involved in oxidative phosphorylation.

Cryo-EM cryoelectron microscopy a technique for the visualization of macromolecules achieved by rapid freezing of a suspension of the biomolecule without the formation of ice.

Cystine the amino acid cysteine can form disulfide bonds and the resulting structure is sometimes called a cystine, particularly in older textbooks.

Denaturation loss of tertiary and secondary structure of a protein leading to a less ordered state that is frequently inactive.

Da Dalton or a unit of atomic mass equivalent to 1/12th the mass of the ^{12}C atom.

Debye–Huckel radius a quantitative expression of the screening effect of counterions on spherical macro ions.

Dihedral angle an angle defined by the bonds between four successive atoms. The backbone dihedral angle ϕ is defined by C'–N–C_a–C'.

Dimer assembly consisting of two subunits.

Dipeptide a molecule containing two amino acids joined by a single peptide bond.

Dipolar ion term synonymous with zwitterions.

Dipole Moment molecules which have an asymmetric distribution of charge are dipoles. The magnitude of the asymmetry is called the dipole moment of the molecule.

Disulfide bond a covalent bond between two sulfur atoms formed from the side chains of cysteine residues, for example.

DNA deoxyribose nucleic acid – the genetic material of almost all systems.

Domain a compact, locally folded region of tertiary structure in a protein.

Elution removal of a molecule from chromatographic matrix.

Edman degradation procedure for systematically sequencing proteins by stepwise removal and identification of N terminal amino acid residue.

EF elongation factor, one of several proteins involved in protein synthesis enhancing ribosomal activity.

Enantiomers also called optical isomers or stereoisomers. The term optical isomers arises from the fact that enantiomers of a compound rotate polarized light in opposite directions.

Endergonic reaction process that has a positive overall free energy process ($\Delta G > 0$).

Enzymes catalytic proteins.

Electrophile literally electron-lover and characterized by atoms with unfilled electron shells.

Eubacteria one of the two major groupings within the prokaryotes.

Eukaryote a cell containing a nucleus that retains the genetic material in the form of chromosomes. Often multicellular and with cells showing compartmentation.

Exons a region in the coding sequence of a gene that is translated into protein (as opposed to introns, which are not). The name comes from the fact that exons are the only parts of an RNA transcript that are seen outside the nucleus of eukaryotic cells.

Fc fragment a proteolytic fragment of an antibody molecule

Fibrous proteins a class of proteins distinguished by a filamentous or elongated three dimensional structure such as collagen.

First order a reaction whose rate is directly proportional to the concentration (activity) of a single reactant.

FT Fourier transform.

FT-IR Fourier transform infrared spectroscopy.

g estimate of centrifugal force so that 5000 ***g*** is 5000 times the force of gravity.

G protein a protein that binds guanine nucleotides such as GTP/GDP.

Gel filtration chromatography also called size exclusion chromatography. Separates biomolecules such as proteins through the use of closely defined pore sizes within an innert matrix according to the molecular mass.

Globular proteins proteins containing polypeptide chains folded into compact structures that are not extended or filamentous and have little free space on the inside.

Heme prosthetic oxygen binding site of globin (and other) proteins. Is a complex of protoporphyrin IX and Fe (II). Carries oxygen in globin proteins

Henderson–Hasselbach equation describes the dissociation of weak acids and bases according to the equation $pH = pK_a + \log([A-]/[HA])$.

Heterodimer a complex of two polypeptide chains in which the two units are non-identical.

HIV human immunodeficiency virus; retrovirus responsible for AIDS.

HLH helix-loop-helix motif found in several eukaryotic DNA binding proteins.

Hormone molecule often but not exclusively protein that is secreted into blood stream and carried systemically where it elicits a physiological response in another tissue.

Hydrolysis cleavage of covalent chemical bond involving water.

Homeobox a DNA binding motif that is widely found in eukaryotic genomes where it encodes a transcription factor whose activity modulates the development, identity and fate of embryonic cell lines.

Homodimer a complex of two units in which both units are identical.

Hsp heat-shock protein – name given to a large group of molecular chaperone proteins.

HTH helix-turn-helix motif found in several DNA binding proteins in which the two helices cross at an angle of $\sim 120°$.

Huntingtin the mutant protein contributing to Huntington's disease.

Hydrogen bond an attractive interaction between the hydrogen atom of a donor group, such as OH or =NH, and a pair of nonbonding electrons on an acceptor group, such as O=C. The donor group atom that carries the hydrogen must be fairly electronegative for the attraction to be significant.

Hydrophilic refers to the ability of an atom or a molecule to engage in attractive interactions with water molecules. Substances that are ionic or can engage in hydrogen bonding are hydrophilic. Hydrophilic substances are either soluble in water or, at least, wettable.

Hydrophobic the molecular property of being unable to engage in attractive interactions with water molecules. Hydrophobic substances are nonionic and nonpolar; they are nonwettable and do not readily dissolve in water.

Hydrophobic effect stabilization of protein structure resulting from association of hydrophobic groups with each other, away from water.

IF initiation factor involved in start of protein synthesis and ribosomal assembly and activation.

Ig immunoglobulin and another name for an antibody group such as IgG.

Importins generic group of proteins with homology to importin α and β that function as heterodimer binding NLS-proteins prior to import into nucleus.

In vitro normally means in the laboratory, but literally in glass.

In vivo in a living organism.

Integral protein a membrane bound protein that can only be removed from the lipid bilayer by extreme treatment. Also called intrinsic protein.

Intron(s) a region in the coding sequence of a gene that is not translated into protein. Introns are common in eukaryotic genes but are rarely found in prokaryotes. They are excised from the RNA transcript before translation.

Ionic strength an expression of the concentration of all ions. In the Debye–Huckel theory $I = 1/2\ \Sigma m_i z_i^2$

Isoelectric focusing a technique for separating ampholytes and polyampholytes based on their pI. Also used to determine pI.

Isoelectric point the pH at which an ampholyte or polyampholyte has a net charge of zero. Same as pI.

Isoenzymes also called isozymes. Represent different proteins from the same species that catalyse identical reactions.

kDa kilodaltons; equivalent to 1000 daltons or approximately 1000 times the mass of a hydrogen atom.

Keratins major fibrous proteins of hair and fingernails. They also compose a major fraction of animal skin.

Kinases enzymes that phosphorylate or transfer phosphorus groups to substrates including other proteins (e.g. protein kinases)

K_m the Michaelis constant. It is the concentration of substrate at which the enzyme-catalysed reaction proceeds with half maximal velocity

Leader sequence a short N-terminal hydrophobic sequence that causes the protein to be translocated into or through a membrane often called a signal sequence.

Mesophile an organism living at normal temperatures in comparison with a thermophile.

Module a sequence motif of between 40–120 residues that occurs in unrelated proteins or as multiple units within proteins.

Molten globule state intermediate structures of a protein in which the overall tertiary framework of the protein is established, but the internal side chains are still free to move about.

Mutation a change in the sequence of DNA.

NES nuclear export signal – a signal consisting of several leucine residues found in proteins and indicating export.

NLS nuclear localization signal – a short stretch of basic amino acid residues that targets proteins for import into the nucleus. The sequence can have bipartite structure consisting of two short sequences of basic residues separated by #10 intervening residues.

NMR spectroscopy acronym for nuclear magnetic resonance spectroscopy, a technique useful in the elucidation of the three-dimensional structure of soluble proteins.

Non-covalent interactions attractive or repulsive forces, such as hydrogen bonds or charge–charge interactions, which are non-covalent in nature, are called non-covalent interactions.

NPC nuclear pore complex – a very large assembly of protein concerned with regulating flux of macromolecules between nucleus and cytoplasm.

Nucleophile atom or group that contains an unshared pair(s) of electrons and is attracted to electrophilic (electron deficient) groups.

Nucleoporins proteins found in the nuclear pore complex.

Oncogene a gene which in a mutated form gives rise to abnormal cell growth or differentiation.

Oncoprotein the product of an oncogene that fails to perform its normal physiological role.

Operator element of DNA at the transcriptional site that binds repressor.

Operon a genetic unit found in prokaryotes that is transcribed as a single mRNA molecule and consisting of several genes of related function.

Oxidative phosphorylation process occurring in mitochondria and bacteria involving oxidation of substrates and the generation of ATP.

Oxidoreductase catalyse redox reactions.

PAGE polyacrylamide gel electrophoresis – a technique for the electrophoretic separation of proteins through polyacrylamide gels.

PCR polymerase chain reaction – a method involving the use of thermostable DNA polymerases to amplify in a cyclic series of primer driven reactions specific DNA sequences from DNA templates.

PDI protein disulfide isomerase – an enzyme catalysing disulfide bond formation or re-arrangement.

Peptide molecules containing peptide bonds are referred to generically as peptides usually less than 40 residues in length.

Peptidase an enzyme that hydrolyses peptide bonds.

Peptide bond the bond that links successive amino acids in a peptide; it consists of an amide bond between the carboxyl group of one amino acid and the amino group of the next.

Peripheral protein also called extrinsic protein and refers to protein weakly associated with membrane.

pH the negative logarithm of the hydrogen ion concentration in an aqueous solution.

Phage shortened version of bacteriophage.

p*I* the isoelectric point or the pH at which an ampholyte or polyampholyte has a net charge of zero.

Pitch the spacing distance between individual adjacent coils of a helix.

p*K* a measure of the tendency of an acid to donate a proton; the negative logarithm of the dissociation constant for an acid. Also called pK_a

Polypeptide a polypeptide is a chain of many amino acids linked by peptide bonds.

Polyprotic acids acids which are capable of donating more than one proton.

Porphyrins a class of compounds found in chlorophyll, the cytochrome proteins, blood, and some natural pigments. They are responsible for the red color of blood and the green color of plants.

Prebiotic era the time span between the origin of the earth and the first appearance of living organism ($\sim 4.6 \times 10^9 - 3.6 \times 10^9$ years ago)

Preinitiation complex the multiprotein complex of transcription factors bound to DNA that facilitates transcription by RNA polymerase.

Preproprotein a protein contain a prosequence in addition to the signal sequence.

Preprotein a protein containing a signal sequence that is cleaved to yield the active form.

Pribnow box prokaryotic promoter region located 10 bases upstream of the transcription start site with a consensus sequence TATAAT.

Primary structure for a nucleic acid or a protein, the sequence of the bases or amino acids in the polynucleotide or polypeptide.

Protoporphyrin IX a tetrapyrrole ring which chelates Fe(II) and other transition metals.

Primary structure or sequence the linear order or sequence of amino acids along a polypeptide chain in a protein.

Prions a class of proteins that causes serious disease without the involvement of DNA/RNA.

Procollagen a newly translated form of collagen in which hydroxylation and addition of sugar residues has occurred, but the triple helix has not formed.

Prokaryote a simple normally unicellular organism that lacks a nucleus. All bacteria are prokaryotes.

Prosequence region of a protein at the N terminal designed to keep the enzyme inactive. The pro sequence is removed in zymogen processing.

Prosthetic group a co-factor such as a metal ion or small molecule such as a heme group. It can be bound covalently or non-covalently to a protein and is usually essential for proper protein function.

Protease a generic group of enzymes that hydrolyse peptide bonds cleaving polypeptide chains into smaller fragments (the term proteinase is also used interchangeably). Often show a specificity for a particular amino acid sequence.

Proteasome assembly of proteins based on a core structure of four heptameric rings that functions to degrade proteins into small peptide fragments.

Proteins biomolecule composed of one or more polypeptide chains containing amino acid residues linked together via peptide bonds.

Proto-oncogene the normal cellular form of an oncogene with the potential to be mutated. Mutation of the gene yields an oncogene and may lead to cancer.

PrP the protein believed to be responsible for transmitting the disease of prions. The protein is encoded by the host's genome and exists in two forms, only one of which causes the disease.

Purine planar, heterocyclic aromatic rings with adenine and guanine being two important bases found in cells.

Quantum a packet of energy

Quaternary structure the level of structure that results between separate, folded polypeptide chains (subunits) to produce the mature or active protein.

R group one of the 20 side chains found attached to the backbone of amino acids.

R state the relaxed state describing the activity of an allosteric enzyme or protein.

Ramachandran plot usually shown as a plot of dihedral angle ϕ against ψ.

Random coil refers to a linear polymer that has no secondary or tertiary structure but instead is wholly flexible with a randomly varying geometry. This is the state of a denatured protein or nucleic acid.

Redox reduction–oxidation reactions.

Renaturation refolding of a denatured protein to assume its active or native state.

RER rough endoplasmic reticulum – characterized by ribosomes attached to this membrane involved in cotranslational targeting.

Residue a name for a monomeric unit with a polymer such as an amino acid within a protein.

Reverse turn a short sequence of 3–5 residues that leads to a polypeptide chain altering direction and characterised by occurrence of certain amino acid residues with distinct dihedral angles. Also called a β bend.

Ribozyme a enzyme based on RNA capable of catalysing a chemical reaction.

S Svedburg unit of sedimentation with the units of 10^{-13} s. An example is the 30S ribosome particle. It is an estimate of how rapidly a protein or protein complex sediments during ultracentrifugation.

Salting in the effect of moderate amounts of ions, which increases the solubility of proteins in solution.

Salting out the effect of an extreme excess of ions which makes proteins precipitate from solution.

Scurvy a condition that occurs with vitamin C deficiency and reflects deficiency in connective tissue and collagen cross linking.

Secondary structure the spatial relationship of amino acid residues in a polypeptide chain that are close together in the primary sequence.

Serpin serine protease inhibitor.

Sheet a fundamental protein secondary structure (ribbon-like) discovered by Linus Pauling. It contains two amino acid residues per turn and forms hydrogen bonds with residues on adjacent chains.

Site-directed mutagenesis technique for altering the sequence of a DNA molecule. If the alteration occurs in a region coding for protein, the amino acid sequence of the protein may be altered as a consequence.

Snurps proteins found in spliceosomes with small nuclear RNAs.

SRP signal recognition particle. A ribonucleoprotein complex involved in cotranslational targeting of nascent polypeptide chains to membranes.

Stereoisomers molecules containing a center of asymmetry that possess same chemical formula but exist with different configuration or arrangement of atoms.

Substrate a reactant in an enzyme catalysed reaction that binds to active site and is converted into product.

T cells cells of the immune system derived from the thymus and concerned with fighting pathogens based on two types killer: T cells and helper T cells.

TATA box A/T rich region of genes that is involved in the binding of RNA polymerase to eukaryotic DNA sequences.

Tertiary structure large-scale folding structure in a linear polymer that is at a higher order than secondary structure. For proteins and RNA molecules, the tertiary structure is the specific three-dimensional shape into which the entire chain is folded.

Thermophile bacteria capable of living at high temperatures sometimes in excess of 90 °C.

Thermosome name given to the proteasome in thermophiles such as *T. acidophilum.*

Tic analogous system to Tim found in chloroplast inner membrane

Tim translocation inner membrane – a collection of proteins forming a protein import pathway in the inner mitochondrial membrane.

TMV tobacco mosaic virus.

Toc analogous system to Tom found in chloroplast outer membrane

Tom translocation outer membrane – a collection of proteins forming a protein import pathway in the outer mitochondrial membrane.

Torsion angle also known as dihedral angle.

Transcription the process of RNA synthesis from a DNA template performed by RNA polymerase and associated proteins known as transcription factors.

Transition state all reactions proceed through a transition state that represents the point of maximum free energy in a reaction coordinate linking reactants and products.

Transition state analogue a stable molecule that resembles closely the transition state complex formed at the active site of enzymes during catalysis.

Translation the process of converting the genetic code as specified by the nucleotide base sequence of mRNA into a corresponding sequence of amino acids within a polypeptide chain.

Transmembrane a protein or helix that completely spans the membrane.

Tropocollagen basic unit of collagen fibre. It is a triple helix of three polypeptide chains, each about 1000 residues in length.

TSE transmissible spongiform encephalopathy – any agent causing spongiform appearance in brain.

UV region of the electromagnetic spectrum extending from ~200 to ~400 nm.

van der Waals interactions weak interactions between uncharged molecular groups that help stabilize a protein's structure.

Variable domain a part of an immunoglobulin that varies in amino acid sequence and tertiary structure from one antibody to another.

vCJD new variant CJD that arose from BSE and is a transmissible spongiform encephalopathy.

V_{max} the maximal velocity in an enzyme-catalysed reaction.

v_0 the initial velocity associated with an enzyme-catalysed reaction

Zwitterion a molecule containing both positively and negatively charged groups but has no overall charge. Amino acids are zwitterionic at ~pH 7.0.

Zymogen an inactive precursor (proenzyme) of a proteolytic enzyme.

Appendices

Appendix 1
The International System (SI) of units related to protein structure

Physical quantity	SI unit	Symbol
Length	Metre	m
Time	Second	s
Temperature	Kelvin	K
Electric potential	Volt	V
Energy	Joule	J
Mass	Kilogram	Kg

Appendix 2
Prefixes associated with SI units

Prefix	Power of 10 (e.g. 10^n)
Tera	12
Giga	9
Mega	6
Kilo	3
Milli	−3
Micro	−6
Nano	−9
Pico	−12
Femto	−15
Atto	−18

Frequently when discussing protein structure bond lengths will be expressed in nanometres (nm). For example, the average distance between two carbon atoms in an aliphatic side chain is ~0.14 nm or 0.14×10^{-9} m. Occasionally a second (non SI) unit is used and is named after the Swedish physicist, Anders J Ångström. It is called the Ångström (Å) and is equivalent to 0.1 nm or 10^{-10} m. Both units are widely and interchangeably used in protein biochemistry and in this textbook.

Appendix 3
Table of important physical constants used in biochemistry

Planck constant	h	$6.6260755 \times 10^{-34}$ J s
	$h/2\pi$	$1.05457266 \times 10^{-34}$ J s
Boltzmann constant	k	1.380658×10^{-23} J K^{-1}
Elementary charge	e	$1.60217733 \times 10^{-19}$ C
Avogadro number	N	6.0221367×10^{23} particles/mol
Speed of light	c	2.99792458×10^{8} ms^{-1}
^{1}H gyromagnetic ratio		2.67515255×10^{8} T s^{-1}
Atomic mass unit	amu	1.66057×10^{-27} kg
Gas constant	R	8.31451 J mol^{-1} K^{-1}
Faraday constant	F	96485.3 C mol^{-1}

Proteins: Structure and Function by David Whitford

Appendix 4
Derivation of the Henderson–Hasselbalch equation concerning the dissociation of weak acids and bases

The Henderson–Hasselbalch equation reflects the logarithmic transformation of the expression for the dissociation of a weak acid or base.

$$\mathrm{HA} \leftrightarrow \mathrm{A}^- + \mathrm{H}^+$$

$$K = [\mathrm{H}^+]\,[\mathrm{A}^-]/[\mathrm{HA}]$$

Rearranging this equation leads to

$$[\mathrm{H}^+] = K\,[\mathrm{HA}]/[\mathrm{A}^-]$$

and by taking negative logarithms this leads to

$$-\log\,[\mathrm{H}^+] = -\log K - \log\,[\mathrm{HA}]/[\mathrm{A}^-]$$

Since $\mathrm{pH} = -\log\,[\mathrm{H}^+]$ and we can define pK as $-\log K$ this leads to the following equation

$$\mathrm{pH} = \mathrm{p}K - \log\,[\mathrm{HA}]/[\mathrm{A}^-]$$

that is related to the Henderson–Hasselbalch equation by a simple changing of signs

$$\mathrm{pH} = \mathrm{p}K + \log\,[\mathrm{A}^-]/[\mathrm{HA}]$$

Expressed more generally and using the Bronsted Lowry definition of an acid as a proton donor and a base as a proton acceptor this equation can be re-written as

$$\mathrm{pH} = \mathrm{pK} + \log\,[\text{proton acceptor}]/[\text{proton donor}]$$

The Henderson–Hasselbalch equation is fundamental to the application of acid–base equilibria in proteins or any other biological system. It is used to calculate the pH formed by mixing known concentrations of proton acceptor and donor whose $\mathrm{p}K_a$'s are known. Alternatively this equation can be used to calculate the molar ratio of donor and acceptor given the pH and pK, or to calculate the pK at a particular pH given the concentrations relative or absolute of proton donor and acceptor.

Appendix 5
Easily accessible molecular graphic software

1. Koradi, R., Billeter, M., and Wüthrich, K. (MOLMOL: a superlative program for display and analysis of macromolecular structures. (obtainable from http://www.mol.biol.ethz.ch/groups/Wuthrichgroups/software/) *J. Mol. Graphics* 1996, **14**, 51–55.
2. Roger Sayle developed Rasmol although no formal citation exists. Rasmol is a very suitable introduction to molecular visualization software. An adaptation of Rasmol for use in web browsers called Chime is available from http://www.mdli.com or http://www.mdli.co.uk.
3. Kraulis, P.J. MOLSCRIPT: A Program to produce both Detailed and Schematic Plots of Protein Structures. *J. Appl. Crystallogr.* 1991, **24**, 946–950.
4. Molecular graphic software such as VMD produced by the Theoretical Biophysics group, an NIH Resource for Macromolecular Modeling and Bioinformatics, at the Beckman Institute, University of Illinois at Urbana-Champaign.
5. Guex, N. and Peitsch, M.C. SWISS-MODEL and the Swiss-PdbViewer: An environment for comparative protein modeling *Electrophoresis* 1997, **18**, 2714–2723. Official site for software http://www.expasy.ch/spdbv.

A wide range of commercial software has also been produced.

Appendix 6 Enzyme nomenclature

Enzyme classes and subclasses	
EC 1 Oxidoreductases	
EC 1.1	Acting on the CH-OH group of donors
EC 1.2	Acting on the aldehyde or oxo group of donors
EC 1.3	Acting on the CH-CH group of donors
EC 1.4	Acting on the CH-NH_2 group of donors
EC 1.5	Acting on the CH-NH group of donors
EC 1.6	Acting on NADH or NADPH
EC 1.7	Acting on other nitrogenous compounds as donors
EC 1.8	Acting on a sulfur group of donors
EC 1.9	Acting on a heme group of donors
EC 1.10	Acting on diphenols and related substances as donors
EC 1.11	Acting on a peroxide as acceptor
EC 1.12	Acting on hydrogen as donor
EC 1.13	Acting on single donors with incorporation of molecular oxygen (oxygenases)
EC 1.14	Acting on paired donors, with incorporation or reduction of molecular oxygen
EC 1.15	Acting on superoxide radicals as acceptor
EC 1.16	Oxidising metal ions
EC 1.17	Acting on CH_2 groups
EC 1.18	Acting on reduced ferredoxin as donor
EC 1.19	Acting on reduced flavodoxin as donor
EC 1.97	Other oxidoreductases
EC 2 Transferases	
EC 2.1	Transferring one-carbon groups
EC 2.2	Transferring aldehyde or ketonic groups
EC 2.3	Acyltransferases
EC 2.4	Glycosyltransferases
EC 2.5	Transferring alkyl or aryl groups, other than methyl groups
EC 2.6	Transferring nitrogenous groups
EC 2.7	Transferring phosphorus-containing groups
EC 2.8	Transferring sulfur-containing groups
EC 2.9	Transferring selenium-containing groups
EC 3 Hydrolases	
EC 3.1	Acting on ester bonds
EC 3.2	Glycosylases
EC 3.3	Acting on ether bonds
EC 3.4	Acting on peptide bonds (peptidases)
EC 3.5	Acting on carbon-nitrogen bonds, other than peptide bonds
EC 3.6	Acting on acid anhydrides
EC 3.7	Acting on carbon-carbon bonds
EC 3.8	Acting on halide bonds
EC 3.9	Acting on phosphorus-nitrogen bonds
EC 3.10	Acting on sulfur-nitrogen bonds
EC 3.11	Acting on carbon-phosphorus bonds
EC 3.12	Acting on sulfur-sulfur bonds
EC 4 Lyases	
EC 4.1	Carbon–carbon lyases
EC 4.2	Carbon–oxygen lyases
EC 4.3	Carbon–nitrogen lyases
EC 4.4	Carbon–sulfur lyases
EC 4.5	Carbon–halide lyases
EC 4.6	Phosphorus–oxygen lyases
EC 4.99	Other lyases
EC 5 Isomerases	
EC 5.1	Racemases and epimerases
EC 5.2	*cis-trans*-isomerases
EC 5.3	Intramolecular isomerases
EC 5.4	Intramolecular transferases (mutases)
EC 5.5	Intramolecular lyases
EC 5.99	Other isomerases
EC 6 Ligases	
EC 6.1	Forming carbon–oxygen bonds
EC 6.2	Forming carbon–sulfur bonds
EC 6.3	Forming carbon–nitrogen bonds
EC 6.4	Forming carbon–carbon bonds
EC 6.5	Forming phosphoric ester bonds

Bibliography

General reading

Alberts B., Bray, D., Lewis, J., Raff, M., Roberts, K. & Watson J.D. *Molecular Biology of the Cell*, 3rd edn. Garland Publishing, New York, 1994.

Barrett, G. C. *Chemistry and Biochemistry of Amino Acids*. Chapman & Hall, London, 1985.

Branden, C. & Tooze, J. *Introduction to Protein Structure*. Garland Publishing, New York, 1991.

Cornish-Bowden, A. *Fundamentals of Enzyme Kinetics*. Butterworths, Oxford, 1979.

Creighton, T. E. *Proteins Structures and Molecular Properties*, 2nd edn. W.H. Freeman, London, 1993.

Darnell, J., Lodish, H. & Baltimore, D. *Molecular Cell Biology*, 2nd edn. Scientific American Books, New York, 1990.

Fersht, A. *Enzyme Structure and Mechanism*, 2nd edn. W.H. Freeman, New York, 1985.

Frausto da Silva, J. J. R. & Williams, R. J. P. *The Biological Chemistry of the Elements*. Oxford University Press, Oxford, 1991.

Gutfreund, H. *Enzymes: Physical Principles*. Wiley-Interscience, New York, 1972

Lehninger, A., Nelson, D. L. & Cox, M. M. (eds) *Principles of Biochemistry*, 3rd edn. Worth Publishers, New York, 2000.

Lippard, S. J. & Berg, J. M. *Principles of Bioinorganic Chemistry*. University Science Books, 1994.

Voet, D. Voet, J. G. & Pratt, C. W. *Fundamentals of Biochemistry*. John Wiley & Sons, Chichester, 1999.

Watson, J. D. *et al. Molecular Biology of the Gene*. Benjamin Cummings, New York, 1988.

Chapter 1

Ingram, V. A case of sickle-cell anemia. *Biochem. Biophys. Acta* 1989, **1000**, 147–150.

Kendrew, J. *The Thread of Life*. G. Bell, 1966.

Mirsky, A. E. The discovery of DNA. *Sci. Am.* 1968, **218**, 78–88.

Olby, R. E. *The Path to the Double Helix: the Discovery of DNA*. Dover Publications, 1995.

Rutherford, N. J. *A Documentary History of Biochemistry 1770–1940*. Fairleigh Dickinson University Press, 1992.

Schrodinger, E. *What is Life?* Cambridge University Press, Cambridge, 1944.

Chapter 2

Barrett, G.C. (ed) *Chemistry and Biochemistry of amino acids*. Chapman & Hall, New York, 1985.

Barrett, G.C. & Elmore, D.T. *Amino Acids and Peptides*. Cambridge University Press, 1998.

Creighton, T.E. *Proteins: Structure and molecular properties*, 2nd edn. Chapters 1–7. W.H. Freeman, New York, 1993.

Creighton, T.E. (ed) *Protein Function: A practical approach*. IRL Press, Oxford, 1989.

Hirs, C.H.W. & Timasheff, S.N. (eds) Enzyme Structure part 1. *Methods Enzymology*, 91. Academic Press, 1983.

Jones, J. *Amino Acids and Peptide Synthesis (Oxford Chemistry Primers)*. Oxford University Press, 2002.

Lamzin, V.S., Dauter, Z. & Wilson, K.S. How nature deals with stereoisomers. *Curr. Opin. Struct. Biol.* 1995, **5**, 830–836.

Means, G.E. & Feeney, R.E. *Chemical Modification of Proteins*. Holden-Day, 1973.

Chapter 3

Berman, H.M., Westbrook, J., Feng, Z., Gilliland, G., Bhat, T.N., Weissig, H., Shindyalov, I.N. & Bourne, P.E. The Protein Data Bank. *Nucl. Acids Res.* 2000, **28**, 235–242.

Davies, D.R., Padlan, E.A. & Sheriff, S. Antibody-antigen complexes. *Annu. Rev. Biochem.* 1990, **59**, 439–473.

Doolittle, R.F. Proteins. *Sci.Am.* 1985, **253**, 88–96.

Goodsell, D.S. & Olson, A.J. Soluble proteins: Size, shape and function. *Trends Biochem. Sci.* 1993, **18**, 65–68.

Kuby, J. *Immunology*. W.H. Freeman, London, 1997.

Lesk, A.M. *Introduction to protein architecture*. Oxford University Press, 2001.

Perutz, M.F. *Hemoglobin structure and respiratory transport*. *Sci. Am.* 1978, **239**, 92–125.

Richardson, J.S. The anatomy and taxonomy of protein structure. *Adv. Prot. Chem.* 1981, **34**, 167–339.

Trabi, M., & Craik, D.J. Circular proteins – no end in sight. *Trends Biochem. Sci.* 2002, **27**, 132–138.

Chapter 4

Baum, J. & Brodsky, B., Folding of peptide models of collagen and misfolding in disease. *Curr. Opin. Struct. Biol.* 1999, **9**, 122–128.

Downing, A. K., Knott, V., Werner, J. M., Cardy, C. M., Campbell, I. D., & Handford, P. A. Solution structure of a pair of Ca^{2+} binding epidermal growth factor-like domains: implications for the Marfan syndrome and other genetic disorders. *Cell* 1996, **85**, 597–605.

Glover, J. N., Harrison, S. C. Crystal structure of the heterodimeric bZIP transcription factor *c-Fos c-Jun* bound to DNA. *Nature* 1995, **373**, 257.

Handford, P. A. Fibrillin-1, a calcium binding protein of extracellular matrix. *Biochim. Biophys. Acta* 2000, **1498**, 84–90.

Kaplan, D., Adams, W. W., Farmer, B. & Viney, C. *Silk Polymers*. American Chemical Society, 1994.

Lupas, A. Coiled coils new structures and new functions. *Trends Biochem. Sci.* 1996, **21** 375–382.

Martin, G. R. Timple, R., Muller, P. K. & Kuhn, K. The genetically distinct collagens. *Trends Biochem. Sci.* 1985, **10**, 285–287.

O'Shea, E. K. Rutkowski, R. & Kim, P. S. Evidence that the leucine zipper is a coiled coil. *Science* 1989, **243**, 538–542.

Parry, D. A. D. The molecular and fibrillar structure of collagen and its relationship to the mechanical properties of connective tissue. *Biophys. Chem.* 1988, **29**, 195–209.

Porter, R. M. & Lane E. B. Phenotypes, genotypes and their contribution to understanding keratin function. *Trends Genet.* 2003, **19**, 278–285.

Prockop, D. J. & Kivirikko, K. I. Collagens – molecular-biology, diseases, and potentials for therapy. *Annu. Rev. Biochem.* 1995, **64**, 403–434.

Royce, P.M. & Steinmann, B. (eds) *Connective Tissue and Its Heritable Disorders: Molecular, Genetic, and Medical Aspects*. John Wiley & Sons, 2002.

Van der Rest, M. & Bruckner, P. Collagens: Diversity at the molecular and supramolecular levels. *Curr. Opin. Struct. Biol.* 1993, **3**, 430–436.

Chapter 5

Benz R. (ed) *Bacterial and Eukaryotic Porins: Structure, Function, Mechanism*. John Wiley & Sons, 2004.

Blankenship, R.E. *Molecular Mechanism of Photosynthesis*. Blackwell Science, 2001.

Capaldi, R.A & Aggeler, R. *Mechanism of the F_1F_0-type ATP synthase, a biological rotary motor*. *Trends Biochem. Sci.* 2002, **27**, 154–160.

Gennis, R.B. *Biomembranes*. Springer Verlag, New York, 1989.

Hamm, H.E. The many faces of G protein signaling. *J. Biol. Chem.* 1998, **273**, 669–672.

Lehninger, A. *Principles of Biochemistry*. 3rd edn. (Nelson, D.L. and Cox, M.M., eds.) *Oxidative Phosphorylation and Photophosphorylation*, pp. 659–690. Worth Publishers

Nicholls, D.G. & Ferguson, S.J. *Bioenergetics 2*, Academic Press, 1992.

Scheffler, I.E. *Mitochondria*. John Wiley & Sons Inc, 1999.

Vance, D.E. & Vance, J.E. *Biochemistry of lipids, lipoproteins and membranes*. 4th edn. Elsevier, 2002.

Torres, J., Stevens, T.J. & Samso, M. *Membrane proteins: the 'Wild West' of structural biology*. *Trends Biochem. Sci.* 2003, **28**, 137–144.

Von Jagow, G., Schaegger, H., & Hunte, C. (eds) *Membrane Protein Purification and Crystallization: A Practical Guide*. Academic Press, 2003.

Chapter 6

Baxevanis, A.D. & Ouellette, B.F.(eds) *Bioinformatics: A practical guide to the analysis of genes and proteins*. John Wiley & Sons Inc, 2004.

Findlay, J.B.C. & Geisow, M.J. (eds) *Protein Sequencing. A practical approach*. IRL press 1989.

Margulis, L & Sagan, C. *What is Life*. Simon & Schuster, New York, 1995.

Miller, S.J. & Orgel, L.E. *The Origins of Life*. Prentice-Hall, New Jersey, 1975.

Mount, D.W. *Bioinformatics: Sequence and Genome Analysis*. Cold Spring Harbor Laboratory Press, 2001.

Orengo, C.A., Thornton, J.M & Jones, D.T. *Bioinformatics (Advanced Texts Series)*. BIOS Scientific Publishers, 2002.

Primrose, S.B. *Principles of Genome Analysis: A Guide to Mapping and Sequencing DNA from Different Organisms*. Blackwell Science, 1998.

Sanger, F. *Sequences, sequences, sequences*. *Annu. Rev. Biochem.* 1988, **57**, 1–28.

Volkenstein, M.V. *Physical Approaches to Biological Evolution*. Springer Verlag, Berlin, 1994.

Webster, D.M. *Protein Structure Prediction: Methods and Protocols*. Humana Press, 2000.

Chapter 7

Bairoch, A. The enzyme databank. *Nucl. Acids. Res.* 1994, **22**, 3626–3627.

Fersht, A.R. *Structure and Mechanism in Protein Science: Guide to Enzyme Catalysis and Protein Folding*. W.H. Freeman and Company, New York, 1999.

Gutfreund, H. *Kinetics for the Life Sciences: Receptors, Transmitters and Catalysts*. Cambridge University Press, Cambridge, 1995.

Kovall, R.A. & Matthews, B.W. Type II restriction endonucleases: structural, functional and evolutionary relationships. *Curr. Opin. Chem. Biol.* 1999, **3**, 578–583.

Kraut, J. How do enzymes work? *Science* 1988, **242**, 533–540.

Martins, L.M. & Earnshaw, W.C. Apoptosis: Alive and kicking in 1997. *Trends Cell Biol.* 1997, **7**, 111–114.

Moore, J.W. & Pearson, R.G. *Kinetics and Mechanism*. John Wiley & Sons, Chichester, 1981.

Vaux, D.L. & Strasser, A. The molecular biology of apoptosis. *Proc. Natl. Acad. Sci. USA* 1996, **93**, 2239–2244.

Wold, F. & Moldave, K. Posttranslational modifications. *Methods Enzymol.* **106** and **107**. Academic Press, 1985.

Chapter 8

Dodson, G. & Wlodawer, A. Catalytic triads and their relatives. *Trends Biochem. Sci.* 1998, **23**, 347–352.

Kornberg, A., & Baker. T.A. *DNA Replication*, 2d ed. W. H. Freeman, San Francisco, 1992.

Murray A.W., & Hunt T, eds. *The Cell Cycle: An Introduction*. Oxford University Press. 1993.

Norbury C, & Nurse P. Animal Cell Cycles and Their Control. *Annu. Rev. Biochem.* 1992, **61**, 441–470.

Spirin, A.S. Ribosomes. Kluwer Academic, New York, 1999.

Stein G. S, (ed.) *The Molecular Basis of Cell Cycle and Growth Control*. John Wiley & Sons Inc, New York, 1999.

Stillman, B. (ed.) *The Ribosome*. Cold Spring Harbor Symp. **66**. Cold Spring Harbor Laboratory Press, 2001.

Trends Biochem. Sci. 1997, **22**, 371–410. Issue devoted to proteasome and proteolytic processes.

Watson, J.D. *The Double Helix: Personal Account of the Discovery of the Structure of DNA*. Penguin 1999.

Zwickl, P. & Wolfgang Baumeister, W. *The Proteasome-Ubiquitin Protein Degradation Pathway*. Springer Verlag, Berlin, 2002.

Chapter 9

Brown, T.A. *Gene Cloning and DNA Analysis: An Introduction*. Blackwell, 4th edn, 2001.

Deutscher, M.P., Colowick, S.P. & Simon, M.I. (eds). Guide to protein purification. *Methods Enzymol.* **182**. Academic Press, London, 1990.

Dunn, B.M. Speicher, D.W., Wingfield, P.T. & John E. Coligan, J.E. (eds) *Short Protocols in Protein Science*. John Wiley & Sons, 2003.

Freifelder, D. *Physical Biochemistry: Applications to Biochemistry and Molecular Biology*. W.H.Freeman, 1982.

Hames, B.D. & Rickwood. (eds) *Gel Electrophoresis of Proteins, A practical approach*, 2nd edn. IRL press, 1990.

Jansen, J-C. & Ryden, L. (eds) *Protein Purification: Principles, High Resolution Methods and Applications*. John Wiley & Sons Inc, 1998.

Meyer, V.R. *Practical High-Performance Liquid Chromatography*, 4th edn. John Wiley & Sons, 2004.

Scopes, R. *Protein purification: Principles and Practice*. Springer-Verlag, Berlin, 1993.

Tanford, C. *The hydrophobic effect; formation of micelles and biological membranes*, 2nd edn. John Wiley & Sons, Chichester, 1980.

Voet, D., Voet, J.G. & Pratt, C.W. *Fundamentals of Biochemistry*, John Wiley & Sons, Ltd, Chichester, 1999.

Chapter 10

Campbell, I.D. & Dwek, R. *Biological Spectroscopy*. Benjamin Cummings, New York, 1984.

Cavanagh, J., Fairbrother, W.J., Palmer III, A.G. & Skelton, N.J. *Protein NMR spectroscopy: Principles and Practice*. Academic Press, 1996.

Chang, R. *Chemistry*. 8th edn. McGraw-Hill Education, 2004.

Drenth, J. *Principles of Protein X-ray Crystallography*. Springer-Verlag, New York, 1994

Fasman, G.D. (ed) *Circular Dichroism and the Conformational Analysis of Biomolecules*. Plenum Publishers, 1996.

Ferry, G. *Dorothy Hodgkin: A life*. Granta Books, 1999.

Harrison, S.C. Whither structural biology? *Nat. Struct. Mol. Biol.* 2004, **11**, 12–15.

Nat. Struct. Biol. 1997, **4**, 841–865 and *Nat. Struct. Biol.* 1998, **5**, 492–522. (Series of short NMR reviews)

Rhodes, G. *Crystallography made crystal clear*. Academic Press 2nd edn, San Diego, 2000.

Wider, G. & Wüthrich, K. NMR spectroscopy of large molecules and multimolecular assemblies in solution. *Curr. Op. Struct. Biol.* 1999, **9**, 594–601.

Chapter 11

Caughey, B. (ed) Prion Proteins. *Adv. Protein Chem.* **57**. Academic Press, 2001

Creighton, T.E. *Protein Folding*. W.H.Freeman, New York, 1992.

Daggett, V. & Fersht, A.R. Is there a unifying mechanism for protein folding? *Trends Biochem. Sci.* 2003, **28**, 18–25.

Horwich, A.L. (ed) Protein folding in the cell. *Adv. Protein Chem.* **59**. Academic Press, 2001.

Dobson, C.M. Protein Misfolding, *Evolution and Disease Trends Biochem. Sci.* 1999, **24**, 329–332.

Ladbury, J.E & Chowdhry, B.Z. (eds) *Biocalorimetry: Applications of Calorimetry in the Biological Sciences*. John Wiley & Sons, Chichester, 1998.

Matthews, B.W. Structural and genetic analysis of protein stability. *Annu. Rev. Biochem.* 1993, **62**, 139–160.

Matthews, C.R. (eds) Protein Folding Mechanisms. *Adv. Protein Chem.* 2000, **53**.

Pain, R. (ed) *Mechanisms of Protein Folding* (Frontiers in Molecular Biology Series). Oxford University Press, 2000.

Saibil, H.R. & Ranson, N.A. The chaperonin folding machine. *Trends Biochem. Sci.* 2002, **27**, 627–632.

Chapter 12

Crystal, R. G. The α_1-antitrypsin gene and its deficiency states. *Trends Genet.* 1989, **5**, 411–417.

Culotta, E. & Koshland, D. E., Jr. p53 sweeps through cancer research. *Science* 1993, **262**, 1958–1959.

Fauci, A. S. HIV and AIDS: 20 years of science. *Nature Medicine* 2003, **9**, 839–843. (and subsequent articles)

Lane, D. & Lain, S. Therapeutic exploitation of the p53 pathway. *Trends Molec. Med.* 2002, **8**, S38–S42.

Gouras, G.K. Current theories for the molecular and cellular pathogenesis of Alzheimer's disease. *Expert Rev. Mol. Med.* 2001. http://www.expertreviews.org/01003167h.htm

Perutz, M.F. *Protein Structure: New Approaches to Disease and Therapy*. W.H. Freeman, 1992.

Rowland-Jones, S.L. AIDS pathogenesis: what have two decades of HIV research taught us? *Nature Rev. Immunol.* 2003, **3**, 343–348.

Prusiner, S. Prion diseases and the BSE crisis. *Science* 1997, **278**, 245–251.

Varmus, H. Retroviruses. *Science* 1988, **240**, 1427–1435.

Wiley, D.C. & Skehel, J.J. The Structure and Function of the Haemagglutinin Membrane Glycoprotein of Influenza Virus. *Annu. Rev. Biochem.* 1987, **56**, 365–394.

References

Abrahams, J. P., Leslie, A. G. W., Lutter, R. & Walker, J. E. Structure at 2.8 Å resolution of F1-ATPase from bovine heart mitochondria. *Nature* 1994, **370**, 621–628.

Abramson, J., Svensson-Ek, M., Byrne, B. & Iwata, S. Structure of cytochrome *c* oxidase: a comparison of the bacterial and mitochondrial enzymes *Biochim. Biophys. Acta* 2001, **1544**, 1–9.

Agarraberes, F. A., & Dice, J. F. Protein translocation across membranes. *Biochim. Biophys. Acta* 2001, **1513**, 1–24.

Agarwal, M. L., Taylor, W. R., Chernov, M. V., Chernova, O. B. & Stark, G. R. The p53 network. *J. Biol. Chem.* 1998, **273**, 1–4.

Aguzzi, A., Glatzel, M., Montrasio, F., Prinz, M. & Heppner, F. L. Interventional strategies against prion diseases. *Nature Rev. Neurosciences* 2001, **2**, 745–749.

Aguzzi, A., Montrasio, F., & Kaeser, P. S. Prions: Health scare and biological challenge. *Nature Reviews Molecular Cell Biology* 2002, **2**, 118–126.

Albright, R. A., & Matthews, B. W. How Cro and λ repressor Distinguish Between Operators: The Structural Basis Underlying a Genetic Switch. *Proc. Natl. Acad. Sci USA* 1996, **95**, 3431–3436.

Allen, T. D., Cronshaw, J. M., Bagley, S. Kiseleva, E. & Goldberg, M. W. The nuclear pore complex: mediator of translocation between nucleus and cytoplasm. *J. Cell Sci.* 2000, **113**, 1651–1659.

Als-Nielsen, J. & McMorrow, D. *Elements of Modern X-ray Physics*. John Wiley & Sons, Chichester, 2001.

Altman, S. & Kirsebom, L. Ribonuclease P. in Gesteland, R. F., Cech, T. R. and Atkins, J. F. (eds), *The RNA World*. Cold Spring Harbor Laboratory Press, Cold Spring Harbor, New York 1999.

Altschul, S. F. Amino acid substitution matrices from an information theoretic perspective. *J. Mol. Biol.* 1991, **219**, 555–665.

Amit, A. G., Mariuzza, R. A., Phillips, S. E. V. & Poljak, R. J. Three dimensional structure of an antigen-antibody complex at 2.8Å resolution. *Science* 1986, **233**, 747–750.

Andrade, C., A peculiar form of peripheral neuropathy: familial atypical generalised amyloidosis with special involvement of peripheral nerves. *Brain* 1952, **75**, 408–427.

Andrews, D. W. & Johnson, A. E. The translocon: more than a hole in the ER membrane? *Trends Biochem. Sci.* 1996, **21**, 365–369.

Anfinsen, C. B. Principles that govern the folding of protein chains. *Science* 1973, **181**, 223–230.

Arakawa, T & Timasheff, S. N. Theory of protein solubility. *Methods Enzymol.* 1985, **114**, 49–77, Academic Press.

Babcock, G. & Wikström, M. Oxygen activation and the conservation of energy in cellular respiration. *Nature* 1992, **356**, 301–309.

Baldwin, R. L. & Rose. G. D, Is protein folding Hierarchic? I. Local structure and peptide folding. *Trends Biochem. Sci.* 1999 **24**, 26–33 II. Folding Intermediates and transition states. *Trends Biochem. Sci.* 1999 **24**, 77–83.

Baltimore. D. Our genome unveiled. *Nature* 2001, **409**, 814–816.

Ban, N. Nissen, P., Hansen, J., Moore, P. B. & Steitz, T. A. The complete atomic structure of the large ribosomal subunit at 2.4 Å resolution. *Science* 2000, **289**, 905–920.

Barber, M & Quinn, G. B. High-Level Expression in *Escherichia coli* of the Soluble, Catalytic Domain of Rat Hepatic Cytochrome b_5 Reductase. *Protein expression and purification* 1996, **8**, 41–47.

Bardwell, J. C. A. & Beckwith, J. The bonds that tie: catalyzed disulfide bond formation *Cell* 1993, **74**, 769–771.

Batey, R. T., Sagar M. B., & Doudna J. A. Structural and energetic analysis of RNA recognition by a universally conserved protein from the signal recognition particle. *J. Mol. Biol.* 2001, **307**, 229–246.

Baum, J. & Brodsky, B. Folding of peptide models of collagen and misfolding in disease. *Curr. Opin. Struct. Biol.* 1999, **9**, 122–128.

Baumeister, W. & Steven, A. C. Macromolecular electron microscopy in the era of structural genomics. *Trends Biochem. Sci.* 2000, **25**, 625–631.

Bax, A. Multi-dimensional nuclear magnetic resonance methods for protein studies. *Curr. Opin. Struct. Biol.* 1994, **4**, 738–744.

Bergfors, T. *Protein Crystallization Techniques, Strategies, and Tips. A Laboratory Manual* 1999.

Berry, E. A., Guergova-Kuras, M., Huang, L. -S. & Antony R. Crofts, A. R. Structure and function of cytochrome bc complexes *Annu. Rev. Biochem.* 2000, **69**, 1005–1075.

Billeter, M., Kline, A. D., Braun, W., Huber, R., & Wüthrich, K. Comparison of the High-Resolution Structures of the α-Amylase Inhibitor Tendamistat Determined by Nuclear Magnetic Resonance in Solution and by X-ray Diffraction in Single Crystals. *J. Mol. Biol.* 1989, **206**, 677–687.

Blobel, G. & Dobberstein, B. Transfer of proteins across membranes. I. Presence of proteolytically processed and unprocessed nascent immunoglobulin light chains on membrane-bound ribosomes of murine myeloma, *J. Cell Biol.* 1975, **67**, 835–851.

Bockaert, J., & Pin, J. P. Molecular tinkering of G protein-coupled receptors: an evolutionary success. *EMBO J.* 1999, **18**, 1723–1729.

Bokman S. H., Ward W. W. Renaturation of *Aequorea* green fluorescent protein. *Biochem. Biophys. Res. Commun.* 1981, **101**, 1372–1380.

Bolton, W. & Perutz. M. F. Three dimensional Fourier synthesis of horse deoxyhaemo-globin at 2.8 Å resolution. *Nature* 1970, **228**, 551–552.

Booth, P., Templer, R. H., Curran, A. R. & Allen, S. J. Can we identify the forces that drive the folding of integral membrane proteins? *Biochem Soc.Trans.* 2001, **29**, 408–413.

Boriack-Sjodin, P. A., Zeitlin, S., Chen, H. -H., Crenshaw, L., Gross, S., Dantanarayana, A., Delgado, P., May, J. A., Dean, T. & Christianson, D. W. Structural analysis of inhibitor binding to human carbonic anhydrase II. *Protein Science* 1998, **7**, 2483–2489.

Bowie, J. U., Clarke, N. D., Pabo, C. O. & Sauer, R. T. Deciphering the message in protein sequence: tolerance to amino acid substitutions. *Science* 1990, **247**, 1306–1310.

Boyer, P. D. The binding change mechanism for ATP synthase – Some probabilities and possibilities. *Biochim. Biophys. Acta* 1993, **1140**, 215–250.

Boyer, P. D., The ATP synthase – a splendid molecular machine, *Annu. Rev. Biochem.* 1997, **66**, 717–749.

Braig, K., Menz R. I., Montgomery M. G., Leslie A. G., & Walker J. E. Structure of bovine mitochondrial F_1-ATPase inhibited by Mg^{2+} ADP and aluminium fluoride. *Structure* 2000, **8**, 567–573.

Braig, K., Otwinowski, Z., Hegde, R., Boisvert, D. C., Joachimiak, A., Horwich, A. L & Sigler, P. B. The crystal structure of the bacterial chaperonin GroEL at 2.8 Å. *Nature* 1994, **371**, 578–586.

Brandts, J. F., Halvorson, H. R., & Brennan, M. Consideration of the possibility that the slow step in protein denaturation reactions is due to *cis-trans* isomerism of proline residues. *Biochemistry* 1975, **14**, 4953–4963.

Brandts, U. Bifurcated ubihydroquinone oxidation in the cytochrome bc 1 complex by protongated charge transfer. *FEBS Lett.* 1996, **387**, 1–6.

Brejc, K., Sixma, T. K., Kitts, P. A., Kain, S. R., Tsien, R. Y., Ormo, M., Remington, S. J. Structural basis for dual excitation and photoisomerization of the Aequorea victoria green fluorescent protein. *Proc. Natl. Acad. Sci U S A* 1997, **94**, 2306–2311.

Brenner, S., Jacob, F. & Meselson, M. An unstable intermediate carrying information from genes to ribosomes for protein synthesis. *Nature* 1961, **190**, 576–581.

Brodersen, D. E., Carter, A. P., Clemons Jr, W. M., Morgan-Warren, R. J., Murphy IV, F. V. Ogle, J. M., Tarry, M. J. Wimberley, B. T. & Ramakrishnan, V. Atomic Structures of the 30S Subunit and Its Complexes with Ligands and Antibiotics. *Cold Spring Harbor Symp.* 2001, **66**, 17–32.

Browner, M. F. & Fletterick, R. J. Phosphorylase: a Biological Transducer. *Trends Biochem. Sci.* 1992, **17**, 66–71.

Burley, S. K. The TATA box binding protein. *Curr. Opin. Struct. Biol.* 1996, **6**, 69–75.

Butler, P. J. G., Klug, A. The assembly of a virus. *Sci. Amer.* Nov. 1978

Byrne, B. & Iwata, S. Membrane protein complexes. *Curr. Opin. Struct. Biol.* 2002, **12**, 239–243.

Cahn, R. S., Ingold, C. K. & Prelog, V. Specification of Molecular Chirality. *Angew. Chem.* 1966, **78**, 413–447.

Cammack, R. & Cooper, C. E. Electron paramagnetic resonance spectroscopy of iron complexes and iron-containing proteins. *Methods Enzymol.* 1993, **22**, 353–384, Academic Press.

Campbell, I. D. & Dwek, R. *Biological Spectroscopy.* Benjamin Cummings, New York, 1984.

Capaldi, R. A. Structure and function of cytochrome oxidase. *Annu. Rev. Biochem.* 1990, **59**, 569–96.

Carter, C. W. Cognition, Mechanism, and Evolutionary Relationships in Aminoacyl-tRNA Synthetases. *Ann. Rev. Biochem.* 1993, **62**, 717–748.

Cavanagh, J., Fairbrother, W. J., Palmer III, A. G. & Skelton, N. J. *Protein NMR spectroscopy: Principles and Practice*. Academic Press, 1996.

Chaddock, J. A., Herbert, M. H., Ling, R. J., Alexander, F. C. G., Fooks, S. J., Revell, D. F., Quinn, C. P., Shone, C. C. & Foster, K. A. Expression and purification of catalytically active, non-toxic endopeptidase derivatives of *Clostridium botulinum* toxin type A. *Protein Expression and Purification* 2002, **25**, 219–228.

Chalfie, M., Tu, Y., Euskirchen, G., Ward W. W., Prasher D. C. Green fluorescent protein as a marker for gene expression. *Science* 1994, **263**, 802–805.

Cheetham, G. M., Jeruzalmi, D. & Steitz, T. A. Structural basis for initiation of transcription from an RNA polymerase- promoter complex *Nature* 1999, **399**, 80–84.

Chen, S., Roseman, A. M., Hunter, A. S., Wood, S. P., Burston, S. G., Ranson, N. A., Clarke, A. R. & Helen R. Saibil, H. R. Location of a folding protein and shape changes in GroEL–GroES complexes imaged by cryo-electron microscopy *Nature* 1994, **371**, 261–264.

Chevet, E., Cameron, P. H., Pelletier, M. F., Thomas D. Y. & Bergeron, J. J. M. The endoplasmic reticulum: integration of protein folding, quality control, signaling and degradation. *Curr. Opin. Struct. Biol.* 2001, **11**, 120–124.

Cho, Y., Gorina, S., Jeffrey, P. D., Pavletich, N. P. Crystal structure of a p53 tumor suppressor-DNA complex: understanding tumorigenic mutations. *Science* 1994, **265**, 346–355.

Chothia, C & Finkelstein, A. V. The classification and origins of protein folding patterns. *Annu. Rev. Biochem.* **59**, 1007–1039, 1990.

Chou, P. Y. & Fasman, G. D. Empirical Predictions of Protein Conformation. *Annu. Rev.Biochem.* 1978, **47**, 251–276.

Cingolani, G., Petosa, C., Weis, K., & Müller, C. W. Structure of importin-β bound to the IBB domain of importin-α. *Nature* 1999, **399**, 221–229.

Cohen, C. & Parry, D. A. D. α helical coiled coils and bundles: how to design an α helical protein. *Prot. Struct. Funct. Genet.* 1990, **7**, 1–15.

Cohen, F. E. & Prusiner, S. B. Pathologic conformations of prion proteins. *Annu. Rev. Biochem.* 1998, **67**, 793–819.

Collinge, J. Human prion diseases and bovine spongiform encephalopathy (BSE). *Hum. Mol. Genet.* 1997, **6**, 1699–1705.

Coux, O., Tanaka, K. & Goldberg, A. L. Structure and Functions of the 20S and 26S Proteasomes. *Annu. Rev. Biochem.* 1996, **65**, 801–847.

Cowan, S. W., Schirmer, T., Rummel, G., Steiert, M., Ghosh, R., Pauptit, R. A., Jansonius N. J. & Rosenbusch, J. P. Crystal structures explain functional properties of two *E. coli* porins. *Nature* 1992, **358**, 727–733.

Crick, F. H. C. The packing of α helices: simple coiled coils. *Acta Cryst.* 1953, **6**, 689–697.

Crick, F. H. C., Barnett, L., Brenner, S. & Watts-Tobin, R. J. General nature of the genetic code for proteins. *Nature* 1961, **192**, 1227–1232.

Crystal, R. G. The α_1-antitrypsin gene and its deficiency states. *Trends Genet.* 1989, **5**, 411–417.

Cserzo, M., Wallin, E., Simon, I., von Heijne, G. & Elofsson, A. Prediction of transmembrane α helices in prokaryotic membrane proteins: the Dense Alignment Surface method. *Prot. Eng.* 1997, **10**, 673–676.

Cullen, B. R, Nuclear RNA export pathways. *Mol. Cell. Biol.* 2000, **20**, 4181–4187.

Dalbey, R. E, Chen, M. Y., Jiang, F. & Samuelson, J. C. *In vivo* Assembly of Transporters and other Membrane Proteins. *Curr. Opin. Cell Biol.* 2000, **12**, 435–442.

Dalbey, R. E. & Robinson, C. Protein translocation into and across the bacterial plasma membrane and the plant thylakoid membrane. *Trends Biochem. Sci.* 1999, **24**, 17–24.

Danna, K. & Nathans, D. Specific cleavage of Simian virus 40 DNA restriction endonuclease of *Haemophilus influenzae*. *Proc. Natl. Acad. Sci. USA* 1971, **68**, 2913–2917.

Davie, E. W. Introduction to the blood clotting cascade and the cloning of blood coagulation factors. *J. Prot. Chem.* 1986, **5**, 247–253.

Davies, D. R. The structure and function of the aspartic proteinases. *Ann. Rev. Biophys. Biophys. Chem.* 1990, **19**, 189–215.

Davies, D. R. & Chacko, S. Antibody structure. *Acc. Chem. Res.* 1993, **26**, 421–427.

Dayhoff, M. The origin and evolution of protein superfamilies. *FASEB J.* 1976, **35**, 2132–2138.

Dayhoff, M. O., R. M. Schwartz and B. C. Orcutt. 1978. A model of evolutionary change in proteins. In *Atlas of Protein Sequence and Structure* Vol. 5 suppl. 2 (ed. M. O. Dayhoff), 345–352. National Biomedical Research Foundation, Washington DC.

DeBondt, H. L., Rosenblatt, J., Jancarik, J., Jones, H. D., Morgan, D. O. & Kim, S. H. Crystal structure of cyclin-dependent kinase 2. *Nature* 1993, **363**, 595–602.

Deisenhofer, J., Epp, O., Miki, K., Huber, R. & Michel, H. Structure of the protein subunits in the photosynthetic reaction centre of R. viridis at 3 Å resolution. *Nature* 1985, **318**, 618–624.

Deisenhofer, J., Epp, O., Miki, K., Huber, R. & Michel, H. X-ray structure analysis of a membrane protein complex. Electron density map at 3 Å resolution and a model of the chromophores of the photosynthetic reaction center from *Rhodopseudomonas viridis*. *J. Mol. Biol.* 1984, **180**, 385–398.

Deisenhofer, J., Epp, O., Sinning, I. & Michel, H. Crystallographic refinement at 2.3 Å resolution and refined model of the photosynthetic reaction centre from *Rhodopseudomonas viridis*. *J. Mol. Biol.* 1995, **246**, 429–457.

DeRose, V. J. & Hoffman, B. M. Protein structure and mechanism studied by electron nuclear double resonance spectroscopy *Methods Enzymol.* 1995, **246**, 554–589 Academic Press.

Dill, K. A. & Chan, H. S. From Levinthal to pathways to funnels. *Nature Struct. Biol.* 1997, **4**, 10–19.

Dill, K. A. Dominant Forces in Protein Folding. *Biochemistry* 1990, **29**, 7133–7155.

Dinner, A. R. & Karplus. M. The roles of stability and contact order in determining protein folding rates. *Nature Struct. Biol.* 2001, **8**, 21–22.

Ditzel, L., Löwe, J. Stock, D., Stetter, K. -O., Huber, H., Huber, R. & Steinbacher, S. Crystal structure of the thermosome, the archaeal chaperonin and homolog of CCT. *Cell* 1998, **93**, 125–138.

Dobson, C. M. Protein Misfolding, Evolution and Disease *Trends Biochem. Sci.* 1999, **24**, 329–332.

Dobson, C. M., Evans, P. A., & Radford, S. E. Understanding How Proteins Fold: The Lysozyme Story so Far, *Trends Biochem. Sci.* 1994, **19**, 31–37.

Dodson, G. & Wlodawer, A. Catalytic triads and their relatives. *Trends Biochem. Sci.* **23**, 347–352, 1998.

Dohlman *et alBiochemistry* 1987, **26**, 2660–2666.

Doolittle, R. F. The multiplicity of domains in proteins. *Annu. Rev. Biochem.* 1995, **64**, 287–314.

Doudna, J. A. & Batey, R. T. Structural insights into the signal recognition particle. *Annu. Rev. Biochem.* 2004, **73**, 539–557.

Downing, A. K., Knott, V., Werner, J. M., Cardy, C. M., Campbell, I. D., & Handford, P. A., Solution structure of a pair of Ca^{2+} binding epidermal growth factor-like domains: implications for the Marfan syndrome and other genetic disorders. *Cell* 1996, **85**, 597–605.

Drenth, J. *Principles of Protein X-ray Crystallography*, Springer-Verlag. New York 1994

Dunbrack, Jr. R. L., & Karplus, M. Conformational analysis of the backbone-dependent rotamer preferences of protein sidechains. *Nature Struct. Biol.* **1**, 334–340, 1994.

Eckert, D. M. & Kim, P. S. Mechanisms of viral membrane fusion and its inhibition. *Annu. Rev. Biochem.* 2001, **70**, 777–810.

Edman, P. & Begg, G. A protein sequenator. *Eur. J. Biochem.* 1967, **1**, 80–91.

Eisenberg, D. The discovery of the α-helix and β-sheet, the principal structural features of proteins. *Proc. Natl. Acad. Sci. USA.* 2003. **100**, 11207–11210.

Ellenberger, T. E., Brandl, C. J., Struhl, K. & Harrison, S. C. The GCN4 basic region leucine zipper binds DNA as a dimer of uninterrupted alpha helices: crystal structure of the protein-DNA complex. *Cell* 1992, **71**, 1223–1237.

Ellis, R. J. Chaperone substrates inside the cell. *Trends Biochem. Sci.* 2000, **25**, 210–212.

Elrod-Erickson, M., Benson, T. E., Pabo, C. O.: High-resolution structures of variant Zif268-DNA complexes: implications for understanding zinc finger-DNA recognition. *Structure* 1998, **6**, 451–464.

Englander S. W., Mayne, L., Bai, Y. & Sosnick T. R. Hydrogen exchange: the modern legacy of Linderström-Lang. *Protein Sci.* 1997, **6**, 1101–1109.

Englander, S. W. & Kallenbach, N. R. Hydrogen exchange and structural dynamics of proteins and nucleic acids. *Quart. Rev. Biophys.* 1984, **16**, 521–655.

Evans, J. N. S. *Biomolecular NMR spectroscopy*. Oxford University Press. 1995.

Evans, P. R. Structural aspects of allostery. *Curr. Opin. Struct. Biol.* 1991, **1**, 773–779.

Farquhar, M. G. Progress in Unraveling Pathways of Golgi Traffic. *Annu. Rev. Cell Biol.* 1985, **1**, 447–488.

Ferguson, M. A. J. & Williams, A. F. Cell surface anchoring of proteins via glycosyl-phosphatidylinositol structures. *Annu. Rev. Biochem.* 1988, **57**, 285–320.

Ferre-D'Amare, A. R., Prendergast, G. C., Ziff, E. B., Burley, S. K. Recognition by Max of its cognate DNA through a dimeric b/HLH/Z domain. *Nature* 1993, **363**, 38–45.

Ferrell, K. Wilkinson, C. R. M., Dubiel, W., & Gordon. C. Regulatory subunit interactions of the 26S proteasome, a complex problem *Trends Biochem. Sci.* 2000, **25**, 83–88.

Fersht, A. R. Protein folding and stability: the pathway of folding of barnase. *FEBS Lett.* 1993, **325**, 5–16.

Fersht, A. R., Knill-Jones, J. W., Bedouelle, H., & Winter G. Reconstruction by site-directed mutagenesis of the transition state for the activation of tyrosine by the tyrosyl-tRNA synthetase: A mobile loop envelopes the transition state in an induced-fit mechanism. *Biochemistry* 1988, **27**, 1581–1587.

Fersht, A. R., Matouschek, A. & Serrano, L Folding of an enzyme: Theory of protein engineering of stability and pathway of protein folding. *J. Mol. Biol.* 1992, **224**, 771–782, 783–804, 805–818, 819–835, 836–846, 847–859.

Findlay, J. B. C. & Geisow, M. J. (eds) *Protein Sequencing. A practical approach* IRL press 1989.

Fischmann, T. O., Bentley, G. A., Bhat, T. N., Boulot, G., Mariuzza, R. A., Phillips, S. E. V., Tello, D., & Poljak, R. J. Crystallographic Refinement of the Three-dimensional Structure of the FabD1.3-Lysozyme Complex at 2.5-Å Resolution. *J. Biol. Chem.* 1991, **266**, 12915–12920.

Frankel, A. D. & Young, J. A. T. HIV-1: Fifteen Proteins and an RNA. *Annu. Rev. Biochem.* 1998, **67**, 1–25.

Gesteland, R. F., Cech, T. R., & Atkins, J. F. *The RNA World*. Cold Spring Harbor Press, New York 1999.

Gether, U. Uncovering Molecular Mechanisms Involved in Activation of G Protein-Coupled Receptors. *Endocrine Rev.* 2000, **21**, 90–113.

Gething, M. J. & Sambrook, J. Protein folding in the cell. *Nature* 1992, **355**, 33–45.

Glover, J. N., & Harrison, S. C., Crystal structure of the heterodimeric bZIP transcription factor *c-Fos c-Jun* bound to DNA. *Nature* 1995, **373**, 257–260.

Gorlich, D. & Rapoport, T. A, Protein translocation into proteoliposomes reconstituted from purified components of the endoplasmic reticulum membrane. *Cell* 1993, **75** 615–630.

Gould, K. L. & Nurse, P. Tyrosine phosphorylation of the fission yeast $cdc2^+$ protein kinase regulates entry into mitosis. *Nature* 1991, **342**, 39–45.

Green, A. A., *J. Biol. Chem.* 95, **47**, 1932.

Griffiths, W. J., Jonsson, A. P., Liu, S., Rai, D. K. & Wang, Y. Electrospray and tandem mass spectrometry in Biochemistry. *Biochem. J.* 2001, **355**, 545–561.

Grigorieff, N., Ceska, T. A., Downing, K. H., Baldwin, J. M. & Henderson, R. Electron-crystallographic refinement of the structure of bacteriorhodopsin. *J. Mol. Biol.* 1996, **259**, 393–421.

Guidotti, G. Membrane proteins. *Annu. Rev. Biochem.* 1972, **41**, 731–752.

Guijarro, J. I. Guijarro, I., Sunde, M., Jones, J. A., Campbell, I. D. & Dobson, C. M. Amyloid fibril formation by an SH3 domain. *Proc. Natl. Acad. Sci. USA.* 1998, **95**, 4224–4228.

Hames, B. D. & Rickwood, D. (Eds.), *Gel Electrophoresis of proteins. A Practical Approach* (2nd Ed). 1990 IRL press.

Handford, P. A. Fibrillin-1, a calcium binding protein of extracellular matrix. *Biochim. Biophys. Acta*, 2000, **1498**, 84–90.

Hansen, J. C., Lebowitz, J. & Demeler, B. Analytical ultracentrifugation of complex macromolecular systems. *Biochemistry* 1994, **33**, 13155–13163.

Hartl F. U. & Hayer-Hartl, M. Molecular chaperones in the cytosol: from nascent chain to folded protein. *Science* 2002, **295**, 1852–1888.

Hartl. F. U. Molecular chaperones in cellular protein folding *Nature* 1996, **381**, 571–580.

Heijne, von G. Patterns of amino acids near signal cleavage sites. *Eur. J. Biochem.* 1983, **133**, 17–21.

Henderson, R. & Unwin, P. N. T. Three-dimensional model of purple membrane obtained by electron microscopy. *Nature* 1975, **257**, 28–32.

Henderson, R., Baldwin, J. M., Ceska, T. A., Zemlin, F., Beckmann, E. & Downing, K. H. Model for the Structure of Bacteriorhodopsin Based on High-Resolution Electron Cryo-microscopy. *J. Mol. Biol.* 1990, **231**, 899–929.

Henikoff, S. & Henikoff, J. G. Amino acid substitution matrices from protein blocks. *Proc. Natl. Acad. Sci. USA* 1992, **89**, 10915–10919.

Hensley, P. Defining the structure and stability of macromolecular assemblies in solution: the re-emergence of analytical ultracentrifugation as a practical tool. *Structure* 1996, **4**, 367–373. 1996.

Hershko, A., & Ciechanover, A. The ubiquitin system. *Annu. Rev. Biochem.* 1998, **67**, 425–479.

Hill, A. F., Desbruslais, M., Joiner, K., Sidle, C. L., Gowland, I., Collinge, J., Doey, L. J. & Lantos, P. The same prion strain cause nvCJD and BSE. *Nature* 1997, **389**, 448–450.

Hochstrasser, M. Ubiquitin-dependent protein degradation. Annu. Rev. Genet. 1996, **30**, 405–439.

Hofmeister, F. On the understanding of the effects of salts. *Arch. Exp. Pathol. Pharmakol. (Leipzig)* 1888, **24**, 247–260.

Holley, R. W., Everett, G. A., Madison, J. T. & Zamir, A. Nucleotide sequences in yeast alanine transfer RNA. *J. Biol. Chem.*, 1965, **240**, 2122–2127.

Homans, S. W. A dictionary of concepts in NMR. Oxford Science Publication. 1992.

Hong, L., Turner, R. T., Koelsch, G., Shin, D., Ghosh, A. K. & Tang, J. Crystal Structure of Memapsin 2 (β-Secretase) in Complex with Inhibitor Om00-3 *Biochemistry* 2002, **41**, 10963–10967.

Hubbard, S. R. & Till, J. R. Protein tyrosine kinase structure and function *Annu. Rev. Biochem.* 2000, **69**, 373–398.

Huffman, J. L. & Brennan, R. G. Prokaryotic transcription regulators: more than just the helix-turn-helix motif. *Curr. Opin. Struct. Biol.* 2002, **12**, 98–106.

Hunkapiller, M. W., Strickler, J. E., & Wilson, K. E. Contemporary methodology for protein structure determination. *Science* 1984, **226**, 304–311.

Ingram, V. A case of sickle-cell anemia. *Biochem. Biophys. Acta* 1989, **1000**, 147–150.

Iwata, S., *et al.* Complete structure of the 11-subunit bovine mitochondrial cytochrome bc_1 complex. *Science* 1998, **281**, 64–71.

Iwata, S., Ostermeier, C., Ludwig, B., & Michel, H. Structure at 2.8 Å resolution of cytochrome c oxidase from *Paracoccus denitrificans*. *Nature* 1998, **376**, 660–669.

Jackson, S. E. & Fersht, A. R. Folding of chymotrypsin inhibitor 2, 1: Evidence for a two-state transition. *Biochemistry* 1991, **30**, 10428–10435.

Jeffrey, P. D., Russo, A. A., Polyak, K., Gibbs, E., Hurwitz, J., Massague, J. & Pavletich, N. P.: Mechanism of CDK activation revealed by the structure of a cyclinA-CDK2 complex. *Nature* 1995, **376**, 313–320.

Jencks, W. P. Economies of enzyme catalysis. *Cold Spring Harbor Symp. Quant. Biol.* 1987, **52**, 65–73.

Jimenez, J. L., Tennent, G., Pepys, M & Saibil, H. R. Structural Diversity of *ex vivo* Amyloid Fibrils Studied by Cryo-electron Microscopy. *J. Mol. Biol.* 2001, **311**, 241–247.

Johnson L. N. Jenkins J. A. Wilson K. S. Stura E. A. & Zanotti G. Proposals for the catalytic mechanism of glycogen phosphorylase b prompted by crystallographic studies on glucose 1-phosphate binding. *J. Mol. Biol.* 1980, **140**, 565–580.

Johnson, W. C. Jr. Protein secondary structure and circular dichroism. A practical guide. *Proteins Struct. Funct. Genet.* 1990, **7**, 205–214.

Jones, D. T., Taylor, W. R. & Thornton. J. M. The rapid generation of mutation data matrices from protein sequences. *Computer Applied Biosciences* 1992, **8**, 275–282.

Jordan, P., Fromme, P., Witt, H. T., Klukas, O., Saenger, W., & Krauss, N. Three-dimensional structure of cyanobacterial photosystem I at 2.5 Å resolution. *Nature* 2001, **411**, 909–917.

Kalies, K. -U. & Hartmann, E. Protein translocation into the endoplasmic reticulum (ER) Two similar routes with different modes. *Eur. J. Biochem.* 1998, **254**, 1–5.

Karwaski, M. F., Wakarchuk, W. W. & Gilbert, M. High-level expression of recombinant *Neisseria* CMP-sialic acid synthetase in *Escherichia coli*. *Protein Expression and Purification* 2002, **25**, 237–240.

Kauzmann, W. Some factors in the interpretation of protein denaturation. *Adv. Prot. Chem.* 1959, **14**, 1–63.

Kay, L. E., Clore, G. M., Bax, A & Gronenborn, A. M. Four-dimensional heteronuclear triple-resonance NMR spectroscopy of interleukin – 1β in solution. *Science* 1990, **249**, 411–414.

Kay, L. E., D. Marion, D. & Bax, A. Practical aspects of three-dimensional heteronuclear NMR of proteins. *J. Magn. Reson.* 1989, **84**, 72–84.

Kay, L. E., Ikura, M., Tschudin, R. & Bax, A. Three-dimensional triple resonance NMR spectroscopy of isotopically enriched proteins. *J. Magn. Reson.* 1990, **89**, 496–514.

Keenan, R. J., Freymann, D. M., Stroud, R. M., & Walter, P. The signal recognition particle. *Annu. Rev. Biochem.* 2001, **70**, 755–775.

Keleti, T. Two rules of enzyme kinetics for reversible Michaelis-Menten mechanisms. *FEBS Lett.* 1986, **208**, 109–112.

Kelly, J. W. Alternative conformations of amyloidogenic proteins govern their behavior, *Curr. Opin. Struct. Biol.* 1996 **6**, 11–17.

Kelly, J. W. Towards an understanding of amyloidosis. *Nature Struct. Biol.* 2002, **5**, 323–324.

Kendrew, J. C., Bodo, G., Dintzis, H. M., Parrish, R. G., Wyckoff, H., and Phillips, D. C. A Three-Dimensional Model of the Myoglobin Molecule Obtained by X-ray Analysis. *Nature*, 1958, **181**, 662.

Kim P. S. & Baldwin R. L. Intermediates in the Folding Reactions of Small Proteins *Annu Rev. Biochem.* 1990, **59**, 631–660.

King, R. W., Deshaies, R. J., Peters, J. M. & Kirschner, M. W. How proteolysis drives the cell cycle. *Science* 1996, **274**, 1652–1659.

Kirby, A. J. The lysozyme mechanism sorted – after 50 years. *Nat. Struct. Biol.* 2001, **8**, 737–739.

Knoll, A. H. The early evolution of eukaryotes: A geological perspective, Science 1992, **256**, 622–627.

Knowles, J. R & Albery, W. J. Perfection in enzyme catalysis, the energetics of triose phosphate isomerase. *Acc. Chem. Res.* 1977, **10**, 105–111.

Knowles, J. R. Tinkering with enzymes: what are we learning? *Science* 1987, **236**, 1252–1257.

Kohlstaedt, L. A., Wang, J., Friedman, J. M., Rice, P. A. & Steitz, T. A. Crystal structure at 3.5 Å resolution of HIV-1 reverse transcriptase complexed with an inhibitor. *Science* 1992, **256**, 1783–1790.

Kraut, J. How do enzymes work? *Science* 1988, **242**, 533–540.

Kwong, P. D., Wyatt, R., Majeed, S., Robinson, J. Sweet, R. W., Sodroski, J. & Hendrickson, W. A. Structures of HIV-1 Gp120 Envelope Glycoproteins from Laboratory-Adapted and Primary Isolates. *Structure* 2000, **8**, 1329–133X.

Kyte J., Doolittle R. F. A simple method for displaying the hydropathic character of a protein. *J. Mol. Biol.* 1982, **157**, 105–132.

Ladbury, J. E & Chowdhry, B. Z. (eds) *Biocalorimetry: Applications of Calorimetry in the Biological Sciences.* John Wiley & Sons Chichester 1998.

Laemmli, U. K. Cleavage of structural proteins during the assembly of the head of bacteriophage T4. *Nature* 1970, **227**, 680–685.

Lander, E. S. Linton, L. M., Birren, B *et al.* International Human Genome Sequencing Consortium. Initial sequencing and analysis of the human genome. *Nature* 2001, **409**, 860–921.

Lander, E. S. The new genomics: global view of biology. *Science* 1996, **274**, 536–539.

Larrabee, J. A. & Choi, S. Fourier transform infrared spectroscopy. *Methods Enzymol.* 1993, **226**, 289–305, Academic Press.

Lashuel, H. A., Lai, Z. & Kelly, J. W. Characterization of the transthyretin acid denaturation pathways by analytical ultracentrifugation: implications for wild-type, V30M, and L55P amyloid fibril formation, *Biochemistry* 1998, **37**, 17851–17864.

Leatherbarrow, R. J., Fersht, A. R., & Winter, G. Transition-state stabilization in the mechanism of tyrosyl-tRNA synthetase revealed by protein engineering. *Proc. Natl. Acad. Sci. USA* 1985, **82**, 7840–7844.

Lee, A. G. A calcium pump made visible. *Curr. Opin. Struct. Biol.* 2002, **12**, 547–554.

Lemmon M. A., Flanagan J. M., Treutlein H. R., Zhang J., & Engelman D. M. Sequence specificity in the dimerization of transmembrane alpha-helices *Biochemistry* 1992, **31**, 12719–12725.

Lin. L. N. & Brandts, J. F. Isomerization of proline-93 during the unfolding and refolding of ribonuclese A. *Biochemistry* 1983, **22**, 559–563.

Lipschutz, R. J. & Fodor, S. P. A. Advanced DNA technologies. *Curr. Opin. Struct. Biol.* 1994, **4**, 376–380.

Lipscomb, W. N. Aspartate Transcarbamylase from *Escherichia Coli*: Activity and Regulation *Adv. Enzymol.* 1994, **73**, 677–751.

Ludwig, S., Pleschka, S., Planz, O., & Wolff, T. Influenza virus induced signalling cascades: targets for antiviral therapy? *Trends Mol. Medicine* 2003, **9**, 46–52.

Luecke, H., Schobert, B., Richter, H. T., Cartailler, P. & Lanyi, J. K. Structure of bacteriorhodopsin at 1.55 Å resolution. *J. Mol. Biol.* 1999, **291**, 899–911.

Luecke, H., Schobert, B., Richter, H. T., Cartailler, P. & Lanyi, J. K. Structural changes in bacteriorhodopsin during ion transport at 2 Å resolution. *Science* 1999, **286**, 255–260.

Luong, C., Miller, A., Barnett, J., Chow J., Ramesha C., & Browner M. F. Flexibility of the NSAID binding site in the structure of human cyclooxygenase-2. *Nature Struct. Biol.* 1996, **3**, 927–933.

Lupas, A. Coiled coils new structures and new functions. *Trends Biochem. Sci.* 1996, **21**, 375–382.

Lynch, D. R. & Synder, S. H. Neuropeptides: multiple molecular forms, metabolic pathways and receptors. *Annu. Rev. Biochem.* 1986, **55**, 773–799.

Malkin, D., Li, F. P., Strong, L. C., Fraumeni, J. F., Jr., Nelson, C. E., Kim, D. H., Kassel, J., Gryka, M. A., Bischoff, F. Z., Tainsky, M. A. & Friend, S. H. Germ line p53 mutations in a familial syndrome of breast cancer, sarcomas, and other neoplasms. *Science* 1990, **250**, 1233–1238.

Mallucci, G. R., Ratte, S., Asante, E. A., Linehan, J., Gowland, I., Jefferys, J. G. R., Collinge, J. Post-natal knockout of prion protein alters hippocampal CA1 properties, but does not result in neurodegeneration. *EMBO J.* 2002, **21**, 202–210.

Mann, M., Hendrickson, R. C & Pandey, A. Analysis of proteins and proteomes by mass spectrometry. *Annu. Rev. Biochem.* 2001, **70**, 437–473.

Margulis, L & Sagan, C. *What is Life*. Simon & Schuster. 1995.

Marquart, M., Walter, J., Deisenhofer, J., Bode, W., & Huber, R. The Geometry of the Reactive Site and of the Peptide Groups in Trypsin, Trypsinogen and its Complexes with Inhibitors *Acta Crystallogr., Sect. B* 1983, **39**, 480–484.

Martin, G. R., Timple, R., Muller, P. K. & Kuhn, K. The genetically distinct collagens. *Trends Biochem. Sci.* 1985, **10**, 285–287.

Martoglio, B. & Dobberstein, B. Snapshots of membrane-translocating proteins. *Trends Cell Biol.* 1996, **6**, 142–147.

Masters, C. L.; Simms, G.; Weinman, N. A.;Multhaup, G.; McDonald, B. L.; Beyreuther, K. Amyloid plaque core protein in Alzheimer disease and Down syndrome. *Proc. Nat. Acad. Sci. USA 1985*, **82**, 4245–4249.

McPherson, A. *Crystallization of Biological Macromolecules*. Cold Spring Harbor Laboratory Press, 1999.

McRee, D. E. *Practical Protein Crystallography*. 2nd edn. Academic Press. 1999.

Meselson, M. & Stahl. F. The replication of DNA in *Escherichia coli*. *Proc. Natl Acad. Sci. USA 1958*, **44**, 671–682.

Michel, H. Three dimensional crystals of a membrane protein complex. The photosynthetic reaction centre from *Rhodopseudomonas viridis*. *J. Mol. Biol.* 1982, **158**, 567–572.

Miller, S. *Cold Spring Harbor Symp. Quant Biol.* 1988, **52**, 17–28.

Miller, S. J. & Orgel, L. E. The Origins of Life, Prentice-Hall, New Jersey, 1975.

Miranker, A., Radford, S. E., Karplus, M., & Dobson, C. M. Demonstration by NMR of Folding Domains in Lysozyme. *Nature*, 1991, **349**, 633–636.

Miranker, A., Robinson, C. V., Radford, S. E., Aplin R. T., Dobson C. M. Detection of Transient Protein Folding Populations by Mass Spectrometry. *Science* 1993, **262**, 896–900.

Moore, P. B. & Steitz, T. A. The structural basis of large ribosomal subunit function. *Annu. Rev. Biochem.* 2003, **72**, 813–850.

Moore, P. B. The ribosome at atomic resolution. *Biochemistry* 2001, **40**, 3243–3250.

Morgan, D. A. Cyclin dependant kinases: Engines, Clocks, and Microprocessors. *Ann Rev. Cell Dev. Biol.* 1997, **13**, 261–291.

Morgan. D. G., Menetret, J. F., Neuhof, A., *et al*, Structure of the Mammalian Ribosome–Channel Complex at 17 Å Resolution. *J. Mol. Biol.* 2002, **324**, 871–886.

Morimoto, R., Tissieres, A & Georgopoulos, C, (eds). *The biology of heat shock proteins and molecular chaperones.* Cold Spring Harbour Laboratory Press 1994.

Morise, H., Shimomura, O., Johnson, F. H., & Winant, J. Intermolecular energy transfer in the bioluminecent system of *Aequorea*. *Biochemistry* 1974, **13**, 2656–2662.

Mullan, M., Crawford, F., Axelman, K., Houlden, H., Lilius, L., Winblad, B. & Lannfelt, L. A pathogenic mutation for probable Alzheimer's disease in the APP gene at the N-terminus of β-amyloid. *Nature Genet. 1992*, **1**, 345–347.

Newman, M., Strzelecka, T., Dorner, L. F., Schildkraut, I. & Aggarwal, A. K. Structure of restriction endonuclease *Bam*HI and its relationship to *Eco*RI, *Nature* 1994, **368**, 660–664.

Nikolov, D. B. & and Burley, S. K. RNA polymerase II transcription initiation: A structural view. *Proc. Natl. Acad Sci.* 1997, **94**, 15–22.

Nissen, P., Hansen, J., Ban, N., Moore, P. B. & Steitz, T. A. The structural basis of ribosome activity in peptide bond synthesis. *Science* 2000, **289**, 920–930.

Noel, J. P. Hamm, H. E. & Sigler, P. B. The 2.2 Å crystal structure of transducin-α complexed with GTPγS. *Nature* 1993, **366**, 654–658.

Nogales, E., Wolf, S. G. & Downing. K. H. Structure of the aβ-tubulin dimer by electron crystallography. *Nature* 1998, **391**, 199–203.

Nogales. E. Structural insight into microtubule function. *Annu. Rev. Biophys. Biomol. Struct.* 2001, **30**, 397–420.

Noiva, R., & Lennarz, W. J. Protein Disulfide Isomerase – A Multifunctional Protein Resident in the Lumen of the Endoplasmic Reticulum. *J. Biol. Chem.* 1992, **267**, 3553–3556.

Nugent, J. H. A. Oxygenic photosynthesis: electron transfer in photosystem I and photosystem II. *Eur. J. Biochem.* 1996, **237**, 519–531.

Nurse, P. Genetic control of cell size at cell division in yeast. *Nature* 1975, **256**, 547–551.

O'Farrell, P. H. High resolution two dimensional electrophoresis. *J. Biol. Chem.* 1975, **250**, 4007–4021.

O'Shea, E. K. Rutkowski, R. & Kim, P. S., Evidence that the leucine zipper is a coiled coil. *Science* 1989, **243**, 538–542.

Oliver, J., Jungnickel, B., Gorlich, D., Rapoport, T., & High S. The Sec61 complex is essential for the insertion of proteins into the membrane of the endoplasmic reticulum. *FEBS Lett.* 1995, **362**, 126–30.

Onuchic, J. N., Wolynes, P. G., Luthey-Schulten, Z. & Socci, N. D. Towards an Outline of the Topography of a Realistic Protein-Folding Funnel. *Proc. Natl. Acad. Sci USA* 1995, **92**, 3626–3630.

Orengo, C. A., Michie, A. D., Jones, S., Jones, D. T., Swindells, M. B., & Thornton, J. M. CATH – A Hierarchic Classification of Protein Domain Structures. *Structure* 1997, **5**, 1093–1108.

Orgel, L. E. Molecular replication. *Nature* 1992, **358**, 203–209.

Orlova, E. V. & Saibil, H. R. Structure determination of macromolecular assemblies by single-particle analysis of cryo-electron micrographs. *Curr. Opin Struct. Biol.* 2004, **14**, 584–590.

Pace, C. N. Conformational stability of proteins. *Trends Biochem. Sci.* 1990, **15**, 14–17.

Pace, C. N., and Scholtz, J. M. A Helix Propensity Scale Based on Experimental Studies of Peptides and Proteins. *Biophys. J.* 1998, **75**, 422–427.

Padlan, E. Anatomy of the Antibody Molecule. *Molecular Immunology* 1994, **31**, 169–178.

Parry, D. A. D. The molecular and fibrillar structure of collagen and its relationship to the mechanical properties of connective tissue. *Biophys. Chem.* 1988, **29**, 195–209.

Passner, J. M., Ryoo, H. D., Shen, L., Mann, R. S. & Aggarwal, A. K. Structure of a DNA-bound Ultrabithorax-Extradenticle homeodomain complex. *Nature* 1999, **397**, 714–719.

Pauling, L. & Corey, R. B. Atomic Coordinates and Structure Factors for Two Helical Configurations of Polypeptide Chains. *Proc. Natl. Acad. Sci. USA* 1951, **37**, 235–240.

Perl, D. Welker, C., Schindler, T., Schroder, K., Marahiel, M. A., Jaenicke, R., and Schmid, F. X. Conservation of rapid two-state folding in mesophilic, thermophilic and hyperthermophilic cold shock proteins. *Nature Structural Biology* 2000, **7**, 380–383.

Perutz, M. F., Rossmann, M. G., Cullis, A. F., Muirhead, H., Will, G. & North, A. C. T. Structure of haemoglobin. A three-dimensional Fourier synthesis at 5.5 Å resolution, obtained by X-ray analysis. *Nature* 1960, **185**, 416–422.

Perutz, M. F., Wilkinson, A. J., Paoli, M., & Dodson, G. The stereochemical mechanism of cooperative effects in hemoglobin revisted. *Annu Rev. Biophys. Biomol. Structure* 1998, **27**, 1–34.

Pfanner, N. & Neupert, W. The mitochondrial protein import apparatus. *Annu. Rev. Biochem.* 1990, **59**, 331–353.

Phillips, M. A. & Fletterick, R. J. Proteases. *Cur. Opinion. Struct. Biol.* 1992, **2**, 713–720.

Picot, D., Loll P. J., & Garavito R. M. The X-ray crystal structure of the membrane protein prostaglandin H2 synthase-l. *Nature* 1994, **367**, 243–249.

Plaxco, K. W., Simons, K. T., & Baker, D. Contact order, transition state placement and the refolding rates of single domain proteins. *J. Mol. Biol.* 1998, **277**, 985–994.

Poignard, P., Saphire, E. O., Parren, P. W. H. I. & Burton, D. R. GP120: Biologic Aspects of Structural Features. *Annu. Rev. Immunol.* 2001, **19**, 253–74.

Ponder, J. W. & F. M. Richards. Tertiary templates for proteins. Use of packing criteria in the enumeration of allowed sequences for different structural classes. *J. Mol. Biol.* 1987, **193**, 775-791.

Popot, J.-L. & Engelman, D. M. Helical membrane protein folding, stability and evolution *Annu. Rev. Biochem.* 2000, **69**, 881–922.

Privalov, P. L. Stability of proteins: Small globular proteins. *Adv. Protein Chem.* 1979, **33**, 167–241.

Privalov, P. L. & Gill, S. J. Stability of protein structure and hydrophobic interaction. *Adv. Prot. Chem.* 1988, **39**, 191–234.

Prusiner, S. B., Novel proteinaceous infectious particles cause scrapie. *Science* 1982, **216**, 136–144.

Radford. S. Protein folding: progress made and promise ahead. *Trends Biochem. Sci.* 2000, **25**, 611–618.

Ramachandran, G. N. & Sasiskharan, V. Conformation of polypeptides and proteins. *Adv. Protein Chem.* 1968, **23**, 283–437.

Ramakrishnan, V. & Moore, P. B. *Curr. Opin. Struct. Biol.* 2001, **11**, 144–154, 2001

Rao, S. T. & Rossmann M. G. Comparison of super-secondary structure in protein *J. Mol. Biol.* 1973, **76**, 241–256.

Rapoport, T. A., Jungnickel, B. & Kutay, U. Protein transport across the eukaryotic endoplasmic reticulum and bacterial inner membranes, *Annu. Rev. Biochem.* 1996, **65**, 271–303.

Rappsilber, J. & Mann, M. What does it mean to identify a protein in proteomics? *Trend Biochem. Sci.* 2002, **27**, 74–78.

Rath, V. L., Silvian, L. F., Beijer, B., Sproat, B. S., Steitz, T. A. How glutaminyl-tRNA synthetase selects glutamine. *Structure* 1997, **6**, 439–449.

Rechsteiner, M. & Rogers, S. W. PEST sequences and regulation by proteolysis. *Trends Biochem. Sci.* 1996, **21**, 267–271.

Riek, R., Hornemann, S., Wider, G., Billeter, M., Glockshuber, R. & Wüthrich, K. NMR structure of the mouse prion protein domain PrP (121–231). *Nature* 1996, **382**, 180–184.

Roder, H., Elöve, G., & Englander, S. W. Structural characterization of folding intermediates in cytochrome c by H-exchange labeling and proton NMR. *Nature* 1888, **335**, 700–704.

Roeder, R. G. The role of general initiation factors in transcription by RNA polymerase II. *Trends Biochem. Sci.* 1996, **21**, 327–335.

Rose, G. N. Turns in peptides and proteins *Adv. Protein Chem.* 1985, **37**, 1–109.

Rosenberg, J. M. Structure and function of restriction endonucleases. *Curr. Opin. Struct. Biol.* 1991, **1**, 104–113.

Rould, M. A., Perona, J. J., Soll, D., & Steitz, T. A. Structure of E. coli glutaminyl-tRNA synthetase complexed with tRNA(Gln) and ATP at 2.8 A resolution. *Science* 1989, **246**, 1135–1141.

Rowland-Jones, S. L. AIDS pathogenesis: what have two decades of HIV research taught us? *Nature Rev. Immunol.* 2003, **3**, 343–348.

Rupp. B., http://www-structure.llnl.gov.

Russell, P. & Nurse, P. cdc25+ functions as an inducer in the mitotic control of fission yeast. *Cell* 1986, **45**, 145–153.

Russo, A. A., *et al.* Nat. Struct. Biol. **3**, 696–XXX 1996

Rutherford, A. W. & Faller, P. The heart of photosynthesis in glorious 3D. *Trends Biochem. Sci.* 2001, **26**, 341–344.

Saibil, H. Molecular chaperones: containers and surfaces for folding, stabilizing or unfolding proteins. *Current Opinion in Struct. Biol.* 2000, **10**, 251–258.

Sambrook, J. & Russell, D. Molecular Cloning: A laboratory manual. Cold Spring Harbor Laboratory Press, 2001

Sambrook, J., Fritsch, E. F. & Maniatis, T. Molecular Cloning. Cold Spring Harbor Laboratory Press, 1989.

Sanger, F. Sequences, sequences, sequences. *Annu. Rev. Biochem.* 1988, **57**, 1–28.

Saraiva, M. J. M. Hereditary transthyretin amyloidosis: molecular basis and therapeutical strategies. *Expert Rev. Mol. Med.* 2002. http://www.expertreviews.org/02004647h.htm

Schekman, R. Dissecting the membrane trafficking system. *Nature Medicine* 2002, **8**, 1055–1058.

Schellman, J. A. The thermodynamic stability of proteins. *Ann. Rev. Biophys. Biophys. Chem.* 1987, **16**, 115–137.

Schmid F. X., Mayr L. M., Mucke, M., & Schonbrunner, E. R. Prolyl isomerases: role in protein folding. *Adv. Prot. Chem.* 1993, **44**, 25–66.

Schopf, J. W. Microfossils of the early Archaen Apex chert: New evidence of the antiquity of life. *Science* 1993, **260**, 640–646.

Schramm, V. L. Enzyme transition states and transition state analog design. *Annu. Rev. Biochem.* 1998, **67**, 693–720.

Schulz, G. E. Structure of porin refined at 1.8 Å resolution. *J. Mol. Biol.* 1992, **227**, 493–509.

Schuster, T. M. & Toedtt, J. M. New revolutions in the evolution of analytical ultracentrifugation. *Curr. Opin. Struct. Biol.* 1996, **6**, 650–658.

Scopes, R. *Protein purification: principles and practice.* Springer-Verlag, Berlin 1993.

Selkoe, D. J. Amyloid β-protein and the genetics of Alzheimer's disease. *J. Biol. Chem.* 1996, **271**, 18295–18298.

Shine, J. & Dalgarno, L. The 3′ terminal sequence of *E.coli* 16S rRNA : Complementary to nonsense triplets and ribosome binding sites. *Proc. Natl. Acad. Sci USA* 1974, **71**, 1342–1346.

Sidransky, D & Hollstein, M. Clinical implications of the p53 gene. *Annu. Rev. Med.* 1996, **47**, 285–30.

Siebert, F. Infrared spectroscopy applied to biochemical and biological problems. *Methods Enzymol.* 1995, **246**, 501–526 Academic Press.

Sigler, P. B., Xu, Z. Rye, H. S. Burston, S. G. , Fenton, W. A & Horwich, A. L. Structure and function in GroEL-mediated protein folding *Annu. Rev. Biochem.* 1998, **67**, 581–608.

Silver, P. A. How proteins enter the nucleus. *Cell* 1991, **64**, 489–497.

Silverman, G. A. *et al.* The Serpins Are an Expanding Superfamily of Structurally Similar but Functionally Diverse Proteins. *J. Biol. Chem.* 2001, **276**, 33293–33296.

Simons K. T., Ruczinski, I., Kooperberg, C., Fox, B., Bystroff, C., & Baker, D. Improved Recognition of Native-like Protein Structures using a Combination of Sequence-dependent and Sequence-independent Features of Proteins. *Proteins* 1999, **34**, 82–95.

Singer, S. J. The molecular organization of membranes. *Annu. Rev. Biochem.* 1974, 805–833.

Singer, S. J. & Nicolson, G. The fluid mosaic model of the structure of cell membranes. *Science* 1972, **175**, 720–731.

Skehel, J. J., Bayley, P. M., Brown, E. B., Martin, S. R., Waterfield, M. D., White, J. M., Wilson, I. A., and Wiley, D. C. Changes in the conformation of influenza virus haemagglutinin at the pH optimum of virus-mediated membrane fusion *Proc. Natl. Acad. Sci. USA* 1982, **79**, 968–972.

Skehel, J. J. & Wiley, D. C. Receptor binding and membrane fusion in virus entry: the influenza haemagglutinin. *Annu. Rev. of Biochem.* 2000, **69**, 531–569.

Smith, W. L., DeWitt, D. L. & Garavito, R. M. Cyclo-oxygenase: Structural, Cellular, and Molecular Biology *Annu. Rev. Biochem.* 2000, **69**, 145–182.

Song, L., Hobaugh, M. R., Shustak, C., Cheley, S., Bayley, H., & Gouaux J. E. Structure of staphylococcal α-hemolysin, a heptameric transmembrane pore, *Science* 1996, **274**, 1859–1866.

Steinhauer, D. A. & Skehel, J. J. Genetics of influenza viruses. *Annu. Rev. Genet.* 2002, **36**, 305–332.

Steitz, T. A. & Schulman, R. G. Crystallographic and NMR studies of the serine proteases. *Annu. Rev. Biophys. Bioenerg.* 1982, **11**, 419–464.

Stock, D., Gibbons, C., Arechaga, I., Leslie, A. G. W. & Walker, J. E. The rotary mechanism of ATP synthase. *Curr. Opin. Struc. Biol.* 2000, **10**, 672–679.

Stock, D., Leslie, A. G. W., & Walker, J. E. Molecular Architecture of the Rotary Motor in ATP Synthase. *Science* 1999, **286**, 1700–1705.

Stoeckenius, W. Bacterial rhodopsins: Evolution of a mechanistic model for the ion pumps Prot. Sci., 1999, **8**, 447–459.

Storey, A., Thomas, M., Kalita, A., Harwood, C., Gardiol, D., Mantovani, F., Breuer, J., Leigh, I. M., Matlashewski, G., & Banks, L. Role of a p53 polymorphism in the development of human papilloma-virus-associated cancer. *Nature* 1998, **393**, 229–234.

Studier, F. W., Rosenberg, A. H., Dunn, J. J., & Dubendorff, J. W. Use of T7 RNA polymerase to direct expression of cloned genes. *Methods in Enzymology* 1990, **185**, 60–89, Academic Press.

Sunde, M., Serpell, L. C., Bartlam, M., Fraser, P. E., Pepys, M. B., & Blake C. C. Common core structure of amyloid fibrils by synchrotron X-ray diffraction. *J. Mol. Biol.* 1997, **273**, 729–739.

Tanford, C. The Hydrophobic Effect; formation of micelles and biological membranes. 2nd ed. Wiley 1980.

Tarn W.-Y. & Steitz. J. A. Pre-mRNA splicing: the discovery of a new spliceosome doubles the challenge *Trends Biochem. Sci.* 1997, **22**, 132–137.

Taylor, K. A. & Glaeser, R. M. Electron diffraction of frozen, hydrated protein crystals *Science* 1974, **186**, 1036–1037.

Toyoshima, C., Nakasako, M., Nomura, H. & Ogawa, H. Crystal structure of the calcium pump of sarcoplasmic reticulum at 2.6 Å resolution. *Nature* 2000, **405** 647

Trabi, M., & Craik, D. J. Circular proteins – no end in sight. *Trends Biochem. Sci.* 2002, **27**, 132–138.

Tsukihara, T. Aoyama, H., Yamashita, E., Tomizaki, T., Yamaguchi, H., Shinzawa-Itoh, K., Nakashima, R., Yaono, R., & Yoshikawa, S. The whole structure of the 13-subunit oxidized cytochrome c oxidase at 2.8 Å. *Science*, 1996, **272**, 1136–1144.

Tugarinov, V., Hwang, P. M. & Kay, L. E. Nuclear magnetic resonance spectroscopy of high-molecular weight proteins. *Annu. Rev. Biochem.* 2004, **73**, 107–146.

Turner, B. G. & Summers, M. F. Structural biology of HIV. *J. Mol. Biol.* 1999, **285**, 1–32.

Unger, V. M. Electron cryomicroscopy methods. *Curr. Opin. Struct. Biol.* 2001, **11**, 548–554.

Vane, J. R. Inhibition of prostaglandin synthesis as a mechanism of action for the aspirin-like drugs. *Nature*, 1971, **231**, 232–235.

Varghese, J. N., Colman, P. M., van Donkelaar, A., Blick, T. J., Sahasrabudhe, A., & McKimm-Breschkin, J. L. Structural evidence for a second sialic acid binding site in avian influenza virus neuraminidases. *Proc Natl Acad Sci USA* 1997, **94**, 11808–11812.

Varghese, J. N., McKimm-Breschkin, J. L., Caldwell, J. B., Kortt, A. A. & Colman, P. M. The structure of the complex between influenza virus neuraminidase and sialic acid, the viral receptor. *Proteins* 1992, **14**, 327–332.

Varghese, J. N. & Colman, P. M. Three-dimensional structure of the neuraminidase of influenza virus A/Tokyo/3/67 at 2.2 Å resolution. *J. Mol. Biol.* 1991, **221**, 473–486.

Varghese, J. N., Laver, W. G. & Colman P. M. Structure of the influenza virus glycoprotein antigen neuraminidase at 2.9 Å resolution. *Nature* 1983, **303**, 35–40.

Varshavsky. A., The ubiquitin system. *Trends Biochem. Sci.* 1997, **22**, 383–387.

Verméglio, A & Joliot, P. The photosynthetic apparatus of *Rhodobacter sphaeroides Trends Microbiol.* 1999, **7**, 435–440.

Viadiu, H., & Aggarwal, A. K. The role of metals in catalysis by the restriction endonuclease BamHI. *Nat. Struct. Biol.* 1998, **5**, 910–6.

Vogelstein, B., Lane, D. & Levine, A. J. Surfing the p53 network. *Nature* 2000, **408**, 307–310.

Voges, D., Zwickl, P. & Baumeister, W. The 26S proteasome: a molecular machine designed for controlled proteolysis *Annu. Rev. Biochem.* 1998, **68**, 1015–1068.

Voos, W. Martin, H., Krimmer, T. & Pfanner, N., Mechanisms of protein translocation into mitochondria *Biochim. Biophys. Acta*, 1999, **1422**, 235–254.

Walter, P. & Johnson, A. E. Signal sequence recogniton and protein targeting to the endoplasmic reticulum membrane. *Annu. Rev. Cell Biol.* 1995, **10**, 87–119.

Wang, J., Smerdon, S. J., Jager, J., Kohlstaedt, L. A., Rice, P. A., Friedman, J. M., Steitz, T. A. Structural basis of asymmetry in the human immunodeficiency virus type 1 reverse transcriptase heterodimer. *Proc Natl Acad Sci USA* 1994, **91**, 7242–7246.

Wang, Y. & van Wart, H. E. Raman and resonance Raman spectroscopy *Methods Enzymol.* 1993, **226**, 319–373, Academic Press.

Warren, A. J. Eukaryotic transcription factors. *Curr. Opin. Struct. Biol.* 2002, **12**. 107–114.

Watson J. D. & Crick. F. H. C. Molecular structure of nucleic acids: a structure for deoxyribose nucleic acid. *Nature* 1953, **171**, 737–738.

Weis, W., Brown, J. H., Cusack, S., Paulson, J. C., Skehel, J. J., & Wiley, D. C. Structure of the influenza virus haemagglutinin complexed with its receptor, sialic acid *Nature* 1988, **333**, 426–431.

Weis. K. Importins and exportins: how to get in and out of the nucleus *Trends Biochem. Sci.* 1998, **23**, 185–189.

Weiss M. S., Olson, R., Nariya, H., Yokota, K., Kamio, Y., & Gouaux, E. Crystal structure of Staphylococcal Lukf delineates conformational changes accompanying formation of a transmembrane channel. *Nat. Struct. Biol.* 1999, **6**, 134–140.

Weissmann, C., Enari, M., Klöhn, P.-C., Rossi, D. & E. Flechsig. E. Transmission of prions. *Proc. Natl. Acad. Sci. USA* 2002, **99**, 16378–16383.

Weissmann, C., Molecular Genetics of Transmissible Spongiform Encephalopathies. *J. Biol. Chem.* 1999, **274**, 3–6.

Wetlaufer, D. B. Ultraviolet spectra of proteins and amino acids. *Adv. Prot. Chem.* 1962, **17**, 303–390.

White, S. H. & Wimley, W. C. Membrane protein folding and stability: Physical principles. *Ann. Rev. Biophys. Biomol. Struct.* 1999, **28**, 319–6.

Wiley D. C., & Skehel J. J. The structure and function of the haemagglutinin membrane glycoprotein of influenza virus. *Annu. Rev. Biochem.* 1987, **56**, 365–94.

Wilmot, C. M. & Thornton, J. M. Analysis and prediction of the different types of beta-turn in proteins. *J. Mol. Biol.* 1988, **203**, 221–232.

Wimberley, B. T., Brodersen, D., Clemons, W., Morgan-Warren, R., Carter, A., Vonrhein, C., Hartsch, T. & Ramakrishnan, V. Structure of the 3OS Ribosomal Subunit. *Nature* 2000, **407**, 327–339.

Wimley, W. C. The versatile β-barrel membrane protein *Curr. Opin. Struct. Biol.* 2003, **13**, 404–411.

Wlodawer. A. & Erickson, J. W. Structure based inhibitors of HIV-1 proteinase. *Annu. Rev. Biochem.* 1993, **62**, 543–585.

Woody, R. W. Circular dichroism. *Methods Enzymol.* 1995, **246**, 34–71, Academic Press.

Wower, J. Rosen, K. V., Hixson, S. S. & Zimmermann, R. A. Recombinant photoreactive tRNA molecules as probes for cross-linking studies. *Biochimie* 1994, **76**, 1235–1246.

Wuthrich, K. NMR of Proteins and Nucleic Acids. John Wiley & Sons. 1984.

Xu, Z., Horwich A. L., & Sigler P. B. The crystal structure of the asymmetric GroEL-GroES-$(ADP)_7$ chaperonin complex. *Nature.* 1997, **388**, 741–750.

Yates. J. R. Mass spectrometry: from genomics to proteomics. *Trends Genet.* 2000, **16**, 5–8.

Yoshida, M., Muneyuki, E., & Hisabori, T. ATP synthase–a marvellous rotary engine of the cell. *Nat. Rev. Mol. Cell Biol.* 2001, **2**, 669–77.

Yoshikawa, S. Beef heart cytochrome oxidase. *Curr. Opin. Struct. Biol.* 1997, **7**, 574–579.

Zahn, R., Liu, A., Luhrs, T., Riek, R., von Schroetter, C., López García, F., Billeter, M., Calzolai, L., Wider, G. & Wüthrich, K. NMR solution structure of the human prion protein. *Proc. Natl. Acad. Sci USA* 2000, **97**, 145–150.

Zhang, D., Kiyatkin, A., Bolin, J. T. & Low, P. S. Crystallographic Structure and Functional Interpretation of the Cytoplasmic Domain of Erythrocyte Membrane Band 3. *Blood* 2000, **96**, 2925–2933.

Zouni, A., Horst-Tobias, W., Kern, J., Fromme, P., Krauss, N., Saenger, W., & Orth, P. Crystal structure of photosystem II from Synechococcus elongatus at 3.8 Å resolution. *Nature* 2001, **409**, 739–743.

INDEX

Entries are arranged alphabetically with page numbers in italic indicating a presence in a figure whilst bold type indicates appearance in a table. Greek letters and numbers are sorted as if they were spelt out; β sandwich becomes beta-sandwich, 5S rRNA appears where *five*S rRNA is normally located in listings. Positional characters are ignored; so 2-phosphoglycolate appears under phosphoglycolate.

Proteins: Structure and Function by David Whitford